Mathematics and Calculus with Applications

Mathematics and Calculus with Applications

Margaret L. Lial/Charles D. Miller
American River College Sacramento, California

Scott, Foresman and Company Glenview, Illinois
Dallas, Tex. Oakland, N.J. Palo Alto, Cal. Tucker, Ga. London

Cover photo: Courtesy of the Chicago Title Insurance Company

Library of Congress Cataloging in Publication Data

Lial, Margaret L
Mathematics and calculus with applications.

Second ed. (c1979) published under title: Mathematics, with applications in the management, natural, and social sciences.
Includes index.
1. Mathematics—1961– I. Miller, Charles David, 1942– joint author. II. Title.
QA37.2.L5 1980 510 79-27939
ISBN 0-673-15352-5

2 3 4 5 6-KPF-85 84 83 82 81

Preface

Mathematics and Calculus with Applications introduces the mathematical ideas needed by students of management, social science, and biology for a combined finite mathematics and calculus course. It can also be used for a separate course in finite mathematics as well as a separate course in calculus, giving you the convenience of having two books in one. This text is an expanded version of *Mathematics with Applications, Second Edition*. It incorporates many of the same features and topics of *Mathematics with Applications,* but expands the calculus portion of the text to include:

Curve Sketching
Implicit Differentiation
Approximation by Differentials
Integration by Parts
Numerical Integration
Improper Integrals
Functions of Several Variables
Economic Lot Size
Probability Density Function
Lagrange Multipliers

The only prerequisite we assume is a previous course in algebra. The fundamentals of algebra are discussed in Chapter 1 for those students who need to review the basics. A diagnostic pretest is included in the *Instructor's Guide* to help identify such students.

Additional features of this text include the following:

- *Mathematical models* are discussed early in the book (Chapter 3) and are emphasized through the use of many applications.
- *A complete chapter on the mathematics of finance* (Chapter 4) includes a section on the important topic of consumer mathematics.
- *A complete chapter on linear programming* (Chapter 7) provides clear explanations and keeps the numerical calculations as simple as possible.
- *Thorough calculus coverage* (Chapters 12–16), together with Chapter 11 on exponential and logarithmic functions and the algebra review (as needed), provide ample material for a complete course in calculus.
- *A large number of examples* show students how to solve problems step-by-step.
- *Problems in the margins of the text* follow each example so that students can check their understanding of the example. Answers are provided beneath the problems so that students can see immediately if their answer is correct.

- *A large number of drill problems* are included to give students plenty of opportunity to practice basic skills.
- *Many applied problems* categorized by subject area show the usefulness of mathematics in many disciplines and accustom students to solving real-world problems.
- *One or more actual case studies* follow most chapters of the book. These cases present the topics of the chapters as they apply to a variety of real-world situations.

This book can be used in a variety of courses:

Finite Mathematics Use as much of Chapters 1–3 as needed, then cover Chapters 4–10 as time and local needs dictate. (At some schools Chapters 9 and 10 on decision theory and statistics are considered optional.)

Calculus Use Chapters 1–3 as needed, then cover Chapters 11–16.

Finite Mathematics and Calculus Use the entire book; cover Chapters 1–3 as needed.

A *Study Guide with Computer Problems* by Margaret Lial is available for use with this book. This study guide offers a brief introduction to the computer terminal and BASIC programming. Twenty-seven programs for solving problems from the text are included. Many additional exercises are designed to provide the student with a deeper insight into the topics of the course.

The *Instructor's Guide* offers two alternate tests for each chapter, two cumulative tests for Chapters 1–10, and two cumulative tests for Chapters 11–16. These tests can be used for preparing midterms or final examinations. The Guide also includes the answers to the even numbered exercises, complete solutions to selected problems, and a description of how some of the mathematical models in the text were constructed.

Many people helped us produce this book, including Jim Buckley, University of Alabama at Birmingham; Robert Eicken, Illinois Central College; Archille J. Laferriere, Boston College; Peter Nicholls, Northern Illinois University; Edward Rozema, University of Tennessee at Chattanooga; David P. Sumner, University of South Carolina; Walter S. McVoy, Wilson Banks, and Larry Spence, Illinois State University.

We also thank James Levy, Arnold Parker, Pamela Carlson, and Tana Vega-Romero of Scott, Foresman for their help in producing the kind of useful, teachable book that we envisioned.

Margaret Lial
Charles Miller

Contents

16 Functions of Several Variables 671

Appendix 696

1
Review of Algebra

This book is about the application of mathematics to various subjects, mainly business, social science, and biology. In almost all of these applications, we begin with some real-world problem that we want to solve. We then try to create a mathematical model of that problem. We discuss mathematical models in more detail later on, but basically a **mathematical model** is an equation (or other mathematical expression) which represents the given problem.

For example, suppose our real-world problem is to find the area of a floor, 12 feet by 18 feet. Here we would construct a mathematical model by using the formula for the area of a rectangle, Area = Length × Width, or $A = LW$. By substituting our numbers, $L = 12$ and $W = 18$, into this formula, we can find the area $A = 12 \times 18 = 216$ square feet.

Most mathematical models involve algebra. We must first use algebra to set up a model and then to simplify the resulting equations. Since algebra is so vital to a study of the applications of mathematics, we begin the book with a review of some of the fundamental ideas of algebra that we will need. Algebra is a sort of advanced arithmetic, so we begin by looking at the real numbers, the basis of arithmetic.

1.1 THE REAL NUMBERS

We begin our study of mathematics and its applications with a look at the basic building blocks of mathematics—numbers. In this section we define the various types of numbers that we use in this book.

The various types of numbers can be explained with the help of a diagram called a **number line.** To draw a number line, choose any point on a horizontal line and label it 0. Then choose any point to the right of 0 and label it 1. The distance between 0 and 1 gives a unit of measure that can be used repeatedly to locate points to the right of 1, which we label 2, 3, 4, and so on, and points to the left of 0, labeled -1, -2, -3, -4, and so on. A number line with several sample numbers located (or **graphed**) on it is shown in Figure 1.1.

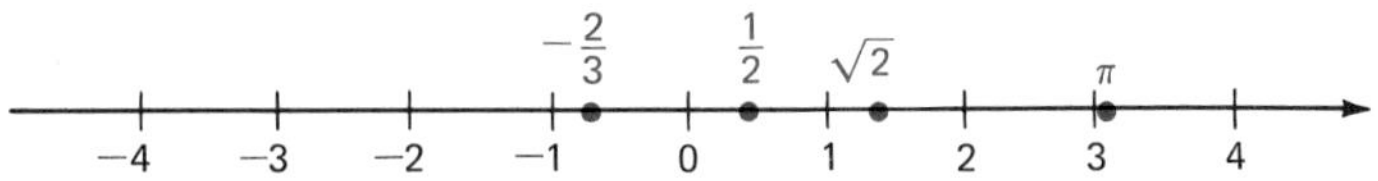

Figure 1.1

1. Draw a number line and graph the numbers -4, -1, 0, 1, 2.5, 13/4 on it.

Answer:

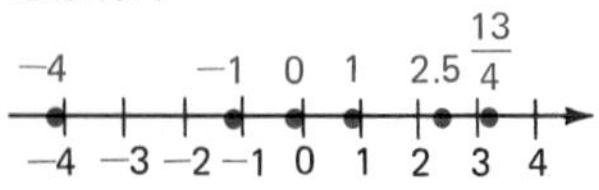

Work Problem 1 at the side.

Any number that can be associated with a point on the number line is called a **real number.*** All the numbers used in this book are real numbers. Often, an application will use only a particular type of real number. The names of the most common types of real numbers are as follows.

Natural (counting) numbers	$1, 2, 3, 4, \cdots$
Whole numbers	$0, 1, 2, 3, 4, \cdots$
Integers	$\cdots, -3, -2, -1, 0, 1, 2, 3, \cdots$
Rational numbers	All numbers of the form p/q, where p and q are integers, with $q \neq 0$
Irrational numbers	Real numbers that are not rational

The three dots in this list show that the numbers continue indefinitely in the same way. Examples of these types of numbers, as well as the relationships among them, are shown in Figure 1.2. Notice, for example, that the integers are also rational numbers and real numbers, but the integers are not irrational numbers.

One example of an irrational number is π, the ratio of the circumference of a circle to its diameter. The number π can be approximated by writing $\pi \approx 3.14159$ or $\pi \approx 22/7$ ($\approx$ means "is approximately equal to"), but there is no rational number that is exactly equal to π. Another irrational number can be found by constructing a triangle having a 90° angle, with the two shortest

* Not all numbers are real numbers. An example of a number that is not a real number is $\sqrt{-1}$.

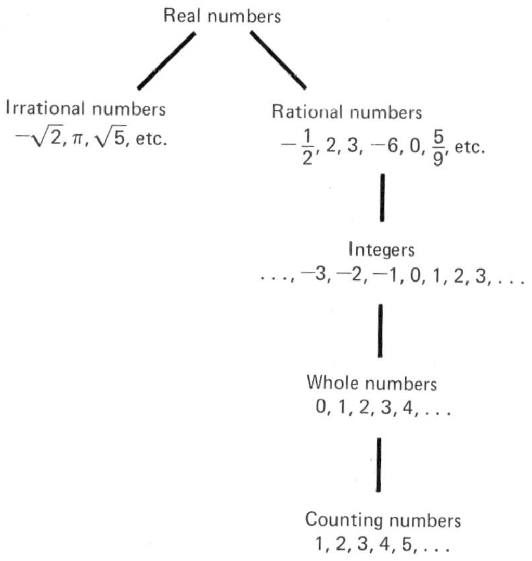

Figure 1.2

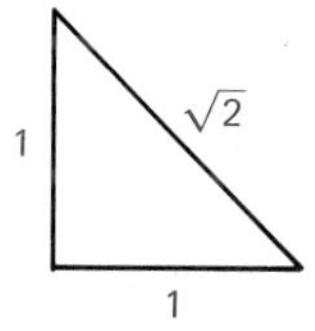

Figure 1.3

sides each 1 unit long, as shown in Figure 1.3. The third side can be shown to have a length which is irrational (the length is $\sqrt{2}$ units).

Many whole numbers have square roots which are irrational numbers; in fact, if a whole number is not the square of an integer, then its square root is irrational.

Example 1 Name all the types of numbers that apply to the following.

(a) 6

Consult Figure 1.2 to see that 6 is a counting number, whole number, integer, rational number, and real number.

(b) 3/4

This number is rational and real.

(c) $\sqrt{8}$

Since 8 is not the square of an integer, $\sqrt{8}$ is irrational and real. ■

2. Name all the types of numbers that apply to the following.
(a) -2
(b) $-5/8$
(c) π

Answer:
(a) Integer, rational, real
(b) Rational, real
(c) Irrational, real

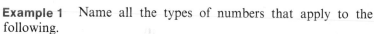

Work Problem 2 at the side.

It is often necessary to write that one number is greater than or less than another number. To do so, we can use the following symbols.

$<$	means *less than*	$\leq$	means *less than or equal to*
$>$	means *greater than*	$\geq$	means *greater than or equal to*

The following definitions tell us how to use the number line to decide which of two given numbers is either greater or less.

If a is to the left of b on a number line, then $a < b$.

If a is to the right of b on a number line, then $a > b$.

Example 2 Write *true* or *false* for each of the following.

(a) $8 < 12$

This statement says that 8 is less than 12, which we know is true.

(b) $-6 > -3$

The graph of Figure 1.4 shows both -6 and -3. Since -6 is to the *left* of -3, $-6 < -3$. Thus, the original statement is false.

Figure 1.4

(c) $-2 \leq -2$

Since $-2 = -2$, this statement is true. ■

3. Write *true* or *false* for the following.
(a) $-9 \leq -2$
(b) $8 > -3$
(c) $-14 \leq -20$

Answer:
(a) True
(b) True
(c) False

Work Problem 3 at the side.

We can use a number line to draw graphs of numbers, as shown in the next few examples.

Example 3 Graph all integers x such that $1 < x < 5$.

The only integers between 1 and 5 are 2, 3, and 4. These integers are graphed on the number line of Figure 1.5. ■

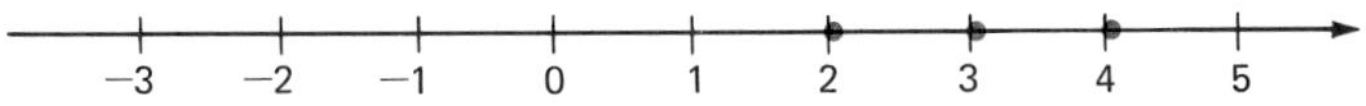

Figure 1.5

4. Graph all integers x such that
(a) $-3 < x < 5$
(b) $1 \leq x \leq 5$.

Answer:

(a) 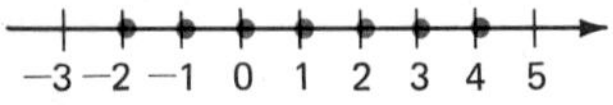

(b) 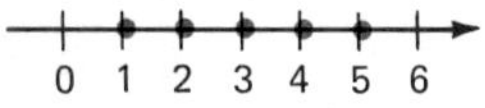

Work Problem 4 at the side.

Example 4 Graph all real numbers x such that $-2 < x < 3$.

Here we need all the real numbers between -2 and 3 and not just the integers. To graph these numbers, draw a heavy line from -2 to 3 on the number line, as in Figure 1.6. Open dots at -2 and 3 show that neither of these points belongs to the graph. ■

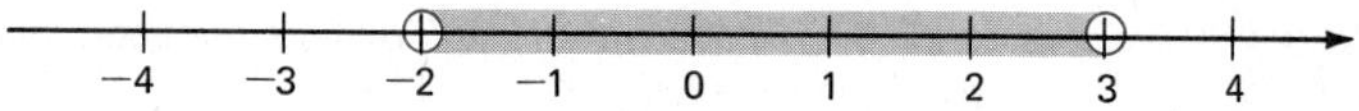

Figure 1.6

Work Problem 5 at the side.

Example 5 Graph all real numbers x such that $x \geq -2$.

Start at -2 and draw a heavy line to the right, as in Figure 1.7. Put a heavy dot at -2 to show that -2 itself is part of the graph. ■

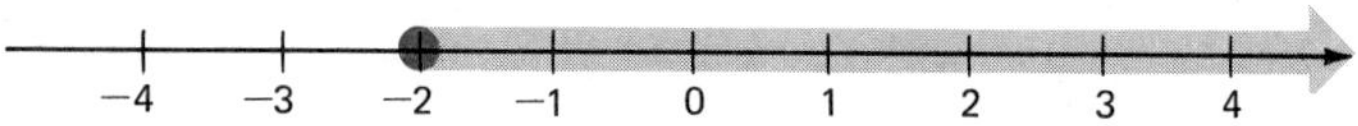

Figure 1.7

Work Problem 6 at the side.

Distance is always given as a nonnegative number. For example, the distance from 0 to -2 on a number line is 2, the same as the distance from 0 to 2. We say that 2 is the absolute value of both numbers, 2 and -2. In general, the **absolute value** of a number a is the distance on the number line from a to 0. We write the absolute value of a as $|a|$. We can state the definition of absolute value more formally as

$$|a| = a \qquad \text{if } a \geq 0$$

$$|a| = -a \qquad \text{if } a < 0.$$

The second part of this definition requires a little care. If a is a *negative* number, then $-a$ is a *positive* number. Thus, for any value of a, we have $|a| \geq 0$.

Example 6 Find each of the following.

(a) $|5|$

Since $5 > 0$, then $|5| = 5$.

(b) $|-5| = 5$

(c) $-|-5| = -(5) = -5$

(d) $|0| = 0$

(e) $|8 - 9|$

First simplify the expression inside the absolute value bars.

$$|8 - 9| = |-1| = 1$$

(f) $|-4 - 7| = |-11| = 11$ ■

5. Graph all real numbers x such that

(a) $-5 < x < 1$

(b) $4 < x < 7$.

Answer:

(a)

(b)

6. Graph all real numbers x such that

(a) $x \geq 4$

(b) $-2 \leq x \leq 1$.

Answer:

(a)

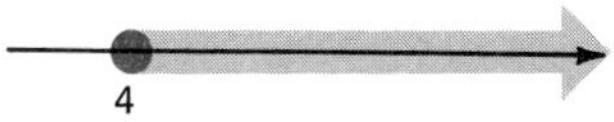

(b)

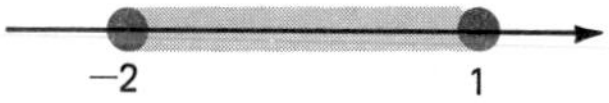

7. Find the following.
(a) $|-6|$
(b) $-|7|$
(c) $-|-2|$
(d) $|-3-4|$
(e) $|2-7|$

Answer:
(a) 6
(b) -7
(c) -2
(d) 7
(e) 5

Work Problem 7 at the side.

1.1 EXERCISES

Name all the types of numbers that apply to the following. (See Example 1.)

1. 6 **2.** -9 **3.** -7 **4.** 0

5. $1/2$ **6.** $-5/11$ **7.** $\sqrt{7}$ **8.** $-\sqrt{11}$

9. π **10.** $1/\pi$

Label each of the following as true *or* false.

11. Every integer is a rational number.

12. Every integer is a whole number.

13. Every whole number is an integer.

14. Some whole numbers are not natural numbers.

15. There is a natural number that is not a whole number.

16. Every rational number is a natural number.

17. Every natural number is a rational number.

18. No whole numbers are rational.

Graph each of the following on a number line. (See Example 3.)

19. All integers x such that $-5 < x < 5$

20. All integers x such that $-4 < x < 2$

21. All whole numbers x such that $x \leq 3$

22. All whole numbers x such that $1 \leq x \leq 8$

23. All natural numbers x such that $-1 < x < 5$

24. All natural numbers x such that $x \leq 2$

Graph all real numbers x satisfying the following conditions. (See Examples 4 and 5.)

25. $x \leq 4$ **26.** $x > 5$

27. $6 \leq x$ **28.** $x \geq 8$

29. $x > -2$ **30.** $-4 < x$

31. $-5 < x < -3$ **32.** $3 < x < 5$

33. $-3 \leq x \leq 6$ **34.** $8 \leq x \leq 14$

35. $1 < x \leq 6$ **36.** $-4 \leq x < 3$

Evaluate each of the following. (See Example 6.)

37. $|-8|$

38. $|9|$

39. $-|-4|$

40. $-|-2|$

41. $|6-4|$

42. $|3-17|$

43. $-|12+(-8)|$

44. $|-6+(-15)|$

45. $|8-(-9)|$

46. $|-3-(-2)|$

47. $|8|-|-4|$

48. $|-9|-|-12|$

49. $-|-4|-|-1-14|$

50. $-|6|-|-12-4|$

In each of the following problems, fill in the blank with either $=$, $<$, *or* $>$.

51. $|5|$ ____ $|-5|$

52. $|3|$ ____ $|-3|$

53. $-|7|$ ____ $|7|$

54. $-|-4|$ ____ $|4|$

55. $|10-3|$ ____ $|3-10|$

56. $|6-(-4)|$ ____ $|-4-6|$

57. $|1-4|$ ____ $|4-1|$

58. $|10-8|$ ____ $|8-10|$

59. $|-2+8|$ ____ $|2-8|$

60. $|3+1|$ ____ $|-3-1|$

61. $|3|\cdot|-5|$ ____ $|3(-5)|$

62. $|3|\cdot|2|$ ____ $|3(2)|$

63. $|3-2|$ ____ $|3|-|2|$

64. $|5-1|$ ____ $|5|-|1|$

65. In general, if a and b are any real numbers having the same sign (both negative or both positive), is it always true that

$$|a+b|=|a|+|b|?$$

66. If a and b are *any* real numbers, is it always true that

$$|a+b|=|a|+|b|?$$

APPLIED PROBLEMS

Management

Use inequality symbols to rewrite each of the following statements which are based on a recent article in Business Week *magazine. Let x represent the unknown in each exercise.*

Example To rewrite "No more than 40% of the employees of Delmar Industries are men," we let x represent the percent of men. Then $x \le 40$. ■

67. Shaklee Products pays its distributors at least 35% of the money received from the sale of its products.

68. Over the last two years, company sales have varied between 102 million dollars and 179 million dollars.

69. The best 5300 of the Shaklee distributors make at least \$12,600 per year. (Hint: let x represent the annual income of the best 5300 distributors.)

70. No distributor made more than \$346,000 last year.

71. Manufacturing costs for Shaklee Products are at least 15% of the retail price.

72. In the last few years the percent of the firm's sales in California has varied between 18% and 57% of total sales.

Social Science

Sociologists measure the status of an individual within a society by evaluating for that individual the number x, which gives the percentage of the population with less income than the given person, and the number y, the percentage of the population with less education. The average status *is defined as* $(x + y)/2$, *while the individual's* status incongruity *is defined by* $|(x - y)/2|$. *People with high status incongruities would include unemployed Ph.D's (low x, high y) and millionaires who didn't make it past the second grade (high x, low y).*

73. What is the highest possible average status for an individual? The lowest?

74. What is the highest possible status incongruity for an individual? The lowest?

75. Jolene Rizzo makes more money than 56% of the population and has more education than 78%. Find her average status and status incongruity.

76. A popular movie star makes more money than 97% of the population and is better educated than 12%. Find the average status and status incongruity for this individual.

1.2 LINEAR EQUATIONS AND INEQUALITIES

One of the main uses of algebra is to solve equations. An **equation** states that two expressions are equal. Examples of equations include $x + 6 = 9$, $4y + 8 = 12$, and $9z = -36$. The letter in each equation, the unknown, is called the **variable.**

A **solution** of an equation is a number which, when substituted for the variable in the equation, produces a true statement. For example, if the number 9 is substituted for x in the equation $2x + 1 = 19$, we get

$$
\begin{aligned}
2x + 1 &= 19 \\
2(9) + 1 &= 19 \qquad \text{Let } x = 9 \\
18 + 1 &= 19,
\end{aligned}
$$

a true statement. Thus, 9 is a solution of the equation $2x + 1 = 19$.

1. Is -4 a solution of the following equations?
(a) $3x + 5 = -7$
(b) $2x - 3 = 5$

Answer:
(a) Yes
(b) No

Work Problem 1 at the side.

Equations which can be written in the form $ax + b = c$, where a, b, and c are real numbers, with $a \neq 0$, are called **linear equations.** Examples of linear equations include $5y + 9 = 16$, $8x = 4$, and $-3p + 5 = -8$. Equations that are *not* linear include absolute value equations, such as $|x| = 4$.

The following properties are used to help solve linear equations. For any real numbers a, b, and c, the following statements are true.

$a(b + c) = ab + ac$	**Distributive property**
If $a = b$, then $a + c = b + c$. (The same number may be added to both sides of an equation.)	**Addition property of equality**
If $a = b$, then $ac = bc$. (The same number may be multiplied on both sides of an equation.)	**Multiplication property of equality**

Example 1 Solve the linear equation $5x - 3 = 12$.

First, use the addition property of equality and add 3 to both sides. We do this to isolate the term containing the variable on one side of the equals sign.

$$5x - 3 = 12$$
$$5x - 3 + 3 = 12 + 3 \qquad \text{Add 3 to both sides}$$
$$5x = 15$$

On the left we have $5x$, but we want just $1x$, or x. To get $1x$, use the fact that $(1/5) \cdot 5 = 1$, and multiply both sides of the equation by $1/5$.

$$5x = 15$$
$$\frac{1}{5}(5x) = \frac{1}{5}(15) \qquad \text{Multiply both sides by } \frac{1}{5}$$
$$1x = 3$$
$$x = 3$$

The solution of the original equation, $5x - 3 = 12$, is thus 3, which can be checked by substituting 3 for x in the original equation. ■

2. Solve the following.
(a) $3p - 5 = 19$
(b) $4y + 3 = -5$
(c) $-2k + 6 = 2$

Answer:
(a) 8
(b) -2
(c) 2

Work Problem 2 at the side.

Example 2 Solve $2k + 3(k - 4) = 2(k - 3)$.

We must first simplify this equation using the distributive property. By this property, $3(k - 4)$ is $3k - 3 \cdot 4$, or $3k - 12$. Also, $2(k - 3)$ is $2k - 2 \cdot 3$, or $2k - 6$. The equation can now be written as

$$2k + 3(k - 4) = 2(k - 3)$$
$$2k + 3k - 12 = 2k - 6.$$

On the left, $2k + 3k = (2 + 3)k = 5k$, again by the distributive property. We have

$$5k - 12 = 2k - 6.$$

One way to proceed is to add $-2k$ to both sides.

$$5k - 12 + (-2k) = 2k - 6 + (-2k)$$
$$3k - 12 = -6$$

Now add 12 to both sides.

$$3k - 12 + 12 = -6 + 12$$
$$3k = 6$$

Finally, multiply both sides by 1/3.

$$\frac{1}{3}(3k) = \frac{1}{3}(6)$$
$$k = 2$$

The solution is 2. ■

3. Solve the following.
(a) $3(m - 6) + 2(m + 4) = 4m - 2$
(b) $-2(y + 3) + 4y = 3(y + 1) - 6$

Answer:
(a) 8
(b) -3

Work Problem 3 at the side.

We can also solve equations involving fractions, as shown in the following example.

Example 3 Solve $\frac{r}{10} - \frac{2}{15} = \frac{3r}{20} - \frac{1}{5}$.

To solve this equation, first eliminate all denominators by multiplying both sides of the equation by a **common denominator,** a number which can be divided (with no remainder) by each denominator in the equation. Here the denominators are 10, 15, 20, and 5. Each of these numbers can be divided into 60, so that 60 is our common denominator. Multiply both sides of the equation by 60 and use the distributive property.

$$\frac{r}{10} - \frac{2}{15} = \frac{3r}{20} - \frac{1}{5}$$

$$60\left(\frac{r}{10}-\frac{2}{15}\right)=60\left(\frac{3r}{20}-\frac{1}{5}\right)$$

$$60\left(\frac{r}{10}\right)-60\left(\frac{2}{15}\right)=60\left(\frac{3r}{20}\right)-60\left(\frac{1}{5}\right)$$

$$6r-8=9r-12$$

Add $-6r$ and 12 to both sides.

$$6r-8+(-6r)+12=9r-12+(-6r)+12$$

$$4=3r$$

Multiply both sides by 1/3, ending up with

$$r=\frac{4}{3}. \quad ■$$

4. Solve the following.

(a) $\frac{x}{2}-\frac{x}{4}=6$

(b) $\frac{2x}{3}+\frac{1}{2}=\frac{x}{4}-\frac{9}{2}$

Answer:
(a) 24
(b) -12

Work Problem 4 at the side.

Example 4 Solve $\frac{4}{3(k+2)}-\frac{k}{3(k+2)}=\frac{5}{3}$.

Multiply both sides of the equation by the common denominator $3(k+2)$.

$$3(k+2)\cdot\frac{4}{3(k+2)}-3(k+2)\cdot\frac{k}{3(k+2)}=3(k+2)\cdot\frac{5}{3}$$

$4-k=5(k+2)$	
$4-k=5k+10$	Distributive property
$4-k+k=5k+10+k$	Add k on both sides
$4=6k+10$	
$4+(-10)=6k+10+(-10)$	Add -10 on both sides
$-6=6k$	
$-1=k$	Multiply by 1/6

The solution is -1. ■

5. Solve the equation

$$\frac{5p+1}{3(p+1)}=\frac{3p-3}{3(p+1)}+\frac{9p-3}{3(p+1)}.$$

Answer: 1

Work Problem 5 at the side.

An equation states that two expressions are equal; an **inequality** states that they are unequal. Examples of inequalities include $3x+1<6$, $5y-8\geq 4$, and $-3x>6$. We can solve inequalities using properties very similar to those used for solving equations. In the following properties, a, b, and c represent real numbers.*

* The following properties can also be stated for $>$, $\geq$, or $\leq$.

If $a < b$, then $a + c < b + c$	**Addition property of inequality**
If $a < b$, and if $c > 0$, then $a\,c < b\,c$. (Both sides of an inequality may be multiplied by the same *positive* number without changing the direction of the inequality symbol.	**Multiplication property of inequality, $c > 0$**
If $a < b$, and if $c < 0$, then $ac > bc$. (Both sides of an inequality may be multiplied by a *negative* number, *if the direction of the inequality symbol is reversed.*)	**Multiplication property of inequality, $c < 0$**

Be careful with the second form of the multiplication property of inequality; if you multiply both sides of an inequality by a negative number, then the direction of the inequality must be reversed. For example, to solve $-3x < 12$, we would multiply both sides by $-1/3$, getting

$$-\frac{1}{3}(-3x) > -\frac{1}{3}(12) \qquad \text{Here we changed } < \text{ to } >.$$

$$x > -4.$$

Example 5 Solve $3x + 5 \geq 11$.

First, add -5 to both sides.

$$3x + 5 + (-5) \geq 11 + (-5)$$

$$3x \geq 6$$

Now multiply both sides by $1/3$.

$$\frac{1}{3}(3x) \geq \frac{1}{3}(6)$$

$$x \geq 2$$

Why did we *not* change the direction of the inequality symbol? ■

Work Problem 6 at the side.

Example 6 Solve $4 - 3y \leq 7 + 2y$.

Add -4 to both sides.

6. Solve the following.
(a) $5x - 11 < 14$
(b) $-3x \leq -12$
(c) $-8x \geq 32$

Answer:
(a) $x < 5$
(b) $x \geq 4$
(c) $x \leq -4$

$$4 - 3y + (-4) \le 7 + 2y + (-4)$$
$$-3y \le 3 + 2y$$

Add $-2y$ to both sides. Remember that *adding* to both sides never changes the direction of the inequality symbol.

$$-3y + (-2y) \le 3 + 2y + (-2y)$$
$$-5y \le 3$$

Multiply both sides by $-1/5$. Since $-1/5$ is negative, we must change the direction of the inequality symbol.

$$-\frac{1}{5}(-5y) \ge -\frac{1}{5}(3)$$
$$y \ge -\frac{3}{5} \quad ■$$

7. Solve the following.
(a) $8 - 6t \ge 2t + 24$
(b) $-4r + 3(r + 1) < 2r$

Answer:
(a) $t \le -2$
(b) $r > 1$

Work Problem 7 at the side.

1.2 EXERCISES

Solve each of the following equations. (See Examples 1–4.)

1. $4x - 1 = 11$

2. $3x + 5 = 23$

3. $-2k + 1 = 19$

4. $-5y - 6 = 14$

5. $4p - 11 + 3p = 2p - 1$

6. $8y - 5y + 4 = 2y - 9$

7. $9x - 2(x - 6) = 10x + 3$

8. $5r - 6(r + 4) = -5r - 4$

9. $2(k - 5) + 4k - 6 = 2k$

10. $2(z - 4) = 7z + 2 - 2z$

11. $4(1 - k) - 2(k + 3) = -6$

12. $3(2r + 1) - 2(r - 2) = 5$

13. $\frac{2f}{5} - \frac{f - 3}{5} = 2$

14. $\frac{3w}{4} - \frac{2w}{3} = \frac{w - 6}{3}$

15. $\frac{y}{3} + \frac{y}{4} = 10 + y$

16. $\frac{p}{5} - \frac{p}{6} = 2$

17. $\frac{m + 1}{8} - \frac{m - 2}{3} = 1$

18. $\frac{y - 6}{3} - \frac{y + 2}{5} = 2$

19. $\frac{2}{r} + \frac{3}{2r} = \frac{7}{6}$

20. $\frac{2}{q} - \frac{3}{4q} = -\frac{5}{12}$

21. $\frac{1}{m - 1} + \frac{2}{3(m - 1)} = -\frac{5}{12}$

22. $\frac{4}{a + 2} - \frac{1}{3(a + 2)} = \frac{11}{9}$

23. $\frac{3}{4(z - 2)} - \frac{2}{3(z - 2)} = \frac{1}{36}$

24. $\frac{2}{3x + 7} - \frac{5}{2(3x + 7)} = -\frac{1}{56}$

Solve each of the following inequalities. (See Examples 5 and 6.)

25. $6x \leq -18$

26. $4m > -32$

27. $-3p < 18$

28. $-5z \leq 40$

29. $-9a < 0$

30. $-4k \geq 0$

31. $2x + 1 \leq 9$

32. $3y - 2 < 10$

33. $-3p - 2 \geq 1$

34. $-5t + 3 \geq -2$

35. $6k - 4 < 3k - 1$

36. $2a - 2 > 4a + 2$

37. $m - (4 + 2m) + 3 < 2m + 2$

38. $2p - (3 - p) \leq -7p - 2$

39. $-2(3y - 8) \geq 5(4y - 2)$

40. $5r - (r + 2) \geq 3(r - 1) + 5$

41. $3p - 1 < 6p + 2(p - 1)$

42. $x + 5(x + 1) > 4(2 - x) + x$

APPLIED PROBLEMS

Natural Science

In the metric system of weights and measures, temperature is measured in degrees Celsius (°C) instead of degrees Fahrenheit (°F). To convert back and forth between the two systems, use the equations

$$\mathrm{C} = \frac{5(\mathrm{F} - 32)}{9} \quad \textit{and} \quad \mathrm{F} = \frac{9}{5}\mathrm{C} + 32.$$

In each of the following exercises, convert to the other system. Round answers to the nearest tenth of a degree if necessary.

43. 20°C

44. 100°C

45. 59°F

46. 86°F

47. 100°F

48. 350°F

49. 40°C

50. 85°C

Management

When a consumer borrows money, the lender must tell the consumer the true annual interest rate of the loan. The method of finding the exact true annual interest rate is given in Section 4.5. However, a quick approximate *rate can be found by using the equation*

$$A = \frac{2pf}{b(q + 1)},$$

where p is the number of payments made in one year, f is the finance charge, b is the balance owed on the loan, and q is the total number of payments. Find the value of the variables not given in each of the following. Round A to the nearest percent and round other variables to the nearest whole numbers.

51. $p = 12, \quad f = \$800, \quad b = \$4000, \quad q = 36;$ find A

52. $p = 12, \quad f = \$60, \quad b = \$740, \quad q = 12;$ find A

53. $A = 14\%$ (or .14), $p = 12, \quad b = \$2000, \quad q = 36;$ find f

54. $A = 11\%, \quad p = 12, \quad b = \$1500, \quad q = 24;$ find f

55. $A = 16\%, \quad p = 12, \quad f = \$370, \quad q = 36;$ find b

56. $A = 10\%, \quad p = 12, \quad f = \$490, \quad q = 48;$ find b

When a loan is paid off early, a portion of the finance charge must be returned to the borrower. By one method of calculating finance charge (called the rule of 78*), the amount of unearned interest (finance charge to be returned) is given by*

$$u = f \cdot \frac{n(n+1)}{q(q+1)}$$

where u represents unearned interest, f is the original finance charge, n is the number of payments remaining when the loan is paid off, and q is the original number of payments. Find the amount of the unearned interest in each of the following.

57. Original finance charge = \$800, loan scheduled to run 36 months, paid off with 18 payments remaining.

58. Original finance charge = \$1400, loan scheduled to run 48 months, paid off with 12 payments remaining.

59. Original finance charge = \$950, loan scheduled to run 24 months, paid off with 6 payments remaining.

60. Original finance charge = \$175, loan scheduled to run 12 months, paid off with 3 payments remaining.

According to a recent article in The Wall Street Journal, *the company that makes the Godzilla movies puts out about \$1.2 million to make each movie. It then receives back about \$5 million from worldwide showings of the film.*

61. Let x represent the number of Godzilla movies that the company makes. Write an equation for C, the cost in millions to make x films.

62. Write an equation for the total income, I, in millions from x films.

63. Use the formula Profit = Income − Cost to write an equation for the profit in millions from x films.

64. Find the company's total profit in millions if 5 films are made.

General

Bill and Cheryl Bradkin went to Portland, Maine, for a week. They needed to rent a car, so they checked out two rental firms. Avis wanted \$14 per day, with no mileage fee. Downtown Toyota wanted \$54 per week and 7¢ per mile.

65. Let x represent the number of miles that the Bradkins would drive in one week. Set up an inequality expressing the rates of the two firms. Then decide how many miles they would have to drive before the Avis car was the better deal.

1.3 POLYNOMIALS

A **polynomial,** one of the key ideas of algebra, is an expression of the form

$$a_n x^n + a_{n-1} x^{n-1} + \cdots + a_1 x + a_0,$$

where $a_0, a_1, a_2, \ldots, a_n$ are real numbers, n is a natural number, and $a_n \neq 0$. Examples of polynomials include

$$5x^4 + 2x^3 + 6x, \quad 8m^3 + 9m^2 - 6m + 3, \quad 10p, \quad \text{and} \quad -9.$$

Expressions that are *not* polynomials include

$$8x^3 + \frac{6}{x}, \quad \frac{9 + x}{2 - x} \quad \text{and} \quad \frac{-p^2 + 5p + 3}{2p - 1}.$$

When we write $9p^4$, the entire expression is called a **term,** the number 9 is called the **coefficient,** p is the **variable,** and 4 is the **exponent.** The expression p^4 means $p \cdot p \cdot p \cdot p$, while p^2 means $p \cdot p$. We discuss exponents in more detail later in this chapter.

Polynomials can be added or subtracted by the distributive property. For example,

$$12y^4 + 6y^4 = (12 + 6)y^4 = 18y^4$$

and

$$-2m^2 + 8m^2 = (-2 + 8)m^2 = 6m^2.$$

Note that the polynomial $8y^4 + 2y^5$ cannot be further simplified. Two terms having the same variable and the same exponent are called **like terms;** other terms are called **unlike terms.** Only like terms may be added or subtracted. To subtract polynomials, we need to know that $-(a + b) = -a - b$. The next example shows how to add and subtract polynomials.

Example 1 Add or subtract as indicated.

(a) $(8x^3 - 4x^2 + 6x) + (3x^3 + 5x^2 - 9x + 8)$

Combine like terms.

$$\begin{aligned}(8x^3 - 4x^2 + 6x) &+ (3x^3 + 5x^2 - 9x + 8) \\ &= (8x^3 + 3x^3) + (-4x^2 + 5x^2) + (6x - 9x) + 8 \\ &= 11x^3 + x^2 - 3x + 8\end{aligned}$$

(b) $(-4x^4 + 6x^3 - 9x^2 - 12) + (-3x^3 + 8x^2 - 11x + 7)$

$$= -4x^4 + 3x^3 - x^2 - 11x - 5$$

(c) $(2x^2 - 11x + 8) - (7x^2 - 6x + 2)$

$$= (2x^2 - 11x + 8) + (-7x^2 + 6x - 2)$$
$$= -5x^2 - 5x + 6 \quad ■$$

Work Problem 1 at the side.

1. Add or subtract.
(a) $(-2x^2 + 7x + 9) + (3x^2 + 2x - 7)$
(b) $(4x + 6) - (13x - 9)$
(c) $(9x^3 - 8x^2 + 2x) - (9x^3 - 2x^2 - 10)$

Answer:
(a) $x^2 + 9x + 2$
(b) $-9x + 15$
(c) $-6x^2 + 2x + 10$

Polynomials may be multiplied using the distributive property. For example, the product of $8x$ and $6x - 4$ is found as follows.

$8x(6x - 4) = 8x(6x) - 8x(4)$	Distributive property
$= 48x^2 - 32x$	$x \cdot x = x^2$

The product of $3p - 2$ and $5p + 1$ can be found by using the distributive property twice:

$$\begin{aligned}(3p-2)(5p+1) &= (3p-2)(5p) + (3p-2)(1)\\ &= 3p(5p) - 2(5p) + 3p(1) - 2(1)\\ &= 15p^2 - 10p + 3p - 2\\ &= 15p^2 - 7p - 2\end{aligned}$$

Work Problem 2 at the side.

2. Find the following products.
(a) $-6r(2r - 5)$
(b) $11m(8m + 3)$
(c) $(5k - 1)(2k + 3)$
(d) $(7z - 3)(2z + 5)$

Answer:
(a) $-12r^2 + 30r$
(b) $88m^2 + 33m$
(c) $10k^2 + 13k - 3$
(d) $14z^2 + 29z - 15$

Multiplication of polynomials can be performed using the distributive property. The reverse process, where we write a polynomial as a product of other polynomials, is called **factoring.** For example, one way to factor the number 18 is to write it as the product $9 \cdot 2$. When 18 is written as $9 \cdot 2$, both 9 and 2 are called **factors** of 18. It is true that $18 = 36 \cdot \frac{1}{2}$, but we do not call 36 and $\frac{1}{2}$ factor of 18; we restrict our attention only to integer factors. Integer factors of 18 are 2, 9; $-2, -9$; 6, 3; $-6, -3$; 18, 1; $-18, -1$.

We will usually be concerned with factoring an algebraic expression such as $15m + 45$. This expression is made up of two terms, $15m$ and 45. Each of these terms can be divided by 15. In fact, $15m = 15 \cdot m$ and $45 = 15 \cdot 3$. We can use the distributive property to write

$$15m + 45 = 15 \cdot m + 15 \cdot 3 = 15(m + 3).$$

Both 15 and $m + 3$ are factors of $15m + 45$. Since 15 divides into all terms of $15m + 45$ (and is the largest number that will do so), it is called the **greatest common factor** for the polynomial $15m + 45$. The process of writing $15m + 45$ as $15(m + 3)$ is called **factoring out** the greatest common factor.

Example 2 Factor out the greatest common factor.

(a) $12p - 18q$

Both $12p$ and $18q$ are divisible by 6. Therefore,

$$12p - 18q = 6 \cdot 2p - 6 \cdot 3q = 6(2p - 3q).$$

(b) $8x^3 - 9x^2 + 15x$

Each of these terms is divisible by x.

$$\begin{aligned}8x^3 - 9x^2 + 15x &= (8x^2) \cdot x - (9x) \cdot x + 15 \cdot x\\ &= x(8x^2 - 9x + 15).\end{aligned}$$ ■

Work Problem 3 at the side.

3. Factor out the greatest common factor.
(a) $12r + 9k$
(b) $75m^2 + 100n^2$
(c) $6m^4 - 9m^3 + 12m^2$

Answer:
(a) $3(4r + 3k)$
(b) $25(3m^2 + 4n^2)$
(c) $3m^2(2m^2 - 3m + 4)$

A polynomial may not have a greatest common factor (other than 1), and yet may still be factorable. For example, the polynomial $x^2 + 5x + 6$ can be factored as $(x + 2)(x + 3)$. To see that this is correct, work out the product $(x + 2)(x + 3)$; you should get $x^2 + 5x + 6$.

If we are given a polynomial such as $x^2 + 5x + 6$, how do we know that it is the product $(x + 2)(x + 3)$? There are two different ways to factor a polynomial of three terms such as $x^2 + 5x + 6$, depending on whether the coefficient of x^2 is 1, or a number other than 1. If the coefficient is 1, proceed as shown in the following example.

Example 3 Factor $y^2 + 8y + 15$.

Since the coefficient of y^2 is understood to be 1, factor by finding two numbers whose *product* is 15, and whose *sum* is 8. Use trial and error to find these numbers. Begin by listing all pairs of integers having a product of 15. As you do this, also form the sum of the numbers.

Products	Sums
$15 \cdot 1 = 15$	$15 + 1 = 16$
$5 \cdot 3 = 15$	$5 + 3 = 8$
$(-1) \cdot (-15) = 15$	$-1 + (-15) = -16$
$(-5) \cdot (-3) = 15$	$-5 + (-3) = -8$

The numbers 3 and 5 have a product of 15 and a sum of 8. Thus, $y^2 + 8y + 15$ factors as

$$y^2 + 8y + 15 = (y + 3)(y + 5).$$

We can also write the answer as $(y + 5)(y + 3)$. ■

Work Problem 4 at the side.

4. Factor the following.
(a) $m^2 + 11m + 30$
(b) $h^2 + 10h + 9$

Answer:
(a) $(m + 5)(m + 6)$
(b) $(h + 9)(h + 1)$

Example 4 Factor $p^2 - 8p - 20$.

We need two numbers whose product is -20 and whose sum is -8. Make a list of all pairs of integers whose product is -20. Choose from this list that pair whose sum is -8. Doing so, you should find the pair -10 and 2; the product of these numbers is -20 and their sum is -8. Therefore,

$$p^2 - 8p - 20 = (p - 10)(p + 2). \quad ■$$

Work Problem 5 at the side.

5. Factor the following.
(a) $a^2 - 7a + 10$
(b) $r^2 - 5r - 14$
(c) $m^2 + 3m - 40$

Answer:
(a) $(a - 5)(a - 2)$
(b) $(r - 7)(r + 2)$
(c) $(m + 8)(m - 5)$

If the coefficient of the squared term is *not* 1, we must use trial and error, as shown in the next example.

Example 5 Factor $2x^2 + 9x - 5$.

The factors of $2x^2$ are $2x$ and x; the possible factors of -5 are -5 and 1, or 5 and -1. We use trial and error, trying various combinations of these factors until we find the one that works (if, indeed, any work). Let's try the product $(2x + 5)(x - 1)$. Multiply it out.

$$\begin{aligned}(2x + 5)(x - 1) &= (2x + 5)(x) + (2x + 5)(-1) \\ &= 2x^2 + 5x - 2x - 5 \\ &= 2x^2 + 3x - 5\end{aligned}$$

This product is not the one we want. So we try another combination

$$\begin{aligned}(2x - 1)(x + 5) &= (2x - 1)(x) + (2x - 1)(5) \\ &= 2x^2 - x + 10x - 5 \\ &= 2x^2 + 9x - 5\end{aligned}$$

This combination led to the correct polynomial; thus

$$2x^2 + 9x - 5 = (2x - 1)(x + 5). \quad ■$$

Work Problem 6 at the side.

6. Factor the following.
(a) $3k^2 + k - 2$
(b) $3m^2 + 5m - 2$
(c) $6p^2 + 13p - 5$

Answer:
(a) $(3k - 2)(k + 1)$
(b) $(3m - 1)(m + 2)$
(c) $(2p + 5)(3p - 1)$

There are two special types of factorizations that occur so often that we list them for future reference.

$$x^2 - y^2 = (x + y)(x - y) \qquad \textbf{Difference of two squares}$$

$$x^2 + 2xy + y^2 = (x + y)^2 \qquad \textbf{Perfect square}$$

Example 6 Factor each of the following.

(a) $x^2 - 25 = x^2 - 5^2 = (x + 5)(x - 5)$
(b) $64p^2 - 49q^2 = (8p)^2 - (7q)^2 = (8p + 7q)(8p - 7q)$
(c) $x^2 + 36$ cannot be factored
(d) $x^2 + 12x + 36 = (x + 6)^2$
(e) $9y^2 - 24yz + 16z^2 = (3y - 4z)^2$ ■

Work Problem 7 at the side.

7. Factor the following.
(a) $r^2 - 81$
(b) $9p^2 - 49$
(c) $y^2 + 100$
(d) $m^2 - 8m + 16$
(e) $100k^2 - 60k + 9$

Answer:
(a) $(r + 9)(r - 9)$
(b) $(3p + 7)(3p - 7)$
(c) cannot be factored
(d) $(m - 4)^2$
(e) $(10k - 3)^2$

Finally, we list two more special types of factorizations that occur from time to time.

$$x^3 - y^3 = (x - y)(x^2 + xy + y^2) \qquad \textbf{Difference of two cubes}$$

$$x^3 + y^3 = (x + y)(x^2 - xy + y^2) \qquad \textbf{Sum of two cubes}$$

Example 7 Factor each of the following.

(a) $y^3 - 8 = y^3 - 2^3 = (y - 2)(y^2 + 2y + 4)$

(b) $m^3 + 125 = m^3 + 5^3 = (m + 5)(m^2 - 5m + 25)$

(c) $8k^3 - 27z^3 = (2k)^3 - (3z)^3 = (2k - 3z)(4k^2 + 6kz + 9z^2)$ ■

Work Problem 8 at the side.

8. Factor the following.

(a) $a^3 + 1000$

(b) $z^3 - 64$

(c) $1000m^3 - 27z^3$

Answer:

(a) $(a + 10)(a^2 - 10a + 100)$

(b) $(z - 4)(z^2 + 4z + 16)$

(c) $(10m-3z)(100m^2+30mz+9z^2)$

1.3 EXERCISES

Add or subtract as indicated. (See Example 1.)

1. $(8m + 9) + (6m - 3)$

2. $(-7p - 11) + (8p + 5)$

3. $(-2k - 3) - (7k - 8)$

4. $(12z + 10) - (3z + 9)$

5. $(2x^2 - 6x + 11) + (-3x^2 + 7x - 2)$

6. $(-3a^2 + 2a - 5) + (7a^2 + 2a + 9)$

7. $(-4y^2 - 3y + 8) - (2y^2 - 6y - 2)$

8. $(7b^2 + 2b - 5) - (3b^2 + 2b - 6)$

9. $(2x^3 - 2x^2 + 4x - 3) - (2x^3 + 8x^2 - 1)$

10. $(3y^3 + 9y^2 - 11y + 8) - (-4y^2 + 10y - 6)$

Find each of the following products.

11. $3p(2p - 5)$

12. $4y(8y + 1)$

13. $-9m(2m^2 + 3m - 1)$

14. $2a(4a^2 - 6a + 3)$

15. $(6k - 1)(2k - 3)$

16. $(8r + 3)(r - 1)$

17. $(3y + 5)(2y - 1)$

18. $(2a - 5)(4a + 3)$

19. $(6y - 2)(6y + 2)$

20. $(8p + 3)(8p - 3)$

Factor out the greatest common factor in each of the following. (See Example 2.)

21. $25k + 30$

22. $6m - 12$

23. $4z + 4$

24. $9y - 9$

25. $8x + 6y + 4z$

26. $15p + 9q + 12s$

27. $6r^2 + 4r + 8$

28. $15z^2 - 10z + 25$

29. $m^3 - 9m^2 + 6m$

30. $y^3 + 6y^2 + 8y$

31. $8a^3 - 16a^2 + 24a$

32. $3y^3 + 24y^2 + 9y$

33. $25p^4 - 20p^3 + 100p^2$

34. $60m^4 - 120m^3 + 50m^2$

Factor each of the following. If a polynomial cannot be factored, write "cannot be factored." (See Examples 3 and 4.)

35. $m^2 + 9m + 14$

36. $p^2 - 2p - 15$

37. $x^2 + 4x - 5$

38. $y^2 + y - 72$

39. $z^2 + 9z + 20$

40. $k^2 + 8k + 15$

41. $b^2 - 8b + 7$

42. $r^2 + r - 20$

43. $a^2 + 4a + 5$

44. $y^2 - 6y + 8$

45. $s^2 + 2s - 35$

46. $n^2 - 12n - 35$

47. $y^2 - 4y - 21$

48. $r^2 + r - 42$

Factor each of the following. If a polynomial cannot be factored, write "cannot be factored." Factor out the greatest common factor as necessary. (See Example 5.)

49. $6a^2 - 48a - 120$

50. $8h^2 - 24h - 320$

51. $3m^3 + 12m^2 + 9m$

52. $3y^4 - 18y^3 + 15y^2$

53. $2x^2 - 5x - 3$

54. $3r^2 - r - 2$

55. $3a^2 + 10a + 7$

56. $4y^2 + y - 3$

57. $2a^2 - 17a + 30$

58. $3k^2 + 2k - 8$

59. $15y^2 + y - 2$

60. $6x^2 + x - 1$

61. $3p^2 - 7p + 10$

62. $8r^2 + r + 6$

63. $5a^2 - 7a - 6$

64. $12s^2 + 11s - 5$

65. $21m^2 + 13m + 2$

66. $20y^2 + 39y - 11$

67. $24a^4 + 10a^3 - 4a^2$

68. $18x^5 + 15x^4 - 75x^3$

69. $32z^5 - 20z^4 - 12z^3$

70. $15x^4 - 7x^3 - 4x^2$

Factor each of the following. (See Examples 6 and 7.)

71. $x^2 - 64$

72. $y^2 - 144$

73. $9m^2 - 25$

74. $4p^2 - 9$

75. $121a^2 - 100$

76. $144m^2 - 169$

77. $9x^2 + 64$

78. $100a^2 + 9$

79. $z^2 + 14z + 49$

80. $y^2 + 20y + 100$

81. $m^2 - 6m + 9$

82. $a^2 - 10a + 25$

83. $9p^2 - 24p + 16$

84. $16m^2 + 40m + 25$

85. $a^3 - 216$

86. $b^3 + 125$

87. $8r^3 - 27$

88. $1000p^3 + 27$

1.4 QUADRATIC EQUATIONS AND INEQUALITIES

A polynomial equation having a highest exponent of 2 is called a **quadratic equation.** Examples of quadratic equations include

$$x^2 + 5x + 6 = 0, \qquad y^2 = 16, \qquad \text{and} \qquad 3p^2 + 11p = 4.$$

Some quadratic equations can be solved by factoring; this method depends on the following property.

Zero-factor property If a and b are real numbers, with $ab = 0$, then $a = 0$ or $b = 0$. (If two numbers have a product of 0, then at least one of the numbers must be 0.)

Example 1 Solve the equation $(x - 4)(3x + 7) = 0$.

Here we are told that the product $(x - 4)(3x + 7)$ is equal to 0. By the zero-factor property, this can only be true if one of the factors is 0. That is, $x - 4 = 0$ or $3x + 7 = 0$. If we solve each of these equations separately, we will find the solutions of the original equation.

$$\begin{aligned} x - 4 &= 0 \qquad \text{or} \qquad & 3x + 7 &= 0 \\ x &= 4 & 3x &= -7 \\ & & x &= -\frac{7}{3} \end{aligned}$$

The solutions of $(x - 4)(3x + 7) = 0$ are 4 and $-7/3$. ■

1. Solve the following equations.
(a) $(y - 6)(y + 2) = 0$
(b) $(5k - 3)(k + 5) = 0$
(c) $(2r - 9)(3r + 5) = 0$

Answer:
(a) $6, -2$
(b) $3/5, -5$
(c) $9/2, -5/3$

Work Problem 1 at the side.

Example 2 Solve the quadratic equation $2m^2 + 5m = 12$.

To begin, rewrite the equation so that we have 0 alone on one side of the equals sign. We do this so that we may use the zero-factor property. Add -12 to both sides, giving

$$2m^2 + 5m - 12 = 0.$$

Use trial and error to factor on the left:

$$2m^2 + 5m - 12 = (2m - 3)(m + 4),$$

so that the given equation becomes

$$(2m - 3)(m + 4) = 0.$$

The product on the left can equal 0 only if at least one of the factors is 0.

$$2m - 3 = 0 \qquad \text{or} \qquad m + 4 = 0$$

Solve each of these equations separately.

$$2m = 3 \qquad\qquad m = -4$$

$$m = \frac{3}{2}$$

The solutions of $2m^2 + 5m = 12$ are $3/2$ and -4. ■

2. Solve the following.
(a) $y^2 + 3y = 10$
(b) $2r^2 + 9r = 5$
(c) $3k^2 = 2k + 8$

Answer:
(a) $2, -5$
(b) $1/2, -5$
(c) $-4/3, 2$

Work Problem 2 at the side.

Not all quadratic equations can be solved by this kind of factoring. For those equations that cannot, we need the more general method given by the **quadratic formula.**

Quadratic formula The solutions of the quadratic equation $ax^2 + bx + c = 0$, where $a \neq 0$, are given by

$$x = \frac{-b \pm \sqrt{b^2 - 4ac}}{2a}$$

Example 3 Solve $x^2 - 4x - 5 = 0$ by the quadratic formula.

First make sure that the equation has 0 alone on one side of the equals sign. Then identify the letters a, b, and c of the quadratic formula. The coefficient of the squared term gives the value of a; here $a = 1$. Also, $b = -4$ and $c = -5$. (Be careful to get the correct signs.) Substitute these values into the quadratic formula.

$$x = \frac{-(-4) \pm \sqrt{(-4)^2 - 4(1)(-5)}}{2(1)} \qquad \text{Let } a = 1,\ b = -4,\ c = -5$$

$$= \frac{4 \pm \sqrt{16 + 20}}{2} \qquad (-4)^2 = (-4)(-4) = 16$$

$$x = \frac{4 \pm 6}{2} \qquad \sqrt{16 + 20} = \sqrt{36} = 6$$

The $\pm$ sign represents the two solutions of the equation. To find each of the solutions, first use $+$ and then use $-$.

$$x = \frac{4 + 6}{2} = \frac{10}{2} = 5 \qquad \text{or} \qquad x = \frac{4 - 6}{2} = \frac{-2}{2} = -1.$$

The two solutions are 5 and -1. ■

3. Use the quadratic formula to solve each of the following equations.
(a) $3x^2 + 11x - 4 = 0$
(b) $2z^2 - 7z - 4 = 0$

Answer:
(a) $1/3, -4$
(b) $-1/2, 4$

Work Problem 3 at the side.

Example 4 Solve $x^2 + 1 = 4x$.
First add $-4x$ to both sides, to get 0 alone on the right side.

$$x^2 - 4x + 1 = 0.$$

Now identify the letters a, b, and c. We have $a = 1$, $b = -4$, and $c = 1$. Substitute these numbers into the quadratic formula.

$$x = \frac{-(-4) \pm \sqrt{(-4)^2 - 4(1)(1)}}{2(1)}$$

$$= \frac{4 \pm \sqrt{16 - 4}}{2}$$

$$= \frac{4 \pm \sqrt{12}}{2}$$

To simplify the solutions, write $\sqrt{12}$ as $\sqrt{4 \cdot 3} = \sqrt{4} \cdot \sqrt{3} = 2\sqrt{3}$. Substituting $2\sqrt{3}$ for $\sqrt{12}$ gives

$$= \frac{4 \pm 2\sqrt{3}}{2}$$

$$= \frac{2(2 \pm \sqrt{3})}{2} \qquad \text{Factor } 4 \pm 2\sqrt{3}$$

$$x = 2 \pm \sqrt{3}.$$

The two solutions are $2 + \sqrt{3}$ and $2 - \sqrt{3}$.

The exact values of the solutions are $2 + \sqrt{3}$ and $2 - \sqrt{3}$. In many cases we need a decimal approximation of these solutions. Use a calculator, or Table 2 in the back of the book, to find that $\sqrt{3} \approx 1.732$, so that (to the nearest thousandth) the solutions are

$$2 + \sqrt{3} \approx 2 + 1.732 = 3.732$$

or

$$2 - \sqrt{3} \approx 2 - 1.732 = .268. \quad \blacksquare$$

4. Find exact and approximate solutions for the following.
(a) $y^2 - 2y = 2$
(b) $x^2 - 6x + 4 = 0$

Answer:
(a) exact: $1 + \sqrt{3}, 1 - \sqrt{3}$; approximate: 2.732, $-.732$
(b) exact: $3 + \sqrt{5}, 3 - \sqrt{5}$; approximate: 5.236, .764

Work Problem 4 at the side.

The above quadratic equations all had two different solutions. This is not always the case, as the following example shows.

Example 5 **(a)** Solve $9x^2 - 30x + 25 = 0$.
We have $a = 9$, $b = -30$, and $c = 25$. By the quadratic formula,

$$x = \frac{-(-30) \pm \sqrt{(-30)^2 - 4(9)(25)}}{2(9)}$$

$$= \frac{30 \pm \sqrt{900 - 900}}{18}$$

$$= \frac{30 \pm 0}{18} = \frac{30}{18} = \frac{5}{3}.$$

The given equation has only one solution.

(b) Solve $x^2 - 6x + 10 = 0$.
Since $a = 1$, $b = -6$, and $c = 10$, we have

$$x = \frac{-(-6) \pm \sqrt{(-6)^2 - 4(1)(10)}}{2(1)} = \frac{6 \pm \sqrt{36 - 40}}{2} = \frac{6 \pm \sqrt{-4}}{2}.$$

There are no real number solutions to this equation since $\sqrt{-4}$ is not a real number. ■

5. Solve the following equations.
(a) $9k^2 - 6k + 1 = 0$
(b) $4m^2 + 28m + 49 = 0$
(c) $2x^2 - 5x + 5 = 0$

Answer:
(a) 1/3
(b) $-7/2$
(c) No real number solutions

Work Problem 5 at the side.

Quadratic Inequalities A **quadratic inequality** is an inequality of the form $ax^2 + bx + c > 0$ (or $<$, or $\leq$, or $\geq$). The highest exponent is always 2. Examples of quadratic inequalities include

$$x^2 - x - 12 < 0, \qquad 3y^2 + 2y \geq 0, \qquad \text{and} \qquad m^2 \leq 4.$$

A method of solving quadratic inequalities is shown in the next few examples.

Example 6 Solve the quadratic inequality $x^2 - x - 12 < 0$.
Since $x^2 - x - 12 = (x - 4)(x + 3)$, our given inequality is really the same as

$$(x - 4)(x + 3) < 0.$$

We want the product of $x - 4$ and $x + 3$ to be negative; this product will be negative if $x - 4$ and $x + 3$ have opposite signs. The factor $x - 4$ is positive when $x - 4 > 0$, or $x > 4$. Thus, $x - 4$ is negative if $x < 4$. In the same way, $x + 3$ is positive when $x > -3$ and negative when $x < -3$. This information is shown in the **sign graph** of Figure 1.8.

We said above that the product $(x - 4)(x + 3)$ will be negative when $x - 4$ and $x + 3$ have opposite signs. From the sign graph of

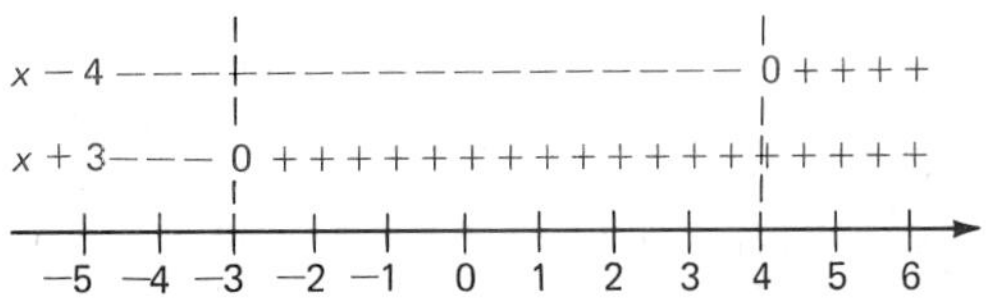

Figure 1.8

Figure 1.8, these expressions have opposite signs whenever x is between -3 and 4. The solution is thus given by $-3 < x < 4$. A graph of this solution is shown in Figure 1.9. ■

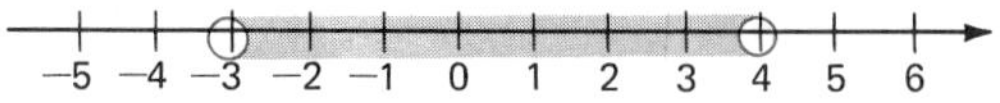

Figure 1.9

6. Solve the following and graph.
(a) $y^2 + 2y - 3 < 0$
(b) $2p^2 + 3p - 2 < 0$

Answer:
(a) $-3 < y < 1$

(b) $-2 < p < 1/2$

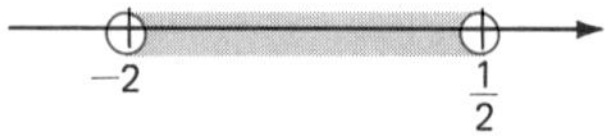

Work Problem 6 at the side.

Example 7 Solve the quadratic inequality $r^2 + 3r \geq 4$.
First rewrite the inequality so that one side is 0.

$$r^2 + 3r \geq 4$$
$$r^2 + 3r - 4 \geq 0 \qquad \text{Add } -4 \text{ to both sides}$$

Factor $r^2 + 3r - 4$ as $(r + 4)(r - 1)$. The factor $r + 4$ is positive when $r > -4$ and negative when $r < -4$. Also, $r - 1$ is positive when $r > 1$ and negative when $r < 1$. We use this information to produce the sign graph in Figure 1.10. The product $(r + 4)(r - 1)$ will be positive when $r + 4$ and $r - 1$ have the same signs, either positive or negative. From Figure 1.10, the solution is seen to be made up of those numbers less than or equal to -4, together with those greater than or equal to 1. This can be written $r \leq -4$ or $r \geq 1$. A graph of the solution is given in Figure 1.11. ■

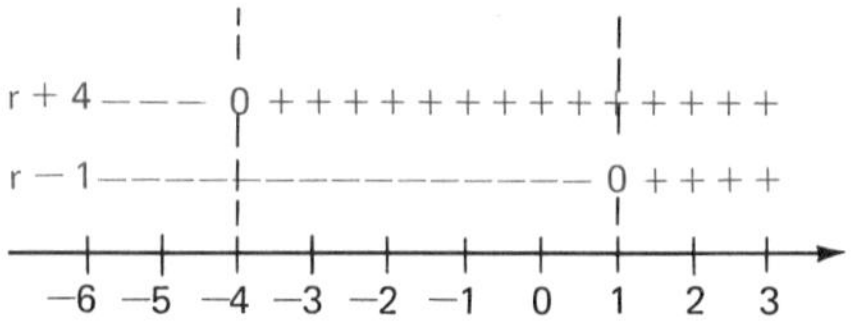

Figure 1.10

7. Solve the following and graph.
(a) $k^2 + 2k - 15 \geq 0$
(b) $3m^2 + 7m \geq 6$

Answer:
(a) $k \leq -5$ or $k \geq 3$

(b) $m \leq -3$ or $m \geq 2/3$

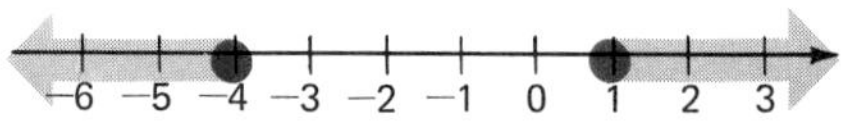

Figure 1.11

Work Problem 7 at the side.

1.4 EXERCISES

Solve each of the following equations. (See Examples 1 and 2.)

1. $(y - 5)(y + 4) = 0$

2. $(m + 3)(m - 1) = 0$

3. $x^2 + 5x + 6 = 0$

4. $y^2 - 3y + 2 = 0$

5. $r^2 - 5r - 6 = 0$

6. $y^2 - y - 12 = 0$

7. $a^2 + 5a = 24$

8. $y^2 = 2y + 15$

9. $x^2 = 3 + 2x$

10. $3m + 4 = m^2$

11. $m^2 + 16 = 8m$

12. $y^2 + 49 = 14y$

13. $2k^2 - k = 10$

14. $6x^2 = 7x + 5$

15. $6x^2 - 5x = 4$

16. $9s^2 + 12s = -4$

17. $m(m - 7) = -10$

18. $z(2z + 7) = 4$

19. $9x^2 - 16 = 0$

20. $25y^2 - 64 = 0$

21. $16x^2 - 16x = 0$

22. $12y^2 - 48y = 0$

Use the quadratic formula to solve each of the following equations. If the solutions involve square roots, give both the exact and approximate solutions. (See Examples 3 and 4.)

23. $3x^2 - 5x + 1 = 0$

24. $2x^2 - 7x + 1 = 0$

25. $2m^2 = m + 4$

26. $p^2 + p - 1 = 0$

27. $k^2 - 10k = -20$

28. $r^2 = 13 - 12r$

29. $2x^2 + 12x + 5 = 0$

30. $5x^2 + 4x = 1$

31. $2r^2 - 7r + 5 = 0$

32. $8x^2 = 8x - 3$

33. $6k^2 - 11k + 4 = 0$

34. $8m^2 - 10m + 3 = 0$

35. $x^2 + 3x = 10$

36. $2x^2 = 3x + 5$

37. $2x^2 - 7x + 30 = 0$

38. $3k^2 + k = 6$

39. $5m^2 + 5m = 0$

40. $8r^2 + 16r = 0$

Solve each of the following quadratic inequalities. Graph each solution. (See Examples 5 and 6.)

41. $(m + 2)(m - 4) < 0$

42. $(k - 1)(k + 2) > 0$

43. $(t + 6)(t - 1) \geq 0$

44. $(y - 2)(y + 3) \leq 0$

45. $y^2 - 3y + 2 < 0$

46. $z^2 - 4z - 5 \leq 0$

47. $2k^2 + 7k - 4 > 0$

48. $6r^2 - 5r - 4 > 0$

49. $q^2 - 7q + 6 \leq 0$

50. $2k^2 - 7k - 15 \leq 0$

51. $6m^2 + m > 1$

52. $10r^2 + r \leq 2$

53. $2y^2 + 5y \leq 3$

54. $3a^2 + a > 10$

55. $x^2 \leq 25$

56. $y^2 \geq 4$

APPLIED PROBLEMS

Management

The commodity market is very unstable; money can be made or lost quickly when invested in soybeans, wheat, pork bellies, and so on. Suppose that an investor kept track of her total profit, P, at time t, measured in months, after she began investing and found that

$$P = 4t^2 - 29t + 30.$$

For example, 4 months after she began investing, her total profits were

$$\begin{aligned} P &= 4 \cdot 4^2 - 29 \cdot 4 + 30 \qquad \textit{Let } t = 4 \\ &= 4 \cdot 16 - 116 + 30 \\ &= 64 - 116 + 30 \\ P &= -22, \end{aligned}$$

so that she was \$22 "in the hole" after 4 months.

57. Find her total profit after 8 months.

58. Find her total profit after 10 months.

59. Find when she has just broken even. (Let $P = 0$, and solve the resulting equation.)

60. Find the time intervals where she has been ahead. (Solve the inequality $4t^2 - 29t + 30 > 0$.)

1.5 RATIONAL EXPRESSIONS

When we get into the later chapters of this book, we will work with algebraic fractions. Examples of these fractions (called **rational expressions**) include

$$\frac{8}{x-1}, \qquad \frac{3x^2 + 4x}{5x - 6} \qquad \text{and} \qquad \frac{2 + \frac{1}{y}}{y}.$$

Methods for working with rational expressions are summarized in the following list.

Properties of rational expressions For all mathematical expressions P, $Q \neq 0$, R, and $S \neq 0$, we have

(a) $\dfrac{P}{Q} = \dfrac{PS}{QS}$ (Fundamental property)

(b) $\dfrac{P}{Q} \cdot \dfrac{R}{S} = \dfrac{PR}{QS}$ (Multiplication)

(c) $\dfrac{P}{Q} + \dfrac{R}{Q} = \dfrac{P+R}{Q}$ (Addition)

(d) $\dfrac{P}{Q} - \dfrac{R}{Q} = \dfrac{P-R}{Q}$ (Subtraction)

(e) $\dfrac{P}{Q} \div \dfrac{R}{S} = \dfrac{P}{Q} \cdot \dfrac{S}{R}$, $R \neq 0$ (Division)

Let us now look at some examples of these properties.

Example 1 Reduce each of the following rational expressions to lowest terms.

(a) $\dfrac{12m}{18}$

Both $12m$ and 18 are divisible by 6. By property (a) above, we have

$$\frac{12m}{18} = \frac{2m \cdot 6}{3 \cdot 6} = \frac{2m}{3}.$$

(b) $\dfrac{8x+16}{4} = \dfrac{8(x+2)}{4} = \dfrac{4 \cdot 2(x+2)}{4} = 2(x+2)$

Here we first factored $8x + 16$. The answer could also be written as $2x + 4$, if desired.

(c) $\dfrac{k^2+7k+12}{k^2+2k-3} = \dfrac{(k+4)(k+3)}{(k-1)(k+3)} = \dfrac{k+4}{k-1}$

The answer cannot be further reduced. ■

Work Problem 1 at the side.

1. Reduce each of the following to lowest terms.

(a) $\dfrac{12k+36}{18}$

(b) $\dfrac{15m+30m^2}{5m}$

(c) $\dfrac{2p^2+3p+1}{p^2+3p+2}$

Answer:

(a) $\dfrac{2(k+3)}{3}$ or $\dfrac{2k+6}{3}$

(b) $3(1 + 2m)$ or $3 + 6m$

(c) $\dfrac{2p+1}{p+2}$

Example 2 Multiply in each of the following.

(a) $\dfrac{2}{3} \cdot \dfrac{y}{5}$

Use property (b) above; multiply on top and multiply on the bottom.

$$\frac{2}{3} \cdot \frac{y}{5} = \frac{2 \cdot y}{3 \cdot 5} = \frac{2y}{15}$$

The result, $2y/15$, cannot be further reduced.

(b) $\frac{3y+9}{6} \cdot \frac{18}{5y+15}$

Factor where possible.

$$\frac{3y+9}{6} \cdot \frac{18}{5y+15} = \frac{3(y+3)}{6} \cdot \frac{18}{5(y+3)}$$

$$= \frac{3 \cdot 18(y+3)}{6 \cdot 5(y+3)} \quad \text{Multiply on top and bottom}$$

$$= \frac{3 \cdot 6 \cdot 3(y+3)}{6 \cdot 5(y+3)} \quad 18 = 6 \cdot 3$$

$$= \frac{3 \cdot 3}{5} \quad \text{Reduce to lowest terms}$$

$$= \frac{9}{5}$$

(c) $\frac{m^2+5m+6}{m+3} \cdot \frac{m}{m^2+3m+2}$

$$= \frac{(m+2)(m+3)}{m+3} \cdot \frac{m}{(m+2)(m+1)} \quad \text{Factor}$$

$$= \frac{m(m+2)(m+3)}{(m+3)(m+2)(m+1)} \quad \text{Multiply}$$

$$= \frac{m}{m+1} \quad \text{Reduce} \ \blacksquare$$

2. Multiply the following.

(a) $\frac{3r^2}{5} \cdot \frac{20}{9r}$

(b) $\frac{y-4}{y^2-2y-8} \cdot \frac{y^2-4}{3y}$

Answer:

(a) $\frac{4r}{3}$

(b) $\frac{y-2}{3y}$

Work Problem 2 at the side.

Example 3 Divide as indicated in each of the following.

(a) $\frac{8x}{5} \div \frac{11x^2}{20}$

As shown in property (e) above, invert the second expression and multiply.

$$\frac{8x}{5} \div \frac{11x^2}{20} = \frac{8x}{5} \cdot \frac{20}{11x^2} \quad \text{Invert and multiply}$$

$$= \frac{8x \cdot 20}{5 \cdot 11x^2} \quad \text{Multiply}$$

$$= \frac{32}{11x} \quad \text{Reduce}$$

(b) $\dfrac{9p - 36}{12} \div \dfrac{5(p - 4)}{18}$

$= \dfrac{9p - 36}{12} \cdot \dfrac{18}{5(p - 4)}$ Invert and multiply

$= \dfrac{9(p - 4)}{12} \cdot \dfrac{18}{5(p - 4)}$ Factor

$= \dfrac{27}{10}$ Multiply and reduce ■

Work Problem 3 at the side.

3. Divide the following.

(a) $\dfrac{5m}{16} \div \dfrac{m^2}{10}$

(b) $\dfrac{2y - 8}{6} \div \dfrac{5y - 20}{3}$

(c) $\dfrac{m^2 - 2m - 3}{m(m + 1)} \div \dfrac{m + 4}{5m}$

Answer:

(a) $\dfrac{25}{8m}$

(b) $\dfrac{1}{5}$

(c) $\dfrac{5(m - 3)}{m + 4}$

Example 4 Add or subtract as indicated.

(a) $\dfrac{4}{5k} - \dfrac{11}{5k}$

As property (d) above shows, when two rational expressions have the same denominators, we subtract by subtracting the numerators.

$$\frac{4}{5k} - \frac{11}{5k} = \frac{4 - 11}{5k} = -\frac{7}{5k}$$

(b) $\dfrac{7}{p} + \dfrac{9}{2p} + \dfrac{1}{3p}$

These three denominators are different; we cannot add until we make them the same. We do this by finding a common denominator into which p, $2p$, and $3p$ all divide. A common denominator here is $6p$.

Rewrite each rational expression, using property (a), so that it has a denominator of $6p$. We have

$$\begin{aligned}\frac{7}{p} + \frac{9}{2p} + \frac{1}{3p} &= \frac{6 \cdot 7}{6 \cdot p} + \frac{3 \cdot 9}{3 \cdot 2p} + \frac{2 \cdot 1}{2 \cdot 3p} \\ &= \frac{42}{6p} + \frac{27}{6p} + \frac{2}{6p} \\ &= \frac{42 + 27 + 2}{6p} \\ &= \frac{71}{6p}.\end{aligned}$$

(c) $\dfrac{3}{k-1} - \dfrac{1}{k}$

The common denominator is $k(k-1)$.

$$\frac{3}{k-1} - \frac{1}{k} = \frac{3 \cdot k}{k(k-1)} - \frac{1(k-1)}{k(k-1)} \quad \text{Property (a)}$$

$$= \frac{3k}{k(k-1)} - \frac{k-1}{k(k-1)}$$

$$= \frac{3k-(k-1)}{k(k-1)} \quad \text{Property (d)}$$

$$= \frac{3k-k+1}{k(k-1)} \quad -(k-1) = -1(k-1) = -k+1$$

$$= \frac{2k+1}{k(k-1)} \quad ■$$

Work Problem 4 at the side.

4. Add or subtract.

(a) $\dfrac{1}{m} - \dfrac{7}{m}$

(b) $\dfrac{3}{4r} + \dfrac{8}{3r}$

(c) $\dfrac{1}{m-2} - \dfrac{3}{2(m-2)}$

Answer:

(a) $-\dfrac{6}{m}$

(b) $\dfrac{41}{12r}$

(c) $\dfrac{-1}{2(m-2)}$

1.5 EXERCISES

Reduce each of the following to lowest terms. Factor as necessary. (See Example 1.)

1. $\dfrac{6m}{24}$

2. $\dfrac{8k}{56}$

3. $\dfrac{7z^2}{14z}$

4. $\dfrac{32y^2}{16y}$

5. $\dfrac{25p^3}{10p^2}$

6. $\dfrac{14z^3}{6z^2}$

7. $\dfrac{8k+16}{9k+18}$

8. $\dfrac{20r+10}{30r+15}$

9. $\dfrac{3(t+5)}{(t+5)(t-3)}$

10. $\dfrac{-8(y-4)}{(y+2)(y-4)}$

11. $\dfrac{8x^2+16x}{4x^2}$

12. $\dfrac{36y^2+72y}{9y}$

13. $\dfrac{m^2-4m+4}{m^2+m-6}$

14. $\dfrac{r^2-r-6}{r^2+r-12}$

15. $\dfrac{x^2+3x-4}{x^2-1}$

16. $\dfrac{z^2-5z+6}{z^2-4}$

17. $\dfrac{8m^2+6m-9}{16m^2-9}$

18. $\dfrac{6y^2+11y+4}{3y^2+7y+4}$

Multiply or divide as indicated in each of the following. Reduce all answers to lowest terms. (See Examples 2 and 3.)

19. $\frac{9k^2}{25} \cdot \frac{5}{3k}$

20. $\frac{21m^3}{9m} \cdot \frac{12m^2}{7m}$

21. $\frac{15p^3}{9p^2} \div \frac{6p}{10p^2}$

22. $\frac{3r^2}{9r^3} \div \frac{8r^3}{6r}$

23. $\frac{a+b}{2p} \cdot \frac{12}{5(a+b)}$

24. $\frac{3(x-1)}{y} \cdot \frac{2y}{7(x-1)}$

25. $\frac{a-3}{16} \div \frac{a-3}{32}$

26. $\frac{9}{2(4-y)} \div \frac{3}{4-y}$

27. $\frac{2k+8}{6} \div \frac{3k+12}{2}$

28. $\frac{5m+25}{10} \cdot \frac{12}{6m+30}$

29. $\frac{9y-18}{6y+12} \cdot \frac{3y+6}{15y-30}$

30. $\frac{12r+24}{36r-36} \div \frac{6r+12}{8r-8}$

31. $\frac{4a+12}{2a-10} \div \frac{a^2-9}{a^2-a-20}$

32. $\frac{6r-18}{9r^2+6r-24} \cdot \frac{12r-16}{4r-12}$

33. $\frac{k^2-k-6}{k^2+k-12} \cdot \frac{k^2+3k-4}{k^2+2k-3}$

34. $\frac{n^2-n-6}{n^2-2n-8} \div \frac{n^2-9}{n^2+7n+12}$

35. $\frac{m^2+3m+2}{m^2+5m+4} \div \frac{m^2+5m+6}{m^2+10m+24}$

36. $\frac{y^2+y-2}{y^2+3y-4} \div \frac{y^2+3y+2}{y^2+4y+3}$

37. $\frac{2m^2-5m-12}{m^2-10m+24} \div \frac{4m^2-9}{m^2-9m+18}$

38. $\frac{6n^2-5n-6}{6n^2+5n-6} \cdot \frac{12n^2-17n+6}{12n^2-n-6}$

Add or subtract as indicated in each of the following. Reduce all answers to lowest terms. (See Example 4.)

39. $\frac{8}{r} + \frac{6}{r}$

40. $\frac{7}{m} - \frac{4}{m}$

41. $\frac{3}{2k} + \frac{5}{3k}$

42. $\frac{8}{5p} + \frac{3}{4p}$

43. $\frac{2}{3y} - \frac{1}{4y}$

44. $\frac{6}{11z} - \frac{5}{2z}$

45. $\frac{a+1}{2} - \frac{a-1}{2}$

46. $\frac{y+6}{5} - \frac{y-6}{5}$

47. $\frac{3}{p} + \frac{1}{2}$

48. $\frac{9}{r} - \frac{2}{3}$

49. $\frac{2}{y} - \frac{1}{4}$

50. $\frac{6}{11} + \frac{3}{a}$

51. $\dfrac{1}{6m}+\dfrac{2}{5m}+\dfrac{4}{m}$

52. $\dfrac{8}{3p}+\dfrac{5}{4p}+\dfrac{9}{2p}$

53. $\dfrac{1}{m-1}+\dfrac{2}{m}$

54. $\dfrac{8}{y+2}-\dfrac{3}{y}$

55. $\dfrac{6}{r}-\dfrac{5}{r-2}$

56. $\dfrac{8}{a-1}-\dfrac{5}{a}$

57. $\dfrac{8}{3(a-1)}+\dfrac{2}{a-1}$

58. $\dfrac{5}{2(k+3)}+\dfrac{2}{k+3}$

59. $\dfrac{2}{5(k-2)}+\dfrac{3}{4(k-2)}$

60. $\dfrac{11}{3(p+4)}-\dfrac{5}{6(p+4)}$

In each of the following exercises, first simplify the numerator and denominator separately, then simplify the final result.

61. $\dfrac{1+\dfrac{1}{x}}{1-\dfrac{1}{x}}$

62. $\dfrac{2-\dfrac{2}{y}}{2+\dfrac{2}{y}}$

63. $\dfrac{\dfrac{1}{x+1}-\dfrac{1}{x}}{\dfrac{1}{x}}$

64. $\dfrac{\dfrac{1}{y+3}-\dfrac{1}{y}}{\dfrac{1}{y}}$

1.6 EXPONENTS AND ROOTS

Earlier in this chapter we saw that $a^2 = a \cdot a$, $a^3 = a \cdot a \cdot a$, and so on. In this section we give a more general meaning to the symbol a^n. First, recall that n is the **exponent** in a^n, and a is called the **base.**

If n is a natural number, then

$$a^n = a \cdot a \cdot a \cdots a,$$

where a appears as a factor n times.

1. Evaluate the following.
(a) 6^3
(b) 5^4
(c) 1^7
(d) $\left(\dfrac{2}{5}\right)^3$

Answer:
(a) 216
(b) 625
(c) 1
(d) 8/125

Example 1 **(a)** $2^3 = 2 \cdot 2 \cdot 2 = 8$
Read 2^3 as "2 cubed."

(b) $5^2 = 5 \cdot 5 = 25$
Read 5^2 as "5 squared."

(c) $4^5 = 4 \cdot 4 \cdot 4 \cdot 4 \cdot 4 = 1024$
Read 4^5 as "4 to the fifth power."

(d) $\left(\dfrac{3}{4}\right)^2 = \dfrac{3}{4} \cdot \dfrac{3}{4} = \dfrac{9}{16}$ ■

Work Problem 1 at the side.

If $n = 0$, then we define $a^n = a^0 = 1$, if $a \neq 0$. (0^0 is a meaningless symbol.) That is, if a is any nonzero real number, then

$$a^0 = 1.$$

Example 2 **(a)** $6^0 = 1$
(b) $(-9)^0 = 1$
(c) $-(4^0) = -(1) = -1$ ■

Work Problem 2 at the side.

2. Evaluate the following.
(a) 17^0
(b) 30^0
(c) $(-10)^0$
(d) $-(12^0)$

Answer:
(a) 1
(b) 1
(c) 1
(d) -1

For problems having negative exponents, we need the following definition.

If n is a natural number, and if $a \neq 0$, then

$$a^{-n} = \frac{1}{a^n}.$$

Example 3 **(a)** $3^{-2} = \frac{1}{3^2} = \frac{1}{9}$

(b) $5^{-4} = \frac{1}{5^4} = \frac{1}{625}$

(c) $9^{-1} = \frac{1}{9^1} = \frac{1}{9}$

(d) $\left(\frac{3}{4}\right)^{-1} = \frac{1}{(\frac{3}{4})^1} = \frac{1}{\frac{3}{4}} = \frac{4}{3}$ ■

Work Problem 3 at the side.

3. Simplify the following.
(a) 6^{-2}
(b) $\left(\frac{5}{8}\right)^{-1}$
(c) 5^{-2}
(d) 10^{-3}
(e) $\left(\frac{1}{2}\right)^{-4}$

Answer:
(a) 1/36
(b) 8/5
(c) 1/25
(d) 1/1000
(e) 16

By using the definitions of exponents given above, we could prove the following.

Properties of exponents For any integers m and n, and any real numbers a and b for which the following exist, we have

(a) $a^m \cdot a^n = a^{m+n}$

(b) $\frac{a^m}{a^n} = a^{m-n}$

(c) $(a^m)^n = a^{mn}$

(d) $(ab)^m = a^m \cdot b^m$

(e) $\left(\frac{a}{b}\right)^m = \frac{a^m}{b^m}$ $(b \neq 0)$.

Example 4 Use the properties of exponents to simplify each of the following. Leave answers with exponents.

(a) $7^4 \cdot 7^6 = 7^{4+6} = 7^{10}$ Property (a)

(b) $\dfrac{9^{14}}{9^6} = 9^{14-6} = 9^8$ Property (b)

(c) $\dfrac{3^9}{3^{17}} = 3^{9-17} = 3^{-8} = \dfrac{1}{3^8}$ Property (b)

(d) $(2^5)^3 = 2^{5 \cdot 3} = 2^{15}$ Property (c)

(e) $(3x)^4 = 3^4 \cdot x^4$ Property (d)

(f) $\left(\dfrac{9}{7}\right)^6 = \dfrac{9^6}{7^6}$ Property (e)

(g) $\dfrac{2^{-3} \cdot 2^5}{2^4 \cdot 2^{-7}} = \dfrac{2^2}{2^{-3}} = 2^{2-(-3)} = 2^5$

(h) $2^{-1} + 3^{-1} = \dfrac{1}{2} + \dfrac{1}{3} = \dfrac{5}{6}$ ■

4. Simplify the following.
(a) $9^6 \cdot 9^4$
(b) $\dfrac{8^7}{8^3}$
(c) $\dfrac{14^9}{14^{12}}$
(d) $(13^4)^3$
(e) $(6y)^4$
(f) $\left(\dfrac{3}{4}\right)^2$
(g) $\dfrac{3^4 \cdot 3^{-6}}{3^5 \cdot 3^{-2}}$
(h) $3^{-1} - 4^{-1}$

Answer:
(a) 9^{10}
(b) 8^4
(c) $1/14^3$
(d) 13^{12}
(e) 6^4y^4
(f) $3^2/4^2$ or $9/16$
(g) 3^{-5} or $1/3^5$
(h) $1/12$

Work Problem 4 at the side.

We have discussed and given meaning to exponentials of the form a^m for all nonzero real numbers a and all *integer* values of m, both positive and negative. In the remainder of this section we define a^m for *rational* values of m. We first look at an exponential of the form $a^{1/n}$, where n is a positive integer. We want any meaning that we assign to $a^{1/n}$ to be consistent with the properties given above. For example, we know that for any real number a, and integers m and n, $(a^m)^n = a^{mn}$. If this property is to hold for the expression $a^{1/n}$, we must have

$$(a^{1/n})^n = a^{(1/n)n} = a^1 = a,$$

or

$$(a^{1/n})^n = a.$$

Thus, the nth power of $a^{1/n}$ must be a. For this reason, $a^{1/n}$ is called an ***n*th root** of a. For example, $a^{1/2}$ denotes a second root, or **square root** of a, while $a^{1/3}$ is the third root, or **cube root** of a.

There are two numbers whose square is 16, namely 4 and -4. Also, there are two possible fourth roots of 81, namely 3 and -3. In these cases we reserve the symbol $a^{1/n}$ for the *positive* root:

$$16^{1/2} = 4 \quad \text{and} \quad 81^{1/4} = 3,$$

and so on.

Example 5 (A calculator, or Tables 1 and 2 in the Appendix will be helpful here.)
(a) $121^{1/2} = 11$, since 11 is positive and $11^2 = 121$
(b) $625^{1/4} = 5$ since $5^4 = 625$
(c) $256^{1/4} = 4$
(d) $64^{1/6} = 2$ ■

Work Problem 5 at the side.

5. Evaluate the following.
(a) $64^{1/2}$
(b) $256^{1/2}$
(c) $4096^{1/4}$
(d) $1296^{1/4}$

Answer:
(a) 8
(b) 16
(c) 8
(d) 6

There is no real number x such that $x^2 = -16$. Therefore, $(-16)^{1/2}$ is not a real number. In general, if $a < 0$ and n is an *even* integer then $a^{1/n}$ is not a real number.

Since $2^3 = 8$, we have $8^{1/3} = 2$. Also, $(-8)^{1/3} = -2$. In general, if a is any real number and n is an *odd* integer, then there is exactly one real number equal to $a^{1/n}$.

Example 6 **(a)** $27^{1/3} = 3$
(b) $(-32)^{1/5} = -2$
(c) $128^{1/7} = 2$
(d) $(-49)^{1/2}$ is not a real number. ■

Work Problem 6 at the side.

6. Evaluate the following.
(a) $125^{1/3}$
(b) $243^{1/5}$
(c) $(-64)^{1/2}$
(d) $(-125)^{1/3}$

Answer:
(a) 5
(b) 3
(c) not a real number
(d) -5

We have now defined a^m for all integer values of m and all exponents of the form $a^{1/n}$, where n is a positive integer. To extend the definition of a^m to include all rational values of m, we make the following definition.

For all real numbers a such that the following roots exist, and for any rational number m/n,

$$a^{m/n} = (a^{1/n})^m.$$

Example 7 **(a)** $27^{2/3} = (27^{1/3})^2 = 3^2 = 9$
(b) $32^{2/5} = (32^{1/5})^2 = 2^2 = 4$
(c) $64^{4/3} = (64^{1/3})^4 = 4^4 = 256$
(d) $25^{3/2} = (25^{1/2})^3 = 5^3 = 125$ ■

Work Problem 7 at the side.

7. Evaluate the following.
(a) $16^{3/4}$
(b) $25^{5/2}$
(c) $32^{7/5}$
(d) $100^{3/2}$

Answer:
(a) 8
(b) 3125
(c) 128
(d) 1000

We now summarize the following properties of rational exponents.

Properties of rational exponents For all rational numbers m and n, and all real numbers a and b for which the following exist, we have

(a) $a^m \cdot a^n = a^{m+n}$

(b) $\dfrac{a^m}{a^n} = a^{m-n}$

(c) $(a^m)^n = a^{mn}$

(d) $(ab)^m = a^m b^m$

(e) $\left(\dfrac{a}{b}\right)^n = \dfrac{a^n}{b^n} \qquad (b \neq 0).$

It is common to express $a^{1/2}$ as $\sqrt{a}$, where the symbol $\sqrt{\ }$ is called a **radical sign.** In general, if n is an integer greater than 1, the symbol $a^{1/n}$ can be written $\sqrt[n]{a}$. Using radical signs, we can write our definition of $a^{m/n}$ as

$$a^{m/n} = (\sqrt[n]{a})^m$$

whenever these roots exist. Using this definition, we can convert back and forth from exponential form to radical form, as the next example shows.

Example 8 Write each of the following in radical form.

(a) $30^{5/2}$

By the definition above,

$$30^{5/2} = (\sqrt{30})^5$$

(b) $7^{-3/4} = \dfrac{1}{7^{3/4}} = \dfrac{1}{(\sqrt[4]{7})^3}$

(c) $2x^{-3/2} = \dfrac{2}{x^{3/2}} = \dfrac{2}{(\sqrt{x})^3}$ ■

8. Write the following in radical form.
(a) $3^{7/2}$
(b) $12^{-2/3}$
(c) $9y^{-3/8}$

Answer:
(a) $(\sqrt{3})^7$
(b) $1/(\sqrt[3]{12})^2$
(c) $9/(\sqrt[8]{y})^3$

Work Problem 8 at the side.

1.6 EXERCISES

Evaluate each of the following. Write all answers without exponents. (See Examples 1–3 and 5–7.)

1. 7^3 **2.** 4^2 **3.** 8^{-1} **4.** 9^{-2}

5. 2^{-3} **6.** 3^{-4} **7.** 5^{-1} **8.** 6^{-3}

9. 8^{-3} **10.** 12^{-2} **11.** $\left(\frac{3}{4}\right)^3$ **12.** $\left(\frac{2}{3}\right)^4$

13. $\left(\frac{5}{8}\right)^2$ **14.** $\left(\frac{6}{7}\right)^3$ **15.** $\left(\frac{1}{2}\right)^{-3}$ **16.** $\left(\frac{1}{5}\right)^{-3}$

17. $\left(\frac{2}{7}\right)^{-2}$ **18.** $\left(\frac{4}{5}\right)^{-3}$ **19.** $81^{1/2}$ **20.** $16^{1/4}$

21. $27^{1/3}$ **22.** $81^{1/4}$ **23.** $8^{2/3}$ **24.** $9^{3/2}$

25. $1000^{2/3}$ **26.** $64^{3/2}$ **27.** $32^{2/5}$ **28.** $32^{6/5}$

29. $-125^{2/3}$ **30.** $-125^{4/3}$ **31.** $\left(\frac{4}{9}\right)^{1/2}$ **32.** $\left(\frac{16}{25}\right)^{1/2}$

33. $\left(\frac{64}{27}\right)^{1/3}$ **34.** $\left(\frac{8}{125}\right)^{1/3}$ **35.** $16^{-5/4}$ **36.** $625^{-1/4}$

37. $\left(\frac{27}{64}\right)^{-1/3}$ **38.** $\left(\frac{121}{100}\right)^{-3/2}$ **39.** $2^{-1} + 4^{-1}$ **40.** $2^{-2} + 3^{-2}$

Simplify each of the following. Write all answers using only positive exponents. (See Example 4.)

41. $\frac{3^8}{3^2}$ **42.** $\frac{4^9}{4^7}$ **43.** $\frac{7^5}{7^9}$ **44.** $\frac{8^6}{8^{12}}$

45. $\frac{3^{-4}}{3^2}$ **46.** $\frac{9^{-2}}{9^5}$ **47.** $\frac{2^{-5}}{2^{-2}}$ **48.** $\frac{14^{-3}}{14^{-8}}$

49. $\frac{6^{-1}}{6}$ **50.** $\frac{15}{15^{-1}}$ **51.** $4^{-3} \cdot 4^6$ **52.** $5^{-9} \cdot 5^{10}$

53. $7^{-5} \cdot 7^{-2}$ **54.** $9^{-1} \cdot 9^{-3}$ **55.** $\frac{8^9 \cdot 8^{-7}}{8^{-3}}$ **56.** $\frac{5^{-4} \cdot 5^6}{5^{-1}}$

57. $\frac{10^8 \cdot 10^{-10}}{10^4 \cdot 10^2}$ **58.** $\frac{2^{-4} \cdot 2^{-3}}{2^6 \cdot 2^{-5}}$

59. $\left(\frac{5^{-6} \cdot 5^3}{5^{-2}}\right)^{-1}$ **60.** $\left(\frac{8^{-3} \cdot 8^4}{8^{-2}}\right)^{-2}$

61. $2^{1/2} \cdot 2^{3/2}$ **62.** $5^{3/8} \cdot 5^{5/8}$

63. $27^{2/3} \cdot 27^{-1/3}$ **64.** $9^{-3/4} \cdot 9^{1/4}$

65. $\frac{4^{2/3} \cdot 4^{5/3}}{4^{1/3}}$ **66.** $\frac{3^{-5/2} \cdot 3^{3/2}}{3^{7/2} \cdot 3^{-9/2}}$

Write each of the following in radical form. (See Example 8.)

67. $7^{3/2}$ **68.** $15^{2/3}$ **69.** $28^{3/5}$ **70.** $80^{3/5}$

71. $60^{-2/3}$ **72.** $3^{-4/3}$ **73.** $6^{-5/4}$ **74.** $22^{-3/5}$

75. $12x^{-3/2}$ **76.** $50y^{-2/3}$ **77.** $8m^{-4/5}$ **78.** $6y^{-2/5}$

79. $(3r)^{-2/3}$ **80.** $(2k)^{-3/2}$

APPLIED PROBLEMS

Management

One important application of mathematics to business and management concerns supply *and* demand. *Usually, as the price of an item increases, the supply increases and the demand decreases. By studying past records of supply and demand at different prices, economists can construct an equation which is an approximate mathematical model of supply and demand for a given item. The next two exercises show examples of this.*

81. The price of a certain type of solar heater is approximated by p, where

$$p = 2x^{1/2} + 3x^{2/3}$$

and x is the number of units supplied. Find the price when the supply is 64 units.

82. The demand for a certain commodity and the price are related by the equation

$$p = 1000 - 200x^{-2/3} \qquad (x > 0)$$

where x is the number of units of the product demanded. Find the price when the demand is 27.

Social Science

In our system of government, the president is elected by the electoral college, *and not by individual voters. Because of this, smaller states have a greater voice in the selection of a president than they would otherwise. Two political scientists have studied the problems of campaigning for president under the current system and have concluded that candidates should allot their money according to the formula*

$$\text{amount for large state} = \left(\frac{E_{\text{large}}}{E_{\text{small}}}\right)^{3/2} \times \text{amount for small state.}$$

Here E_{large} *represents the electoral vote of the large state, and* E_{small} *represents the electoral vote of the small state. Find the amount that should be spent in each of the following larger states if* \$1,000,000 *is spent in the small state and the following statements are true.*

83. The large state has 48 electoral votes and the small state has 3.

84. The large state has 36 electoral votes and the small state has 4.

Natural Science

A Delta Airlines map gives a formula for calculating the visible distance from a jet plane to the horizon. On a clear day, this distance is approximated by

$$D = 1.22x^{1/2},$$

where x is altitude in feet, and D is distance to the horizon in miles. Use a calculator or Table 2 to find D for an altitude of

85. 5000 feet; **86.** 10,000 feet;

87. 30,000 feet; **88.** 40,000 feet.

(This problem requires a calculator with an x^y key.) The Galápagos Islands are a chain of islands ranging in size from .2 to 2249 square miles. A biologist has shown that the number of different land-plant species on an island in this chain is related to the size of the island by

$$S = 28.6A^{0.32},$$

where A is the area of an island in square miles and S is the number of different plant species on that island. Estimate S (rounding to the nearest whole number) for islands of area

89. 1 square mile;

90. 25 square miles;

91. 300 square miles;

92. 2000 square miles.

CHAPTER 1 TEST

[1.1] *Name all the types of numbers that apply to the following.*

1. -9

2. $\sqrt{5}$

Graph each of the following on a number line.

3. $x \geq -3$

4. $-4 < x \leq 6$

Evaluate each of the following.

5. $-|-6|$

6. $|-8 - (-4)|$

[1.2] *Solve each of the following equations.*

7. $4k - 11 + 3k = 2k - 1$

8. $2(k + 3) - 4(1 - k) = 6$

9. $\dfrac{2x}{5} - \dfrac{x - 3}{5} = 2$

Solve each of the following inequalities.

10. $-5k \leq 20$

11. $3m + 5 > 4m - 9$

12. Find the unearned interest if the original finance charge is $1200, the loan was scheduled for 24 payments, and the loan is paid off with 6 payments left.

[1.3] *Factor each of the following as completely as possible.*

13. $8y + 16$

14. $p^2 - 9p + 14$

15. $k^2 + 4k - 45$

16. $2a^2 + 7a - 15$

17. $3m^2 - 8m - 35$

18. $16p^2 + 24p + 9$

[1.4] *Solve each of the following quadratic equations.*

19. $x^2 + 3x = 18$

20. $6z^2 = 11z + 10$

21. $y^2 - 4y + 2 = 0$

22. $9k^2 - 6k = 2$

Solve each of the following quadratic inequalities.

23. $(x-1)(x+3) \le 0$

24. $3y^2 > 5y + 2$

[1.5] *Work each of the following problems as indicated.*

25. $\frac{6p}{7} \cdot \frac{28p}{15}$

26. $\frac{9r^2}{16} \div \frac{27r}{32}$

27. $\frac{m^2 - 2m - 3}{m(m-3)} \div \frac{m+1}{4m}$

28. $\frac{5}{2r} + \frac{6}{r}$

29. $\frac{3}{8y} + \frac{1}{2y} + \frac{7}{9y}$

30. $\frac{8}{r-1} - \frac{3}{r}$

[1.6] *Simplify each of the following. Write all answers without exponents.*

31. 4^{-2}

32. $6^{-5} \cdot 6^{7}$

33. $\frac{8^{-2}}{8^{-1}}$

34. $64^{3/2}$

35. $9^{-5/2}$

36. $\left(\frac{144}{49}\right)^{-1/2}$

CASE 1 ESTIMATING SEED DEMANDS—THE UPJOHN COMPANY*

The Upjohn Company has a subsidiary which buys seeds from farmers and then resells them. Each spring the firm contracts with farmers to grow the seeds. The firm must decide on the number of acres that it will contract for. The problem faced by the company is that the demand for seeds is not constant, but fluctuates from year to year. Also, the number of tons of seed produced per acre varies, depending on weather and other factors. In an attempt to decide the number of acres that should be planted in order to maximize profits, a company mathematician created a model of the variables involved in determining the number of acres to plant.

The analysis of this model required advanced methods that we will not go into. We can, however, give the conclusion; the number of acres that will maximize profit in the long run is found by solving the equation

$$F(AX + Q) = \frac{(S - C_p)X - C_A}{(S - C_p + C_c)X} \qquad \textbf{(1)}$$

for A. The function $F(z)$ represents the chances that z tons of seed will be demanded by the marketplace. The variables in the equation are:

A = number of acres of land contracted by the company
X = quantity of seed produced per acre of land
Q = quantity of seed in inventory from previous years
S = selling price per ton of seed
C_p = variable cost (production, marketing, etc.) per ton of seed
C_c = cost to carry over one ton of seed from previous year
C_A = variable cost per acre of land

To advise management of the number of acres of seed to contract for, the mathematician studied past records to find the values of the various variables. From these records and from predictions of future trends, it was concluded that $S = \$10{,}000$ per ton, $X = .1$ ton per acre (on the average), $Q = 200$ tons, $C_p = \$5000$ per ton, $C_A = \$100$ per acre, $C_c = \$3000$ per ton.

The function $F(z)$ is found by the same process to be approximated by

$$F(z) = \frac{z}{1000} - \frac{1}{2}, \qquad \text{if} \qquad 500 \le z \le 1500 \text{ tons.} \qquad \textbf{(2)}$$

EXERCISES

1. Evaluate $AX + Q$ using the values of X and Q given above.

2. Find $F(AX + Q)$, using equation (2) and your results from Exercise 1.

3. Solve equation (1) for A.

4. How many acres should be planted?

5. How many tons of seed will be produced?

6. Find the total revenue that will be received from the sale of the seeds.

---

* Based on work by David P. Rutten, Senior Mathematician, The Upjohn Company, Kalamazoo, Michigan.

2 Sets and Functions

One idea that comes up time and again in mathematics, particularly in applications, is that of a function. We will see examples where the price of a commodity is a function of the supply, the number of people in a city is related to time, and so on. We discuss functions at the end of this chapter, and then use these ideas throughout the book.

However, before we can discuss functions, we need to look at some topics basic to an understanding of them. A brief discussion of sets and set notation begins the chapter. This is followed by a section showing applications of sets to surveys. Then we define functions. The chapter ends with a section on a special kind of function, the sequence function.

2.1 SETS

A **set** is a collection of objects. We could form a set containing one of each type of coin now put out by the government. Another set might be made up of all the students in your class. In mathematics, sets are usually made up of numbers. The set containing the numbers 3, 4, and 5 could be written

$$\{3, 4, 5\}$$

where we use set braces, $\{\ \ \}$, to enclose the numbers belonging to the set. The numbers 3, 4, and 5 are called the **elements** or **members** of this set. To show that 4 is an element of the set $\{3, 4, 5\}$, we use the symbol $\in$ and write

$$4 \in \{3, 4, 5\}.$$

Also, $5 \in \{3, 4, 5\}$. To show that 8 is *not* an element of this set, place a slash through the symbol.

$$8 \notin \{3, 4, 5\}$$

We often name sets with capital letters, so that if

$$B = \{5, 6, 7\},$$

we have, for example, $6 \in B$ and $10 \notin B$.

1. Write *true* or *false*.
(a) $9 \in \{8, 4, -3, -9, 6\}$
(b) $4 \notin \{3, 9, 7\}$
(c) If $M = \{0, 1, 2, 3, 4\}$, then $0 \in M$.

Answer:
(a) False
(b) True
(c) True

Work Problem 1 at the side.

Two sets are **equal** if they contain exactly the same elements. The sets $\{5, 6, 7\}$, $\{7, 6, 5\}$, and $\{6, 5, 7\}$ all contain exactly the same elements and are equal. In symbols,

$$\{5, 6, 7\} = \{7, 6, 5\} = \{6, 5, 7\}.$$

Sets which do not contain exactly the same elements are *not equal.* For example, the sets $\{5, 6, 7\}$ and $\{5, 6, 7, 8\}$ do not contain exactly the same elements and are not equal. This is written as follows:

$$\{5, 6, 7\} \neq \{5, 6, 7, 8\}.$$

Sometimes we are more interested in a common property of a set rather than in a list of the elements in the set. We can express this common property using **set-builder notation.** Write

$$\{x | x \text{ has property } P\}$$

to represent the set of all elements having some property P.

Example 1 Write the elements belonging to each of the following sets.

(a) $\{x | x \text{ is a counting number less than } 5\}$
The counting numbers less than 5 make up the set $\{1, 2, 3, 4\}$.

(b) $\{x | x \text{ is a state that touches Florida}\} = \{\text{Alabama, Georgia}\}$ ■

2. List the elements in the following sets.
(a) $\{x | x \text{ is a counting number more than 5 and less than 8}\}$
(b) $\{x | x \text{ is an integer}, -3 < x \leq 1\}$

Answer:
(a) $\{6, 7\}$
(b) $\{-2, -1, 0, 1\}$

Work Problem 2 at the side.

When discussing a particular situation or problem, we can usually identify a **universal set** (whether expressed or implied) which contains all the elements appearing in any set used in the given problem. The letter U is used to represent the universal set.

For example, when discussing the set of company employees who favor a certain pension proposal, we might choose the universal set to be the set of all company employees. In discussing the types of species found by Charles Darwin on the Galápagos Islands, the universal set might be the set of all species on all Pacific islands. The choice of a universal set is often arbitrary and depends on the problem under discussion.

Sometimes every element of one set also belongs to another

set. For example, if

$$A = \{3, 4, 5, 6\}$$

and

$$B = \{2, 3, 4, 5, 6, 7, 8\},$$

then every element of A is also an element of B. This means that A is a **subset** of B, written $A \subset B$. For example, the set of all presidents of corporations is a subset of the set of all executives of corporations.

Example 2 Decide whether the following statements are true or false.

(a) $\{3, 4, 5, 6\} = \{4, 6, 3, 5\}$

Both sets contain exactly the same elements; the sets are equal. The statement is true. (The fact that the elements are in a different order doesn't matter.)

(b) $\{5, 6, 9, 10\} \subset \{5, 6, 7, 8, 9, 10, 11\}$

Every element of the first set is also an element of the second. This statement is also true. ■

3. Write *true* or *false*.
(a) $\{3, 4, 5\} \subset \{2, 3, 4, 6\}$
(b) $\{1, 2, 5, 8\} \subset \{1, 2, 5, 10, 11\}$
(c) $\{3, 6, 9, 10\} \subset \{3, 9, 11, 13\}$

Answer:
(a) False
(b) False
(c) False

Work Problem 3 at the side.

Figure 2.1 shows a drawing which represents a set A which is a subset of set B. The rectangle of the drawing represents the universal set, U. Diagrams like this are called **Venn diagrams.** We use Venn diagrams as an aid in clarifying and discussing sets.

By the definition of subset, the **empty set** (which contains no elements) is a subset of every set. That is, if A is a set, and we use $\varnothing$ to represent the empty set, then

$$\varnothing \subset A.$$

Example 3 List all possible subsets for each of the following sets.

(a) $\{7, 8\}$

There are four subsets of $\{7, 8\}$:

$$\varnothing, \quad \{7\}, \quad \{8\}, \quad \{7, 8\}$$

(b) $\{a, b, c\}$

There are eight subsets of $\{a, b, c\}$:

$$\varnothing, \quad \{a\}, \quad \{b\}, \quad \{c\}, \quad \{a, b\}, \quad \{a, c\}, \quad \{b, c\}, \quad \{a, b, c\}$$ ■

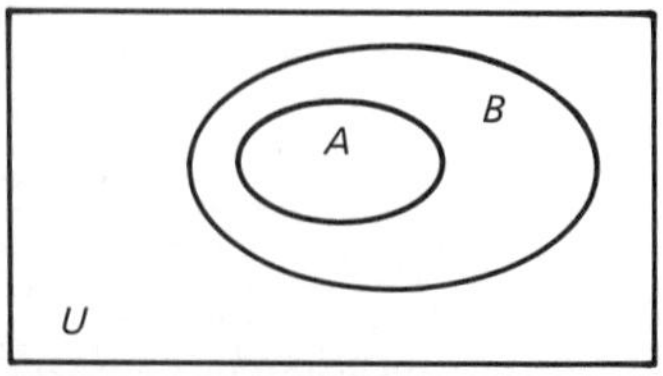

$A \subset B$

Figure 2.1

In Example 3, we found all subsets of $\{7, 8\}$ and all subsets of $\{a, b, c\}$ by trial and error. An alternate method uses a **tree diagram.** A tree diagram is a systematic way of listing all the subsets of a given set. Figures 2.2(a) and (b) show tree diagrams for finding the subsets of $\{7, 8\}$ and $\{a, b, c\}$.

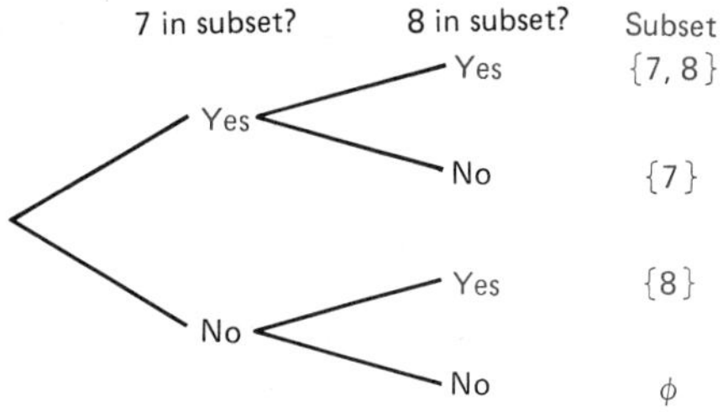

(a) Find the subsets of {7, 8}

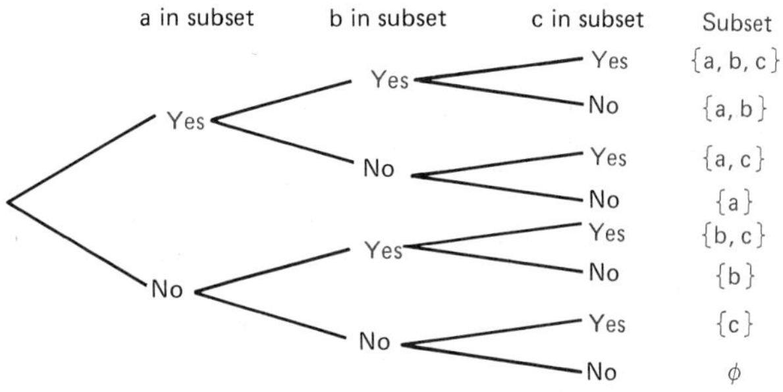

(b) Find the subsets of {a, b, c}

Figure 2.2

By studying examples and tree diagrams similar to the ones above, we are led to the following rule.

A set containing n elements has 2^n subsets.

Example 4 Find the number of subsets for each of the following sets.

(a) {3, 4, 5, 6, 7}
This set has five elements; thus it has 2^5 or 32 subsets.

(b) {−1, 2, 3, 4, 5, 6, 12, 14}
This set has 8 elements and therefore has 2^8 or 256 subsets. ■

Work Problem 4 at the side.

4. Find the number of subsets for each of the following sets.
(a) {1, 2, 3}
(b) {−6, −5, −4, −3, −2, −1, 0}
(c) {6}

Answer:
(a) 8
(b) 128
(c) 2

2.1 EXERCISES

Write true *or* false *for each of the following.*

1. $3 \in \{2, 5, 7, 9, 10\}$

2. $6 \in \{-2, 6, 9, 5\}$

3. $1 \in \{3, 4, 5, 1, 11\}$
4. $12 \in \{19, 17, 14, 13, 12\}$
5. $9 \notin \{2, 1, 5, 8\}$
6. $3 \notin \{7, 6, 5, 4\}$
7. $\{2, 5, 8, 9\} = \{2, 5, 9, 8\}$
8. $\{3, 0, 9, 6, 2\} = \{2, 9, 0, 3, 6\}$
9. $\{5, 8, 9\} = \{5, 8, 9, 0\}$
10. $\{3, 7, 12, 14\} = \{3, 7, 12, 14, 0\}$
11. {all counting numbers less than 6} $= \{1, 2, 3, 4, 5, 6\}$
12. {all whole numbers greater than 7 and less than 10} $= \{8, 9\}$
13. {all whole numbers not greater than 4} $= \{0, 1, 2, 3\}$
14. {all counting numbers not greater than 3} $= \{0, 1, 2\}$
15. $\{x|x$ is a whole number, $x \leq 5\} = \{0, 1, 2, 3, 4, 5\}$
16. $\{x|x$ is an integer, $-3 \leq x < 4\} = \{-3, -2, -1, 0, 1, 2, 3, 4\}$
17. $\{x|x$ is an odd integer, $6 \leq x \leq 18\} = \{7, 9, 11, 15, 17\}$
18. $\{x|x$ is an even counting number, $x \leq 9\} = \{0, 2, 4, 6, 8\}$

Let

$$A = \{2, 4, 6, 8, 10, 12\} \qquad D = \{2, 10\}$$
$$B = \{2, 4, 8, 10\} \qquad U = \{2, 4, 6, 8, 10, 12, 14\}$$
$$C = \{4, 10, 12\}$$

Write true *or* false *for each of the following. (See Examples 1 and 3.)*

19. $A \subset U$
20. $C \subset U$
21. $D \subset B$
22. $D \subset A$
23. $A \subset B$
24. $B \subset C$
25. $\varnothing \subset A$
26. $\varnothing \subset \varnothing$
27. $\{4, 8, 10\} \subset B$
28. $\{0, 2\} \subset D$
29. $D \not\subset B$
30. $A \not\subset C$
31. There are exactly 32 subsets of A.
32. There are exactly 16 subsets of B.
33. There are exactly 6 subsets of C.
34. There are exactly 4 subsets of D.

Find the number of subsets for each of the following sets. (See Example 3.)

35. $\{4, 5, 6\}$ **36.** $\{3, 7, 9, 10\}$

37. $\{5, 9, 10, 15, 17\}$ **38.** $\{6, 9, 1, 4, 3, 2\}$

39. $\varnothing$ **40.** $\{0\}$

41. $\{x|x$ is a counting number between 6 and $12\}$

42. $\{x|x$ is a whole number between 8 and $12\}$

APPLIED PROBLEMS

Natural Science

A Hershey bar of a certain size contains 220 calories. Suppose you eat two of these candy bars and then decide to exercise and get rid of the calories. A list of possible exercises shows the following information.

Exercise	Abbreviation	Calories per hour
Sitting around	s	100
Light exercise	l	170
Moderate exercise	m	300
Severe exercise	e	450
Very severe exercise	v	600

The universal set here is $U = \{s, l, m, e, v\}$. Find all subsets of U (with no element listed twice) that will burn off the calories from the candy bars in

43. one hour; **44.** two hours.

2.2 SET OPERATIONS

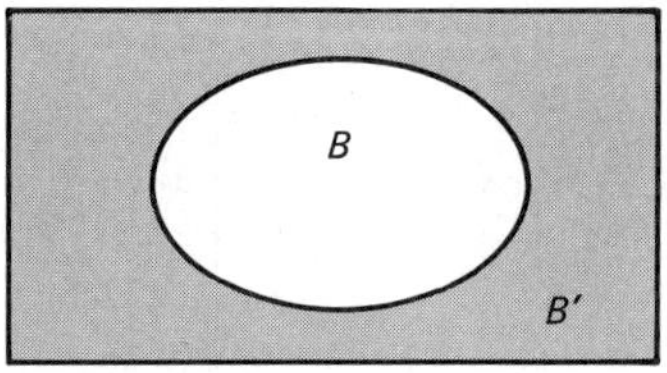

Figure 2.3

If we are given a set A, and a universal set U, we can form the set of all elements which do *not* belong to set A. This set is called the **complement** of set A. For example, if set A is the set of all the female students in your class, and U is the set of all students, then the complement of A would be the set of all the male students in the class. The complement of set A is written A'. (Read: "A-prime.") The Venn diagram of Figure 2.3 shows a set B. Its complement, B', is shown in color.

Example 1 Let $U = \{1, 2, 3, 4, 5, 6, 7\}$, $A = \{1, 3, 5, 7\}$ and $B = \{3, 4, 6\}$. Find each of the following sets.

1. Let $U = \{a, b, c, d, e, f, g\}$, with $K = \{c, d, f, g\}$ and $R = \{a, c, d, e, g\}$. Find
(a) K'
(b) R'.

Answer:
(a) $\{a, b, e\}$
(b) $\{b, f\}$

(a) A'
Set A' contains the elements of U that are not in A.

$$A' = \{2, 4, 6\}$$

(b) $B' = \{1, 2, 5, 7\}$
(c) $\varnothing' = U$ and $U' = \varnothing$ ■

Work Problem 1 at the side.

Given two sets A and B, the set of all elements belonging to both set A and set B is called the **intersection** of the two sets, written $A \cap B$. For example, the elements that belong to both $A = \{1, 2, 4, 5, 7\}$ and $B = \{2, 4, 5, 7, 9, 11\}$ are 2, 4, 5, and 7, so that

$$A \cap B = \{1, 2, 4, 5, 7\} \cap \{2, 4, 5, 7, 9, 11\} = \{2, 4, 5, 7\}.$$

The Venn diagram of Figure 2.4 shows two sets A and B; their intersection, $A \cap B$, is shown in color.

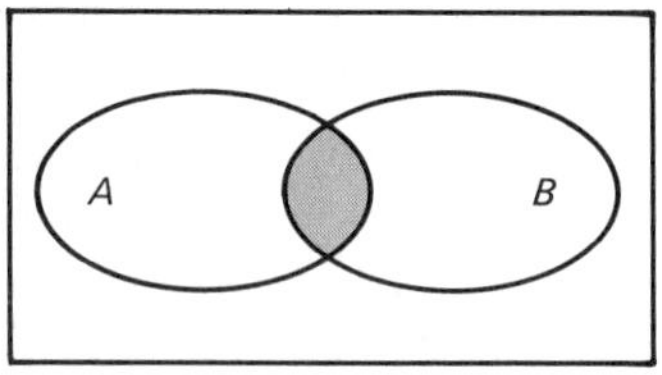

Figure 2.4

Example 2 **(a)** $\{9, 15, 25, 36\} \cap \{15, 20, 25, 30, 35\} = \{15, 25\}$
The elements 15 and 25 are the only ones belonging to both sets.
(b) $\{-2, -3, -4, -5, -6\} \cap \{-4, -3, -2, -1, 0, 1, 2\} = \{-4, -3, -2\}$ ■

2. Find the following.
(a) $\{1, 2, 3, 4\} \cap \{3, 5, 7, 9\}$
(b) $\{a, b, c, d, e, g\} \cap \{a, c, e, g, h, j\}$

Answer:
(a) $\{3\}$
(b) $\{a, c, e, g\}$

Work Problem 2 at the side.

Two sets that have no elements in common are called **disjoint sets.** For example, there are no elements common to both $\{50, 51, 54\}$ and $\{52, 53, 55, 56\}$, so that these two sets are disjoint, and

$$\{50, 51, 54\} \cap \{52, 53, 55, 56\} = \varnothing.$$

The result of this example can be generalized: if A and B are any two disjoint sets, then $A \cap B = \varnothing$.

The set of all elements belonging to set A or set B is called the **union** of the two sets, written $A \cup B$. For example,

$$\{1, 3, 5\} \cup \{3, 5, 7, 9\} = \{1, 3, 5, 7, 9\}.$$

The Venn diagram of Figure 2.5 shows two sets A and B; their union, $A \cup B$, is shown in color.

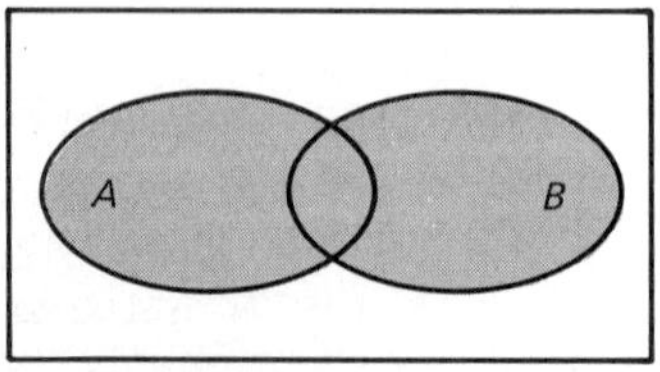

Figure 2.5

Example 3 Find the union of $\{1, 2, 5, 9, 14\}$ and $\{1, 3, 4, 8\}$.
Begin by listing the elements of the first set, $\{1, 2, 5, 9, 14\}$.

Then include any elements from the second set that are not already listed. Doing this gives us

$$\{1, 2, 5, 9, 14\} \cup \{1, 3, 4, 8\} = \{1, 2, 3, 4, 5, 8, 9, 14\}. \quad ■$$

Example 4 $\{1, 3, 5, 7\} \cup \{2, 4, 6\} = \{1, 2, 3, 4, 5, 6, 7\}$. ■

3. Find the following.
(a) $\{a, b, c\} \cup \{a, c, e\}$
(b) $\{a, c, d, e, f\} \cup \{a, b, c, d, g\}$

Answer:
(a) $\{a, b, c, e\}$
(b) $\{a, b, c, d, e, f, g\}$

Work Problem 3 at the side.

Finding the complement of a set, the intersection of two sets, or the union of two sets are examples of **set operations.** These are similar to operations on numbers, such as addition, subtraction, multiplication, and division.

Venn Diagrams We included Venn diagrams above to help in understanding set union and intersection. The rectangular region in a Venn diagram represents the universal set, U. If we include only a single set A inside the universal set, as in Figure 2.6, we divide the total region of U into two regions. Region 1 represents those elements outside of set A, while region 2 represents those elements belonging to set A.

If we include two sets A and B inside U, we get the Venn diagram of Figure 2.7. Two sets divide the universal set into four regions. As labeled in Figure 2.7, region 1 includes those elements outside of both set A and set B. Region 2 includes those elements belonging to A and not to B. Region 3 includes those elements belonging to both A and B. Which elements belong to region 4?

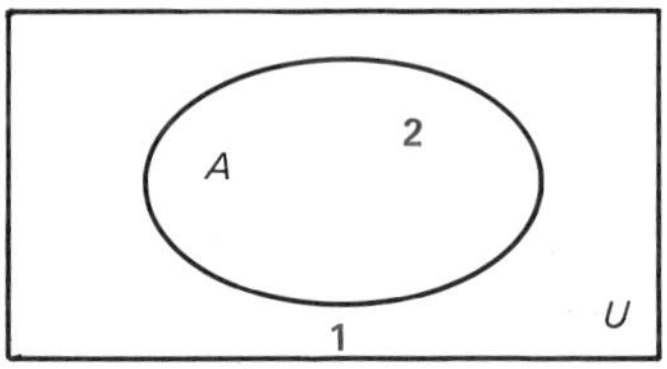

One set leads to 2 regions

Figure 2.6

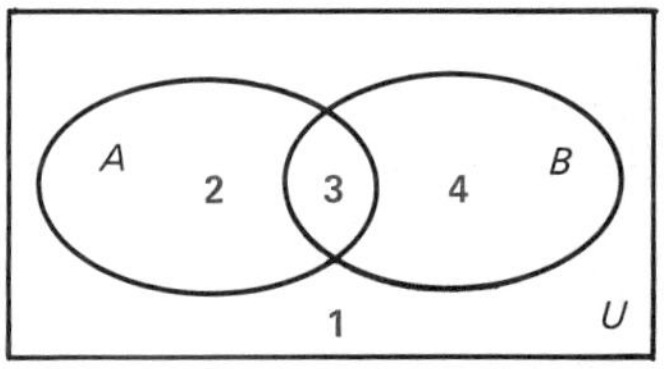

Two sets lead to 4 regions

Figure 2.7

Example 5 Draw Venn diagrams similar to Figure 2.7 and shade the regions representing the following sets.

(a) $A' \cap B$

Set A' contains all the elements outside of set A. As labeled in Figure 2.7, A' is made up of regions 1 and 4. Set B is made up of the elements in regions 3 and 4. The intersection of sets A' and B, the set $A' \cap B$, is made up of the elements in the region common to regions 1 and 4 and regions 3 and 4. The result, region 4, is shaded in Figure 2.8.

(b) $A' \cup B'$

Again, set A' is represented by regions 1 and 4, while B' is made up of regions 1 and 2. To find $A' \cup B'$, we need the elements belonging to either regions 1 and 4 or to regions 1 and 2. The result, regions 1, 2, and 4, is shaded in Figure 2.9. ■

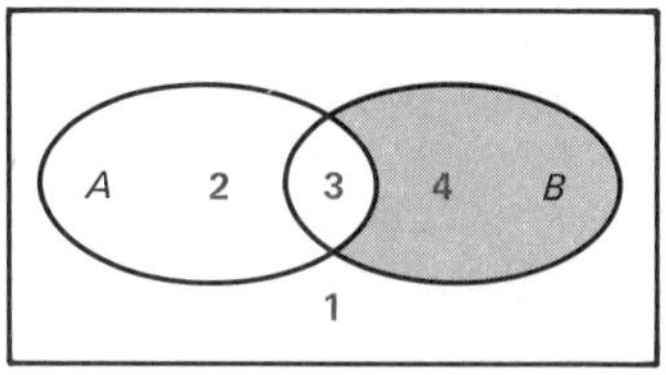

Figure 2.8

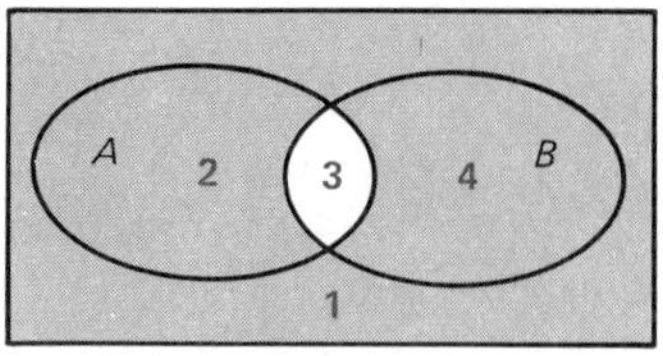

Figure 2.9

Work Problem 4 at the side.

We can draw Venn diagrams with three sets inside U. Three sets divide the universal set into eight regions, which are numbered in Figure 2.10.

Example 6 Shade $A' \cup (B \cap C')$ on a Venn diagram.

We first find $B \cap C'$. Set B is made up of regions 3, 4, 7, and 8, while C' is made up of regions 1, 2, 3, and 8. The overlap of these regions, the set $B \cap C'$, is made up of regions 3 and 8. Set A' is made up of regions 1, 6, 7, and 8. The union of regions 3 and 8 and regions 1, 6, 7, 8 is regions 1, 3, 6, 7, 8, which are shaded in Figure 2.11. ■

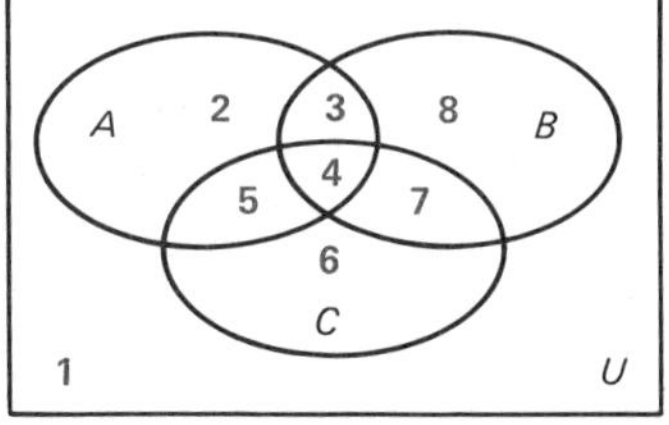

Three sets lead to 8 regions

Figure 2.10

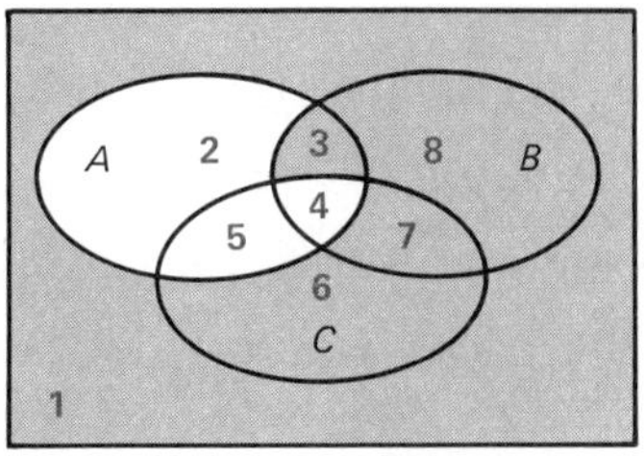

Figure 2.11

Work Problem 5 at the side.

4. Draw Venn diagrams for the following.
(a) $A \cup B'$
(b) $A' \cap B'$

Answer:
(a)

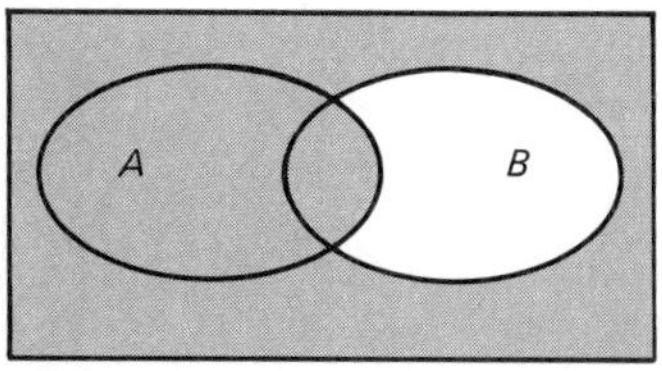

$A \cup B'$

(b)

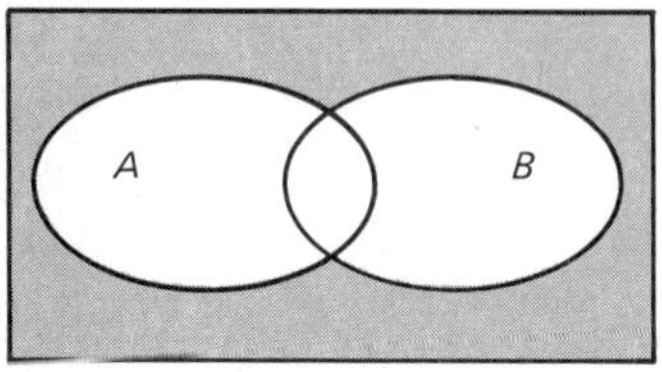

$A' \cap B'$

5. Draw a Venn diagram for $(A \cup B)' \cap C$.

Answer:

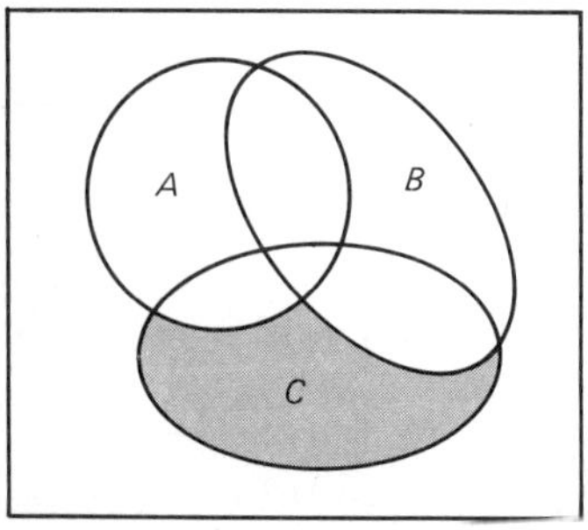

$(A \cup B)' \cap C$

2.2 EXERCISES

Write true *or* false *for each of the following. (See Examples 2–4.)*

1. $\{5, 7, 9, 19\} \cap \{7, 9, 11, 15\} = \{7, 9\}$

2. $\{8, 11, 15\} \cap \{8, 11, 19, 20\} = \{8, 11\}$

3. $\{2, 1, 7\} \cup \{1, 5, 9\} = \{1\}$

4. $\{6, 12, 14, 16\} \cup \{6, 14, 19\} = \{6, 14\}$

5. $\{3, 2, 5, 9\} \cap \{2, 7, 8, 10\} = \{2\}$

6. $\{8, 9, 6\} \cup \{9, 8, 6\} = \{8, 9\}$

7. $\{3, 5, 9, 10\} \cap \varnothing = \{3, 5, 9, 10\}$

8. $\{3, 5, 9, 10\} \cup \varnothing = \{3, 5, 9, 10\}$

9. $\{1, 2, 4\} \cup \{1, 2, 4\} = \{1, 2, 4\}$

10. $\{1, 2, 4\} \cap \{1, 2, 4\} = \varnothing$

11. $\varnothing \cup \varnothing = \varnothing$

12. $\varnothing \cap \varnothing = \varnothing$

Let $U = \{2, 3, 4, 5, 7, 9\}$, $X = \{2, 3, 4, 5\}$, $Y = \{3, 5, 7, 9\}$, *and* $Z = \{2, 4, 5, 7, 9\}$. *Find each of the following sets.*

13. $X \cap Y$

14. $X \cup Y$

15. $Y \cup Z$

16. $Y \cap Z$

17. $X \cup U$

18. $Y \cap U$

19. X'

20. Y'

21. $X' \cap Y'$

22. $X' \cap Z$

23. $Z' \cap \varnothing$

24. $Y' \cup \varnothing$

25. $X \cup (Y \cap Z)$

26. $Y \cap (X \cup Z)$

Let $U = \{\text{all students in this school}\}$
$M = \{\text{all students taking this course}\}$
$N = \{\text{all students taking accounting}\}$
$P = \{\text{all students taking zoology}\}$

Describe each of the following sets in words.

27. M'

28. $M \cup N$

29. $N \cap P$

30. $N' \cap P'$

31. $M \cup P$

32. $P' \cup M'$

Use a Venn diagram similar to Figure 2.7 to show each of the following sets. (See Example 5.)

33. $B \cap A'$

34. $A \cup B'$

35. $A' \cup B$

36. $A' \cap B'$

37. $B' \cup (A' \cap B')$

38. $(A \cap B) \cup B'$

39. U'

40. $\varnothing'$

Use a Venn diagram similar to Figure 2.10 to show each of the following sets. (See Example 6.)

41. $(A \cap B) \cap C$

42. $(A \cap C') \cup B$

43. $A \cap (B \cup C')$

44. $A' \cap (B \cap C)$

45. $(A' \cap B') \cap C$

46. $(A \cap B') \cup C$

47. $(A \cap B') \cap C$

48. $A' \cap (B' \cup C)$

APPLIED PROBLEMS

Natural Science

The lists below show some symptoms of an overactive thyroid and an underactive thyroid.

Underactive thyroid	Overactive thyroid
Sleepiness, s	Insomnia, i
Dry hands, d	Moist hands, m
Intolerance of cold, c	Intolerance of heat, h
Goiter, g	Goiter, g

49. Find the smallest possible universal set U that includes all the symptoms listed.

Let N be the set of symptoms for an underactive thyroid, and let O be the set of symptoms for an overactive thyroid. Find each of the following sets.

50. O'

51. N'

52. $N \cap O$

53. $N \cup O$

54. $N \cap O'$

2.3 SURVEYS: AN APPLICATION OF SETS

We can use Venn diagrams to solve problems in surveying groups of people. Suppose a group of 60 freshman business students at a large university was surveyed, with the following results.

19 of the students read *Business Week*;
18 read *The Wall Street Journal*;
50 read *Playboy*;
13 read *Business Week* and *The Journal*;
11 read *The Journal* and *Playboy*;
13 read *Business Week* and *Playboy*;
9 read all three.

Let us use this data to help answer the following questions.

(a) How many students read none of the publications?
(b) How many read only *Playboy*?
(c) How many read *Business Week* and *The Journal*, but not *Playboy*?

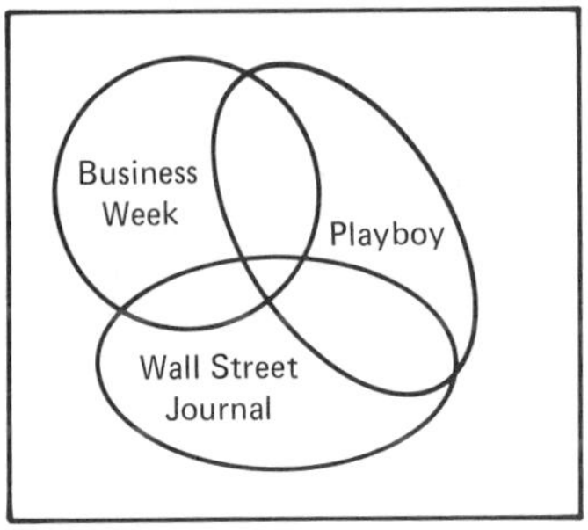

Figure 2.12

Many of the students are listed more than once in the data above. For example, some of the 50 students who read *Playboy* also read *Business Week*. The 9 students who read all three are counted in the 13 who read *Business Week* and *Playboy*, and so on.

We can use a Venn diagram, as shown in Figure 2.12, to better illustrate this data. Since 9 students read all three publications, we begin by placing 9 in the region that belongs to all three circles, as shown in Figure 2.13. We know that 13 students read *Business Week* and *Playboy*. However, 9 of these 13 also read *The Journal*. Therefore, only $13 - 9 = 4$ read just *Business Week* and *Playboy*. Place the number 4 in the area of Figure 2.13 common to *Business Week* and *Playboy* readers. In the same way, place 4 in the region common only to *Business Week* and *The Journal*, and 2 in the region common only to *Playboy* and *The Journal*.

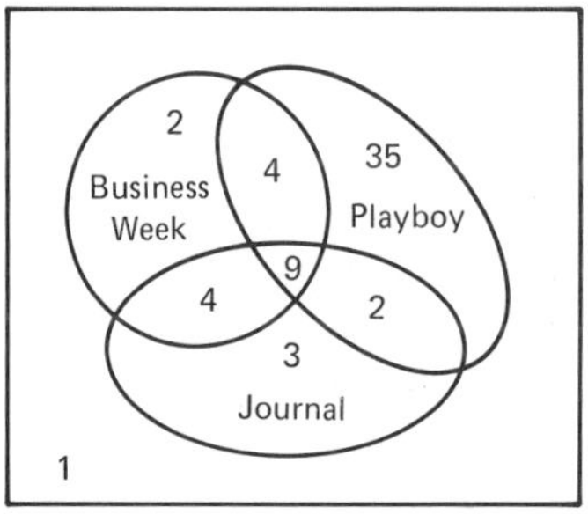

Figure 2.13

We know 19 students read *Business Week*. However, we have already placed $4 + 9 + 4 = 17$ readers in the region representing *Business Week*. Thus, the rest of this region will contain only $19 - 17 = 2$ students. These 2 students read *Business Week* only—not *Playboy* and not *The Journal*. In the same way, 3 students read only *The Journal* and 35 read only *Playboy*.

We have placed $2 + 4 + 3 + 4 + 9 + 2 + 35 = 59$ students in the three regions of Figure 2.13. We know that 60 students were surveyed; thus, $60 - 59 = 1$ student reads none of these three publications and so is placed outside all three regions.

We can now use Figure 2.13 to answer the questions asked above.

(a) Only 1 student reads none of the three publications.
(b) From Figure 2.13, 35 students read only *Playboy*.
(c) The overlap of the regions representing *Business Week* and *The Journal* shows that 4 students read *Business Week* and *The Journal* but not *Playboy*.

1. In the example about the three publications, how many students read exactly
(a) 1
(b) 2
of the publications?

Answer:
(a) 40
(b) 10

Work Problem 1 at the side.

Example 1 Jeff Friedman is a section chief for an electric utility company. The employees in his section cut down tall trees, climb poles, and splice wire. Friedman reported the following information to the management of the utility.

"Out of the 100 employees in my section,
45 can cut tall trees;
50 can climb poles;
57 can splice wire;
28 can cut trees and climb poles;
20 can climb poles and splice wire;
25 can cut trees and splice wire;
11 can do all three;
9 can't do any of the three (management trainees)."

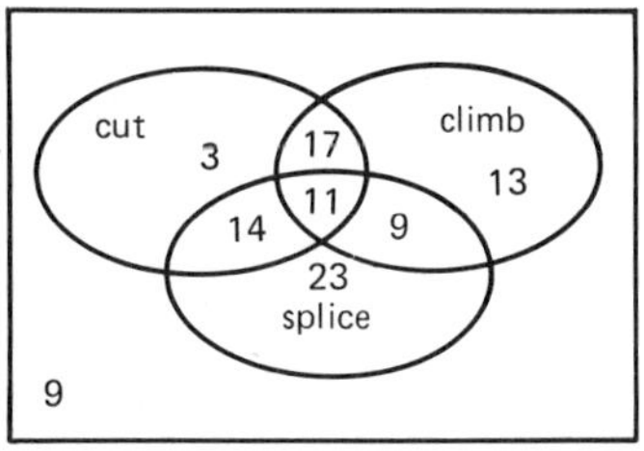

Figure 2.14

From the data supplied by Friedman we can find the numbers shown in Figure 2.14. By adding all the numbers from the regions, we find the total number of Friedman's employees to be

$$9 + 3 + 14 + 23 + 11 + 9 + 17 + 13 = 99.$$

Friedman claimed to have 100 employees, but his data indicates only 99. The management decided that Friedman didn't qualify as a section chief, and reassigned him as a nightshift meter reader in Guam. (Moral: he should have taken this course.) ■

2.3 APPLIED PROBLEMS

Management

Use Venn diagrams to answer the following questions. (See Example 1.)

1. Jeff Friedman, of Example 1 in the text, was again reassigned, this time to the home economics department of the electric utility. He interviewed 140 people in a suburban shopping center to find out some of their cooking habits. He obtained the following results. Should he be reassigned yet one more time?

58 use microwave ovens;
63 use electric ranges;
58 use gas ranges;
19 use microwave ovens and electric ranges;
17 use microwave ovens and gas ranges;
4 use both gas and electric ranges;
1 uses all three;
2 cook only with solar energy.

2. Toward the middle of the harvesting season, peaches for canning come in three types: earlies, lates, and extra lates, depending on the expected date of ripening. During a certain week, the following data was recorded at a fruit delivery station.

   34 trucks went out carrying early peaches;
   61 had late peaches;
   50 had extra lates;
   25 had earlies and lates;
   30 had lates and extra lates;
   8 had earlies and extra lates;
   6 had all three;
   9 had only figs (no peaches at all).

   (a) How many trucks had only late variety peaches?
   (b) How many had only extra lates?
   (c) How many had only one type of peaches?
   (d) How many trucks in all went out during the week?

3. A chicken farmer in Wappingers Falls, New York, surveyed his flock with the following results. The farmer had

   9 fat red roosters;
   2 fat red hens;
   37 fat chickens;
   26 fat roosters;
   7 thin brown hens;
   18 thin brown roosters;
   6 thin red roosters;
   5 thin red hens.

   Answer the following questions about the flock. Hint: you need a Venn diagram with regions for fat, for male (a rooster is a male, a hen is a female), and for red (assume that brown and red are opposites in the chicken world). How many chickens were
   (a) fat?
   (b) red?
   (c) male?
   (d) fat, but not male?
   (e) brown, but not fat?
   (f) red and fat?

General

4. Country-western songs emphasize three basic themes: love, prison, and trucks. A survey of the local country-western radio station produced the following data.

   12 songs were about a truck driver who was in love while in prison;
   13 about a prisoner in love;
   28 about a person in love;

18 about a truck driver in love;

3 about a truck driver in prison who was not in love;

2 about a prisoner who was not in love and did not drive a truck;

8 about a person out of jail who was not in love, and did not drive a truck;

16 about truck drivers who were not in prison.

(a) How many songs were surveyed?

Find the number of songs about

(b) truck drivers;

(c) prisoners;

(d) truck drivers in prison;

(e) people not in prison;

(f) people not in love.

Natural Science

5. After a genetics experiment, the number of pea plants having certain characteristics was tallied, with the results as follows.

22 were tall;

25 had green peas;

39 had smooth peas;

9 were tall and had green peas;

17 were tall and had smooth peas;

20 had green peas and smooth peas;

6 had all three characteristics;

4 had none of the characteristics.

(a) Find the total number of plants counted.

(b) How many plants were tall and had peas which were neither smooth nor green?

(c) How many plants were not tall but had peas which were smooth and green?

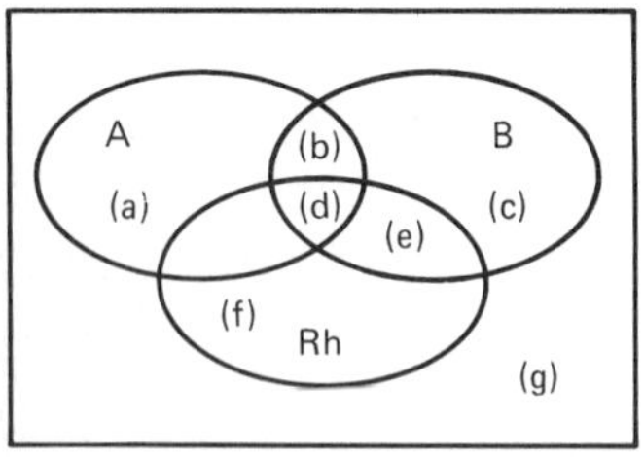

6. Human blood can contain either no antigens, the A antigen, the B antigen, or both the A and B antigens. A third antigen, called the Rh antigen, is important in human reproduction, and again may or may not be present in an individual. Blood is called type A-positive if the individual has the A and Rh, but not the B antigen. A person having only the A and B antigens is said to have type AB-negative blood. A person having only the Rh antigen has type O-positive blood. Other blood types are defined in a similar manner. Identify the blood type of the individuals in regions (a)–(g) of the Venn diagram.

7. (Use the diagram from Exercise 6.) In a certain hospital, the following data was recorded.

25 patients had the A antigen;

17 had the A and B antigens;

27 had the B antigen;

22 had the B and Rh antigens;

30 had the Rh antigen;
12 had none of the antigens;
16 had the A and Rh antigens;
15 had all three antigens.

How many patients
(a) were represented?
(b) had exactly one antigen?
(c) had exactly two antigens?
(d) had O-positive blood?
(e) had AB-positive blood?
(f) had B-negative blood?
(g) had O-negative blood?
(h) had A-positive blood?

Social Science

8. A survey of 80 sophomores at a western college showed that

36 take English;
32 take history;
32 take political science;
16 take political science and history;
16 take history and English;
14 take political science and English;
6 take all three.

How many students:
(a) take English and neither of the other two?
(b) take none of the three courses?
(c) take history, but neither of the other two?
(d) take political science and history, but not English?
(e) do not take political science?

9. The following table shows the number of people in a certain small town in Georgia who fit in the given categories.

Age	Drink vodka (V)	Drink bourbon (B)	Drink gin (G)	Totals
20–25 (Y)	40	15	15	70
26–35 (M)	30	30	20	80
over 35 (O)	10	50	10	70
Totals	80	95	45	220

Using the letters given in the table, find the number of people in each of the following sets.
(a) $Y \cap V$
(b) $M \cap B$
(c) $M \cup (B \cap Y)$
(d) $Y' \cap (B \cup G)$
(e) $O' \cup G$
(f) $M' \cap (V' \cap G')$

10. The following table shows the results of a survey in a medium-sized town in Tennessee. The survey asked questions about the investment habits of local citizens.

Age	Stocks (S)	Bonds (B)	Savings accounts (A)	Totals
18–29 (Y)	6	2	15	23
30–49 (M)	14	5	14	33
50 or over (O)	32	20	12	64
Totals	52	27	41	120

Using the letters given in the table, find the number of people in each of the following sets.
(a) $Y \cap B$
(b) $M \cup A$
(c) $Y \cap (S \cup B)$
(d) $O' \cup (S \cup A)$
(e) $(M' \cup O') \cap B$

2.4 FUNCTIONS

A common problem in many real-life situations is to describe relationships between quantities. For example, assuming that the number of hours a student studies each day is related to the grade he or she receives in the course, how can the relationship be expressed? One way is to set up pairs of symbols representing hours of study and the corresponding grades that resulted. For example, we might have the pairs

$$(3, \text{A}), \quad (2\tfrac{1}{2}, \text{B}), \quad (2, \text{C}), \quad (1, \text{D}), \quad (0, \text{F})$$

for the relationship in a particular case. This is an example of a relation.

In a more complex relationship, a formula of some sort can often be used to describe how one quantity changes with respect to another. For example, if a bank pays 6% interest per year, then we can write the interest, I, that a deposit of P dollars would earn in one year as

$$I = .06 \times P, \qquad \text{or} \qquad I = .06P,$$

to describe the relationship between interest and the amount deposited. Since situations which can be described in this way occur so frequently, the idea of a relation, and the special kind of relation called a function are very important.

We call (a, b), an **ordered pair** of numbers, where a is the first element or first **component,** of the ordered pair, and b is the second element or second **component.** A **relation** is defined as any set of ordered pairs.

Usually, there is some formula or rule that shows how the second component of an ordered pair is obtained from the first.

Example 1 List all ordered pairs belonging to the relation

$$\{(x, y)|x = 1, 2, 3, 4; y = x + 3\}.$$

Here x can take on any of the values 1, 2, 3, or 4. To find y, use the equation $y = x + 3$. For example, if $x = 1$, we have

$$y = x + 3$$
$$y = 1 + 3 \qquad \text{Let } x = 1$$
$$y = 4.$$

"If $x = 1$, then $y = 4$" is abbreviated as the ordered pair (1, 4). In the same way, if $x = 2$, then $y = 2 + 3 = 5$; this gives the ordered pair (2, 5). When $x = 3$, we have $y = 6$, producing (3, 6). Finally, when $x = 4$, we get (4, 7). The set of ordered pairs belonging to the given relation is $\{(1, 4), (2, 5), (3, 6), (4, 7)\}$. ■

1. Find the ordered pairs belonging to the relation

$\{(x, y)|x = 0, 1, 2, 3, 4; y = 2x - 5\}$.

Answer: $\{(0, -5), (1, -3), (2, -1), (3, 1), (4, 3)\}$

Work Problem 1 at the side.

The set of first components of the ordered pairs of a relation is called the **domain** of the relation, while the set of second components is called the **range** of the relation.

Example 2 Find the domain and range of the relation

$$\{(x, y)|x = 1, 2, 3, 4; y = x + 3\}.$$

In Example 1 above, we found that the ordered pairs of this relation form the set $\{(1, 4), (2, 5), (3, 6), (4, 7)\}$. The set of first components of the ordered pairs, $\{1, 2, 3, 4\}$, makes up the domain, while the range is the set of second components, $\{4, 5, 6, 7\}$. ■

2. Find the domain and range of the relation

$\{(x, y)|x = 0, 1, 2, 3, 4; y = 2x - 5\}$.

Answer: Domain is $\{0, 1, 2, 3, 4\}$; range is $\{-5, -3, -1, 1, 3\}$

Work Problem 2 at the side.

It is often useful to draw a graph of the ordered pairs of a relation. To do this, we use the perpendicular crossed number lines of a **Cartesian coordinate system,** as shown in Figure 2.15. We let the horizontal number line, or ***x*-axis,** represent the first component of the ordered pairs of the relation, while the vertical or ***y*-axis** represents the second component. The point where the number

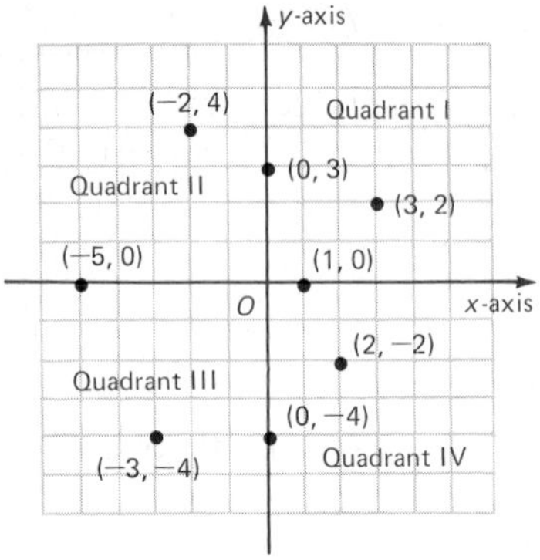

Figure 2.15

lines cross is the zero point on both lines; this point is called the **origin.**

To plot points on the graph corresponding to ordered pairs, proceed as in the following example. Locate the point $(-2, 4)$ by starting at the origin and counting 2 units to the left on the horizontal axis, and 4 units upward and parallel to the vertical axis. This point is shown in Figure 2.15, along with several other sample points. The number -2 is the *x-coordinate* and 4 is the *y-coordinate* of the point $(-2, 4)$.

The x-axis and y-axis divide the graph into four parts or **quadrants.** For example, Quadrant I includes all those points whose x- and y-coordinates are both positive. The quadrants are numbered as shown in Figure 2.15. The points of the axes themselves belong to no quadrant. The set of points corresponding to the ordered pairs of a relation is called the **graph** of the relation.

3. Locate $(-1, 6)$, $(-3, -5)$, $(4, -3)$, $(0, 2)$, and $(-5, 0)$ on a coordinate system.

Answer:

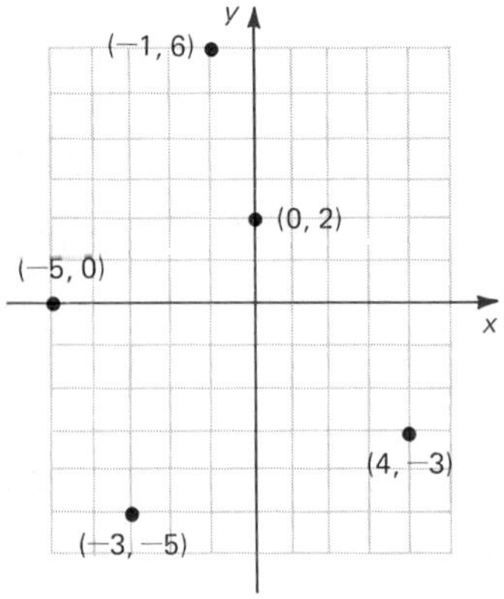

Work Problem 3 at the side.

Example 3 Find the set of all ordered pairs belonging to the relation

$$\{(x, y) | y \le 2x + 1;\ x = -1, 0;\ y = -2, -1, 0, 1\}.$$

Graph the ordered pairs of the relation.

When $x = -1$, we have

$$\begin{aligned} y &\le 2x + 1 \\ y &\le 2(-1) + 1 && \text{Let } x = -1 \\ y &\le -2 + 1 \\ y &\le -1. \end{aligned}$$

The possible y-values from the relation are -2, -1, 0, and 1. Of these, only -2 and -1 satisfy $y \le -1$. This leads to two ordered pairs, $(-1, -2)$ and $(-1, -1)$.

When $x = 0$, we have

$$\begin{aligned} y &\le 2x + 1 \\ y &\le 2(0) + 1 && \text{Let } x = 0 \\ y &\le 1. \end{aligned}$$

Any of the numbers -2, -1, 0, or 1 will satisfy $y \le 1$, giving the four ordered pairs $(0, -2)$, $(0, -1)$, $(0, 0)$, and $(0, 1)$. In summary, the set of ordered pairs belonging to the given relation is

$$\{(-1, -2), (-1, -1), (0, -2), (0, -1), (0, 0), (0, 1)\}.$$

A graph of the ordered pairs of this relation is shown in Figure 2.16. ■

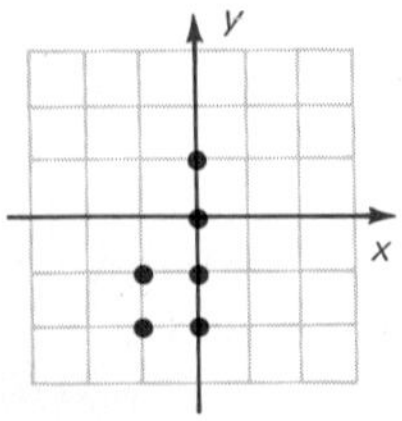

Figure 2.16

4. Identify the ordered pairs belonging to $\{(x, y) | y \leq x + 6; x = 1, 2; y = 5, 6, 7, 8\}$.

Graph the ordered pairs.

Answer: $\{(1, 5), (1, 6), (1, 7), (2, 5), (2, 6), (2, 7), (2, 8)\}$

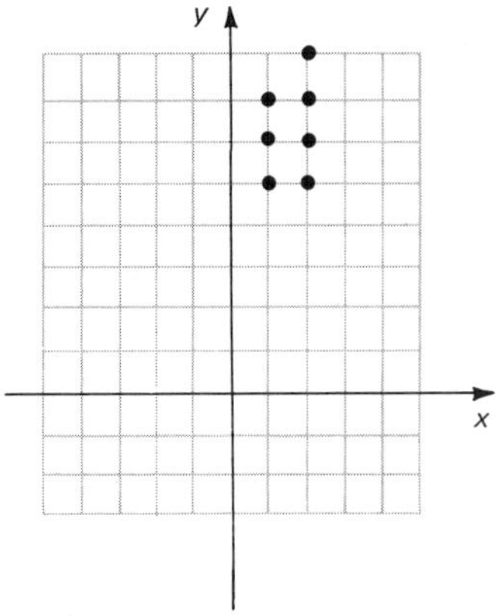

Work Problem 4 at the side.

The most important relations are those which are also functions. A **function** is a special type of relation with the property that each domain element x corresponds to exactly one range element y. We call x the **independent variable** and y the **dependent variable.** For example, the relation

$$\{(-1, 2), (2, 3), (3, 3)\}$$

is a function, since to each value of x we assign exactly one value of y. On the other hand,

$$\{(-1, 2), (-1, 3), (3, 3)\}$$

is not a function. Here we assign *two* range elements, 2 and 3, to the one domain element -1.

Letters, such as f, g, or h, are often used to name functions. For example,

$$f = \{(x, y) | y = 2x + 3\}$$

is a function named f. When the domain of a function is not specified, as in f, we assume it to be the set of all real numbers that are meaningful replacements for the independent variable. The range of f is thus the set of all real numbers.

Given the function f above, we can select any real number x and use the equation $y = 2x + 3$ to find exactly one value of y. For example, if we choose the value $x = 4$, then $y = 2(4) + 3 = 11$, so that (4, 11) belongs to function f. If we choose $x = -5$, we can find that $(-5, -7)$ also belongs to f. In the same way, $(-3, -3)$, $(-1, 1)$, and (0, 3) also belong to the function.

We saw that (4, 11) belongs to f. This can be abbreviated by writing

$$f(4) = 11. \qquad \text{(Read "}f\text{ of 4 equals 11.")}$$

In the same way, $f(-5) = -7$, $f(-3) = -3$, $f(-1) = 1$, and $f(0) = 3$.

Example 4 Let $g = \{(x, y) | y = -4x + 5\}$. Find each of the following.

(a) $g(3)$

Replace x with 3 in the equation $y = -4x + 5$.

$$y = -4(3) + 5 \qquad \text{Let } x = 3$$
$$y = -12 + 5 = -7$$

Therefore, $g(3) = -7$.

(b) $g(-8) = -4(-8) + 5 \qquad \text{Let } x = -8$

$= 37$ ■

5. Let $f = \{(x, y) | y = -3x - 8\}$. Find
(a) $f(0)$
(b) $f(-1)$
(c) $f(5)$.

Answer:
(a) -8
(b) -5
(c) -23

Work Problem 5 at the side.

Writing all the symbols in $f = \{(x, y) | y = 2x + 3\}$ takes a lot of time. To save time, we often abbreviate

$$f = \{(x, y) | y = 2x + 3\}$$

as

$$f(x) = 2x + 3$$

or even

$$y = 2x + 3.$$

This last way of writing the function f is especially useful since it makes it clear that x is the independent variable and y is the dependent variable. Another way of saying that x is the independent variable is to say that y is a function of x. For example, sales are often a function of advertising, the number of bacteria is a function of time, and so on.

Example 5 Let $f(x) = x^2 - 4x + 5$. Find $f(3)$, $f(0)$, and $f(a)$.

To find $f(3)$, substitute 3 for x.

$$f(3) = 3^2 - 4(3) + 5 = 9 - 12 + 5 = 2$$

We can find $f(0)$ and $f(a)$ in the same way.

$$f(0) = 0^2 - 4(0) + 5 = 5$$
$$f(a) = a^2 - 4a + 5 \quad \blacksquare$$

6. Let $f(x) = 5x^2 - 2x + 1$. Find
(a) $f(1)$
(b) $f(3)$
(c) $f(m)$.

Answer:
(a) 4
(b) 40
(c) $5m^2 - 2m + 1$

Work Problem 6 at the side.

Example 6 Suppose the sales of a small company are approximated by

$$S(x) = 100 + 80x,$$

where $S(x)$ represents the total sales in thousands of dollars in year x, with $x = 0$ representing 1978. Find the sales in each of the following years.

(a) 1978

In the equation for $S(x)$, x represents years, where $x = 0$ represents 1978. We can find the sales for 1978 if we find $S(0)$, that is, if we substitute 0 for x.

$$S(0) = 100 + 80(0) \qquad \text{Let } x = 0$$
$$= 100$$

Since $S(x)$ represents sales in thousands of dollars, sales totaled \$100,000 in 1978.

(b) 1982
To estimate sales in 1982, let $x = 4$.

$$S(4) = 100 + 80(4) = 100 + 320 = 420,$$

so that sales are about \$420,000 in 1982. ■

7. A developer estimates that the total cost of building x large apartment complexes in a year is approximated by

$$A(x) = x^2 + 80x + 60,$$

where $A(x)$ represents the cost in hundred thousands of dollars. Find the cost of building
(a) 4 complexes
(b) 10 complexes.

Answer:
(a) \$39,600,000
(b) \$96,000,000

Work Problem 7 at the side.

For a relation to be a function, we must be able to assign exactly one value of y to each value of x in the domain of the function. Figure 2.17 shows the graph of a relation. For the value $x = x_1$, the graph gives the two y-values, y_1 and y_2. Since the x-value corresponds to two different y-values, the relation is not a function. Based on this is the **vertical line test** for a function:

If a vertical line cuts the graph of a relation at more than one point, then the relation is not a function.

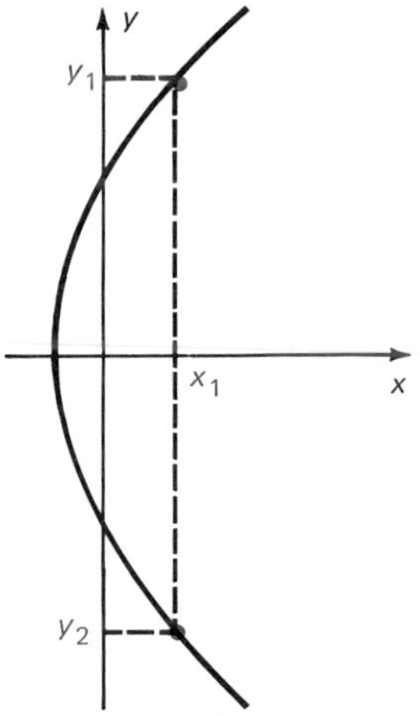

Figure 2.17

8. Does this graph represent a function?

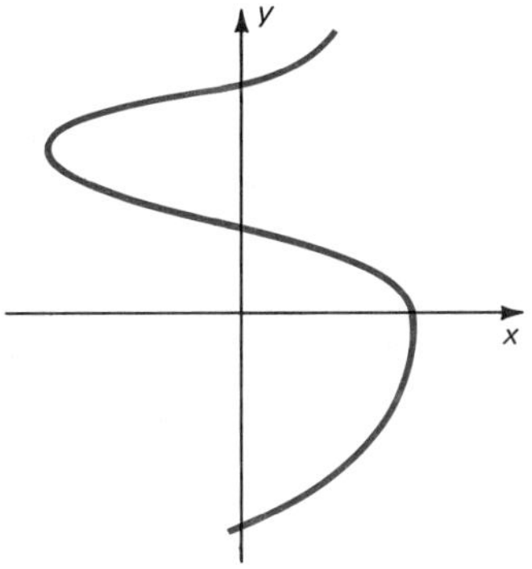

Answer:
No

Work Problem 8 at the side.

2.4 EXERCISES

List the ordered pairs belonging to each of the following relations. Assume that the domain of x for each exercise is $\{-2, -1, 0, 1, 2, 3\}$. Graph each relation and give the range of each. (See Examples 1 and 2.)

1. $y = x - 1$

2. $y = 2x + 3$

3. $y = -4x + 9$

4. $y = -6x + 12$

5. $y = -x - 5$

6. $y = -2x - 3$

7. $2x + y = 9$

8. $3x + y = 16$

9. $2y - x = 5$

10. $6x - y = -3$

11. $y = x(x + 1)$

12. $y = (x - 2)(x - 3)$

13. $y = x^2$

14. $y = -2x^2$

15. $y = 3 - 4x^2$

16. $y = 5 - x^2$

17. $y = \dfrac{1}{x + 3}$

18. $y = \dfrac{-2}{x + 4}$

19. $y = \dfrac{3x - 3}{x + 5}$

20. $y = \dfrac{2x + 1}{x + 3}$

List the ordered pairs belonging to each of the following relations. Assume that the domain of x in each exercise is {0, 1}. The range is given in set braces. Graph each relation. (See Example 3.)

21. $y \leq x + 4$; $\{2, 3, 4, 5\}$

22. $y \leq 2x - 1$; $\{-2, -1, 0, 1\}$

23. $y > 3x + 1$; $\{1, 2, 3, 4, 5\}$

24. $y > -x + 2$; $\{2, 3, 4\}$

Identify any of the following that represent functions.

25.

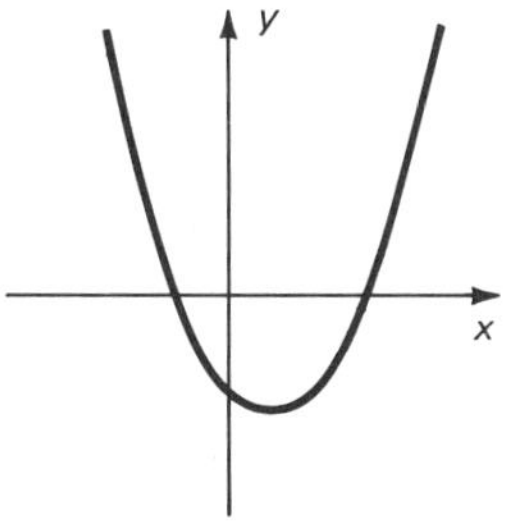

26.

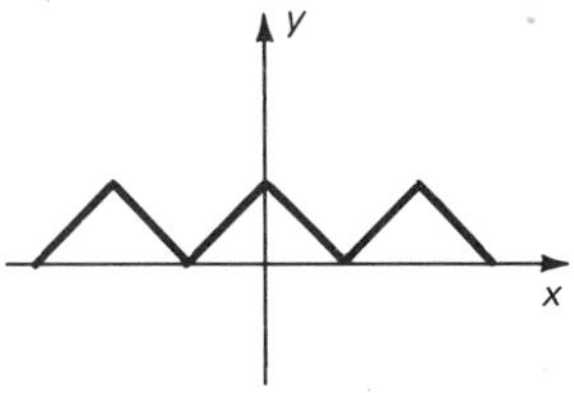

27.

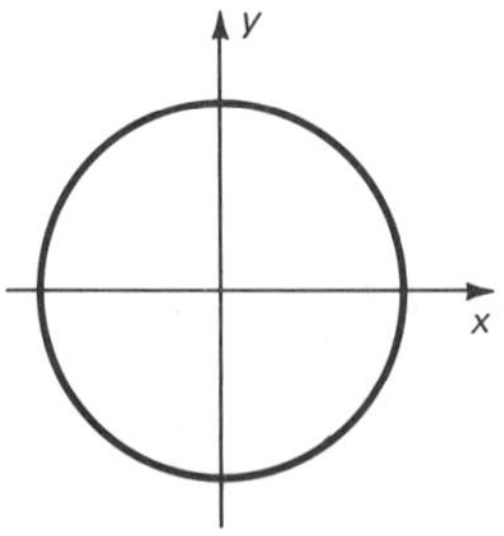

28.

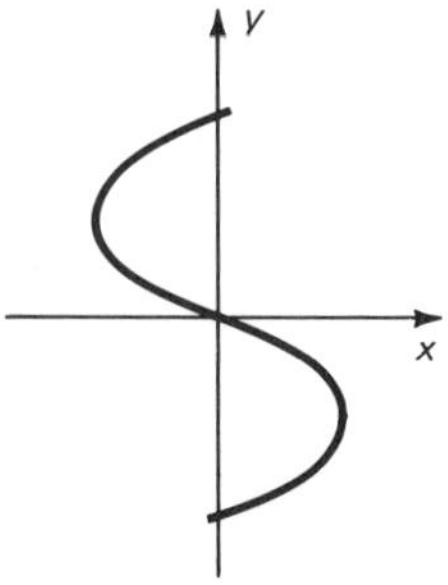

29.

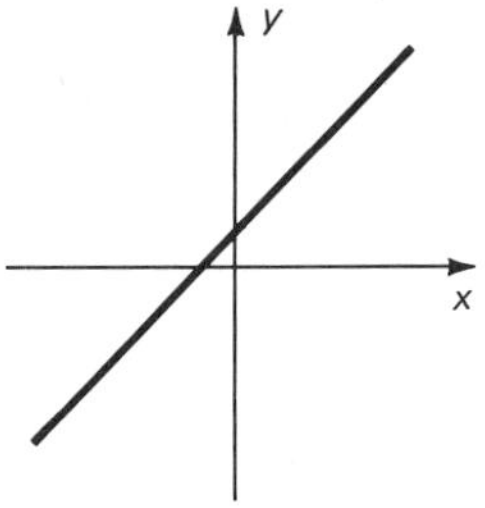

30.

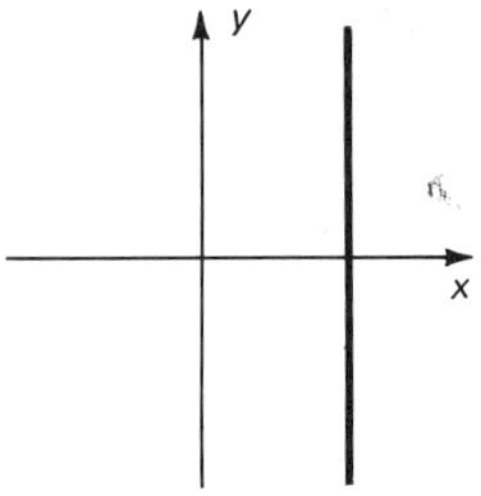

For each of the following functions, find (a) $f(4)$ (b) $f(-3)$ (c) $f(0)$ (d) $f(a)$.

31. $f(x) = 3x + 2$

32. $f(x) = 5x - 6$

33. $f(x) = -2x - 4$

34. $f(x) = -3x + 7$

35. $f(x) = 2x^2 + 4x$

36. $f(x) = x^2 - 2x$

37. $f(x) = -x^2 + 5x + 1$

38. $f(x) = -x^2 - x + 5$

39. $f(x) = (x + 1)(x + 2)$

40. $f(x) = (x + 3)(x - 4)$

Let $f = \{(x, y) | y = 2x - 3\}$. Find each of the following. In Exercise 49, first find $f(2)$. (See Example 4.)

41. $f(0)$ **42.** $f(-1)$ **43.** $f(-6)$ **44.** $f(4)$

45. $f(a)$ **46.** $f(-r)$ **47.** $f(m + 3)$ **48.** $f(p - 2)$

49. $f[f(2)]$ **50.** $f[f(-3)]$

APPLIED PROBLEMS

Management

Suppose the sales of a small company that sells by mail are approximated by

$$S(t) = 1000 + 50(t + 1),$$

where $S(t)$ represents sales in thousands of dollars. Here t is time in years, with $t = 0$ representing the year 1977. Find the sales in each of the following years.

51. 1977 **52.** 1978 **53.** 1980 **54.** 1982

A chain-saw rental firm charges \$7 per day or fraction of a day to rent a saw, plus a fixed fee of \$4 for resharpening the blade. Let $S(x)$ represent the cost of renting a saw for x days. Find each of the following.

55. $S\left(\frac{1}{2}\right)$ **56.** $S(1)$ **57.** $S\left(1\frac{1}{4}\right)$ **58.** $S\left(3\frac{1}{2}\right)$

59. $S(4)$ **60.** $S\left(4\frac{1}{10}\right)$ **61.** $S\left(4\frac{9}{10}\right)$

62. A portion of the graph of $y = S(x)$ is shown here. Explain how the graph could be continued.

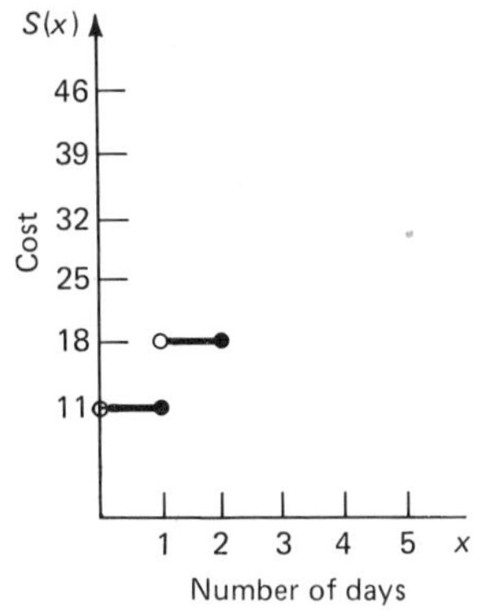

General

To rent a midsized car from Avis costs \$18 per day or fraction of a day. If you pick up the car in Boston and drop it off in Utica, there is a fixed \$40 charge. Let $C(x)$ represent the cost of renting the car for x days, taking it from Boston to Utica. Find each of the following.

63. $C\left(\frac{3}{4}\right)$ **64.** $C\left(\frac{9}{10}\right)$ **65.** $C(1)$ **66.** $C\left(1\frac{5}{8}\right)$

67. $C\left(2\frac{1}{9}\right)$ **68.** Graph the function $y = C(x)$.

2.5 SEQUENCE FUNCTIONS (OPTIONAL)

Throughout much of this book we look at various types of functions which are useful in writing mathematical descriptions of real-world situations. In this section we begin our study of various types of useful functions by looking at sequence functions.

A **sequence function** (or **sequence**) is a function whose domain is the set of positive integers. For example,

$$a(n) = 2n, \qquad n = 1, 2, 3, 4, \ldots$$

is a sequence. The letter n is used as a variable instead of x to emphasize the fact that the domain includes only positive integers. For the same reason, a is used to name the function instead of f.

The range values of a sequence function, such as

$$a(1) = 2, \qquad a(2) = 4, \qquad a(3) = 6, \qquad \text{and so on}$$

from the sequence given above, are called the **terms** of the sequence. Instead of writing $a(5)$ for the fifth term of a sequence, it is customary to write

$$a_5 = 10.$$

In the same way, for the sequence above, $a_1 = 2$, $a_2 = 4$, $a_8 = 16$, $a_{20} = 40$, and $a_{51} = 102$.

We often write a_n for the **general** or ***n*th** term of a sequence. For example, for the sequence 4, 7, 10, 13, 16, . . . the general term is given by $a_n = 1 + 3n$.

This formula for a_n can be used to find any term of the sequence that we might need. For example, the first three terms of the sequence are

$$a_1 = 1 + 3(1) = 4, \quad a_2 = 1 + 3(2) = 7, \quad \text{and} \quad a_3 = 1 + 3(3) = 10.$$

Also, $a_8 = 25$ and $a_{12} = 37$.

Example 1 Find the first four terms for the sequences having the following general terms.

(a) $a_n = -4n + 2$

Replace n, in turn, with 1, 2, 3, and 4. If $n = 1$, we have

$$a_1 = -4(1) + 2 = -4 + 2 = -2. \qquad \text{Let } n = 1$$

Also, $$a_2 = -4(2) + 2 = -6. \qquad \text{Let } n = 2$$

When $n = 3$, we have $a_3 = -10$; finally, $a_4 = -4(4) + 2 = -14$. The first four terms of this sequence are -2, -6, -10, and -14.

(b) $a_n = \dfrac{n+1}{n+2}$

When $n = 1$, we have $a_1 = \dfrac{1+1}{1+2} = \dfrac{2}{3}$.

Also,

$$a_2 = \frac{2+1}{2+2} = \frac{3}{4}, \qquad a_3 = \frac{3+1}{3+2} = \frac{4}{5}, \qquad \text{and} \qquad a_4 = \frac{5}{6}. \quad \blacksquare$$

Work Problem 1 at the side.

1. Find the first four terms for each of the following sequences.

(a) $a_n = -9 + 7n$

(b) $a_n = (2n + 3)(n - 1)$

Answer:

(a) $-2, 5, 12, 19$

(b) $0, 7, 18, 33$

A sequence in which each term after the first is found by adding the same number to the preceding term is called an **arithmetic sequence.** The sequence of Example 1(a) above is an arithmetic sequence; -4 is added to any term to get the next term.

The sequence

$$8, 13, 18, 23, 28, \ldots$$

is an arithmetic sequence since each term after the first is found by adding 5 to the previous term. The number 5, the difference between any two adjacent terms, is called the **common difference.**

If a_1 is the first term of an arithmetic sequence and d is the common difference, then we can get the second term by adding d to the first term: $a_2 = a_1 + d$. The third term is found by adding d to the second term.

$$a_3 = a_2 + d = (a_1 + d) + d = a_1 + 2d$$

In the same way, $a_4 = a_1 + 3d$ and $a_5 = a_1 + 4d$. In general, the nth term of an arithmetic sequence is given by $a_n = a_1 + (n - 1)d$. This result can be summarized as follows.

Theorem 2.1 If an arithmetic sequence has first term a_1 and common difference d, then a_n, the nth term of the sequence, is given by

$$\mathbf{a_n = a_1 + (n - 1)d.}$$

Example 2 A company had sales of \$50,000 during its first year of operation. If the sales increase by \$6000 per year, find its total sales in the eleventh year.

Since the sales for each year after the first are found by adding \$6000 to the sales of the previous year, we have an arithmetic sequence with $a_1 = 50{,}000$ and $d = 6000$. Using the formula for the nth term of an arithmetic sequence, we find that the sales during the eleventh year are given by

$$a_{11} = 50{,}000 + (11 - 1)6000 = 50{,}000 + 60{,}000 = 110{,}000,$$

or \$110,000. ■

Work Problem 2 at the side.

2. Suppose a field has 10 rabbits in it at the end of the first month of an experiment, with the number of rabbits then increasing by 8 per month. Find the number of rabbits present after
(a) 3 months
(b) 8 months
(c) 12 months.

Answer:
(a) 26
(b) 66
(c) 98

In Theorem 2.1 we gave a formula for the nth term of an arithmetic sequence. We can also find a formula for the *sum* of the first n terms of an arithmetic sequence. This formula can be developed by a method used by the mathematician C. F. Gauss (1777–1855) when he was a small child.

According to an old story, one of Gauss' teachers was a tyrant in the classroom. One day, he demanded that the students find the sum of the first hundred counting numbers, the sum

$$1 + 2 + 3 + 4 + 5 + \cdots + 98 + 99 + 100.$$

The other students in the class began by adding two numbers at a time, but Gauss found a quicker way. He first wrote the sum twice.

$$\begin{array}{l} 1 + 2 + 3 + 4 + \ldots + 99 + 100 \\ 100 + 99 + 98 + 97 + \ldots + 2 + 1 \end{array}$$

He then noticed that when added vertically, each pair of numbers added up to 101.

$$\begin{array}{r} 1 + 2 + 3 + 4 + \ldots + 99 + 100 \\ 100 + 99 + 98 + 97 + \ldots + 2 + 1 \\ \hline 101 + 101 + 101 + 101 + \ldots + 101 + 101 \end{array}$$

There are 100 of these sums of 101, for a total of $100 \cdot 101 = 10{,}100$. However, when he found this total, he had added the numbers 1 through 100 *twice*. To compensate for this, he took half of 10,100.

$$1 + 2 + 3 + 4 + \ldots + 99 + 100 = \frac{10{,}100}{2} = 5050$$

We can use a method very similar to the one Gauss used to find a formula for the sum of the first n terms of an arithmetic sequence. Two versions of such a formula are given in the next theorem.

Theorem 2.2 Suppose an arithmetic sequence has first term a_1, common difference d, and general term a_n. Then the sum of the first n terms of the sequence, S_n, is given by

$$\boldsymbol{S_n = \frac{n}{2}[2a_1 + (n - 1)d]}$$

or

$$\boldsymbol{S_n = \frac{n}{2}(a_1 + a_n).}$$

Either of these two formulas can be used to find the sum of the first n terms of an arithmetic sequence.

Example 3 Find the sum of the first 25 terms of the following arithmetic sequence.

$$3, 8, 13, 18, 23, \ldots$$

In this sequence, $a_1 = 3$ and $d = 5$. If we use the first formula of Theorem 2.2, we have

$$\begin{aligned} S_{25} &= \frac{25}{2}[2(3) + (25 - 1)5] \qquad \text{Let } n = 25,\ a_1 = 3,\ d = 5 \\ &= \frac{25}{2}[6 + 120] \\ &= 1575. \quad \blacksquare \end{aligned}$$

Work Problem 3 at the side.

3. For the sequence 12, 15, 18, 21, . . . , find the sum of
(a) the first 10 terms
(b) the first 31 terms.

Answer:
(a) 255
(b) 1767

Example 4 Find the sum of the first thousand counting numbers.

The first thousand counting numbers, 1, 2, 3, 4, . . . , 1000, can be thought of as an arithmetic sequence with $a_1 = 1$, $a_{1000} = 1000$, and $n = 1000$. We can find S_{1000} most quickly by using the second formula of Theorem 2.2.

$$S_{1000} = \frac{1000}{2}(1 + 1000) \qquad \text{Let } n = 1000,\ a_1 = 1,\ a_{1000} = 1000$$
$$= 500(1001)$$
$$= 500{,}500 \quad \blacksquare$$

Work Problem 4 at the side.

4. Find the sum of the first 500 counting numbers.

Answer: 125,250

Example 5 Find the total sales of the company of Example 2 during its first 11 years.

We know from Example 2 that $a_1 = 50{,}000$, $a_{11} = 110{,}000$, and $n = 11$. By the second formula of Theorem 2.2, we have

$$S_{11} = \frac{11}{2}(50{,}000 + 110{,}000) = 880{,}000.$$

The total sales for 11 years are \$880,000. ■

Work Problem 5 at the side.

5. A new firm sells 60 solar heating systems during its first year in business. Each year after that, it sells 40 more systems than in the preceding year. Find the number of systems that will be sold during the firm's first 10 years in business.

Answer: 2400

In an arithmetic sequence, each term after the first is found by adding the same number to the preceding term. In a **geometric sequence,** each term after the first is found by *multiplying* the preceding term by the same number. Thus,

$$3, -6, 12, -24, 48, -96, \ldots$$

is a geometric sequence in which each term after the first is found by multiplying the preceding term by the number -2. The number -2 is called the **common ratio.**

If a_1 is the first term of a geometric sequence and r is the common ratio, then the second term is given by $a_2 = a_1 r$ and the third term by $a_3 = a_2 r = a_1 r^2$. Also, $a_4 = a_1 r^3$, and $a_5 = a_1 r^4$. In general, $a_n = a_1 r^{n-1}$, as stated in the following theorem.

Theorem 2.3 If a geometric sequence has first term a_1 and common ratio r, then

$$a_n = a_1 r^{n-1}.$$

Example 6 Find the indicated term for each of the following geometric sequences.

(a) 6, 24, 96, 384, . . . ; find a_7.

Here $a_1 = 6$. To find r, choose any term except the first and divide it by the preceding term. If we choose 96, we have

$$r = \frac{96}{24} = 4.$$

Now use the formula of Theorem 2.3.

$$\begin{aligned} a_7 &= 6(4)^{7-1} && \text{Let } n = 7, a_1 = 6, r = 4 \\ &= 6(4)^6 \\ &= 6(4096) && 4^6 = 4096 \\ &= 24{,}576 \end{aligned}$$

(b) 8, -16, 32, -64, 128, . . . ; find a_6.

In this sequence $a_1 = 8$ and $r = -2$.

$$\begin{aligned} a_6 &= 8(-2)^{6-1} && \text{Let } n = 6, a_1 = 8, r = -2 \\ &= 8(-2)^5 \\ &= 8(-32) \\ &= -256 \quad \blacksquare \end{aligned}$$

6. Find a_4 and a_7 for the geometric sequence having $a_1 = 5$ and $r = -2$.

Answer: $a_4 = -40$; $a_7 = 320$

Work Problem 6 at the side.

Example 7 The number of bacteria in a culture triples every four hours. If 1000 bacteria are present initially, how many bacteria will be present at the end of 24 hours?

The number of bacteria at the end of the first four hours is 3 times the initial number of 1000. Thus, $a_1 = 3000$. At the end of the next four hours, the 3000 is tripled, so that $a_2 = 9000$. Thus, the number of bacteria at the end of each four-hour period is a geometric sequence with first term equal to 3000 (the number present at the end of the initial four-hour period), and common ratio 3.

Since there are 6 four-hour periods in a 24-hour day, the number of bacteria present after 24 hours will be given by the sixth term of the geometric sequence.

$$a_6 = 3000(3)^{6-1} = 3000(3)^5 = 3000(243) = 729{,}000 \text{ bacteria} \quad \blacksquare$$

7. In Example 7, how many bacteria will be present after

(a) 8 hours

(b) 12 hours?

Answer:

(a) 9000

(b) 27,000

Work Problem 7 at the side.

Often, we need to find the sum of the first n terms of a geometric sequence. Just as with arithmetic sequences, we can find a formula for this sum. This formula is given in the next theorem.

Theorem 2.4 If a geometric sequence has first term a_1 and common ratio r, then the sum of the first n terms, S_n, is given by

$$S_n = \frac{a_1(r^n - 1)}{r - 1}, \qquad r \neq 1.$$

Example 8 Find the sum of the first six terms of the geometric sequence 3, 12, 48, ...

Here $a_1 = 3$ and $r = 4$. We want S_6, which we find by the formula of Theorem 2.4.

$$S_6 = \frac{3(4^6 - 1)}{4 - 1} \qquad \text{Let } n = 6,\ a_1 = 3,\ r = 4$$

$$= \frac{3(4096 - 1)}{3}$$

$$= 4095 \quad \blacksquare$$

8. Find S_4 and S_7 for the geometric sequence 5, 15, 45, 135, ...

Answer: $S_4 = 200$; $S_7 = 5465$

Work Problem 8 at the side.

Example 9 Find S_6 for the geometric sequence having $a_1 = 72$ and $r = -1/2$.

Use Theorem 2.4.

$$S_6 = \frac{72\left[\left(-\frac{1}{2}\right)^6 - 1\right]}{-\frac{1}{2} - 1} \qquad \text{Let } n = 6,\ a_1 = 72, \text{ and } r = -\frac{1}{2}$$

$$= \frac{72\left[\frac{1}{64} - 1\right]}{-\frac{3}{2}}$$

$$= \frac{72\left[-\frac{63}{64}\right]}{-\frac{3}{2}}$$

$$= 72\left[-\frac{63}{64}\right]\left(-\frac{2}{3}\right) \qquad \text{Invert and multiply}$$

$$= \frac{189}{4} \quad \blacksquare$$

9. Find S_5 for the geometric sequence having $a_1 = -3$ and $r = -5$.

Answer: -1563

Work Problem 9 at the side.

Several applications of geometric sequences are included in Chapters 3 and 4.

2.5 EXERCISES

In the following exercises, a formula for the general term of a sequence is given. Use the formula to find the first five terms of the sequence. Identify each sequence as arithmetic, geometric, *or* neither. *(See Examples 1 and 6.)*

1. $a_n = 6n + 5$

2. $a_n = 12n - 3$

3. $a_n = 3n - 7$

4. $a_n = 5n - 12$

5. $a_n = -6n + 4$

6. $a_n = -11n + 10$

7. $a_n = 2^n$

8. $a_n = 3^n$

9. $a_n = (-2)^n$

10. $a_n = (-3)^n$

11. $a_n = 3(2^n)$

12. $a_n = -4(2^n)$

13. $a_n = \dfrac{n+1}{n+5}$

14. $a_n = \dfrac{2n}{n+1}$

15. $a_n = \dfrac{1}{n+1}$

16. $a_n = \dfrac{1}{n+8}$

Identify each of the following sequences as arithmetic, geometric, *or* neither. *For an arithmetic sequence, give the common difference. For a geometric sequence, give the common ratio.*

17. 6, 14, 22, 30, 38, 46, . . .

18. 40, 46, 52, 58, 64, . . .

19. 5, 8, 11, 14, 17, 20, 23, . . .

20. 23, 34, 45, 54, 63, 72, . . .

21. 4, 12, 36, 108, . . .

22. 7, 14, 28, 56, 112, . . .

23. 2, 5, 9, 14, 20, 27, . . .

24. 1, 4, 9, 16, 25, 36, . . .

25. 12, 9, 6, 3, 0, −3, −6, . . .

26. 37, 31, 25, 19, 13, 7, . . .

27. −18, −15, −12, −9, −6, . . .

28. −21, −17, −13, −9, −5, . . .

29. 3, −6, 12, −24, 48, −96, . . .

30. −5, 10, −20, 40, −80, . . .

31. −5, 6, −7, 8, −9, 10, −11, . . .

32. −12, 9, −6, 3, . . .

Find the indicated term for each of the following arithmetic sequences.

33. $a_1 = 10, d = 5$; find a_{13}

34. $a_1 = 6, d = 9$; find a_8

35. $a_1 = 8, d = 3$; find a_{20}

36. $a_1 = 13, d = 7$; find a_{11}

37. 6, 9, 12, 15, 18, . . . ; find a_{25}

38. 14, 17, 20, 23, 26, 29, . . . ; find a_{13}

39. $-9, -13, -17, -21, -25, \ldots$; find a_{15}

40. $-4, -11, -18, -25, -32, \ldots$; find a_{18}

Find the sum of the first six terms for each of the following arithmetic sequences. (See Example 3.)

41. 3, 6, 9, 12, . . .

42. 11, 13, 15, 17, . . .

43. 88, 98, 108, . . .

44. 92, 95, 98, . . .

45. $a_1 = 8, d = 9$

46. $a_1 = 12, d = 6$

47. $a_1 = 7, d = -4$

48. $a_1 = 13, d = -5$

Find a_5 for each of the following geometric sequences. (See Example 6.)

49. $a_1 = 3, r = 2$

50. $a_1 = 5, r = 3$

51. $a_1 = -8, r = 3$

52. $a_1 = -6, r = 2$

53. $a_1 = 1, r = -3$

54. $a_1 = 12, r = -2$

55. $a_1 = 1024, r = 1/2$

56. $a_1 = 729, r = 1/3$

Find the sum of the first four terms for each of the following geometric sequences. (See Example 8.)

57. $a_1 = 1, r = 2$

58. $a_1 = 3, r = 3$

59. $a_1 = 5, r = 1/5$

60. $a_1 = 6, r = 1/2$

61. $a_1 = 128, r = -3/2$

62. $a_1 = 81, r = -2/3$

Sums of the terms of a sequence are often written in sigma notation *(which is also used in statistics). For example, to evaluate*

$$\sum_{i=1}^{4} (3i + 7)$$

(where $\sum$ *is the capital Greek letter* sigma), *first evaluate* $3i + 7$ *for* $i = 1$, *then for* $i = 2$, $i = 3$, *and finally,* $i = 4$. *Then add the results. We have*

$$\sum_{i=1}^{4} (3i + 7) = [3(1) + 7] + [3(2) + 7] + [3(3) + 7] + [3(4) + 7]$$

$$= 10 + 13 + 16 + 19 = 58.$$

Evaluate each of the following sums.

63. $\sum_{i=1}^{5} (3 - 2i)$

64. $\sum_{i=1}^{4} (2i - 5)$

65. $\sum_{i=1}^{8} (2i + 1)$

66. $\sum_{i=1}^{4} (-6i + 8)$

67. $\sum_{i=1}^{5} i(2i + 1)$

68. $\sum_{i=1}^{4} (2i - 1)(i + 2)$

69. $\sum_{i=1}^{4} (3i + 1)(i + 1)$

70. $\sum_{i=1}^{5} (i - 3)(i + 5)$

A sum in the form

$$\sum_{i=1}^{n} (ai + b),$$

where a and b are real numbers, represents the sum of the first n terms of an arithmetic sequence. The first term is $a_1 = a(1) + b = a + b$ and the common difference is a. The nth term is $a_n = an + b$. Using these numbers, the sum can be found by applying Theorem 2.2. Use this method to find each of the following sums.

71. $\sum_{i=1}^{5} (2i + 8)$

72. $\sum_{i=1}^{6} (4i - 5)$

73. $\sum_{i=1}^{4} (-8i + 6)$

74. $\sum_{i=1}^{9} (2i - 3)$

75. $\sum_{i=1}^{500} i$

76. $\sum_{i=1}^{1000} i$

Sigma notation can also be used for geometric sequences. A sum in the form

$$\sum_{i=1}^{n} a(b^i)$$

represents the sum of the first n terms of a geometric sequence having first term $a_1 = ab$ and common ratio b. These sums can thus be found by applying Theorem 2.4. Use this method to find each of the following sums.

77. $\sum_{i=1}^{4} 3(2^i)$

78. $\sum_{i=1}^{3} 2(3^i)$

79. $\sum_{i=1}^{4} \frac{1}{2}(4^i)$

80. $\sum_{i=1}^{4} \frac{3}{2}(2^i)$

81. $\sum_{i=1}^{4} \frac{4}{3}(3^i)$

82. $\sum_{i=1}^{4} \frac{5}{3}(3^i)$

APPLIED PROBLEMS

General

83. Joy is hired for \$11,400 per year, with annual raises of \$600. What will she earn during her eighth year with the company?

84. What will be the total income received by Joy (see Exercise 83) during her first eight years with the company?

Management

85. A certain machine annually loses 30% of the value it had at the beginning of that year. If its initial value is \$10,000, use Theorem 2.3 and a calculator to find its value (a) at the end of the fifth year, (b) at the end of the eighth year.

Natural Science

86. A certain colony of bacteria increases in number by 10% per hour. After five hours, what is the percentage increase in the population over the initial population?

CHAPTER 2 TEST

[2.1]

Write true *or* false *for each of the following statements.*

1. $5 \in \{2, 3, 4, 5, 6\}$
2. $\{7, 9\} \subset \{8, 9, 10, 11\}$
3. $\varnothing \subset \{1, 2, 3, 4\}$
4. There are exactly 16 subsets of $\{1, 2, 3, 4, 5\}$.
5. $\{9, 10, 11, 12\} \cap \{7, 8, 9, 10\} = \{7, 8, 9, 10, 11, 12\}$
6. $\{3, 4, 5, 6\} \cup \{1, 2, 3, 4, 5\} = \{1, 2, 3, 4, 5, 6\}$
7. $\{5, 9, 14\} \cap \{-1, 3, 5, 7, 9\} = \{5, 9\}$
8. $\{1, 2, 3\} \cap \{4, 5, 6\} = \varnothing$

[2.2]

Let $U = \{a, b, c, d, e, f, g, h\}$, $X = \{a, c, e\}$, *and* $Y = \{a, e, f, g\}$. *Find each of the following sets.*

9. $X \cap Y$ **10.** $X \cup Y$ **11.** X' **12.** $X' \cap Y'$

Draw a Venn diagram for each of the following sets.

13. $A \cap B'$ **14.** $A' \cup B$

15. $A' \cap (B \cup C')$ **16.** $(A \cup B)' \cap C$

[2.3]

The local department store inventoried its Levis department with the following results.

59 pairs were new;

34 had a button fly;

40 were blue;

22 were new and had a button fly;
7 had a button fly and were blue;
32 were new and blue;
4 were new and blue, with a button fly;
7 were none of these.

Find the number of pairs which were

17. blue, but not new;

18. new, but not blue;

19. new and blue, but had no button fly;

20. not new and not blue.

[2.4] **21.** List the ordered pairs belonging to the relation

$$\{(x, y) | y = -x + 6;\ x = -2, -1, 0, 1, 2, 3\}.$$

Give the domain and range.

Let $f(x) = -2x^2 + 3x + 1$. *Find each of the following.*

22. $f(-2)$ **23.** $f(3)$ **24.** $f[f(0)]$ **25.** $f(p)$

[2.5] *Write the first five terms for each of the following sequences. Identify any which are arithmetic or geometric.*

26. $a_n = -4n + 2$ **27.** $a_n = (-2)^n$ **28.** $a_n = \dfrac{n+2}{n+5}$

29. Find a_{12} for the arithmetic sequence having $a_1 = 6$ and $d = 5$.

30. Find the sum of the first 20 terms for the arithmetic sequence having $a_1 = -6$ and $d = 8$.

31. Find a_4 for the geometric sequence with $a_1 = -3$ and $r = 2$.

32. Find the sum of the first 6 terms for the geometric sequence with $a_1 = 8000$ and $r = -1/2$.

CASE 2 ESTIMATING ITEM CATALOG SALES—MONTGOMERY WARD AND COMPANY*

When preparing a new catalog for publication, buyers at Montgomery Ward must estimate the total sales of each item in the catalog for the life of the catalog. This is often done by using the actual sales of the same (or a similar) item from past catalogs. For example, the actual sales of six items in one line of goods for the 20-week life of one past catalog were as shown in the following table. (A "line of goods" is one group of items, such as men's sport shirts, plastic dinnerware, etc.)

	Beginning of week											
Item	2	3	4	5	6	7	8	9	10	11	12	13
1	20	39	80	172	251	384	558	850	1231	1697	2232	3066
2	6	34	97	182	318	480	680	998	1359	1857	2409	3108
3	4	20	50	80	122	186	273	420	584	785	1026	1361
4	3	7	9	17	31	41	53	77	101	135	163	214
5	2	3	11	23	40	60	100	151	197	286	372	487
6	4	23	51	102	182	261	380	620	876	1223	1550	2181
Total	39	126	298	576	944	1412	2044	3116	4348	5983	7752	10,417

	Beginning of week							
Item	14	15	16	17	18	19	20	Final
1	3867	4331	4419	4468	4543	4569	4605	4646
2	3889	4209	4258	4282	4304	4317	4328	4341
3	1705	1892	1913	1927	1942	1952	1959	1961
4	267	295	307	316	324	328	329	331
5	589	660	675	686	691	694	699	700
6	2655	2861	2883	2896	2904	2911	2916	2924
Total	12,972	14,248	14,502	14,575	14,708	14,771	14,836	14,903

The numbers in this table represent **cumulative sales.** For example, 20 units of item 1 were sold by the beginning of week 2, with 39 units sold by the beginning of week 3, and 80 units sold by the beginning of week 4. A total of $80 - 39 = 41$ units were sold during week 3, for example.

After gathering the data of the table, company analysts calculate the terms of a sequence as follows. Term $a_1 = 0$ and represents the fraction of total sales at the beginning of week 1. To get a_2, find the total of all sales before the beginning of the second week. From the "Total" row of the table above, this number is 39. Divide this number by the total sales of all items in this line of merchandise for the entire life of the catalog, 14,903 units.

---

* Reprinted by permission from P. H. Hartung, "A Simple Style Goods Inventory Model," *Management Science*, Volume 19, #12, August, 1973, Copyright 1973 The Institute of Management Sciences.

Then convert this answer into a percent.

$$a_2 = 39/14,903 = .003 = .3\%$$

A total of .3% of all business should come before the second week begins. Also,

$$a_6 = 944/14,903 = .063 = 6.3\%.$$

Thus, 6.3% of all business should come before the sixth week begins.

EXERCISES

Use the table above to find the number of units of item 3 sold during:

1. 2 weeks **2.** 4 weeks **3.** 12 weeks **4.** 16 weeks

Find each of the following terms of the sequence discussed above.

5. a_4 **6.** a_8 **7.** a_{12} **8.** a_{16}

CASE 3 DETERMINING THE OPTIMUM NUMBER OF CHRISTMAS CARDS*

The Christmas card market is highly volatile—designs that are popular this year might not sell at all in other years. The total market is declining slightly every year, mainly because of ever higher postal rates. It is thus difficult to be a profitable producer of Christmas cards, especially for the smaller firms in the business.

The firm involved in this analysis is a small firm. It hired a mathematical analyst to help decide how many of each type of Christmas card should be produced for a given season. Producing a large number of cards would satisfy all possible demand, but costs would be too high. Producing fewer cards would cut costs but lead to lost revenue. The job of the analyst was to develop a rule that would tell the company the number of cards to produce in order to maximize profit.

To begin, the analyst studied company operations carefully. He found that each year the company receives many design proposals, from which it chooses 500 designs. In January a few hundred copies of each design are printed and shown to wholesalers and retailers at conventions. Orders start coming in around the first of February, and by mid-April it is possible to get a fairly accurate projection of total sales for a particular design.

With these projections in hand, the company has to decide how many of each design to print. Printing is subject to three restrictions or **constraints:**

1. Cards are printed ten at a time on large sheets from one set of plates. The ten cards can be of the same or different designs.

---

* Based on "A Decision Rule for Producing Christmas Greeting Cards," S. K. Goyal, *Operations Research* (July 1974): 795–801. Copyright © 1974, ORSA. By permission of S. K. Goyal. No further reproduction permitted without consent of the copyright owner.

2. A design cannot be placed on more than one set of plates because of unacceptable differences in printing quality.
3. It is not possible to make a reprinting from a given set of plates.

Cards left over after Christmas are destroyed at no additional cost. After cards are printed, they are sent to a warehouse for cutting and creasing.

The following variables were used in arriving at a mathematical model.

C = set-up cost for a particular set of plates
N = number of designs on the set of plates
P = variable cost of printing one sheet (paper, ink, printing)
Q = total number of sheets printed
$B(Q)$ = total gross profit from producing Q sheets
For design R on a given set of plates:
$D(R)$ = estimated demand (made in April)
$F(R)$ = cost of assembling a card in the warehouse
$Q(R)$ = number of cards sent out for assembly (it is understood that $Q(R) \leq D(R)$)
$S(R)$ = wholesale price per card
$Z(R)$ = number of times a design is laid out on a set of plates

The total revenue to be received from cards of design 1 on a sheet ($R = 1$) is given by $S(1) \cdot Q(1)$. The total revenue from cards of design 2 is $S(2) \cdot Q(2)$. Thus, the total revenue from all the cards on a given sheet is

$$S(1) \cdot Q(1) + S(2) \cdot Q(2) + \cdots + S(N) \cdot Q(N).$$

This sum can be abbreviated, using **summation notation,** as

$$\text{total revenue} = \sum_{R=1}^{N} S(R) \cdot Q(R).$$

In the same way, the total production cost is given by

$$\text{total production cost} = C + Q \cdot P + \sum_{R=1}^{N} F(R) \cdot Q(R).$$

The total profit, $B(Q)$, from producing Q sheets is thus

$$B(Q) = \text{revenue} - \text{cost}$$

$$B(Q) = \sum_{R=1}^{N} S(R) \cdot Q(R) - \left[C + Q \cdot P + \sum_{R=1}^{N} F(R) \cdot Q(R) \right].$$

We want to find a value of Q that will make $B(Q)$ as large as possible. The necessary value of Q must satisfy the condition that the cost to produce one more card (called the *marginal cost*) must equal the revenue from the sale of that extra card (the *marginal revenue*). In other words, $B(Q)$ must not be less than $B(Q + 1)$ (or else $Q + 1$ sheets should be printed to get the larger profit). Also, $B(Q)$ must be greater than $B(Q - 1)$ (or else only $Q - 1$ sheets need be printed).

Using these marginal cost considerations, the analyst produced the following procedure for determining a value of Q that maximizes $B(Q)$. A typical set of data is shown in the table on page 84.

1. For each possible design R on a given set of plates, calculate $D(R)$, the estimated demand, and $Z(R)$, the number of times that the design is laid out on a given set of plates.

2. Calculate the quotient $D(R)/Z(R)$, which gives the number of sheets that would have to be printed to meet the total estimated demand for this design. Then list all designs in *descending order* using this quotient.

3. Find P, the cost of printing one sheet.

4. Find $S(R)$, the wholesale price per card, and $F(R)$, the cost of assembling a card.

5. Determine the gross profit per card as the result of selling one more card, $S(R) - F(R)$.

6. In Step 2 we found the number of sheets that would have to be printed in order to satisfy all estimated demand. If one less sheet were to be printed, profit would be lost from each design having sales estimates higher than this number.

   For example, look at line 5 in the table. For design number 5, $D(R) = D(5) = 16{,}440$, $Z(5) = 1$, and $D(5)/Z(5) = 16{,}440$. Thus, a total of 16,440 sheets must be printed to satisfy all estimated demand for design 5. If only 16,440 sheets were printed, the demand for designs 1, 2, 3, and 4 would not be fully met. If one less sheet were to be printed, the profit from selling one card each of designs 1, 2, and 4 and the profit from selling two cards of design 3 would be lost.

   Thus, at line 5 the loss of profit from printing one less sheet than the value given by $D(5)/Z(5)$ is

$$0.380 + 0.530 + 0.377 + 0.377 + 0.520 + 0.572 = 2.756.$$

   At Step 6, this "loss of profit" column should be completed.

7. In a similar way, complete the column showing the additional profit if one extra sheet is printed. Again, let us use line 5 of the table as an example. At line 5, we would print 16,440 sheets. If one extra sheet were to be printed, additional profit would be earned from sales of designs 1, 2, 3, and 4. (Why would no additional profit be earned from the design 5 card that the sheet contains?) This additional profit amounts to

$$0.380 + 0.530 + 0.377 + 0.377 + 0.520 = 2.184.$$

8. In this example, P, the cost of printing one sheet, is 1.035. Look down the "loss of profit" column and the "additional profit" column for two numbers in the same row such that $P = 1.035$ is between them. Here these numbers occur in row 3: 1.664 and 0.910, with

$$0.910 < 1.035 < 1.664.$$

   (Here we have used the idea of marginal cost, discussed above.) This row of the table shows the number of sheets to print: from row 3, a total of 17,500 sheets should be printed. Since only 17,500 sheets will be printed, the demand for designs 1 and 2 will not be met. Exactly enough cards will be printed to satisfy the demand for design 3, while too many cards will be printed of all the other designs.

R	Estimated demand $D(R)$	$Z(R)$	Number of sheets $D(R)/Z(R)$	Wholesale price $S(R)$	Assembly cost $F(R)$	Gross profit $S(R) - F(R)$	Loss of profit	Additional profit
1	21,000	1	21,000	0.50	0.120	0.380	0.380	0
2	19,000	1	19,000	0.75	0.220	0.530	0.910	0.380
3	35,000	2	17,500	0.50	0.123	0.377	1.664	0.910
4	16,842	1	16,842	0.75	0.230	0.520	2.184	1.664
5	16,440	1	16,440	0.75	0.178	0.572	2.756	2.184
6	15,280	1	15,280	0.50	0.130	0.370	3.120	2.756
7	14,380	1	14,380	0.50	0.123	0.377	3.497	3.120
8	14,340	1	14,340	0.75	0.210	0.540	4.037	3.497
9	27,144	2	13,572	0.50	0.130	0.370	4.777	4.037
10	13,044	1	13,044	0.75	0.186	0.564	5.341	4.777
11	13,000	1	13,000	1.00	0.278	0.722	6.063	5.341
12	12,004	1	12,004	0.50	0.128	0.372	6.435	6.063
13	11,968	1	11,968	0.50	0.158	0.342	6.777	6.435
14	11,500	1	11,500	1.00	0.310	0.690	7.467	6.777

Management tested this model by using data from the past five years. It was found that the model would have improved results greatly. In actual use, the model resulted in a \$30,000 saving in the value of cards destroyed, with an additional profit of \$14,000.

EXERCISES

1. Calculate the gross revenue if 21,000 sheets are printed. (Multiply the estimated demand for each design and the gross profit per card. Then add these products for all 14 designs.)

2. From the answer in Exercise 1 subtract the total cost of printing 21,000 sheets ($21{,}000 \times 1.035$ dollars) and the fixed plate-preparation cost of \$609. The result is the total gross profit.

Find the total gross profit for each of the following numbers of printed sheets.

3. 19,000 (If 19,000 sheets are printed, the estimated demand will be met for all designs except design 1.)

4. 17,500

5. 16,842

6. Which number led to maximum total gross profit? Does this agree with the numbers given in the text?

3 Linear and Polynomial Models

As we have seen, a **mathematical model** is a mathematical description of a real-world situation. To construct a mathematical model of a given situation, we need a solid understanding of the situation to be modeled. We also need a good, solid knowledge of the possible mathematical functions that can be used to build the model.

Much mathematical theory has been developed over the years, with a large fraction of it being useful for model building. Yet the very richness and diversity of contemporary mathematics too often serves as a barrier between a person in another field and the mathematical tools that person needs. There are so many useful functions that it is often hard to know which one to pick for a particular model.

One way to get around this problem is to have a thorough understanding of the basic and most useful functions that are available for model building. In this chapter and later ones, several such functions are discussed, along with models constructed from each. In this chapter, we first discuss linear functions, used for writing models for data whose graphs can be approximated by a straight line. Then we discuss several functions which have curves for graphs—quadratic, polynomial, and rational.

3.1 LINEAR FUNCTIONS

Any relation of the form

$$ax + by = c,$$

where a, b, and c are real numbers, with not both a and b equal to 0, is called a **linear relation.** Examples of linear relations include $4x + 5y = 9$, $x - 6y = 8$, $x = 3$, $y = 5$, and so on. To see that $x = 3$ is a linear relation, rewrite it as $x + 0y = 3$.

If we are given a particular linear relation, such as $y = x + 1$, and a particular value of x, say $x = 2$, we can find the corresponding value of y.

$$y = x + 1$$
$$y = 2 + 1 \qquad \text{Let } x = 2$$
$$y = 3$$

Therefore, the ordered pair (2, 3) belongs to the relation. Check that (0, 1), (4, 5), (−2, −1), (−5, −4), (−3, −2), among many others, are also ordered pairs that satisfy the relation $y = x + 1$.

To graph $y = x + 1$, we can begin by drawing a graph containing all the ordered pairs listed above. This graph is shown in Figure 3.1(a). All the points of this graph appear to be on one straight line, which can be drawn through the points plotted, as in Figure 3.1(b). This straight line is the graph of the relation $y = x + 1$. Since any vertical line will cut the graph of Figure 3.1(b) in only one point, $y = x + 1$ is a function. A linear relation that is also a function is called a **linear function.**

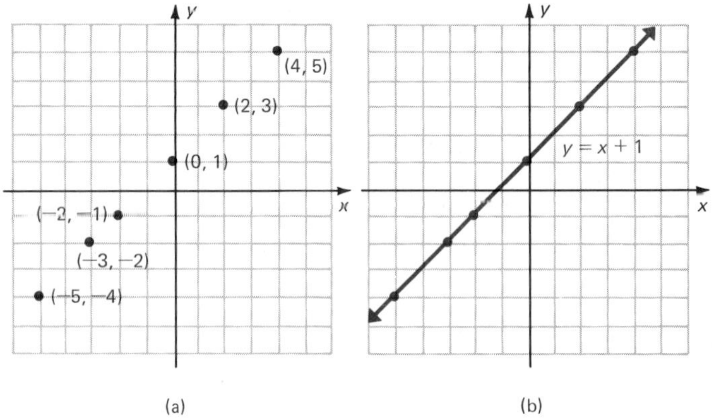

Figure 3.1

Example 1 Use the relation $x + 2y = 6$ to complete the ordered pairs (−6,), (−4,), (−2,), (0,), (2,), (4,). Graph these points and then draw a straight line through them.

To complete the ordered pair (−6,), we let $x = -6$ in the equation $x + 2y = 6$.

$$x + 2y = 6$$
$$-6 + 2y = 6 \qquad \text{Let } x = -6$$
$$2y = 12 \qquad \text{Add 6 to both sides}$$
$$y = 6$$

Ordered pair: (−6, 6).

1. Use the relation $2x - 3y = 12$ to complete the ordered pairs $(-6,\ \ )$, $(-3,\ \ )$, $(0,\ \ )$, $(3,\ \ )$, $(6,\ \ )$. Graph these points and then draw a straight line through them.

Answer: $(-6, -8)$, $(-3, -6)$, $(0, -4)$, $(3, -2)$, $(6, 0)$.

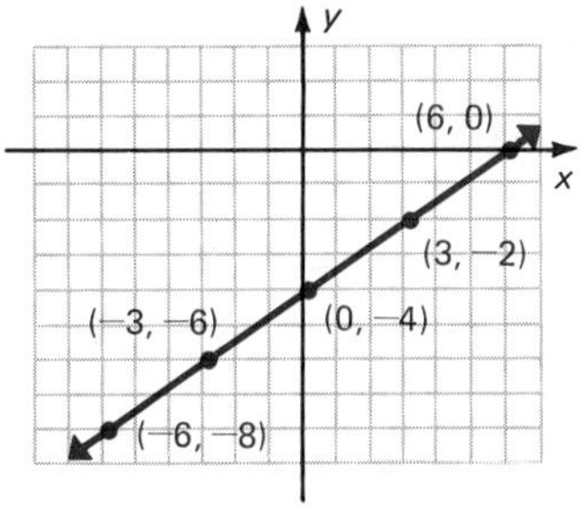

In the same way, if $x = -4$, then $y = 5$, giving $(-4, 5)$. Check that the remaining ordered pairs are as graphed in Figure 3.2(a). A line is drawn through the points in Figure 3.2(b). ■

Work Problem 1 at the side.

It is shown in more advanced courses that the graph of any linear relation is a straight line. Since a straight line is completely determined if we know any two distinct points that the line passes through, we really need to locate only two distinct points to draw the graph. Two points that are often useful for this purpose are the ***x*-intercept** and the ***y*-intercept.** The x-intercept is the x value (if one exists) at which the graph of the relation crosses the x-axis. The y-intercept in turn is the y value (if one exists) at which the graph crosses the y-axis. At the point where the graph crosses the y-axis, $x = 0$. Also, $y = 0$ at the x-intercept. (See Figure 3.3.)

Example 2 Use intercepts to draw the graph of $y = -2x + 5$.

To find the y-intercept, the point where the line crosses the y-axis, we let $x = 0$.

$$y = -2x + 5$$
$$y = -2(0) + 5 \qquad \text{Let } x = 0$$
$$y = 5$$

Thus the y-intercept is the ordered pair $(0, 5)$. In the same way, we find the x-intercept by letting $y = 0$.

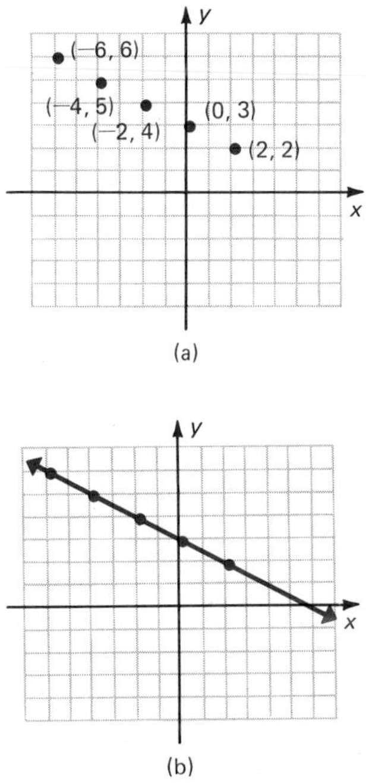

Figure 3.2

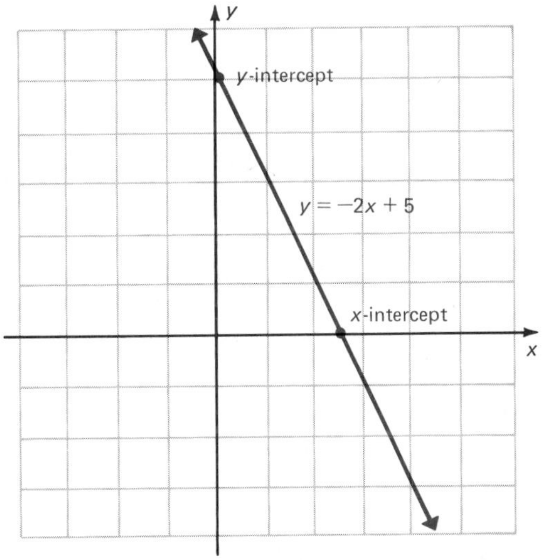

Figure 3.3

$0 = -2x + 5$	Let $y = 0$
$2x = 5$	Add $2x$ on both sides
$x = 5/2 = 2\frac{1}{2}$	Multiply both sides by $1/2$

The graph goes through $(2\frac{1}{2}, 0)$.

Using these two intercepts, we get the graph of Figure 3.3. To check the work, we can find a third point on the line. We do this by choosing another value of x (or y) and finding the corresponding value of the other variable. Check that $(1, 3)$, $(2, 1)$, $(3, -1)$, and $(4, -3)$ satisfy the relation $y = -2x + 5$ and also lie on the line of Figure 3.3. Does $y = -2x + 5$ represent a function? ■

Work Problem 2 at the side.

2. Find the intercepts of each of the following. Graph the lines.
(a) $3x + 4y = 12$
(b) $5x - 2y = 8$

Answer:
(a) x-intercept 4
y-intercept 3

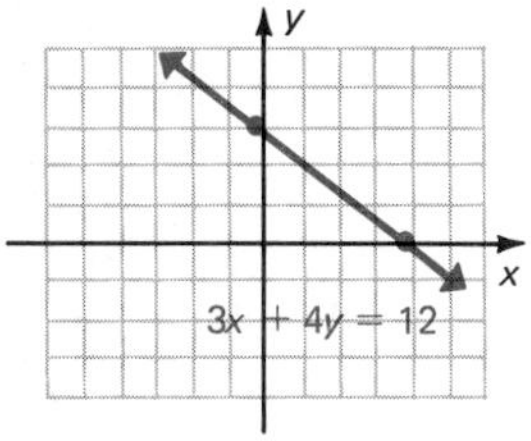

(b) x-intercept $8/5$
y-intercept -4

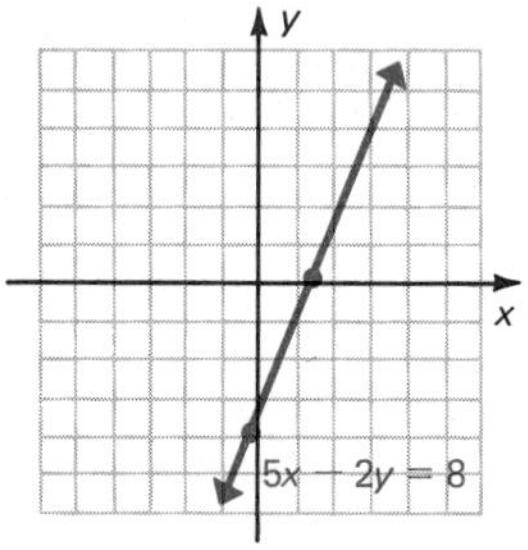

In the discussion of intercepts given above, we added the phrase "if one exists" when talking about the place where a graph crosses the x-axis. The next example shows that not every linear relation has an x-intercept.

Example 3 Graph $y = -3$.

Using $y = -3$, or equivalently, $y = 0x - 3$, we always get the same y value, -3, for any value of x. Therefore, there is no value of x that will make $y = 0$, so that the graph has no x-intercept. Since $y = -3$ is a linear relation, with a straight-line graph, and since the graph cannot cross the x-axis, the line must be parallel to the x-axis. For any value of x, we have $y = -3$. Therefore, the graph is the horizontal line which is parallel to the x-axis with y-intercept -3, as shown in Figure 3.4. Note that the graph is the graph of a function. In general, the graph of $y = k$, where k is a real number, is the horizontal line having y-intercept k. ■

Work Problem 3 at the side.

3. Graph $y = -5$. Does this graph represent a function?

Answer: Yes

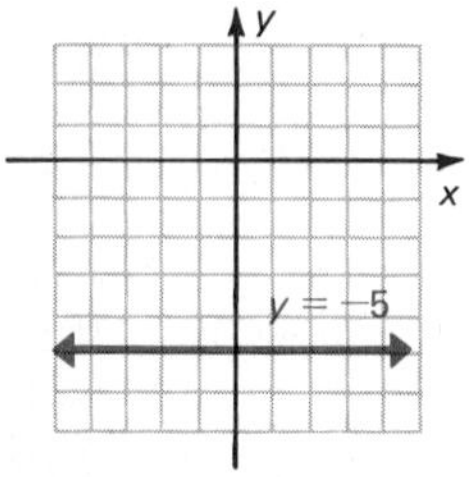

Example 4 Graph $x = -1$.

To obtain the graph of $x = -1$, we can complete some ordered pairs using the equivalent form, $x = 0y - 1$. For example, $(-1, 0)$, $(-1, 1)$, $(-1, 2)$, and $(-1, 4)$ are some ordered pairs satisfying the relation $x = -1$. (The first component of these ordered pairs is always -1, which is what $x = -1$ means.) Here, more than one second component corresponds to the same first component, -1. As the graph of Figure 3.5 shows, a vertical line can cut this graph in more than one point. (In fact, a vertical line cuts the graph in an infinite number of points.) For this reason, $x = -1$ is not a function. ■

4. Graph $x = 4$. Is this a function?

Answer: No

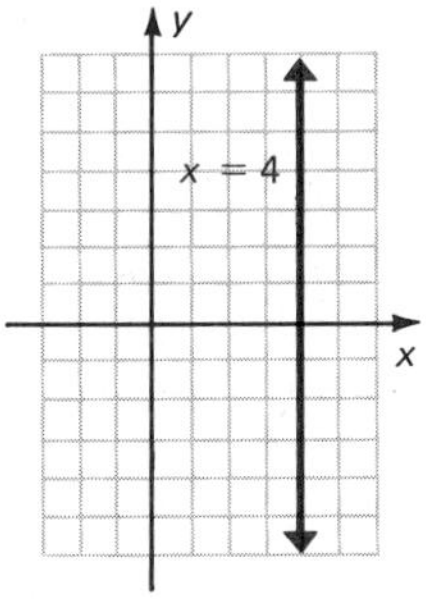

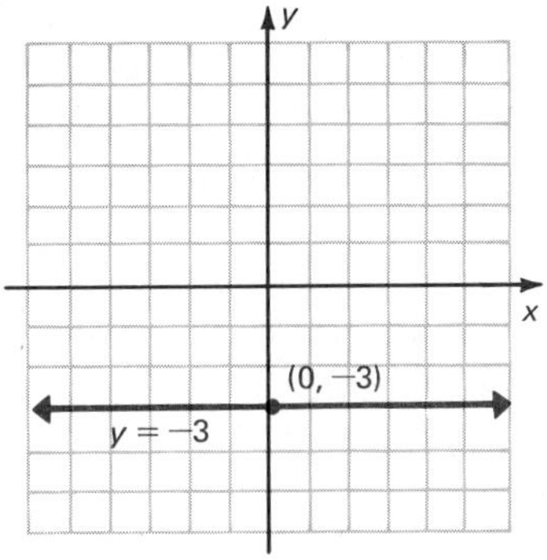

Figure 3.4

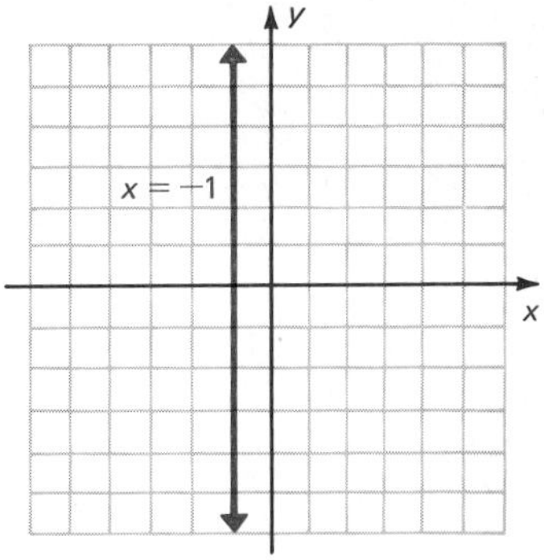

Figure 3.5

Work Problem 4 at the side.

As shown in Example 4 above, $x = -1$ is a linear relation that is not a linear function. Linear relations of this form, $x = k$, where k is a real number, are the only linear relations that are not also linear functions.

5. Graph
(a) $x = 5y$
(b) $5x = y$

Answer:
(a)

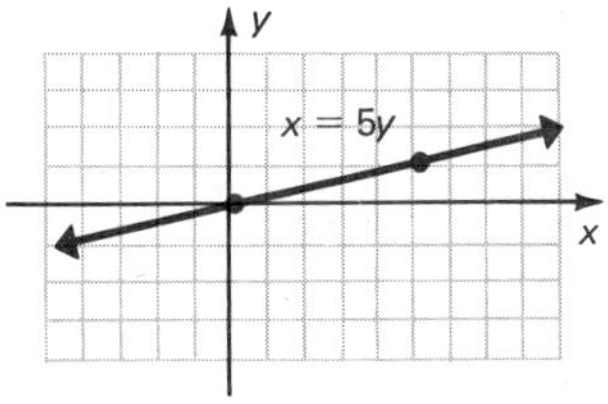

(b)

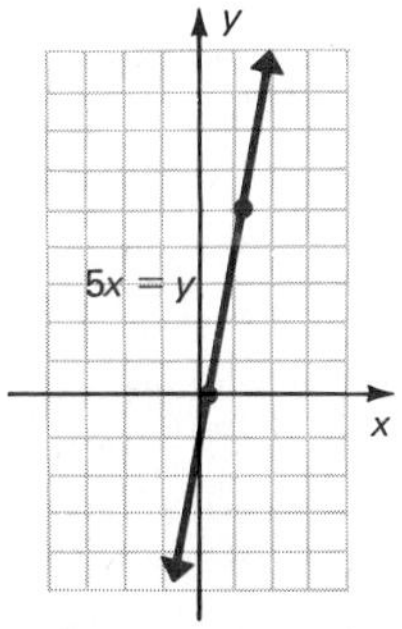

Example 5 Graph $y = -3x$.
Begin by looking for the x-intercept. If $y = 0$, then

$$y = -3x$$
$$0 = -3x \qquad \text{Let } y = 0$$
$$0 = x. \qquad \text{Multiply both sides by } -1/3$$

We have the ordered pair (0, 0). If we let $x = 0$, we end up with exactly the same ordered pair, (0, 0). We need two points to determine a straight line, and the intercepts have led to only one point. To get a second point, we can choose some other value of x (or y). For example, if $x = 2$, we have

$$y = -3x = -3(2) = -6, \qquad \text{Let } x = 2$$

giving the ordered pair $(2, -6)$. The two ordered pairs that we have found, (0, 0) and $(2, -6)$, were used to get the line shown in Figure 3.6 on the following page. ■

Work Problem 5 at the side.

Linear functions can be very useful in setting up a mathematical model for a real-life situation. In almost every case, linear (or any other reasonably simple) functions provide only approximations to real-world situations. However, such functions can often provide remarkably useful approximations.

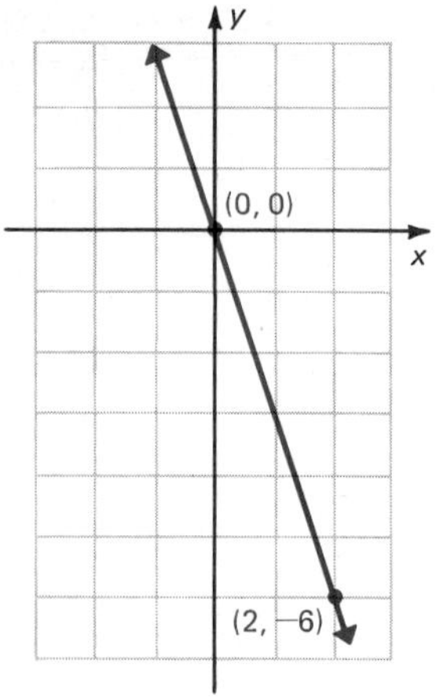

Figure 3.6

In particular, linear functions are often good choices for **supply and demand curves.** Typically, as the price of an item increases, the demand for the item decreases, while the supply increases. The following example shows this. (Note: In these examples we write price as a function of the number of units demanded or supplied. Some people would rather reverse this and write the number of units demanded as a function of the price, which is certainly valid. We have chosen the method used here since it is the approach favored by economists.)

Example 6 Suppose that Greg Odjakjian, an economist, has studied the supply and demand for aluminum siding and has come up with the conclusion that price, p, and demand, x, in appropriate units and for an appropriate domain, are related by the linear function

$$p = 60 - \frac{3}{4}x.$$

(a) Find the demand at a price of \$40.
When $p = 40$, we have

$$p = 60 - \frac{3}{4}x$$

$$40 = 60 - \frac{3}{4}x \qquad \text{Let } p = 40$$

$$-20 = -\frac{3}{4}x \qquad \text{Add } -60 \text{ on both sides}$$

$$\frac{80}{3} = x. \qquad \text{Multiply both sides by } -\frac{4}{3}$$

At a price of \$40, 80/3 units will be demanded; this gives the ordered pair (80/3, 40). (It is customary to write the ordered pairs so that price comes second.)

(b) Find the price if the demand is 32 units.
Let $x = 32$.

$$p = 60 - \frac{3}{4}x$$

$$p = 60 - \frac{3}{4}(32) \qquad \text{Let } x = 32$$

$$p = 60 - 24$$

$$p = 36$$

With a demand of 32 units, we get a price of \$36. This gives the ordered pair (32, 36).

(c) Graph $p = 60 - \frac{3}{4}x$.

Use the ordered pairs (80/3, 40) and (32, 36) to get the graph shown in black in Figure 3.7. Only the portion of the graph in Quadrant I is shown, since our function is only meaningful for positive values of p and x. ■

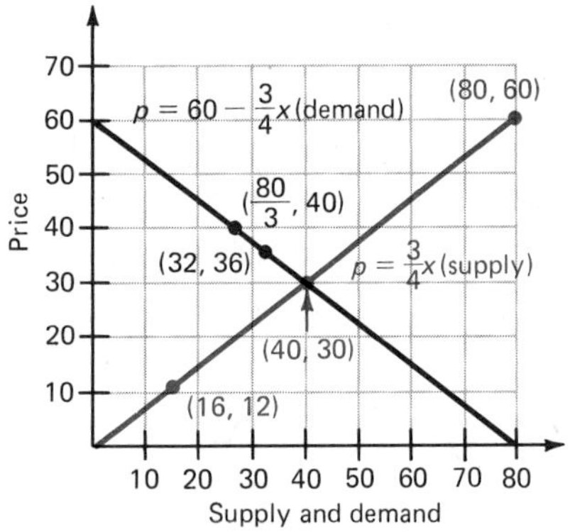

Figure 3.7

Work Problem 6 at the side.

6. Suppose price and demand are related by $p = 100 - 4x$.
(a) Find the price if the demand is 10 units.
(b) Find the demand if the price is \$80.

Answer:
(a) \$60
(b) 5 units

Example 7 Suppose now that the economist of Example 6 concludes that the price and supply of siding are related by

$$p = \frac{3}{4}x.$$

(a) Find the supply if the price is \$60.

We have $\quad 60 = \frac{3}{4}x \quad$ Let $p = 60$

$$80 = x.$$

If the price is \$60, then 80 units will be supplied to the marketplace. This gives the ordered pair (80, 60).

(b) Find the price if the supply is 16 units.

$$p = \frac{3}{4}(16) = 12 \qquad \text{Let } x = 16$$

If the supply is 16 units, then the price is \$12. This gives the ordered pair (16, 12).

(c) Graph $p = \frac{3}{4}x$.

Use the ordered pairs (80, 60) and (16, 12) to get the graph shown in color in Figure 3.7. ■

As shown in the graphs of Figure 3.7, both the supply and the demand functions pass through the point (40, 30). If the price of the siding is more than \$30, the supply will exceed the demand. At a price less than \$30, the demand will exceed the supply. Only at a price of \$30 will demand and supply be equal. For this reason, \$30 is called the **equilibrium price.** When the price is \$30, demand and supply both equal 40 units. This second number is called the **equilibrium supply** or **equilibrium demand.**

Example 8 Use algebra to find the equilibrium supply for the aluminum siding. (See Examples 6 and 7.)

The equilibrium supply is found when the prices from both supply and demand are equal. From Example 6 we have $p = 60 - \frac{3}{4}x$; in Example 7 we have $p = \frac{3}{4}x$. Thus,

$$60 - \frac{3}{4}x = \frac{3}{4}x$$

$$240 - 3x = 3x \qquad \text{Multiply both sides by 4}$$

$$240 = 6x \qquad \text{Add } 3x \text{ to both sides}$$

$$40 = x.$$

The equilibrium supply is 40 units, the same answer that we found above. ■

7. The demand for a certain commodity is related to the price by $p = 80 - \frac{2}{3}x$. The supply is related to the price by $p = \frac{4}{3}x$. Find
(a) the equilibrium demand
(b) the equilibrium price.

Answer:
(a) 40
(b) 160/3

Work Problem 7 at the side.

3.1 EXERCISES

Graph each of the following linear relations. Identify any which are not *linear functions. (See Examples 1–5.)*

1. $y = 2x + 1$
2. $y = 3x - 1$
3. $y = 4x$
4. $y = x + 5$
5. $3y + 4x = 12$
6. $4y + 5x = 10$
7. $y = -2$
8. $x = 4$
9. $6x + y = 12$
10. $x + 3y = 9$
11. $x - 5y = 4$
12. $2y + 5x = 20$
13. $x + 5 = 0$
14. $y - 4 = 0$
15. $5y - 3x = 12$
16. $2x + 7y = 14$
17. $8x + 3y = 10$
18. $9y - 4x = 12$
19. $y = 2x$
20. $y = -5x$

21. $y = -4x$

22. $y = x$

23. $x + 4y = 0$

24. $x - 3y = 0$

APPLIED PROBLEMS

Management

Suppose that the demand and price for a certain model of electric can opener are related by

$$p = 16 - \frac{5}{4}x,$$

where p is price and x is demand, in appropriate units. (See Example 6.) Find the price for a demand of

25. 0 units;

26. 4 units;

27. 8 units.

Find the demand for the electric can opener at a price of

28. \$6;

29. \$11;

30. \$16.

31. Graph $p = 16 - \frac{5}{4}x$.

Suppose the price and supply of the item above are related by

$$p = \frac{3}{4}x.$$

(See Example 7.) Find the supply when the price is

32. \$0;

33. \$10;

34. \$20.

35. Graph $p = \frac{3}{4}x$ on the same axes used for Exercise 31.

36. Find the equilibrium supply. (See Example 8.)

37. Find the equilibrium price.

Let the supply and demand functions for strawberry-flavored licorice be

$$\textit{supply:}\quad p = \frac{3}{2}x$$

and

$$\textit{demand:}\quad p = 81 - \frac{3}{4}x.$$

38. Graph these on the same axes.

39. Find the equilibrium demand.

40. Find the equilibrium price.

Let the supply and demand functions for butter pecan ice cream be given by

$$\textit{supply:}\quad p = \frac{2}{5}x$$

and

$$\textit{demand:}\quad p = 100 - \frac{2}{5}x.$$

41. Graph these on the same axes.

42. Find the equilibrium demand.

43. Find the equilibrium price.

Let the supply and demand functions for sugar be given by

$$\textit{supply:}\quad p = 1.4x - .6$$

and

$$\textit{demand:}\quad p = -2x + 3.2.$$

44. Graph these on the same axes.

45. Find the equilibrium demand.

46. Find the equilibrium price.

In a recent issue of Business Week, *the president of Insta-Tune, a chain of franchised automobile tune-up shops, says that people who buy a franchise and open a shop pay a weekly fee of*

$$y = .07x + \$135$$

to company headquarters. Here y is the fee and x is the total amount of money taken in during the week by the tune-up center. Find the weekly fee if x is

47. \$0; **48.** \$1000; **49.** \$2000; **50.** \$3000.

51. Graph the function.

Social Science

In a recent issue of The Wall Street Journal, *we are told that the relationship between the amount of money that an average family spends on food, x, and the amount of money it spends on eating out, y, is approximated by the model*

$$y = .36x.$$

Find y if x is

52. \$40; **53.** \$80; **54.** \$120.

55. Graph the function.

3.2 SLOPE AND THE EQUATION OF A LINE

As we said in the previous section, the graph of a straight line is completely determined if we know two different points that the line goes through. We can also draw the graph of a straight line if we know only *one* point that the line goes through, and, in addition, the "steepness" of the line. To get a number which represents the "steepness" of a line, we define the **slope** of a line, as in the next paragraph.

Figure 3.8 shows a line passing through the two different points $(x_1, y_1) = (-3, 5)$ and $(x_2, y_2) = (2, -4)$. The difference in the two x values,

$$x_2 - x_1 = 2 - (-3) = 5$$

in this example, is called the **change in x.** The symbol Δx (read "delta x") is used to represent the change in x. In the same way, Δy represents the **change in y.** In our example,

$$\Delta y = y_2 - y_1 = -4 - 5 = -9.$$

The slope of a line through the two different points (x_1, y_1) and (x_2, y_2) is defined as the change in y divided by the change in x or

$$\textbf{slope} = \frac{\textbf{change in } y}{\textbf{change in } x} = \frac{\Delta y}{\Delta x} = \frac{y_2 - y_1}{x_2 - x_1}.$$

The slope of the line in Figure 3.8 is

$$\text{slope} = \frac{\Delta y}{\Delta x} = \frac{-4 - 5}{2 - (-3)} = \frac{-9}{5}.$$

Using similar triangles from geometry, it can be shown that the slope is independent of the choice of points on the line. That is, the same slope will be obtained for *any* choice of two different points on the line.

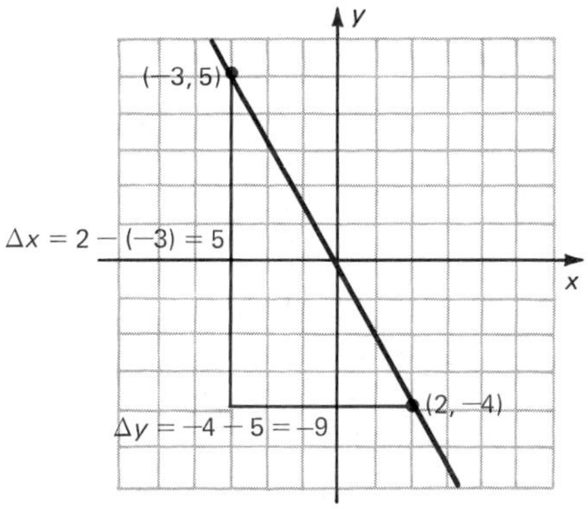

Figure 3.8

Example 1 Find the slope of the line through the points $(-7, 6)$ and $(4, 5)$.

Let $(x_1, y_1) = (-7, 6)$. Then $(x_2, y_2) = (4, 5)$. Use the definition of slope:

$$\text{slope} = \frac{\Delta y}{\Delta x} = \frac{5 - 6}{4 - (-7)} = \frac{-1}{11}.$$

We could have let $(x_1, y_1) = (4, 5)$ and $(x_2, y_2) = (-7, 6)$. In that case,

$$\text{slope} = \frac{6 - 5}{-7 - 4} = \frac{1}{-11} = \frac{-1}{11},$$

the same answer. The order in which coordinates are subtracted does not matter, as long as it is done consistently. ■

1. Find the slope of the line through
(a) $(6, 11), (-4, -3)$
(b) $(-3, 5), (-2, 8)$.

Answer:
(a) 7/5
(b) 3

Work Problem 1 at the side.

Example 2 Find the slope of the line $3x - 4y = 12$. To find the slope, we need two different points that the line goes through. We can find two such points here by first letting $x = 0$ and then letting $y = 0$.

If $x = 0$,		If $y = 0$,	
$3x - 4y = 12$		$3x - 4y = 12$	
$3(0) - 4y = 12$	Let $x = 0$	$3x - 4(0) = 12$	Let $y = 0$
$-4y = 12$		$3x = 12$	
$y = -3$		$x = 4$	

Ordered pairs: $(0, -3)$ and $(4, 0)$.

Use the two ordered pairs to find the slope.

$$\text{slope} = \frac{0 - (-3)}{4 - 0} = \frac{3}{4} \quad ■$$

2. Find the slope of
(a) $8x + 5y = 9$
(b) $2x - 7y = 6$
(c) $8y = x$.

Answer:
(a) $-8/5$
(b) 2/7
(c) 1/8

Work Problem 2 at the side.

In Example 2, we found that the slope of $3x - 4y = 12$ is 3/4. We found this slope by getting two different points that the line goes through. We can get the slope more quickly if we use a shortcut. Solve the equation for y.

$$3x - 4y = 12$$

$$-4y = -3x + 12 \qquad \text{Add } -3x \text{ on both sides}$$

$$y = \frac{3}{4}x - 3 \qquad \text{Multiply both sides by } -\frac{1}{4}$$

In this final equation, the slope is the number in front of x (the **coefficient** of x). Also, from this last equation, if we let $x = 0$, we find the y-intercept, -3. An equation that is solved for y always gives the slope and y-intercept, as stated in the next theorem.

Theorem 3.1 *Slope-intercept form* If a linear function is expressed in the form

$$y = mx + b,$$

then m is the slope of the line and b is the y-intercept.

The equation $y = mx + b$ is called the **slope-intercept form** of the equation of a line since the slope, m, and the y-intercept, b, can be read directly from the equation.

Example 3 Find the slope and y-intercept for each of the following lines.

(a) $5x - 3y = 1$

Solve for y:

$$\begin{aligned} 5x - 3y &= 1 \\ -3y &= -5x + 1 \\ y &= \frac{5}{3}x - \frac{1}{3}. \end{aligned}$$

The slope is 5/3 and the y-intercept is $-1/3$.

(b) $-9x + 6y = 2$

Solve for y:

$$\begin{aligned} -9x + 6y &= 2 \\ 6y &= 9x + 2 \\ y &= \frac{3}{2}x + \frac{1}{3}. \end{aligned}$$

The slope is 3/2 and the y-intercept is 1/3. ■

3. Find the slope and y-intercept for

(a) $x + 4y = 6$

(b) $3x - 2y = 1$

Answer:

(a) $-1/4$; 6/4 or 3/2

(b) 3/2; $-1/2$

Work Problem 3 at the side.

The slope and y-intercept of a line can be used to draw the graph of the line as shown in the next example.

Example 4 Use the slope and y-intercept to graph $3x - 2y = 2$.

Solve for y:

$$\begin{aligned} 3x - 2y &= 2 \\ -2y &= -3x + 2 \\ y &= \frac{3}{2}x - 1 \end{aligned}$$

The slope is 3/2 and the y-intercept is -1.

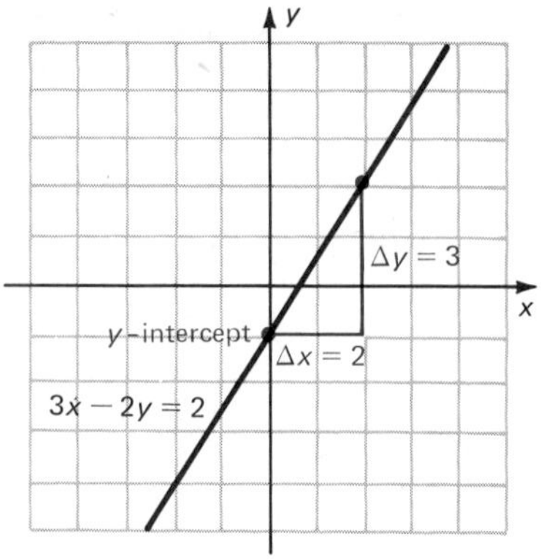

Figure 3.9

To draw the graph, first locate the y-intercept, as shown in Figure 3.9. To find a second point on the graph, we use the slope. If m represents the slope, then we know that

$$m = \frac{\Delta y}{\Delta x} = \frac{3}{2}$$

in our example. If x changes by 2 units ($\Delta x = 2$), then y will change by 3 units ($\Delta y = 3$). We find our second point by starting at the y-intercept graphed in Figure 3.9 and moving 2 units to the right and 3 units up. Once we find this second point, we can draw a line through it and the y-intercept. ■

Work Problem 4 at the side.

4. Use the slope and y-intercept to graph
(a) $2x + 5y = 10$
(b) $-x + y = 4$.

Answer:
(a)

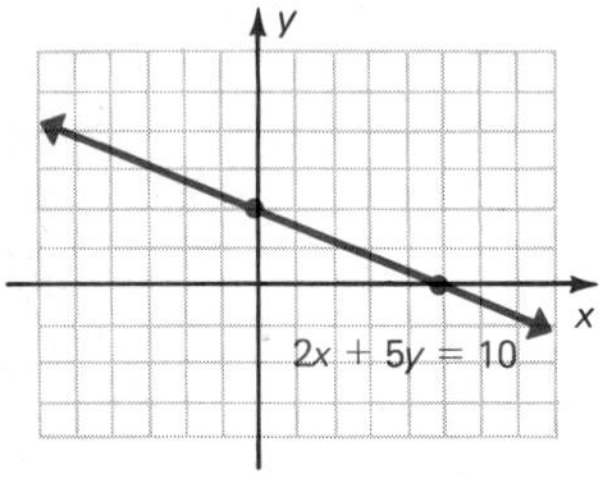

(b)

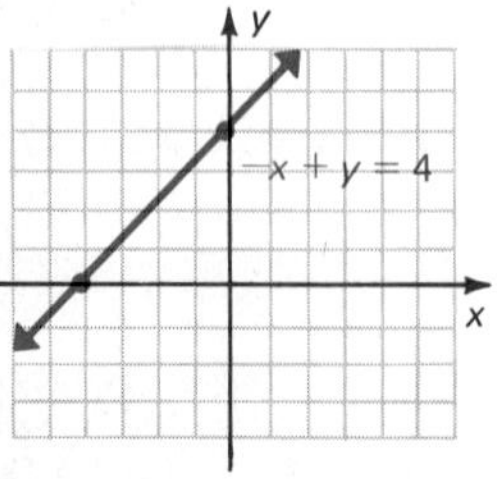

The slope-intercept form of the equation of a line involves the slope and the y-intercept. Sometimes, however, we know the slope of a line, together with one point (perhaps *not* the y-intercept) that the line passes through. To find the equation of a line in this case, we use the point-slope form of the equation of a line.

To obtain the **point-slope form** of the equation of a line, let (x_1, y_1) be any fixed point on the line and let (x, y) represent any other point on the line. If m is the slope of the line, we can use the definition of slope to get

$$\frac{y - y_1}{x - x_1} = m$$

or

$$y - y_1 = m(x - x_1).$$

The next theorem summarizes this result.

Theorem 3.2 *Point-slope form* If a line has slope m and passes through the point (x_1, y_1), then the equation of the line is given by

$$y - y_1 = m(x - x_1).$$

Example 5 Find the equation of the line going through the following point and having the given slope.

(a) $(-4, 1)$, $m = -3$

Here $x_1 = -4$, $y_1 = 1$, and $m = -3$. Substitute these values into the point-slope form of the equation of a line.

$$\begin{aligned} y - y_1 &= m(x - x_1) \\ y - 1 &= -3[x - (-4)] && \text{Let } y_1 = 1, m = -3, x_1 = -4 \\ y - 1 &= -3(x + 4) \\ y - 1 &= -3x - 12 && \text{Distributive property} \\ y &= -3x - 11 \end{aligned}$$

(b) $(3, -7)$, $m = 5/4$

$$\begin{aligned} y - y_1 &= m(x - x_1) \\ y - (-7) &= \frac{5}{4}(x - 3) && \text{Let } y_1 = -7, m = \frac{5}{4}, x_1 = 3 \\ y + 7 &= \frac{5}{4}(x - 3) \\ 4y + 28 &= 5(x - 3) && \text{Multiply both sides by 4} \\ 4y + 28 &= 5x - 15 \\ 4y &= 5x - 43 \quad \blacksquare \end{aligned}$$

Work Problem 5 at the side.

5. Find the equation of the line having the given slope and going through the given point
(a) $m = -3/5$, $(5, -2)$
(b) $m = 1/3$, $(6, 8)$.

Answer:
(a) $5y = -3x + 5$
(b) $3y = x + 18$

The point-slope form can also be used to find the equation of a line if we know two different points that the line goes through. The procedure for doing this is shown in the next example.

Example 6 Find the equation of the line through $(5, 4)$ and $(-10, -2)$.

Begin by using the definition of slope to find the slope of the line which passes through the two points.

$$\text{slope} = m = \frac{-2 - 4}{-10 - 5} = \frac{-6}{-15} = \frac{2}{5}$$

Use $m = 2/5$ and either of the given points in the point-slope form. If we let $(x_1, y_1) = (5, 4)$, we have

$$\begin{aligned} y - y_1 &= m(x - x_1) \\ y - 4 &= \frac{2}{5}(x - 5) && \text{Let } y_1 = 4,\ m = \frac{2}{5},\ x_1 = 5 \\ 5y - 20 &= 2(x - 5) && \text{Multiply both sides by 5} \\ 5y - 20 &= 2x - 10 \\ 5y &= 2x + 10. \end{aligned}$$

Check that we get the same result if we let $(x_1, y_1) = (-10, -2)$. ■

6. Find the equation of the line through
(a) (2, 3) and (−4, 6)
(b) (−8, 2) and (3, −6).

Answer:
(a) $2y = -x + 8$
(b) $11y = -8x - 42$

Work Problem 6 at the side.

Example 7 Find the equation of the line through (8, −4) and (−2, −4).

Find the slope.

$$m = \frac{-4 - (-4)}{-2 - 8} = \frac{0}{-10} = 0$$

Choose, say, (8, −4) as (x_1, y_1).

$$\begin{aligned} y - y_1 &= m(x - x_1) \\ y - (-4) &= 0(x - 8) && \text{Let } y_1 = -4,\ m = 0,\ x_1 = 8 \\ y + 4 &= 0 && 0(x - 8) = 0 \\ y &= -4 \quad ■ \end{aligned}$$

As we saw in the previous section, $y = -4$ represents a horizontal line, with y-intercept −4. In general, every horizontal line has a slope of 0.

7. Find the equation of the line through (−2, 5) and (7, 5).

Answer: $y = 5$

Work Problem 7 at the side.

Example 8 Find the equation of the line through (4, 3) and (4, −6).

Find the slope.

$$m = \frac{-6 - 3}{4 - 4} = \frac{-9}{0}$$

Division by 0 is impossible, so there is no slope here. If we graph the given ordered pairs (4, 3) and (4, −6) and draw a line through them, we find that the line is vertical. In the last section, we saw that vertical lines have equations of the form $x =$ some number. By looking at the two ordered pairs given above, we find that the desired equation is

$$x = 4. \quad ■$$

In general,

a vertical line has no slope.

8. Find the equation of the line through $(-5, 1)$ and $(-5, 7)$.

Answer: $x = -5$

Work Problem 8 at the side.

Following is a summary of the types of equations of lines that we have studied.

Equation	Description
$ax + by = c$	if $a \neq 0$ and $b \neq 0$, line has x-intercept c/a and y-intercept c/b
$x = k$	**vertical line,** x-intercept k, no y-intercept, no slope
$y = k$	**horizonal line,** y-intercept k, no x-intercept, slope 0
$y = mx + b$	**slope-intercept form,** slope m, y-intercept b
$y - y_1 = m(x - x_1)$	**point-slope form,** slope m, line passes through (x_1, y_1)

3.2 EXERCISES

In each of the following exercises, find the slope, if it exists, of the line going through the given pair of points. (See Example 1.)

1. $(-8, 6), (2, 4)$

2. $(-3, 2), (5, 9)$

3. $(-1, 4), (2, 6)$

4. $(3, -8), (4, 1)$

5. The origin and $(-4, 6)$

6. The origin and $(8, -2)$

7. $(-2, 9), (-2, 11)$

8. $(7, 4), (7, 12)$

9. $(3, -6), (-5, -6)$

10. $(5, -11), (-9, -11)$

Find the slope and y-intercept of each of the following lines. (See Examples 2 and 3.)

11. $y = 3x + 4$

12. $y = -3x + 2$

13. $y + 4x = 8$

14. $y - x = 3$

15. $3x + 4y = 5$

16. $2x - 5y = 8$

17. $3x + y = 0$

18. $y - 4x = 0$

19. $2x + 5y = 0$

20. $3x - 4y = 0$

21. $y = 8$

22. $y = -4$

23. $y + 2 = 0$

24. $y - 3 = 0$

25. $x = -8$

26. $x = 3$

Graph the line going through the given point and having the given slope.

27. $(-4, 2)$, $m = 2/3$

28. $(3, -2)$, $m = 3/4$

29. $(-5, -3)$, $m = -2$

30. $(-1, 4)$, $m = 2$

31. $(8, 2)$, $m = 0$

32. $(2, -4)$, $m = 0$

33. $(6, -5)$, no slope

34. $(-8, 9)$, no slope

35. $(0, -2)$, $m = 3/4$

36. $(0, -3)$, $m = 2/5$

37. $(5, 0)$, $m = 1/4$

38. $(-9, 0)$, $m = 5/2$

Find the equation for each line having the given y-intercept and slope. (See Example 5.)

39. 4, $m = -3/4$

40. -3, $m = 2/3$

41. -2, $m = -1/2$

42. $3/2$, $m = 1/4$

43. $5/4$, $m = 3/2$

44. $-3/8$, $m = 3/4$

Find equations for each of the following lines. (See Examples 5–8.)

45. Through $(-4, 1)$, $m = 2$

46. Through $(5, 1)$, $m = -1$

47. Through $(0, 3)$, $m = -3$

48. Through $(-2, 3)$, $m = 3/2$

49. Through $(3, 2)$, $m = 1/4$

50. Through $(0, 1)$, $m = -2/3$

51. Through $(-1, 1)$ and $(2, 5)$

52. Through $(4, -2)$ and $(6, 8)$

53. Through $(9, -6)$ and $(12, -8)$

54. Through $(-5, 2)$ and $(7, 5)$

55. Through $(-8, 4)$ and $(-8, 6)$

56. Through $(2, -5)$ and $(4, -5)$

57. Through $(-1, 3)$ and $(0, 3)$

58. Through $(2, 9)$ and $(2, -9)$

APPLIED PROBLEMS

Many real-world situations can be approximately described by a straight-line graph. One way to find the equation of such a straight line is to use two typical data points from the graph and the point-slope form of the equation of a line. In each of the following problems, assume that the data can be approximated fairly closely by a straight line. Use the given information to find the equation of the line. Find the slope of each of the lines.

Management

59. A company finds that it can make a total of 20 hot tubs for \$13,900, while 10 tubs cost \$7500. Let y be the total cost to produce x hot tubs.

60. The sales of a small company were \$27,000 in its second year of operation and \$63,000 in its fifth year. Let y represent sales in year x.

Natural Science

61. When a certain industrial pollutant is introduced into a river, the reproduction of catfish declines. In a given period of time, three tons of the pollutant results in a fish population of 37,000. Also, 12 tons of pollutant produce a fish population of 28,000. Let y be the fish population when x tons of pollutant are introduced into the river.

62. In the snake *Lampropelbis polyzona*, total length, y, is related to tail length, x, in the domain 30 mm $\leq x \leq$ 200 mm, by a linear function. Find such a linear function, if a snake 455 mm long has a 60 mm tail, and a 1050 mm snake has a 140 mm tail.

Social Science

63. According to research done by the political scientist James March, if the Democrats win 45% of the two-party vote for the House of Representatives, they win 42.5% of the seats. If the Democrats win 55% of the vote, they win 67.5% of the seats. Let y be the percent of seats won, and x the percent of the two-party vote.

64. If the Republicans win 45% of the two-party vote, they win 32.5% of the seats (see Exercise 63.) If they win 60% of the vote, they get 70% of the seats. Let y represent the percent of the seats, and x the percent of the vote.

3.3 LINEAR MATHEMATICAL MODELS

Throughout this book, we have been setting up mathematical models—mathematical descriptions of real-world situations. In this section, we discuss mathematical models in more detail, then we look at some mathematical models using the linear equations we have discussed in the last two sections.

If we completely understand the principles causing a certain event to happen, then the mathematical model describing that event can often be very accurate. As a rule, the mathematical models constructed in the physical sciences are excellent at predicting events. For example, if a body falls in a vacuum, then d, the distance in feet that the body will fall in t seconds, is given by

$$d = \frac{1}{2}gt^2,$$

where g is a constant representing gravity (as an approximation,

$g = 32$ feet per second per second). Using this equation, d can be predicted exactly for a known value of t. In the same way, astronomers have formulated very precise mathematical models of the movements of our solar system. Predictions about the future positions of heavenly bodies can be made with great accuracy. Thus, when astronomers say that Halley's comet will next appear in 1986, few dispute the statement.

The situation is different in the fields of management and social science. Mathematical models in these fields tend to be less accurate approximations, and often gross approximations at that. While an astronomer can say with almost certainty that Halley's comet will reappear in 1986, no corporation vice-president would dare predict that company sales will be \$185,000 higher in 1986 than in 1985. So many variables enter into determining sales of a company that no present-day mathematical model can ever hope to produce results comparable to those produced in physical science.

In spite of these limitations of mathematical models, such models have found a large and increasing acceptance in management and economic decision making. There is one main reason for this—mathematical models produce very useful results.

The former vice-president for corporate planning for Anheuser-Busch, Mr. Robert S. Weinberg, offers the following reasons for the wide use of mathematical models in management and economics.

(1) A mathematical model leads to a well-defined statement of the problem. For example, decisions about new products have often been made strictly on the intuition of a top manager. By studying past records, by surveying potential consumers, and by carefully analyzing costs at various possible levels of production, it is possible to produce a mathematical model of the factors affecting a new product. This model can then be used to produce results of the following type:

There is a 15% chance of making a profit of \$10 million.
There is a 20% chance of making a profit of \$2 million.
There is a 25% chance of breaking even.
There is a 40% chance of losing \$3 million.

With this information at hand, management can make a much more informed decision.

(2) A model makes clear the assumptions behind the problem and its possible solutions.

(3) A model can be used to identify areas in which further research is needed.

Let us now look at some mathematical models of real-world situations.

Example 1 *Sales analysis* It is common to compare the change in sales of two companies by comparing the rates at which these sales change. If the sales of the two companies can be approximated by linear functions, we can use the work of the last section to find rates of change. For example, the chart below shows sales in two different years for two different companies.

Company	Sales in 1978	Sales in 1981
A	\$10,000	\$16,000
B	5000	14,000

The sales of Company A increased from \$10,000 to \$16,000 over this 3-year period, for a total increase of \$6000. Thus, the average rate of change of sales is

$$\frac{\$6000}{3} = \$2000.$$

If we assume that the sales of Company A have increased linearly (that is, that the sales can be closely approximated by a linear function), then we can find the equation describing the sales by finding the equation of the line through $(-3, 10{,}000)$ and $(0, 16{,}000)$, where $x = 0$ represents 1981 and $x = -3$ represents 1978. The slope of the line is

$$\frac{16{,}000 - 10{,}000}{0 - (-3)} = 2000,$$

which is the same as the annual rate of change we found above. Using the point-slope form of the equation of a line, we find that

$$y - 16{,}000 = 2000(x - 0),$$
$$y = 2000x + 16{,}000$$

gives the equation describing sales. ■

1. Assume that the sales of Company B (see the chart above) have also increased linearly. Find the equation giving its sales and the average rate of change of sales.

Answer: $y = 3000x + 14{,}000$; 3000

Work Problem 1 at the side.

As both the example and the problem show, the average rate of change is the same as the slope of the line. This is always true for data that can be modeled with a linear function.

Example 2 Suppose that a researcher has concluded that a dosage of x grams of a certain stimulant causes a rat to gain y grams of weight, where

$$y = 2x + 50.$$

If the researcher administers 30 grams of the stimulant, how much weight will the rat gain?

Let $x = 30$. Then the rat will gain

$$y = 2(30) + 50 = 110$$

grams of weight. ■

2. Find the weight gain if the researcher administers 50 grams of the stimulant.

Answer: 150 grams

Work Problem 2 at the side.

Based on Example 2 above, the average rate of change of weight gain with respect to the amount of stimulant is given by the slope of the line. The slope of $y = 2x + 50$ is 2. If the researcher increases the dose of stimulant by 1 gram, the rat will gain 2 grams of weight.

3. A certain new anticholesterol drug is related to the blood cholesterol level by the linear model

$$y = 280 - 3x,$$

when x is the dosage of the drug (in grams) and y is the blood cholesterol level.

(a) Find the blood cholesterol level if 12 grams of the drug are administered.

(b) In general, an increase of 1 gram in the dose causes what change in the blood cholesterol level?

Answer:

(a) 244

(b) a decrease of 3 units.

Work Problem 3 at the side.

Example 3 *Cost analysis* Suppose that the cost of producing clock-radios can be approximated by the linear model

$$C(x) = 12x + 100,$$

where $C(x)$ is the cost in dollars to produce x radios.

The cost to produce 0 radios is

$$C(0) = 12(0) + 100 = 100 \qquad \text{Let } x = 0$$

dollars. This sum, \$100, represents the costs involved in designing the radio, setting up a plant, training the workers, and so on. The value of $C(0)$ is called the **fixed cost.**

Once the company has invested the fixed cost into the clock-radio project, what will then be the additional cost per radio? To find out, let's first find the cost of a total of 5 radios:

$$C(5) = 12(5) + 100 = 160,$$

or \$160. The cost of 6 radios is

$$C(6) = 12(6) + 100 = 172,$$

or \$172.

The sixth radio itself thus costs \$172 − \$160 = \$12 to produce. In the same way, the 81st radio costs $C(81) - C(80) = \$1072 -$

$1060 = $12 to produce. In fact, the $(n + 1)$st radio costs

$$C(n + 1) - C(n) = [12(n + 1) + 100] - [12n + 100] = 12$$

dollars to produce. Since each additional radio costs $12 to produce, $12 is called the **variable cost** per radio. Note that 12 is also the slope of the cost function, $C(x) = 12x + 100$.

In economics, the cost of producing an additional item is called the **marginal cost** of that item. In the clock-radio example, the marginal cost of each radio is $12. ■

4. The cost in dollars to produce x kilograms of chocolate candy is given by $C(x)$, where in dollars

$$C(x) = 3.5x + 800.$$

Find each of the following:
(a) the fixed cost
(b) the total cost for 12 kilograms
(c) the variable cost
(d) the marginal cost of the 40th kilogram.

Answer:
(a) $800
(b) $842
(c) $3.50 per kilogram
(d) $3.50

Work Problem 4 at the side.

In general,

if a cost function is given by a linear function of the form $C(x) = mx + b$, then m represents the variable cost per item and b the fixed cost. Conversely, if the fixed cost of producing an item is b and the variable cost is m, then the cost function, $C(x)$, for producing x items, is given by $C(x) = mx + b$.

Example 4 In a midwest city, a taxi company charges riders a fixed charge of 50¢, plus 60¢ per mile. Write a cost function, $C(x)$, which is a mathematical model for the cost of a ride of x miles.

Here the fixed cost is $b = 50$ cents, with a variable cost of $m = 60$ cents. The cost function, $C(x)$, is thus given by

$$C(x) = 60x + 50.$$

For example, a taxi ride of 4 miles will cost $C(4) = 60(4) + 50 = 290$, or $2.90. For each additional mile, the cost increases by 60¢. ■

5. Avis charges $18 for a one-day rental of a certain model car, plus 17¢ per mile. Write a cost function, $C(x)$, giving the cost of driving the car x miles in one day.

Answer: $C(x) = .17x + 18$

Work Problem 5 at the side.

Example 5 The variable cost for raising a certain type of frog for laboratory study is $12 per unit of frogs, while the cost to produce 100 units is $1500. Find the cost function, $C(x)$, if we know it is linear.

Since the cost function is linear, it can be expressed in the form $C(x) = mx + b$. We are told that the variable cost is $12 per unit, which gives the value for m in the model. The model can thus be written $C(x) = 12x + b$. To find b, we use the fact that the cost of producing 100 units of frogs is $1500, or $C(100) = 1500$. Substituting $x = 100$ and $C(x) = 1500$ into $C(x) = 12x + b$, we have

$$C(x) = 12x + b$$
$$1500 = 12 \cdot 100 + b$$
$$1500 = 1200 + b$$
$$b = 300.$$

The desired model is given by $C(x) = 12x + 300$. The fixed cost is \$300. ■

6. The total cost of producing 10 units of a business calculator is \$220. The variable cost per calculator is \$14. Find the cost function, $C(x)$, if we know that it is linear.

Answer: $C(x) = 14x + 80$

Work Problem 6 at the side.

Depreciation Since machines and equipment wear out or become obsolete over a period of time, business firms must take into account a number which represents the amount of value that the equipment has lost during each year of its useful life. This lost value, called **depreciation,** may be calculated in several ways. The simplest way is to use **straight-line,** or **linear,** depreciation, in which an item having a useful life of n years is assumed to lose $1/n$ of its value each year. Thus, a typewriter with a ten-year life would be assumed to lose 1/10 of its value each year.

A machine may have some **scrap value** at the end of its useful life. For this reason, depreciation is calculated on **net cost**—the difference between purchase price and scrap value. To find the annual straight-line depreciation on an item having a net cost of I dollars and a useful life of n years, multiply the net cost by the fraction of the value lost each year, $1/n$. This gives

$$\textbf{annual straight-line depreciation} = \frac{1}{n} \cdot I.$$

For example, in 4 years the depreciation will total

$$4\left(\frac{1}{n} \cdot I\right) = \frac{4}{n} \cdot I.$$

The total undepreciated balance after 4 years will thus be given by the net cost minus the total depreciation thus far, or

$$I - \frac{4}{n} \cdot I = I\left(1 - \frac{4}{n}\right).$$

To generalize this result, first note that the total depreciation at the end of year j, where j is a whole number between 1 and n inclusive, or $1 \le j \le n$, is

$$j\left(\frac{1}{n} \cdot I\right) = \frac{j}{n} \cdot I,$$

while the amount undepreciated after j years is

$$I - \frac{j}{n} \cdot I = I\left(1 - \frac{j}{n}\right).$$

Example 6 An asset has a purchase price of \$100,000 and a scrap value of \$40,000. The useful life of the asset is 10 years. Find each of the following for this asset.

(a) the net cost

Since net cost = purchase price − scrap value, we have

$$\text{net cost} = \$100{,}000 - \$40{,}000 = \$60{,}000.$$

(b) the annual depreciation

The useful life of the asset is 10 years. Therefore, 1/10 of the net cost is depreciated each year. The annual depreciation by the straight-line method is

$$\frac{1}{n} \cdot I = \frac{1}{10} \cdot 60{,}000 = 6000$$

dollars.

(c) the undepreciated balance after 4 years

Let $j = 4$ in the formula above.

$$\begin{aligned} I\left(1 - \frac{j}{n}\right) &= 60{,}000\left(1 - \frac{4}{10}\right) && \text{Let } I = 60{,}000,\ j = 4,\ n = 10 \\ &= 60{,}000\left(\frac{6}{10}\right) \\ &= 36{,}000 \end{aligned}$$

dollars. The total amount that will be depreciated over the life of this asset is \$60,000. After 4 years, \$36,000 of this amount has not yet been claimed as depreciation. ■

Work Problem 7 at the side.

Straight-line depreciation is the easiest method of depreciation to use, but it often does not accurately reflect the rate at which assets actually lose value. Some assets, such as new cars, lose more value annually at the beginning of their useful life than at the end. For this reason, two other methods of depreciation, the **sum-of-the-years'-digits** method (discussed in Exercises 33 and 34 below) and **double declining balance** (discussed in Section 3.6) are often used.

7. A backhoe (used for digging basements, swimming pools, and other holes) costs \$65,000, has a useful life of 7 years, and a scrap value of \$16,000. Find the
(a) net cost
(b) annual depreciation
(c) amount undepreciated after 5 years.

Answer:
(a) \$49,000
(b) \$7000
(c) \$14,000

Example 7 *Break-even analysis* A firm producing chicken feed finds that the total cost, $C(x)$, of producing x units is given by

$$C(x) = 20x + 100.$$

The feed sells for \$24 per unit, so that the revenue, $R(x)$, from selling x units is given by the product of the price per unit and the number of units sold,

$$R(x) = 24x.$$

The firm will just break even (no profit and no loss), as long as revenue just equals cost, or $R(x) = C(x)$. This is true whenever

$$R(x) = C(x)$$

$$24x = 20x + 100 \qquad \text{Substitute for } R(x) \text{ and } C(x)$$

$$4x = 100 \qquad \text{Add } -20x \text{ to both sides}$$

$$x = 25.$$

The point at which revenue just equals sales (here, $x = 25$) is called the **break-even point.**

The graphs of $C(x) = 20x + 100$ and $R(x) = 24x$ are shown in Figure 3.10. The break-even point is shown on the graph. If the company produces more than 25 units (if $x > 25$), it makes a profit; if $x < 25$ it loses money. ■

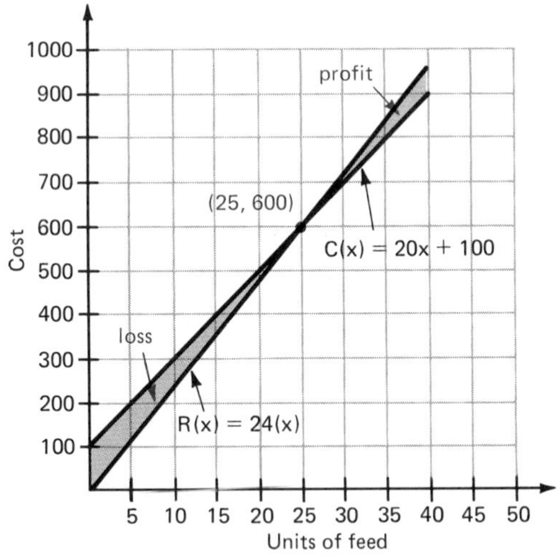

Figure 3.10

8. For a certain magazine, $C(x) = .20x + 1200$, while $R(x) = .50x$, where x is the number of magazines sold. Find the break-even point.

Answer: 4000 magazines

Work Problem 8 at the side.

3.3 APPLIED PROBLEMS

Management

1. Suppose the sales of a particular brand of electric guitar satisfy the relationship

$$S(x) = 300x + 2000,$$

where $S(x)$ represents the number of guitars sold in year x, with $x = 0$ corresponding to 1979. Find the sales in each of the following years.
(a) 1981 (b) 1982 (c) 1983 (d) 1979
(e) Find the annual rate of change of the sales.

Natural Science

2. If the population of ants in an anthill satisfies the relationship

$$A(x) = 1000x + 6000,$$

where $A(x)$ represents the number of ants present at the end of month x, find the number of ants present at the end of each of the following months. Let $x = 0$ represent June.
(a) June
(b) July
(c) August
(d) December
(e) What is the monthly rate of change of the number of ants?

3. Let $N(x) = -5x + 100$ represent the number of bacteria (in thousands) present in a certain tissue culture at time x, measured in hours, after an antibacterial spray is introduced into the environment. Find the number of bacteria present at each of the following times.
(a) $x = 0$
(b) $x = 6$
(c) $x = 20$
(d) What is the hourly rate of change in the number of bacteria? Interpret the negative sign in the answer.

Social Science

4. Let $R(x) = -8x + 240$ represent the number of students present in a large business mathematics class, where x represents the number of hours of study required weekly. Find the number of students present at each of the following levels of required study.
(a) $x = 0$
(b) $x = 5$
(c) $x = 10$
(d) What is the rate of change of the number of students in the class with respect to the number of hours of study? Interpret the negative sign in the answer.
(e) The professor in charge of the class likes to have exactly 16 students. How many hours of study must he require in order to have exactly 16 students? (Hint: Find a value of x such that $R(x) = 16$.)

Management

5. Assume that the sales of a certain appliance dealer are given by a linear function. Suppose the sales were \$85,000 in 1968 and \$150,000 in 1978. Let $x = 0$ represent 1968, with $x = 10$ representing 1978.

(a) Find the equation giving the dealer's yearly sales.
(b) What were the dealer's sales in 1976?
(c) Estimate sales in 1981.

6. Assume the sales of a certain automotive parts company are given by a linear function. Suppose the sales were \$200,000 in 1972 and \$1,000,000 in 1979. Let $x = 0$ represent 1972, and $x = 7$ represent 1979.
(a) Find the equation giving the company's yearly sales.
(b) Find the sales in 1974.
(c) Estimate the sales in 1981.

7. Suppose the stock of rubber dog bones, $B(x)$, on hand at the beginning of the day in a pet store is given by

$$B(x) = 600 - 20x,$$

where $x = 1$ corresponds to June 1, and x is measured in days. If the store is open every day of the month, find the supply of dog bones on hand at the beginning of each of the following days.
(a) June 6
(b) June 12
(c) June 24
(d) When will the last bone from this stock be sold?
(e) What is the daily rate of change of this stock?

Social Science

8. In psychology, the just-noticeable-difference (JND) for some stimulus is defined as the amount by which the stimulus must be increased so that a person will perceive it as having just barely been increased. For example, suppose a research study indicates that a line 40 centimeters in length must be increased to 42 cm before a subject thinks that it is longer. In this case, the JND would be $42 - 40 = 2$ cm. In a particular experiment, the JND (y) is given by

$$y = 0.03x,$$

where x represents the original length of the line. Find the JND for lines having the following lengths.
(a) 10 cm
(b) 20 cm
(c) 50 cm
(d) 100 cm
(e) Find the rate of change in the JND with respect to the original length of the line.

Management

Write a cost function for each of the following. Identify all variables used. (See Example 4.)

9. A chain-saw rental firm charges \$12 plus \$1 per hour.

10. A trailer-hauling service charges \$45 plus \$2 per mile.

11. A parking garage charges 35¢ plus 30¢ per half-hour.

12. For a one-day rental, a car rental firm charges \$14 plus 6¢ per mile.

Assume that each of the following can be expressed as a linear cost function. Find the appropriate cost function in each case. (See Example 5.)

13. Fixed cost, \$100; 50 items cost \$1600 to produce.

14. Fixed cost, \$400; 10 items cost \$650 to produce.

15. Fixed cost, \$1000; 40 items cost \$2000 to produce.

16. Fixed cost, \$8500; 75 items cost \$11,875 to produce.

17. Variable cost, \$50; 80 items cost \$4500 to produce.

18. Variable cost, \$120; 100 items cost \$15,800 to produce.

19. Variable cost, \$90; 150 items cost \$16,000 to produce.

20. Variable cost, \$120; 700 items cost \$96,500 to produce.

The manager of a McDonald's restaurant in Sacramento (an ex-student) told us that his cost function for producing coffee is $C(x) = .048x$, *where* $C(x)$ *is the total cost in dollars of producing* x *cups. (He is ignoring the cost of the coffee pot and the cost of labor.) Find the total cost of producing the following numbers of cups.*

21. 500 cups **22.** 1000 cups **23.** 1001 cups

24. Find the marginal cost of the 1001st cup.

For each of the assets in Exercises 25–30 find the straight-line depreciation in year four, and the amount undepreciated after 4 years. (See Example 6.)

25. Cost: \$50,000; scrap value: \$10,000; life: 20 years

26. Cost: \$120,000; scrap value: \$0; life: 10 years

27. Cost: \$80,000; scrap value: \$20,000; life: 30 years

28. Cost: \$720,000; scrap value: \$240,000; life: 12 years

29. Cost: \$1,400,000; scrap value: \$200,000; life: 8 years

30. Cost: \$2,200,000; scrap value: \$400,000; life: 12 years

31. Suppose an asset has a net cost of \$80,000 and a four-year life.
(a) Find the straight-line depreciation in each of years 1, 2, 3, and 4 of the item's life.
(b) Find the sum of all depreciation for the four-year life.

32. A forklift truck has a net cost of \$12,000, with a useful life of 5 years.
(a) Find the straight-line depreciation in each of years 1, 2, 3, 4, and 5 of the forklift's life.
(b) Find the sum of all depreciation for the five-year life.

Some assets, such as new cars, lose more value annually at the beginning of their useful life than at the end. By one method of depreciation for such assets, called the **sum-of-the-years'-digits** *the depreciation in year j, which we call A_j, is given by*

$$A_j = \frac{n - j + 1}{n(n + 1)} \cdot 2I,$$

where n is the useful life of the item, $1 \leq j \leq n$, and I is the net cost of the item. (Another approach to this method of depreciation is shown in the exercises of Section 5.2.)

33. For a certain asset, $n = 4$ and $I = \$10{,}000$.
  (a) Use the sum-of-the-years'-digits method to find the depreciation for each of the four years covering the useful life of the asset.
  (b) What would be the annual depreciation by the straight-line method?

34. A machine tool costs $105,000 and has a scrap value of $25,000, with a useful life of 4 years.
  (a) Use the sum-of-the-years'-digits method to find the depreciation in each of the four years of the machine tool's life.
  (b) Find the total depreciation by this method.

For each of the following assets, use the sum-of-the-years'-digits method to find the depreciation in year 1 and year 4.

35. Cost: $36,500; scrap value: $8500; life: 10 years

36. Cost: $6250; scrap value: $250; life: 5 years

37. Cost: $18,500; scrap value: $3900; life: 6 years

38. Cost: $275,000; scrap value: $25,000; life: 20 years

39. The cost to produce x units of wire is $C(x) = 50x + 5000$, while the revenue is $R(x) = 60x$. Find the break-even point and the revenue at the break-even point.

40. The cost to produce x units of squash is $C(x) = 100x + 6000$, while the revenue is $R(x) = 500x$. Find the break-even point.

You are the manager of a firm. You are considering the manufacture of a new product, so you ask the accounting department to produce cost estimates and the sales department to produce sales estimates. After you receive the data, you must decide whether or not to go ahead with production of the new product. Analyze the following data (find a break-even point) and then decide what you would do. (See Example 7.)

41. $C(x) = 85x + 900$; $R(x) = 105x$; not more than 38 units can be sold.

42. $C(x) = 105x + 6000$; $R(x) = 250x$; not more than 400 units can be sold.

43. $C(x) = 70x + 500$; $R(x) = 60x$ (Hint: what does a negative break-even point mean?)

44. $C(x) = 1000x + 5000$; $R(x) = 900x$.

x year	y sales
0	48
1	59
2	66
3	75
4	80
5	90

The sales of a certain furniture company in thousands of dollars are shown in the chart at the side.

45. Graph this data, plotting years on the x-axis and sales on the y-axis. (Note that the data points can be closely approximated by a straight line.)

46. Draw a line through the points (2, 66) and (5, 90). The other four points should be close to this line. (These two points were selected as "best" representing the line that could be drawn through the data points.)

47. Use the two points of Exercise 46 to find an equation for the line that approximates the data.

48. Complete the following chart.

Year	Sales (actual)	Sales (predicted from equation of Exercise 47)	Difference, actual minus predicted
0			
1			
2			
3			
4			
5			

(We will obtain a formula for the "best" line through the points in Section 10.6.)

49. Use the result of Exercise 47 to predict sales in year 7.

50. Do the same for year 9.

Social Science

Most people are not very good at estimating the passage of time. Some people estimate it too fast, and others, too slow. One psychologist has constructed a mathematical model for actual time as a function of estimated time: if y represents actual time and x estimated time, then

$$y = mx + b,$$

where m and b are constants that must be determined experimentally for each person. Suppose that for a particular person, $m = 1.25$ and $b = -5$. Find y if x is

51. 30 minutes;

52. 60 minutes;

53. 120 minutes;

54. 180 minutes.

Suppose that for another person, $m = .85$ and $b = 1.2$. Find y if x is

55. 15 minutes;

56. 30 minutes;

57. 60 minutes;

58. 120 minutes.

For this same person, find x if y is

59. 60 minutes;

60. 90 minutes.

3.4 QUADRATIC MODELS

While linear functions provide mathematical models for many different types of problems, not all real-world situations can be adequately approximated by these functions. In many cases, we need a *curve*, not a straight line.

One curve which has proven useful in many different mathematical models is obtained from the quadratic function. A function of the form

$$y = ax^2 + bx + c,$$

where $a \neq 0$, and where a, b, and c are real numbers, is called a **quadratic function.** Without the restriction $a \neq 0$, we might have $y = bx + c$, a linear function. In general, unless restricted for some reason, domains of quadratic functions include all real numbers, but ranges are more restricted, as we will see.

Example 1 Graph the quadratic function $y = x^2$.

The function $y = x^2$ is perhaps the simplest of all quadratic functions. (To see that $y = x^2$ is quadratic, let $a = 1$, $b = 0$, and $c = 0$ in the general quadratic function $y = ax^2 + bx + c$.) To graph $y = x^2$, we can choose some values for x, and then find the corresponding values for y, as in the table of Figure 3.11. These points can then be plotted, with a smooth curve drawn through them, as in Figure 3.11. The domain of the resulting graph is the set of all real numbers, while the range is the set of all nonnegative real numbers. ■

The curve in Figure 3.11 is called a **parabola.** The lowest point on this parabola, (0, 0), is called the **vertex.** It can be proved that every quadratic function has a graph which is a parabola.

Parabolas have many useful properties. Cross sections of radar dishes and spotlights form parabolas. Discs often visible on the sidelines of televised football games are microphones having reflectors with parabolic cross sections. These microphones are used by the television networks to pick up the shouted signals of the quarterbacks.

Example 2 Graph the quadratic function $y = x^2 + 2$.

For any value of x that we choose, the corresponding value of y will be 2 more than for the parabola graphed in Figure 3.11 above. The graph of the parabola $y = x^2 + 2$, as shown in Figure 3.12, is shifted 2 units upward in comparison to the graph of $y = x^2$. The point (0, 2) is the vertex of the parabola. ■

Work Problem 1 at the side.

1. Graph each of the following parabolas.
(a) $y = x^2 - 4$
(b) $y = x^2 + 5$

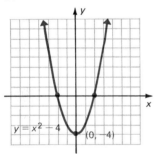

Answer:
(a)

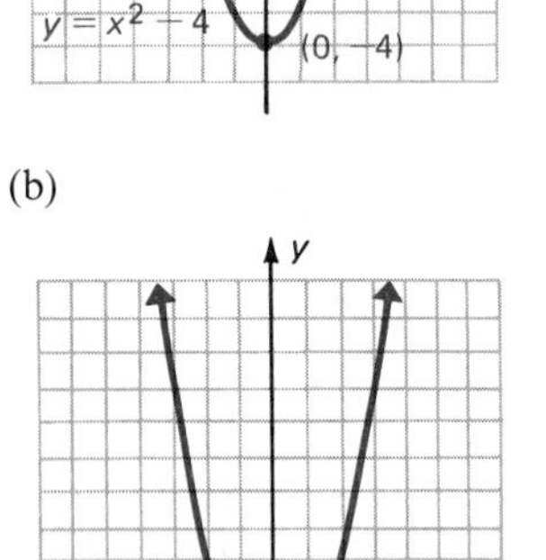

(b)

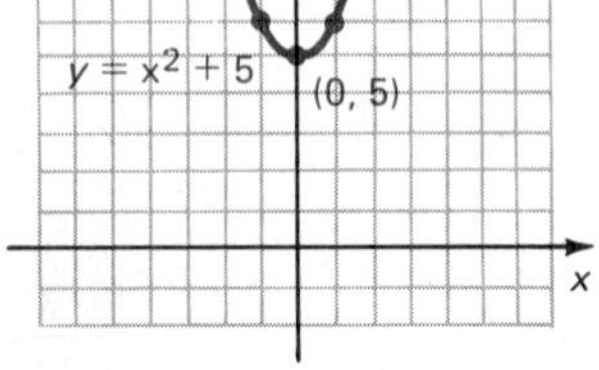

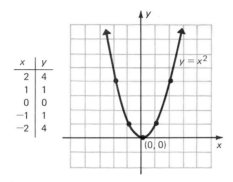

x	y
2	4
1	1
0	0
−1	1
−2	4

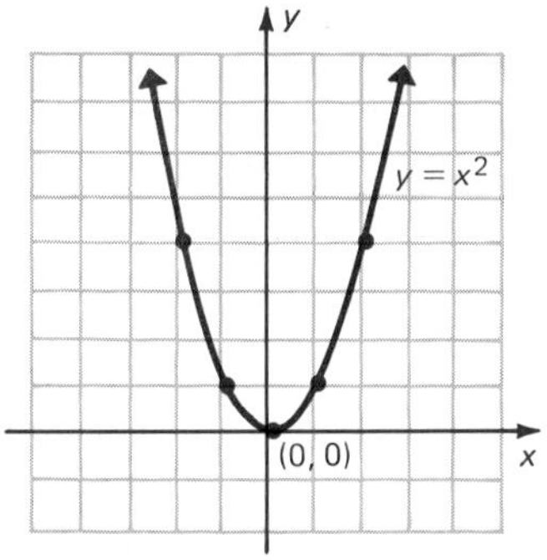

Figure 3.11

x	y
2	6
1	3
0	2
−1	3
−2	6

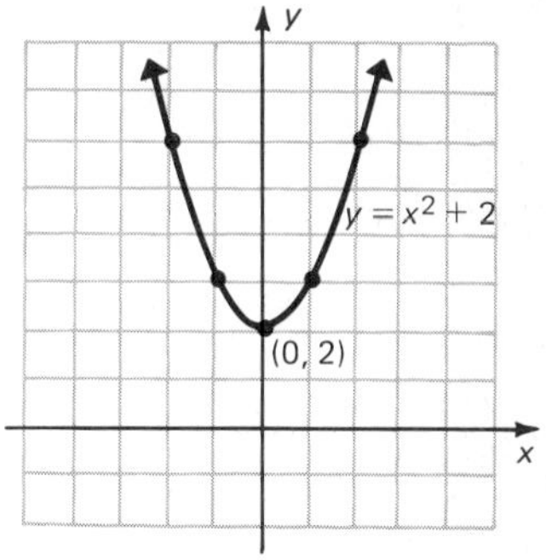

Figure 3.12

Example 3 Graph $y = -(x + 2)^2$.

Check that the ordered pairs shown in thc table in Figure 3.13 belong to the function. These ordered pairs show that the graph of $y = -(x + 2)^2$ is "upside down" in comparison to $y = x^2$, and also shifted 2 units to the left. Here the vertex, $(-2, 0)$, is the *highest* point on the graph. ■

Work Problem 2 at the side.

As shown in Example 2, the graph of $y = x^2 + 2$ was shifted upward 2 units in comparison with the graph of $y = x^2$. In Example 3, the graph of $y = -(x + 2)^2$ is shifted 2 units to the left when compared to $y = x^2$, and opens downward. Both upward

x	y
0	−4
−1	−1
−2	0
−3	−1
−4	−4

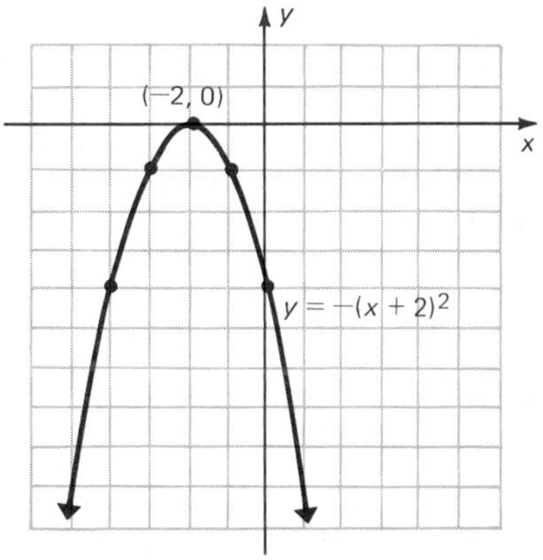

Figure 3.13

2. Graph the following.
(a) $y = (x + 4)^2$
(b) $y = -(x - 3)^2$

Answer:
(a)

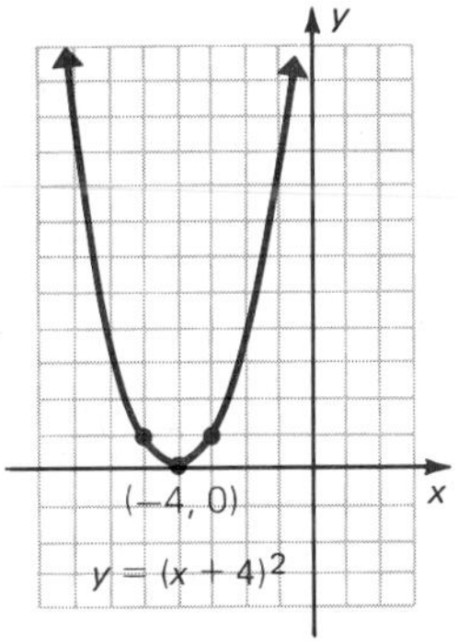

(b)

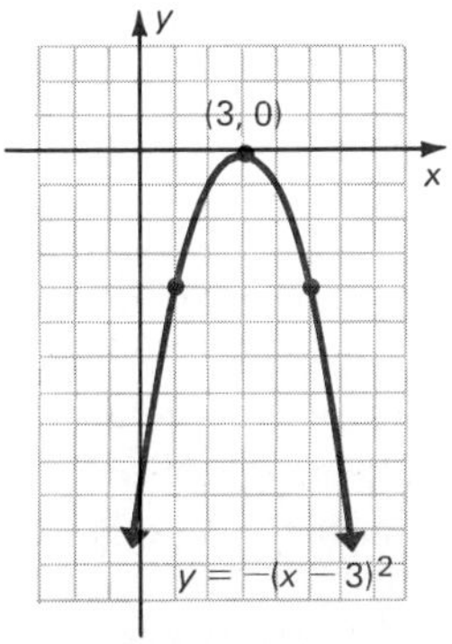

(or downward) and side-to-side shifts can occur in the same parabola, as shown in the next example.

Example 4 Graph $y = 2(x - 3)^2 - 5$.

Using ideas similar to those of the examples above, show that this parabola has vertex $(3, -5)$ and opens upward. The coefficient 2 causes the values of y to increase more rapidly than in the parabola $y = x^2$, so that the graph in this example is "narrower" than the graph of $y = x^2$. Figure 3.14 shows a table of ordered pairs satisfying $y = 2(x - 3)^2 - 5$, as well as the graph of the function. ■

x	y
1	3
2	−3
3	−5
4	−3
5	3

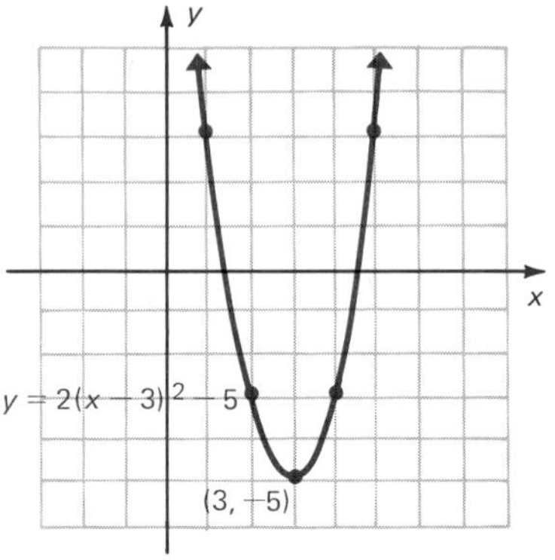

Figure 3.14

Work Problem 3 at the side.

3. Graph the following.
(a) $y = (x + 4)^2 - 3$
(b) $y = -2(x - 3)^2 + 1$

Answer:
(a)

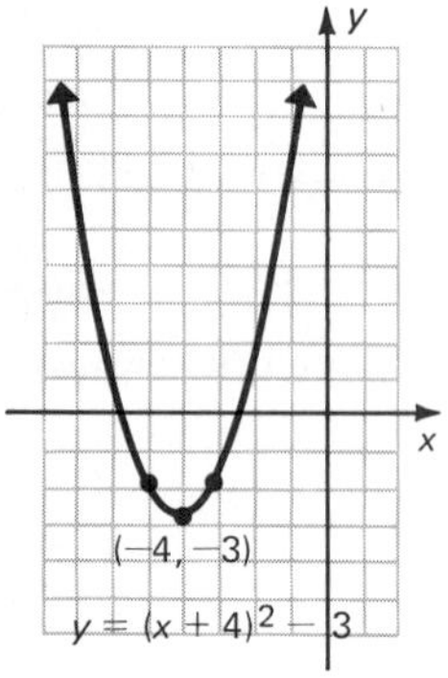

(b)

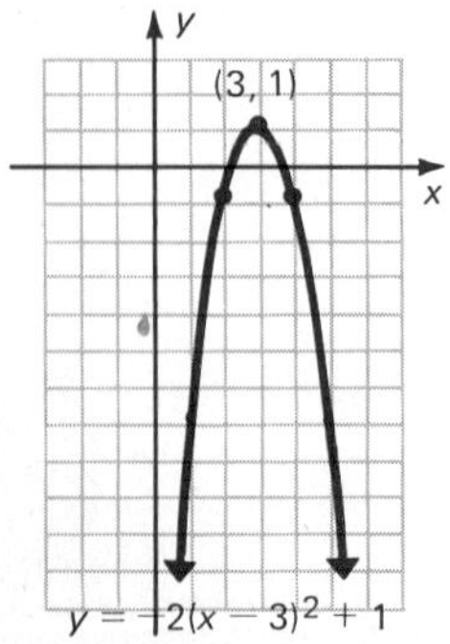

In summary,

if a quadratic function is written in the form $y = a(x - h)^2 + k$, then the graph of the function is a **parabola** having its vertex at (h, k). If $a > 0$, the parabola opens **upward;** if $a < 0$, it opens **downward.** If $0 < |a| < 1$, the parabola is **"broader"** than $y = x^2$, while if $|a| > 1$, the parabola is **"narrower"** than $y = x^2$.

If a quadratic function is not given in the form $y = a(x - h)^2 + k$, it is helpful to convert it to this form, as shown by the next example.

Example 5 Graph $y = x^2 - 2x + 3$.

To graph this parabola, we can use the method of **completing the square** to first rewrite the equation in the form $y = a(x - h)^2 + k$. To use this method, first write $y = x^2 - 2x + 3$ as

$$y = (x^2 - 2x \qquad) + 3.$$

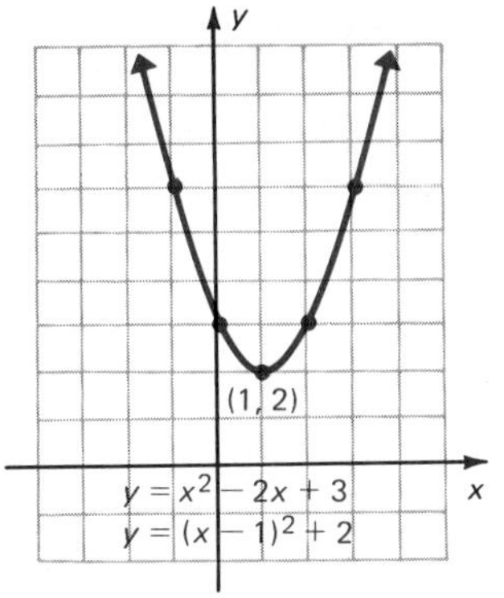

Figure 3.15

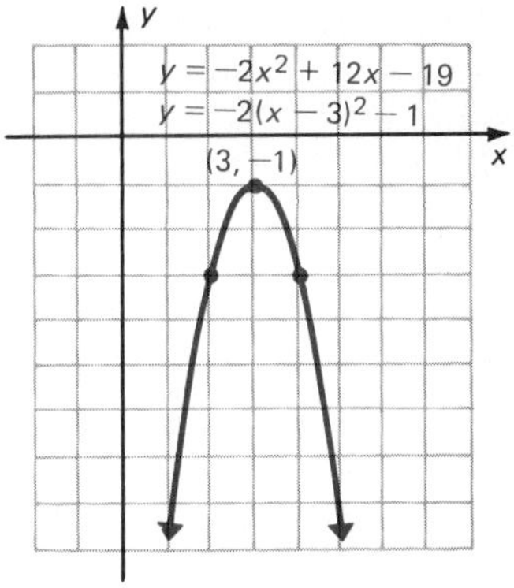

Figure 3.16

We want to write the expression inside the parentheses as the square of some quantity. To do this, take half the coefficient of x, $\frac{1}{2}(-2) = -1$, and square this result: $(-1)^2 = 1$.* Now add 1 inside the parentheses and compensate by subtracting 1 on the outside. This gives

$$y = (x^2 - 2x + 1) + 3 - 1.$$

Factor $x^2 - 2x + 1$ as $(x - 1)(x - 1)$, or $(x - 1)^2$. This gives

$$y = (x - 1)^2 + 2.$$

By this result, we see that the given function has a graph which is a parabola which opens upward and has a vertex at (1, 2). By plotting a few points, we get the graph shown in Figure 3.15. ■

Work Problem 4 at the side.

Example 6 Graph $y = -2x^2 + 12x - 19$.

To complete the square, first factor -2 from $-2x^2 + 12x$.

$$y = -2x^2 + 12x - 19 = -2(x^2 - 6x \qquad) - 19$$

Take half of -6: $\frac{1}{2}(-6) = -3$. Square this result: $(-3)^2 = 9$. Add 9 inside the parentheses. We added 9 inside the parentheses; this 9 is multiplied by the -2 outside the parentheses: $-2(9) = -18$. To compensate for this -18, add 18 outside the parentheses.

$$y = -2(x^2 - 6x + 9) - 19 + 18$$
$$y = -2(x - 3)^2 - 1$$

The parabola opens downward, has its vertex at $(3, -1)$, and is narrower than $y = x^2$. By plotting a few points, we get the graph shown in Figure 3.16. ■

* We take half the coefficient of x since $(a + b)^2 = a^2 + 2ab + b^2$.

4. Complete the square for each of the following. Then graph the parabola.

(a) $y = x^2 - 6x + 11$
(b) $y = x^2 + 8x + 18$

Answer:

(a) $y = (x - 3)^2 + 2$

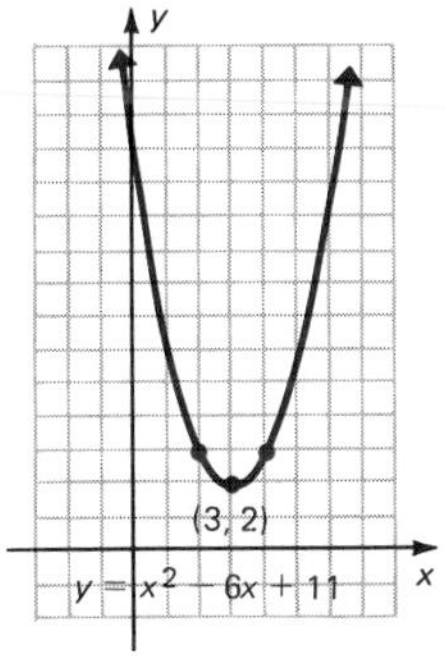

(b) $y = (x + 4)^2 + 2$

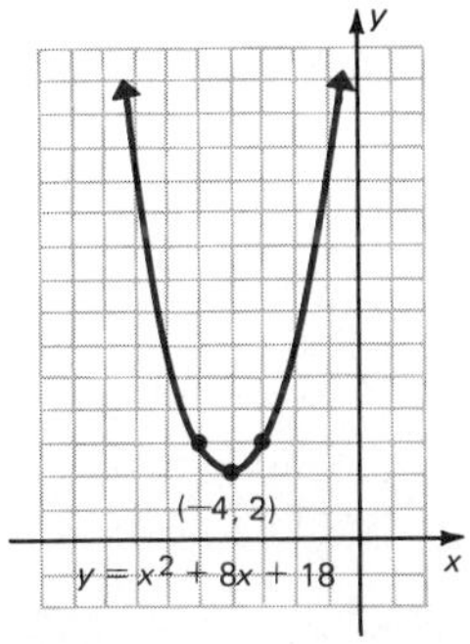

5. Complete the square and graph the following.
(a) $y = 3x^2 - 12x + 14$
(b) $y = -x^2 + 6x - 12$

Answer:
(a) $y = 3(x - 2)^2 + 2$

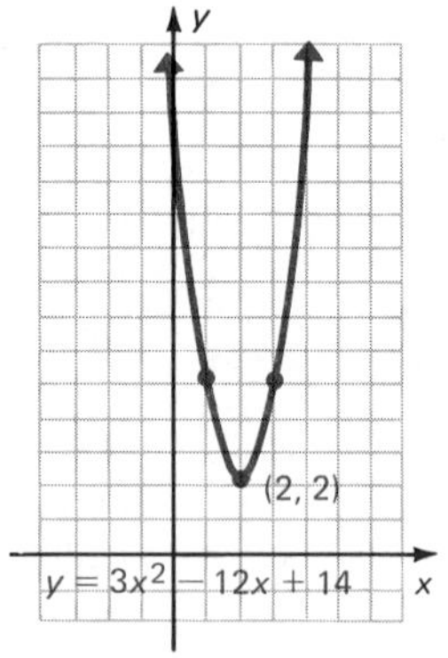

(b) $y = -(x - 3)^2 - 3$

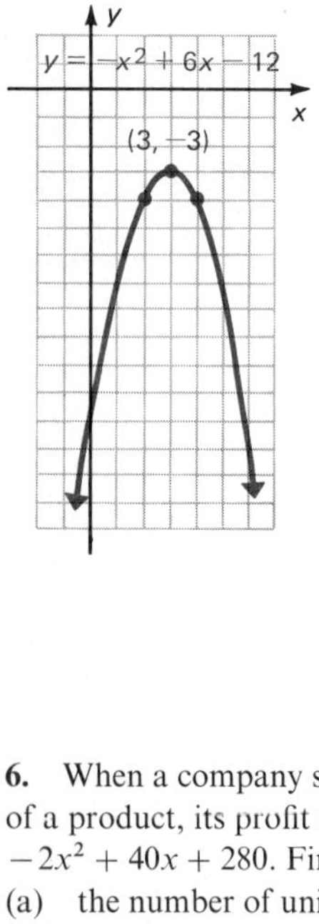

6. When a company sells x units of a product, its profit is $P(x) = -2x^2 + 40x + 280$. Find
(a) the number of units that should be sold so that maximum profit is received
(b) the maximum profit.

Answer:
(a) 10 units
(b) \$480

Work Problem 5 at the side.

Example 7 Elmyra Weintraub owns and operates Aunt Elmyra's Blueberry Pies. Always wanting to maximize her profits, she hires a consultant to analyze her business operations. The consultant tells Weintraub that her profits, $P(x)$, are given by

$$P(x) = 120x - x^2,$$

where x is the number of units of pies that she makes. How many units of pies should she make in order to maximize profit? What is the maximum profit?

Complete the square on the profit function as follows.

$$\begin{aligned} P(x) &= -x^2 + 120x \\ &= -(x^2 - 120x \qquad\qquad) \\ &= -(x^2 - 120x + 3600) + 3600 \\ P(x) &= -(x - 60)^2 + 3600 \end{aligned}$$

Figure 3.17 shows a graph of that portion of the profit function in Quadrant I. The maximum profit can be found from the vertex of the parabola, (60, 3600). The maximum profit of \$3600 is obtained when 60 units of pies are made. Here the profit increases as more and more pies are made, up to the point $x = 60$, and then decreases as more and more pies are made past this point. ■

Work Problem 6 at the side.

Example 8 Suppose that the price and demand for an item are related by

$$p = 150 - 6x^2,$$

where p is the price and x is the number of items demanded (in hundreds). The price and supply are related by

$$p = 10x^2 + 2x,$$

where x is the supply of the item (in hundreds).

Both of these functions have graphs which are parabolas; these parabolas can be graphed by completing the square, giving the results shown in Figure 3.18.

We include only those portions of the graphs that lie in the first quadrant, since neither supply, demand, nor price can be negative. To find the equilibrium demand, we solve the equation

$$\begin{aligned} 150 - 6x^2 &= 10x^2 + 2x \\ 0 &= 16x^2 + 2x - 150 && \text{Add } -150 \text{ and } 6x^2 \text{ to both sides} \\ 0 &= 8x^2 + x - 75. && \text{Multiply both sides by } 1/2 \end{aligned}$$

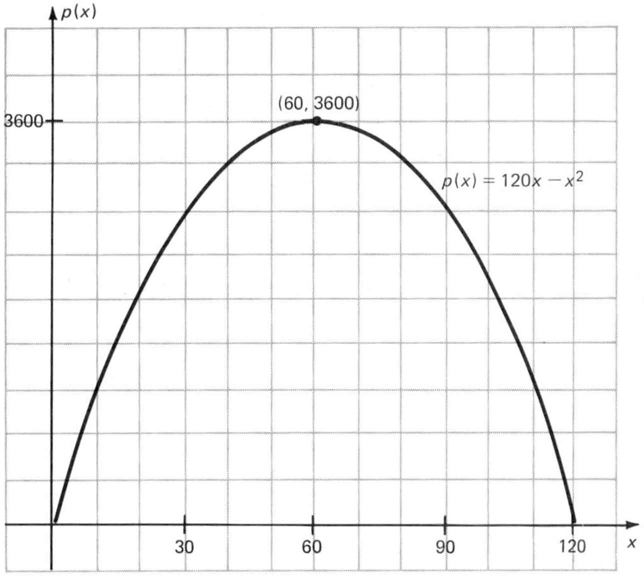

Figure 3.17

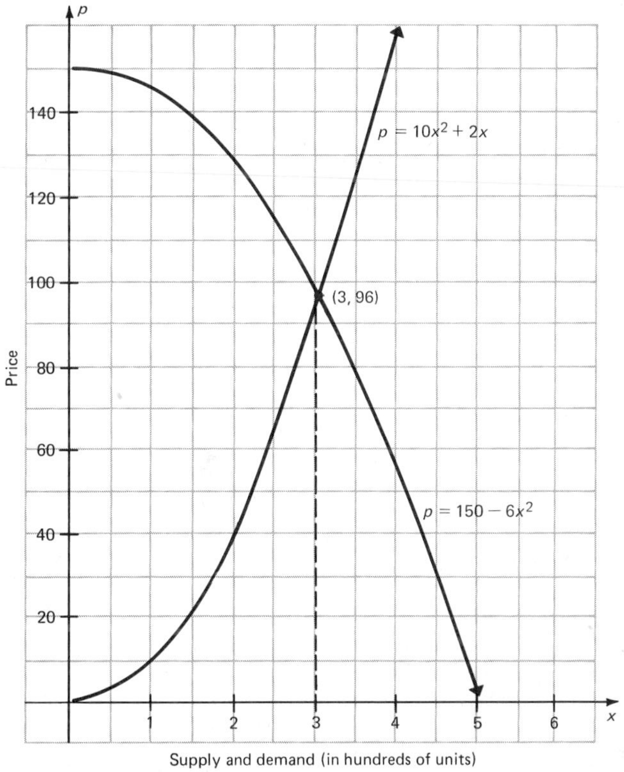

Figure 3.18

This equation can be solved by the quadratic formula of Chapter 1. Here $a = 8$, $b = 1$, and $c = -75$.

$$x = \frac{-1 \pm \sqrt{1 - 4(8)(-75)}}{2(8)}$$

$$= \frac{-1 \pm \sqrt{1 + 2400}}{16} \qquad -4(8)(-75) = 2400$$

$$= \frac{-1 \pm 49}{16} \qquad \sqrt{1 + 2400} = \sqrt{2401} = 49$$

$$x = \frac{-1 + 49}{16} = \frac{48}{16} = 3 \quad \text{or} \quad x = \frac{-1 - 49}{16} = -\frac{50}{16} = -\frac{25}{8}$$

We cannot make $-25/8$ units, so we discard that answer, and use only $x = 3$. Since x represents supply and demand (in hundreds), we see that equilibrium demand (and supply) is 300 units. To find the equilibrium price, substitute 3 for x in either the supply or the demand function. If we use the supply function we have

$$p = 10x^2 + 2x$$

$$p = 10 \cdot 3^2 + 2 \cdot 3 \qquad \text{Let } x = 3$$

$$= 10 \cdot 9 + 6$$

$$p = 96. \quad \blacksquare$$

Work Problem 7 at the side.

7. The price and demand for an item are related by $p = 32 - x^2$, while price and supply are related by $p = x^2$. Find
(a) the equilibrium supply
(b) the equilibrium price.

Answer:
(a) 4
(b) 16

3.4 EXERCISES

Graph the following parabolas. Find the vertex of each. (See Examples 1–6.)

1. $y = 2x^2$

2. $y = -\frac{1}{2}x^2$

3. $y = -x^2 + 1$

4. $y = x^2 - 3$

5. $y = 3x^2 - 2$

6. $y = -2x^2 + 4$

7. $y = (x + 2)^2$

8. $y = (x - 3)^2$

9. $y = -(x - 4)^2$

10. $y = -(x + 5)^2$

11. $y = -2(x - 3)^2$

12. $y = -3(x + 4)^2$

13. $y = (x - 1)^2 - 3$

14. $y = (x - 2)^2 + 1$

15. $y = -(x + 4)^2 + 2$

16. $y = -(x + 1)^2 - 3$

17. $y = x^2 - 4x + 6$

18. $y = x^2 + 6x + 3$

19. $y = x^2 + 12x + 1$

20. $y = x^2 - 10x + 3$

21. $y = 2x^2 + 4x + 1$

22. $y = 3x^2 + 6x + 5$

23. $y = -x^2 + 6x - 6$

24. $y = -x^2 + 2x + 5$

APPLIED PROBLEMS

Management

25. George Lobell runs a sandwich shop. By studying data for his past costs, he has found that a mathematical model describing the cost of operating his shop is given by

$$C(x) = x^2 - 10x + 40,$$

where $C(x)$ is the daily cost to make x units of sandwiches.
(a) Complete the square for $C(x)$.
(b) Graph the parabola resulting from part (a).
(c) From the vertex of the parabola, find the number of units of sandwiches George must sell to produce minimum cost.
(d) What is the minimum cost?

26. Josette Skelnik runs a taco stand. She has found that her profits are approximated by

$$P(x) = -x^2 + 60x - 400,$$

where $P(x)$ is the profit from selling x units of tacos.
(a) Complete the square on $P(x)$.
(b) Graph the parabola.
(c) Find the number of units of tacos that Josette should make to produce maximum profit.
(d) What is the maximum profit?

27. French fries produce a tremendous profit (150% to 300%) for many fast food restaurants. Management, therefore, desires to maximize the number of bags sold. Suppose that a mathematical model connecting p, the profit per day from french fries (in hundreds of dollars), and x, the price per bag (in dimes), is

$$p = -x^2 + 6x - 1.$$

(a) Find the price per bag that leads to maximum profit.
(b) What is the maximum profit?

Natural Science

28. A researcher in physiology has decided that a good mathematical model for the number of impulses fired after a nerve has been stimulated is given by the quadratic function

$$y = -x^2 + 20x - 60,$$

where y is the number of responses per millisecond, and x is the number of milliseconds since the nerve was stimulated.
(a) When will the maximum firing rate be reached?
(b) What is the maximum firing rate?

Management

29. Suppose the price p, is related to the demand x, where x is measured in hundreds of items, by

$$p = 640 - 5x^2.$$

Find p when the demand is
(a) 0 (b) 5 (c) 10.
(d) Graph the function.

30. Suppose that the price and supply of the item in Exercise 29 are related by

$$p = 5x^2.$$

(a) Graph $p = 5x^2$ on the axes used in Exercise 29.
(b) Find the equilibrium supply.
(c) Find the equilibrium price.

Suppose that the supply and demand of a certain textbook are related to price by

$$\textit{supply:}\quad p = \frac{1}{5}x^2 \qquad \textit{demand:}\quad p = -\frac{1}{5}x^2 + 40.$$

31. Use the square root table in the back of the book, or a calculator, to estimate the demand at a price of
(a) 10 (b) 20 (c) 30 (d) 40.

(e) Graph $p = -\frac{1}{5}x^2 + 40$.

32. Approximate the supply at a price of
(a) 5 (b) 10 (c) 20 (d) 30

(e) Graph $p = \frac{1}{5}x^2$ on the axes used in Exercise 31.

33. Find the equilibrium demand in Exercises 31 and 32.

34. Find the equilibrium price in Exercises 31 and 32.

35. A charter flight charges a fare of \$200 per person, plus \$4 per person for each unsold seat on the plane. If the plane holds 100 passengers, and if x represents the number of unsold seats, find the following.
(a) An expression for the total revenue received for the flight. (Hint: Multiply the number of people flying, $100 - x$, by the price per ticket.)
(b) The graph for the expression of part (a).
(c) The number of unsold seats that will produce the maximum revenue.
(d) The maximum revenue.

36. The demand for a certain type of cosmetic is given by

$$p = 500 - x,$$

where p is the price when x units are demanded.
(a) Find the revenue, $R(x)$, that would be obtained at a demand of x. (Hint: revenue = demand × price.)
(b) Graph the revenue function, $R(x)$.
(c) From the graph of the revenue function, estimate the price that will produce the maximum revenue.
(d) What is the maximum revenue?

Natural Science

37. Between the months of June and October, the percent of maximum possible chlorophyl production in a leaf is approximated by $C(x)$, where

$$C(x) = 10x + 50.$$

Here x is time in months, with $x = 1$ representing June. From October through December, $C(x)$ is approximated by

$$C(x) = -20(x - 5)^2 + 100,$$

where x is as above. Find the percent of maximum possible chlorophyl production in each of the following months.
(a) June
(b) July
(c) September
(d) October
(e) November
(f) December

38. Use your results in Exercise 37 to sketch a graph of $y = C(x)$, from June through December. In which month is chlorophyl production a maximum?

3.5 POLYNOMIAL AND RATIONAL MODELS

A **polynomial function of degree *n*** is a function of the form

$$y = a_n x^n + a_{n-1} x^{n-1} + \cdots + a_1 x + a_0,$$

where $a_n \neq 0$. We have already discussed polynomial functions of degree $n = 1$ (linear functions) and $n = 2$ (quadratic functions). In this section we discuss polynomial functions of degree $n = 3$ and $n = 4$. Polynomial functions of higher degree are usually discussed using the methods of calculus. In fact, the methods we present are really useful only for fairly simple polynomials. For more complicated ones, calculus or a computer plotter are necessary.

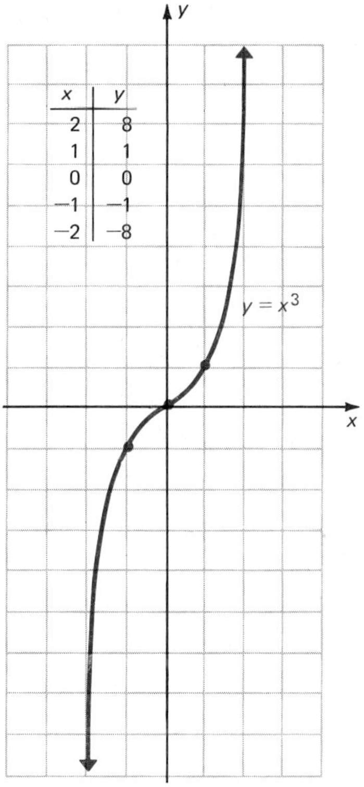

Figure 3.19

Example 1 Graph $y = x^3$.

First, find several ordered pairs belonging to the graph, as shown in the table beside Figure 3.19. Then plot the resulting ordered pairs, and draw a smooth curve through them, getting the graph shown in Figure 3.19. ■

Work Problem 1 at the side.

1. Graph $y = x^3 - 5$.

Answer:

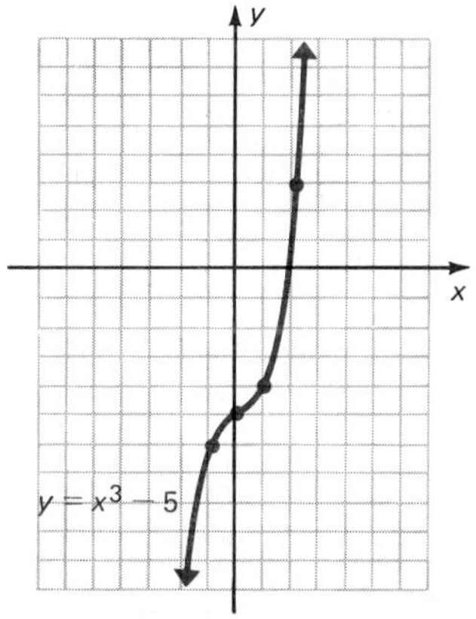

Example 2 Graph $y = x^3 - 2x^2 - x + 2$.

Once again, we start by completing several different ordered pairs, as in the table of Figure 3.20. (A calculator is helpful in finding these.) Plot the points and draw a smooth curve through them, with a result as shown in Figure 3.20.

This graph is typical of the graphs of polynomial functions of degree $n = 3$ where the coefficient of the third degree term is positive. As shown by the graph, both the domain and range of this function are the set of all real numbers. ■

x	y
3	8
2	0
1	0
0	2
−1	0
−2	−12

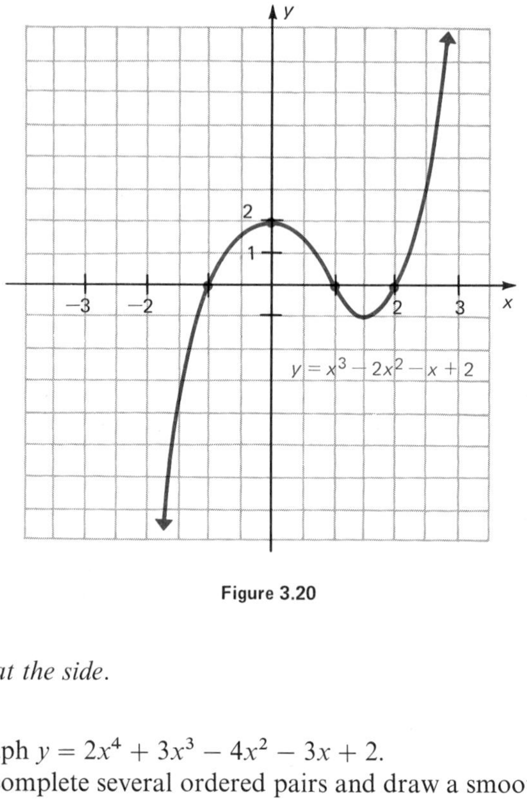

Figure 3.20

2. Graph $y = x^3 - 2x^2 - 5x + 6$.

Answer:

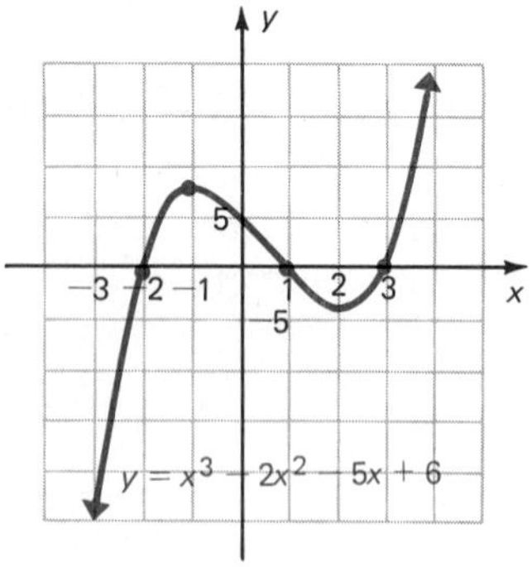

Work Problem 2 at the side.

Example 3 Graph $y = 2x^4 + 3x^3 - 4x^2 - 3x + 2$.

If we again complete several ordered pairs and draw a smooth curve through them, we get the graph shown in Figure 3.21.

This graph is typical of polynomial functions of degree $n = 4$, where the coefficient of the fourth degree term is positive. The domain is the set of all real numbers and, for this polynomial, the range is of the form $\{y | y \geq k,\ k \text{ is a real number}\}$. ■

Rational Functions Functions which can be expressed as the quotient of two polynomials, such as

$$y = \frac{1}{x}, \qquad y = \frac{x+1}{x-1}, \qquad \text{and} \qquad y = \frac{3x^2 + 4x - 1}{x - 1}$$

are called **rational functions.**

x	y
2	36
1	0
1/2	0
0	2
−1	0
−2	0
−3	56

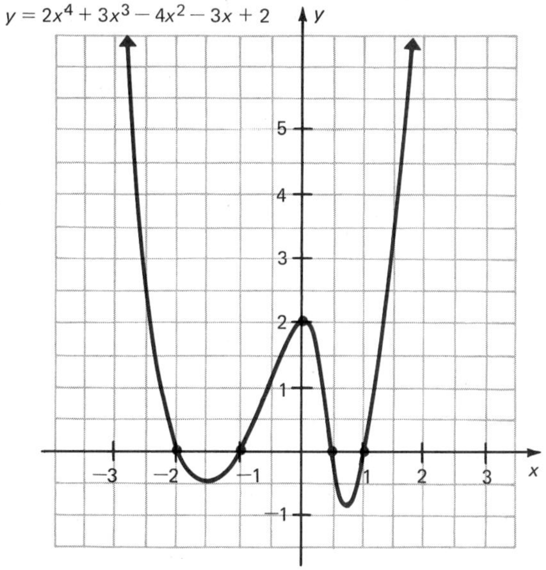

Figure 3.21

Example 4 Graph the rational function $y = \dfrac{2}{1+x}$.

First note that the function is undefined for $x = -1$, since the denominator would then become $1 + (-1) = 0$. Therefore, the graph of this function will not cross the vertical line $x = -1$. We can replace x with any number we wish, except -1, so that we can let x approach -1 as closely as we wish.

From the following table, check that as x gets closer to -1, then $1 + x$ gets closer to 0, and $|2/(1 + x)|$ gets larger. When the resulting ordered pairs are graphed, as in Figure 3.22, we see that the graph gets closer and closer to the vertical line $x = -1$. This line is called a **vertical asymptote.** (Roughly, an **asymptote** is a straight line that a graph approaches but never reaches.)

x	$1+x$	$\dfrac{2}{1+x}$	Ordered pair
$-.5$	$.5$	4	$(-.5, 4)$
$-.8$	$.2$	10	$(-.8, 10)$
$-.9$	$.1$	20	$(-.9, 20)$
$-.99$	$.01$	200	$(-.99, 200)$
-1.01	$-.01$	-200	$(-1.01, -200)$
-1.1	$-.1$	-20	$(-1.1, -20)$
-1.5	$-.5$	-4	$(-1.5, -4)$

x	y
3	1/2
2	2/3
1	1
0	2
−1/2	4
−3/2	−4
−2	−2
−3	−1
−4	−2/3

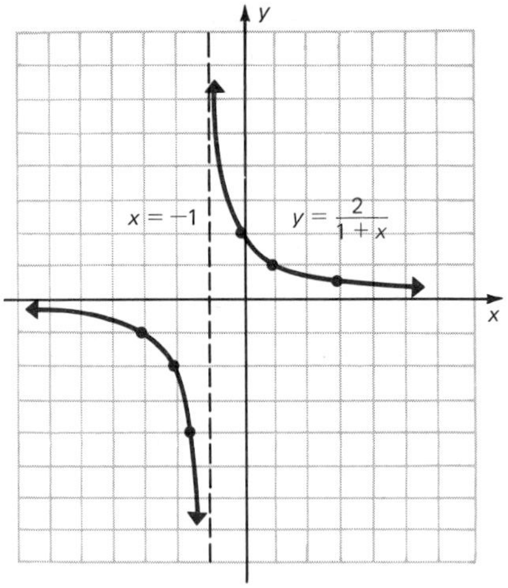

Figure 3.22

As $|x|$ gets larger, then $2/(1 + x)$ gets closer to 0. Because of this, the graph has a **horizontal asymptote** on the x-axis. By using the two asymptotes and by plotting some points, we get the graph shown in Figure 3.22. ■

Work Problem 3 at the side.

Example 5 Graph $y = \dfrac{3x+2}{2x+4}$.

The value $x = -2$ makes the denominator 0. This means that the graph has a vertical asymptote at $x = -2$.* To find a horizontal asymptote, let x get larger, as shown in the following chart.

x	$\dfrac{3x+2}{2x+4}$	Ordered pair
10	$\dfrac{32}{24} = 1.33$	(10, 1.33)
20	$\dfrac{62}{44} = 1.41$	(20, 1.41)
100	$\dfrac{302}{204} = 1.48$	(100, 1.48)
100,000	$\dfrac{300{,}002}{200{,}004} = 1.49998$	(100,000, 1.49998)

* Actually it is also necessary to make sure that $x = -2$ does not make the numerator 0. If both numerator and denominator are 0, then we cannot be sure we have an asymptote.

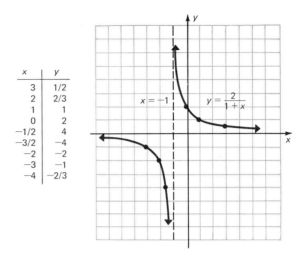

3. Graph the following.

(a) $y = \dfrac{3}{5 - x}$

(b) $y = \dfrac{-4}{x + 4}$

Answer:

(a)

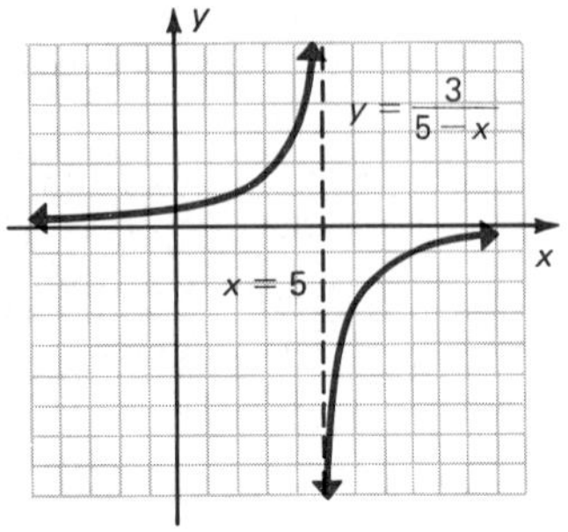

(b)

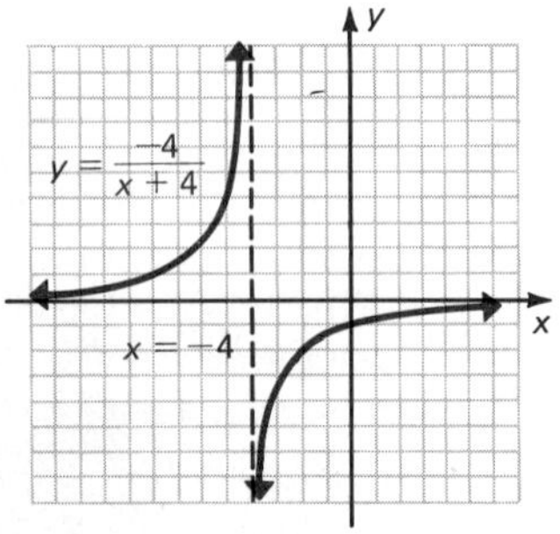

We see that as x gets larger, $(3x + 2)/(2x + 4)$ gets closer to 1.5, or 3/2. Therefore, the line $y = 3/2$ is a horizontal asymptote. Use your calculator to show that as x gets more negative, and takes on values of -10, -100, -1000, $-100{,}000$, and so on, the graph again approaches the line $y = 3/2$. Using these asymptotes and plotting several points, we get the graph of Figure 3.23. ■

x	y
3	1.1
2	1
1	5/6
0	1/2
−1	−1/2
−3/2	−5/2
−5/2	11/2
−3	7/2
−4	5/2
−6	2

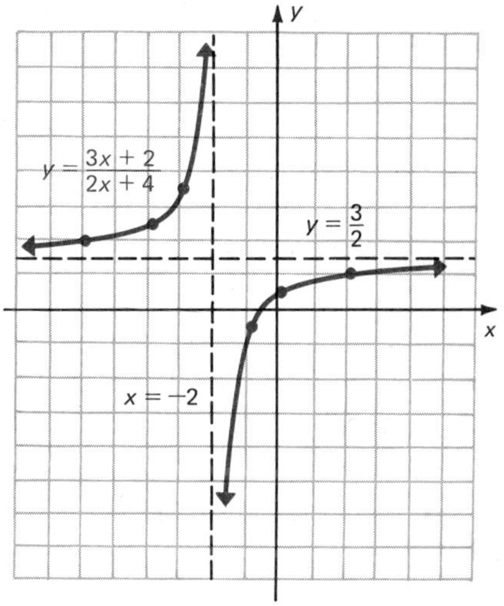

Figure 3.23

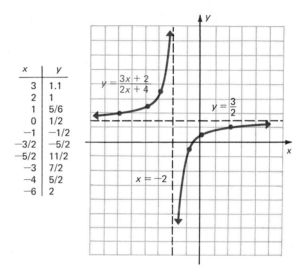

In Example 5 above, we found that $y = 3/2$ is a horizontal asymptote for the rational function $y = (3x + 2)/(2x + 4)$. The horizontal asymptote can also be found by solving the equation $y = (3x + 2)/(2x + 4)$ for x. To do this, first multiply both sides of the equation by $2x + 4$. This gives

$$y(2x + 4) = 3x + 2$$

or

$$2xy + 4y = 3x + 2.$$

Collect all terms containing x on one side of the equation:

$$2xy - 3x = 2 - 4y.$$

Factor out x on the left and solve for x.

$$x(2y - 3) = 2 - 4y.$$

$$x = \frac{2 - 4y}{2y - 3}$$

From this form of the equation, we see that y cannot take on the value $y = 3/2$. This means that $y = 3/2$ is a horizontal asymptote.

4. Graph the following.

(a) $y = \dfrac{3x + 5}{x - 3}$

(b) $y = \dfrac{2 - x}{x + 3}$

Answer:

(a)

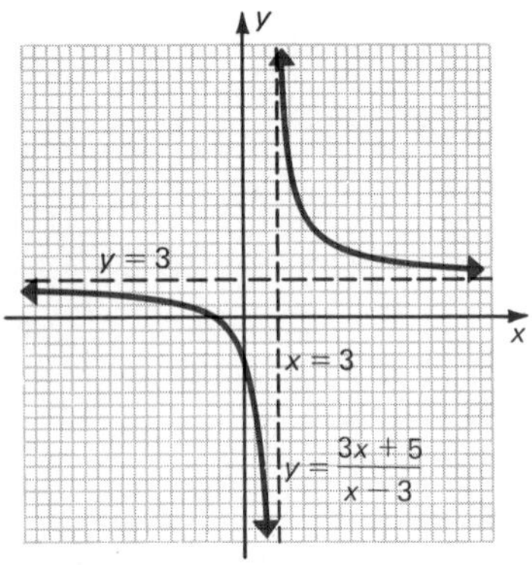

(b)

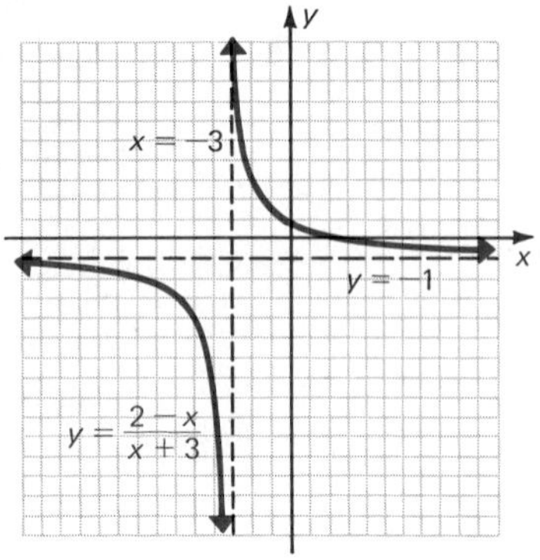

Work Problem 4 at the side.

The final two examples of this section show mathematical models using rational functions.

Example 6 The U.S. Maritime Administration estimated that in a recent year the cost of building an oil tanker of 50,000 deadweight tons in the United States was \$409 per ton. The cost per ton for a 100,000-ton tanker was \$310, while the cost per ton for a 400,000-ton tanker was \$178.

Figure 3.24 shows these values plotted on a graph, where x represents tons (in thousands) and y represents the cost per ton. There is a gap in the information presented by the government agency: the data skips from $x = 100$ to $x = 400$, with no intermediate values. If we could fit a curve through the data points we do have, we could then approximate any desired intermediate values.

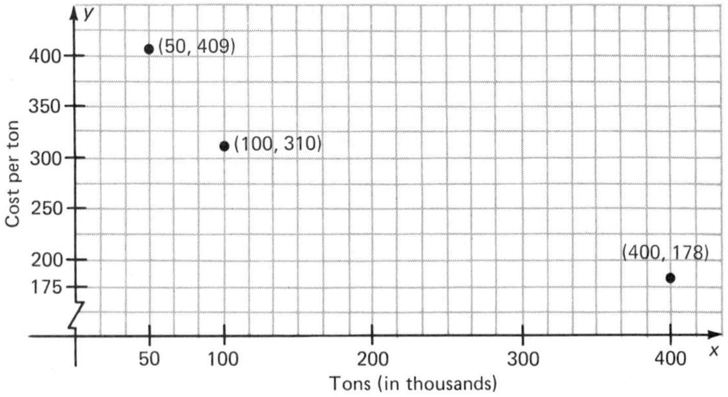

Figure 3.24

Many different functions have graphs which could give a good mathematical model for the given data. By studying the graph of Figure 3.24, we might decide that a rational function, such as the one graphed in Figure 3.22, is one such function. Using methods which we do not discuss here, we can get

$$y = \frac{110{,}000}{x + 225}$$

as a mathematical model which approximates the given data. We can substitute the known values $x = 50$, $x = 100$, and $x = 400$ to test the "goodness of fit" of the model. Doing this, we obtain the following results.

x	y (given)	y (from function)
50	409	400
100	310	338
400	178	176

As the chart shows, the rational function is a good mathematical model for the data, so that we can use it to approximate y for intermediate values of x. If we let $x = 150$, we get

$$y = \frac{110{,}000}{150 + 225} = 293.$$

Also, $y = \$259$ when $x = 200$, while if $x = 300$, then $y = \$210$. We can use these points to graph the function, as shown in Figure 3.25. ■

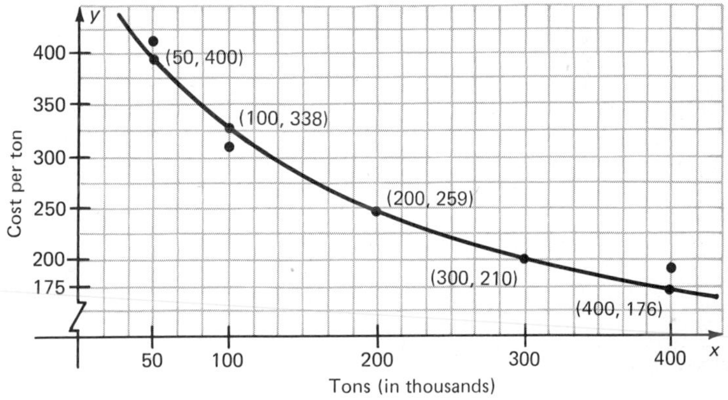

Figure 3.25

Work Problem 5 at the side.

5. Using the result of Example 6, find the cost per ton to build a ship of
(a) 75,000 tons
(b) 350,000 tons.

Answer:
(a) $367
(b) $191

Example 7 In many situations involving environmental pollution, it turns out that much of the pollutant can be removed from the air or water at a fairly reasonable cost, but the last, small part of the pollutant can get increasingly expensive to remove.

Cost as a function of the percentage of pollutant removed from the environment can be calculated for various percentages of removal, with a curve fitted through the resulting data points. This curve then leads to a mathematical model of the situation. Rational functions often are a good choice for these **cost-benefit models.**

For example, suppose a cost-benefit model is given by

$$y = \frac{18x}{106 - x},$$

where y is the cost (in thousands of dollars) of removing x percent of a certain pollutant. The domain of x is the set $\{x|0 \le x \le 100\}$; any amount of pollutant from 0% to 100% can be removed. To remove 100% of the pollutant here would cost

$$y = \frac{18(100)}{106 - 100} = 300,$$

or \$300,000. Check that 95% of the pollutant can be removed for \$155,000, 90% for \$101,000, and 80% for \$55,000. Using these points, as well as others that we could obtain from the function above, we get the graph shown in Figure 3.26. ■

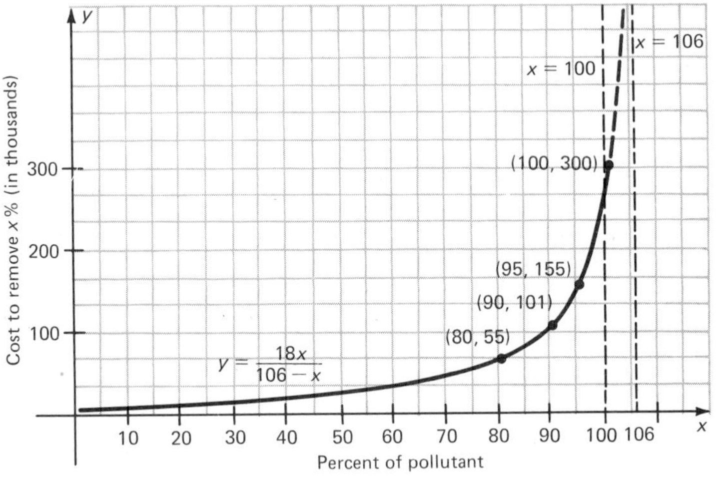

Figure 3.26

3.5 EXERCISES

Graph each of the following polynomial functions. (See Examples 1–3.)

1. $y = x^3 + 2$
2. $y = x^3 - 1$
3. $y = x^4$
4. $y = x^4 - 2$
5. $y = (x - 2)^3$
6. $y = (x + 1)^3$
7. $y = (x + 3)^4$
8. $y = (x - 5)^4$
9. $y = x^3 - 2x^2 - 5x + 6$
10. $y = x^3 + 3x^2 - 4x - 12$
11. $y = x^4 + 2x^3 - 7x^2 - 8x + 12$
12. $y = x^4 - x^3 - 7x^2 + x + 6$

Graph each of the following rational functions. (See Examples 4 and 5.)

13. $y = \dfrac{1}{x + 2}$
14. $y = \dfrac{1}{x - 1}$

15. $y = \dfrac{-4}{x-3}$

16. $y = \dfrac{-1}{x+3}$

17. $y = \dfrac{2}{x}$

18. $y = \dfrac{-3}{x}$

19. $y = \dfrac{2}{3+2x}$

20. $y = \dfrac{4}{5+3x}$

21. $y = \dfrac{3x}{x-1}$

22. $y = \dfrac{4x}{3-2x}$

23. $y = \dfrac{x+1}{x-4}$

24. $y = \dfrac{x-3}{x+5}$

25. $y = \dfrac{2x-1}{4x+2}$

26. $y = \dfrac{3x-6}{6x-1}$

27. $y = \dfrac{1-2x}{5x+20}$

28. $y = \dfrac{6-3x}{4x+12}$

29. $y = \dfrac{-x-4}{3x+6}$

30. $y = \dfrac{-x+8}{2x+5}$

APPLIED PROBLEMS

Natural/Social Science

Based on an article we recently read, we constructed the mathematical model

$$A(x) = \frac{-7}{480}x^3 + \frac{127}{120}x$$

as the approximate alcohol concentration (in tenths of a percent) in an average person's bloodstream x hours after drinking about eight ounces of 100 proof whiskey. The function is approximately valid for $0 \le x < 9$. Use a calculator to find each of the following values.

31. $A(0)$ **32.** $A(1)$ **33.** $A(2)$ **34.** $A(4)$

35. $A(6)$ **36.** $A(7)$ **37.** $A(8)$

38. Graph $y = A(x)$.

39. Using the graph you drew for Exercise 38, estimate the time of maximum alcohol concentration.

40. In one state, a person is legally drunk if the blood alcohol concentration exceeds .15%. Use the graph of Exercise 38 and estimate the period in which this average person is legally drunk.

Management

Suppose that the average cost per unit, $C(x)$, of producing x units of margarine is given by

$$C(x) = \frac{500}{x+30}.$$

Find the cost per unit of producing each of the following quantities.

41. 10 units **42.** 20 units **43.** 50 units

44. 70 units **45.** 100 units **46.** Graph $y = C(x)$.

Using the information given in Example 6 of the text, estimate the cost per ton for building ships weighing

47. 25,000 tons; **48.** 125,000 tons;

49. 250,000 tons; **50.** 275,000 tons.

Natural Science

Suppose a cost-benefit model (see Example 7) is given by

$$y = \frac{6.5x}{102 - x},$$

where y is the cost in thousands of dollars of removing x percent of a certain pollutant. Find the cost of removing the following percents of pollutants.

51. 0% **52.** 50% **53.** 80% **54.** 90%

55. 95% **56.** 99% **57.** 100%

58. Graph the function.

Suppose a cost-benefit model is given by

$$y = \frac{6.7x}{100 - x},$$

where y is the cost in thousands of dollars of removing x percent of a given pollutant. Find the cost of removing each of the following percents of pollutants.

59. 50% **60.** 70% **61.** 80% **62.** 90%

63. 92% **64.** 95% **65.** 98% **66.** 99%

67. Is it possible, according to this function, to remove *all* the pollutant?

68. Graph the function.

What percent of pollutant can be removed for

69. \$10,000? **70.** \$38,000?

General

Antique car fans often enter their cars in a concours d'elegance *in which a maximum of 100 points can be awarded to a particular car. Points are awarded for the general attractiveness of the car. Based on a recent article in* Business Week, *we constructed the following mathematical model for the cost, in thousands of dollars, of restoring a car so that it will win x points.*

$$C(x) = \frac{10x}{49(101 - x)}$$

Find the cost of restoring a car so that it will win

71. 99 points; **72.** 100 points.

Management

In management, **product-exchange functions** *give the relationship between quantities of two items that can be produced by the same machine or factory. For example, an oil refinery can produce gasoline, heating oil, or a combination of the two; a winery can produce red wine, white wine, or a combination of the two. Sketch the portion of the graph of each of the following functions in Quadrant I, and then estimate the maximum quantities of each product that can be produced. (Hint: look at the intercepts.)*

73. The product exchange function for gasoline, x, and heating oil, y, in hundreds of gallons per day, is

$$y = \frac{125{,}000 - 25x}{125 + 2x}.$$

74. A drug factory found the product exchange function for a red tranquilizer, x, and a blue tranquilizer, y, is

$$y = \frac{900{,}000{,}000 - 30{,}000x}{x + 90{,}000}.$$

3.6 CONSTRUCTING MATHEMATICAL MODELS (OPTIONAL)

In this section we actually construct two different mathematical models—one for double declining balance depreciation and one showing sales and prices for Volvo automobiles for the last few years.

As we develop these two models, notice the fundamental difference between them. To develop the model for depreciation, we start with the basic principles, apply some of the mathematics we have learned in this course, and come up with a model which gives the depreciation of an item *exactly*.

On the other hand, the model we develop for Volvo sales cannot go back to basic principles. (What are the basic principles for car sales? How do we write equations for them?) Rather, we construct a mathematical model for Volvo sales by gathering data on past sales and using it to predict future sales. Such methods can never give exact answers. The best we can expect is an approximation; if we are careful and lucky we can get a good approximation.

Depreciation When a business buys an asset (such as a machine or building) it doesn't treat the cost of the asset as an expense immediately. Instead, it *depreciates* the cost of the asset over the lifetime of the asset.

For example, a machine costing \$10,000 and having a useful life of 8 years, at which time it is worthless, might be depreciated at the rate of $10,000/8 = $1250 per year. Depreciation by this method is called straight-line depreciation, which we discussed in Section 3.3. When we use straight-line depreciation, we are assuming that the asset loses an equal amount of value during each year of its life.

This assumption of equal loss of value is not valid for many assets, such as new cars. A new car may lose 30% of its value during the first year. For assets that lose value quickly at first, and then less rapidly in later years, the Internal Revenue Service permits the use of alternate methods of depreciation. In Exercise Set 3.3, we discussed the sum-of-the-years'-digits method of depreciation, one alternate method.

Another common method is **double declining balance** depreciation. To find the depreciation in the first year, multiply the cost of the asset, C, by the fraction $2/n$, where n is the life of the asset in years. That is, the depreciation in year 1 is

$$\textbf{depreciation in year 1} = \frac{2}{n} \cdot C.$$

The number 2 in this formula shows the origin of the word "double" in the name of this method.

Example 1 A forklift costs $9600 and has a useful life of 8 years. Find the depreciation in year 1. Use the double declining balance method.

Using the formula above,

$$\begin{aligned}\text{depreciation in year 1} &= \frac{2}{n} \cdot C \\ &= \frac{2}{8} \cdot 9600 \qquad \text{Let } n = 8 \text{ and } C = 9600 \\ &= 2400\end{aligned}$$

dollars. ■

1. An asset costs $5000 and has a life of 5 years. Find the depreciation in year 1 by the double declining balance method.

Answer: $2000

Work Problem 1 at the side.

The depreciation in later years of an asset's life can be found by multiplying the depreciation from the previous year by $1 - 2/n$.

For example, in Problem 1 at the side, we found that an asset costing $5000 with a life of 5 years would lead to a depreciation of $2000 during the first year of its life. To find the depreciation in year 2, multiply the depreciation in year 1 by $1 - \frac{2}{5}$, as follows:

$$\textbf{depreciation in year 2} = \textbf{(depreciation in year 1)} \times \left(1 - \frac{2}{n}\right)$$

$$= 2000\left(1 - \frac{2}{5}\right) \qquad \text{Let } n = 5$$
$$= 2000\left(\frac{3}{5}\right)$$
$$= 1200$$

dollars. To find the depreciation in year 3, multiply this result by $1 - \frac{2}{5} = \frac{3}{5}$.

2. Find the depreciation in year 3.

Answer: $720

Work Problem 2 at the side.

The depreciation by the double declining balance method in each year of an asset's life forms a sequence:

Year	1	2	3	4
Amount of depreciation	$\frac{2}{n} \cdot C$	$\frac{2}{n} \cdot C \cdot \left(1 - \frac{2}{n}\right)$	$\frac{2}{n} \cdot C \cdot \left(1 - \frac{2}{n}\right)^2$	$\frac{2}{n} \cdot C \cdot \left(1 - \frac{2}{n}\right)^3$

This sequence is geometric, since each term is found by multiplying the preceding term by $1 - 2/n$. By the results shown in Section 2.6, where we first discussed sequences, the depreciation in year j is found from the formula for the general term of a geometric sequence. In our sequence the first term is $a_1 = \frac{2}{n} \cdot C$ and the common ratio is $r = 1 - \frac{2}{n}$. Thus, the general term, a_j, is

$$a_j = a_1 \cdot r^{j-1}$$
$$a_j = \frac{2}{n} \cdot C \cdot \left(1 - \frac{2}{n}\right)^{j-1}.$$

Here j represents a particular year in the life of the asset having a cost of C dollars and a total life of n years. This formula is a mathematical model for double declining balance depreciation.

If double declining balance depreciation is used for each year of an asset's life, then the total depreciation will be less than the net cost of the asset. For this reason, it is permissible to switch to straight-line depreciation toward the end of the useful life of the asset.

Example 2 Oxford Typo, Inc., buys a new printing press for $39,000. The press has a life of 6 years. Find the depreciation in years 1, 2, and 3. Use the double declining balance method.

Use the mathematical model given above for a_j; replace j in turn by 1, 2, and 3.

$$a_j = \frac{2}{n} \cdot C \cdot \left(1 - \frac{2}{n}\right)^{j-1}$$

$$a_1 = \frac{2}{6} \cdot (39{,}000) \cdot \left(1 - \frac{2}{6}\right)^{1-1} \qquad \text{Let } j = 1,\ n = 6,\ C = 39{,}000$$

$$= 13{,}000\left(1 - \frac{2}{6}\right)^0 \qquad \frac{2}{6} \cdot 39{,}000 = 13{,}000$$

$$= 13{,}000(1) \qquad \left(1 - \frac{2}{6}\right)^0 = 1$$

$$a_1 = 13{,}000$$

Find a_2 and a_3 in the same way.

$$a_2 = \frac{2}{6} \cdot (39{,}000) \cdot \left(1 - \frac{2}{6}\right)^{2-1} \qquad a_3 = \frac{2}{6} \cdot (39{,}000) \cdot \left(1 - \frac{2}{6}\right)^{3-1}$$

$$= 13{,}000 \cdot \left(1 - \frac{1}{3}\right)^{2-1} \qquad = 13{,}000 \cdot \left(1 - \frac{1}{3}\right)^2$$

$$= 13{,}000 \cdot \left(\frac{2}{3}\right)^1 \qquad = 13{,}000 \cdot \left(\frac{2}{3}\right)^2$$

$$= 13{,}000 \cdot \left(\frac{2}{3}\right) \qquad = 13{,}000 \cdot \left(\frac{4}{9}\right)$$

$$a_2 = 8667 \qquad a_3 = 5778 \ \blacksquare$$

3. An asset costs $27,000 and has a useful life of 6 years. Use the double declining balance method and find the depreciation in years 1, 2, and 3.

Answer: $9000, $6000, $4000

Work Problem 3 at the side.

Supply and Demand In the past few years, Volvo automobile sales have declined in the United States. Much of the cause of this decline is price increases—in the last five years the cost of a basic Volvo sedan has increased by 78%. In the rest of this example, let us set up a mathematical model showing the relationship between sales and price. To begin, we gather data on past sales and prices. Typical prices and sales data are shown in the following chart.* (The numbers have been rounded for simplicity.)

Total Volvo sales	Price of basic sedan
54,000	$4100
61,000	$4350
53,000	$5250
60,000	$5900
44,000	$7100

* Most information in this chart was supplied by Nancy Fiesler of Volvo of America.

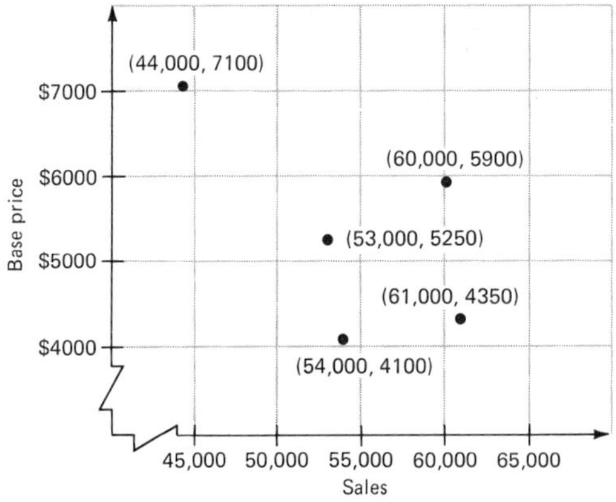

Figure 3.27

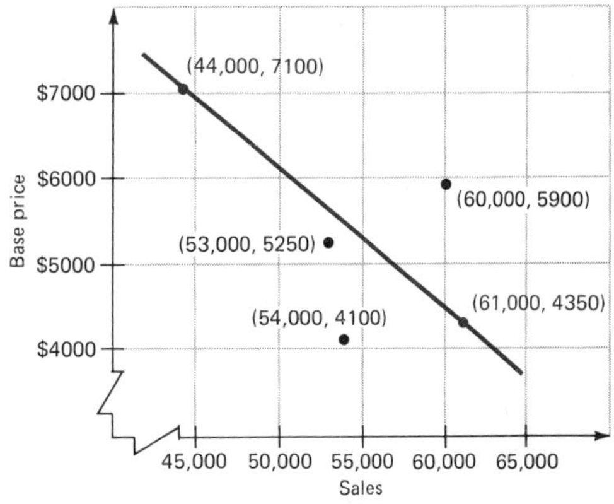

Figure 3.28

Clearly, we cannot use this data to obtain a precise and exact mathematical model, such as the one for depreciation. There are too many variables here; the state of the economy, gasoline prices, and other factors all affect Volvo sales. For example, the chart shows that when the price increased from \$5250 to \$5900, sales *increased* from 53,000 to 60,000. However, a careful model can at least approximate the degree to which sales have fallen as the price increases. A graph of the data in the chart is shown in Figure 3.27.

None of the curves that we have studied will fit exactly through the data points of the graph. However, a straight line provides a reasonable fit to the points. In Chapter 10 we discuss methods for finding the best possible line through these data points. For now, we can just choose two typical points and draw a line through them. Let's choose (44,000, 7100) and (61,000, 4350). A line through these points is shown in Figure 3.28. This line is the mathematical model of the relationship between price and demand.

We can use the two points (44,000, 7100) and (61,000, 4350), along with the point-slope form of the equation of a line to find the equation of the line of Figure 3.28. First, find the slope.

$$m = \frac{\text{change in } y}{\text{change in } x} = \frac{4350 - 7100}{61{,}000 - 44{,}000} = \frac{-2750}{17{,}000} = -.16$$

We can use $m = -.16$ and the point (44,000, 7100) to find the equation of the line of Figure 3.28.

$$\begin{aligned} p - p_1 &= m(x - x_1) && \text{Point-slope form} \\ p - 7100 &= -.16(x - 44{,}000) && \\ p - 7100 &= -.16x + 7040 && -.16(-44{,}000) = 7040 \\ p &= -.16x + 14{,}140 && \end{aligned}$$

This is the mathematical model of the relationship between price, p, and demand, x.

Example 3 Suppose the factory dictates a price of \$8000. Estimate the number of cars that will be sold.

Let $p = 8000$ in the model above.

$$p = -.16x + 14{,}140$$
$$8000 = -.16x + 14{,}140 \qquad \text{Let } p = 8000$$
$$-6140 = -.16x \qquad \text{Add } -14{,}140 \text{ to both sides}$$
$$x = 38{,}375 \qquad \text{Use a calculator to divide both sides by } -.16$$

At a price of \$8000, about 38,000 cars will be sold. ■

4. Find the sales at a price of \$8500.

Answer: About 35,000

Work Problem 4 at the side.

Example 4 Suppose the manager would like to know the price to charge so that 50,000 cars can be sold. What advice should be given?

Use the model developed above, with $x = 50{,}000$.

$$p = -.16x + 14{,}140$$
$$p = -.16(50{,}000) + 14{,}140 \qquad \text{Let } x = 50{,}000$$
$$p = -8000 + 14{,}140$$
$$p = 6140$$

To sell 50,000 cars, set the price at \$6140 each. ■

5. Find the price so that 45,000 cars could be sold.

Answer: \$6940

Work Problem 5 at the side.

3.6 APPLIED PROBLEMS

Management

An apartment house costs \$180,000. The owners decide to depreciate it over 6 years (it isn't built very well). Use the double declining balance method to find the depreciation in each of the following years. (See Examples 1 and 2.)

1. year 1 **2.** year 2 **3.** year 3 **4.** year 4

A new airplane costs \$600,000 and has a useful life of 10 years. Use the double declining balance method of depreciation to find the depreciation in each of the following years. (See Examples 1 and 2.)

5. year 1 **6.** year 2 **7.** year 3 **8.** year 4

Complete the following tables, which compare the three methods of depreciation. [Recall: I is the net cost of an asset; $I = C -$ scrap value. Also, the formula for depreciation in year j, by the sum-of-the-years'-digits method, is

$$a_j = \frac{n - j + 1}{n(n+1)}(2I).]$$

9. Cost: \$1400; life: 3 years; scrap value: \$275

Depreciation in year	Straight-line	Double declining	Sum-of-years' digits
1			
2			
3			
Totals			

10. Cost: \$55,000; life: 5 years; scrap value: \$0

Depreciation in year	Straight-line	Double declining	Sum-of-years' digits
1			
2			
3			
4			
5			
Totals			

The following problems refer to the mathematical model developed in the text for Volvo sales.

Find the price that would produce the following levels of sales. (See Example 4.)

11. 40,000 **12.** 48,000 **13.** 54,000 **14.** 60,000

Find the number of cars that would be sold at the following prices. (See Example 3.)

15. \$6000 **16.** \$7000 **17.** \$7400 **18.** \$8100

*The following table shows the percent profit at typical McDonald's restaurants in a recent year.**

Annual store sales in thousands, x	250	375	450	500	600	650
Percent pretax profit, y	9.3	14.8	14.2	15.9	19.2	21.0

19. Draw a graph for this data that is similar to Figure 3.27.

20. Pick two points that lead to a "typical" straight line through the data. Draw the line.

21. Find the slope of your line.

22. Find the equation of your line.

Use your line to estimate profit for sales of

23. \$300,000; **24.** \$400,000; **25.** \$475,000; **26.** \$700,000.

* This data was supplied by Harvey Lubelchek, McDonald's Systems, Inc.

CHAPTER 3 TEST

Graph each of the following functions.

[3.1] **1.** $x + 3y = 6$

2. $4x - y = 8$

3. $x + 2 = 0$

4. $y = 3$

5. $2x = y$

[3.4] **6.** $y = -x^2$

7. $y = (x + 2)^2$

8. $y = (x - 1)^2 - 3$

[3.5] **9.** $y = x^3 - 2$

10. $y = \dfrac{1}{x - 3}$

11. $y = \dfrac{-3}{2x - 4}$

12. $y = \dfrac{5x + 5}{3x - 5}$

[3.1] *The supply and demand of a certain commodity are related by*

supply: $p = 6x + 3$ *demand:* $p = 19 - 2x.$

13. Find the equilibrium supply.

14. Find the equilibrium price.

Find the slope of each of the following lines.

15. through $(-2, 5)$ and $(4, 7)$

16. $8x - 3y = 7$

[3.2] *Find the equation of each of the following lines.*

17. through $(-2, 3)$, $m = 1/2$.

18. through $(8, -1)$ and $(3, -2)$.

19. through $(3, -4)$ and is vertical.

[3.3] **20.** A company can make 8 units of paper for \$300; the fixed cost is \$60. Find the cost function.

The cost of producing x units of vodka is $C(x)$, where $C(x) = 20x + 100$. The revenue from the sale of x units is $R(x) = 40x$.

21. Find the break-even point.

22. What revenue will the company receive if it sells just that number of units?

[3.4] *Find the vertex of each of the following parabolas.*

23. $y = x^2 - 6x + 5$

24. $y = -2x^2 + 8x - 9$

25. $y = -x^2 - 6x + 1$

Suppose that the cost, $C(x)$, of producing x units of saddles is approximated by

$$C(x) = x^2 - 14x + 59.$$

26. How many units of saddles should be manufactured so as to produce the minimum total cost?

27. What is the minimum total cost?

[3.5] *A cost-benefit curve for pollution control is given by*

$$y = \frac{9.2x}{106 - x},$$

where y is the cost in thousands of dollars of removing x percent of a specific industrial pollutant. Find y for each of the following values of x.

28. $x = 50$

29. $x = 98$

30. What percent of the pollutant can be removed for $22,000?

[3.3] and [3.6] *An asset costs $21,000 and has a 6-year life. Find the depreciation in year 1 by each of the following methods. Assume that the asset has no scrap value.*

31. straight-line

32. sum-of-the-years' digits

33. double declining balance

CASE 4 MARGINAL COST—BOOZ, ALLEN AND HAMILTON*

Booz, Allen and Hamilton is a large management consulting firm. One of the services it provides to client companies is profitability studies, which show ways in which the client can increase profit levels. The client company requesting the analysis presented in this case is a large producer of a staple food. The company buys from farmers, and then processes the food in its mills, resulting in a finished product. The company sells both at retail under its own brands, and in bulk to other companies who use the product in the manufacture of convenience foods.

The client company has been reasonably profitable in recent years, but the management retained Booz, Allen and Hamilton to see whether its consultants could suggest ways of increasing company profits. The management of the company had long operated with the philosophy of trying to process and sell as much of its product as possible, since they felt this would lower the average processing cost per unit sold. However, the consultants found that the client's fixed mill costs were quite low, and that, in fact, processing extra units made the cost per unit start to increase. (There are several reasons for this: the company must run three shifts, machines break down more often, and so on.)

In this case, we shall discuss the marginal cost of two of the company's products. The marginal cost (cost of producing an extra unit) of production for product A was found by the consultants to be approximated by the linear function

$$y = .133x + 10.09,$$

where x is the number of units produced (in millions) and y is the marginal cost.

For example, at a level of production of 3.1 million units, an additional unit of product A would cost about

$$\begin{aligned} y &= .133(3.1) + 10.09 \\ &= \$10.50. \end{aligned}$$

At a level of production of 5.7 million units, an extra unit costs \$10.85. Figure 1 shows a graph of the marginal cost function from $x = 3.1$ to $x = 5.7$, the domain to which the function above was found to apply.

The selling price for product A is \$10.73 per unit, so that, as shown on the graph of Figure 1, the company was losing money on many units of the product that it sold. Since the selling price could not be raised if the company was to remain competitive, the consultants recommended that production of product A be cut.

For product B, the Booz, Allen and Hamilton consultants found a marginal cost function given by

$$y = .0667x + 10.29,$$

---

* This case was supplied by John R. Dowdle of the Chicago office of Booz, Allen and Hamilton.

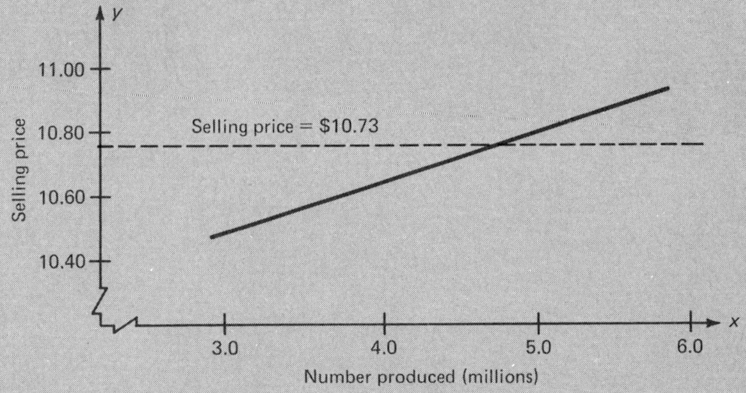

Figure 1

with x and y as defined above. Verify that at a production level of 3.1 million units, the marginal cost is \$10.50, while at a production level of 5.7 million units, the marginal cost is \$10.67. Since the selling price of this product is \$9.65, the consultants again recommended a cutback in production.

The consultants ran similar cost analyses of other products made by the company, and then issued their recommendation: The company should reduce total production by 2.1 million units. The analysts predicted that this would raise profits for the products under discussion from \$8.3 million annually to \$9.6 million, which is very close to what actually happened when the client took this advice.

EXERCISES

1. At what level of production, x, was the marginal cost of a unit of product A equal to the selling price?

2. Graph the marginal cost function for product B from $x = 3.1$ million units to $x = 5.7$ million units.

3. Find the number of units for which marginal cost equals the selling price for product B.

4. For product C, the marginal cost of production is

$$y = .133x + 9.46.$$

(a) Find the marginal cost at a level of production of 3.1 million units; of 5.7 million units.
(b) Graph the marginal cost function.
(c) For a selling price of \$9.57, find the level of production for which the cost equals the selling price.

4 Mathematics of Finance

Not too many years ago, money could be borrowed by the largest and most secure corporations for 3% (and home mortgages could be had for $4\frac{1}{2}$%). Today, however, even the largest corporations must pay at least 7% for their money, and as much as 12% at times. Thus it is important that the management of corporations and consumers (who pay 18% to Sears or Wards) alike have a good understanding of the cost of borrowing money.

The cost of borrowing money is called **interest.** The formulas for interest are developed in this chapter.

4.1 COMPOUND INTEREST AND PRESENT VALUE

In business, interest is normally charged on interest (**compound interest**). To find a formula for compound interest, first suppose that P dollars is deposited at a certain rate of yearly interest (i). The interest earned during the first year is given by

$$P \cdot i \cdot 1 = Pi,$$

where we used the formula for simple interest,

simple interest = principal · rate · time.

At the end of one year, the amount on deposit will be the sum of the original principal and the interest earned, or

$$P + Pi = P(1 + i). \qquad \textbf{(1)}$$

If the deposit earns compound interest, we must calculate the interest earned during the second year by finding the interest on the total amount on deposit at the end of the first year. Thus, the interest earned during the second year (again found by the formula for simple interest) is given by

$$P(1 + i)(i)(1) = P(1 + i)i, \qquad \textbf{(2)}$$

so that the total amount on deposit at the end of the second year is given by the sum of equation (1) and equation (2), or

$$\begin{aligned} P(1 + i) + P(1 + i)i &= P(1 + i) \cdot (1 + i) \\ &= P(1 + i)^2. \end{aligned}$$

In the same way, it can be shown that the total on deposit at the end of three years is given by $P(1 + i)^3$. We can generalize these results in the following theorem.

Theorem 4.1 If P dollars is deposited for n years in an account paying a rate of interest i compounded annually, the account will contain a total of

$$\boldsymbol{A = P(1 + i)^n}$$

dollars at the end of n years.

Example 1 Suppose $1000 is deposited for 6 years in an account paying 5% per year compounded annually.

(a) Find the final amount in the account.

In the formula of Theorem 4.1, we have $P = 1000$, $i = 5\% = .05$, and $n = 6$. The final amount on deposit is given by

$$\begin{aligned} A &= P(1 + i)^n \\ A &= 1000(1 + .05)^6 \\ A &= 1000(1.05)^6 \end{aligned}$$

dollars. We could find $(1.05)^6$ by using a calculator, or by using common logarithms, or by using special compound interest tables. Table 6 at the back of this book is such a table. To find $(1.05)^6$, look for 5% across the top and 6 down the side. You should find 1.34010; thus $(1.05)^6 = 1.34010$, and

$$A = 1000(1.34010) = 1340.10$$

dollars, which represents the final amount on deposit.

(b) Find the actual amount of interest earned.

To find the amount of interest that was earned, take the final amount on deposit and subtract the initial deposit. We get

amount of interest = \$1340.10 − \$1000 = \$340.10. ■

Example 2 Find the final amount on deposit if $6500 is deposited for 14 years in an account paying 6% compounded annually.

Look in Table 6 for 6% across the top and 14 down the side. You should find the number 2.26090. Therefore, A, the final amount on deposit, is given by

$$A = 6500(2.26090) = 14{,}695.85$$

dollars. The amount of interest earned is \$14,695.85 − \$6500 = \$8195.85. ■

1. Find the final amount on deposit if \$12,000 is deposited for 7 years in an account paying 5% compounded annually. Also, find the amount of interest earned.

Answer: \$16,885.20; \$4885.20

Work Problem 1 at the side.

In the examples above, we assumed that interest was compounded annually. In actual practice interest often is compounded more frequently. Common periods of compounding include every six months (semiannually), every three months (quarterly), every month, or every day. When interest is compounded more often than annually, we must use the more general formula given in the next theorem. This theorem is derived in a manner very similar to that which we used above.

Theorem 4.2 If P dollars is deposited for n years in an account paying a rate of interest i compounded m times a year (m *periods*), then the account will contain a total of

$$A = P\left(1 + \frac{i}{m}\right)^{mn}$$

dollars at the end of n years.

Example 3 Find the amount of interest earned by a deposit of \$1000 for 6 years at 8% compounded quarterly.

Use the formula of Theorem 4.2. Here we have $P = 1000$, $i = 8\% = .08$, $n = 6$, and $m = 4$ (4 quarters in one year).

$$\begin{aligned} A &= 1000\left(1 + \frac{.08}{4}\right)^{4(6)} \\ &= 1000(1 + .02)^{24} \\ A &= 1000(1.02)^{24} \end{aligned}$$

The value of $(1.02)^{24}$ can be found in Table 6; locate 2% across the top and 24 periods at the left. You should find the number 1.60844. Thus,

$$A = 1000(1.60844) = 1608.44$$

dollars. The account contains a total of \$1608.44 and the interest is \$1608.44 − \$1000 = \$608.44. ■

2. Find the final amount on deposit and the interest earned if \$2500 is deposited in an account paying 6% compounded quarterly. The money is deposited for 5 years.

Answer: $A = \$3367.15$; \$867.15

Work Problem 2 at the side.

Example 4 Find the final amount on deposit if \$900 is deposited at 6% compounded semiannually for 8 years.

In 8 years, there are $2 \cdot 8 = 16$ semiannual periods. If interest is 6% per year, then $6\%/2 = 3\%$ is earned per semiannual period. Look in Table 6 for 3% and 16 periods, finding the number 1.60471. Finally, we can find the amount on deposit,

$$A = 900(1.60471) = 1444.24$$

dollars. ■

3. Suppose \$4000 is deposited at 4% compounded semiannually for 8 years. Find the final amount on deposit.

Answer: $A = \$5491.16$

Work Problem 3 at the side.

The more often interest is compounded, the more interest that will be earned. Using a compound interest table more complete than the one in this text, and using the formula above, we can get the results shown in the chart below.

Interest on \$1000 at 6% per Year for 10 Years

Compounded	Interest
Not at all (simple interest)	\$600
Yearly	790.85
Semiannually	806.11
Quarterly	814.02
Monthly	819.40
Daily	822.03

Note the pattern shown by the chart. It makes a big difference whether interest is compounded or not. Interest differs by \$190.85 when simple interest is compared to interest compounded annually. However, increasing the frequency of compounding makes smaller and smaller differences in the amount of interest earned. In fact, it can be shown that even if interest is compounded *every instant* the total amount of interest earned will be only slightly more than for daily compounding. Compounding every instant is discussed in more detail in Chapter 11 of this book.

If we deposit \$1 at 4% compounded quarterly, we can use Table 6 to find that at the end of one year the account will contain \$1.0406, an increase of 4.06% over the original \$1. The actual increase of 4.06% in our money is somewhat higher than the stated increase of 4%. To differentiate between these two numbers, 4% is called the **nominal** or **stated rate** of interest, while 4.06% is called the **effective rate.**

Example 5 Find the effective rate corresponding to a nominal rate of 6% compounded semiannually.

Look in Table 6 for 3% and 2 periods. You should find the number 1.06090. Thus, \$1 will increase to \$1.06090, an actual increase of 6.09%. The effective rate is 6.09%. ■

4. Find the effective rate corresponding to a nominal rate of 12% per year, compounded monthly.

Answer: 12.683%

Work Problem 4 at the side.

The equation for interest compounded annually, $A = P(1 + i)^n$, has four variables, A, P, i, and n. If we know any three of them, we have enough information to find the fourth. In particular, if we know A, the amount of money we wish to end up with, and also know i and n, then we can find P. Here P is the amount that we should deposit today to produce A dollars in n years. The next example shows how this works.

Example 6 Joan must pay a lump sum of \$6000 in 8 years. What lump sum deposited today at 4% compounded annually will amount to \$6000 in 8 years?

Here $A = 6000$, $i = .04$, $n = 8$, and P is unknown. Substituting these values into the formula for compound interest gives

$$6000 = P(1 + .04)^8$$

or

$$6000 = P(1.04)^8.$$

To solve for P, we can multiply both sides of this last equation by $(1.04)^{-8}$. Recall: $(1.04)^8 \cdot (1.04)^{-8} = (1.04)^{8+(-8)} = (1.04)^0 = 1$. Doing so gives

$$6000(1.04)^{-8} = P.$$

It is not easy to evaluate $(1.04)^{-8}$. To avoid this problem, we use Table 7. Look for 4% across the top and 8 periods down the side, finding .73069. Therefore,

$$P = 6000(.73069) = 4384.14$$

dollars. If Joan deposits \$4384.14 for 8 years in an account paying 4% compounded annually, she will have \$6000 when she needs it. ■

As this last example shows, \$6000 in 8 years is the same as

\$4384.14 today (if money can be deposited at 4% compounded annually). An amount that can be deposited today to yield a given sum in the future is called the **present value** of this given sum.

Example 7 Find the present value of \$16,000 in 9 years if money can be deposited at 6% compounded annually.

Look in Table 7 for 6% across the top and 9 periods down the side. You should find .59190. The present value is

$$16{,}000(.59190) = 9470.40.$$

A deposit of \$9470.40 today, at 6% compounded annually, will produce a total deposit of \$16,000 in 9 years. ■

5. Find the present value in Example 7 if money can be deposited at only 4% compounded annually.

Answer: \$11,241.44

Work Problem 5 at the side.

We can also solve $A = P(1 + i)^n$ for n, as the following example shows.

Example 8 Suppose the general level of inflation in the economy averages 3% per year. Find the number of years it would take for the general level of prices to double.

We want to find the number of years it will take for \$1 worth of goods and services to cost \$2. That is, we want to find n in the equation

$$2 = 1(1.03)^n \qquad A = 2,\ P = 1,\ i = .03$$

or

$$2 = (1.03)^n.$$

We could find n by using logarithms or certain calculators, but we can get a reasonable approximation by reading down the 3% column of Table 6. Read down the column until you come to the number closest to 2. The number closest to 2 is 1.97359; this number corresponds to 23 periods. Thus, the general level of prices will double in about 23 or 24 years. ■

6. Find the number of years it takes for prices to double if inflation averages
(a) 4% per year
(b) 8% per year

Answer:
(a) about 18 years
(b) about 9 years

Work Problem 6 at the side.

4.1 EXERCISES

Find the final amounts when the following deposits are made. (See Examples 1, 2, and 4.)

1. \$1000 at 4% compounded annually for 8 years

2. \$1000 at 4% compounded annually for 10 years
3. \$4500 at 6% compounded annually for 20 years
4. \$810 at 5% compounded annually for 12 years
5. \$470 at 4% compounded semiannually for 12 years
6. \$15,000 at 6% compounded semiannually for 11 years
7. \$46,000 at 8% compounded semiannually for 5 years
8. \$1050 at 10% compounded semiannually for 13 years
9. \$7500 at 4% compounded quarterly for 9 years
10. \$8000 at 4% compounded quarterly for 4 years
11. \$6500 at 8% compounded quarterly for 6 years
12. \$9100 at 8% compounded quarterly for 4 years

Find the amount of interest earned by the following deposits. (See Example 3.)

13. \$6000 at 5% compounded annually for 8 years
14. \$21,000 at 6% compounded annually for 5 years
15. \$43,000 at 6% compounded semiannually for 10 years
16. \$750 at 8% compounded semiannually for 5 years
17. \$2000 at 8% compounded quarterly for 4 years
18. \$50,000 at 12% compounded quarterly for 3 years

Find the present value of the following sums. (See Examples 6 and 7.)

19. \$4500 at 6% compounded annually for 9 years
20. \$11,500 at 4% compounded annually for 12 years
21. \$3800 at 5% compounded annually for 7 years
22. \$9100 at 6% compounded annually for 11 years
23. \$2000 at 4% compounded semiannually for 11 years
24. \$2000 at 6% compounded semiannually for 11 years
25. \$8800 at 6% compounded semiannually for 4 years
26. \$7500 at 4% compounded quarterly for 9 years
27. If money can be invested at 4% compounded quarterly, which is larger, \$1000 now or \$1210 in 5 years? (Hint: find the present value of the \$1210.)
28. If money can be invested at 4% compounded annually, which is larger, \$10,000 now or \$15,000 in 10 years?

Find the effective rate corresponding to each of the following nominal rates. (See Example 5.)

29. 4% compounded semiannually

30. 8% compounded quarterly

31. 8% compounded semiannually

32. 10% compounded semiannually

33. 12% compounded semiannually

34. 12% compounded quarterly

APPLIED PROBLEMS

Management

Use the ideas of Example 8 in the text to answer the following questions.

35. The consumption of electricity has increased historically at 6% per year. If it continued to increase at this rate indefinitely, find the number of years before the electric utilities would need to double generating capacity.

36. Suppose a conservation campaign coupled with higher rates caused demand for electricity to increase at only 2% per year. (See Exercise 35.) Find the number of years before the utilities would need to double generating capacity.

4.2 ANNUITIES

Suppose you decide to set aside money for your newborn child's education by depositing \$1000 at the end of each year for 18 years. Suppose also that the bank pays 5% interest compounded annually. How much money altogether would you have at the end of the 18-year period?

A sequence of payments such as this is called an **annuity.** If the payments are made at the end of the time period, and if the frequency of payments is the same as the frequency of compounding, the annuity is called an **ordinary annuity.**

To find the total amount of money in the above account after 18 years, we proceed as follows. The initial deposit, at the end of the first year, earns interest for 17 years. Thus, the initial \$1000 will amount to $1000(1 + .05)^{17} = 1000(1.05)^{17}$. The second \$1000 deposit earns interest for 16 years and finally amounts to $1000(1.05)^{16}$, and so on. The final deposit of \$1000 earns no interest at all. These results are shown in the table.

Periodic Payments of \$1000

Year	Final amount of each payment
1	$\$1000(1.05)^{17}$
2	$1000(1.05)^{16}$
3	$1000(1.05)^{15}$
4	$1000(1.05)^{14}$
⋮	⋮
16	$1000(1.05)^{2}$
17	$1000(1.05)^{1}$
18	1000, or $1000(1.05)^{0}$

By adding the numbers in the right column of the chart, we get the total amount on deposit after 18 years.

$$1000 + 1000(1.05)^1 + 1000(1.05)^2 + \cdots + 1000(1.05)^{14} + 1000(1.05)^{15} + 1000(1.05)^{16} + 1000(1.05)^{17}$$

The numbers in this sum are the terms of a geometric sequence (recall: we discussed sequences in Chapter 2) with $a_1 = 1000$, $r = 1.05$, and $n = 18$. Thus, if we use V to represent the **amount of the annuity** (the final amount on deposit), we can find V by the formula for the sum of the terms of a geometric sequence (given in Theorem 2.4).

Using this formula, we have

$$V = \frac{1000[(1.05)^{18} - 1]}{.05}.$$

The number $[(1.05)^{18} - 1]/.05$ is hard to evaluate. Normally a special table (or calculator) is used. Table 8 in this book gives these values. Find 5% across the top and 18 periods down the side.

$$\frac{(1.05)^{18} - 1}{.05} = 28.13238$$

Finally, we can find V, the total amount on deposit after 18 years.

$$V = \frac{1000[(1.05)^{18} - 1]}{.05}$$

$$= 1000(28.13238)$$

$$V = \$28{,}132.38$$

Since a total of $18 \times \$1000 = \$18{,}000$ was deposited over the years, a total of

$$\$28{,}132.38 - \$18{,}000 = \$10{,}132.38$$

in interest was earned.

Now let us write a general formula for the amount of an ordinary annuity. Suppose R dollars is deposited at the end of each year for n years, at a rate of interest i compounded annually. The first payment of R dollars will earn interest for $n - 1$ years and will finally amount to

$$R(1+i)^{n-1}$$

dollars. The R dollars deposited at the end of the second year will finally amount to $R(1+i)^{n-2}$ dollars, and so on. The last payment earns no interest at all. If we let V represent the amount of the annuity, then

$$V = R + R(1+i)^1 + R(1+i)^2 + \cdots + R(1+i)^{n-1}.$$

This sum is the sum of the first n terms of the geometric sequence having first term R and common ratio $1 + i$. Using the formula for the sum of the first n terms of a geometric sequence, we have

$$V = \frac{R[(1+i)^n - 1]}{(1+i) - 1} = \frac{R[(1+i)^n - 1]}{i} = R\left[\frac{(1+i)^n - 1}{i}\right].$$

In summary, we have the following result.

Theorem 4.3 If R dollars is deposited at the end of each year in an account paying a rate of interest i compounded annually, then after n years the account will contain a total of V dollars, where

$$V = R\left[\frac{(1+i)^n - 1}{i}\right].$$

The expression in brackets is often written with the symbol $s_{\overline{n}|i}$ (read "s-angle-n-at-i"); that is

$$s_{\overline{n}|i} = \frac{(1+i)^n - 1}{i}.$$

With this symbol, the conclusion of Theorem 4.3 becomes

$$V = R \cdot s_{\overline{n}|i}.$$

Values of $s_{\overline{n}|i}$ are given in Table 8.

Example 1 Tom Bleser is an athlete who feels that his playing career will last exactly 7 years. To prepare for his future, he deposits \$22,000 at the end of each year for 7 years in an account paying 6% compounded annually. How much will he have on deposit after 7 years?

His payments form an ordinary annuity with $R = 22{,}000$,

$n = 7$, and $i = .06$. The amount of this annuity is (by Theorem 4.3)

$$V = 22{,}000\left[\frac{(1.06)^7 - 1}{.06}\right].$$

From Table 8, the number in brackets, $s_{\overline{7}|.06}$, is found to be 8.39384, so that

$$V = 22{,}000(8.39384) = 184{,}664.48$$

dollars. ■

Work Problem 1 at the side.

1. Johnson Building Materials deposits \$2500 at the end of each year in an account paying 5% per year compounded annually. Find the total amount on deposit after
(a) 6 years
(b) 10 years.

Answer:
(a) \$17,004.78
(b) \$31,444.73

An ordinary annuity involves payments at the end of a given time period, with the frequency of payments equal to the period of compounding. The examples above have involved annual payments and interest compounded annually. If the payment period and the frequency of compounding are not annual, the problem is changed in much the same manner as for compound interest.

Example 2 Suppose \$1000 is deposited at the end of each 6-month period for 8 years in an account paying 6% compounded semi-annually. Find the final amount in the account.

Interest of $6\%/2 = 3\%$ is earned semiannually. In 8 years there are $8 \cdot 2 = 16$ semiannual periods. We need to find $s_{\overline{16}|.03}$ from Table 8. By looking in the 3% column and down 16 periods, we find that $s_{\overline{16}|.03} = 20.15688$. Thus, the \$1000 deposits will lead to a final total of

$$V = 1000(20.15688) = 20{,}156.88$$

dollars. ■

Work Problem 2 at the side.

2. Helen's Dry Goods deposits \$5800 at the end of each quarter for 4 years. Find the final amount on deposit if the interest rate is
(a) 4% compounded quarterly
(b) 8% compounded quarterly.

Answer:
(a) \$100,095.59
(b) \$108,107.88

Just as we can solve the formula $A = P(1 + i)^n$ for its various parts, we can also use the formula of Theorem 4.3 to find the value of variables other than V. In Example 3 below, we are given V, the amount of money we want at the end, and we need to find R, the amount of each payment.

Example 3 Betsy wants to buy an expensive movie camera three years from now. She wants to deposit an equal amount at the end of each quarter for 3 years in order to accumulate enough money to pay for the camera. The camera costs \$1200, and the bank pays 4% interest compounded quarterly. Find the amount of each of the twelve deposits she will make.

This example describes an ordinary annuity with $V = 1200$, $i = .01$ ($4\%/4 = 1\%$ per quarter), and $n = 3 \cdot 4 = 12$ periods. The unknown here is the amount of each payment, R. If we use the formula for the amount of an annuity from Theorem 4.3, we have

$$V = R\left[\frac{(1+i)^n - 1}{i}\right]$$

$$1200 = R\left[\frac{(1.01)^{12} - 1}{.01}\right].$$

From Table 8 (1%, 12 periods) we find that the expression in brackets, $s_{\overline{12}|.01}$, is 12.68250, so that

$$1200 = R(12.68250)$$

or

$$R = 94.62$$

dollars (divide both sides by 12.68250). ■

3. Francisco needs $8000 in 6 years so that he can go on an archeological dig. He wants to deposit equal payments at the end of each quarter so that he will have enough to go on the dig. Find the amount of each payment if the bank pays
(a) 4% compounded quarterly
(b) 8% compounded quarterly.

Answer:
(a) $296.59
(b) $262.97

Work Problem 3 at the side.

In Example 3, we had to divide 1200 by 12.68250, a difficult task if no calculator is available. Dividing by 12.68250 gives the same result as multiplying by the *reciprocal* of 12.68250. (Recall: the reciprocal of a number x is $1/x$). Table 10 in the back of the book gives the values of $1/s_{\overline{n}|i}$. Look in Table 10 for 1%, 12 periods, and find that $1/s_{\overline{12}|.01} = .07885$. Using the numbers of Example 3, we find that

$$R = 1200 \cdot \frac{1}{s_{\overline{12}|.01}} = 1200(.07885) = 94.62,$$

the same quarterly payment as Example 3.

4. Use Table 10 to rework the the previous problem.

Answer:
(a) $296.56
(b) $262.96
(The answers differ by a few cents from the previous answers—this is due to rounding errors.)

Work Problem 4 at the side.

An annuity in which payments are made at the *beginning* of each period is called an **annuity due.** To find the amount of an annuity due, proceed as follows.

1. Use a row in Table 8 corresponding to one more period than in the annuity. (For an annuity of 12 periods, use row 13 of the table.)
2. Using the correct interest rate per period, find the necessary

number from the table. Multiply this number and the amount of each payment.

3. Subtract the amount of one payment from the product found in Step 2.

Example 4 Find the amount of an annuity due if payments of \$500 are made at the beginning of each quarter for 7 years, in an account paying 4% compounded quarterly.

Follow the steps above.

1. In 7 years, there are $4 \cdot 7 = 28$ periods. Look in row 29 $(28 + 1)$ of the table.

2. Use the $4\%/4 = 1\%$ column of the table. You should find the number 33.45039. Multiply this number by 500, the amount of each payment.

$$500(33.45039) = 16{,}725.20$$

3. Subtract the amount of one payment from this result.

$$\$16{,}725.20 - \$500 = \$16{,}225.20$$

The account will contain a total of \$16,225.20 after 7 years. ■

5. Ms. Black deposits \$800 at the beginning of each six-month period for 5 years. Find the final amount in the account if the bank pays
(a) 6% compounded semi-annually
(b) 8% compounded semi-annually.

Answer:
(a) \$9446.24
(b) \$9989.08

Work Problem 5 at the side.

4.2 EXERCISES

Find each of the following values. Use Table 8 or Table 10.

1. $s_{\overline{12}|.05}$ **2.** $s_{\overline{20}|.06}$ **3.** $s_{\overline{16}|.04}$ **4.** $s_{\overline{40}|.01}$

5. $s_{\overline{20}|.01}$ **6.** $s_{\overline{18}|.015}$ **7.** $1/s_{\overline{15}|.04}$ **8.** $1/s_{\overline{30}|.015}$

9. $1/s_{\overline{19}|.06}$ **10.** $1/s_{\overline{24}|.05}$

Find the value of the following ordinary annuities. Interest is compounded annually. (See Example 1.)

11. $R = 100, \quad i = .04, \quad n = 10$

12. $R = 1000, \quad i = .06, \quad n = 12$

13. $R = 10{,}000, \quad i = .05, \quad n = 19$

14. $R = 100{,}000, \quad i = .04, \quad n = 23$

15. $R = 8500, \quad i = .05, \quad n = 30$

16. $R = 11{,}200, \quad i = .04, \quad n = 25$

17. $R = 46{,}000, \quad i = .06, \quad n = 32$

18. $R = 29{,}500, \quad i = .05, \quad n = 15$

Find the value of each of the following ordinary annuities. Payments are made and interest is compounded as given. (See Examples 2 and 3.)

19. $R = 800$, $i = .06$, 10 years, semiannually

20. $R = 4600$, $i = .08$, 9 years, quarterly

21. $R = 15{,}000$, $i = .04$, 6 years, quarterly

22. $R = 42{,}000$, $i = .06$, 12 years, semiannually

Find the amount of each of the following annuities due. Interest is compounded annually. (See Example 4.)

23. $R = 600$, $i = .06$, $n = 8$

24. $R = 1400$, $i = .05$, $n = 10$

25. $R = 20{,}000$, $i = .05$, $n = 6$

26. $R = 4000$, $i = .06$, $n = 11$

Find the amounts of each of the following annuities due.

27. Payments of \$1000 made at the beginning of each year for 9 years at 6% compounded annually

28. \$750 deposited at the beginning of each year for 15 years at 5% compounded annually

29. \$100 deposited at the beginning of each quarter for 9 years at 6% compounded quarterly

30. \$1500 deposited at the beginning of each semiannual period for 11 years at 4% compounded semiannually

Find the periodic payment (R) that will amount to the following sums under the given conditions.

31. $V = \$10{,}000$, interest is 5% compounded annually, payments made at the end of each year for 12 years

32. $V = \$100{,}000$, interest is 6% compounded annually, payments made at the end of each year for 9 years

33. $V = \$50{,}000$, interest is 4% compounded quarterly, payments made at the end of each quarter for 8 years

APPLIED PROBLEMS

Solve each of the following problems involving ordinary annuities. In Problems 37 and 38 you will also need to use the ideas on compound interest from Section 4.1.

Management

34. Pat Dillon deposits \$12,000 at the end of each year for 23 years in an account paying 5% compounded annually. Find the final amount she will have on deposit.

35. Pat's brother-in-law works in a bank which pays 4% compounded annually. If she deposits her money in this bank (instead of the one of Exercise 34), how much would she have in her account?

36. How much would Pat lose over 23 years by using her brother-in-law's bank? (See Exercises 34 and 35.)

37. Pam Parker deposits \$1000 at the end of each year for 8 years in an account paying 5% compounded annually. She then leaves the money alone, with no further deposits, for an additional 6 years. Find the final amount on deposit after the entire 14-year period.

38. Chuck deposits \$10,000 at the end of each year for 12 years in an account paying 6% compounded annually. He then puts the total amount on deposit in another account paying 6% compounded semiannually for another 9 years. Find the final amount on deposit after the entire 21-year period.

39. Ray Berkowitz needs \$10,000 in 8 years. What amount R can he deposit at the end of each year, at 5% compounded annually, so that he will have his \$10,000?

40. Find Berkowitz's annual deposit (See Exercise 39) if the money is deposited at 6% compounded annually.

41. Barb Silverman wants to buy a \$12,000 car in 6 years. How much money must she deposit at the end of each quarter in an account paying 4% compounded quarterly, so that she will have enough to pay for her car?

42. Harv's Meats knows that it must buy a new deboner machine in 4 years. The machine costs \$12,000. In order to accumulate enough money to be able to pay for the machine, Harv decides to deposit a sum of money at the end of each quarter, in an account paying 4% compounded quarterly. How much should each payment be?

4.3 PRESENT VALUE OF AN ANNUITY

As we saw in Theorem 4.3, if deposits of R dollars are made at the end of each year for n years at a rate of interest i compounded annually, then the account will contain

$$V = R \cdot s_{\overline{n}|i} = R\left[\frac{(1+i)^n - 1}{i}\right]$$

dollars after the n years. Let us now find the lump sum A that can can be deposited today at a rate of interest i compounded annually, which will amount to the same V dollars in n years.

First recall that A dollars deposited today will amount to $A(1+i)^n$ dollars after n years at a rate of interest i compounded annually. We want this amount, $A(1+i)^n$, to be the same as V, the amount of the annuity. Substituting $A(1 + i)^n$ for V in the formula above, we have

$$A(1 + i)^n = R\left[\frac{(1 + i)^n - 1}{i}\right].$$

We want to solve this equation for A. To do so, divide both sides of the equation by $(1 + i)^n$.

$$A = \frac{R}{(1 + i)^n}\left[\frac{(1 + i)^n - 1}{i}\right]$$

Replace $1/(1 + i)^n$ by $(1 + i)^{-n}$.

$$A = R(1 + i)^{-n}\left[\frac{(1 + i)^n - 1}{i}\right]$$

Finally, use the distributive property, and the fact that $(1 + i)^{-n} \cdot (1 + i)^n = 1$.

$$A = R\frac{(1 + i)^{-n}(1 + i)^n - (1 + i)^{-n}}{i} = R\left[\frac{1 - (1 + i)^{-n}}{i}\right]$$

The amount A is called the **present value of the annuity.** In summary, we have the following theorem.

Theorem 4.4 The present value of an annuity made up of payments of R dollars at the end of each year for n years at a rate of interest i compounded annually is given by A, where

$$\boldsymbol{A = R\left[\frac{1 - (1 + i)^{-n}}{i}\right].}$$

The expression in brackets is abbreviated as $\boldsymbol{a_{\overline{n}|i}}$, so that

$$\boldsymbol{A = R \cdot a_{\overline{n}|i}}.$$

Values of $a_{\overline{n}|i}$ are given in Table 9.

Example 1 What lump sum deposited today at 6% compounded annually will yield the same total amount as payments of \$1500 at the end of each year for 12 years, also at 6% compounded annually?

We want to find the present value of an annuity of \$1500 per year for 12 years at 6% compounded annually. Using Table 9, we have $a_{\overline{12}|.06} = 8.38384$, so that

$$A = 1500(8.38384) = 12{,}575.76$$

dollars. A lump sum deposit of \$12,575.76 today at 6% compounded annually will yield the same total after 12 years as deposits of \$1500 at the end of each year for 12 years, at 6% compounded annually.

Let's check this result. The final amount in 12 years of a deposit of \$12,575.76 today at 6% compounded annually can be found by using the formula for compound interest. From Table 6, we see that \$12,575.76 will give a total of

$$(12{,}575.76)(2.01220) = 25{,}304.94$$

dollars. On other hand, from Table 8 we see that deposits of \$1500 at the end of each year for 12 years, at 6% compounded annually, give an amount of

$$1500(16.86994) = 25{,}304.91$$

dollars. (The difference of 3¢ is due to rounding errors.)

In summary, there are two ways we can have \$25,304.91 in 12 years at 6% compounded annually—a single deposit of \$12,575.76 today, or payments of \$1500 at the end of each year for 12 years. ■

Example 2 Mr. Jones and Ms. Gonsalez are both alumni of the Forestview Institute of Technology. They both agree to contribute to the endowment fund of FIT. Mr. Jones says that he will give \$500 at the end of each year for 9 years. Ms. Gonsalez would rather give a lump sum today. What lump sum can she give that will be equivalent to Mr. Jones' annual gifts, if the endowment fund earns 5% compounded annually?

Here $R = 500$, $n = 9$, and $i = .05$. The necessary number from Table 9 is $a_{\overline{9}|.05} = 7.10782$. Ms. Gonsalez must therefore donate a lump sum of

$$500(7.10782) = 3553.91$$

dollars today. ■

Work Problem 1 at the side.

1. What lump sum today is equivalent to an annuity of \$1000 at the end of each quarter for 4 years, if the money is deposited at 8% compounded quarterly?

Answer: \$13,577.71

We can also use the formula of Theorem 4.4 if we know the lump sum and want to find the periodic payment of the annuity. This is shown in the next example.

Example 3 Chuck's house cost \$52,000. His down payment was \$8000, leaving a balance of \$44,000 to be financed by a mortgage. The bank gives Chuck a 30-year mortgage at an interest rate of 8%. What payments must Chuck make at the end of each year for 30 years in order to pay off the \$44,000 balance of the mortgage?

A single lump sum payment of \$44,000 today would pay off the mortgage. However, Chuck wishes to make annual payments for 30 years. This makes \$44,000 the present value of an annuity of 30 annual payments with interest at 8%, compounded annually.

The formula for the present value of an annuity that we developed above is $A = R \cdot a_{\overline{n}|i}$. In our example, $A = 44{,}000$, $i = .08$, $n = 30$, and R is unknown. From Table 9, $a_{\overline{30}|.08} = 11.25778$, and

$$44{,}000 = R(11.25778).$$

To solve this equation, we need to use a calculator and divide both sides by 11.25778. Doing so gives

$$R = 3908.41.$$

Chuck must make payments of \$3908.41 at the end of each year for 30 years. ■

As a practical matter, payments on a mortgage are made monthly, and at a higher interest rate. We show in the next section how to calculate house payments with more realistic numbers.

2. Roberta Runck owes \$20,000, which she has agreed to pay off by making payments at the end of each year for 11 years. Interest is at 6% compounded annually. Find the amount of each payment.

Answer: \$2535.86

Work Problem 2 at the side.

The calculation in Example 3 would have been easier if we knew values of $1/a_{\overline{n}|i}$. These can be found from Table 10, which gives values of $1/s_{\overline{n}|i}$. To see how, first recall that

$$a_{\overline{n}|i} = \frac{1-(1+i)^{-n}}{i}, \qquad \text{so that} \qquad \frac{1}{a_{\overline{n}|i}} = \frac{i}{1-(1+i)^{-n}}.$$

Add $-i$ to both sides of the equation on the right.

$$\frac{1}{a_{\overline{n}|i}} - i = \frac{i}{1-(1+i)^{-n}} - i$$

Simplify on the right.

$$= \frac{i - i + i(1+i)^{-n}}{1-(1+i)^{-n}}$$

$$= \frac{i(1+i)^{-n}}{1-(1+i)^{-n}}$$

Multiply the numerator and denominator by $(1 + i)^n$.

$$= \frac{i(1+i)^{-n} \cdot (1+i)^n}{[1-(1+i)^{-n}](1+i)^n}$$

$$\frac{1}{a_{\overline{n}|i}} - i = \frac{i}{(1+i)^n - 1}$$

The fraction on the right is $1/s_{\overline{n}|i}$, so that

$$\frac{1}{a_{\overline{n}|i}} - i = \frac{1}{s_{\overline{n}|i}}$$

or, finally,

$$\frac{1}{a_{\overline{n}|i}} = \frac{1}{s_{\overline{n}|i}} + i.$$

Example 4 **(a)** Find $1/a_{\overline{12}|.03}$.

From Table 10, $1/s_{\overline{12}|.03} = .07046$. By the formula above,

$$\begin{aligned}\frac{1}{a_{\overline{12}|.03}} &= \frac{1}{s_{\overline{12}|.03}} + i \\ &= .07046 + .03 \\ &= .10046.\end{aligned}$$

(b) $$\frac{1}{a_{\overline{20}|.06}} = \frac{1}{s_{\overline{20}|.06}} + .06 = .08718 \quad ■$$

Work Problem 3 at the side.

3. Find each of the following.
(a) $1/a_{\overline{12}|.02}$
(b) $1/a_{\overline{25}|.06}$

Answer:
(a) .09456
(b) .07823

4.3 EXERCISES

Find each of the following values. Use Table 9 or Table 10.

1. $a_{\overline{15}|.06}$ **2.** $a_{\overline{10}|.03}$ **3.** $a_{\overline{18}|.04}$ **4.** $a_{\overline{30}|.015}$

5. $a_{\overline{16}|.01}$ **6.** $a_{\overline{32}|.02}$ **7.** $1/a_{\overline{6}|.05}$ **8.** $1/a_{\overline{18}|.03}$

9. $1/a_{\overline{12}|.015}$ **10.** $1/a_{\overline{32}|.01}$

Find the present value of each of the following ordinary annuities. (See Example 1.)

11. Payments of \$1000 are made annually for 9 years at 5% compounded annually.

12. Payments of \$5000 are made annually for 11 years at 6% compounded annually.

13. Payments of \$890 are made annually for 16 years at 5% compounded annually.

14. Payments of \$1400 are made annually for 8 years at 5% compounded annually.

15. Payments of \$10,000 are made semiannually for 15 years at 6% compounded semiannually.

16. Payments of \$50,000 are made quarterly for 10 years at 4% compounded quarterly.

17. Payments of \$105,000 are made quarterly for 4 years at 6% compounded quarterly.

18. Payments of \$18,500 are made every six months for 18 years at 8% compounded semiannually.

Find the lump sum deposited today that will yield the same total amount as payments of \$10,000 at the end of each year for 15 years, at each of the following interest rates. Interest is compounded annually. (See Example 2.)

19. 3% **20.** 4% **21.** 5% **22.** 6%

APPLIED PROBLEMS

Management

23. In his will the late Mr. Hudspeth said that each child in his family could have an annuity of \$2000 at the end of each year for 9 years, or the equivalent present value. If money can be deposited at 5% compounded annually, what is the present value?

24. In the "Million Dollar Lottery" a winner is paid a million dollars at the rate of \$50,000 per year for 20 years. Assume that these payments form an ordinary annuity, and that the lottery managers can invest money at 6% compounded annually. Find the lump sum that the management must put away today to pay off the "million dollar" winner.

25. Lynn Meyers buys a new car costing \$6000. She agrees to make payments at the end of each six-month period for 4 years. If she pays 6% interest, compounded semiannually, what is the amount of each payment?

26. Find the total amount of interest that Lynn will pay. (See Exercise 25.)

27. A new house costs \$80,000 after a down payment. The house will be paid off with payments at the end of each year for 25 years. The interest rate is 8% compounded annually. Find the amount of each payment.

28. How much interest, in total, will be paid in Exercise 27?

29. What lump sum deposited today at 5% compounded annually for 8 years will provide the same amount as \$1000 deposited at the end of each year for 8 years at 6% compounded annually? (Hint: first find the total amount of the annual deposits and then find the present value of that amount.)

30. What lump sum deposited today at 4% compounded quarterly for 10 years will yield the same final amount as deposits of \$4000 at the end of each six-month period for 10 years at 6% compounded semiannually?

4.4 AMORTIZATION AND SINKING FUNDS

To **amortize** a loan is to pay it off. The methods for calculating the amount of the various payments necessary to pay off the loan can be found with the formulas of this chapter. Example 1 shows this.

Example 1 A speculator agrees to pay \$15,000 for a parcel of land; this amount will be paid off in 3 years with semiannual payments, at an interest rate of 8% compounded semiannually.

(a) Find the amount of each payment.

If the speculator paid \$15,000 immediately, there would be no need for any payments at all. Thus, \$15,000 is the present value of an annuity of R dollars, $2 \cdot 3 = 6$ periods, and $i = 8\%/2 = .04$. If A is the present value of an annuity, we know that

$$A = R \cdot a_{\overline{n}|i}.$$

In our example, $A = 15{,}000$, with

$$15{,}000 = R \cdot a_{\overline{6}|.04}$$

or

$$R = 15{,}000\left(\frac{1}{a_{\overline{6}|.04}}\right).$$

By the formula of the previous section, $1/a_{\overline{6}|.04} = 1/s_{\overline{6}|.04} + .04$. From Table 10, we have $1/s_{\overline{6}|.04} = .15076$, so that $1/a_{\overline{6}|.04} = .15076 + .04 = .19076$, and

$$R = 15{,}000(.19076) = 2861.40.$$

Each payment is \$2861.40. (With a calculator, much of this work could have been avoided.)

(b) Find the portion of the first payment that represents interest.

Interest is at 8% per year, compounded semiannually, or 4% per semiannual period. During the first period, the entire \$15,000 is owed. Interest on this amount for 6 months (1/2 year) is found by the formula for simple interest.

$$\begin{aligned}\text{Interest} &= \text{principal} \cdot \text{rate} \cdot \text{time}\\ &= 15{,}000(.08)(\tfrac{1}{2})\\ &= 600\end{aligned}$$

dollars. At the end of 6 months, the speculator makes a payment of \$2861.40; since \$600 of this represents interest, a total of

$$\$2861.40 - \$600 = \$2261.40$$

is applied to the reduction of the original debt.

(c) Find the balance due after 6 months.

The original balance due is \$15,000. After 6 months, \$2261.40 is applied to the reduction of this debt. The debt after 6 months is thus

$$\$15{,}000 - \$2261.40 = \$12{,}738.60.$$

(d) How much interest is due for the second 6-month period?

A total of \$12,738.60 is owed for the second 6 months. Interest on this amount is

$$\text{Interest} = (12{,}738.60)(.08)(\tfrac{1}{2}) = 509.54$$

dollars. A payment of \$2861.40 is made at the end of this period; a total of

$$\$2861.40 - \$509.54 = \$2351.86$$

is applied to the debt. ■

1. Find the balance on the loan of Example 1 after the second 6-month period.

Answer: \$12,738.60 − \$2351.86 = \$10,386.74

2. Find the amount of the interest due for the third 6-month period in Example 1.

Answer: $(10{,}386.74)(.08)(\tfrac{1}{2}) =$ 415.47

Work Problem 1 at the side.

Work Problem 2 at the side.

By continuing this process, we can get the **amortization schedule** shown here. Note that the payment is always the same, except perhaps for a small adjustment in the final one. Payment 0 represents the original amount of the loan.

Amortization Schedule

Payment number	Amount of payment	Interest for period	Portion to principal	Principal at end of period
0	—	—	—	\$15,000
1	\$2861.40	\$600	\$2261.40	\$12,738.60
2	\$2861.40	\$509.54	\$2351.86	\$10,386.74
3	\$2861.40	\$415.47	\$2445.93	\$7940.81
4	\$2861.40	\$317.63	\$2543.77	\$5397.04
5	\$2861.40	\$215.88	\$2645.52	\$2751.52
6	\$2861.58	\$110.06	\$2751.52	\$0

In the example above, a portion of the principal is paid off with each payment. In other debts, only interest is paid periodically, with the entire principal amount of the loan repaid at some time in the future. In a case such as this, periodic payments can be deposited into an account, called a **sinking fund.** When the principal is due, the sinking fund will have just enough in it to pay off the debt. The next example shows how to calculate the payments into a sinking fund.

Example 2 The Stockdales are close to retirement. They agree to sell an antique urn to a local museum for \$17,000. Their tax adviser tells them to defer the receipt of this money until they retire, 5 years in the future. The museum agrees to pay them 6% per year for 5 years, and then pay them the \$17,000 in a lump sum.

(a) What annual interest payment will the museum make? Use the formula for simple interest.

$$\text{Interest} = (17{,}000)(.06)(1) = 1020$$

The museum makes annual interest payments of \$1020 to the Stockdales. (This amount does not play any part in the rest of the example.)

(b) Find the amount of each payment the museum must make into a sinking fund so that it will have \$17,000 in 5 years. Assume that the museum can earn 5% compounded annually on its money. Also, assume that the payments will be made annually.

These payments are the periodic payments into an ordinary annuity. The annuity will amount to \$17,000 in 5 years, at 5% compounded annually. From the formulas of Section 4.2, we have

$$17{,}000 = R \cdot s_{\overline{5}|.05}$$

$$R = 17{,}000\left(\frac{1}{s_{\overline{5}|.05}}\right)$$

$$= 17{,}000(.18097) \qquad \text{Table 10}$$

$$R = 3076.49.$$

If the museum deposits \$3076.49 at the end of each year for 5 years in an account paying 5% compounded annually, it will have the total amount that it needs. This is shown in the following table. Again, the last payment differs slightly from the others.

Payment number	Amount of deposit	Interest earned	Total
1	\$3076.49	\$0.00	\$3076.49
2	\$3076.49	\$153.82	\$6306.80
3	\$3076.49	\$315.34	\$9698.63
4	\$3076.49	\$484.93	\$13,260.05
5	\$3076.95	\$663.00	\$17,000.00

■

Work Problem 3 at the side.

3. A firm must pay off \$40,000 worth of bonds in 7 years. Find the amount of each annual payment to be made into a sinking fund if money earns 6% compounded annually.

Answer: \$4765.60

4.4 EXERCISES

Find the payment necessary to amortize each of the following loans. (See Example 1.)

1. \$1000, 6% compounded annually, 9 annual payments
2. \$2500, 8% compounded quarterly, 6 quarterly payments
3. \$41,000, 8% compounded semiannually, 10 semiannual payments

4. \$90,000, 5% compounded annually, 12 annual payments
5. \$140,000, 12% compounded quarterly, 15 quarterly payments
6. \$7400, 10% compounded semiannually, 18 semiannual payments
7. \$5500, 18% compounded monthly, 24 monthly payments
8. \$45,000, 18% compounded monthly, 36 monthly payments

Find the amount of each payment to be made into a sinking fund so that enough will be present to pay off the indicated loans. (See Example 2(b).)

9. loan \$2000, money earns 6% compounded annually, 5 annual payments
10. loan \$8500, money earns 5% compounded annually, 7 annual payments
11. loan \$11,000, money earns 8% compounded semiannually, 12 semiannual payments
12. loan \$75,000, money earns 6% compounded semiannually, 9 semiannual payments
13. loan \$50,000, money earns 6% compounded quarterly, 18 quarterly payments
14. loan \$25,000, money earns 8% compounded quarterly, 12 quarterly payments
15. loan \$6000, money earns 18% compounded monthly, 36 monthly payments
16. loan \$9000, money earns 18% compounded monthly, 24 monthly payments

APPLIED PROBLEMS

Management

17. An insurance firm pays \$4000 for a new printer for its computer. It amortizes this loan in 4 annual payments at 6% compounded annually. Prepare an amortization schedule for this machine.
18. Large semitrailer trucks cost \$52,000 each. Ace Trucking buys such a truck and agrees to pay for it by a loan which will be amortized with 9 annual payments, at 6% compounded annually. Prepare an amortization schedule for this truck.
19. IBM charges \$900 for a Self-Correcting Selectric Typewriter. A firm of tax accountants buys 8 of these machines. They make a down payment of \$1200 and agree to amortize the balance with semiannual payments at 8% compounded semiannually for 4 years. Prepare an amortization schedule for these machines.
20. When Denise Sullivan opened her law office, she bought \$14,000 worth of law books and \$7200 worth of office furniture. She paid \$1200 down and agreed to amortize the balance with semiannual payments for 5 years, at 6% compounded semiannually. Prepare an amortization schedule for this purchase.

Helen Spence sells some land in Nevada. She will be paid a lump sum of \$60,000 in 7 years. Until then, the buyer pays 8% interest, compounded quarterly.

21. Find the amount of each quarterly interest payment.

22. The buyer sets up a sinking fund so that enough money will be present to pay off the \$60,000. The buyer wants to make semiannual payments into the sinking fund; the account pays 6% compounded semiannually. Find the amount of each payment into the fund.

23. Prepare a table similar to the one of Example 2(b), showing the amount in the sinking fund after each deposit.

Jeff Reschke bought a rare stamp for his collection. He agreed to pay \$1500 for the stamp. He will pay a lump sum of \$1500 after 5 years. Until then, he pays 6% interest, compounded semiannually.

24. Find the amount of each semiannual interest payment.

25. Jeff sets up a sinking fund so that enough money will be present to pay off the \$1500. Jeff wants to make annual payments into the fund. The accounts pays 5% compounded annually. Find the amount of each payment into the fund.

26. Prepare a table similar to the one of Example 2(b), showing the amount in the sinking fund after each deposit.

When Ms. Thompson died, she left \$25,000 for her husband. He deposits the money at 5% compounded annually. He wants to make annual withdrawals from the account so that the money (principal and interest) is gone in 8 years.

27. Find the amount of each withdrawal. (Hint: you will need $1/a_{\overline{8}|.05}$.)

28. Find the amount of each withdrawal in Exercise 27 if the money must last 12 years.

The trustees of St. Albert's College have accepted a gift of \$150,000. The donor has directed the trustees to deposit the money into an account paying 6% per year, compounded semiannually. The trustees may withdraw money at the end of each 6-month period; the money must last 5 years.

29. Find the amount of each withdrawal.

30. Find the amount of each withdrawal in Exercise 29 if the money must last 7 years.

4.5 CONSUMER MATHEMATICS

In this section we use mathematics to solve problems that are important to consumers and to the businesses that sell to them. Many of these answers concern monthly payments. Typically, a consumer buys an item and pays for it over a period of time. We can think of the cash price of the item after any down payments as the present value of the annuity represented by the monthly payment.

We saw in Theorem 4.4 that the present value of an annuity, A, made up of payments of R dollars at the end of each year for n years at a rate of interest i compounded annually is given by

$$A = Ra_{\overline{n}|i}$$

This result can also be used when payments are made more often than annually. For example, suppose you buy a \$50,000 house and pay \$10,000 down. You must then borrow the \$40,000 still owed on the house. To do this, you might take out a mortgage from the bank. The interest rate charged by the bank varies from time to time, but let us assume that you are charged 9%, with the mortgage due to be paid off in 30 years. What is the monthly payment?

In 30 years, there are $30 \cdot 12 = 360$ months. An interest rate of 9% per year is the same as $9\%/12 = \frac{3}{4}\% = .0075$ per month. If we substitute these values into the formula above, we have

$$40{,}000 = R \cdot a_{\overline{360}|.0075}.$$

It is not at all easy to evaluate $a_{\overline{360}|.0075}$ unless you have a good financial calculator, or access to a computer. A common way out of this problem is to use a special table, such as the one below. This table gives the monthly payment per \$1000 of the amount financed.

Monthly Payments Required to Repay a \$1,000 Loan

No. Years	6%	6 1/2%	6.9%	7%	7.2%	7 1/4%	7 1/2%	7 5/8%	7 3/4%	8%	8 1/4%	8.4%	8 1/2%	8 3/4%	9%	9 1/4%	9 1/2%	9 3/4%	10%
5	\$19.34	\$19.57	\$19.76	\$19.81	\$19.90	\$19.92	\$20.04	\$20.10	\$20.16	\$20.28	\$20.40	\$20.47	\$20.52	\$20.64	\$20.76	\$20.89	\$21.01	\$21.13	\$21.25
6	16.58	16.81	17.01	17.05	17.15	17.17	17.30	17.36	17.42	17.54	17.66	17.73	17.78	17.91	18.03	18.16	18.28	18.41	18.53
7	14.61	14.85	15.05	15.10	15.20	15.22	15.34	15.41	15.47	15.59	15.72	15.79	15.84	15.97	16.09	16.22	16.35	16.48	16.61
8	13.15	13.39	13.59	13.64	13.74	13.76	13.89	13.95	14.01	14.14	14.27	14.34	14.40	14.53	14.65	14.78	14.92	15.05	15.18
9	12.01	12.26	12.46	12.51	12.61	12.64	12.77	12.83	12.89	13.02	13.15	13.23	13.28	13.42	13.54	13.67	13.81	13.95	14.08
10	11.11	11.36	11.56	11.62	11.72	11.75	11.88	11.94	12.01	12.14	12.27	12.35	12.40	12.54	12.67	12.80	12.94	13.08	13.22
11	10.37	10.63	10.84	10.89	10.99	11.02	11.15	11.22	11.29	11.42	11.56	11.63	11.69	11.83	11.97	12.09	12.24	12.38	12.52
12	9.76	10.02	10.24	10.29	10.40	10.42	10.56	10.62	10.69	10.83	10.97	11.05	11.11	11.25	11.39	11.52	11.67	11.81	11.95
13	9.25	9.52	9.73	9.79	9.89	9.92	10.06	10.13	10.20	10.34	10.48	10.56	10.62	10.75	10.90	11.04	11.19	11.34	11.48
14	8.82	9.09	9.30	9.36	9.47	9.50	9.64	9.71	9.78	9.92	10.06	10.14	10.20	10.35	10.49	10.63	10.79	10.94	11.09
15	8.44	8.72	8.94	8.99	9.11	9.13	9.28	9.35	9.42	9.56	9.71	9.79	9.85	10.00	10.15	10.29	10.45	10.60	10.75
16	8.12	8.40	8.62	8.68	8.79	8.82	8.96	9.04	9.11	9.25	9.40	9.49	9.55	9.70	9.85	9.99	10.15	10.31	10.46
17	7.84	8.12	8.34	8.40	8.52	8.55	8.69	8.77	8.84	8.99	9.14	9.22	9.29	9.44	9.59	9.74	9.90	10.06	10.22
18	7.59	7.87	8.10	8.16	8.28	8.31	8.45	8.53	8.60	8.75	8.91	9.00	9.06	9.21	9.37	9.52	9.68	9.84	10.00
19	7.37	7.65	7.89	7.95	8.07	8.10	8.25	8.32	8.40	8.55	8.70	8.79	8.86	9.02	9.17	9.32	9.49	9.65	9.82
20	7.17	7.46	7.70	7.76	7.88	7.91	8.06	8.14	8.21	8.37	8.53	8.62	8.68	8.84	9.00	9.15	9.33	9.49	9.66
25	6.45	6.76	7.01	7.07	7.20	7.23	7.39	7.48	7.56	7.72	7.89	7.99	8.06	8.23	8.40	8.56	8.74	8.92	9.09
30	6.00	6.33	6.59	6.66	6.79	6.83	7.00	7.08	7.17	7.34	7.52	7.62	7.69	7.87	8.05	8.22	8.41	8.60	8.78

Example 1 Find the monthly payment necessary to pay off a home loan of $55,000, if the interest rate is 9%. Assume that the loan is for 25 years.

Look in the table for 9% and 25 years. You should find the number 8.40. The loan balance is $55,000, which represents 55 thousands. The monthly payment is thus

$$\text{monthly payment} = 55(8.40) = 462$$

dollars. This amount represents the sum necessary for paying principal and interest. In many areas, it is necessary to add an amount for property taxes and fire insurance to get the final payment. ■

1. Find the monthly payment necessary to pay off a 9%, 30-year home loan of
(a) $35,000
(b) $72,000
(c) $62,500.

Answer:
(a) $281.75
(b) $579.60
(c) $503.13

Work Problem 1 at the side.

In 1969, the federal government passed the Truth-in-Lending Act, which requires that lenders notify borrowers as to the **true annual interest rate** that they will pay. The calculation of the true annual interest rate also uses $a_{\overline{n}|i}$ and is quite complicated. Once again, in practice this rate is found with a table. A portion of a table is shown here.

Number of monthly payments	True Annual Interest Rate						
	14%	$14\frac{1}{2}$%	15%	$15\frac{1}{2}$%	16%	$16\frac{1}{2}$%	17%
	(Finance charge per $100 of amount financed)						
6	$ 4.12	4.27	4.42	4.57	4.72	4.87	5.02
12	7.74	8.03	8.31	8.59	8.88	9.16	9.45
18	11.45	11.87	12.29	12.72	13.14	13.57	13.99
24	15.23	15.80	16.37	16.94	17.51	18.09	18.66
30	19.10	19.81	20.54	21.26	21.99	22.72	23.45
36	23.04	23.92	24.80	25.68	26.57	27.46	28.35
42	27.06	28.10	29.15	30.19	31.25	32.31	33.37
48	31.17	32.37	33.59	34.81	36.03	37.27	38.50

To use the table, find the actual finance charge being assessed and the total amount being financed. Then divide the finance charge by the amount financed and multiply by 100. Look in the table above. Find the number of payments at the left. Read across, and find the number closest to the answer you found. The true annual interest rate is then given at the top of the column.

Example 2 Jenny Toms bought a new car for $4200. She paid $1600 down, and agreed to payments of $89.50 per month for 36 months. Find the true annual interest rate that she paid.

The amount financed is the difference of the purchase price and the down payment, or \$4200 − \$1600 = \$2600. Over 36 months she will repay a total of 36(89.50) = \$3222. The finance charge is thus

$$\$3222 - \$2600 = \$622.$$

Now divide the finance charge by the amount financed and multiply by 100.

$$\frac{622}{2600} \times 100 = 23.92$$

Find 36 at the left in the table above. Follow across to the right until the number nearest to 23.92 is found. Here we find exactly 23.92. Read up the column containing 23.92 to find that the true annual interest rate is 14.50%. ■

2. Find the true annual interest rate if the amount financed is \$3000 and the finance charge is \$675. Assume that the loan is paid off in 30 months.

Answer: $16\frac{1}{2}\%$

Work Problem 2 at the side.

4.5 EXERCISES

Find the monthly payment necessary to pay off the following home loans. (See Example 1.)

	Loan amount	Interest rate	Term of loan
1.	\$40,000	9%	30 years
2.	\$35,000	$8\frac{1}{2}\%$	25 years
3.	\$55,000	8%	25 years
4.	\$44,000	9%	30 years
5.	\$27,000	$8\frac{1}{2}\%$	30 years
6.	\$64,000	9%	25 years
7.	\$37,500	8%	25 years
8.	\$26,500	$8\frac{1}{2}\%$	30 years

In some areas, you must prepay taxes and insurance by adding 1/12 of the annual total to your monthly payment. Find the total monthly payment for each of the following loans.

	Amount of loan	Interest rate	Term of loan	Annual taxes	Annual insurance
9.	\$40,000	8%	30 years	\$1200	\$180
10.	\$50,000	$8\frac{1}{2}\%$	25 years	\$1440	\$240

11.	$29,000	8%	25 years	$540	$110
12.	$41,000	9%	30 years	$1000	$200
13.	$35,500	$8\frac{1}{2}\%$	25 years	$750	$105
14.	$38,500	$9\frac{1}{2}\%$	25 years	$590	$180

Find the true annual percentage rate for each of the following loans. (See Example 2.)

	Amount financed	Finance charge	Number of monthly payments
15.	$1194	$106.00	12
16.	$680	$187.00	36
17.	$200	$35.00	24
18.	$3780	$1292.00	48
19.	$115	$14.61	18
20.	$500	$87.50	24

APPLIED PROBLEMS

In each of the following problems, assume that equal payments are made monthly.

Management

21. At a recent appliance sale, Gil Eckern bought a portable color television for $222. He decided to charge the set, paying nothing down, with payments of $38.69 per month for 6 months. Find the true annual percentage rate of this loan.

22. On a 12-month loan of $1000, the A Sharp Guitar Store advertises only $92 interest. What is the annual percentage rate of this loan?

23. As a result of rising gasoline costs, Lottie Salzman decides to buy a new small car. The total purchase price is $6300, with terms of $800 down and the balance payable at $191 per month for 36 months. Find (a) the total amount financed and (b) the annual percentage rate.

24. Ernie Lopez borrows $4200 for remodeling his home. The total interest he pays is $1310 over a 4-year period. His payments are made monthly. What is the annual percentage rate he is charged?

25. Swanson's Television and Appliance wants to advertise a complete home stereo tape system for $300, nothing down and monthly payments of $19 a month for 18 months. They must also include the true annual percentage rate in the ad; what is it?

26. Donna Sharp bought a new IBM typewriter. The total price was $900, with $140 down. Sharp pays $37 per month for 24 months. Find the true annual interest rate.

CHAPTER 4 TEST

[4.1] *Find the final amounts when each of the following deposits are made.*

1. \$1000 at 5% compounded annually for 9 years
2. \$2800 at 6% compounded semiannually for 10 years
3. \$5000 at 8% compounded quarterly for 5 years

Find the amount of interest earned by the following deposits.

4. \$6000 at 6% compounded annually for 8 years
5. \$12,500 at 8% compounded quarterly for 3 years

Find the present value of the following sums.

6. \$2500 at 5% compounded annually for 14 years
7. \$1800 at 6% compounded semiannually for 12 years

8. In four years, Mr. Heeren must pay a \$5000 pledge to his church's building fund. What lump sum can he invest today, at 6% compounded semiannually, so that he will have enough to pay his pledge?

[4.2] *Find the value of each of the following annuities.*

9. Ordinary annuity of \$500 every 6 months, at 6% compounded semiannually for 8 years
10. Annuity due of \$900 every quarter, at 8% compounded quarterly for 5 years

11. Georgette Dahl deposits \$350 at the end of each quarter for 9 years. If the account pays 4% compounded quarterly, find the final amount in the account.
12. J. Euclid deposits \$1000 at the beginning of each six-month period in an account paying 6% compounded semiannually. How much will be in the account after 5 years?

[4.3] *Find the present value for each of the following ordinary annuities.*

13. Payments of \$850 are made annually for 4 years at 5% compounded annually.
14. Payments of \$1500 are made quarterly for 7 years at 6% compounded quarterly.

15. Vicky Manchester borrows \$20,000 from the bank. The money is to help expand her business. She agrees to repay the money in equal payments at the end of each year for 9 years. Interest is at 6% compounded annually. Find the amount of each payment.

[4.4] *Find the amount of the indicated payments necessary to amortize each of the following loans.*

16. \$80,000 loan, 5% compounded annually, 9 annual payments

17. \$3200 loan, 8% compounded quarterly, 10 quarterly payments

Find the amount of each payment to be made into a sinking fund so that enough money will be present to pay off the indicated loan.

18. \$6500 loan, money earns 4% compounded annually, 9 annual payments

19. \$37,000 loan, money earns 6% compounded semiannually, 17 semi-annual payments

[4.5] *Find each of the following monthly house payments. Use the table in Section 4.5.*

20. Amount owed is \$45,000, interest is 9%, term of loan is 25 years

21. Amount owed is \$51,000, interest is $8\frac{1}{2}$%, term of loan is 30 years.

22. Two roommates buy a house. The mortgage is for \$32,000 at $8\frac{1}{2}$% for 30 years. Annual taxes are \$1800 with an annual fire insurance premium of \$270. Find the total monthly payment for the house including taxes and insurance.

Find the true annual interest rate for each of the following.

23. Finance charge \$1040, amount financed \$4000, 36 monthly payments

24. Finance charge \$148, amount financed \$870, 24 monthly payments

CASE 5 PRESENT VALUE*

The Southern Pacific Railroad, with lines running from Oregon to Louisiana, is one of the country's most profitable railroads. The railroad has vast landholdings (granted by the government in the last half of the nineteenth century) and is diversified into trucking, pipelines, and data transmission.

The railroad was recently faced with a decision on the fate of an old bridge which crosses the Kings River in central California. The bridge is on a minor line which carries very little traffic. Just north of the Southern Pacific bridge is another bridge, owned by the Santa Fe Railroad; it too carries little traffic. The Southern Pacific had two alternatives: it could replace its own bridge or it could negotiate with the Santa Fe for the rights to run trains over its bridge. In the second alternative a yearly fee would be paid to the Santa Fe, and new connecting tracks would be built. The situation is shown in Figure 1.

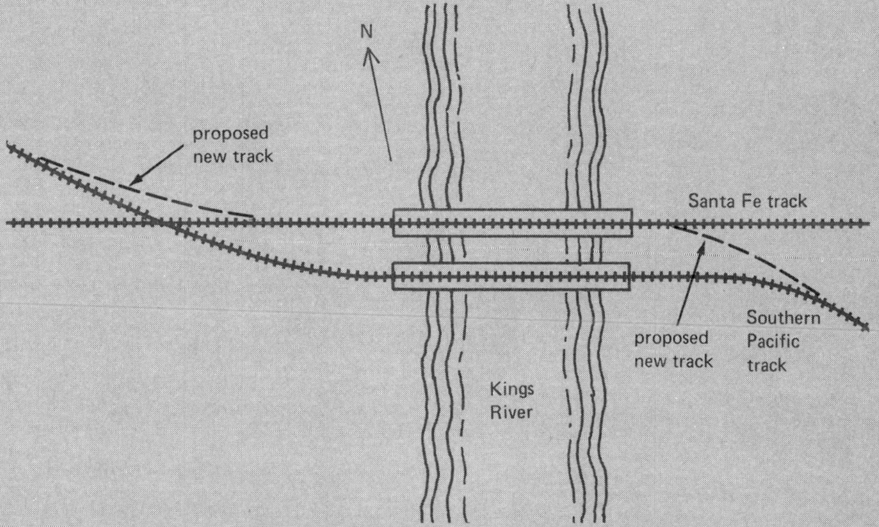

To find the better of these two alternatives, the railroad used the following approach.

1. Calculate estimated expenses for each alternative.
2. Calculate annual cash flows in after-tax dollars. At a 48% corporate tax rate, \$1 of expenses costs \$.52, and \$1 of revenue can bring a maximum of \$.52 in profit. Cash flow for a given year is found by the following formula.

$$\text{cash flow} = -.52\,(\text{operating and maintenance expenses}) + .52\,(\text{savings and revenue}) + .48\,(\text{depreciation})$$

* Based on information supplied by The Southern Pacific Transportation Company, San Francisco.

3. Calculate the net present values of all cash flows for future years. The formula used is

$$\text{net present value} = \sum(\text{cash flow in year } i)(1 + k)^{1-i}$$

where i is a particular year in the life of the project and k is the assumed annual rate of interest. (Recall: $\sum$ indicates a sum.) The net present value is found for interest rates from 0% to 20%.

4. The interest rate that leads to a net present value of $0 is called the **rate of return** on the project.

Let us now see how these steps worked out in practice.

Alternative 1: Operate over the Santa Fe bridge First, estimated expenses were calculated.

1976	Work done by Southern Pacific on Santa Fe track	$27,000
1976	Work by Southern Pacific on its own track	11,600
1976	Undepreciated portion of cost of current bridge	97,410
1976	Salvage value of bridge	12,690
1977	Annual maintenance of Santa Fe track	16,717
1977	Annual rental fee to Santa Fe	7,382

From these figures and others not given here, annual cash flows and net present values were calculated. The following table was then prepared.

Interest rate, %	Net present value
0	$85,731
4	67,566
8	53,332
12	42,008
16	32,875
20	25,414

Although the table does not show a net present value of $0, the interest rate that leads to that value is 44%. This is the rate of return for this alternative.

Alternative 2: Build a new bridge Again, estimated expenses were calculated.

1976	Annual maintenance	$2,870
1976	Annual inspection	120
1976	Repair trestle	17,920
1977	Install bridge	189,943
1977	Install walks and handrails	15,060
1978	Repaint steel (every 10 years)	10,000
1978	Repair trusses (every 10 years, increases by $200)	2,000
1981	Replace ties	31,000
1988	Repair concrete (every 10 years)	400
2021	Replace ties	31,000

After cash flows and net present values were calculated, the following table was prepared.

Interest rate, %	Net present value
0	$399,577
4	96,784
7.8	0
8	−3,151
12	−43,688
16	−62,615
20	−72,126

In this alternative the net present value is $0 at 7.8%, the rate of return.

Based on this analysis, the first alternative, renting from the Santa Fe, is clearly preferable.

EXERCISES

Find the cash flow in each of the following years. Use the formula given above.

1. Alternative 1, year 1977, operating expenses $6228, maintenance expenses $2976, savings $26,251, depreciation $10,778

2. Alternative 1, year 1984, same as Exercise 1, but depreciation is only $1347

3. Alternative 2, year 1976, maintenance $2870, operating expenses $6386, savings $26,251, no depreciation

4. Alternative 2, year 1980, operating expenses $6228, maintenance expenses $2976, savings $10,618, depreciation $6736

5 Matrix Theory

The study of matrices has been of interest to mathematicians for some time. Recently, however, the use of matrices has assumed greater importance in the fields of management, natural science, and social science because it provides a natural way to organize data. The theory of matrices is yet another type of mathematical model which we can use to solve many of the problems that arise in these fields, from inventory control to models of a nation's economy.

5.1 BASIC OPERATIONS ON MATRICES

A **matrix** (plural: matrices) is a rectangular array of numbers. Each number is called an **element** or entry. We indicate a matrix by enclosing an array of numbers in brackets, and we refer to the matrix by a capital letter. For example,

$$A = \begin{bmatrix} 2 & -4 & 1 \\ 7 & 8 & 6 \end{bmatrix}$$

is a matrix with two (horizontal) rows and three (vertical) columns.

We can use matrices to help us organize data, as we will show here:

The EZ Life Company manufactures sofas and armchairs in three models, A, B, and C. The company has regional warehouses in New York, Chicago, and San Francisco. In its August shipment, the company sends 10 model A sofas, 12 model B sofas, 5 model C sofas, 15 model A chairs, 20 model B chairs, and 8 model C chairs to each warehouse.

To organize this data, we might first list it as follows.

sofas	10 model A	12 model B	5 model C
chairs	15 model A	20 model B	8 model C

Alternatively, we might tabulate the data in a chart.

		Model		
		A	B	C
Furniture	Sofa	10	12	5
	Chair	15	20	8

With the understanding that the numbers in each row refer to the furniture type (sofa, chair) and the numbers in each column refer to the model (A, B, C), the same information can be given by a matrix, as follows.

$$M = \begin{bmatrix} 10 & 12 & 5 \\ 15 & 20 & 8 \end{bmatrix}$$

Matrices are classified by their **dimension** or **order,** that is, by the number of rows and columns that they contain. For example, the matrix

$$\begin{bmatrix} 2 & 7 & 5 \\ 4 & 6 & 9 \end{bmatrix}$$

has two rows and three columns. This matrix is said to be of **dimension** 2×3 (read "2 by 3") or **order** 2×3. In general, a matrix with m rows and n columns is of order $m \times n$. The number of rows is always given first.

Example 1

(a) The matrix $\begin{bmatrix} 6 & 5 \\ 3 & 4 \\ 5 & -1 \end{bmatrix}$ is of order 3×2.

(b) $\begin{bmatrix} 5 & 8 & 9 \\ 0 & 5 & -3 \\ -4 & 0 & 5 \end{bmatrix}$ is of order 3×3.

(c) $[1 \quad 6 \quad 5 \quad -2 \quad 5]$ is of order 1×5.

(d) $\begin{bmatrix} 3 \\ -5 \\ 0 \\ 2 \end{bmatrix}$ is of order 4×1. ■

Work Problem 1 at the side.

A matrix having the same number of rows as columns is called a **square matrix.** The matrix given in Example 1(b) above is a

1. Give the order of each of the following matrices.

(a) $\begin{bmatrix} 2 & 1 & -5 & 6 \\ 3 & 0 & 7 & -4 \end{bmatrix}$

(b) $\begin{bmatrix} 1 & 2 & 3 \\ 4 & 5 & 6 \\ 9 & 8 & 7 \end{bmatrix}$

Answer:
(a) 2×4
(b) 3×3

square matrix, as are

$$\begin{bmatrix} -5 & 6 \\ 8 & 3 \end{bmatrix} \quad \text{and} \quad \begin{bmatrix} 0 & 0 & 0 & 0 \\ -2 & 4 & 1 & 3 \\ 0 & 0 & 0 & 0 \\ -5 & -4 & 1 & 8 \end{bmatrix}.$$

If a matrix contains only one row, it is called a **row matrix** or row vector. The matrix in Example 1(c) is a row matrix, as are

$$[5 \quad 8] \qquad [6 \quad -9 \quad 2] \qquad \text{and} \qquad [-4 \quad 0 \quad 0 \quad 0].$$

A matrix of only one column, such as in Example 1(d), is a **column matrix** or column vector.

2. Use the numbers 2, 5, −8, 4 to write
(a) a row matrix
(b) a column matrix
(c) a square matrix.

Answer:
(a) $[2 \quad 5 \quad -8 \quad 4]$
(b) $\begin{bmatrix} 2 \\ 5 \\ -8 \\ 4 \end{bmatrix}$
(c) $\begin{bmatrix} 2 & 5 \\ -8 & 4 \end{bmatrix}$ or $\begin{bmatrix} 2 & -8 \\ 5 & 4 \end{bmatrix}$

(Other answers are possible.)

Work Problem 2 at the side.

Two matrices are **equal** if they are of the same order and if each corresponding element, position by position, is equal. Using this definition, the matrices

$$\begin{bmatrix} 2 & 1 \\ 3 & -5 \end{bmatrix} \quad \text{and} \quad \begin{bmatrix} 1 & 2 \\ -5 & 3 \end{bmatrix}$$

are not equal (even though they contain the same elements and are of the same order) since the corresponding elements differ.

Example 2 **(a)** From the definition of equality given above, the only way that the statement

$$\begin{bmatrix} 2 & 1 \\ p & q \end{bmatrix} = \begin{bmatrix} x & y \\ -1 & 0 \end{bmatrix}$$

can be true is if $2 = x$, $1 = y$, $p = -1$, and $q = 0$.

(b) The statement

$$\begin{bmatrix} x \\ y \end{bmatrix} = \begin{bmatrix} 1 \\ 4 \\ 0 \end{bmatrix}$$

can never be true, since the two matrices are of different order. (One is 2×1 and the other is 3×1.) ■

3. Give the values of the variables that make each of the following statements true.

(a) $\begin{bmatrix} x & 2 \\ 5 & y \end{bmatrix} = \begin{bmatrix} 6 & p \\ q & -1 \end{bmatrix}$

(b) $[1 \quad 2 \quad x] = \begin{bmatrix} y \\ 2 \\ 8 \end{bmatrix}$

Answer:
(a) $x = 6, y = -1, p = 2, q = 5$
(b) Can never be true

Work Problem 3 at the side.

The matrix given earlier in this section,

$$M = \begin{bmatrix} 10 & 12 & 5 \\ 15 & 20 & 8 \end{bmatrix}$$

shows the August shipment from the EZ Life plant to its New York warehouse. If matrix N below gives the September shipment to the same warehouse, what is the total shipment for each item of furniture for these two months?

$$N = \begin{bmatrix} 45 & 35 & 20 \\ 65 & 40 & 35 \end{bmatrix}$$

If 10 model A sofas were shipped in August and 45 in September, altogether 55 model A sofas were shipped in the two months. Similarly, the other corresponding entries may be added, to get a new matrix, call it Q, which represents the total shipment for the two months.

$$Q = \begin{bmatrix} 55 & 47 & 25 \\ 80 & 60 & 43 \end{bmatrix}$$

It is convenient to refer to Q as the "sum" of M and N.

The way we added the two matrices above illustrates the following definition of addition of matrices. The **sum** of two $m \times n$ matrices X and Y is the $m \times n$ matrix $X + Y$ in which each element is the sum of the corresponding elements of X and Y. It is important to remember that

only matrices with the same dimensions can be added.

Example 3 Find each of the following sums.

(a)

$$\begin{bmatrix} 5 & -6 \\ 8 & 9 \end{bmatrix} + \begin{bmatrix} -4 & 6 \\ 8 & -3 \end{bmatrix} = \begin{bmatrix} 5 + (-4) & -6 + 6 \\ 8 + 8 & 9 + (-3) \end{bmatrix}$$

$$= \begin{bmatrix} 1 & 0 \\ 16 & 6 \end{bmatrix}$$

(b) The matrices

$$A = \begin{bmatrix} 5 & 8 \\ 6 & 2 \end{bmatrix} \quad \text{and} \quad B = \begin{bmatrix} 3 & 9 & 1 \\ 4 & 2 & 5 \end{bmatrix}$$

are of different orders. Therefore, the sum $A + B$ does not exist. ■

Work Problem 4 at the side.

4. Find each of the following sums.

(a) $\begin{bmatrix} 2 & 5 & 7 \\ 3 & -1 & 4 \end{bmatrix} + \begin{bmatrix} -1 & 2 & 0 \\ 10 & -4 & 5 \end{bmatrix}$

(b) $\begin{bmatrix} 1 \\ 2 \\ 3 \end{bmatrix} + \begin{bmatrix} 2 & -1 \\ 4 & 5 \\ 6 & 0 \end{bmatrix}$

Answer:

(a) $\begin{bmatrix} 1 & 7 & 7 \\ 13 & -5 & 9 \end{bmatrix}$

(b) Not possible

Example 4 The September shipments from the EZ Life Company to the San Francisco and Chicago warehouses are given in matrices S and C below.

$$S = \begin{bmatrix} 30 & 32 & 28 \\ 43 & 47 & 30 \end{bmatrix}, \qquad C = \begin{bmatrix} 22 & 25 & 38 \\ 31 & 34 & 35 \end{bmatrix}$$

What was the total amount shipped to the three warehouses in September? (See matrix N above for New York.)

The total of the September shipments is represented by the sum of the three matrices N, S, and C.

$$\begin{aligned} N + S + C &= \begin{bmatrix} 45 & 35 & 20 \\ 65 & 40 & 35 \end{bmatrix} + \begin{bmatrix} 30 & 32 & 28 \\ 43 & 47 & 30 \end{bmatrix} + \begin{bmatrix} 22 & 25 & 38 \\ 31 & 34 & 35 \end{bmatrix} \\ &= \begin{bmatrix} 97 & 92 & 86 \\ 139 & 121 & 100 \end{bmatrix} \end{aligned}$$

From the resulting matrix above, we see, for example, that the total number of model C sofas shipped to the three warehouses in September was 86. ■

The **additive inverse** (or **negative**) of a matrix X is the matrix $-X$ in which each element is the additive inverse of the corresponding element of X.

If

$$A = \begin{bmatrix} 1 & 2 & 3 \\ 0 & -1 & 5 \end{bmatrix} \quad \text{and} \quad B = \begin{bmatrix} -2 & 3 & 0 \\ 1 & -7 & 2 \end{bmatrix},$$

then by the definition of the additive inverse of a matrix, we have

$$-A = \begin{bmatrix} -1 & -2 & -3 \\ 0 & 1 & -5 \end{bmatrix} \quad \text{and} \quad -B = \begin{bmatrix} 2 & -3 & 0 \\ -1 & 7 & -2 \end{bmatrix}.$$

By the definition of matrix addition, for each matrix X, the sum $X + (-X)$ is a **zero matrix,** O, whose elements are all zeros. There is an $m \times n$ zero matrix for each pair of values of m and n. Zero matrices have the following **identity property:** If O is an $m \times n$ zero matrix, and A is any $m \times n$ matrix, then

$$A + O = O + A = A.$$

We can now define **subtraction** of matrices in a manner comparable to subtraction for real numbers. That is, for matrices X and Y, we define

$$X - Y = X + (-Y).$$

Using A, B, and $-B$ as defined above, we have

$$A - B = A + (-B) = \begin{bmatrix} 1 & 2 & 3 \\ 0 & -1 & 5 \end{bmatrix} + \begin{bmatrix} 2 & -3 & 0 \\ -1 & 7 & -2 \end{bmatrix}$$
$$= \begin{bmatrix} 3 & -1 & 3 \\ -1 & 6 & 3 \end{bmatrix}.$$

This definition means that we can perform matrix subtraction by simply subtracting corresponding elements.

Example 5 **(a)** $[8 \quad 6 \quad -4] - [3 \quad 5 \quad -8] = [5 \quad 1 \quad 4]$
(b) The matrices

$$\begin{bmatrix} -2 & 5 \\ 0 & 1 \end{bmatrix} \quad \text{and} \quad \begin{bmatrix} 3 \\ 5 \end{bmatrix}$$

have different orders and cannot be subtracted. ■

5. Find each of the following differences.

(a) $\begin{bmatrix} 2 & 5 \\ -1 & 0 \end{bmatrix} - \begin{bmatrix} 6 & 4 \\ 3 & -2 \end{bmatrix}$

(b) $\begin{bmatrix} 1 & 5 & 6 \\ 2 & 4 & 8 \end{bmatrix} - \begin{bmatrix} 2 & 1 \\ 10 & 3 \end{bmatrix}$

Answer:

(a) $\begin{bmatrix} -4 & 1 \\ -4 & 2 \end{bmatrix}$

(b) Not possible

Work Problem 5 at the side.

Example 6 During September the Chicago warehouse of the EZ Life Company shipped out the following number of each model.

$$K = \begin{bmatrix} 5 & 10 & 8 \\ 11 & 14 & 15 \end{bmatrix}$$

What was the Chicago warehouse inventory on October 1, taking into account only the number of items received and sent out during the month?

The number of each kind of item received during September is given by matrix C from Example 4; the number of each model sent out during September is given by matrix K. Thus the October 1 inventory will be represented by the matrix $C - K$ as shown below.

$$\begin{bmatrix} 22 & 25 & 38 \\ 31 & 34 & 35 \end{bmatrix} - \begin{bmatrix} 5 & 10 & 8 \\ 11 & 14 & 15 \end{bmatrix} = \begin{bmatrix} 17 & 15 & 30 \\ 20 & 20 & 20 \end{bmatrix}$$ ■

5.1 EXERCISES

Mark each of the following statements as true *or* false. *If false, tell why.*

1. $\begin{bmatrix} 1 & 3 \\ 5 & 7 \end{bmatrix} = \begin{bmatrix} 1 & 5 \\ 3 & 7 \end{bmatrix}$

2. $\begin{bmatrix}1\\2\\3\end{bmatrix} = [1 \quad 2 \quad 3]$

3. $\begin{bmatrix}x\\y\end{bmatrix} = \begin{bmatrix}3\\5\end{bmatrix}$ if $x = 3$ and $y = 5$.

4. $\begin{bmatrix}3 & 5 & 2 & 8\\1 & -1 & 4 & 0\end{bmatrix}$ is a 4×2 matrix.

5. $\begin{bmatrix}1 & 9 & -4\\3 & 7 & 2\\-1 & 1 & 0\end{bmatrix}$ is a square matrix.

6. $\begin{bmatrix}2 & 4 & -1\\3 & 7 & 5\\0 & 0 & 0\end{bmatrix} = \begin{bmatrix}2 & 4 & -1\\3 & 7 & 5\end{bmatrix}$

Find the order of each of the following. Identify any square, column, or row matrices. (See Example 1.)

7. $\begin{bmatrix}-4 & 8\\2 & 3\end{bmatrix}$

8. $\begin{bmatrix}-9 & 6 & 2\\4 & 1 & 8\end{bmatrix}$

9. $\begin{bmatrix}-6 & 8 & 0 & 0\\4 & 1 & 9 & 2\\3 & -5 & 7 & 1\end{bmatrix}$

10. $[8 \quad -2 \quad 4 \quad 6 \quad 3]$

11. $\begin{bmatrix}2\\4\end{bmatrix}$

12. $[-9]$

Find the values of the variables in each of the following. (See Example 2.)

13. $\begin{bmatrix}2 & 1\\4 & 8\end{bmatrix} = \begin{bmatrix}x & 1\\y & z\end{bmatrix}$

14. $\begin{bmatrix}-5\\y\end{bmatrix} = \begin{bmatrix}-5\\8\end{bmatrix}$

15. $\begin{bmatrix}x+6 & y+2\\8 & 3\end{bmatrix} = \begin{bmatrix}-9 & 7\\8 & k\end{bmatrix}$

16. $\begin{bmatrix}9 & 7\\r & 0\end{bmatrix} = \begin{bmatrix}m-3 & n+5\\8 & 0\end{bmatrix}$

17. $\begin{bmatrix}-7+z & 4r & 8s\\6p & 2 & 5\end{bmatrix} + \begin{bmatrix}-9 & 8r & 3\\2 & 5 & 4\end{bmatrix} = \begin{bmatrix}2 & 36 & 27\\20 & 7 & 12a\end{bmatrix}$

18. $\begin{bmatrix}a+2 & 3z+1 & 5m\\4k & 0 & 3\end{bmatrix} + \begin{bmatrix}3a & 2z & 5m\\2k & 5 & 6\end{bmatrix} = \begin{bmatrix}10 & 14 & 80\\10 & 5 & 9\end{bmatrix}$

Perform the indicated operations where possible.

19. $\begin{bmatrix} 1 & 2 & 5 & -1 \\ 3 & 0 & 2 & -4 \end{bmatrix} + \begin{bmatrix} 8 & 10 & -5 & 3 \\ -2 & -1 & 0 & 0 \end{bmatrix}$

20. $\begin{bmatrix} 1 & 5 \\ 2 & -3 \\ 3 & 7 \end{bmatrix} + \begin{bmatrix} 2 & 3 \\ 8 & 5 \\ -1 & 9 \end{bmatrix}$

21. $\begin{bmatrix} 1 & 5 & 7 \\ 2 & 2 & 3 \end{bmatrix} + \begin{bmatrix} 4 & 8 & -7 \\ 1 & -1 & 5 \end{bmatrix}$

22. $\begin{bmatrix} 2 & 4 \\ -8 & 1 \end{bmatrix} + \begin{bmatrix} 9 & -3 \\ 8 & 5 \end{bmatrix}$

23. $\begin{bmatrix} 1 & 3 & -2 \\ 4 & 7 & 1 \end{bmatrix} + \begin{bmatrix} 3 & 0 \\ 6 & 4 \\ -5 & 2 \end{bmatrix}$

24. $\begin{bmatrix} 1 & 3 & -2 \\ 4 & 7 & 1 \end{bmatrix} - \begin{bmatrix} 3 & 6 & -5 \\ 0 & 4 & 2 \end{bmatrix}$

25. $\begin{bmatrix} 2 & 8 & 12 & 0 \\ 7 & 4 & -1 & 5 \\ 1 & 2 & 0 & 10 \end{bmatrix} - \begin{bmatrix} 1 & 3 & 6 & 9 \\ 2 & -3 & -3 & 4 \\ 8 & 0 & -2 & 17 \end{bmatrix}$

26. $\begin{bmatrix} 2 & 1 \\ 5 & -3 \\ -7 & 2 \\ 9 & 0 \end{bmatrix} + \begin{bmatrix} 1 & -8 & 0 \\ 5 & 3 & 2 \\ -6 & 7 & -5 \\ 2 & -1 & 0 \end{bmatrix}$

27. $\begin{bmatrix} -4x+2y & -3x+y \\ 6x-3y & 2x-5y \end{bmatrix} + \begin{bmatrix} -8x+6y & 2x \\ 3y-5x & 6x+4y \end{bmatrix}$

28. $\begin{bmatrix} 4k-8y \\ 6z-3x \\ 2k+5a \\ -4m+2n \end{bmatrix} - \begin{bmatrix} 5k+6y \\ 2z+5x \\ 4k+6a \\ 4m-2n \end{bmatrix}$

Using matrices $O = \begin{bmatrix} 0 & 0 \\ 0 & 0 \end{bmatrix}$, $P = \begin{bmatrix} m & n \\ p & q \end{bmatrix}$, $T = \begin{bmatrix} r & s \\ t & u \end{bmatrix}$, *and* $X = \begin{bmatrix} x & y \\ z & w \end{bmatrix}$, *verify that the following statements are true.*

29. $X + T$ is a 2×2 matrix (Closure property)

30. $X + T = T + X$ (Commutative property of addition of matrices)

31. $X + (T + P) = (X + T) + P$ (Associative property of addition of matrices)

32. $X + (-X) = 0$ (Inverse property of addition of matrices)

33. $P + 0 = P$ (Identity property of addition of matrices)

34. Which of the above properties are valid for matrices that are not square?

APPLIED PROBLEMS

General

35. When John inventoried his screw collection, he found that he had 7 flathead long screws, 9 flathead medium, 8 flathead short, 2 roundhead long, no roundhead medium, and 6 roundhead short. Write this information first as a 3×2 matrix and then as a 2×3 matrix.

36. At the grocery store, Miguel bought 4 quarts of milk, 2 loaves of bread, 4 chickens, and an apple. Mary bought 2 quarts of milk, a loaf of bread, 5 chickens, and 4 apples. Write this information first as a 2×4 matrix and then as a 4×2 matrix.

Natural Science

37. A dietician prepares a diet specifying the amounts a patient should eat of four basic food groups: group I, meats; group II, fruits and vegetables; group III, breads and starches; group IV, milk products. Amounts are given in "exchanges" which represent 1 ounce (meat), 1/2 cup (fruits and vegetables), 1 slice (bread), 8 ounces (milk), or other suitable measurements.
   (a) The number of "exchanges" for breakfast for each of the four food groups respectively are 2, 1, 2, and 1; for lunch, 3, 2, 2, and 1; and for dinner, 4, 3, 2, and 1. Write a 3×4 matrix using this information.
   (b) The amounts of fat, carbohydrates, and protein in each food group respectively are as follows.
      Fat: 5, 0, 0, 10
      Carbohydrates: 0, 10, 15, 12
      Protein: 7, 1, 2, 8
   Use this information to write a 4×3 matrix.
   (c) There are 8 calories per exchange of fat, 4 calories per exchange of carbohydrates, and 5 calories per exchange of protein; summarize this data in a 3×1 matrix.

38. At the beginning of a laboratory experiment, five baby rats measured 5.6, 6.4, 6.9, 7.6, and 6.1 centimeters in length, and weighed 144, 138, 149, 152, and 146 grams respectively.
   (a) Write a 2×5 matrix using this information.
   (b) At the end of two weeks, their lengths were 10.2, 11.4, 11.4, 12.7, and 10.8 centimeters, and they weighed 196, 196, 225, 250, and 230 grams. Write a 2×5 matrix with this information.
   (c) Use matrix subtraction and the matrices found in (a) and (b) to write a matrix which gives the amount of change in length and weight for each rat.
   (d) The following week the rats gained as shown in the matrix below.

$$\begin{matrix} \text{Length} \\ \text{Weight} \end{matrix} \begin{bmatrix} 1.8 & 1.5 & 2.3 & 1.8 & 2.0 \\ 25 & 22 & 29 & 33 & 20 \end{bmatrix}$$

   What were their lengths and weights at the end of this week?

5.2 MULTIPLICATION OF MATRICES

In work with matrices, a real number is called a **scalar.** The product of a scalar k and a matrix X is the matrix kX, each of whose elements is a number equal to k times the corresponding element of X. Thus

$$-3\begin{bmatrix} 2 & -5 \\ 1 & 7 \end{bmatrix} = \begin{bmatrix} -6 & 15 \\ -3 & -21 \end{bmatrix}.$$

Finding the product of two matrices is more involved. However, this kind of multiplication is important in applications which are useful in solving practical problems. To understand the reasoning behind matrix multiplication, it may be helpful to consider another example concerning the EZ Life Company discussed in Section 5.1. Sofas and chairs of the same model are often sold as sets. Matrix W below shows the number of each model set in each warehouse.

$$\begin{array}{l} \\ \text{New York} \\ \text{Chicago} \\ \text{San Francisco} \end{array} \begin{array}{c} \begin{array}{ccc} A & B & C \end{array} \\ \begin{bmatrix} 10 & 7 & 3 \\ 5 & 9 & 6 \\ 4 & 8 & 2 \end{bmatrix} \end{array} = W$$

Suppose the selling price of model A sets is \$400, of model B sets is \$500, and of model C sets is \$600. To find the total value of the sets in the New York warehouse, we multiply as follows.

Type	Number of sets	Price of set	Total
A	10	\$400	\$4000
B	7	\$500	\$3500
C	3	\$600	\$1800
			\$9300

(Total for New York)

Thus, the total value of the three kinds of sets in New York is \$9300. In the same way, the Chicago sets have a total value of

$$5(\$400) + 9(\$500) + 6(\$600) = \$10{,}100,$$

and in San Francisco, the total value of the sets is

$$4(\$400) + 8(\$500) + 2(\$600) = \$6800.$$

We can write the selling prices as a column matrix, P.

$$\begin{bmatrix} 400 \\ 500 \\ 600 \end{bmatrix} = P$$

Now we can indicate the product of matrices

$$W = \begin{bmatrix} 10 & 7 & 3 \\ 5 & 9 & 6 \\ 4 & 8 & 2 \end{bmatrix} \quad \text{and} \quad P = \begin{bmatrix} 400 \\ 500 \\ 600 \end{bmatrix}$$

as

$$WP = \begin{bmatrix} 10 & 7 & 3 \\ 5 & 9 & 6 \\ 4 & 8 & 2 \end{bmatrix} \begin{bmatrix} 400 \\ 500 \\ 600 \end{bmatrix} = \begin{bmatrix} 9300 \\ 10{,}100 \\ 6800 \end{bmatrix}.$$

Recall that we found the product by multiplying the elements of the *rows* of the matrix on the left and the corresponding elements of the *column* of the matrix on the right, and then finding the sum of these elements. Notice that the product of a 3×3 matrix and a 3×1 matrix is a 3×1 matrix.

Example 1 Find the product AB given

$$A = \begin{bmatrix} 2 & 3 & -1 \\ 4 & 2 & 2 \end{bmatrix} \quad \text{and} \quad B = \begin{bmatrix} 1 \\ 8 \\ 6 \end{bmatrix}.$$

Step 1 Multiply the elements of the first row of A and the corresponding elements of the column of B.

$$\begin{bmatrix} \mathbf{2} & \mathbf{3} & \mathbf{-1} \\ 4 & 2 & 2 \end{bmatrix} \begin{bmatrix} \mathbf{1} \\ \mathbf{8} \\ \mathbf{6} \end{bmatrix} \qquad \mathbf{2}(\mathbf{1}) + \mathbf{3}(\mathbf{8}) + (\mathbf{-1})(\mathbf{6}) = 20$$

Therefore, 20 is the first row entry of the product matrix AB.

Step 2 Multiply the elements of the second row of A with the corresponding elements of B.

$$\begin{bmatrix} 2 & 3 & -1 \\ \mathbf{4} & \mathbf{2} & \mathbf{2} \end{bmatrix} \begin{bmatrix} \mathbf{1} \\ \mathbf{8} \\ \mathbf{6} \end{bmatrix} \qquad \mathbf{4}(\mathbf{1}) + \mathbf{2}(\mathbf{8}) + \mathbf{2}(\mathbf{6}) = 32$$

The second row entry of the product is 32.

Step 3 Write the product using the two entries we found above.

$$AB = \begin{bmatrix} 2 & 3 & -1 \\ 4 & 2 & 2 \end{bmatrix} \begin{bmatrix} 1 \\ 8 \\ 6 \end{bmatrix} = \begin{bmatrix} 20 \\ 32 \end{bmatrix} \quad \blacksquare$$

Work Problem 1 at the side.

1. Find the product AB given

$$A = \begin{bmatrix} 2 & 4 \\ 5 & 6 \end{bmatrix} \quad \text{and} \quad B = \begin{bmatrix} -3 \\ 4 \end{bmatrix}.$$

Answer: $AB = \begin{bmatrix} 10 \\ 9 \end{bmatrix}$

Example 2 Find the product CD given

$$C = \begin{bmatrix} -3 & 4 & 2 \\ 5 & 0 & 4 \end{bmatrix} \quad \text{and} \quad D = \begin{bmatrix} -6 & 4 \\ 2 & 3 \\ 3 & -2 \end{bmatrix}.$$

Step 1

$$\begin{bmatrix} \mathbf{-3} & \mathbf{4} & \mathbf{2} \\ 5 & 0 & 4 \end{bmatrix}\begin{bmatrix} \mathbf{-6} & 4 \\ \mathbf{2} & 3 \\ \mathbf{3} & -2 \end{bmatrix} \qquad (-3)(-6) + 4(2) + 2(3) = 32$$

Step 2

$$\begin{bmatrix} \mathbf{-3} & \mathbf{4} & \mathbf{2} \\ 5 & 0 & 4 \end{bmatrix}\begin{bmatrix} -6 & \mathbf{4} \\ 2 & \mathbf{3} \\ 3 & \mathbf{-2} \end{bmatrix} \qquad (-3)(4) + 4(3) + 2(-2) = -4$$

Step 3

$$\begin{bmatrix} -3 & 4 & 2 \\ \mathbf{5} & \mathbf{0} & \mathbf{4} \end{bmatrix}\begin{bmatrix} \mathbf{-6} & 4 \\ \mathbf{2} & 3 \\ \mathbf{3} & -2 \end{bmatrix} \qquad 5(-6) + 0(2) + 4(3) = -18$$

Step 4

$$\begin{bmatrix} -3 & 4 & 2 \\ \mathbf{5} & \mathbf{0} & \mathbf{4} \end{bmatrix}\begin{bmatrix} -6 & \mathbf{4} \\ 2 & \mathbf{3} \\ 3 & \mathbf{-2} \end{bmatrix} \qquad 5(4) + 0(3) + 4(-2) = 12$$

Step 5 The product is

$$CD = \begin{bmatrix} -3 & 4 & 2 \\ 5 & 0 & 4 \end{bmatrix}\begin{bmatrix} -6 & 4 \\ 2 & 3 \\ 3 & -2 \end{bmatrix} = \begin{bmatrix} 32 & -4 \\ -18 & 12 \end{bmatrix}.$$

Here the product of a 2×3 matrix and a 3×2 matrix is a 2×2 matrix. ■

2. Find the product CD given

$$C = \begin{bmatrix} 1 & 3 & 5 \\ 2 & -4 & -1 \end{bmatrix} \quad \text{and}$$

$$D = \begin{bmatrix} 2 & -1 \\ 4 & 3 \\ 1 & -2 \end{bmatrix}.$$

Answer: $CD = \begin{bmatrix} 19 & -2 \\ -13 & -12 \end{bmatrix}$

Work Problem 2 at the side.

In general, the **product** AB of an $m \times n$ matrix A and an $n \times k$ matrix B is found as follows. Multiply each element of the first row of A by the corresponding element of the first column of B. The sum of these n products is the first row, first column element of AB.

Similarly, the sum of the products found by multiplying the

elements of the first row of A times the corresponding elements of the second column of B gives the first row, second column element of AB, and so on.

In general, to find the ith row, jth column element of AB, multiply each element in the ith row of A by the corresponding element in the jth column of B. The sum of these products will give the elements of row i, column j of AB.

As this definition shows,

> two matrices A and B can be multiplied only if the number of columns of A is the same as the number of rows of B.

The final product will have as many rows as A and as many columns as B.

Example 3 Suppose matrix A is 2×2 and matrix B is 2×4. Can the product AB be calculated? What is the order of the product? The following diagram helps decide the answers to these questions.

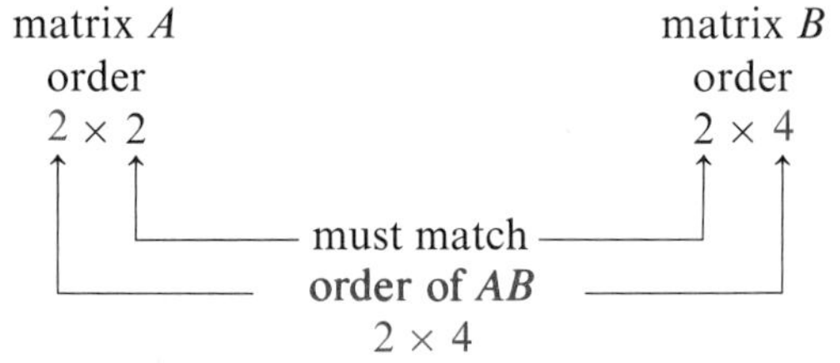

■

Work Problem 3 at the side.

3. Give the order of each of the following products that can be found.

(a) $\begin{bmatrix} 2 & 4 \\ 6 & 8 \end{bmatrix}\begin{bmatrix} 1 & 2 & 3 \\ 0 & -1 & 2 \end{bmatrix}$

(b) $\begin{bmatrix} 1 & 2 \\ 5 & 10 \\ 12 & 7 \end{bmatrix}\begin{bmatrix} 2 & 4 \\ 3 & 6 \\ 9 & 1 \end{bmatrix}$

(c) $\begin{bmatrix} 5 \\ 2 \\ 4 \end{bmatrix}[1 \quad 0 \quad 6]$

Answer:
(a) 2×3
(b) Not possible
(c) 3×3

Example 4 Let

$$A = \begin{bmatrix} 1 & -3 \\ 7 & 2 \end{bmatrix} \quad \text{and} \quad B = \begin{bmatrix} 1 & 0 & -1 \\ 3 & 1 & 4 \end{bmatrix}.$$

(a) Find AB.
Use the definition of matrix multiplication.

$$AB = \begin{bmatrix} 1 & -3 \\ 7 & 2 \end{bmatrix}\begin{bmatrix} 1 & 0 & -1 \\ 3 & 1 & 4 \end{bmatrix}$$

$$= \begin{bmatrix} 1(1) + (-3)(3) & 1(0) + (-3)1 & 1(-1) + (-3)4 \\ 7(1) + 2(3) & 7(0) + 2(1) & 7(-1) + 2(4) \end{bmatrix}$$

$$= \begin{bmatrix} -8 & -3 & -13 \\ 13 & 2 & 1 \end{bmatrix}$$

(b) Find BA.
Since B is a 2×3 matrix and A is a 2×2 matrix, we cannot find the product BA. ■

Example 5 A contractor builds three kinds of houses, models A, B, and C, with a choice of two styles, Spanish or contemporary. Matrix P shows the number of each kind of house he is planning to build for a new 100-home subdivision. The amounts for each of the exterior materials he uses depend primarily on the style of the house. These amounts are shown in matrix Q. (Concrete is in cubic yards, lumber in units of 1000 board feet, brick in 1000's, and shingles in units of 100 square feet.) Matrix R gives the cost for each kind of material.

$$\begin{array}{l} \\ \text{Model } A \\ \text{Model } B \\ \text{Model } C \end{array} \begin{array}{c} \begin{array}{cc} \text{Spanish} & \text{Contemporary} \end{array} \\ \begin{bmatrix} 0 & 30 \\ 10 & 20 \\ 20 & 20 \end{bmatrix} \end{array} = P$$

$$\begin{array}{l} \\ \text{Spanish} \\ \text{Contemporary} \end{array} \begin{array}{c} \begin{array}{cccc} \text{Concrete} & \text{Lumber} & \text{Brick} & \text{Shingles} \end{array} \\ \begin{bmatrix} 10 & 2 & 0 & 2 \\ 50 & 1 & 20 & 2 \end{bmatrix} \end{array} = Q$$

$$\begin{array}{l} \\ \text{Concrete} \\ \text{Lumber} \\ \text{Brick} \\ \text{Shingles} \end{array} \begin{array}{c} \text{Cost per unit} \\ \begin{bmatrix} 20 \\ 180 \\ 60 \\ 25 \end{bmatrix} \end{array} = R$$

(a) What is the total cost for each model house?

To find the cost for each model, we must first find PQ, which will show the amount of each material needed for each model house.

$$PQ = \begin{bmatrix} 0 & 30 \\ 10 & 20 \\ 20 & 20 \end{bmatrix} \begin{bmatrix} 10 & 2 & 0 & 2 \\ 50 & 1 & 20 & 2 \end{bmatrix}$$

$$= \begin{array}{c} \begin{array}{cccc} \text{Concrete} & \text{Lumber} & \text{Brick} & \text{Shingles} \end{array} \\ \begin{bmatrix} 1500 & 30 & 600 & 60 \\ 1100 & 40 & 400 & 60 \\ 1200 & 60 & 400 & 80 \end{bmatrix} \end{array} \begin{array}{l} \\ \text{Model } A \\ \text{Model } B \\ \text{Model } C \end{array}$$

If we now multiply PQ times R, the cost matrix, we will get the total cost for each model house.

$$\begin{bmatrix} 1500 & 30 & 600 & 60 \\ 1100 & 40 & 400 & 60 \\ 1200 & 60 & 400 & 80 \end{bmatrix} \begin{bmatrix} 20 \\ 180 \\ 60 \\ 25 \end{bmatrix} = \begin{array}{c} \text{Cost} \\ \begin{bmatrix} 72{,}900 \\ 54{,}700 \\ 60{,}800 \end{bmatrix} \end{array} \begin{array}{l} \\ \text{Model } A \\ \text{Model } B \\ \text{Model } C \end{array}$$

(b) How much of each of the four kinds of material must he order?

The totals of the columns of matrix PQ will give a matrix whose elements represent the total amounts of each material needed for the subdivision. Let us call this matrix T, and write it as a row matrix.

$$T = [3800 \quad 130 \quad 1400 \quad 200]$$

(c) What is the total cost for material?

To find the total cost of all the materials, we need the product of matrix T, the matrix showing the total amounts of each material, and matrix R, the cost matrix. (To multiply these and get a 1×1 matrix, representing total cost, we must multiply a 1×4 matrix by a 4×1 matrix. This is why we wrote T as a row matrix in (b) above.)

$$TR = [3800 \quad 130 \quad 1400 \quad 200] \begin{bmatrix} 20 \\ 180 \\ 60 \\ 25 \end{bmatrix} = [188{,}400].$$

(d) Suppose the contractor builds the same number of homes in five subdivisions. Calculate the total amount of each material for each model for all five subdivisions.

Multiply PQ by the scalar 5, as follows.

$$5\begin{bmatrix} 1500 & 30 & 600 & 60 \\ 1100 & 40 & 400 & 60 \\ 1200 & 60 & 400 & 80 \end{bmatrix} = \begin{bmatrix} 7500 & 150 & 3000 & 300 \\ 5500 & 200 & 2000 & 300 \\ 6000 & 300 & 2000 & 400 \end{bmatrix} \quad \blacksquare$$

We can introduce a notation to help us keep track of what quantities a matrix represents. For example, we can say that matrix P, from Example 5, represents models/styles, matrix Q represents styles/materials, and matrix R represents materials/cost. In each case we write the meaning of the rows first and the columns second. When we found the product PQ in Example 5, the rows of the matrix represented models and the columns represented materials. Therefore, we can say the matrix product PQ represents models/materials. Note that the common quantity, styles, in both P and Q was eliminated in the product PQ. Do you see that the product $(PQ)R$ represents models/cost?

In practical problems this notation helps us decide in which order to multiply matrices so that the results are meaningful. In Example 5(c) we could have found either product RT or product TR. However, since T represents subdivisions/materials and R represents materials/cost, we multiplied T times R to get subdivisions/cost.

4. Let

$$A = \begin{array}{c} \textit{Vitamin} \\ \begin{array}{ccc} \text{C} & \text{E} & \text{K} \end{array} \\ \begin{bmatrix} 2 & 7 & 5 \\ 4 & 6 & 9 \end{bmatrix} \end{array} \begin{array}{c} \text{X} \\ \text{Y} \end{array} \ \textit{Brand}$$

and

$$B = \begin{array}{c} \textit{Cost} \\ \begin{array}{cc} \text{X} & \text{Y} \end{array} \\ \begin{bmatrix} 12 & 14 \\ 18 & 15 \\ 9 & 10 \end{bmatrix} \end{array} \begin{array}{c} \text{C} \\ \text{E} \\ \text{K} \end{array} \ \textit{Vitamin.}$$

(a) What quantities do matrices A and B represent?
(b) What quantities does the product AB represent?
(c) What quantities does the product BA represent?

Answer:
(a) A = brand/vitamin, B = vitamin/cost
(b) AB = brand/cost
(c) Not meaningful, although the product BA can be found

Work Problem 4 at the side.

5.2 EXERCISES

In each of the following exercises, the dimensions of two matrices A and B are given. Find the dimensions of the product AB and the product BA, whenever these products exist. (See Example 3.)

1. A is 2×2, B is 2×2

2. A is 3×3, B is 3×3

3. A is 4×2, B is 2×4

4. A is 3×1, B is 1×3

5. A is 3×5, B is 5×2

6. A is 4×3, B is 3×6

7. A is 4×2, B is 3×4

8. A is 7×3, B is 2×7

Let

$$A = \begin{bmatrix} -2 & 4 \\ 0 & 3 \end{bmatrix} \quad \textit{and} \quad B = \begin{bmatrix} -6 & 2 \\ 4 & 0 \end{bmatrix}.$$

Find each of the following.

9. $2A$

10. $-3B$

11. $-4B$

12. $5A$

13. $-4A + 5B$

14. $3A - 10B$

Find each of the following matrix products where possible. (See Examples 1 and 2.)

15. $\begin{bmatrix} 1 & 2 \\ 3 & 4 \end{bmatrix} \begin{bmatrix} -1 \\ 7 \end{bmatrix}$

16. $\begin{bmatrix} -1 & 5 \\ 7 & 0 \end{bmatrix} \begin{bmatrix} 6 \\ 2 \end{bmatrix}$

17. $\begin{bmatrix} 2 & 2 & -1 \\ 3 & 0 & 1 \end{bmatrix} \begin{bmatrix} 0 & 2 \\ -1 & 4 \\ 0 & 2 \end{bmatrix}$

18. $\begin{bmatrix} -9 & 2 & 1 \\ 3 & 0 & 0 \end{bmatrix} \begin{bmatrix} 2 \\ -1 \\ 4 \end{bmatrix}$

Compute the following products where possible.

19. $\begin{bmatrix} 1 & 2 \\ 3 & 4 \end{bmatrix} \begin{bmatrix} -1 & 5 \\ 7 & 0 \end{bmatrix}$

20. $\begin{bmatrix} -1 & 5 \\ 7 & 0 \end{bmatrix} \begin{bmatrix} 1 & 1 \\ 3 & 4 \end{bmatrix}$

21. $\begin{bmatrix} -2 & -3 & 7 \\ 1 & 5 & 6 \end{bmatrix} \begin{bmatrix} 1 \\ 2 \\ 3 \end{bmatrix}$

22. $\begin{bmatrix} 6 \\ 5 \\ 4 \end{bmatrix} [-1 \quad 1 \quad 1]$

23. $\begin{bmatrix} 4 & 3 \\ 1 & 2 \\ 0 & -5 \end{bmatrix} \begin{bmatrix} 2 & -2 \\ 1 & -1 \end{bmatrix}$

24. $[-1 \quad 1 \quad 1] \begin{bmatrix} 6 \\ 5 \\ 4 \end{bmatrix}$

25. $\begin{bmatrix} 2 & -2 \\ 1 & -1 \end{bmatrix} \begin{bmatrix} 4 & 3 \\ 1 & 2 \\ 0 & -5 \end{bmatrix}$

26. $\begin{bmatrix} 1 & 0 \\ 0 & 1 \end{bmatrix} \begin{bmatrix} 1 & 5 & 2 \\ -3 & 4 & -1 \end{bmatrix} \begin{bmatrix} 1 & 0 & 0 \\ 0 & 1 & 0 \\ 0 & 0 & 1 \end{bmatrix}$

Let

$$A = \begin{bmatrix} -2 & 4 \\ 1 & 3 \end{bmatrix} \quad \textit{and} \quad B = \begin{bmatrix} -2 & 1 \\ 3 & 6 \end{bmatrix}.$$

27. Find AB.

28. Find BA.

29. Did you get the same answer in Exercises 27 and 28? Do you think that matrix multiplication is commutative?

30. In general, for matrices A and B such that AB and BA both exist, does AB always equal BA?

Given matrices

$$P = \begin{bmatrix} m & n \\ p & q \end{bmatrix}, \quad X = \begin{bmatrix} x & y \\ z & w \end{bmatrix}, \quad T = \begin{bmatrix} r & s \\ t & u \end{bmatrix},$$

verify that the following statements are true.

31. $(PX)T = P(XT)$ (Associative property)

32. $P(X + T) = PX + PT$ (Distributive property)

33. PX is a 2×2 matrix (Closure property)

34. $P(X - T) = PX - PT$ (Distributive property)

35. $k(X + T) = kX + kT$ for any real number k

36. $(k + h)P = kP + hP$ for any real numbers k and h

37. Let I be the matrix $I = \begin{bmatrix} 1 & 0 \\ 0 & 1 \end{bmatrix}$, and let matrices P, X, and T be defined as above.

(a) Find IP, PI, IX.
(b) Without calculating, guess what the matrix IT might be.
(c) Suggest a reason for naming a matrix such as I an identity matrix.

APPLIED PROBLEMS

Management

38. The Bread Box, a small neighborhood bakery, sells four main items: sweet rolls, bread, cake, and pie. The amount of certain major ingredients required to make these items is given in matrix A.

$$A = \begin{bmatrix} 1 & 4 & \frac{1}{4} & \frac{1}{4} & 1 \\ 0 & 3 & 0 & \frac{1}{4} & 0 \\ 4 & 3 & 2 & 1 & 1 \\ 0 & 1 & 0 & \frac{1}{3} & 0 \end{bmatrix}$$

	Eggs	Flour*	Sugar*	Shortening*	Milk*
Sweet rolls (dozen)	1	4	1/4	1/4	1
Bread (loaves)	0	3	0	1/4	0
Cake (1)	4	3	2	1	1
Pie (1)	0	1	0	1/3	0

The cost (in cents) for each ingredient when purchased in large lots and in small lots is given by matrix B.

$$B = \begin{bmatrix} 5 & 5 \\ 8 & 10 \\ 10 & 12 \\ 12 & 15 \\ 5 & 6 \end{bmatrix}$$

	Cost (in cents) Large lot	Small lot
Eggs	5	5
Flour†	8	10
Sugar†	10	12
Shortening†	12	15
Milk†	5	6

(a) Use matrix multiplication to find a matrix representing the comparative costs per item under the two purchase options.

Suppose a day's orders consist of 20 dozen sweet rolls, 200 loaves of bread, 50 cakes, and 60 pies.

(b) Represent these orders as a 1×4 matrix and use matrix multiplication to write as a matrix the amount of each ingredient required to fill the day's orders.
(c) Use matrix multiplication to find a matrix representing the costs under the two purchase options to fill the day's orders.

39. In Exercise 37, Section 5.1, label the matrices found in parts (a), (b), and (c) respectively X, Y, and Z.
(a) Find the product matrix XY. What do the entries of this matrix represent?
(b) Find the product matrix YZ. What do the entries represent?

40. The EZ Life Company buys three machines costing \$90,000, \$60,000, and \$120,000 respectively. They plan to depreciate the machines over a three-year period using the sum-of-the-years'-digits method, where the company assumes that the machines will have 20% of their value left at the end of three years. Thus they will compute the depreciation

* Measured in cups.
† Cost per cup.

of 80% of the cost using rates of 1/2, 1/3, and 1/6 respectively for the three-year period.

(a) Write a column matrix C to represent the cost of these three assets.

(b) To get a matrix which represents 80% of the costs, let

$$P = \begin{bmatrix} .8 & 0 & 0 \\ 0 & .8 & 0 \\ 0 & 0 & .8 \end{bmatrix}.$$

Then PC is the desired matrix. Calculate PC.

(c) Write a row matrix R representing the three rates given above.

(d) Find $(PC)R$. This product matrix (which should be 3×3) represents the depreciation charge for each asset for each year of the three-year period.

5.3 MULTIPLICATIVE INVERSES

In Section 5.1, we defined a zero matrix which has properties similar to those of the real number zero. Another real number which has special properties is the number 1, the identity element of the real numbers. We know that for any real number a, $a \cdot 1 = 1 \cdot a = a$. In this section, we define an identity matrix I which has properties similar to those of the number 1. We then use this identity matrix to find the multiplicative inverse of any square matrix which has an inverse.

We know that if I is to be the identity matrix, the products AI and IA must both equal A. This means that we can find an identity matrix only for square matrices. Otherwise, IA and AI would not both exist. The 2×2 **identity matrix** which satisfies these conditions is

$$\mathbf{I} = \begin{bmatrix} \mathbf{1} & \mathbf{0} \\ \mathbf{0} & \mathbf{1} \end{bmatrix}.$$

To check that I, as defined above, is really the 2×2 identity matrix, let

$$A = \begin{bmatrix} a & b \\ c & d \end{bmatrix}.$$

Then AI and IA should both equal A.

$$AI = \begin{bmatrix} a & b \\ c & d \end{bmatrix} \begin{bmatrix} 1 & 0 \\ 0 & 1 \end{bmatrix} = \begin{bmatrix} a & b \\ c & d \end{bmatrix} = A$$

$$IA = \begin{bmatrix} 1 & 0 \\ 0 & 1 \end{bmatrix} \begin{bmatrix} a & b \\ c & d \end{bmatrix} = \begin{bmatrix} a & b \\ c & d \end{bmatrix} = A$$

This verifies that we have defined I correctly. (It can also be shown that I is the only 2×2 identity matrix.)

In a similar way, we have the following identity matrices for 3×3 matrices and 4×4 matrices respectively:

$$I = \begin{bmatrix} 1 & 0 & 0 \\ 0 & 1 & 0 \\ 0 & 0 & 1 \end{bmatrix}$$

$$I = \begin{bmatrix} 1 & 0 & 0 & 0 \\ 0 & 1 & 0 & 0 \\ 0 & 0 & 1 & 0 \\ 0 & 0 & 0 & 1 \end{bmatrix}$$

By generalizing, we can find an identity matrix for any $n \times n$ matrix.

Recall that the multiplicative inverse of a number a is defined to be $1/a$. The inverse $1/a$ exists for all real numbers a except 0. Because of the way matrix multiplication is defined, *only square matrices can have inverses*. The multiplicative inverse of a matrix A is written A^{-1}. This matrix must satisfy the statements

$$AA^{-1} = I \qquad \textit{and} \qquad A^{-1}A = I.$$

In the rest of this section, we develop a method for finding the multiplicative inverse of any $n \times n$ matrix which has an inverse. The method we will use depends on three fundamental **row operations** which may be performed on any row of any given matrix. (These row operations are also useful in solving systems of equations, which we will do in Section 6.2.) The necessary row operations are:

1. Interchanging any two rows of a matrix;
2. Multiplying the elements of any row of a matrix by the same nonzero scalar k;
3. Adding a multiple of one row to another row.

For example, the first row operation allows us to change the matrix

$$\begin{bmatrix} 1 & 3 & 5 \\ 0 & 1 & 2 \\ 2 & 1 & -2 \end{bmatrix} \quad \text{to} \quad \begin{bmatrix} 0 & 1 & 2 \\ 1 & 3 & 5 \\ 2 & 1 & -2 \end{bmatrix}$$

by interchanging the first two rows. Note that row three is left unchanged.

The second row operation allows us to change

$$\begin{bmatrix} 1 & 3 & 5 \\ 0 & 1 & 2 \\ 2 & 1 & -2 \end{bmatrix} \quad \text{to} \quad \begin{bmatrix} -2 & -6 & -10 \\ 0 & 1 & 2 \\ 2 & 1 & -2 \end{bmatrix}$$

by multiplying the elements of the first row of the original matrix by -2. Note that rows two and three are left unchanged.

The third row operation allows us to change

$$\begin{bmatrix} 1 & 3 & 5 \\ 0 & 1 & 2 \\ 2 & 1 & -2 \end{bmatrix} \quad \text{to} \quad \begin{bmatrix} -1 & 2 & 7 \\ 0 & 1 & 2 \\ 2 & 1 & -2 \end{bmatrix}$$

by first multiplying each element in the third row of the original matrix by -1 and then adding the results to the corresponding elements in the first row of that matrix. We write this out as

$$\begin{bmatrix} 1+2(-1) & 3+1(-1) & 5+(-2)(-1) \\ 0 & 1 & 2 \\ 2 & 1 & -2 \end{bmatrix} = \begin{bmatrix} -1 & 2 & 7 \\ 0 & 1 & 2 \\ 2 & 1 & -2 \end{bmatrix}.$$

Note that rows two and three are left unchanged.

Work Problem 1 at the side.

1. Let

$$A = \begin{bmatrix} 6 & 2 & 1 \\ -1 & 3 & 4 \end{bmatrix}.$$

(a) Interchange rows one and two.

(b) Multiply row one by $-1/2$.

(c) Add 3 times row one to row two.

Answer:

(a) $\begin{bmatrix} -1 & 3 & 4 \\ 6 & 2 & 1 \end{bmatrix}$

(b) $\begin{bmatrix} -3 & -1 & -\frac{1}{2} \\ -1 & 3 & 4 \end{bmatrix}$

(c) $\begin{bmatrix} 6 & 2 & 1 \\ 17 & 9 & 7 \end{bmatrix}$

Before we can find inverses, we define an **augmented matrix** to be a combination of two matrices written as one, such as

$$\left[\begin{array}{cc|c} 1 & 2 & 2 \\ 3 & 4 & 5 \end{array}\right].$$

The vertical bar separates the two original matrices

$$\begin{bmatrix} 1 & 2 \\ 3 & 4 \end{bmatrix} \quad \text{and} \quad \begin{bmatrix} 2 \\ 5 \end{bmatrix},$$

but does not prevent us from performing row operations on the augmented matrix.

To obtain A^{-1} for any $n \times n$ matrix for which A^{-1} exists, we first form the augmented matrix $[A|I]$ where A is the $n \times n$ matrix and I is the $n \times n$ multiplicative identity matrix. We perform row operations on $[A|I]$ until we have an augmented matrix of the form $[I|B]$. Matrix B is then the desired matrix A^{-1}.

Unfortunately, it is not readily apparent that this method should produce the desired matrix; however, in any particular case, it is easy to verify that the matrix B from $[I|B]$ is indeed A^{-1} by verifying that $AB = BA = I$.

Work Problem 2 at the side.

2. Given

$$A = \begin{bmatrix} 1 & 2 \\ 4 & 6 \end{bmatrix}$$

and

$$B = \begin{bmatrix} -3 & 1 \\ 2 & -\frac{1}{2} \end{bmatrix},$$

decide if they are inverses.

Answers: Yes, since $AB = BA = I$.

Example 1 Find A^{-1} if $A = \begin{bmatrix} 2 & 4 \\ 1 & -1 \end{bmatrix}$.

Step 1 Use the 2×2 identity matrix I to form the augmented matrix $[A|I]$.

$$[A|I] = \left[\begin{array}{cc|cc} 2 & 4 & 1 & 0 \\ 1 & -1 & 0 & 1 \end{array}\right]$$

Step 2 Now we perform row operations on $[A|I]$ until we obtain a new matrix of the form $[I|B]$, B is then the inverse we want. We begin by getting a 1 in the upper left-hand corner by multiplying the top row of $[A|I]$ by 1/2.

$$\left[\begin{array}{cc|cc} 1 & 2 & \frac{1}{2} & 0 \\ 1 & -1 & 0 & 1 \end{array}\right]$$

Step 3 To get a 0 for the first element in row two, multiply the new row one by -1 and add the results to row two.

$$\left[\begin{array}{cc|cc} 1 & 2 & \frac{1}{2} & 0 \\ 0 & -3 & -\frac{1}{2} & 1 \end{array}\right]$$

Step 4 To get a 1 for the second element in row two, multiply row two by $-1/3$.

$$\left[\begin{array}{cc|cc} 1 & 2 & \frac{1}{2} & 0 \\ 0 & 1 & \frac{1}{6} & -\frac{1}{3} \end{array}\right]$$

Step 5 To get a 0 for the second element in row one, multiply the new row two by -2 and add the results to row one.

$$\left[\begin{array}{cc|cc} 1 & 0 & \frac{1}{6} & \frac{2}{3} \\ 0 & 1 & \frac{1}{6} & -\frac{1}{3} \end{array}\right]$$

We have now transformed $[A|I]$ into $[I|B]$. The transformation gives us

$$B = \begin{bmatrix} \frac{1}{6} & \frac{2}{3} \\ \frac{1}{6} & -\frac{1}{3} \end{bmatrix}$$

which should equal A^{-1}. To check, multiply A times B. The result should be I. We have

$$AB = \begin{bmatrix} 2 & 4 \\ 1 & -1 \end{bmatrix} \begin{bmatrix} \frac{1}{6} & \frac{2}{3} \\ \frac{1}{6} & -\frac{1}{3} \end{bmatrix} = \begin{bmatrix} \frac{1}{3} + \frac{2}{3} & \frac{4}{3} - \frac{4}{3} \\ \frac{1}{6} - \frac{1}{6} & \frac{2}{3} + \frac{1}{3} \end{bmatrix} = \begin{bmatrix} 1 & 0 \\ 0 & 1 \end{bmatrix} = I.$$

Also, $BA = I$.

Thus, $A^{-1} = \begin{bmatrix} \frac{1}{6} & \frac{2}{3} \\ \frac{1}{6} & -\frac{1}{3} \end{bmatrix}$. ■

In Step 2 above we could have simply exchanged rows one and two to get a 1 in the desired position. This can be done whenever there is a row which begins with a 1. It simplifies the work somewhat and is a good shortcut. The final result, of course, is the same.

Work Problem 3 at the side.

3. (a) Find A^{-1} if

$$A = \begin{bmatrix} 2 & 2 \\ 4 & 1 \end{bmatrix}.$$

(b) Check your answer by finding AA^{-1} and $A^{-1}A$.

Answer:

(a) $\begin{bmatrix} -\frac{1}{6} & \frac{1}{3} \\ \frac{2}{3} & -\frac{1}{3} \end{bmatrix}$

(b) $\begin{bmatrix} 1 & 0 \\ 0 & 1 \end{bmatrix}$

Example 2 Find A^{-1} if $A = \begin{bmatrix} 1 & 0 & 1 \\ 2 & -2 & -1 \\ 3 & 0 & 0 \end{bmatrix}$.

Step 1 Write the augmented matrix $[A|I]$.

$$[A|I] = \left[\begin{array}{ccc|ccc} 1 & 0 & 1 & 1 & 0 & 0 \\ 2 & -2 & -1 & 0 & 1 & 0 \\ 3 & 0 & 0 & 0 & 0 & 1 \end{array}\right]$$

Step 2 Since 1 is already in the upper left-hand corner as desired, we begin by selecting the row operation which will result in a 0 for the first element in row two. We multiply row one by -2 and add the result to row two. This gives

$$\left[\begin{array}{ccc|ccc} 1 & 0 & 1 & 1 & 0 & 0 \\ 0 & -2 & -3 & -2 & 1 & 0 \\ 3 & 0 & 0 & 0 & 0 & 1 \end{array}\right].$$

Step 3 To get 0 for the first element in row three, multiply row one by -3 and add to row three. The new matrix is

$$\left[\begin{array}{ccc|ccc} 1 & 0 & 1 & 1 & 0 & 0 \\ 0 & -2 & -3 & -2 & 1 & 0 \\ 0 & 0 & -3 & -3 & 0 & 1 \end{array}\right].$$

Step 4 To get 1 for the second element in row two, multiply row two by $-1/2$, obtaining the new matrix

$$\left[\begin{array}{ccc|ccc} 1 & 0 & 1 & 1 & 0 & 0 \\ 0 & 1 & \frac{3}{2} & 1 & -\frac{1}{2} & 0 \\ 0 & 0 & -3 & -3 & 0 & 1 \end{array}\right].$$

Step 5 To get 1 for the third element in row three, multiply row three by $-1/3$, with the result

$$\left[\begin{array}{ccc|ccc} 1 & 0 & 1 & 1 & 0 & 0 \\ 0 & 1 & \frac{3}{2} & 1 & -\frac{1}{2} & 0 \\ 0 & 0 & 1 & 1 & 0 & -\frac{1}{3} \end{array}\right].$$

Step 6 To get 0 for the third element in row one, multiply row three by -1 and add to row one, which gives

$$\left[\begin{array}{ccc|ccc} 1 & 0 & 0 & 0 & 0 & \frac{1}{3} \\ 0 & 1 & \frac{3}{2} & 1 & -\frac{1}{2} & 0 \\ 0 & 0 & 1 & 1 & 0 & -\frac{1}{3} \end{array}\right].$$

Step 7 To get 0 for the third element in row two, multiply row three by $-3/2$ and add to row two.

$$\left[\begin{array}{ccc|ccc} 1 & 0 & 0 & 0 & 0 & \frac{1}{3} \\ 0 & 1 & 0 & -\frac{1}{2} & -\frac{1}{2} & \frac{1}{2} \\ 0 & 0 & 1 & 1 & 0 & -\frac{1}{3} \end{array}\right]$$

From the last transformation, we get the desired inverse.

$$A^{-1} = \begin{bmatrix} 0 & 0 & \frac{1}{3} \\ -\frac{1}{2} & -\frac{1}{2} & \frac{1}{2} \\ 1 & 0 & -\frac{1}{3} \end{bmatrix}$$

Confirm this by forming the product $A^{-1}A$, which should equal I. ■

4. (a) Find A^{-1} if

$$A = \begin{bmatrix} 2 & 8 \\ -1 & -5 \end{bmatrix}.$$

(b) Check your answer by finding AA^{-1}.

Answer:

(a) $\begin{bmatrix} \frac{5}{2} & 4 \\ -\frac{1}{2} & -1 \end{bmatrix}$

(b) $\begin{bmatrix} 1 & 0 \\ 0 & 1 \end{bmatrix}$

Work Problem 4 at the side.

We can see from the examples above that the most efficient order is to work column by column from left to right. For each column a 1 in the proper position should be the result of the first change. Next, perform the steps that obtain zeros in the remainder of the column. Then proceed to the next column. Since it is tedious to find an inverse with paper and pencil, these same steps can be adapted for a computer program. A computer can produce the inverse of a large matrix, even a 40×40 matrix, in a few seconds.*

Example 3 Find A^{-1} if $A = \begin{bmatrix} 2 & -4 \\ 1 & -2 \end{bmatrix}$.

Using row operations to transform the first column of the augmented matrix

$$\left[\begin{array}{cc|cc} 2 & -4 & 1 & 0 \\ 1 & -2 & 0 & 1 \end{array}\right]$$

* See, for example, Margaret Lial, *Study Guide with Computer Problems* (Glenview, Ill.: Scott, Foresman and Company, 1979).

results in the following matrices.

$$\left[\begin{array}{rr|rr} 1 & -2 & \frac{1}{2} & 0 \\ 1 & -2 & 0 & 1 \end{array}\right]$$

$$\left[\begin{array}{rr|rr} 1 & -2 & \frac{1}{2} & 0 \\ 0 & 0 & -\frac{1}{2} & 1 \end{array}\right]$$

At this point, we wish to change the matrix so that the second element of row two will be 1. Since that element is now 0, there is no way to complete the desired transformation.

What is wrong? Remember, near the beginning of this section we mentioned that some matrices do not have inverses. Matrix A is an example of a matrix that has no inverse. Thus, in this case there is no matrix A^{-1} such that $AA^{-1} = A^{-1}A = A$. ■

5.3 EXERCISES

Use the third row operation to change each of the following matrices as indicated.

1. $\begin{bmatrix} 2 & 4 \\ 4 & 7 \end{bmatrix}$; $-\frac{3}{4}$ times row two added to row one.

2. $\begin{bmatrix} -1 & 4 \\ 7 & 0 \end{bmatrix}$; 7 times row one added to row two.

3. $\begin{bmatrix} 1 & \frac{1}{5} \\ 5 & 2 \end{bmatrix}$; -5 times row one added to row two.

4. $\begin{bmatrix} 5 & 7 \\ 2 & -4 \end{bmatrix}$; $\frac{4}{7}$ times row one added to row two.

5. $\begin{bmatrix} 1 & 5 & 6 \\ -2 & 3 & -1 \\ 4 & 7 & 0 \end{bmatrix}$; 2 times row one added to row two.

6. $\begin{bmatrix} 2 & 5 & 6 \\ 4 & -1 & 2 \\ 3 & 7 & 1 \end{bmatrix}$; -6 times row three added to row one.

7. $\begin{bmatrix} -3 & 1 & -4 \\ 2 & 1 & 3 \\ -7 & 5 & 2 \end{bmatrix}$; -5 times row two added to row three.

8. $\begin{bmatrix} 4 & 10 & -8 \\ 7 & 4 & 3 \\ -1 & 1 & 0 \end{bmatrix}$; -4 times row three added to row two.

Decide whether or not the given 2 × 2 matrices are inverses of each other. (Check to see if their product is I.)

9. $\begin{bmatrix} 2 & 3 \\ 1 & 1 \end{bmatrix}$ and $\begin{bmatrix} -1 & 3 \\ 1 & -2 \end{bmatrix}$

10. $\begin{bmatrix} 5 & 7 \\ 2 & 3 \end{bmatrix}$ and $\begin{bmatrix} 3 & -7 \\ -2 & 5 \end{bmatrix}$

11. $\begin{bmatrix} 2 & 1 \\ 3 & 2 \end{bmatrix}$ and $\begin{bmatrix} 2 & 1 \\ -3 & 2 \end{bmatrix}$

12. $\begin{bmatrix} -1 & 2 \\ 3 & -5 \end{bmatrix}$ and $\begin{bmatrix} -5 & -2 \\ -3 & -1 \end{bmatrix}$

Decide whether or not the given 3 × 3 matrices are inverses of each other.

13. $\begin{bmatrix} 1 & 2 & 0 \\ 0 & 1 & 0 \\ 0 & 1 & 0 \end{bmatrix}$ and $\begin{bmatrix} 1 & -2 & 0 \\ 0 & 1 & 0 \\ 0 & -1 & 1 \end{bmatrix}$

14. $\begin{bmatrix} 0 & 1 & 0 \\ 0 & 0 & -2 \\ 1 & -1 & 0 \end{bmatrix}$ and $\begin{bmatrix} 1 & 0 & 1 \\ 1 & 0 & 0 \\ 0 & -1 & 0 \end{bmatrix}$

15. $\begin{bmatrix} 1 & 3 & 3 \\ 1 & 4 & 3 \\ 1 & 3 & 4 \end{bmatrix}$ and $\begin{bmatrix} 7 & -3 & -3 \\ -1 & 1 & 0 \\ -1 & 0 & 1 \end{bmatrix}$

16. $\begin{bmatrix} -1 & 0 & 2 \\ 3 & 1 & 0 \\ 0 & 2 & -3 \end{bmatrix}$ and $\begin{bmatrix} -\frac{1}{5} & \frac{4}{15} & -\frac{2}{15} \\ \frac{3}{5} & \frac{1}{5} & \frac{2}{5} \\ \frac{2}{5} & \frac{2}{15} & -\frac{1}{15} \end{bmatrix}$

Find the inverse, if it exists, for each of the following matrices. (See Examples 1–3.)

17. $\begin{bmatrix} 1 & -1 \\ 2 & 0 \end{bmatrix}$

18. $\begin{bmatrix} -1 & 2 \\ -2 & -1 \end{bmatrix}$

19. $\begin{bmatrix} 3 & -1 \\ -5 & 2 \end{bmatrix}$

20. $\begin{bmatrix} -1 & -2 \\ 3 & 4 \end{bmatrix}$

21. $\begin{bmatrix} -6 & 4 \\ -3 & 2 \end{bmatrix}$

22. $\begin{bmatrix} 5 & 10 \\ -3 & -6 \end{bmatrix}$

23. $\begin{bmatrix} 1 & 0 & 0 \\ 0 & -1 & 0 \\ 1 & 0 & 1 \end{bmatrix}$

24. $\begin{bmatrix} 1 & 0 & 1 \\ 0 & -1 & 0 \\ 2 & 1 & 1 \end{bmatrix}$

25. $\begin{bmatrix} -1 & -1 & -1 \\ 4 & 5 & 0 \\ 0 & 1 & -3 \end{bmatrix}$

26. $\begin{bmatrix} 2 & 0 & 4 \\ 3 & 1 & 5 \\ -1 & 1 & -2 \end{bmatrix}$

27. $\begin{bmatrix} 1 & 2 & 3 \\ -3 & -2 & -1 \\ -1 & 0 & 1 \end{bmatrix}$

28. $\begin{bmatrix} 2 & 0 & 4 \\ 1 & 0 & -1 \\ 3 & 0 & -2 \end{bmatrix}$

29. $\begin{bmatrix} 2 & 4 & 6 \\ -1 & -4 & -3 \\ 0 & 1 & -1 \end{bmatrix}$

30. $\begin{bmatrix} 2 & 2 & -4 \\ 2 & 6 & 0 \\ -3 & -3 & 5 \end{bmatrix}$

31. $\begin{bmatrix} 1 & -2 & 3 & 0 \\ 0 & 1 & -1 & 1 \\ -2 & 2 & -2 & 4 \\ 0 & 2 & -3 & 1 \end{bmatrix}$

32. $\begin{bmatrix} 1 & 1 & 0 & 2 \\ 2 & -1 & 1 & -1 \\ 3 & 3 & 2 & -2 \\ 1 & 2 & 1 & 0 \end{bmatrix}$

Let $A = \begin{bmatrix} a & b \\ c & d \end{bmatrix}$. *Show that each of the following statements is true.*

33. $IA = A$

34. $AI = A$

35. $A \cdot O = O$

36. Find A^{-1}. (Assume $ad - bc \neq 0$.) Show that $AA^{-1} = I$.

37. Show that $A^{-1}A = I$.

38. Using the definitions and properties listed in this section, show that for square matrices A and B of the same order, if $AB = O$ and if A^{-1} exists, then $B = O$.

5.4 APPLICATIONS OF MATRICES

Applications of matrices sometimes require the solution of a matrix equation. Matrix equations can be solved in much the same way as algebraic equations, by using the properties of addition and multiplication. Some of these properties were presented in the exercise sets of the first two sections of this chapter. We also need to use the multiplication property of equality, which states that we can multiply both sides of a matrix equation by any appropriate matrix. The next example shows how the method works.

Example 1 Given $A = \begin{bmatrix} 2 & 2 \\ -1 & -2 \end{bmatrix}$ and $B = \begin{bmatrix} 2 & 4 \\ 3 & -1 \end{bmatrix}$, find a matrix X such that $AX = B$.

To solve the matrix equation $AX = B$, we first see if A^{-1} exists. If so, we then use the facts that $A^{-1}A = I$ and $IX = X$, as follows.

$$AX = B$$

$$A^{-1}(AX) = A^{-1}B \qquad \text{Multiply both sides by } A^{-1}$$

$$(A^{-1}A)X = A^{-1}B$$

$$IX = A^{-1}B$$

$$X = A^{-1}B$$

(Note that when multiplying by matrices on both sides of a matrix equation, we must be careful to multiply in the same order on both sides of the equation, since (unlike multiplication of real numbers) multiplication of matrices is noncommutative.)

Since we now know $X = A^{-1}B$, we must first find A^{-1} using row operations.

Step 1 $[A|I] = \left[\begin{array}{cc|cc} 2 & 2 & 1 & 0 \\ -1 & -2 & 0 & 1 \end{array}\right]$

Step 2 $\left[\begin{array}{cc|cc} -1 & -2 & 0 & 1 \\ 2 & 2 & 1 & 0 \end{array}\right]$ Exchange rows 1 and 2

Step 3 $\left[\begin{array}{cc|cc} 1 & 2 & 0 & -1 \\ 2 & 2 & 1 & 0 \end{array}\right]$ Multiply row 1 by -1

Step 4 $\left[\begin{array}{cc|cc} 1 & 2 & 0 & -1 \\ 0 & -2 & 1 & 2 \end{array}\right]$ Multiply row 1 by -2; add to row 2

Step 5 $\left[\begin{array}{cc|cc} 1 & 2 & 0 & -1 \\ 0 & 1 & -\frac{1}{2} & -1 \end{array}\right]$ Multiply row 2 by $-\frac{1}{2}$

Step 6 $\left[\begin{array}{cc|cc} 1 & 0 & 1 & 1 \\ 0 & 1 & -\frac{1}{2} & -1 \end{array}\right] = [I|A^{-1}]$ Multiply row 2 by -2; add to row 1

Thus, $A^{-1} = \begin{bmatrix} 1 & 1 \\ -\frac{1}{2} & -1 \end{bmatrix}$. (Check by showing that $AA^{-1} = I$.)

Now we can find X by multiplying A^{-1} and B.

$$X = A^{-1}B = \begin{bmatrix} 1 & 1 \\ -\frac{1}{2} & -1 \end{bmatrix}\begin{bmatrix} 2 & 4 \\ 3 & -1 \end{bmatrix} = \begin{bmatrix} 5 & 3 \\ -4 & -1 \end{bmatrix}.$$

Now verify the result.

$$AX = \begin{bmatrix} 2 & 2 \\ -1 & -2 \end{bmatrix}\begin{bmatrix} 5 & 3 \\ -4 & -1 \end{bmatrix} = \begin{bmatrix} 2 & 4 \\ 3 & -1 \end{bmatrix} = B. \quad \blacksquare$$

Work Problem 1 at the side.

1. (a) Solve the matrix equation $AX = B$ for X, given

$$A = \begin{bmatrix} 1 & 2 \\ 4 & 7 \end{bmatrix} \quad \text{and} \quad B = \begin{bmatrix} 3 \\ 5 \end{bmatrix}.$$

(b) Check your answer.

Answer:

(a) $\begin{bmatrix} -11 \\ 7 \end{bmatrix}$

(b) $\begin{bmatrix} 1 & 2 \\ 4 & 7 \end{bmatrix}\begin{bmatrix} -11 \\ 7 \end{bmatrix} = \begin{bmatrix} 3 \\ 5 \end{bmatrix}$

In the next chapter we use matrix equations to solve systems of linear equations. This is one of the most important general applications of matrices.

We now present two specific real-world applications of matrices.

Leontief Models An interesting application of matrix theory was developed by W. W. Leontief* as a model of a simplified economy

* Leontief won the Nobel Prize for Economics in 1973. (See *Newsweek*, October 29, 1973, page 94.)

with n basic commodities. The production of each commodity uses some (perhaps all) of the other commodities in the economy. The amounts of each commodity used in the production of 1 unit of each commodity can be written as an $n \times n$ matrix A, called the **technological matrix** of the economy.

Example 2 Suppose a simple economy is based on just three commodities: grain, fuel, and transportation. Production of 1 unit of grain requires 1/4 unit of fuel and 1/3 unit of transportation, production of 1 unit of fuel requires 1/2 unit of grain and 1/4 unit of transportation, while production of 1 unit of transportation requires 1/4 unit of grain and 1/4 unit of fuel. Find the technological matrix of this economy.

The technological matrix of the economy would be

$$\begin{array}{l} \\ \text{Grain} \\ \text{Fuel} \\ \text{Transportation} \end{array} \begin{array}{c} \begin{array}{ccc} \text{Grain} & \text{Fuel} & \text{Transportation} \end{array} \\ \begin{bmatrix} 0 & \frac{1}{4} & \frac{1}{3} \\ \frac{1}{2} & 0 & \frac{1}{4} \\ \frac{1}{4} & \frac{1}{4} & 0 \end{bmatrix} \end{array} = A.$$

The top row represents the amount of each of the three commodities consumed in the production of 1 unit of grain. The second row gives the corresponding amounts required to produce 1 unit of fuel, and the last row gives the amounts needed to produce 1 unit of transportation. (Although it is perhaps unrealistic that production of a unit of each commodity requires none of that commodity, it allows us to consider a simpler matrix than would otherwise be possible.) ■

2. Write a 2×2 technological matrix in which 1 unit of electricity requires 1/2 unit of water and 1/3 unit of electricity, while one unit of water requires no water but 1/4 unit of electricity.

Answer: $\begin{bmatrix} \frac{1}{3} & \frac{1}{2} \\ \frac{1}{4} & 0 \end{bmatrix}$

Work Problem 2 at the side.

Let the number of units of each commodity produced, which is called the **gross production,** be represented by a row matrix,

$$X = [x_1 \quad x_2 \quad x_3 \quad \cdots \quad x_n].$$

In Example 2, suppose $X = [60\ 52\ 40]$. Then 60 units of grain, 52 units of fuel, and 40 units of transportation are produced. Since matrix A gives the amount of each commodity used to produce 1 unit of the various commodities, the matrix product XA gives the amount of each commodity used up in the production process.

$$XA = [60 \quad 52 \quad 40] \begin{bmatrix} 0 & \frac{1}{4} & \frac{1}{3} \\ \frac{1}{2} & 0 & \frac{1}{4} \\ \frac{1}{4} & \frac{1}{4} & 0 \end{bmatrix} = [36 \quad 25 \quad 33].$$

Thus 36 units of grain, 25 units of fuel, and 33 units of transportation are used up to produce 60 units of grain, 52 units of fuel, and 40 units of transportation.

The differences between gross production and the amount used up in the production process is the **net production,** Y. Thus

$$Y = X - XA.$$

In our example,

$$\begin{aligned} Y &= [60 \quad 52 \quad 40] - [36 \quad 25 \quad 33] \\ &= [24 \quad 27 \quad 7]. \end{aligned}$$

Thus, net production is 24 units of grain, 27 of fuel, and 7 of transportation.

3. (a) Write a 1×2 matrix to represent gross production of 9000 units of electricity and 12,000 units of water.
(b) Find XA using A from the last problem.
(c) Find Y using $Y = X - XA$.

Answer:
(a) [9000 12,000]
(b) [6000 4500]
(c) [3000 7500]

Work Problem 3 at the side.

In practical problems, A and Y usually are known and we need to find X. That is, we need to decide what gross production X is necessary in order to get a required amount of net production Y. We can use matrix algebra to decide on X.

By the distributive property, we have

$$Y = X - XA = X(I - A).$$

We want to solve the matrix equation

$$Y = X(I - A)$$

for X. If the matrix $I - A$ has an inverse, then

$$X = Y(I - A)^{-1}.$$

In our example, suppose we want a net production of 516 units of grain, 258 units of fuel, and 129 units of transportation. Then

$$Y = [516 \quad 258 \quad 129].$$

(Why did we write Y as a 1×3 rather than a 3×1 matrix?)

To find the gross production X, we must first get $I - A$.

$$I - A = \begin{bmatrix} 1 & 0 & 0 \\ 0 & 1 & 0 \\ 0 & 0 & 1 \end{bmatrix} - \begin{bmatrix} 0 & \frac{1}{4} & \frac{1}{3} \\ \frac{1}{2} & 0 & \frac{1}{4} \\ \frac{1}{4} & \frac{1}{4} & 0 \end{bmatrix} = \begin{bmatrix} 1 & -\frac{1}{4} & -\frac{1}{3} \\ -\frac{1}{2} & 1 & -\frac{1}{4} \\ -\frac{1}{4} & -\frac{1}{4} & 1 \end{bmatrix}$$

Using row operations, we find the inverse of $I - A$.

$$(I - A)^{-1} = \begin{bmatrix} \frac{60}{43} & \frac{64}{129} & \frac{76}{129} \\ \frac{36}{43} & \frac{176}{129} & \frac{80}{129} \\ \frac{24}{43} & \frac{20}{43} & \frac{56}{43} \end{bmatrix}$$

We know $X = Y(I - A)^{-1}$. Thus

$$X = [516 \quad 258 \quad 129]\begin{bmatrix} \frac{60}{43} & \frac{64}{129} & \frac{76}{129} \\ \frac{36}{43} & \frac{176}{129} & \frac{80}{129} \\ \frac{24}{43} & \frac{20}{43} & \frac{56}{43} \end{bmatrix} = [1008 \quad 668 \quad 632].$$

From this last result, we see that a gross production of 1008 units of grain, 668 units of fuel, and 632 units of transportation is required to produce net amounts of 516, 258, and 129 units respectively.

Example 3 An economy depends on two basic products, oil and wheat. To produce 1 metric ton of wheat requires 1/4 metric ton of wheat and 1/12 metric ton of oil. Production of 1 metric ton of oil consumes 1/3 metric ton of wheat and 1/9 metric ton of oil. Find the gross production necessary for a net production of 500 metric tons of wheat and 1000 metric tons of oil.

The technological matrix is

$$A = \begin{bmatrix} \frac{1}{4} & \frac{1}{12} \\ \frac{1}{3} & \frac{1}{9} \end{bmatrix}.$$

We first find $I - A$.

$$I - A = \begin{bmatrix} \frac{3}{4} & -\frac{1}{12} \\ -\frac{1}{3} & \frac{8}{9} \end{bmatrix}$$

Next, we calculate the inverse of $I - A$.

$$(I - A)^{-1} = \begin{bmatrix} \frac{32}{23} & \frac{3}{23} \\ \frac{12}{23} & \frac{27}{23} \end{bmatrix}$$

To find the gross production X, we use the equation $X = Y(I - A)^{-1}$. Since we want a net production of $Y = [500\ 1000]$, the gross production is given by

$$X = [500\ 1000]\begin{bmatrix} \frac{32}{23} & \frac{3}{23} \\ \frac{12}{23} & \frac{27}{23} \end{bmatrix} \approx [1217\ 1239].$$

Both numbers in the final result have been rounded off to the nearest whole number. The result shows that a gross production of 1217 metric tons of wheat and 1239 metric tons of oil are necessary to produce a net of 500 metric tons of wheat and 1000 metric tons of oil. ■

Work Problem 4 at the side.

4. A simple economy depends on just two products, beer and pretzels.
(a) Suppose 1/2 unit of beer and 1/2 unit of pretzels are needed to make 1 unit of beer and 3/4 unit of beer is needed to make 1 unit of pretzels. Write the technological matrix A for the economy.
(b) Find $I - A$.
(c) Find $(I - A)^{-1}$.
(d) Find the gross production X that will be needed to get a net production of $Y = [100 \quad 1000]$.

Answer:

(a) $\begin{bmatrix} \frac{1}{2} & \frac{1}{2} \\ \frac{3}{4} & 0 \end{bmatrix}$

(b) $\begin{bmatrix} \frac{1}{2} & -\frac{1}{2} \\ -\frac{3}{4} & 1 \end{bmatrix}$

(c) $\begin{bmatrix} 8 & 4 \\ 6 & 4 \end{bmatrix}$

(d) $[6800 \quad 4400]$

Code Theory Governments need sophisticated methods of coding and decoding messages. One example of such an advanced code uses matrix theory. Such a code takes the letters in the words and divides them into groups. (Each space between words is treated as a letter; punctuation is disregarded.) Then, numbers are assigned to the letters of the alphabet. For our purposes, let the letter *a* correspond to 1, *b* to 2, and so on. We let the number 27 correspond to a space between words.

For example, the message

mathematics is for the birds

can be divided into groups of three letters each.

mat *hem* *ati* *cs–* *is–* *for* *–th* *e–b* *ird* *s––*

Note that we used – to represent a space between words. We now write a column matrix for each group of three symbols using the corresponding numbers, as determined above, instead of letters. For example, the letters *mat* can be encoded as

$$\begin{bmatrix} 13 \\ 1 \\ 20 \end{bmatrix}.$$

The coded message then is the set of 3×1 column matrices

$$\begin{bmatrix} 13 \\ 1 \\ 20 \end{bmatrix} \begin{bmatrix} 8 \\ 5 \\ 13 \end{bmatrix} \begin{bmatrix} 1 \\ 20 \\ 9 \end{bmatrix} \begin{bmatrix} 3 \\ 19 \\ 27 \end{bmatrix} \begin{bmatrix} 9 \\ 19 \\ 27 \end{bmatrix} \begin{bmatrix} 6 \\ 15 \\ 18 \end{bmatrix} \begin{bmatrix} 27 \\ 20 \\ 8 \end{bmatrix} \begin{bmatrix} 5 \\ 27 \\ 2 \end{bmatrix} \begin{bmatrix} 9 \\ 18 \\ 4 \end{bmatrix} \begin{bmatrix} 19 \\ 27 \\ 27 \end{bmatrix}.$$

5. Write the message "*when*" using 2×1 matrices.

Answer: $\begin{bmatrix} 23 \\ 8 \end{bmatrix} \begin{bmatrix} 5 \\ 14 \end{bmatrix}$

Work Problem 5 at the side.

We can further complicate the code by choosing a matrix which has an inverse, in this case a 3×3 matrix, call it M, and find the products of this matrix and each of the above column matrices. (Note that the size of each group, the assignment of numbers to letters, and the choice of matrix M must all be predetermined.)

Suppose

$$M = \begin{bmatrix} 1 & 3 & 3 \\ 1 & 4 & 3 \\ 1 & 3 & 4 \end{bmatrix}.$$

If we find the products of M and the column matrices above, we have a new set of column matrices,

$$\begin{bmatrix} 76 \\ 77 \\ 96 \end{bmatrix} \begin{bmatrix} 62 \\ 67 \\ 75 \end{bmatrix} \quad \text{and so on.}$$

The entries of these matrices can then be transmitted to an agent as the message *76, 77, 96, 62, 67, 75*, and so on.

6. Use the matrix given below to find the 2×1 matrices to be transmitted for the message you encoded in Problem 5.

$$\begin{bmatrix} 2 & 1 \\ 5 & 0 \end{bmatrix}$$

Answer: $\begin{bmatrix} 54 \\ 115 \end{bmatrix} \begin{bmatrix} 24 \\ 25 \end{bmatrix}$

Work Problem 6 at the side.

When the agent receives the message, he divides it into groups of numbers and forms each group into a column matrix. After multiplying each column matrix by the matrix M^{-1}, the message can be read.

Although this type of code is relatively simple, it is actually difficult to break. Many ramifications are possible. For example, a long message might be placed in groups of 20, thus requiring a 20×20 matrix for coding and decoding. Finding the inverse of such a matrix would require an impractical amount of time if calculated by hand. For this reason some of the largest computers are used by government agencies involved in coding.

5.4 EXERCISES

For each of the following, solve the matrix equation $AX = B$ for X.

1. $A = \begin{bmatrix} 1 & 3 \\ -2 & 4 \end{bmatrix}, \quad B = \begin{bmatrix} 15 \\ 10 \end{bmatrix}$

2. $A = \begin{bmatrix} 2 & -2 \\ 1 & 0 \end{bmatrix}, \quad B = \begin{bmatrix} -4 \\ 3 \end{bmatrix}$

3. $A = \begin{bmatrix} -2 & 4 \\ 3 & -1 \end{bmatrix}, \quad B = \begin{bmatrix} 40 & -20 \\ 80 & 20 \end{bmatrix}$

4. $A = \begin{bmatrix} 5 & 10 \\ 8 & 14 \end{bmatrix}, \quad B = \begin{bmatrix} -15 & 10 \\ 20 & 5 \end{bmatrix}$

5. $A = \begin{bmatrix} 1 & 0 & 2 \\ -1 & 1 & 0 \\ 3 & 0 & 4 \end{bmatrix}, \quad B = \begin{bmatrix} 8 \\ 4 \\ -6 \end{bmatrix}$

6. $A = \begin{bmatrix} 2 & 4 & 0 \\ 1 & -2 & 0 \\ 0 & 0 & 3 \end{bmatrix}, \quad B = \begin{bmatrix} 72 \\ -24 \\ 48 \end{bmatrix}$

7. $A = \begin{bmatrix} -3 & 0 & 6 \\ 1 & 1 & 0 \\ 0 & 2 & 5 \end{bmatrix}, \quad B = \begin{bmatrix} 12 & 3 \\ -6 & 0 \\ 0 & -3 \end{bmatrix}$

8. $A = \begin{bmatrix} -2 & 2 & 0 \\ 3 & 1 & 0 \\ 0 & 0 & 1 \end{bmatrix}, \quad B = \begin{bmatrix} 8 & -16 \\ 0 & 32 \\ -11 & 5 \end{bmatrix}$

Use matrix algebra to solve the following matrix equations for X. Then use the given matrices to find X and check your work.

9. $N = X - MX$ $\quad N = \begin{bmatrix} 8 \\ -12 \end{bmatrix}, \quad M = \begin{bmatrix} 0 & 1 \\ -2 & 1 \end{bmatrix}$

10. $A = BX + X$ $\quad A = \begin{bmatrix} 4 & 6 \\ -2 & 2 \end{bmatrix}, \quad B = \begin{bmatrix} -2 & -2 \\ 3 & 3 \end{bmatrix}$

11. Use the methods of the text to encode the message

Arthur is here.

Break the message into groups of two letters and use the matrix

$$M = \begin{bmatrix} -1 & 2 \\ 2 & 5 \end{bmatrix}.$$

12. Encode the following message by breaking it into groups of two letters and using the matrix of Exercise 11.

Attack at dawn unless too cold.

13. Finish encoding the message given in the text.

APPLIED PROBLEMS

In Exercises 14 and 15, refer to Example 3.

Management

14. If the net production requirements are changed to 690 metric tons of wheat and 920 metric tons of oil, what gross production will be needed?

15. Change the technological matrix so that production of 1 ton of wheat requires 1/5 metric ton of oil (and no wheat), and 1/3 metric ton of wheat (and no oil) is required to produce 1 metric ton of oil. To get the same net production, what gross production will be needed?

In Exercises 16–19, refer to Example 2.

16. If the net production requirements are changed to 516 units of each commodity, what gross production will be needed?

17. Suppose 1/3 unit of fuel (no grain or transportation) is required to produce 1 unit of grain, 1/4 unit of transportation is required to produce 1 unit of fuel, and 1/2 unit of grain is required to produce 1 unit of transportation. What gross production is needed to get a net production of 1000 units of each commodity?

18. Suppose 1/4 unit of fuel and 1/2 unit of transportation are required to produce 1 unit of grain, 1/2 unit of grain and 1/4 unit of transportation to produce 1 unit of fuel, and 1/4 unit of grain and 1/4 unit of fuel to produce one unit of transportation. What gross production is needed to get a net production of 1000 units of each commodity?

19. If the technological matrix is changed so that 1/4 unit of fuel and 1/2 unit of transportation are required to produce 1 unit of grain, 1/2 unit of grain and 1/4 unit of transportation are required to produce 1 unit of fuel, and 1/4 unit each of grain and fuel are required to produce 1 unit of transportation, find the gross production needed to get a net production of 500 units of each commodity.

20. A primitive economy depends on two basic goods, yams and pork. Production of 1 bushel of yams requires 1/4 bushel of yams and 1/2 of a pig. To produce 1 pig requires 1/6 bushel of yams. Find the necessary gross production for a net production of
(a) 1 bushel of yams and 1 pig;
(b) 100 bushels of yams and 70 pigs.

21. The Bread Box Bakery sells three types of cakes, each requiring the amounts of the basic ingredients shown in the following matrix.

$$\begin{array}{l} \\ \\ \text{Flour (cups)} \\ \text{Sugar (cups)} \\ \text{Eggs} \end{array} \begin{array}{c} \text{Types of cake} \\ \begin{array}{ccc} \text{I} & \text{II} & \text{III} \end{array} \\ \begin{bmatrix} 2 & 4 & 2 \\ 2 & 1 & 2 \\ 2 & 1 & 3 \end{bmatrix} \end{array}$$

To fill its daily orders for these three kinds of cake, the bakery uses 72 cups of flour, 48 cups of sugar, and 60 eggs.
(a) Write a 3×1 matrix for the amounts used daily.
(b) Let the number of daily orders for cakes be a 3×1 matrix X with entries x_1, x_2, and x_3. Write a matrix equation which you can solve for X, using the given matrix and the matrix from part (a).
(c) Solve the equation you wrote in part (b) to find the number of daily orders for each type of cake.

CHAPTER 5 TEST

[5.1] *For each of the following, find the order of the matrices, find the values of any variables, and identify any square, row, or column matrices.*

1. $\begin{bmatrix} 2 & 3 \\ 5 & q \end{bmatrix} = \begin{bmatrix} a & b \\ c & 9 \end{bmatrix}$

2. $\begin{bmatrix} 2 & x \\ y & 6 \\ 5 & z \end{bmatrix} = \begin{bmatrix} a & -1 \\ 4 & 6 \\ p & 7 \end{bmatrix}$

3. $[m \quad 4 \quad z \quad -1] = [12 \quad k \quad -8 \quad r]$

4. The activities of a grazing animal can be classified roughly into three categories: grazing, moving, and resting. Suppose horses spend 8 hours grazing, 8 moving, and 8 resting; cattle spend 10 grazing, 5 moving, and 9 resting; sheep spend 7 grazing, 10 moving, and 7 resting; and goats spend 8 grazing, 9 moving, and 7 resting. Write this information as a 4×3 matrix.

Given the matrices

$$A = \begin{bmatrix} 4 & 10 \\ -2 & -3 \\ 6 & 9 \end{bmatrix}, \quad B = \begin{bmatrix} 2 & 3 & -2 \\ 2 & 4 & 0 \\ 0 & 1 & 2 \end{bmatrix}, \quad C = \begin{bmatrix} 5 & 0 \\ -1 & 3 \\ 4 & 7 \end{bmatrix}, \quad D = \begin{bmatrix} 6 \\ 1 \\ 0 \end{bmatrix},$$

$$E = [1 \quad 3 \quad -4], \quad F = \begin{bmatrix} -1 & 4 \\ 3 & 7 \end{bmatrix}, \quad G = \begin{bmatrix} 2 & 5 \\ 1 & 6 \end{bmatrix},$$

find each of the following which exists.

[5.2] **5.** $A + C$ **6.** $2G - 4F$

7. $B - A$ **8.** AF

9. AC **10.** DE

[5.3] **11.** F^{-1} **12.** B^{-1}

[5.4] **13.** Solve the matrix equation $AX = B$ for X, if

$$A = \begin{bmatrix} 2 & 4 \\ -1 & -3 \end{bmatrix}, \quad B = \begin{bmatrix} 8 \\ 3 \end{bmatrix}, \quad \text{and} \quad X = \begin{bmatrix} x \\ y \end{bmatrix}.$$

14. An economy depends on two commodities, goats and cheese. It takes 2/3 of a unit of goats to produce 1 unit of cheese and 1/2 unit of cheese to produce 1 unit of goats.
(a) Write the technological matrix.
(b) What gross production will be required to get a net production of 400 units of cheese and 800 units of goats?

15. Let $M = \begin{bmatrix} 4 & -1 \\ 2 & 6 \end{bmatrix}$.

(a) Use M to encode the message: *Meet at the cave*. Use 2×1 matrices.
(b) What matrix should be used to decode the message?

6
Linear Systems

Many applications of mathematics require the solution of a large number of equations or inequalities having many variables. Any set of equations is a **system of equations.** The solution of a system is the intersection of the solution sets of the individual equations. If the system includes any inequalities, it is called a **system of inequalities.** It is customary to write a system by listing the equations or inequalities without set braces. For example, the system of equations $2x + y = 4$ and $x - y = 6$ is written as

$$\begin{aligned} 2x + y &= 4 \\ x - y &= 6. \end{aligned}$$

In this chapter, we discuss methods of solving systems of equations or inequalities. Use of the computer has led to increased application of linear systems since large systems with many variables can now be quickly solved. Some of the methods we present here are well suited to computer solution of linear systems.

6.1 SOLUTION OF SYSTEMS OF LINEAR EQUATIONS BY ELIMINATION

A first-degree equation in n unknowns is any equation of the form

$$a_1x_1 + a_2x_2 + \ldots + a_nx_n = k,$$

where $a_1, a_2, \ldots, a_n$ and k are all real numbers. For example,

$2x+3y=6$ is a first-degree equation in two unknowns,

$4x-3y+2z=8$ is a first-degree equation in three unknowns,

$6x-y+7z+2w=1$ is a first-degree equation in four unknowns,

and so on. As discussed in Section 1.2, first degree equations are also called linear equations. We confine our discussion of linear equations to those with two or three variables, although the methods of solution can be extended to systems with more variables.

Recall from Chapter 2 that the graph of the solution set of a linear equation in two variables, such as $5x - 2y = 8$, is a straight line. Thus, there are three possibilities for the solution set of a system of two linear equations in two variables.

1. The two graphs are lines intersecting at a single point. The coordinates of this point give the solution of the system. [See Figure 6.1(a).]

2. The graphs are distinct parallel lines. When this is the case, the system is said to be **inconsistent**; that is, there is no solution common to both equations. The solution set of such a system is $\varnothing$. [See Figure 6.1(b).]

3. The graphs are the same line. In this case, the equations are said to be **dependent,** since any solution of one equation is also a solution of the other. Here the solution set of the system can be written as either of the two identical sets $\{(x, y) | a_1x + b_1y = c_1\}$ or $\{(x, y) | a_2x + b_2y = c_2\}$.[See Figure 6.1(c).]

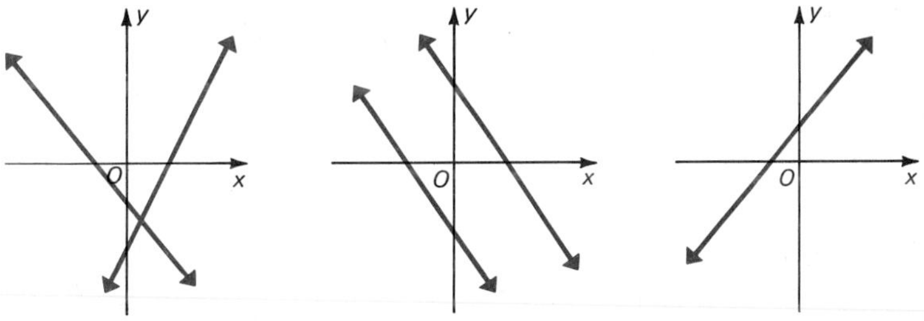

(a) (b) (c)

Figure 6.1

An equation formed by adding the corresponding sides of a pair of equations is called a **linear combination** of the two equations. For example, given $5x - y = 3$ and $2x + 7y = 9$, a linear combination of these equations is their sum,

$$7x + 6y = 12.$$

We can use linear combinations to solve systems of equations by eliminating variables from a system until only one variable is left. Once the value of this variable is found, the values of the remaining variables are then found by substitution. The following example illustrates the method.

Example 1 Solve the system

$$3x - 4y = 1 \quad (1)$$
$$2x + 3y = 12. \quad (2)$$

We need to find a way to eliminate either x or y from this system. Notice that if we multiply both sides of equation (1) by 2 and both sides of equation (2) by -3, we obtain two new equations

$$6x - 8y = 2 \qquad \textbf{(3)}$$

and

$$-6x - 9y = -36. \qquad \textbf{(4)}$$

By taking the linear combination of equations (3) and (4), we can now eliminate the x variable.

$$\begin{array}{r} 6x - 8y = \quad 2 \\ -6x - 9y = -36 \\ \hline -17y = -34 \end{array}$$

This gives us the value of y. $\qquad y = 2$

To find the value of x, we substitue 2 for y in one of the given equations. By substituting 2 for y in equation (1), we have

$$\begin{aligned} 3x - 4(2) &= 1 \\ 3x &= 9 \\ x &= 3. \end{aligned}$$

Thus, the solution set of the given system is $\{(3, 2)\}$, which can be checked by substituting 3 for x and 2 for y in equation (2). The graph of the system is shown in Figure 6.2. ∎

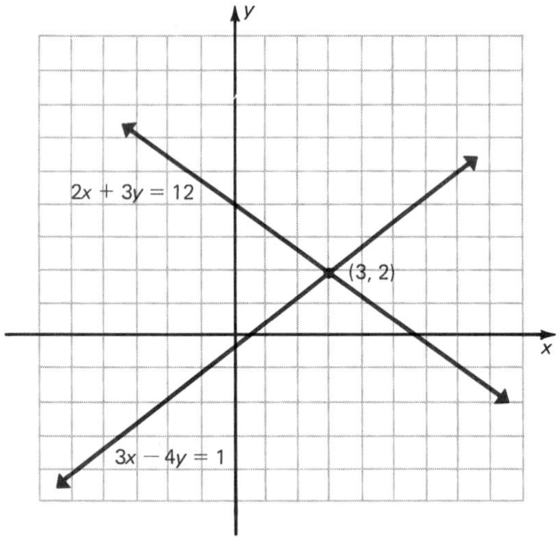

Figure 6.2

Since the method of solution shown in Example 1 results in the elimination of one variable from the system, it is called the **elimination method** for solving a system.

1. Solve the system of equations

$$3x + 2y = -1$$
$$5x - 3y = 11.$$

Answer: $\{(1, -2)\}$

Work Problem 1 at the side.

Example 2 Solve the system

$$3x - 2y = 4$$
$$-6x + 4y = 7.$$

We can eliminate the variable x if we multiply both sides of the top equation by 2 and then form a linear combination by adding.

$$\begin{aligned} 6x - 4y &= 8 \\ -6x + 4y &= 7 \\ \hline 0 &= 15 \end{aligned}$$

Both variables are eliminated and we end up with the false statement $0 = 15$. This is a signal that these two equations have no solutions at all in common. Thus the system of equations is inconsistent and the solution set is $\varnothing$. (See Figure 6.3.) ■

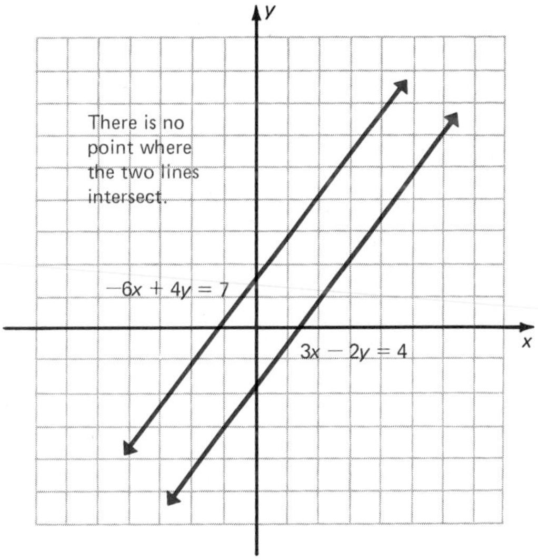
y
x
There is no point where the two lines intersect.
−6x + 4y = 7
3x − 2y = 4

Figure 6.3

Example 3 Solve the system

$$-4x + y = 2$$
$$8x - 2y = -4.$$

Multiply both sides of the top equation by 2 and then form a linear equation by adding.

$$\begin{aligned} -8x + 2y &= 4 \\ 8x - 2y &= -4 \\ \hline 0 &= 0 \end{aligned}$$

This true statement, $0 = 0$, tells us that the two equations have the same graph. Thus, there is an infinite number of solutions for this system. (See Figure 6.4.) The solution set can be written as either

$$\{(x, y) | -4x + y = 2\} \quad \text{or} \quad \{(x, y) | 8x - 2y = -4\}. \blacksquare$$

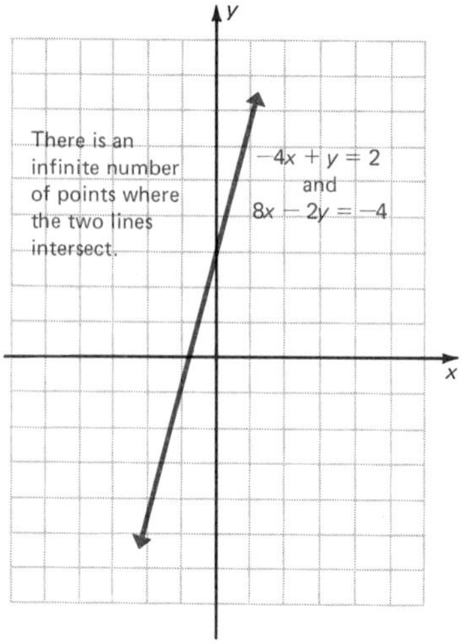
y
x
There is an infinite number of points where the two lines intersect.
−4x + y = 2
and
8x − 2y = −4

Figure 6.4

2. Solve each of the following systems.

(a) $\begin{aligned} 2x + 7y &= 8 \\ -4x - 14y &= 12 \end{aligned}$

(b) $\begin{aligned} 3x - 4y &= 13 \\ 12x - 16y &= 52 \end{aligned}$

Answers:

(a) $\varnothing$

(b) $\{(x, y) | 3x - 4y = 13\}$ or $\{(x, y) | 12x - 16y = 52\}$

Work Problem 2 at the side.

Any one solution of a linear equation in three variables $ax + by + cz = k$ is an **ordered triple** (x, y, z). The solution set of such an equation is comprised of an infinite set of ordered triples. It is shown in geometry that the graph of a linear equation in three variables is a plane. By considering the possible intersections of the planes representing three equations in three unknowns, we can see that the solution set of such a system may be either a single ordered triple (a *point*), an infinite set of ordered triples (a *line* or *plane*), or $\varnothing$.

We again use the elimination method to solve a system in three or more variables, as shown in the next example.

Example 4 Solve the system

$$2x + y - z = 2 \quad \textbf{(5)}$$
$$x + 3y + 2z = 1 \quad \textbf{(6)}$$
$$x + y + z = 2. \quad \textbf{(7)}$$

We need to use several steps to obtain an equation in one variable. First, we must eliminate the *same variable* from each of two pairs of equations. Then we must eliminate another variable from the two resulting equations. Suppose we wish to eliminate x first. Multiplying both sides of (5) by -1 and both sides of (6) by 2, we have

$$\begin{array}{r} -2x - y + z = -2 \\ 2x + 6y + 4z = 2 \\ \hline 5y + 5z = 0 \end{array}$$

or

$$y + z = 0. \qquad \textbf{(8)}$$

We must now eliminate x again from a different pair of equations, say (6) and (7). To do this, multiply both sides of (6) by -1 and both sides of (7) by 1.

$$\begin{array}{r} -x - 3y - 2z = -1 \\ x + y + z = 2 \\ \hline -2y - z = 1 \end{array} \qquad \textbf{(9)}$$

Equations (8) and (9) can now be added to find y.

$$\begin{array}{r} -2y - z = 1 \\ y + z = 0 \\ \hline -y \quad = 1 \end{array} \qquad \text{or} \qquad y = -1$$

By substituting -1 for y in equation (8), we find z. [We could also use equation (9).]

$$\begin{aligned} y + z &= 0 \\ -1 + z &= 0 \\ z &= 1 \end{aligned}$$

Now, using equation (7) and substituting -1 for y and 1 for z, we find x. [We could also use equation (5) or (6).]

$$\begin{aligned} x + y + z &= 2 \\ x - 1 + 1 &= 2 \\ x &= 2 \end{aligned}$$

Thus the solution set of the system is $\{(2, -1, 1)\}$. ■

Work Problem 3 at the side.

3. Solve the system

$$\begin{aligned} 2x - y + 3z &= 2 \\ x + 2y - z &= 6 \\ -x - y + z &= -5. \end{aligned}$$

Answer: $\{(3, 1, -1)\}$

The systems of equations discussed so far have had the same number of equations as variables. When this is the case, there is either *one* solution, *no* solution, or an *infinite number* of solutions. However, sometimes a model leads to a system of equations with fewer equations than variables. Such systems always have an infinite number of solutions or no solution.

Example 5 Solve

$$\begin{aligned} 2x + 3y + 4z &= 6 \\ x - 2y + z &= 9 \\ 3x - 6y + 3z &= 27. \end{aligned}$$

Note that the third equation here is a multiple of the second equation. Thus, this system really has only the two equations,

$$\begin{aligned} 2x + 3y + 4z &= 6 \\ x - 2y + z &= 9. \end{aligned} \qquad \textbf{(10)}$$

We can eliminate x (we could have chosen y or z instead) by first multiplying equation (10) by -2.

$$\begin{aligned} 2x + 3y + 4z &= 6 \\ -2x + 4y - 2z &= -18 \\ \hline 7y + 2z &= -12 \end{aligned}$$

We can now solve for either y in terms of z or z in terms of y. Let's solve for z.

$$\begin{aligned} 2z &= -12 - 7y \\ z &= -\frac{7}{2}y - 6 \end{aligned}$$

Substituting for z in equation (10) gives x in terms of y.

$$x - 2y + z = 9 \qquad \textbf{(10)}$$

$$x - 2y + \left(-\frac{7}{2}y - 6\right) = 9$$

$$x = \frac{11}{2}y + 15$$

We now have both x and z in terms of y. Each choice of a value for y leads to values for x and z. (For this reason, y in this system is called a **parameter.**) For example, verify the following.

If $y = -2$, $x = 4$ and $z = 1$.
If $y = 0$, $x = 15$ and $z = -6$.
If $y = 2$, $x = 26$ and $z = -13$.

An infinite number of ordered triples are possible. This infinite set is the solution to the system. This kind of solution is sometimes written as

$$\begin{aligned} &y \text{ arbitrary} \\ &x = \frac{11}{2}y + 15 \\ &z = -\frac{7}{2}y - 6. \end{aligned}$$ ■

4. Solve the system of Example 5 by eliminating z instead of x. Then let x be the arbitrary variable. Verify that the solution is the same by checking the ordered triples found in Example 5.

Answer: x arbitrary, $y = \frac{2}{11}x - \frac{30}{11}$, $z = -\frac{7}{11}x + \frac{39}{11}$

Work Problem 4 at the side.

Example 5 discussed a system with one more variable than equation. If there are two more variables than equations, there will be two parameters (arbitrary variables), and so on. What about the case where there are fewer variables than equations? For example, consider the system

$$\begin{aligned} 2x + 3y &= 8 \\ x - y &= 4 \\ 5x + y &= 7. \end{aligned}$$

Since each equation has a line as its graph, the possibilities are three lines which intersect at a common point, three lines which cross at three different points, three lines of which two are the same line so that the intersection would be a point, three lines of which two are parallel so that the intersection would be two different points, three lines which are all parallel so that there would be no intersection, and so on. Verify that the point (1, 2) satisfies both the first and last equations but not the second equation of the system above. Therefore, the system in the example has no solution.

5. Solve

$$\begin{aligned} 3x + y &= 12 \\ x - 4y &= 17 \\ 2x + 5y &= -5. \end{aligned}$$

Answer: $\{(5, -3)\}$

Work Problem 5 at the side.

Example 6 The Busy Bee Bakery sells two special cakes: Chocolate Dream Cake which requires 1/2 hour of the baker's time and costs \$3 for ingredients, and Coconut Surprise Cake which requires 1/2 hour to prepare and costs \$2 for ingredients. How many of each type can the bakery turn out each day if the baker spends 4 hours per day making these cakes and the bakery spends \$18 per day on their ingredients?

Let x represent the number of Chocolate Dream Cakes and y the number of Coconut Surprise Cakes the bakery can make each day. The values of x and y are restricted by the amount of time and money available for these cakes.

$$\tfrac{1}{2}x + \tfrac{1}{2}y = 4 \qquad \text{(Time restriction)}$$
$$3x + 2y = 18 \qquad \text{(Cost restriction)}$$

This is a system of equations which we can solve, using the methods of this section, to get $x = 2$ and $y = 6$. Thus, the bakery can make 2 Chocolate Dream Cakes and 6 Coconut Surprise Cakes each day under the conditions of the problem. ■

6. Write a system of equations to solve Example 6 if the baker spends 1/3 hour per cake on the Dream Cakes and 1/2 hour per cake on the Surprise Cakes, and the costs are changed to \$2.50 and \$1.50 respectively.

Answer: $\frac{1}{3}x + \frac{1}{2}y = 4$

$2.5x + 1.5y = 18$

Work Problem 6 at the side.

Example 7 Find the equation of the parabola $y = ax^2 + bx + c$ which goes through the points $(0, 3)$, $(2, 15)$, and $(-3, 45)$.

We need to find suitable values for a, b, and c. Since the given points satisfy the equation $y = ax^2 + bx + c$, we can write

$$3 = a(0)^2 + b(0) + c \quad \text{or} \quad 3 = c$$
$$15 = a(2)^2 + b(2) + c \quad \text{or} \quad 15 = 4a + 2b + c$$
$$45 = a(-3)^2 + b(-3) + c \quad \text{or} \quad 45 = 9a - 3b + c.$$

This is a system of equations which can be solved by the methods of this section to give $a = 4$, $b = -2$, and $c = 3$. The equation of the parabola through the given points, thus, is

$$y = 4x^2 - 2x + 3. \quad ■$$

7. Find the equation of the parabola $y = ax^2 + bx + c$ which goes through the points $(0, 4)$, $(2, 0)$ and $(1, 1)$.

Answer: $y = x^2 - 4x + 4$

Work Problem 7 at the side.

6.1 EXERCISES

Use the elimination method to solve each of the following systems of two equations in two unknowns. (See Examples 1–3.)

1. $x + y = 9$
$2x - y = 0$

2. $4x + y = 9$
$3x - y = 5$

3. $5x + 3y = 7$
$7x - 3y = -19$

4. $2x + 7y = -8$
$-2x + 3y = -12$

5. $3x + 2y = -6$
$5x - 2y = -10$

6. $-6x + 2y = 8$
$5x - 2y = -8$

7. $2x - 3y = -7$
$5x + 4y = 17$

8. $4m + 3n = -1$
$2m + 5n = 3$

9. $5p + 7q = 6$
$10p - 3q = 46$

10. $12s - 5t = 9$
$3s - 8t = -18$

11. $6x + 7y = -2$
$7x - 6y = 26$

12. $2a + 9b = 3$
$5a + 7b = -8$

13. $\begin{aligned} 3x + 2y &= 5 \\ 6x + 4y &= 8 \end{aligned}$

14. $\begin{aligned} 9x - 5y &= 1 \\ -18x + 10y &= 1 \end{aligned}$

15. $\begin{aligned} 4x - y &= 9 \\ -8x + 2y &= -18 \end{aligned}$

16. $\begin{aligned} 3x + 5y + 2 &= 0 \\ 9x + 15y + 6 &= 0 \end{aligned}$

In Exercises 17–20, first multiply both sides of each equation by its common denominator to eliminate the fractions. Then use the elimination method to solve.

17. $\begin{aligned} \frac{x}{2} + \frac{y}{3} &= 8 \\ \frac{2x}{3} + \frac{3y}{2} &= 17 \end{aligned}$

18. $\begin{aligned} \frac{x}{5} + 3y &= 31 \\ 2x - \frac{y}{5} &= 8 \end{aligned}$

19. $\begin{aligned} \frac{x}{2} + y &= \frac{3}{2} \\ \frac{x}{3} + y &= \frac{1}{3} \end{aligned}$

20. $\begin{aligned} x + \frac{y}{3} &= -6 \\ \frac{x}{5} + \frac{y}{4} &= -\frac{7}{4} \end{aligned}$

Use the elimination method to solve each of the following systems of three equations in three unknowns. Check your answers. (See Example 4.)

21. $\begin{aligned} x + y + z &= 2 \\ 2x + y - z &= 5 \\ x - y + z &= -2 \end{aligned}$

22. $\begin{aligned} 2x + y + z &= 9 \\ -x - y + z &= 1 \\ 3x - y + z &= 9 \end{aligned}$

23. $\begin{aligned} x + 3y + 4z &= 14 \\ 2x - 3y + 2z &= 10 \\ 3x - y + z &= 9 \end{aligned}$

24. $\begin{aligned} 4x - y + 3z &= -2 \\ 3x + 5y - z &= 15 \\ -2x + y + 4z &= 14 \end{aligned}$

25. $\begin{aligned} x + 2y + 3z &= 8 \\ 3x - y + 2z &= 5 \\ -2x - 4y - 6z &= 5 \end{aligned}$

26. $\begin{aligned} 3x - 2y - 8z &= 1 \\ 9x - 6y - 24z &= -2 \\ x - y + z &= 1 \end{aligned}$

27. $\begin{aligned} 2x - 4y + z &= -4 \\ x + 2y - z &= 0 \\ -x + y + z &= 6 \end{aligned}$

28. $\begin{aligned} 4x - 3y + z &= 9 \\ 3x + 2y - 2z &= 4 \\ x - y + 3z &= 5 \end{aligned}$

29. $\begin{aligned} x + 4y - z &= 6 \\ 2x - y + z &= 3 \\ 3x + 2y + 3z &= 16 \end{aligned}$

30. $\begin{aligned} 3x + y - z &= 7 \\ 2x - 3y + z &= -7 \\ x - 4y + 3z &= -6 \end{aligned}$

31. $\begin{aligned} 5m + n - 3p &= -6 \\ 2m + 3n + p &= 5 \\ -3m - 2n + 4p &= 3 \end{aligned}$

32. $\begin{aligned} 2r - 5s + 4t &= -35 \\ 5r + 3s - t &= 1 \\ r + s + t &= 1 \end{aligned}$

33. $\begin{aligned} a - 3b - 2c &= -3 \\ 3a + 2b - c &= 12 \\ -a - b + 4c &= 3 \end{aligned}$

34. $\begin{aligned} 2x + 2y + 2z &= 6 \\ 3x - 3y - 4z &= -1 \\ x + y + 3z &= 11 \end{aligned}$

Solve each of the following systems of equations. Let x be the arbitrary variable. (See Example 5.)

35. $\begin{aligned} 5x + 3y + 4z &= 19 \\ 3x - y + z &= -4 \end{aligned}$

36. $\begin{aligned} 3x + y - z &= 0 \\ 2x - y + 3z &= -7 \end{aligned}$

37. $\begin{aligned} x + 2y + 3z &= 11 \\ 2x - y + z &= 2 \end{aligned}$

38. $\begin{aligned} -x + y - z &= -7 \\ 2x + 3y + z &= 7 \end{aligned}$

39. $\begin{aligned} x + y - z + 2w &= -20 \\ 2x - y + z + w &= 11 \\ 3x - 2y + z - 2w &= 27 \end{aligned}$

40. $\begin{aligned} 4x + 3y + z + 2w &= 1 \\ -2x - y + 2z + 3w &= 0 \\ x + 4y + z - w &= 12 \end{aligned}$

Solve each of the following systems of equations.

41. $\begin{aligned} 5x + 2y &= 7 \\ -2x + y &= -10 \\ x - 3y &= 15 \end{aligned}$

42. $\begin{aligned} 9x - 2y &= -14 \\ 3x + y &= -4 \\ -6x - 2y &= 8 \end{aligned}$

43. $\begin{aligned} x + 7y &= 5 \\ 4x - 3y &= 2 \\ -x + 2y &= 10 \end{aligned}$

44. $\begin{aligned} -3x - 2y &= 11 \\ x + 2y &= -14 \\ 5x + y &= -9 \end{aligned}$

45. $\begin{aligned} x + y &= 2 \\ y + z &= 4 \\ x + z &= 3 \\ y - z &= 8 \end{aligned}$

46. $\begin{aligned} 2x + y &= 7 \\ x + 3z &= 5 \\ y - 2z &= 6 \\ x + 4y &= 10 \end{aligned}$

Write each of the following as a system of equations and then solve. (See Example 7.)

47. Find a and b so that the line $ax + by = 5$ contains the points $(-2, 1)$ and $(-1, -2)$.

48. Find m and b so that the line $y = mx + b$ contains the points $(4, 6)$ and $(-5, -3)$.

49. Find a, b, and c so that the graph of $y = ax^2 + bx + c$ contains the points $(2, 3)$, $(-1, 0)$, and $(-2, 2)$.

50. Find a, b, and c so that the points $(2, 14)$, $(0, 0)$, and $(-1, -1)$ lie on the graph of $y = ax^2 + bx + c$.

APPLIED PROBLEMS

Write a system of equations for each of the following; then solve the system. (See Example 6.)

Social Science

51. A working couple earned a total of \$2176. The wife earned \$32 per day; the husband earned \$4 per day less. Find the number of days each worked if the total number of days worked by both was 72.

Management

52. Midtown Manufacturing Company makes two products, plastic plates and plastic cups. Both require time on two machines: plates—one hour on machine A and two hours on machine B; cups—three hours on machine A and one hour on machine B. Both machines operate 15 hours a day. How many of each product can be produced in a day under these conditions?

53. A company produces two models of bicycles, model 201 and model 301. Model 201 requires 2 hours of assembly time and model 301 requires 3 hours of assembly time. The parts for model 201 cost \$25 per bike and the parts for model 301 cost \$30 per bike. If the company has a total of 34 hours of assembly time and \$365 available per day for these two models, how many of each can be made in a day?

54. Juanita invests \$10,000, received from her grandmother, in three ways. With one part, she buys mutual funds which offer a return of 8% per year. The second part, which amounts to twice the first, is used to buy government bonds at 9% per year. She puts the rest in the bank at 5% annual interest. The first year her investments bring a return of \$830. How much did she invest in each way?

55. To get the necessary funds for a planned expansion, a small company took out three loans totaling \$25,000. The company was able to borrow some of the money at 8%. They borrowed \$2000 more than one-half the amount of the 8% loan at 10%, and the rest at 9%. The total annual interest was \$2220. How much did they borrow at each rate?

56. The business analyst for Midtown Manufacturing wants to find an equation which can be used to project sales of a relatively new product. For the years 1976, 1977, and 1978, sales were \$15,000, \$32,000, and \$123,000 respectively.

(a) Graph the sales for the years 1976, 1977, and 1978, letting the year 1976 equal 0 on the x-axis. Let the values on the vertical axis be in thousands. [For example, the point (1977, 32,000) will be graphed as (1, 32).]

(b) Find the equation of the straight line $ax + by = c$ through the points for 1976 and 1978.

(c) Find the equation of the parabola $y = ax^2 + bx + c$ through the three given points. (See Example 7.)

(d) Find the projected sales for 1981, first by using the equation from part (b) and second, by using the equation from part (c). If you were to estimate sales of the product in 1981, which result would you choose? Why?

6.2 SOLUTION OF SYSTEMS OF LINEAR EQUATIONS BY MATRICES

We have learned how to solve linear systems of equations by the elimination method. In this section, we see how to solve these systems using matrices. (Matrix methods are particularly suitable for computer solution of large systems of equations having many unknowns.) To begin, we write a system of three equations and three unknowns in the form

$$\begin{aligned} a_1x + b_1y + c_1z &= d_1 \\ a_2x + b_2y + c_2z &= d_2 \\ a_3x + b_3y + c_3z &= d_3. \end{aligned}$$

Now we write the coefficients as a 3×3 matrix,

$$\begin{bmatrix} a_1 & b_1 & c_1 \\ a_2 & b_2 & c_2 \\ a_3 & b_3 & c_3 \end{bmatrix},$$

which we call the **coefficient matrix** of the system. If we then take the coefficient matrix and adjoin the column of constants, we have the **augmented matrix** of the system,

$$\left[\begin{array}{ccc|c} a_1 & b_1 & c_1 & d_1 \\ a_2 & b_2 & c_2 & d_2 \\ a_3 & b_3 & c_3 & d_3 \end{array}\right].$$

(The bar is used to separate the constants from the coefficients.)

We can transform the rows of this augmented matrix just as we would the equations of a system, since the augmented matrix is actually a shortened form of the system. Operations which produce valid transformations are given in the following theorem.*

Theorem 6.1 For any nonzero real number k and any augmented matrix of a system of linear equations, the following operations will result in the matrix of an equivalent system.

(a) Interchanging any two rows of a matrix.

(b) Multiplying the elements of any row of a matrix by the same nonzero real number k.

(c) Adding a multiple of one row of a matrix to another row.

The method of solving systems of equations based on Theorem 6.1 is called the **Gaussian method.** This procedure was developed by the mathematician Karl F. Gauss (1777–1855). The following example illustrates how we can use the Gaussian method to solve a system of equations.

Example 1 Solve the system

$$3x - 4y = 1 \quad \textbf{(1)}$$
$$5x + 2y = 19. \quad \textbf{(2)}$$

* These operations are the same as the row operations used in Chapter 5 to find matrix inverses.

First we write the augmented matrix for this system.

$$\left[\begin{array}{cc|c} 3 & -4 & 1 \\ 5 & 2 & 19 \end{array}\right]$$

Now if we rewrite equation (1) so that x has a coefficient of 1, it becomes

$$x - \frac{4}{3}y = \frac{1}{3}. \qquad \textbf{(3)}$$

If we replace equation (1) in the system by equation (3), the augmented matrix would then be

$$\left[\begin{array}{cc|c} 1 & -\frac{4}{3} & \frac{1}{3} \\ 5 & 2 & 19 \end{array}\right].$$

This matrix gives us a 1 for the first element in row one. Notice that it also could have been obtained from the original augmented matrix by multiplying each element in row one by 1/3 [using Theorem 6.1(b)].

Now if we solve equation (3) for x, we have

$$x = \frac{4}{3}y + \frac{1}{3}.$$

Substituting this for x in equation (2) gives

$$5\left(\frac{4}{3}y + \frac{1}{3}\right) + 2y = 19$$

$$\frac{20}{3}y + \frac{5}{3} + 2y = 19$$

$$\frac{26}{3}y = \frac{52}{3}. \qquad \textbf{(4)}$$

If we replace equation (2) in the system by equation (4), the augmented matrix would then be

$$\left[\begin{array}{cc|c} 1 & -\frac{4}{3} & \frac{1}{3} \\ 0 & \frac{26}{3} & \frac{52}{3} \end{array}\right].$$

This matrix gives us a 0 for the first element in row two. Notice that it also could have been obtained from the matrix directly above by multiplying each element of row one by -5 and adding the results to the corresponding elements in row two [using Theorem 6.1(c)].

If we now solve equation (4), we get $y = 2$. By letting $y = 2$ replace equation (4), the augmented matrix can be written as

$$\left[\begin{array}{cc|c} 1 & -\frac{4}{3} & \frac{1}{3} \\ 0 & 1 & 2 \end{array}\right].$$

This matrix gives us a 1 for the second element in row two. Notice that it also could have been obtained from the matrix above by multiplying each element of row two by 3/26 [using Theorem 6.1(b)].

Finally, if we substitute $y = 2$ into equation (3), we get

$$x - \frac{4}{3}y = \frac{1}{3}$$
$$x - \frac{4}{3}(2) = \frac{1}{3}$$
$$x = 3.$$

By letting $x = 3$ replace equation (3), the augmented matrix can be written as

$$\left[\begin{array}{cc|c} 1 & 0 & 3 \\ 0 & 1 & 2 \end{array}\right].$$

This matrix gives us a 0 for the second element in row one. This also could have been obtained from the matrix above by multiplying each element of row two by 4/3 and adding the results to the corresponding elements in row one [using Theorem 6.1(c)].

This last augmented matrix represents the system

$$\begin{aligned} x \quad &= 3 \\ y &= 2, \end{aligned}$$

which gives the solution set $\{(3, 2)\}$. Notice that this solution can be read directly from the third column of the final matrix. ■

In Example 1 we began with the augmented matrix $[A|B]$ and changed it to the augmented matrix $[I|C]$, where I represents the identity matrix. The column matrix C gave us the values of the variables in the solution. When using Theorem 6.1 to transform the augmented matrix, it is best to work column by column from left to right. For each column, a 1 in the proper position should be the result of the first change. Next, perform the steps that obtain zeros in the remainder of the column. Then proceed to the next column. (Recall that these are the same procedures used in Chapter 5 when using row operations to find the inverse of a matrix.)

Work Problem 1 at the side.

1. Use the Gaussian method to solve the system

$$\begin{aligned} 2x + 4y &= 22 \\ -4x + y &= -8. \end{aligned}$$

Answer: $\{(3, 4)\}$

Example 2 Use the Gaussian method to solve the system

$$\begin{aligned} x - y + 5z &= -6 \\ 3x + 3y - z &= 10 \\ x + 3y + 2z &= 5. \end{aligned}$$

Begin by writing the augmented matrix of the linear system.

$$\left[\begin{array}{ccc|c} 1 & -1 & 5 & -6 \\ 3 & 3 & -1 & 10 \\ 1 & 3 & 2 & 5 \end{array}\right]$$

We want the final matrix to be of the form

$$\left[\begin{array}{ccc|c} 1 & 0 & 0 & m \\ 0 & 1 & 0 & n \\ 0 & 0 & 1 & p \end{array}\right],$$

where m, n, and p are real numbers. From this final form of the matrix, we can read the solution: $x = m$, $y = n$, and $z = p$.

We already have 1 for the first element in row one. To get 0 for the first element in row two, multiply each element in the first row by -3 and add the results to the corresponding elements in the second row [using Theorem 6.1(c)].

$$\left[\begin{array}{ccc|c} 1 & -1 & 5 & -6 \\ 0 & 6 & -16 & 28 \\ 1 & 3 & 2 & 5 \end{array}\right]$$

Now, to change the first element in row three to 0, multiply each element of the first row by -1 and add the results to the corresponding elements of the third row [again using Theorem 6.1(c)].

$$\left[\begin{array}{ccc|c} 1 & -1 & 5 & -6 \\ 0 & 6 & -16 & 28 \\ 0 & 4 & -3 & 11 \end{array}\right]$$

This transforms the first column. We transform the second and third columns similarly.

$$\left[\begin{array}{ccc|c} 1 & -1 & 5 & -6 \\ 0 & 1 & -\frac{8}{3} & \frac{14}{3} \\ 0 & 4 & -3 & 11 \end{array}\right]$$ Second row multiplied by $\frac{1}{6}$ [Theorem 6.1(b)]

$$\left[\begin{array}{ccc|c} 1 & 0 & \frac{7}{3} & -\frac{4}{3} \\ 0 & 1 & -\frac{8}{3} & \frac{14}{3} \\ 0 & 4 & -3 & 11 \end{array}\right]$$ Second row added to first row [Theorem 6.1(c)]

$$\left[\begin{array}{ccc|c} 1 & 0 & \frac{7}{3} & -\frac{4}{3} \\ 0 & 1 & -\frac{8}{3} & \frac{14}{3} \\ 0 & 0 & \frac{23}{3} & -\frac{23}{3} \end{array}\right]$$ -4 times second row added to third row [Theorem 6.1(c)]

$$\begin{bmatrix} 1 & 0 & \frac{7}{3} & \Big| & -\frac{4}{3} \\ 0 & 1 & -\frac{8}{3} & \Big| & \frac{14}{3} \\ 0 & 0 & 1 & \Big| & -1 \end{bmatrix}$$ Third row multiplied by $\frac{3}{23}$ [Theorem 6.1(b)]

$$\begin{bmatrix} 1 & 0 & 0 & \Big| & 1 \\ 0 & 1 & -\frac{8}{3} & \Big| & \frac{14}{3} \\ 0 & 0 & 1 & \Big| & -1 \end{bmatrix}$$ $-\frac{7}{3}$ times third row added to first row [Theorem 6.1(c)]

$$\begin{bmatrix} 1 & 0 & 0 & \Big| & 1 \\ 0 & 1 & 0 & \Big| & 2 \\ 0 & 0 & 1 & \Big| & -1 \end{bmatrix}$$ $\frac{8}{3}$ times third row added to second row [Theorem 6.1(c)]

The linear system associated with this last augmented matrix is

$$\begin{aligned} x \qquad &= 1 \\ y \quad &= 2 \\ z &= -1, \end{aligned}$$

and the solution set is $\{(1, 2, -1)\}$. ■

2. Use the Gaussian method to solve

$$\begin{aligned} x + y - z &= 6 \\ 2x - y + z &= 3 \\ -x + y + z &= -4. \end{aligned}$$

Answer: (3, 1, −2)

Work Problem 2 at the side.

Example 3 Use the Gaussian method to solve the system

$$\begin{aligned} x + y &= 2 \\ 2x + 2y &= 5. \end{aligned}$$

Begin by writing the augmented matrix.

$$\begin{bmatrix} 1 & 1 & \Big| & 2 \\ 2 & 2 & \Big| & 5 \end{bmatrix}$$

To get a 0 for the first element in row two, multiply the numbers in row one by −2 and add the results to the corresponding elements in row two.

$$\begin{bmatrix} 1 & 1 & \Big| & 2 \\ 0 & 0 & \Big| & 1 \end{bmatrix}$$

The next step is to get a 1 for the second element in row two. Since this is impossible, we cannot go further.

If we put the matrix back into equation form, we have

$$\begin{aligned} x + y &= 2 \\ 0x + 0y &= 1. \end{aligned}$$

Since the second equation is $0 = 1$, we see that the system is inconsistent and has no solution. ■

Example 4 Use the Gaussian method to solve

$$\begin{aligned} x + 2y - z &= 0 \\ 3x - y + z &= 6 \\ -2x - 4y + 2z &= 0. \end{aligned}$$

The augmented matrix is

$$\left[\begin{array}{rrr|r} 1 & 2 & -1 & 0 \\ 3 & -1 & 1 & 6 \\ -2 & -4 & 2 & 0 \end{array}\right].$$

The first element in row one is a 1. We use Theorem 6.1 to get zeros in the rest of column one.

$$\left[\begin{array}{rrr|r} 1 & 2 & -1 & 0 \\ 0 & -7 & 4 & 6 \\ -2 & -4 & 2 & 0 \end{array}\right]$$

$$\left[\begin{array}{rrr|r} 1 & 2 & -1 & 0 \\ 0 & -7 & 4 & 6 \\ 0 & 0 & 0 & 0 \end{array}\right]$$

At this point we cannot continue since all entries in the last row are zero. (Why?) Converting back to the system of equations, we have

$$\begin{aligned} x + 2y - z &= 0 \\ -7y + 4z &= 6 \\ 0 &= 0. \end{aligned}$$

To complete the solution, select an arbitrary parameter, say z. Verify that the solution can then be written as

$$\begin{aligned} &z \text{ arbitrary} \\ &y = -\frac{6}{7} + \frac{4}{7}z \\ &x = \frac{12}{7} - \frac{1}{7}z. \quad \blacksquare \end{aligned}$$

Work Problem 3 at the side.

3. Use the Gaussian method to solve the following.

(a) $\begin{aligned} 3x + 9y &= -6 \\ -x - 3y &= 2 \end{aligned}$

(b) $\begin{aligned} 2x + 9y &= 12 \\ 4x + 18y &= 5 \end{aligned}$

Answer:

(a) y arbitrary, $x = -3y - 2$

(b) $\varnothing$

6.2 EXERCISES

Write the augmented matrix for each of the following systems. ***Do not solve.***

1. $\begin{aligned} 2x + 3y &= 11 \\ x + 2y &= 8 \end{aligned}$

2. $\begin{aligned} 3x + 5y &= -13 \\ 2x + 3y &= -9 \end{aligned}$

3. $\begin{aligned} x + 5y &= 6 \\ y &= 1 \end{aligned}$

4. $\begin{aligned} 2x + 7y &= 1 \\ 5x \quad &= -15 \end{aligned}$

5. $\begin{aligned} 2x + y + z &= 3 \\ 3x - 4y + 2z &= -7 \\ x + y + z &= 2 \end{aligned}$

6. $\begin{aligned} 4x - 2y + 3z &= 4 \\ 3x + 5y + z &= 7 \\ 5x - y + 4z &= 7 \end{aligned}$

7. $\begin{aligned} x + y \quad &= 2 \\ 2y + z &= -4 \\ z &= 2 \end{aligned}$

8. $\begin{aligned} x \quad &= 6 \\ y + 2z &= 2 \\ x \quad - 3z &= 6 \end{aligned}$

9. $\begin{aligned} x \quad &= 5 \\ y \quad &= -2 \\ z &= 3 \end{aligned}$

10. $\begin{aligned} x \quad &= 8 \\ y + z &= 6 \\ z &= 2 \end{aligned}$

Write the system of equations associated with each of the following augmented matrices. ***Do not solve.***

11. $\left[\begin{array}{cc|c} 1 & 0 & 2 \\ 0 & 1 & 3 \end{array}\right]$

12. $\left[\begin{array}{cc|c} 1 & 0 & 5 \\ 0 & 1 & -3 \end{array}\right]$

13. $\left[\begin{array}{cc|c} 2 & 1 & 1 \\ 3 & -2 & -9 \end{array}\right]$

14. $\left[\begin{array}{cc|c} 1 & -5 & -18 \\ 6 & 2 & 20 \end{array}\right]$

15. $\left[\begin{array}{ccc|c} 1 & 0 & 0 & 2 \\ 0 & 1 & 0 & 3 \\ 0 & 0 & 1 & -2 \end{array}\right]$

16. $\left[\begin{array}{ccc|c} 1 & 0 & 1 & 4 \\ 0 & 1 & 0 & 2 \\ 0 & 0 & 1 & 3 \end{array}\right]$

Use the Gaussian method to solve each of the following systems of equations. (See Examples 1–3.)

17. $\begin{aligned} x + y &= 5 \\ x - y &= -1 \end{aligned}$

18. $\begin{aligned} x + 2y &= 5 \\ 2x + y &= -2 \end{aligned}$

19. $\begin{aligned} x + y &= -3 \\ 2x - 5y &= -6 \end{aligned}$

20. $\begin{aligned} 3x - 2y &= 4 \\ 3x + y &= -2 \end{aligned}$

21. $\begin{aligned} 2x - 3y &= 10 \\ 2x + 2y &= 5 \end{aligned}$

22. $\begin{aligned} 4x + y &= 5 \\ 2x + y &= 3 \end{aligned}$

23. $\begin{aligned} 2x - 5y &= 10 \\ 4x - 5y &= 15 \end{aligned}$

24. $\begin{aligned} 4x - 2y &= 3 \\ -2x + 3y &= 1 \end{aligned}$

25. $\begin{aligned} 2x - 3y &= 2 \\ 4x - 6y &= 1 \end{aligned}$

26. $\begin{aligned} x + 2y &= 1 \\ 2x + 4y &= 3 \end{aligned}$

27. $\begin{aligned} 6x - 3y &= 1 \\ -12x + 6y &= -2 \end{aligned}$

28. $\begin{aligned} x - y &= 1 \\ -x + y &= -1 \end{aligned}$

29. $\begin{aligned} x + y \quad &= -1 \\ y + z &= 4 \\ x \quad + z &= 1 \end{aligned}$

30. $\begin{aligned} x \quad - z &= -3 \\ y + z &= 9 \\ x \quad + z &= 7 \end{aligned}$

31. $\begin{aligned} x + y - z &= 6 \\ 2x - y + z &= -9 \\ x - 2y + 3z &= 1 \end{aligned}$

32. $\begin{aligned} x + 3y - 6z &= 7 \\ 2x - y + 2z &= 0 \\ x + y + 2z &= -1 \end{aligned}$

33. $\begin{aligned} -x + y \quad &= -1 \\ y - z &= 6 \\ x \quad + z &= -1 \end{aligned}$

34. $\begin{aligned} x + y \quad &= 1 \\ 2x \quad - z &= 0 \\ y + 2z &= -2 \end{aligned}$

35.
$$\begin{aligned} x - 2y + z &= 5 \\ 2x + y - z &= 2 \\ -2x + 4y - 2z &= 2 \end{aligned}$$

36.
$$\begin{aligned} 3x + 5y - z &= 0 \\ 4x - y + 2z &= 1 \\ -6x - 10y + 2z &= 0 \end{aligned}$$

37.
$$\begin{aligned} x + 2y - w &= 3 \\ 2x + 4z + 2w &= -6 \\ x + 2y - z &= 6 \\ 2x - y + z + w &= -3 \end{aligned}$$

38.
$$\begin{aligned} x + 3y - 2z - w &= 9 \\ 2x + 4y + 2w &= 10 \\ -3x - 5y + 2z - w &= -15 \\ x - y - 3z + 2w &= 6 \end{aligned}$$

APPLIED PROBLEMS

Management

At rush hours, substantial traffic congestion is encountered at the traffic intersections shown in the figure.

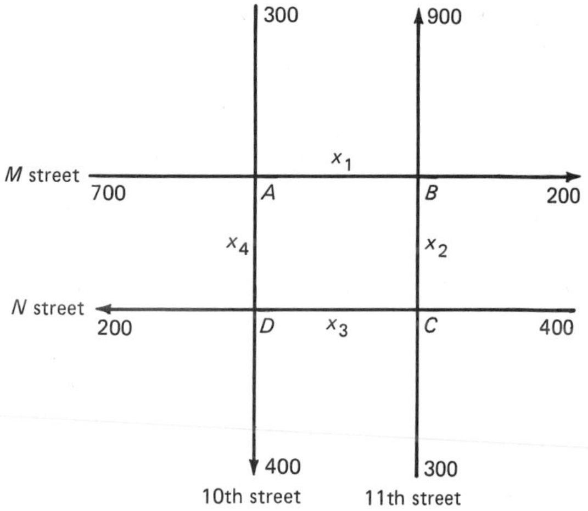

The city wishes to improve the signals at these corners so as to speed the flow of traffic. The traffic engineers first gather data. As the figure shows, 700 cars per hour come down M Street to intersection A; 300 cars per hour come to intersection A on 10th Street. A total of x_1 of these cars leave A on M Street, while x_4 cars leave A on 10th Street. The number of cars entering A must equal the number leaving, so that

$$x_1 + x_4 = 700 + 300$$

or

$$x_1 + x_4 = 1000.$$

For intersection B, x_1 cars enter B on M Street, and x_2 cars enter B on 11th Street. The figure shows that 900 cars leave B on 11th while 200 leave on M. We have

$$x_1 + x_2 = 900 + 200$$
$$x_1 + x_2 = 1100.$$

At intersection C, 400 cars enter on N Street, 300 on 11th Street, while x_2 leave on 11th Street and x_3 leave on N Street. This gives

$$x_2 + x_3 = 400 + 300$$
$$x_2 + x_3 = 700.$$

Finally, intersection D has x_3 cars entering on N and x_4 entering on 10th. There are 400 leaving D on 10th and 200 leaving on N, so that

$$x_3 + x_4 = 400 + 200$$
$$x_3 + x_4 = 600.$$

39. Use the four equations to set up an augmented matrix, and then use the Gaussian method to solve it. (Hint: keep going until you get a row of all zeros.)

40. Since you got a row of all zeros, the system of equations does not have a unique solution. Write three equations, corresponding to the three nonzero rows of the matrix.

41. Solve each of the equations for x_4.

42. One of your equations should have been $x_4 = 1000 - x_1$. What is the largest possible value of x_1 so that x_4 is not negative? What is the largest value of x_4 so that x_1 is not negative?

43. Your second equation should have been $x_4 = x_2 - 100$. Find the smallest possible value of x_2 so that x_4 is not negative.

44. For the third equation, $x_4 = 600 - x_3$, find the largest possible values of x_3 and x_4 so that neither variable is negative.

45. Give the maximum possible values of x_1, x_2, x_3, and x_4 so that all the equations are satisfied and all variables are nonnegative.

6.3 SOLUTION OF LINEAR SYSTEMS BY INVERSES

Another way to use matrices to solve linear systems is to write the system as a matrix equation $AX = B$, where X is the matrix of the variables in the system. As shown in Chapter 5, multiplying both sides of the matrix equation $AX = B$ by A^{-1} gives

$$X = A^{-1}B.$$

Thus, to find X, we first find A^{-1} and then find the product $A^{-1}B$. By letting the corresponding elements of X and $A^{-1}B$ be equal, we get the solution of the system.

Example 1 Use the inverse of the coefficient matrix to solve the linear system

$$2x - 3y = 4$$
$$x + 5y = 2.$$

To represent the system as a matrix equation, we use one matrix for the coefficients, one for the variables, and one for the constants.

$$A = \begin{bmatrix} 2 & -3 \\ 1 & 5 \end{bmatrix}, \quad X = \begin{bmatrix} x \\ y \end{bmatrix}, \quad \text{and} \quad B = \begin{bmatrix} 4 \\ 2 \end{bmatrix}$$

The system can then be written in matrix form as the equation $AX = B$ since

$$AX = \begin{bmatrix} 2 & -3 \\ 1 & 5 \end{bmatrix}\begin{bmatrix} x \\ y \end{bmatrix} = \begin{bmatrix} 2x - 3y \\ x + 5y \end{bmatrix} = \begin{bmatrix} 4 \\ 2 \end{bmatrix} = B.$$

To solve the system, we must find A^{-1}. To do this, we use row operations on augmented matrix $[A|I]$ to get

$$\left[\begin{array}{cc|cc} 1 & 0 & \frac{5}{13} & \frac{3}{13} \\ 0 & 1 & -\frac{1}{13} & \frac{2}{13} \end{array}\right].$$

Therefore, $$A^{-1} = \begin{bmatrix} \frac{5}{13} & \frac{3}{13} \\ -\frac{1}{13} & \frac{2}{13} \end{bmatrix}.$$

Next, find the product $A^{-1}B$.

$$A^{-1}B = \begin{bmatrix} \frac{5}{13} & \frac{3}{13} \\ -\frac{1}{13} & \frac{2}{13} \end{bmatrix}\begin{bmatrix} 4 \\ 2 \end{bmatrix} = \begin{bmatrix} 2 \\ 0 \end{bmatrix}$$

Since $X = A^{-1}B$,

$$X = \begin{bmatrix} x \\ y \end{bmatrix} = \begin{bmatrix} 2 \\ 0 \end{bmatrix}.$$

Thus, the solution set of the system is $\{(2, 0)\}$. ■

Work Problem 1 at the side.

Example 2 Use the inverse of the coefficient matrix to solve the system

$$\begin{aligned} -x - 2y + 2z &= 9 \\ 2x + y - z &= -3 \\ 3x - 2y + z &= -6. \end{aligned}$$

The matrices we need are

$$A = \begin{bmatrix} -1 & -2 & 2 \\ 2 & 1 & -1 \\ 3 & -2 & 1 \end{bmatrix}, \quad X = \begin{bmatrix} x \\ y \\ z \end{bmatrix}, \quad \text{and} \quad B = \begin{bmatrix} 9 \\ -3 \\ -6 \end{bmatrix}.$$

To find A^{-1}, we write the augmented matrix

$$[A|I] = \left[\begin{array}{ccc|ccc} -1 & -2 & 2 & 1 & 0 & 0 \\ 2 & 1 & -1 & 0 & 1 & 0 \\ 3 & -2 & 1 & 0 & 0 & 1 \end{array}\right]$$

1. (a) Write the matrix of coefficients, the matrix of variables, and the matrix of constants for the system

$$\begin{aligned} 2x + 6y &= -14 \\ -x - 2y &= 3. \end{aligned}$$

(b) Solve the system in part (a) by using the inverse of the coefficient matrix.

Answer:

(a) $A = \begin{bmatrix} 2 & 6 \\ -1 & -2 \end{bmatrix}, X = \begin{bmatrix} x \\ y \end{bmatrix}$

$B = \begin{bmatrix} -14 \\ 3 \end{bmatrix}$

(b) $\{(5, -4)\}$

and use row operations to get $[I|A^{-1}]$, from which

$$A^{-1} = \begin{bmatrix} \frac{1}{3} & \frac{2}{3} & 0 \\ \frac{5}{3} & \frac{7}{3} & -1 \\ \frac{7}{3} & \frac{8}{3} & -1 \end{bmatrix}.$$

We now find $A^{-1}B$.

$$A^{-1}B = \begin{bmatrix} \frac{1}{3} & \frac{2}{3} & 0 \\ \frac{5}{3} & \frac{7}{3} & -1 \\ \frac{7}{3} & \frac{8}{3} & -1 \end{bmatrix} \begin{bmatrix} 9 \\ -3 \\ -6 \end{bmatrix} = \begin{bmatrix} 1 \\ 14 \\ 19 \end{bmatrix}$$

Since $X = A^{-1}B$,

$$X = \begin{bmatrix} x \\ y \\ z \end{bmatrix} = \begin{bmatrix} 1 \\ 14 \\ 19 \end{bmatrix}.$$

Then $x = 1$, $y = 14$, $z = 19$ and the solution set is $\{(1, 14, 19)\}$. ■

6.3 EXERCISES

Solve each of the following systems of equations by using the inverse of the coefficient matrix. (See Example 1.) The inverses for the first four problems can be found in the text and exercises of Section 5.3.

1. $2x + 3y = 10$
$x - y = -5$

2. $-x + 2y = 15$
$-2x - y = 20$

3. $2x + y = 5$
$5x + 3y = 13$

4. $-x - 2y = 8$
$3x + 4y = 24$

5. $-x + y = 1$
$2x - y = 1$

6. $3x - 6y = 1$
$-5x + 9y = -1$

7. $2x - 2y = -8$
$3x - y = 4$

8. $x + 3y = -14$
$2x - y = 7$

9. $3x + 3y = 2$
$x + 2y = 3$

10. $x + 2y = 4$
$3x + 2y = -2$

11. $-x - 8y = 12$
$3x + 24y = -36$

12. $4x + 12y = 24$
$2x + 6y = 15$

Solve each of the following systems of equations by using the inverse of the coefficient matrix. (See Example 2.) The inverses for the first four problems can be found in the text and exercises of Section 5.3.

13. $-x - y - z = 1$
$4x + 5y = -2$
$y - 3z = 3$

14. $2x + 4z = -8$
$3x + y + 5z = 2$
$-x + y - 2z = 4$

15. $2x + 4y + 6z = 4$
$-x - 4y - 3z = 8$
$y - z = -4$

16. $2x + 2y - 4z = 12$
$2x + 6y = 16$
$-3x - 3y + 5z = -20$

17. $x + 2y + 3z = 5$
$2x + 3y + 2z = 2$
$-x - 2y - 4z = -1$

18. $x + y - 3z = 4$
$2x + 4y - 4z = 8$
$-x + y + 4z = -3$

19.
$$\begin{aligned} x + 2y \quad &= -10 \\ -x \quad + 4z &= 8 \\ -y + z &= 4 \end{aligned}$$

20.
$$\begin{aligned} x + \quad z &= 3 \\ y + 2z &= 8 \\ -x + y \quad &= 4 \end{aligned}$$

21.
$$\begin{aligned} 2x - 2y \quad &= 5 \\ 4y + 8z &= 7 \\ x \quad + 2z &= 1 \end{aligned}$$

22.
$$\begin{aligned} 6x + 5y \quad &= 10 \\ x \quad - z &= 3 \\ -12x - 10y \quad &= -20 \end{aligned}$$

Solve each of the following systems of equations by using the inverse of the coefficient matrix. The inverses were found in the exercises of Section 5.3.

23.
$$\begin{aligned} x - 2y + 3z \quad &= 4 \\ y - z + w &= -8 \\ -2x + 2y - 2z + 4w &= 12 \\ 2y - 3z + w &= -4 \end{aligned}$$

24.
$$\begin{aligned} x + y \quad + 2w &= 3 \\ 2x - y + z - w &= 3 \\ 3x + 3y + 2z - 2w &= 5 \\ x + 2y + z \quad &= 3 \end{aligned}$$

6.4 SYSTEMS OF LINEAR INEQUALITIES

Many mathematical models of real situations are best expressed as inequalities, rather than as equations. A **system of inequalities** may include some equations but must include at least one inequality. Since the solution set of an inequality is usually infinite, the clearest and most meaningful description of the solution set is its graph. The solution set of a system of inequalities is the graph of the intersection of the solution sets of the members of the system.

We use the idea of a **half-plane** to graph linear inequalities. A line divides the plane into three parts, the line itself and two half-planes, one on either side of the line. In Figure 6.5, the line divides

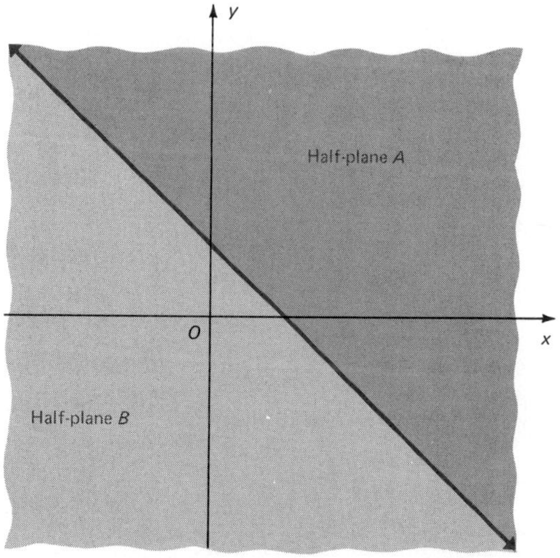

Figure 6.5

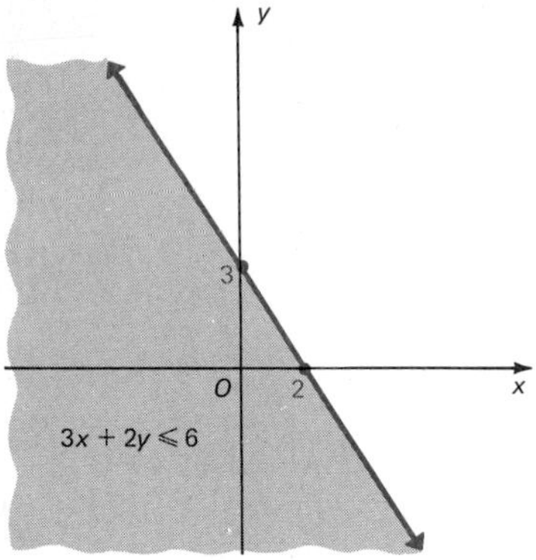

Figure 6.6

the plane into half-planes A and B, and the line itself. The line belongs to neither half-plane, and the half-planes themselves have no points in common.

A linear inequality is a relation such as $3x + 2y \leq 6$. The graph of $3x + 2y \leq 6$ is the union of the graphs of the equation $3x + 2y = 6$ and the inequality $3x + 2y < 6$. To graph the solution set of $3x + 2y \leq 6$, we first graph the line $3x + 2y = 6$. This line divides the plane into two half-planes, as shown in Figure 6.6.

It can be shown that the points in one of these half-planes will satisfy $3x + 2y < 6$, while the points in the other do not. To find which half-plane satisfies the inequality portion of our example, $3x + 2y < 6$, we choose any point in the plane that is not on the line $3x + 2y = 6$. If the coordinates of the chosen point satisfy $3x + 2y < 6$, then the half-plane containing the chosen point, together with the line $3x + 2y = 6$, gives the desired graph. The point (0, 0) is often a good point to use for the test.

If we substitute 0 for x and 0 for y in the inequality $3x + 2y < 6$, we have

$$3(0) + 2(0) < 6$$
$$0 + 0 < 6,$$

a true statement. Because of this, the point (0, 0) is located in the desired half-plane. The shaded region of Figure 6.6, together with the line, is the graph of $3x + 2y \leq 6$.

Work Problem 1 at the side.

1. Graph $2x + 5y \leq 10$.

Answer:

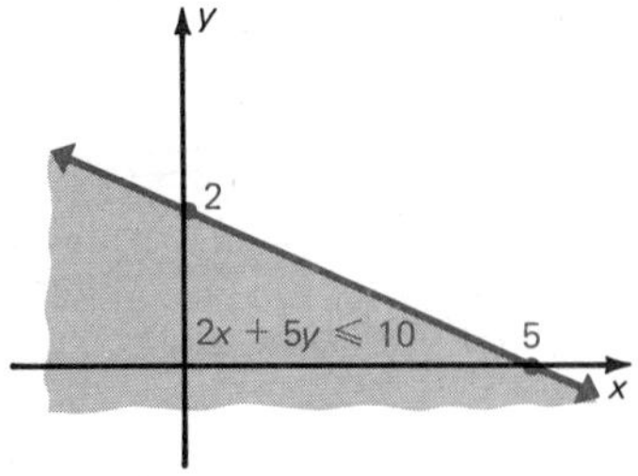

Example 1 Graph $x + 4y > 4$.

Here the points of the line $x + 4y = 4$ are not included in the graph of $x + 4y > 4$. (Why?) However, we still graph the line $x + 4y = 4$ to serve as a boundary for the graph of the inequality. It is customary to make the line dashed to show that the points of the line itself are omitted from the solution. If we try (0, 0) as a test point in the original inequality, we have

$$0 + 4(0) > 4$$
$$0 > 4,$$

a false statement. Thus, we shade the half-plane not including (0, 0), as shown in Figure 6.7. ■

Work Problem 2 at the side.

2. Graph $2x + 3y > 12$.

Answer:

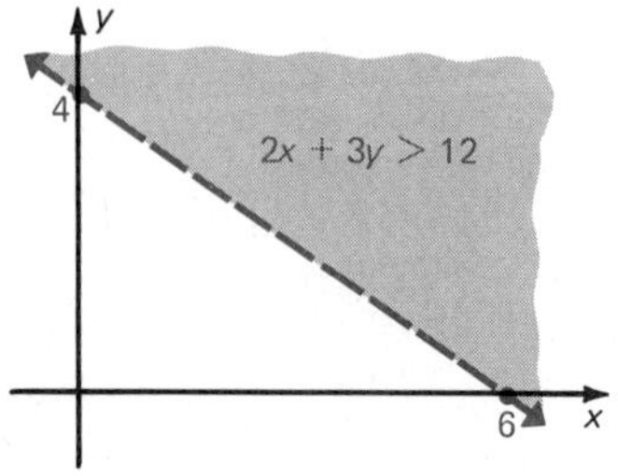

Example 2 Graph $x \leq -1$.

Recall from Chapter 3 that the straight line $x = -1$ is vertical.

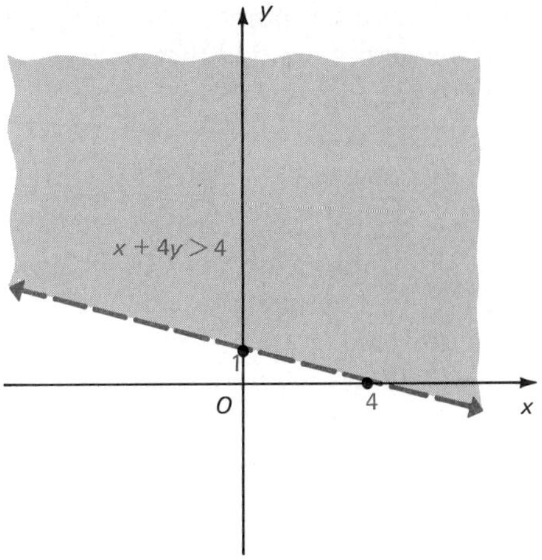

Figure 6.7

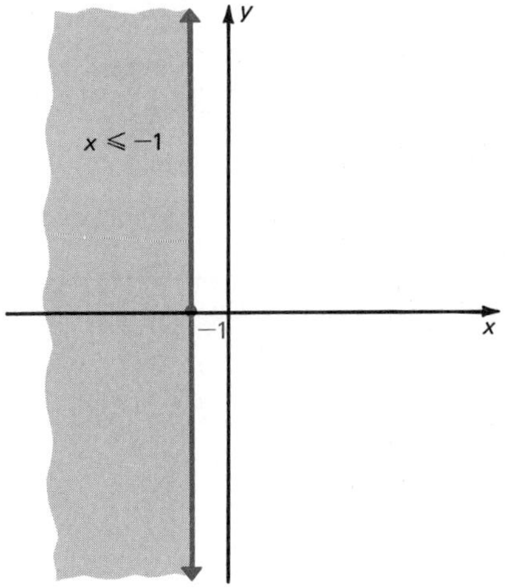

Figure 6.8

3. Graph each of the following.
(a) $x \geq 3$
(b) $y - 3 \leq 0$

Answer:
(a)

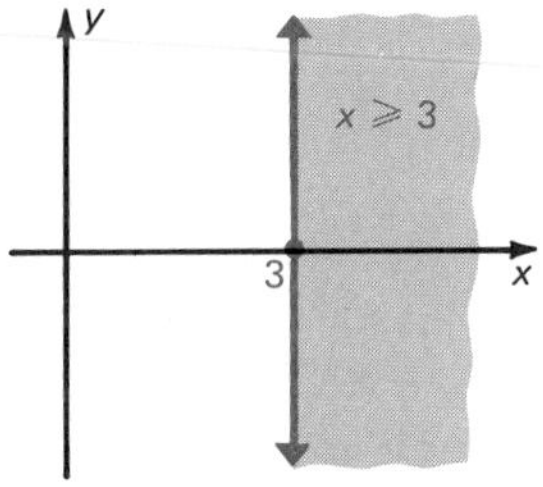

(b)

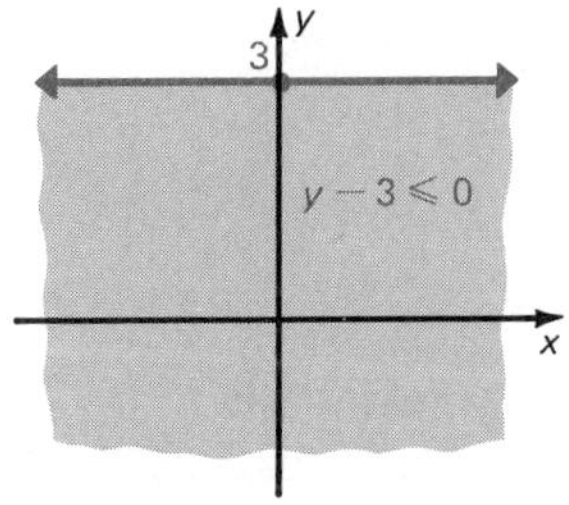

If we use (0, 0) as a test point, we find that we must shade the half-plane not including (0, 0). (See Figure 6.8.) Similarly, the graph of $y \leq -1$ would include the horizontal line $y = -1$ and the region below that line. ■

Work Problem 3 at the side.

Example 3 Graph $x \geq 3y$.

First, graph $x = 3y$, as in Figure 6.9. We cannot use (0, 0) as a test point here, since it is on the line $x = 3y$. Choose some other

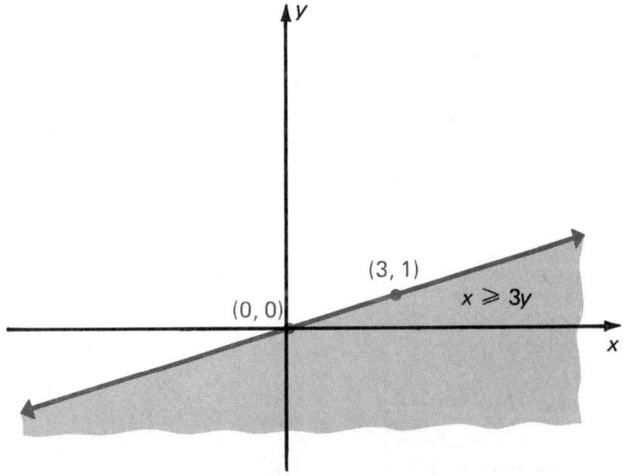

Figure 6.9

4. Graph $2x < y$.

Answer:

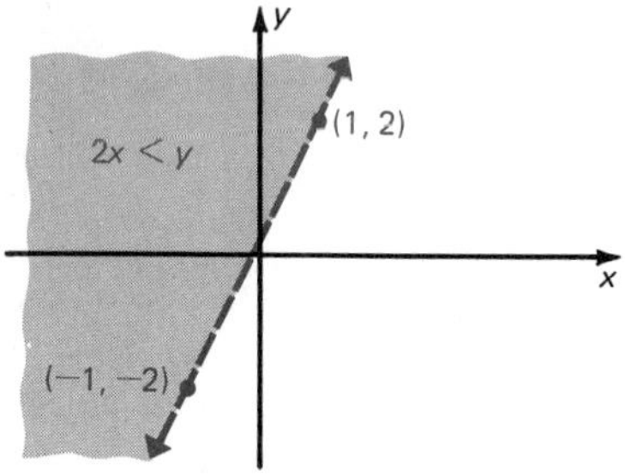

5. Graph the system

$$x + y \leq 6$$
$$2x + y \geq 4.$$

Answer:

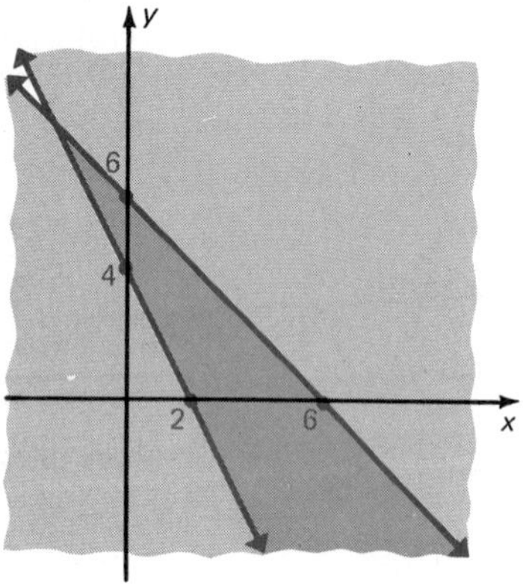

point such as (6, 1) as a test point. By substituting 6 for x and 1 for y in the original inequality, we find that we must shade the half-plane containing (6, 1). See Figure 6.9. ■

Work Problem 4 at the side.

To graph the solution of a system of inequalities, we graph all the inequalities on the same axes and identify, by heavy shading, the region common to all graphs (the intersection). The next example shows how this is done.

Example 4 Graph the system

$$y < -3x + 12$$
$$x < 2y.$$

The heavily shaded region of Figure 6.10 is the intersection of the graphs of each inequality of the system. Since the points on the boundary lines are not in the solution, the lines are dashed. ■

Work Problem 5 at the side.

Example 5 Graph the system

$$2x - 5y \leq 10$$
$$x + 2y \leq 8$$
$$x \geq 0$$
$$y \geq 0.$$

The graph is shown in Figure 6.11. ■

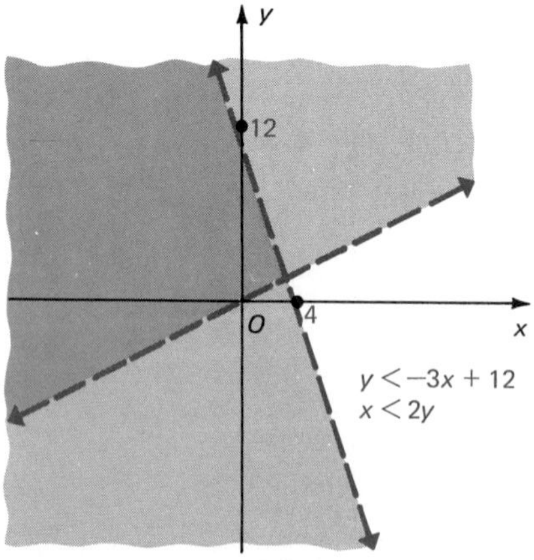

Figure 6.10

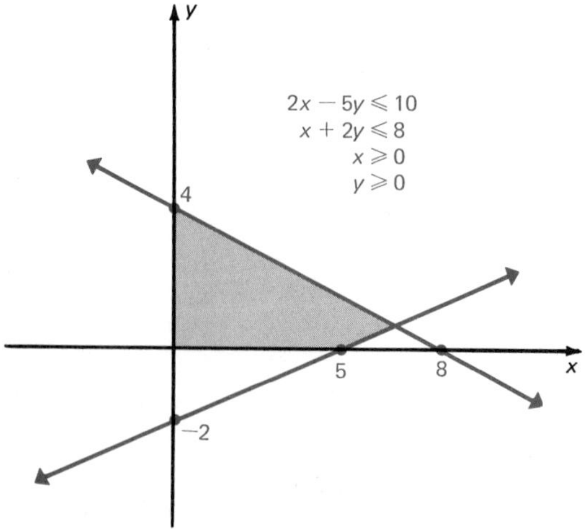

Figure 6.11

6. Graph the system

$$x + 4y \le 8$$
$$x - y \ge 3$$
$$x \ge 0$$
$$y \ge 0.$$

Answer:

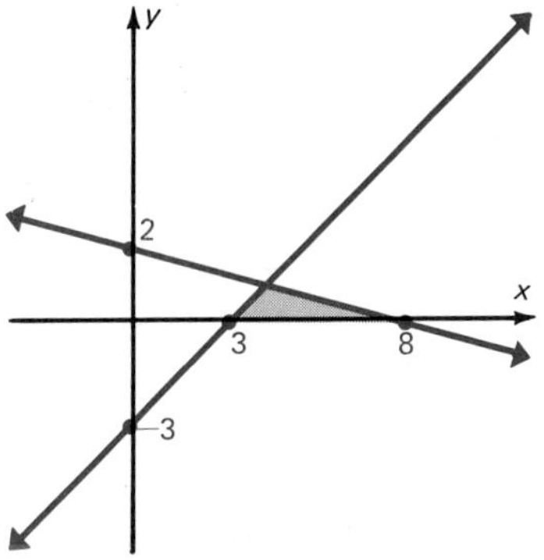

Work Problem 6 at the side.

6.4 EXERCISES

Graph each of the following linear inequalities. (See Examples 1–3.)

1. $x + y \le 2$
2. $y \le x + 1$
3. $x \ge 3 + y$
4. $y \ge x - 3$
5. $4x - y < 6$
6. $3y + x > 4$
7. $3x + y < 6$
8. $2x - y > 2$
9. $x + 3y \ge -2$
10. $2x + 3y \le 6$
11. $4x + 3y > -3$
12. $5x + 3y > 15$
13. $2x - 4y < 3$
14. $4x - 3y < 12$
15. $x \le 5y$
16. $2x \ge y$
17. $-3x < y$
18. $-x > 6y$
19. $x + y \le 0$
20. $3x + 2y \ge 0$
21. $y < x$
22. $y > -2x$
23. $x < 4$
24. $y > 5$
25. $y \le -2$
26. $x \ge 3$

Graph each of the following systems of inequalities. (See Examples 4 and 5.)

27. $x + y \le 1$; $x - y \ge 2$

28. $3x - 4y < 6$; $2x + 5y > 15$

29. $2x - y < 1$; $3x + y < 6$

30. $x + 3y \le 6$; $2x + 4y \ge 7$

31. $-x - y < 5$; $2x - y < 4$

32. $6x - 4y > 8$; $3x + 2y > 4$

33. $x + y \le 4$; $x - y \le 5$; $4x + y \le -4$

34. $3x - 2y \ge 6$; $x + y \le -5$; $y \le 4$

35. $-2 < x < 3$; $-1 \le y \le 5$; $2x + y < 6$

36. $-2 < x < 2$; $y > 1$; $x - y > 0$

37. $2y + x \ge -5$; $y \le 3 + x$; $x \ge 0$; $y \ge 0$

38. $2x + 3y \le 12$; $2x + 3y > -6$; $3x + y < 4$; $x \ge 0$; $y \ge 0$

39. $3x + 4y > 12$; $2x - 3y < 6$; $0 \le y \le 2$; $x \ge 0$

40. $0 \le x \le 9$; $x - 2y \ge 4$; $3x + 5y \le 30$; $y \ge 0$

APPLIED PROBLEMS

Management

41. The California Almond Growers have 2400 boxes of almonds to be shipped from their plant in Sacramento to Des Moines and San Antonio. The Des Moines market needs at least 1000 boxes, while the San Antonio market must have at least 800 boxes. Let x_1 = the number

of boxes to be shipped to Des Moines and x_2 = the number of boxes to be shipped to San Antonio.
(a) Write a system of inequalities to express the conditions of the problem. Remember that the variables must be nonnegative.
(b) Graph the system.

42. A cement manufacturer produces at least 3.2 million barrels of cement annually. He is told by the Environmental Protection Agency that his operation emits 2.5 pounds of dust for each barrel produced. The EPA has ruled that annual emissions must be reduced to 1.8 million pounds. To do this the manufacturer plans to replace the present dust collectors with two types of electronic precipitators. One type would reduce emissions to .5 pounds per barrel and would cost 16¢ per barrel. The other would reduce the dust to .3 pounds per barrel and would cost 20¢ per barrel. The manufacturer does not want to spend more than .8 million dollars on the precipitators. He needs to know how many barrels he should produce with each type.
(a) Let x = the number of barrels in millions produced with the first type and y = the number of barrels in millions produced with the second type. Write inequalities to express the manufacturer's restrictions. Remember that the number of barrels cannot be a negative number.
(b) Graph the system.

CHAPTER 6 TEST

[6.1] *Solve each of the following systems by the elimination method.*

1.
$$\begin{aligned} 2x + 3y &= 10 \\ -3x + y &= 18 \end{aligned}$$

2.
$$\begin{aligned} \frac{x}{2} + \frac{y}{4} &= 3 \\ \frac{x}{4} - \frac{y}{2} &= 4 \end{aligned}$$

3.
$$\begin{aligned} 2x - 3y + z &= -5 \\ x + 4y + 2z &= 13 \\ 5x + 5y + 3z &= 14 \end{aligned}$$

Write each of the following problems as a system of equations and solve.

4. Find the values for a and b so that the graph of the equation $ax + by = 4$ contains the points $(-2, 1)$ and $(4, 3)$.

5. The Waputi Indians make woven blankets, rugs, and skirts. Each blanket requires 24 hours for spinning the yarn, 4 hours for dying the yarn, and 15 hours for weaving. Rugs require 30, 5, and 18 hours, and skirts 12, 3, and 9 hours respectively. If there are 306, 59, and 201 hours available for spinning, dying, and weaving respectively, how many of each item can be made? (Hint: Simplify the equations you write, if possible, before solving the system.)

[6.2] *Solve each of the following systems of equations by the Gaussian method.*

6. $$\begin{aligned} 2x + 4y &= -6 \\ -3x - 5y &= 12 \end{aligned}$$

7. $$\begin{aligned} x - y + 3z &= 13 \\ 4x + y + 2z &= 17 \\ 3x + 2y + 2z &= 1 \end{aligned}$$

[6.3] *Solve each of the following systems of equations by inverses.*

8. $$\begin{aligned} 2x + y &= 5 \\ 3x - 2y &= 4 \end{aligned}$$

9. $$\begin{aligned} x + y + z &= 1 \\ 2x + y &= -2 \\ 3y + z &= 2 \end{aligned}$$

[6.4] *Graph the solution of each of the following systems of inequalities.*

10. $$\begin{aligned} x + y &\le 6 \\ 2x - y &\ge 3 \end{aligned}$$

11. $$\begin{aligned} x + 3y &\ge 6 \\ 4x - 3y &\le 12 \\ x &\ge 0 \\ y &\ge 0 \end{aligned}$$

7
Linear Programming

Of the applications presented in this book, linear programming is perhaps the single most useful one. It has gained importance as a quantitative tool in management in a relatively short time. While linear algebra (matrix theory) developed over several hundred years, linear programming developed from a practical problem faced by George B. Dantzig in 1947. Dantzig was concerned with the problem of allocating supplies for the United States Air Force in the least expensive way.

Many mathematical models representing problems in business, biology, and economics involve either maximizing or minimizing a linear function subject to certain constraints. In particular, a linear programming problem either maximizes or minimizes a linear function (called the **objective function**) subject to linear **constraints** which can be expressed as linear equations or inequalities. For example, we may wish to maximize a linear (objective) function representing profit which is restricted by the constraint that the number of items manufactured cannot be more than some given number. Such problems can be solved by linear programming methods.

In this chapter we first discuss a *graphical method* for solving linear programming problems involving only two variables. We then introduce the *simplex method* for solving problems with more variables or with many constraints.

7.1 THE GRAPHICAL SOLUTION

Many practical problems involve **optimizing** (that is, either maximizing or minimizing) some linear (objective) function subject to one or more linear constraints. The following examples illustrate how we can optimize an objective function by graphing a system of linear inequalities.

Example 1 Find the values of x and y which will maximize the value of the objective function $z = 2x + 5y$ subject to the linear constraints

$$\begin{aligned} 3x + 2y &\leq 6 \\ -2x + 4y &\leq 8 \\ x &\geq 0 \\ y &\geq 0. \end{aligned}$$

The graph of the solution set of these four inequalities is shown in Figure 7.1. This bounded region is called the **region of feasible solutions.** Any point in the interior or on the boundary of this region will satisfy all constraints.

Now we must choose one point (x, y) of the feasible region that gives the maximum possible value of z. To locate that point, let us add to the graph of Figure 7.1 lines which represent the objective function at possible values of z. Figure 7.2 shows the lines representing the objective function when $z = 0$, 5, 10, and 15. These lines have equations

$$\begin{aligned} 0 &= 2x + 5y \qquad & 10 &= 2x + 5y \\ 5 &= 2x + 5y \qquad & 15 &= 2x + 5y. \end{aligned}$$

(Why are all the lines parallel?) From the figure, we see that z cannot be as large as 15 because the graph for $z = 15$ is outside the

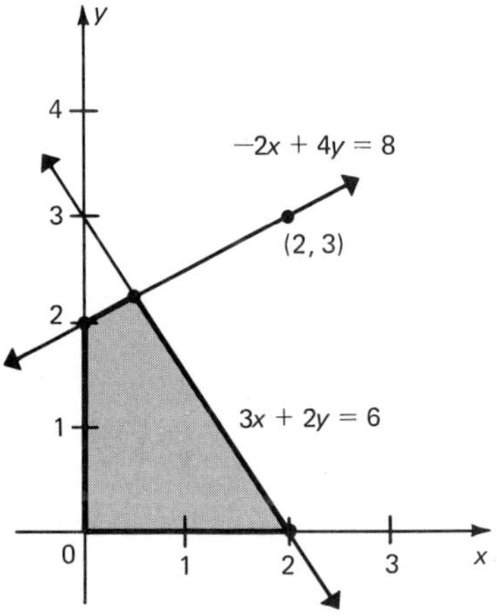

Figure 7.1

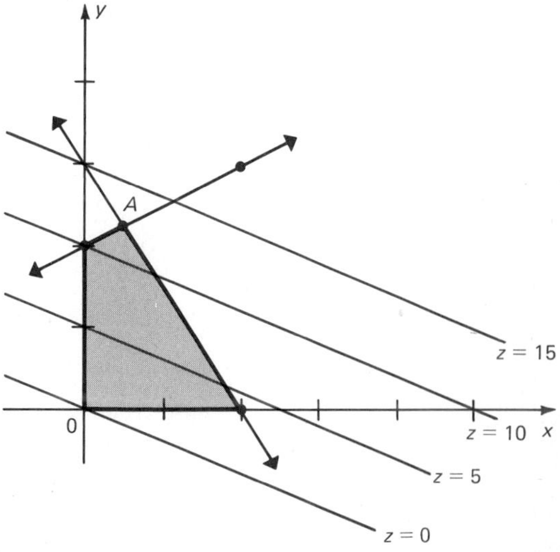

Figure 7.2

feasible region. The maximum possible value of z will be obtained if we draw another line, parallel to the others and between the lines representing the objective function when $z = 10$ and $z = 15$. We will maximize z and still satisfy all constraints if this intermediate line just "touches" the feasible region.

This occurs at point A, a **vertex** (or **corner**) **point** of the region of feasible solutions. To find the coordinates of this point A, note in Figure 7.1 that the point lies on the two lines with equations

$$3x + 2y = 6$$
$$-2x + 4y = 8.$$

By solving this system of two equations, we find that A has coordinates (1/2, 9/4). Thus, the value of z at this point is

$$z = 2x + 5y$$
$$z = 2\left(\frac{1}{2}\right) + 5\left(\frac{9}{4}\right) \qquad \text{Substitute } x = \frac{1}{2}, y = \frac{9}{4}$$
$$z = 12\frac{1}{4}.$$

The maximum possible value of z under the given conditions is thus $12\frac{1}{4}$. Out of all the feasible solutions, each of which satisfies the constraints of the problem, the vertex point (1/2, 9/4) gives us the optimum value. ■

As we can see from the parallel lines in Figure 7.2,

the optimum value will always be at a vertex point if the region of feasible solutions is bounded.

Work Problem 1 at the side.

1. Suppose the objective function above is changed to $z = 5x + 2y$.
(a) Sketch the graphs of the objective function when $z = 0$, $z = 5$, and $z = 10$ on the region of feasible solutions given in Figure 7.1.
(b) From the graph, decide what values of x and y will maximize the objective function.

Answer:
(a)

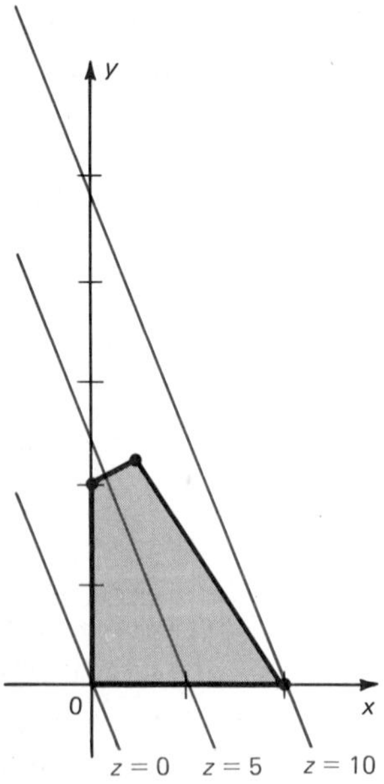

(b) (2, 0)

Example 2 Find the values of x and y which will minimize the value of the objective function $z = 2x + 4y$ subject to the linear constraints

$$x + 2y \geq 10$$
$$3x + y \geq 10$$
$$x \geq 0$$
$$y \geq 0.$$

Figure 7.3 shows the region of feasible solutions and the lines which represent the objective function when $z = 10$, $z = 20$, $z = 40$, and $z = 50$. The line representing the objective function touches the region of feasible solutions when $z = 20$. Two vertex points (2, 4) and (10, 0) lie on this line. In this case, both points (2, 4) and (10, 0) as well as all the points on the boundary line between them give the same optimum value of z. Thus, in this example, there is an infinite number of equally "good" values of x and y which give the same minimum value of the objective function $z = 2x + 4y$. ■

The feasible region of Figure 7.3 illustrates an important point about linear programming problems: *not all problems have a solution.* For example, we cannot find a maximum value of $z = 2x + 4y$. As the graph shows, the feasible region is unbounded and goes indefinitely to the upper right. Thus, there is no point that could ever lead to a maximum value of $z = 2x + 4y$.

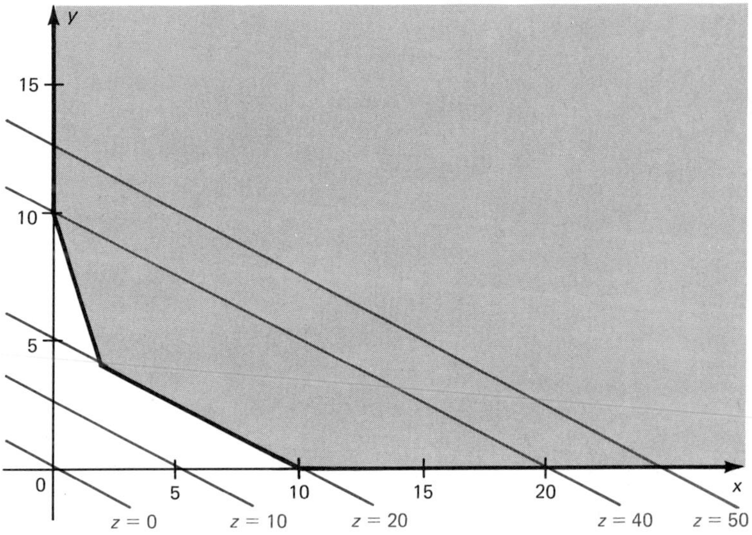

Figure 7.3

Work Problem 2 at the side.

By considering Figures 7.2 and 7.3, we can see that the following theorem is true for cases with two variables. The proof of the general case is beyond the scope of this text.

Theorem 7.1 *Fundamental Theorem of Linear Programming* If an optimum value (either a maximum or a minimum) of the objective function exists, it will occur at one or more of the vertex points.

Theorem 7.1 simplifies the job of finding an optimum value.

2. The sketch below shows a region of feasible solutions with a line representing the graph of the objective function for a particular value of z. From the sketch decide what ordered pair would minimize z.

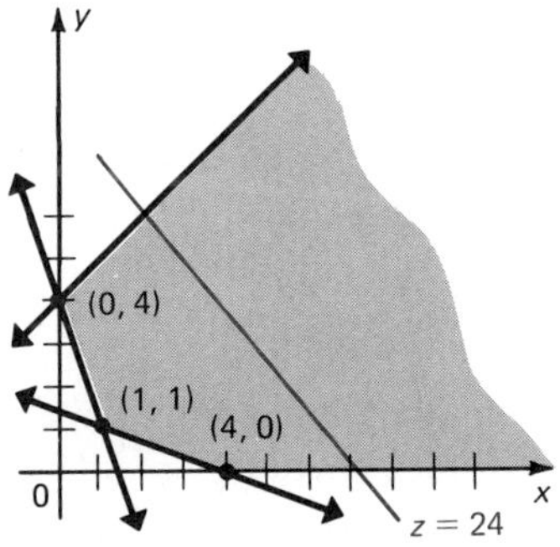

Answer: (1, 1)

All we need to do is sketch the feasible region and identify all vertex points. Then test each point in the objective function and choose the one which produces the optimum value for the objective function. For unbounded regions, we must consider whether the required optimum can be found, as illustrated in Example 2 above. For bounded regions, we no longer need to consider the parallel lines used above.

For example, we could have solved the problem in Example 1 by looking at Figure 7.1 and identifying the four vertex points as (0, 0), (0, 2), (1/2, 9/4), and (2, 0). To find each vertex point, identify the equations of any two lines which intersect. Solving this system of two equations gives the coordinates of a vertex point. Substitute each of the four points into the objective function, $z = 2x + 5y$, and then choose the vertex point that produces the maximum value of z.

Vertex point	Value of $z = 2x + 5y$
(0, 0)	$2(0) + 5(0) = 0$
(0, 2)	$2(0) + 5(2) = 10$
$(\frac{1}{2}, \frac{9}{4})$	$2(\frac{1}{2}) + 5(\frac{9}{4}) = 12\frac{1}{4}$ (maximum)
(2, 0)	$2(2) + 5(0) = 4$

From these results, we see that the vertex point (1/2, 9/4) yields the maximum value of $12\frac{1}{4}$. This is the same result that we obtained above.

3. (a) Identify the vertex points in the graph from Problem 2.
(b) Which vertex point would minimize $z = 2x + 3y$?

Answer:
(a) (0, 4), (1, 1), (4, 0)
(b) (1, 1)

Work Problem 3 at the side.

Example 3 Sketch the feasible region for the following set of constraints.

$$3y - 2x \geq 0$$
$$y + 8x \leq 52$$
$$y - 2x \leq 2$$
$$x \geq 3$$

Then find the maximum and minimum values of the objective function

$$z = 5x + 2y.$$

From the graph in Figure 7.4, we see that the feasible region is bounded. We use the vertex points from the graph to find the maximum and minimum values of the objective function.

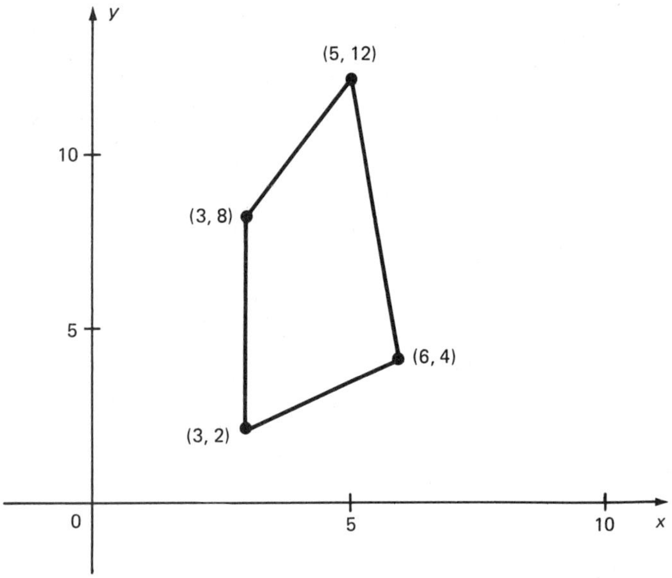

Figure 7.4

Vertex point	Value of $z = 5x + 2y$	
(3, 2)	$5(3) + 2(2) = 19$	(minimum)
(6, 4)	$5(6) + 2(4) = 38$	
(5, 12)	$5(5) + 2(12) = 49$	(maximum)
(3, 8)	$5(3) + 2(8) = 31$	

The minimum value of $z = 5x + 2y$ is 19 at the vertex point (3, 2). The maximum value is 49 at (5, 12). ■

Work Problem 4 at the side.

Example 4 An office manager needs to purchase new filing cabinets. He knows that Ace cabinets cost \$10 each, require 6 square feet of floor space, and hold 8 cubic feet of files. On the other hand, each Excello cabinet costs \$20, requires 8 square feet of floor space, and holds 12 cubic feet. His budget permits him to spend no more than \$140 on files, while the office has room for no more than 72 square feet of cabinets. The manager desires the greatest storage capacity within the limitations imposed by funds and space. How many of each type cabinet should he buy?

It is helpful to organize the information given in the problem as follows.

4. Use the region of feasible solutions in the sketch to find the following.

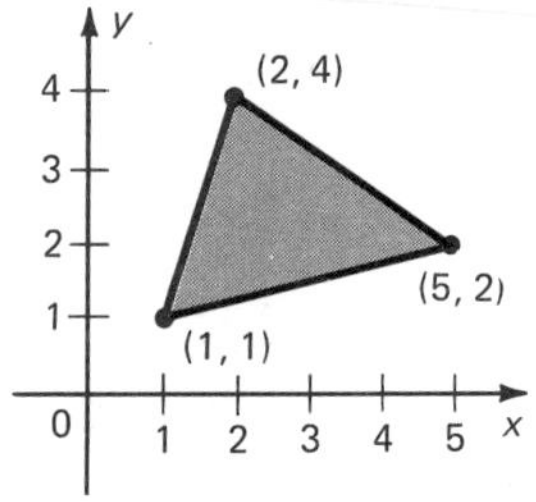

(a) The values of x and y which maximize $z = 2x - y$.
(b) The maximum value of $z = 2x - y$.
(c) The values of x and y which minimize $z = 4x + 3y$.
(d) The minimum value of $z = 4x + 3y$.

Answer:
(a) (5, 2)
(b) 8
(c) (1, 1)
(d) 7

	Cost	Space required	Storage capacity
1 Ace cabinet	\$10	6 sq. ft.	8 cu. ft.
1 Excello cabinet	\$20	8 sq. ft.	12 cu. ft.

5. Midtown Manufacturing Company makes plastic plates and cups, both of which require time on two machines. Plates require one hour on machine A and two on machine B, while cups require three hours on machine A and one on machine B. Each machine is operated for at most 15 hours per day. The profit on each plate is 25¢ and on each cup is 40¢. How many plates and how many cups should be manufactured to maximize the total profit?

(a) Make a chart to organize the information given in the problem.
(b) Assign variables to the quantities to be found.
(c) Write an equation for the objective function.
(d) Write 4 inequalities for the constraints.
(e) Find the maximum profit.

Answer:

(a)

	Machine A	Machine B	Profit
1 plate	1 hour	2 hours	25¢
1 cup	3 hours	1 hour	40¢

(b) x = number of plates
y = number of cups
z = maximum profit

(c) $z = .25x + .40y$

(d) $x + 3y \le 15$
$2x + y \le 15$
$x \ge 0$
$y \ge 0$

(e) \$2.70 at (6, 3)

The objective function to be maximized is the amount of storage capacity provided by some combination of Ace and Excello cabinets. If we let x represent the number of Ace cabinets to be bought and y the number of Excello cabinets, then the objective function is

$$\text{storage space} = z = 8x + 12y.$$

From the chart, we see that the constraints imposed by cost and space can be expressed as follows.

$$10x + 20y \le 140 \quad \text{(cost)}$$
$$6x + 8y \le 72 \quad \text{(floor space)}$$

Also, the number of cabinets cannot be negative. Thus, $x \ge 0$ and $y \ge 0$.

The graph of the feasible region is the shaded area in Figure 7.5. By Theorem 7.1, we know that the maximum value of the objective function will occur at one of the vertex points or on the boundary. From the figure, we can identify three of the vertex points as (0, 0), (0, 7), and (12, 0). The coordinates of the other point, labeled Q in the figure, can be found by solving the system of equations

$$10x + 20y = 140$$
$$6x + 8y = 72.$$

By solving this system, we find that Q is the point (8, 3). We must now test the coordinates of these four points in the objective function z to determine which point maximizes it. The results are shown below.

Vertex point	Value of $z = 8x + 12y$
(0, 0)	0
(0, 7)	84
(12, 0)	96
(8, 3)	100 (maximum)

Thus, the objective function, storage space, is maximized when $x = 8$ and $y = 3$. The manager should buy 8 Ace cabinets and 3 Excello cabinets. ■

Work Problem 5 at the side.

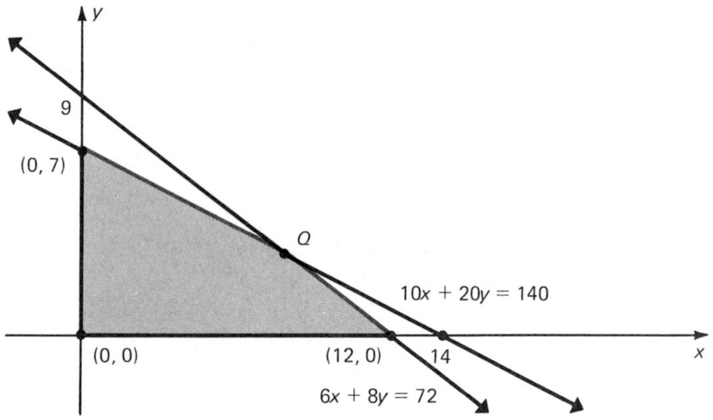

Figure 7.5

7.1 EXERCISES

Exercises 1–4 show regions of feasible solutions. Use these regions to find maximum and minimum values of each given objecitve function.

1. $z = 3x + 5y$

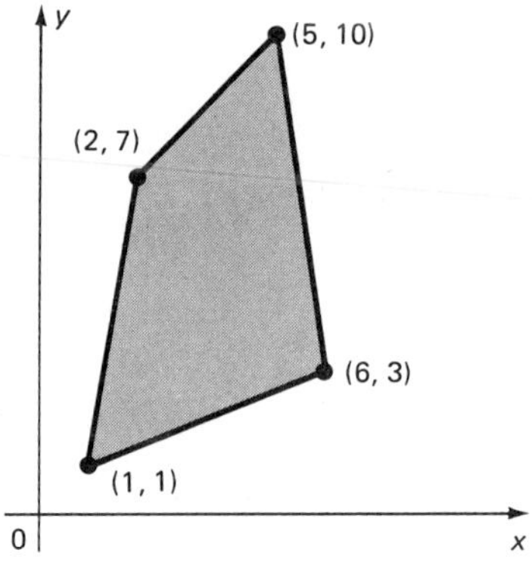

2. $z = 6x + y$

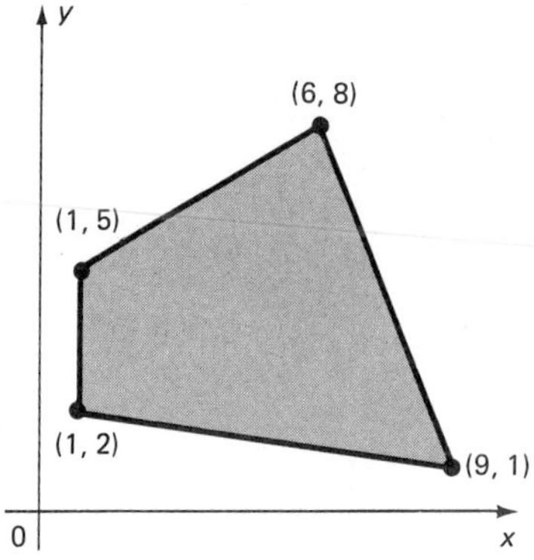

3. $z = .40x + .75y$

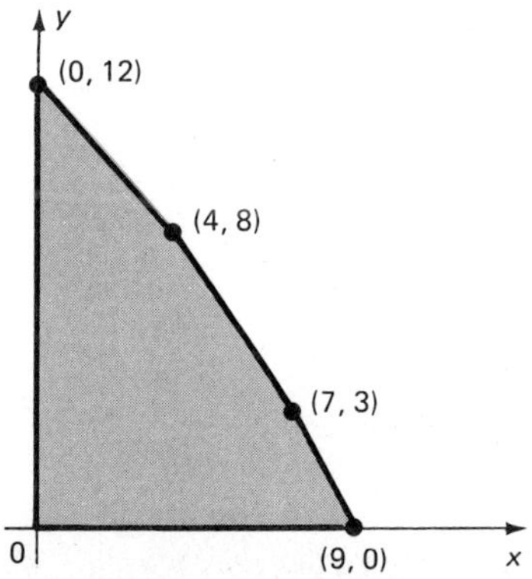

4. $z = .35x + 1.25y$

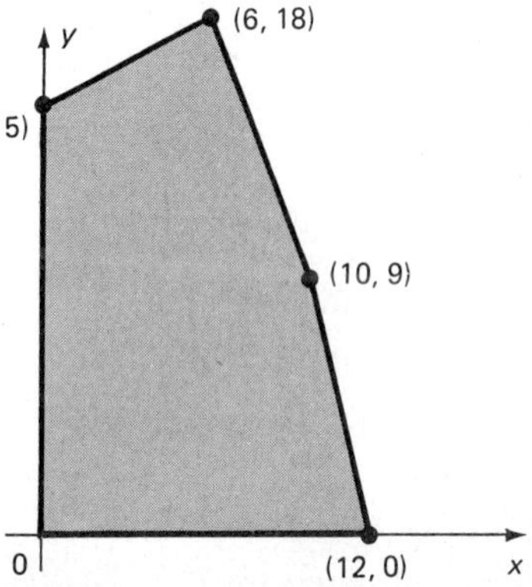

Use graphical methods to solve each of the following problems. (See Example 3.)

5. Find $x \geq 0$ and $y \geq 0$ such that

$$2x + 3y \leq 6$$
$$4x + y \leq 6$$

and $z = 5x + 2y$ is maximized.

6. Find $x \geq 0$ and $y \geq 0$ such that

$$x + y \leq 10$$
$$5x + 2y \geq 20$$
$$-x + 2y \geq 0$$

and $z = x + 3y$ is minimized.

7. Find $x \geq 2$ and $y \geq 5$ such that

$$3x - y \geq 12$$
$$x + y \leq 15$$

and $z = 2x + y$ is minimized.

8. Find $x \geq 10$ and $y \geq 20$ such that

$$2x + 3y \leq 100$$
$$5x + 4y \leq 200$$

and $z = x + 3y$ is maximized.

9. Find $x \geq 0$ and $y \geq 0$ such that

$$x - y \leq 10$$
$$5x + 3y \leq 75$$

and $z = 4x + 2y$ is maximized.

10. Find $x \geq 0$ and $y \geq 0$ such that

$$10x - 5y \leq 100$$
$$20x + 10y \geq 150$$

and $z = 4x + 5y$ is minimized.

11. Find values of $x \geq 0$ and $y \geq 0$ which maximize $z = 10x + 12y$ subject to each of the following sets of constraints.

(a) $x + y \leq 20$, $x + 3y \leq 24$

(b) $3x + y \leq 15$, $x + 2y \leq 18$

(c) $2x + 5y \geq 22$, $4x + 3y \leq 28$, $2x + 2y \leq 17$

12. Find values of $x \geq 0$ and $y \geq 0$ which minimize $z = 3x + 2y$ subject to each of the following sets of constraints.

(a) $10x + 7y \leq 42$, $4x + 10y \geq 35$

(b) $6x + 5y \geq 25$, $2x + 6y \geq 15$

(c) $x + 2y \geq 10$, $2x + y \geq 12$, $x - y \leq 8$

APPLIED PROBLEMS

Solve each of the following linear programming problems by the graphical method. (See Example 4.)

Management

13. Farmer Jones raises only pigs and geese. He wants to raise no more than 16 animals, including no more than 12 geese. He spends \$5 to raise a pig and \$2 to raise a goose. He has \$50 available for this purpose. Pigs sell for \$10 and geese for \$5. How many of each should he raise for maximum profit? What is the maximum profit?

14. Mark, who is ill, takes vitamin pills. Each day he must have at least 16 units of vitamin A, 5 units of vitamin B_1, and 20 units of vitamin C. He can choose between pill #1 which contains 8 units of A, 1 of B_1, and 2 of C, and pill #2 which contains 2 units of A, 1 of B_1, and 7 of C. Pill #1 costs 15¢ and pill #2 costs 30¢. How many of each pill should he buy in order to minimize his cost? What is the minimum cost?

15. A machine shop manufactures two types of bolts. Each can be made on any of three groups of machines, but the time required on each group differs, as shown in the table below.

		Machine groups I	II	III
Bolts	Type 1	.4 hour	.5 hour	.2 hour
	Type 2	.3 hour	.2 hour	.4 hour

Production schedules are made up for one week at a time. In this period there are 1200 hours of machine time available for each machine group. Type 1 bolts sell for 10¢ and type 2 bolts for 12¢. How many of each type of bolt should be manufactured per week to maximize revenue? What is the maximum revenue?

16. Seall Manufacturing Company makes color television sets. It produces a bargain set that sells for \$100 profit and a deluxe set that sells for \$150 profit. On the assembly line the bargain set requires 3 hours, while the deluxe set takes 5 hours. The cabinet shop spends one hour on the cabinet for the bargain set and 3 hours on the cabinet for the deluxe set. Both sets require 2 hours of time for testing and packing. On a particular production run the Seall Company has available 3900 work hours on the assembly line, 2100 work hours in the cabinet shop, and 2200 work hours in the testing and packing department. How many sets of each type should it produce to make maximum profit? What is the maximum profit?

17. A manufacturer of refrigerators must ship at least 100 refrigerators to its two West coast warehouses. Each warehouse holds a maximum of 100 refrigerators. Warehouse A holds 25 refrigerators already, while warehouse B has 20 on hand. It costs \$12 to ship a refrigerator to warehouse A and \$10 to ship one to warehouse B. How many refrigerators should be shipped to each warehouse to minimize cost? What is the minimum cost?

18. In a small town in South Carolina, zoning rules require that the window space (in square feet) in a house be at least one-sixth of the space used up by solid walls. The cost to heat the house is 2¢ for each square foot of solid walls and 8¢ for each square foot of windows. Find the maximum total area (windows plus walls) if \$16 is available to pay for heat.

19. A candy company has 100 kilograms of chocolate-covered nuts and 125 kilograms of chocolate-covered raisins to be sold as two different mixes. One mix will contain half nuts and half raisins and will sell for \$1 per kilogram. The other mix will contain 1/3 nuts and 2/3 raisins and will sell for \$.80 per kilogram. How many kilograms of each mix should the company prepare for maximum revenue?

Natural Science

20. Ms. Oliveras was given the following advice. She should supplement her daily diet with at least 6000 USP units of vitamin A, at least 195 milligrams of vitamin C, and at least 600 USP units of vitamin D. Ms. Oliveras finds that Mason's Pharmacy carries Upjohn vitamin pills at

5¢ each and Lilly vitamins at 4¢ each. Each Upjohn pill contains 3000 USP units of A, 45 milligrams of C, and 75 USP units of D, while the Lilly pills contain 1000 USP units of A, 50 milligrams of C, and 200 USP units of D. What combination of vitamin pills should she buy to obtain the least possible cost? What is the least possible cost per day?

21. Sam, who is dieting, requires two food supplements, I and II. He can get these supplements from two different products, A and B, as shown in the following table.

		Supplement (grams per serving) I	II
Product	A	3	2
	B	2	4

Sam's physician has recommended that he include at least 15 grams of each supplement in his daily diet. If product A costs 25¢ per serving and product B costs 40¢ per serving, how can he satisfy his requirements most economically?

Management

22. A small country can grow only two crops for export, coffee and cocoa. The country has 500,000 hectares* of land available for the crops. Long-term contracts require that at least 100,000 hectares be devoted to coffee and at least 200,000 hectares to cocoa. Cocoa must be processed locally, and production bottlenecks limit cocoa to 270,000 hectares. Coffee requires two workers per hectare, with cocoa requiring five. No more than 1,750,000 people are available for these crops. Coffee produces a profit of $220 per hectare, and cocoa a profit of $310 per hectare. How many hectares should the country devote to each crop in order to maximize the profit?

7.2 THE SIMPLEX METHOD—SLACK VARIABLES

In the previous section, we solved linear programming problems using the graphical method. The graphical method illustrates the basic ideas of linear programming, but is suitable only for problems with two variables. For problems with more than two variables, or with two variables and many constraints, we use the simplex method.

In using the **simplex method,** we proceed systematically from one feasible solution (on a boundary) to another in such a way that the objective function is automatically improved with each new feasible solution. Finally, we reach an optimum solution or determine that none exists.

The simplex method requires a number of steps. We have

* A hectare is a metric unit of land measure; one hectare is equivalent to approximately 2.47 acres.

divided the presentation of these steps into three sections of this chapter in which we discuss maximizing problems only. We show how to set up the problem in this section, how to begin the method in Section 7.3, and how to complete the method in Section 7.4.

Since the simplex method is used for problems with a large number of variables it is not always possible to use letters such as x, y, z, or w as variable names. We use instead the symbols x_1 (read "x-sub-one"), x_2, x_3, and so on, as variables. These variable names lend themselves easily to use on the computer.*

In a linear programming problem, we must have a linear function which we want to optimize and a set of constraints which are expressed as linear equations or inequalities. To use the simplex method, all constraints must be expressed in the linear form

$$a_1x_1 + a_2x_2 + a_3x_3 + \cdots \leq b$$

where $x_1, x_2, x_3, \ldots$ are variables and $a_1, a_2, \ldots, b$ are constants. That is, we want all terms with variables on the left and the constant term on the right. We restrict our discussion to those problems which meet the following conditions.

1. All variables are nonnegative. $(x_i \geq 0)$

2. The constant term is nonnegative. $(b \geq 0)$

3. The problems must be maximization problems in which all constraints are $\leq$ inequalities. (We discuss minimization problems in Section 7.5.)

We begin the simplex method by converting the constraints, which are linear inequalities, into linear equations. (We then have a system of linear equations which can be solved by the matrix methods of Chapter 6.) The linear inequalities are converted into linear equations by adding a nonnegative variable, called a **slack variable,** to each constraint. For example, we can convert the inequality $x_1 + x_2 \leq 10$ into an equation by adding the slack variable x_3:

$$x_1 + x_2 \leq 10$$

becomes

$$x_1 + x_2 + x_3 = 10, \qquad \text{where } x_3 \geq 0.$$

The inequality $x_1 + x_2 \leq 10$ says that the sum $x_1 + x_2$ is less than or perhaps equal to 10. The variable x_3 "takes up any slack" and represents the amount by which $x_1 + x_2$ fails to equal 10. (x_3 is 0 if $x_1 + x_2 = 10$.)

---

* A BASIC computer program for linear programming is given in *Study Guide with Computer Problems*, by Margaret L. Lial, (Glenview, Ill.: Scott, Foresman and Company, 1979).

Example 1 A farmer has 100 acres of available land which he wishes to plant with a mixture of potatoes, corn, and cabbage. It costs him \$100 to produce an acre of potatoes, \$40 to produce an acre of corn, and \$70 to produce an acre of cabbage. He has a maximum of \$5000 to spend. He makes a profit of \$30 per acre of potatoes, \$10 per acre of corn, and \$15 per acre of cabbage. How many acres of each crop should he plant to maximize his profit?

We can summarize the information in the problem as follows.

	Cost	Profit
1 acre of potatoes	\$100	\$30
1 acre of corn	40	10
1 acre of cabbage	70	15

If the number of acres allotted to each of the three crops is represented by x_1, x_2, and x_3, respectively, then the constraints of the example can be expressed as

$$x_1 + x_2 + x_3 \leq 100 \quad \text{(number of acres)}$$
$$100x_1 + 40x_2 + 70x_3 \leq 5000 \quad \text{(production costs)}$$

where x_1, x_2, and x_3 are all nonnegative. The first of these constraints says that $x_1 + x_2 + x_3$ is less than or perhaps equal to 100. Since it is possible that $x_1 + x_2 + x_3$ might not equal 100, but only be less than 100, there might be some "slack" in this inequality. We use x_4 to represent any slack, and express $x_1 + x_2 + x_3 \leq 100$ as the equation

$$x_1 + x_2 + x_3 + x_4 = 100.$$

Here the slack variable x_4 represents the amount of the farmer's 100 acres that will not be used. (x_4 may well be 0.)

In the same way, the constraint $100x_1 + 40x_2 + 70x_3 \leq 5000$ can be converted into an equation by adding a slack variable, x_5:

$$100x_1 + 40x_2 + 70x_3 + x_5 = 5000.$$

The slack variable x_5 represents any unused portion of the farmer's \$5000 in capital. (Again, x_5 may be zero.)

The objective function here represents the profit. The farmer wants to maximize

$$z = 30x_1 + 10x_2 + 15x_3,$$

subject to the constraints

$$x_1 + x_2 + x_3 \leq 100$$
$$100x_1 + 40x_2 + 70x_3 \leq 5000.$$

We can now restate this problem as follows. Find $x_1 \geq 0$, $x_2 \geq 0$,

$x_3 \geq 0$, $x_4 \geq 0$, and $x_5 \geq 0$ such that

$$\begin{aligned} x_1 + x_2 + x_3 + x_4 \quad &= 100 \\ 100x_1 + 40x_2 + 70x_3 \quad + x_5 &= 5000, \end{aligned}$$

and $z = 30x_1 + 10x_2 + 15x_3$ is maximized.

Here we added slack variables x_4 and x_5. The problem now is in a form suitable for solution by the simplex method. We continue with this example in the next section. ■

This example shows how to begin the simplex method by converting each linear inequality into a linear equation. This is done using slack variables.

1. Rewrite the following set of constraints as equations by adding slack variables.

$$\begin{aligned} x_1 + x_2 + x_3 &\leq 12 \\ 2x_1 + 4x_2 \quad &\leq 15 \\ x_2 + 3x_3 &\leq 10 \end{aligned}$$

Answer:

$$\begin{aligned} x_1 + x_2 + x_3 + x_4 \quad &= 12 \\ 2x_1 + 4x_2 \quad + x_5 \quad &= 15 \\ x_2 + 3x_3 \quad + x_6 &= 10 \end{aligned}$$

Work Problem 1 at the side.

Example 2 Restate the following linear programming problem in a form suitable for solution by the simplex method. Maximize $z = 3x_1 + 2x_2 + x_3$ subject to the constraints

$$\begin{aligned} 2x_1 + x_2 + x_3 &\leq 150 \\ x_1 + x_2 + 5x_3 &\leq 100 \\ 2x_1 + 3x_2 + x_3 &\leq 320 \end{aligned}$$

and $x_1 \geq 0$, $x_2 \geq 0$, $x_3 \geq 0$.

We must restate each of these constraints as an equation. To do this, we need three nonnegative slack variables, x_4, x_5, and x_6, one for each constraint. By adding slack variables we can restate the problem as follows: Find $x_1 \geq 0$, $x_2 \geq 0$, $x_3 \geq 0$, $x_4 \geq 0$, $x_5 \geq 0$, and $x_6 \geq 0$ such that

$$\begin{aligned} 2x_1 + x_2 + x_3 + x_4 \quad &= 150 \\ x_1 + x_2 + 5x_3 \quad + x_5 \quad &= 100 \\ 2x_1 + 3x_2 + x_3 \quad + x_6 &= 320 \end{aligned}$$

and $z = 3x_1 + 2x_2 + x_3$ is maximized. ■

7.2 EXERCISES

Convert each of the following inequalities into equations by adding a slack variable. (See Example 2.)

1. $x_1 + 2x_2 \leq 6$

2. $3x_1 + 5x_2 \leq 100$

3. $2x_1 + 4x_2 + 3x_3 \leq 100$

4. $8x_1 + 6x_2 + 5x_3 \leq 250$

For each of the following problems,

(*a*) *determine the number of slack variables needed;*

(*b*) *name them;*

(*c*) *use slack variables to convert each constraint into a linear equation.*

5. Maximize $z = 10x_1 + 12x_2$ subject to

$$\begin{aligned} 4x_1 + 2x_2 &\leq 20 \\ 5x_1 + x_2 &\leq 50 \\ 2x_1 + 3x_2 &\leq 25 \end{aligned}$$

and $x_1 \geq 0$, $x_2 \geq 0$.

6. Maximize $z = 1.2x_1 + 3.5x_2$ subject to

$$\begin{aligned} 2.4x_1 + 1.5x_2 &\leq 10 \\ 1.7x_1 + 1.9x_2 &\leq 15 \end{aligned}$$

and $x_1 \geq 0$, $x_2 \geq 0$.

7. Maximize $z = 8x_1 + 3x_2 + x_3$ subject to

$$\begin{aligned} 7x_1 + 6x_2 + 8x_3 &\leq 118 \\ 4x_1 + 5x_2 + 10x_3 &\leq 220 \end{aligned}$$

and $x_1 \geq 0$, $x_2 \geq 0$, $x_3 \geq 0$.

8. Maximize $z = 12x_1 + 15x_2 + 10x_3$ subject to

$$\begin{aligned} 2x_1 + 2x_2 + x_3 &\leq 8 \\ x_1 + 4x_2 + 3x_3 &\leq 12 \end{aligned}$$

and $x_1 \geq 0$, $x_2 \geq 0$, $x_3 \geq 0$.

9. Maximize $z = x_1 + 2x_2 + x_3 + 5x_4$ subject to

$$\begin{aligned} x_1 + x_2 + x_3 + x_4 &\leq 50 \\ 3x_1 + x_2 + 2x_3 + x_4 &\leq 100 \end{aligned}$$

and $x_1 \geq 0$, $x_2 \geq 0$, $x_3 \geq 0$, $x_4 \geq 0$.

10. Maximize $z = x_1 + x_2 + 4x_3 + 5x_4$ subject to

$$\begin{aligned} x_1 + 2x_2 + 3x_3 + x_4 &\leq 115 \\ 2x_1 + x_2 + 8x_3 + 5x_4 &\leq 200 \\ x_1 \qquad + x_3 \qquad &\leq 50 \end{aligned}$$

and $x_1 \geq 0$, $x_2 \geq 0$, $x_3 \geq 0$, $x_4 \geq 0$.

APPLIED PROBLEMS

Management

Set up each of the following word problems for solution by the simplex method. (See Example 1.)

(a) Write the objective function.

(b) List the constraints.

(c) Convert each constraint into a linear equation.

11. A candy company has 100 kilograms of chocolate-covered nuts and 125 kilograms of chocolate-covered raisins to be sold as two different mixtures. One mix will contain half nuts and half raisins and will sell for \$1 per kilogram. The other mix will contain 1/3 nuts and 2/3 raisins, and will sell for \$.80 per kilogram. How many kilograms of each mix should the company prepare for maximum revenue? (See Exercise 19, Section 7.1.)

12. Seall Manufacturing Company makes color television sets. It produces a bargain set that sells for \$100 profit and a deluxe set that sells for \$150 profit. On the assembly line the bargain set requires 3 hours' work, while the deluxe set takes 5 hours. The cabinet shop spends one hour on the cabinet for the bargain set and 3 hours on the cabinet for the deluxe set. Both sets require 2 hours of time for testing and packing. On a particular production run the Seall Company has available 3900 work hours on the assembly line, 2100 work hours in the cabinet shop, and 2200 work hours in the testing and packing department. How many sets of each type should it produce to make maximum profit? What is the maximum profit? (See Exercise 16, Section 7.1.)

13. A cat breeder has the following amounts of cat food: 90 units of tuna, 80 units of liver, and 50 units of chicken. To raise a Siamese cat, the breeder must use 2 units of tuna, 1 of liver, and 1 of chicken per day, while raising a Persian cat requires 1, 2, and 1 units respectively per day. If a Siamese cat sells for \$12, while a Persian cat sells for \$10, how many of each should be raised in order to obtain maximum gross income? What is the maximum gross income?

7.3 THE SIMPLEX METHOD—PIVOTS

We know from Theorem 7.1 that any optimum solution for a linear programming problem must occur at one or more vertex points of the region of feasible solutions. We could find the optimum by checking every possible vertex, but it is not possible in most practical problems. Instead, we use the simplex method to go systematically from one vertex to another in such a way that the value of the objective function is improved, or at least remains the same, each time.

In the last section we saw how to prepare a linear programming problem for solution by the simplex method, converting linear inequalities into linear equations by adding slack variables. For example, given the linear constraints

$$\begin{aligned} 2x_1 + x_2 + x_3 &\le 150 \\ 2x_1 + 2x_2 + 8x_3 &\le 200 \\ 2x_1 + 3x_2 + x_3 &\le 320, \end{aligned}$$

we introduce slack variables to rewrite the inequalities as the system of linear equations

$$\begin{aligned} 2x_1 + x_2 + x_3 + x_4 \qquad\qquad &= 150 \\ 2x_1 + 2x_2 + 8x_3 \qquad + x_5 \qquad &= 200 \\ 2x_1 + 3x_2 + x_3 \qquad\qquad + x_6 &= 320. \end{aligned}$$

One solution for this system of linear equations is found when $x_1 = 0, x_2 = 0, x_3 = 0, x_4 = 150, x_5 = 200$, and $x_6 = 320$. We write this solution as (0, 0, 0, 150, 200, 320).

We can identify such a solution more readily if we write the system of equations as an augmented matrix. For example, the system of equations above leads to the augmented matrix

$$\begin{array}{c} \begin{matrix} x_1 & x_2 & x_3 & x_4 & x_5 & x_6 \end{matrix} \\ \left[\begin{array}{cccccc|c} 2 & 1 & 1 & 1 & 0 & 0 & 150 \\ 2 & 2 & 8 & 0 & 1 & 0 & 200 \\ 2 & 3 & 1 & 0 & 0 & 1 & 320 \end{array}\right]. \end{array}$$

The three columns headed x_4, x_5, and x_6 form a 3×3 identity matrix. If all the other variables have the value 0, the system will reduce to

$$\begin{aligned} 1 \cdot x_4 + 0 \cdot x_5 + 0 \cdot x_6 &= 150 \\ 0 \cdot x_4 + 1 \cdot x_5 + 0 \cdot x_6 &= 200 \\ 0 \cdot x_4 + 0 \cdot x_5 + 1 \cdot x_6 &= 320, \end{aligned}$$

or simply $x_4 = 150$, $x_5 = 200$, and $x_6 = 320$. These results lead to the same solution (0, 0, 0, 150, 200, 320) that we found above. Thus, by identifying the columns which form an identity matrix, the nonzero values of a solution can be read directly from the matrix.

Example 1 Read the solution from the matrix shown below.

$$\begin{array}{c} \begin{matrix} x_1 & x_2 & x_3 & x_4 & x_5 \end{matrix} \\ \left[\begin{array}{ccccc|c} 2 & 1 & 8 & 5 & 0 & 27 \\ 9 & 0 & 3 & 12 & 1 & 45 \end{array}\right] \end{array}$$

Here the columns of the identity matrix are under the variables x_2 and x_5. Thus the other three variables are zero in the solution. Since the column of the identity matrix under x_2 has a 1 in the *first* row, then $x_2 = 27$, where 27 is the constant in the *first* row. Similarly, $x_5 = 45$. The solution is (0, 27, 0, 0, 45). ■

Work Problem 1 at the side.

In the discussion at the beginning of this section, we arbitrarily let $x_1 = 0$, $x_2 = 0$, and $x_3 = 0$. This seems to contradict common sense. These three variables came from the given linear programming problem, while the variables x_4, x_5, and x_6 were merely introduced to help find the final answer to the given problem. However what we are doing does have a purpose, since we start with a solution and then use the simplex method to improve the solution until we finally reach an optimum. In addition, the following theorem tells us that three variables in the optimum solution of this problem must be zero.

1. Read the solution from the matrix shown below.

$$\begin{array}{c} \begin{matrix} x_1 & x_2 & x_3 & x_4 & x_5 \end{matrix} \\ \left[\begin{array}{ccccc|c} 1 & 2 & 0 & 1 & 10 & 25 \\ 0 & 6 & 1 & 3 & 1 & 50 \end{array}\right] \end{array}$$

Answer: (25, 0, 50, 0, 0)

Theorem 7.2 Suppose an objective function has k variables subject to m linear inequalities, and suppose m slack variables are introduced. Then, if an optimum solution exists, there is always one optimum solution for which at least k of the $m + k$ variables are 0.

Because of this theorem, we need only consider solutions for which at least k variables are 0. To get started, we choose the simplest of these solutions, the one in which the original variables are all 0.

Example 2 In the example about the farmer (Example 1, Section 7.2), we found the objective function

$$z = 30x_1 + 10x_2 + 15x_3,$$

which has three variables. Thus, applying Theorem 7.2, we have $k = 3$. This objective function is subject to two constraints (number of acres and production costs). We added two slack variables to change the inequalities into equations. Thus, $m = 2$ and $m + k = 2 + 3 = 5$. By Theorem 7.2, if there is an optimal solution, there will be one where at least three of the five variables are 0. ■

2. (a) Find k and m of Theorem 7.2 for the linear programming problem given in Exercise 9 of the last section.
(b) Write a solution to the problem which could be used to start the simplex method.

Answer:
(a) $k = 4$, $m = 2$
(b) (0, 0, 0, 0, 50, 100)

Work Problem 2 at the side.

Example 3 In the example at the beginning of this section, we used the augmented matrix

$$\begin{array}{c} \begin{matrix} x_1 & x_2 & x_3 & x_4 & x_5 & x_6 & \end{matrix} \\ \left[\begin{array}{cccccc|c} 2 & 1 & 1 & 1 & 0 & 0 & 150 \\ 2 & 2 & 8 & 0 & 1 & 0 & 200 \\ 2 & 3 & 1 & 0 & 0 & 1 & 320 \end{array}\right] \begin{matrix} = x_4 \\ = x_5 \\ = x_6 . \end{matrix} \end{array}$$

From this matrix we read the solution (0, 0, 0, 150, 200, 320). We have written the nonzero variables of the solution at the right of the matrix.

Now, suppose we want x_2 to be nonzero and x_5 to be zero in the solution. We say that x_2 is the **entering variable** and x_5 the **departing variable.** We show this using the matrix below by having one arrow point to the second row to indicate that x_5 departs and another arrow point to the second column to indicate that x_2 is to enter. The number in both the second row and the second column is called the **pivot.** Here the pivot, 2, is circled.

$$\begin{array}{c} \begin{matrix} x_1 & x_2 & x_3 & x_4 & x_5 & x_6 & \end{matrix} \\ \left[\begin{array}{cccccc|c} 2 & 1 & 1 & 1 & 0 & 0 & 150 \\ 2 & \textcircled{2} & 8 & 0 & 1 & 0 & 200 \\ 2 & 3 & 1 & 0 & 0 & 1 & 320 \end{array}\right] \begin{matrix} = x_4 & \\ = x_5 & \leftarrow \\ = x_6 & \end{matrix} \\ \begin{matrix} & \uparrow & & & & & \end{matrix} \end{array}$$

3. Find and circle the pivot in the matrix shown below.

$$\begin{array}{cccccc|c} x_1 & x_2 & x_3 & x_4 & x_5 & x_6 & \\ 1 & 2 & 5 & 3 & 1 & 0 & 40 \\ 3 & 1 & 0 & 3 & 0 & 1 & 28 \end{array} \leftarrow$$

(↑ under the x_3 column)

Answer: 5

Work Problem 3 at the side.

Starting where we left off in Example 3, we now use matrix row operations (see Chapter 5) to obtain a new matrix where the x_2 column, instead of the x_5 column, will become

$$\begin{matrix} 0 \\ 1 \\ 0. \end{matrix}$$

To change the pivot from a 2 to a 1, multiply the second row by 1/2, which gives the following matrix.

$$\begin{array}{cccccc|c} x_1 & x_2 & x_3 & x_4 & x_5 & x_6 & \\ 2 & 1 & 1 & 1 & 0 & 0 & 150 \\ 1 & 1 & 4 & 0 & \frac{1}{2} & 0 & 100 \\ 2 & 3 & 1 & 0 & 0 & 1 & 320 \end{array}$$

Now we need to get a 0 both above and below this 1. To get a 0 above, multiply the second row of the new matrix above by -1 and add the result to the top row.

$$\begin{array}{cccccc|c} x_1 & x_2 & x_3 & x_4 & x_5 & x_6 & \\ 1 & 0 & -3 & 1 & -\frac{1}{2} & 0 & 50 \\ 1 & 1 & 4 & 0 & \frac{1}{2} & 0 & 100 \\ 2 & 3 & 1 & 0 & 0 & 1 & 320 \end{array}$$

Now multiply the second row by -3 and add the result to the third row.

$$\begin{array}{cccccc|cl} x_1 & x_2 & x_3 & x_4 & x_5 & x_6 & & \\ 1 & 0 & -3 & 1 & -\frac{1}{2} & 0 & 50 & = x_4 \\ 1 & 1 & 4 & 0 & \frac{1}{2} & 0 & 100 & = x_2 \\ -1 & 0 & -11 & 0 & -\frac{3}{2} & 1 & 20 & = x_6 \end{array}$$

From this matrix we can read the solution (0, 100, 0, 50, 0, 20). ■

When we use row operations to change the matrix in this way, we are simply replacing the original system of equations with an equivalent system of equations. This is exactly what we did when we used the row operations to solve systems of equations in Section 6.2.

4. Change the following matrix using row operations so that the x_2 column replaces the x_4 column and becomes

$$\begin{matrix} 1 \\ 0. \end{matrix}$$

$$\begin{array}{ccccc|cl} x_1 & x_2 & x_3 & x_4 & x_5 & & \\ 4 & \textcircled{2} & 0 & 1 & 0 & 200 & = x_4 \leftarrow \\ 3 & 1 & 1 & 0 & 1 & 300 & = x_5 \end{array}$$

(↑ under the x_2 column)

Answer:

$$\begin{array}{ccccc|cl} x_1 & x_2 & x_3 & x_4 & x_5 & & \\ 2 & 1 & 0 & \frac{1}{2} & 0 & 100 & = x_2 \\ 1 & 0 & 1 & -\frac{1}{2} & 1 & 200 & = x_5 \end{array}$$

Work Problem 4 at the side.

In the simplex method, this process is repeated until an optimum solution is found, if one exists. In the next section, we discuss how to decide which variables to use as entering and departing variables in order to improve the value of the objective function. We also learn to know when an optimum solution has been reached, or that one does not exist.

7.3 EXERCISES

Write the solution that can be read from each of the following. (See Example 1.)

1.
$$\begin{array}{c} \begin{array}{cccccc} x_1 & x_2 & x_3 & x_4 & x_5 & \end{array} \\ \left[\begin{array}{ccccc|c} 2 & 2 & 0 & 3 & 1 & 15 \\ 3 & 4 & 1 & 6 & 0 & 20 \end{array}\right] \end{array}$$

2.
$$\begin{array}{c} \begin{array}{cccccc} x_1 & x_2 & x_3 & x_4 & x_5 & \end{array} \\ \left[\begin{array}{ccccc|c} 0 & 2 & 1 & 1 & 3 & 5 \\ 1 & 5 & 0 & 1 & 2 & 8 \end{array}\right] \end{array}$$

3.
$$\begin{array}{c} \begin{array}{ccccccc} x_1 & x_2 & x_3 & x_4 & x_5 & x_6 & \end{array} \\ \left[\begin{array}{cccccc|c} 6 & 2 & 1 & 3 & 0 & 0 & 8 \\ 2 & 2 & 0 & 1 & 0 & 1 & 7 \\ 2 & 1 & 0 & 3 & 1 & 0 & 6 \end{array}\right] \end{array}$$

4.
$$\begin{array}{c} \begin{array}{ccccccc} x_1 & x_2 & x_3 & x_4 & x_5 & x_6 & \end{array} \\ \left[\begin{array}{cccccc|c} 0 & 2 & 0 & 1 & 2 & 2 & 3 \\ 0 & 3 & 1 & 0 & 1 & 2 & 2 \\ 1 & 4 & 0 & 0 & 3 & 5 & 5 \end{array}\right] \end{array}$$

For each of the following matrices, the entering and departing variables are indicated by arrows.

(a) Determine the pivot and perform the indicated transformation.

(b) State the resulting solution. (See Example 3.)

5.
$$\begin{array}{c} \begin{array}{cccccc} x_1 & x_2 & x_3 & x_4 & x_5 & \end{array} \\ \left[\begin{array}{ccccc|c} 1 & 2 & 4 & 1 & 0 & 56 \\ 2 & 2 & 1 & 0 & 1 & 40 \end{array}\right] \begin{array}{l} \\ \leftarrow \end{array} \\ \begin{array}{cccccc} & \uparrow & & & & \end{array} \end{array}$$

6.
$$\begin{array}{c} \begin{array}{cccccc} x_1 & x_2 & x_3 & x_4 & x_5 & \end{array} \\ \left[\begin{array}{ccccc|c} 5 & 4 & 1 & 1 & 0 & 50 \\ 3 & 3 & 2 & 0 & 1 & 40 \end{array}\right] \begin{array}{l} \leftarrow \\ \end{array} \\ \begin{array}{cccccc} & & \uparrow & & & \end{array} \end{array}$$

7.
$$\begin{array}{c} \begin{array}{ccccccc} x_1 & x_2 & x_3 & x_4 & x_5 & x_6 & \end{array} \\ \left[\begin{array}{cccccc|c} 2 & 2 & 1 & 1 & 0 & 0 & 12 \\ 1 & 2 & 3 & 0 & 1 & 0 & 45 \\ 3 & 1 & 1 & 0 & 0 & 1 & 20 \end{array}\right] \begin{array}{l} \leftarrow \\ \\ \\ \end{array} \\ \begin{array}{ccccccc} & & \uparrow & & & & \end{array} \end{array}$$

8.
$$\begin{array}{c} \begin{array}{ccccccc} x_1 & x_2 & x_3 & x_4 & x_5 & x_6 & \end{array} \\ \left[\begin{array}{cccccc|c} 4 & 2 & 3 & 1 & 0 & 0 & 22 \\ 2 & 2 & 5 & 0 & 1 & 0 & 28 \\ 1 & 3 & 2 & 0 & 0 & 1 & 45 \end{array}\right] \begin{array}{l} \\ \leftarrow \\ \\ \end{array} \\ \begin{array}{ccccccc} & & \uparrow & & & & \end{array} \end{array}$$

Restate the following linear programming problems, adding slack variables as necessary, and then write the resulting systems of equations as augmented matrices.

9. Find $x_1 \geq 0$ and $x_2 \geq 0$ such that

$$\begin{aligned} 2x_1 + 3x_2 &\leq 6 \\ 4x_1 + x_2 &\leq 6 \end{aligned}$$

and $z = 5x_1 + x_2$ is maximized.

10. Find $x_1 \geq 50$, $x_2 \geq 50$ such that

$$\begin{aligned} 2x_1 + 3x_2 &\leq 100 \\ 5x_1 + 4x_2 &\leq 200 \end{aligned}$$

and $z = x_1 + 3x_2$ is maximized.

11. Find $x_1 \geq 0$, $x_2 \geq 0$ such that

$$\begin{aligned} x_1 + x_2 &\leq 10 \\ 5x_1 + 2x_2 &\leq 20 \\ x_1 + 2x_2 &\leq 0 \end{aligned}$$

and $z = x_1 + 3x_2$ is maximized.

12. Find $x_1 \geq 0$, $x_2 \geq 0$ such that

$$\begin{aligned} x_1 + x_2 &\leq 10 \\ 5x_1 + 3x_2 &\leq 75 \end{aligned}$$

and $z = 4x_1 + 2x_2$ is maximized.

13. Find $x_1 \geq 0$, $x_2 \geq 0$ such that

$$\begin{aligned} 3x_1 + x_2 &\leq 12 \\ x_1 + x_2 &\leq 15 \end{aligned}$$

and $z = 2x_1 + x_2$ is maximized.

14. Find $x_1 \geq 0$, $x_2 \geq 0$ such that

$$\begin{aligned} 10x_1 + 4x_2 &\leq 100 \\ 20x_1 + 10x_2 &\leq 150 \end{aligned}$$

and $z = 4x_1 + 5x_2$ is maximized.

APPLIED PROBLEMS

Management

Set up each of the following problems for solution by the simplex method; that is, express the linear constraints and objective function, add slack variables, and set up the augmented matrix.

15. A small boat manufacturer builds three types of fiberglass boats: prams, runabouts, and trimarans. The pram sells at a profit of \$75, the runabout at a profit of \$90, and the trimaran at a profit of \$100. The factory is divided into two sections. Section A does the molding and construction work, while section B does the painting, finishing and equipping. The pram takes 1 hour in section A and 2 hours in section B. The runabout takes 2 hours in A and 5 hours in B. The trimaran takes 3 hours in A and 4 hours in B. Section A has a total of 6240 hours available and section B has 10,800 hours available for the year. The manufacturer has ordered a supply of fiberglass that will build at most 3000 boats, figuring the average amount used per boat. How many of each type of boat should be made to produce maximum profit? What is the maximum profit?

16. Caroline's Quality Candy Confectionery is famous for fudge, chocolate cremes, and pralines. Its candy-making equipment is set up to make 100-pound batches at a time. Currently there is a chocolate shortage and the company can get only 120 pounds of chocolate in the next shipment. On a week's run, the confectionery's cooking and processing equipment is available for a total of 42 machine hours. During the same period the employees have a total of 56 work hours available for packaging. A batch of fudge requires 20 pounds of chocolate while a batch of cremes uses 25 pounds of chocolate. The cooking and processing take 120 minutes for fudge, 150 minutes for chocolate cremes, and 200 minutes for pralines. The packaging times measured in minutes per one pound box are 1, 2, and 3 respectively, for fudge, cremes, and pralines. Determine how many batches of each type of candy the confectionery should make, assuming that the profit per pound box is 50¢ on fudge, 40¢ on chocolate cremes, and 45¢ on pralines.

7.4 THE SIMPLEX METHOD—SOLVING MAXIMIZATION PROBLEMS

We have learned how to prepare a linear programming problem for solution by first converting the constraints to linear equations with slack variables. We then wrote the coefficients of the variables from the linear equations as an augmented matrix. Finally, we used the pivot to go from one vertex of the region of feasible solutions to another.

Now we are ready to put all this together and produce an optimum value for the objective function. To see how this is done, let us finally complete the example about the farmer. (Recall Example 1 from Section 7.2.) In this example, we are to maximize the profit

$$z = 30x_1 + 10x_2 + 15x_3$$

subject to the constraints

$$\begin{aligned} x_1 + x_2 + x_3 &\le 100 \\ 100x_1 + 40x_2 + 70x_3 &\le 5000, \end{aligned}$$

with $x_1 \ge 0$, $x_2 \ge 0$, and $x_3 \ge 0$.

We first added slack variables x_4 and x_5, where $x_4 \ge 0$ and $x_5 \ge 0$.

$$x_1 + x_2 + x_3 + x_4 = 100 \qquad (1)$$

$$100x_1 + 40x_2 + 70x_3 + x_5 = 5000 \qquad (2)$$

The objective function here is $z = 30x_1 + 10x_2 + 15x_3$. We can rewrite this function to include the slack variables x_4 and x_5 as follows.

$$z = 30x_1 + 10x_2 + 15x_3 + 0x_4 + 0x_5$$

In Section 7.2, we noted that we would require all equations used in the simplex method to be in linear form with variable terms on the left and a constant term on the right. Furthermore, that constant and all variables must be nonnegative. Therefore, for the simplex method, the objective function must be rewritten as

$$-30x_1 - 10x_2 - 15x_3 - 0x_4 - 0x_5 + z = 0. \qquad \textbf{(3)}$$

To begin the simplex method, form an augmented matrix using the coefficients of equations (1), (2), and (3).

$$\begin{array}{c} \\ \\ \\ \end{array}\begin{array}{cccccc|c} x_1 & x_2 & x_3 & x_4 & x_5 & z & \\ 1 & 1 & 1 & 1 & 0 & 0 & 100 \\ 100 & 40 & 70 & 0 & 1 & 0 & 5000 \\ \hline -30 & -10 & -15 & 0 & 0 & 1 & 0 \end{array}\begin{array}{l} \\ = x_4 \\ = x_5 \\ = z \end{array}$$

We used a horizontal line to separate the objective function from the constraints. This augmented matrix is called the first **simplex tableau.** The z column in the tableau above will never change and will be left off later tableaus.

From the tableau, we read off the solution (0, 0, 0, 100, 5000). Notice that we have ignored the z column. However, from the lower right corner of the tableau, we see that this solution leads to a value of \$0 for z. A profit of \$0 for the farmer is certainly not an optimum, so we try to improve this profit.

Since the coefficients of x_1, x_2, and x_3 of the objective function are nonzero, we can improve the profit by making any one of these variables take on a nonzero value in a solution. To decide which variable to use, look in the bottom row in the above matrix. The coefficient of x_1 (-30) is the most negative in this row. This means that x_1 has the largest coefficient in the objective function, and we can most reasonably increase profit by increasing x_1.

To make x_1 nonzero, either x_4 or x_5 must become 0. To decide whether x_4 or x_5 should be used, solve equations (1) and (2) above for x_4 and x_5, respectively.

$$x_4 = 100 - x_1 - x_2 - x_3$$
$$x_5 = 5000 - 100x_1 - 40x_2 - 70x_3$$

We are changing x_1 to a nonzero value. Thus, x_2 and x_3 still equal 0. Replacing x_2 and x_3 with 0 gives

$$x_4 = 100 - x_1$$
$$x_5 = 5000 - 100x_1.$$

Since both x_4 and x_5 must remain nonnegative, there is a limit to how much we can increase x_1. From the equation $x_4 = 100 - x_1$ we see that x_1 cannot exceed 100/1 or 100. In the second equation,

$x_5 = 5000 - 100x_1$, we see that x_1 cannot exceed 5000/100 or 50. To satisfy both these conditions, x_1 cannot exceed 50 (the smaller of 50 and 100). If we let $x_1 = 50$, $x_2 = 0$, $x_3 = 0$, and $x_5 = 0$, we find that $x_4 = 50$, leading to the solution (50, 0, 0, 50, 0).

This solution gives a profit of

$$z = 30x_1 + 10x_2 + 15x_3 + 0x_4 + 0x_5$$
$$z = 30(50) + 10(0) + 15(0) + 0(50) + 0(0)$$
$$z = 1500.$$

We could have found the same result by using row operations on the first simplex tableau given above. First, choose the most negative number in the bottom row of the tableau. (If there is no negative number, we cannot further improve the value of the objective function.) The most negative number identifies the entering variable. This variable (x_1 here) is the one whose value is to be increased in the next solution. The entering variable is shown with an arrow on the tableau below.

To find the departing variable, the variable which will become 0 in place of x_1, compute the quotients that we found above. This is done automatically by dividing each number from the last column of the tableau by the corresponding coefficient of x_1 (the entering variable).

$$\begin{array}{r} \text{Quotients} \\ 100/1 = 100 \\ 5000/100 = 50 \\ \\ \\ \end{array} \qquad \begin{array}{c} \begin{array}{ccccc} x_1 & x_2 & x_3 & x_4 & x_5 \end{array} \\ \left[\begin{array}{ccccc|c} 1 & 1 & 1 & 1 & 0 & 100 \\ \boxed{100} & 40 & 70 & 0 & 1 & 5000 \\ \hline -30 & -10 & -15 & 0 & 0 & 0 \end{array}\right] \begin{array}{l} = x_4 \\ = x_5 \leftarrow \\ = z \end{array} \\ \uparrow \qquad\qquad\qquad\qquad \end{array}$$

The smaller quotient is 50. This quotient identifies x_5 as the departing variable. (See the arrow at the right above.) The entering and departing variables locate 100 as the pivot. Now use row transformations to obtain the second simplex tableau. First multiply the second row by 1/100. Then add -1 times the new second row to the old first row to get the new first row. How do we obtain the new third row?

$$\begin{array}{c} \begin{array}{ccccc} x_1 & x_2 & x_3 & x_4 & x_5 \end{array} \\ \left[\begin{array}{ccccc|c} 0 & .6 & .3 & 1 & -.01 & 50 \\ 1 & .4 & .7 & 0 & .01 & 50 \\ \hline 0 & 2 & 6 & 0 & .3 & 1500 \end{array}\right] \begin{array}{l} = x_4 \\ = x_1 \\ = z \end{array} \end{array}$$

We can now read the new solution from this tableau. Above the horizontal line, the columns that form the 2×2 identity matrix (x_1 and x_4 here) identify the variables which have nonzero values. From the first row, we get $x_4 = 50$ and from the second row, we get $x_1 = 50$; the other variables are all zero. Thus, the solution from this tableau is (50, 0, 0, 50, 0), as we found above. The entry 1500 (in color above) gives the value of the objective function for this solution: $z = \$1500$.

Since all the entries in the last row of the tableau are positive, we cannot improve the value of z beyond \$1500. Remember that the last row represents the coefficients of the objective function. If we add the coefficient of 1 for the z column which was dropped, we have

$$0x_1 + 2x_2 + 6x_3 + 0x_4 + .3x_5 + z = 1500.$$

Solving for z, we get

$$z = 1500 - 0x_1 - 2x_2 - 6x_3 - 0x_4 - .3x_5.$$

If the variables x_2, x_3, or x_5 are nonzero, z will decrease. Thus, we have the optimal solution as soon as all coefficients in the last row of the tableau are positive.

If the last row contained a negative number, we could improve the solution. We would repeat the entire procedure, locating new entering and departing variables and a pivot. We would then transform the matrix again. This process can be repeated as long as negative numbers appear in the last row.

We can finally state the solution to the problem about the farmer: the optimum value of z is 1500 when $x_1 = 50$, $x_2 = 0$, $x_3 = 0$, $x_4 = 50$, and $x_5 = 0$. That is, the farmer will make a maximum profit of \$1500 by planting 50 acres of potatoes. Another 50 acres should be left unplanted. It may seem strange that leaving assets unused can produce a maximum profit, but such results occur often in practical applications.

Example 1 Given the simplex tableau below, find the entering and departing variables and the pivot.

$$\begin{array}{c} \begin{matrix} x_1 & x_2 & x_3 & x_4 & x_5 & \end{matrix} \\ \left[\begin{array}{rrrrr|r} 1 & -2 & 1 & 0 & 0 & 100 \\ 3 & 4 & 0 & 1 & 0 & 200 \\ 5 & 0 & 0 & 0 & 1 & 150 \\ \hline -10 & -25 & 0 & 0 & 0 & 0 \end{array}\right] \begin{matrix} = x_3 \\ = x_4 \\ = x_5 \\ = z \end{matrix} \end{array}$$

To find the entering variable, we look for the most negative number in the last row, which is -25. Thus, x_2 is the entering variable. To find the departing variable, we must find the quotients formed by the entries in the last column and those in the x_2 column. Here the three quotients are $100/-2$, $200/4$, and $150/0$. We can't divide by 0, so we must disregard the last quotient. Since the quotients predict the value of the entering variable in the new solution, they cannot be negative. Remember, all variables must be nonnegative. The only usable quotient is $200/4 = 50$. Therefore, x_4 is the departing variable and 4 is the pivot. *If all the quotients are either negative or have zero denominators, there will be no optimal solution.* The quotients, then, determine whether or not an optimal solution exists. ■

Example 2 The matrix given below is the first simplex tableau of a linear programming problem. Complete the solution.

$$\begin{array}{c} \begin{matrix} x_1 & x_2 & x_3 & x_4 & x_5 & \end{matrix} \\ \left[\begin{array}{rrrrr|r} 1 & 2 & 4 & 1 & 0 & 38 \\ 2 & 1 & 1 & 0 & 1 & 16 \\ \hline -2 & -3 & -1 & 0 & 0 & 0 \end{array}\right] \begin{matrix} = x_4 \\ = x_5 \\ \end{matrix} \end{array}$$

Since the most negative number in the last row is -3, x_2 is the entering variable. The quotients, shown below on the left of the tableau, indicate that x_5 is the departing variable. Thus the pivot is 1.

$$\begin{matrix} 38/2 = 19 \\ 16/1 = 16 \\ \end{matrix} \quad \begin{array}{c} \begin{matrix} x_1 & x_2 & x_3 & x_4 & x_5 & \end{matrix} \\ \left[\begin{array}{rrrrr|r} 1 & 2 & 4 & 1 & 0 & 38 \\ 2 & \textcircled{1} & 1 & 0 & 1 & 16 \\ \hline -2 & -3 & -1 & 0 & 0 & 0 \end{array}\right] \begin{matrix} = x_4 \\ = x_5 \leftarrow \\ \end{matrix} \\ \uparrow \end{array}$$

Performing row transformations, we get the second tableau.

$$\begin{array}{c} \begin{matrix} x_1 & x_2 & x_3 & x_4 & x_5 & \end{matrix} \\ \left[\begin{array}{rrrrr|r} -3 & 0 & 2 & 1 & -2 & 6 \\ 2 & 1 & 1 & 0 & 1 & 16 \\ \hline 4 & 0 & 2 & 0 & 3 & 48 \end{array}\right] \begin{matrix} = x_4 \\ = x_2 \\ = z \end{matrix} \end{array}$$

All numbers in the last row are nonnegative. Therefore, the solution cannot be improved. The optimum value of 48 occurs when $x_1 = 0$, $x_2 = 16$, $x_3 = 0$, $x_4 = 6$, and $x_5 = 0$. ■

Work Problem 1 at the side.

1. For the simplex tableau below, find the following.
(a) the entering variable
(b) the departing variable
(c) the pivot

$$\begin{array}{c} \begin{matrix} x_1 & x_2 & x_3 & x_4 & x_5 & \end{matrix} \\ \left[\begin{array}{rrrrr|r} 1 & 2 & 6 & 1 & 0 & 16 \\ 1 & 3 & 0 & 0 & 1 & 25 \\ \hline -1 & -4 & -3 & 0 & 0 & 0 \end{array}\right] \begin{matrix} = x_4 \\ = x_5 \\ = z \end{matrix} \end{array}$$

Answer:
(a) x_2
(b) x_4
(c) 2

Example 3 Given the first simplex tableau of a linear programming problem, complete its solution.

$$\begin{matrix} \text{Quotients} \\ 30/2 = 15 \\ 10/1 = 10 \\ 20/4 = 5 \\ \end{matrix} \quad \begin{array}{c} \begin{matrix} x_1 & x_2 & x_3 & x_4 & x_5 & x_6 & \end{matrix} \\ \left[\begin{array}{rrrrrr|r} 2 & 5 & 3 & 1 & 0 & 0 & 30 \\ 1 & \frac{9}{4} & 5 & 0 & 1 & 0 & 10 \\ \textcircled{4} & 1 & 4 & 0 & 0 & 1 & 20 \\ \hline -3 & -2 & -2 & 0 & 0 & 0 & 0 \end{array}\right] \begin{matrix} = x_4 \\ = x_5 \\ = x_6 \leftarrow \\ = z \end{matrix} \\ \uparrow \end{array}$$

We begin by determining the entering and departing variables. Since -3 is the most negative entry in the last row, x_1 is the entering variable. Now compute the quotients as shown at the left above. Each number from the far right column of the matrix is divided by the corresponding coefficient from the x_1 column. The smallest of these quotients is 5, so that x_6 is the departing variable. Thus, the pivot is 4.

Performing row transformations, we get the second tableau.

$$\begin{array}{l} \text{Quotients} \\ 20/\frac{9}{2} = 40/9 \\ 5/2 \\ 5/\frac{1}{4} = 20 \\ \\ \end{array} \quad \begin{array}{c} \begin{array}{cccccc} x_1 & x_2 & x_3 & x_4 & x_5 & x_6 \end{array} \\ \left[\begin{array}{cccccc|c} 0 & \frac{9}{2} & 1 & 1 & 0 & -\frac{1}{2} & 20 \\ 0 & \textcircled{2} & 4 & 0 & 1 & -\frac{1}{4} & 5 \\ 1 & \frac{1}{4} & 1 & 0 & 0 & \frac{1}{4} & 5 \\ \hline 0 & -\frac{5}{4} & 1 & 0 & 0 & \frac{3}{4} & 15 \end{array}\right] \\ \uparrow \end{array} \begin{array}{l} = x_4 \\ = x_5 \quad \leftarrow \\ = x_1 \\ = z \end{array}$$

There is still a negative number in the bottom row, so we must repeat this process. The most negative number in the bottom row is $-5/4$. Thus, x_2 is the entering variable. Compute the quotients shown at the left above to find that x_5 is the departing variable (5/2 is the smallest quotient). The pivot is 2. Using 2 as the pivot we get the third simplex tableau.

$$\begin{array}{c} \begin{array}{cccccc} x_1 & x_2 & x_3 & x_4 & x_5 & x_6 \end{array} \\ \left[\begin{array}{cccccc|c} 0 & 0 & -8 & 1 & -\frac{9}{4} & \frac{1}{16} & \frac{35}{4} \\ 0 & 1 & 2 & 0 & \frac{1}{2} & -\frac{1}{8} & \frac{5}{2} \\ 1 & 0 & \frac{1}{2} & 0 & -\frac{1}{8} & \frac{9}{32} & \frac{35}{8} \\ \hline 0 & 0 & \frac{7}{2} & 0 & \frac{5}{8} & \frac{19}{32} & \frac{145}{8} \end{array}\right] \end{array} \begin{array}{l} = x_4 \\ = x_2 \\ = x_1 \\ = z \end{array}$$

All entries in the bottom row are positive or zero, so that we cannot further improve the value of z. The optimum value, 145/8 or $18\frac{1}{8}$, is found for z when $x_1 = 35/8$, $x_2 = 5/2$, $x_3 = 0$, $x_4 = 35/4$, $x_5 = 0$, and $x_6 = 0$. ■

Work Problem 2 at the side.

We can now summarize the steps involved in the simplex method for finding the maximum value of an objective function.

1. Set up the objective function.
2. Write all necessary constraints.
3. Convert each constraint into an equation by adding the necessary slack variables.
4. Set up the first simplex tableau.

2. For the simplex tableau from problem 1, find the following.
(a) the second tableau
(b) the solution shown in the second tableau
(c) Can the solution be improved?

Answer:

(a)

$$\begin{array}{c} \begin{array}{ccccc} x_1 & x_2 & x_3 & x_4 & x_5 \end{array} \\ \left[\begin{array}{ccccc|c} \frac{1}{2} & 1 & 3 & \frac{1}{2} & 0 & 8 \\ -\frac{1}{2} & 0 & -9 & -\frac{3}{2} & 1 & 1 \\ \hline 1 & 0 & 9 & 2 & 0 & 32 \end{array}\right] \end{array} \begin{array}{l} = x_2 \\ = x_5 \\ \\ \end{array}$$

(b) (0, 8, 0, 0, 1)
(c) no

5. Locate the most negative number in the bottom row. This number establishes the entering variable.

6. Form the necessary quotients to find the departing variable. Disregard any negative quotients or quotients with a 0 denominator. The smallest nonnegative quotient indicates the departing variable. If all quotients must be disregarded, no maximum solution exists.*

7. Locate the pivot by using the entering and departing variables.

8. Transform the tableau so that the pivot becomes 1 and all other numbers in that column become 0.

9. If the numbers in the bottom row are all positive or zero, you are through. If not, go back to step 5 above and transform the latest tableau. Keep doing this until a tableau is obtained with no negative numbers in the bottom row.

10. The maximum value of the objective function is given by the number in the lower right corner of the final tableau.

7.4 EXERCISES

Given the following first tableaus, use the simplex method to solve the indicated linear programming problems. (See Examples 2 and 3.)

1.

$$\begin{array}{c} \begin{matrix} x_1 & x_2 & x_3 & x_4 & x_5 & \end{matrix} \\ \left[\begin{array}{rrrrr|r} 1 & 2 & 4 & 1 & 0 & 8 \\ 2 & 2 & 1 & 0 & 1 & 10 \\ \hline -2 & -5 & -1 & 0 & 0 & 0 \end{array}\right] \begin{matrix} = x_4 \\ = x_5 \\ = z \end{matrix} \end{array}$$

2.

$$\begin{array}{c} \begin{matrix} x_1 & x_2 & x_3 & x_4 & x_5 & \end{matrix} \\ \left[\begin{array}{rrrrr|r} 2 & 2 & 1 & 1 & 0 & 10 \\ 1 & 2 & 3 & 0 & 1 & 15 \\ \hline -3 & -2 & -1 & 0 & 0 & 0 \end{array}\right] \begin{matrix} = x_4 \\ = x_5 \\ = z \end{matrix} \end{array}$$

3.

$$\begin{array}{c} \begin{matrix} x_1 & x_2 & x_3 & x_4 & x_5 & \end{matrix} \\ \left[\begin{array}{rrrrr|r} 1 & 3 & 1 & 0 & 0 & 12 \\ 2 & 1 & 0 & 1 & 0 & 10 \\ 1 & 1 & 0 & 0 & 1 & 4 \\ \hline -2 & -1 & 0 & 0 & 0 & 0 \end{array}\right] \begin{matrix} = x_3 \\ = x_4 \\ = x_5 \\ = z \end{matrix} \end{array}$$

4.

$$\begin{array}{c} \begin{matrix} x_1 & x_2 & x_3 & x_4 & x_5 & x_6 & \end{matrix} \\ \left[\begin{array}{rrrrrr|r} 2 & 2 & 1 & 1 & 0 & 0 & 50 \\ 1 & 1 & 3 & 0 & 1 & 0 & 40 \\ 4 & 2 & 5 & 0 & 0 & 1 & 80 \\ \hline -2 & -3 & -5 & 0 & 0 & 0 & 0 \end{array}\right] \begin{matrix} = x_4 \\ = x_5 \\ = x_6 \\ = z \end{matrix} \end{array}$$

* Some special circumstances are noted at the end of Section 7.5.

5.

$$
\begin{array}{cccccc|c}
x_1 & x_2 & x_3 & x_4 & x_5 & x_6 & \\
2 & 2 & 8 & 1 & 0 & 0 & 40 \\
4 & -5 & 6 & 0 & 1 & 0 & 60 \\
2 & -2 & 6 & 0 & 0 & 1 & 24 \\
\hline
-14 & -10 & -12 & 0 & 0 & 0 & 0
\end{array}
\quad
\begin{array}{l}
\\ = x_4 \\ = x_5 \\ = x_6 \\ = z
\end{array}
$$

6.

$$
\begin{array}{ccccc|c}
x_1 & x_2 & x_3 & x_4 & x_5 & \\
3 & 2 & 4 & 1 & 0 & 18 \\
2 & 1 & 5 & 0 & 1 & 8 \\
\hline
-1 & -4 & -2 & 0 & 0 & 0
\end{array}
\quad
\begin{array}{l}
\\ = x_4 \\ = x_5 \\ = z
\end{array}
$$

Use the simplex method to solve each of the following problems.

7. Maximize $z = 10x_1 + 12x_2$ subject to

$$
\begin{aligned}
4x_1 + 2x_2 &\le 20 \\
5x_1 + x_2 &\le 50 \\
2x_1 + 2x_2 &\le 24
\end{aligned}
$$

and $x_1 \ge 0$, $x_2 \ge 0$.

8. Maximize $z = 1.2x_1 + 3.5x_2$ subject to

$$
\begin{aligned}
2.4x_1 + 1.5x_2 &\le 10 \\
1.7x_1 + 1.9x_2 &\le 15
\end{aligned}
$$

and $x_1 \ge 0$, $x_2 \ge 0$.

9. Maximize $z = 8x_1 + 3x_2 + x_3$ subject to

$$
\begin{aligned}
x_1 + 6x_2 + 8x_3 &\le 118 \\
x_1 + 5x_2 + 10x_3 &\le 220
\end{aligned}
$$

and $x_1 \ge 0$, $x_2 \ge 0$, $x_3 \ge 0$.

10. Maximize $z = 12x_1 + 15x_2 + 5x_3$ subject to

$$
\begin{aligned}
2x_1 + 2x_2 + x_3 &\le 8 \\
x_1 + 4x_2 + 3x_3 &\le 12
\end{aligned}
$$

and $x_1 \ge 0$, $x_2 \ge 0$, $x_3 \ge 0$.

11. Maximize $z = x_1 + 2x_2 + x_3 + 5x_4$ subject to

$$
\begin{aligned}
x_1 + 2x_2 + x_3 + x_4 &\le 50 \\
3x_1 + x_2 + 2x_3 + x_4 &\le 100
\end{aligned}
$$

and $x_1 \ge 0$, $x_2 \ge 0$, $x_3 \ge 0$, $x_4 \ge 0$.

12. Maximize $z = x_1 + x_2 + 4x_3 + 5x_4$ subject to

$$
\begin{aligned}
x_1 + 2x_2 + 3x_3 + x_4 &\le 115 \\
2x_1 + x_2 + 8x_3 + 5x_4 &\le 200 \\
x_1 \qquad + x_3 \qquad &\le 50
\end{aligned}
$$

and $x_1 \ge 0$, $x_2 \ge 0$, $x_3 \ge 0$, $x_4 \ge 0$.

APPLIED PROBLEMS

Solve the following problems using the simplex method.

Natural Science

13. A biologist has 500 kilograms of nutrient A, 600 kilograms of nutrient B, and 300 kilograms of nutrient C. These nutrients will be used to make 4 types of food, whose contents (in percent of nutrient per kilogram of food) and whose "growth values" are as shown below.

Food	Nutrient (%) A	B	C	Growth value
P	0	0	100	90
Q	0	75	25	70
R	37.5	50	12.5	60
S	62.5	37.5	0	50

How many kilograms of each food should be produced in order to maximize total growth value?

Social Science

14. A political party is planning a half-hour television show. The show will have 3 minutes of direct requests for money from viewers. Three of the party's politicians will be on the show—a senator, a congresswoman, and a governor. The senator, a party "elder statesman," demands that he be on at least twice as long as the governor. The total time taken by the senator and the governor must be at least twice the time taken by the congresswoman. Based on a pre-show survey, it is believed that 40, 60, and 50 (in thousands) viewers will watch the program for each minute the senator, congresswoman, and governor, respectively are on the air. Find the time that should be allotted to each politician in order to get the maximum number of viewers. Find the maximum number of viewers.

Management

15. A candy company has 100 kilograms of chocolate-covered nuts and 125 kilograms of chocolate-covered raisins to be sold as two different mixtures. One mix will contain half nuts and half raisins and will sell for \$1 per kilogram. The other mix will contain 1/3 nuts and 2/3 raisins, and will sell for \$.80 per kilogram. How many kilograms of each mix should the company prepare for maximum revenue? (See Exercise 11, Section 7.2.)

16. A baker has 150 units of flour, 90 units of sugar, and 150 of raisins. A loaf of raisin bread requires 1 unit of flour, 1 of sugar, and 2 of raisins, while a raisin cake needs 5, 2, and 1 units, respectively. If raisin bread sells for 35¢ a loaf and raisin cake for 80¢ each, how many of each should be baked so that gross income is maximized? What is the maximum gross income?

17. A manufacturer of bicycles builds one-, three-, and ten-speed models. The bicycles need both aluminum and steel. The company has available 91,800 units of steel and 42,000 units of aluminum. The one-, three-, and ten-speed models need respectively 17, 27, and 34 units of steel, and 12, 21, and 15 units of aluminum. How many of each type of bicycle should be made in order to maximize profit if the company makes \$8 per one-speed bike, \$12 per three-speed, and \$22 per ten-speed?

18. Caroline's Quality Candy Confectionery is famous for fudge, chocolate cremes, and pralines. Its candy-making equipment is set up to make 100-pound batches at a time. Currently there is a chocolate shortage and the company can get only 120 pounds of chocolate in the next shipment. On a week's run, the confectionery's cooking and processing equipment is available for a total of 42 machine hours. During the same period the employees have a total of 56 work hours available for packaging. A batch of fudge requires 20 pounds of chocolate while a batch of cremes uses 25 pounds of chocolate. The cooking and processing take 120 minutes for fudge, 150 minutes for chocolate cremes, and 200 minutes for pralines. The packaging times measured in minutes per one pound box are 1, 2, and 3, respectively for fudge, cremes and pralines. Determine how many batches of each type of candy the confectionery should make, assuming that the profit per pound box is 50¢ on fudge, 40¢ on chocolate cremes, and 45¢ on pralines. Also, find the maximum profit for the week. (See Exercise 16, Section 7.3.)

7.5 THE SIMPLEX METHOD—MINIMIZATION PROBLEMS (OPTIONAL)

Up to this point, we have discussed the simplex method only for maximizing problems. As we show in this section, the steps we have presented can also be used to solve minimizing problems.

Example 1 Minimize $w = 100y_1 + 5000y_2$ where $y_1 \geq 0$, $y_2 \geq 0$, and

$$\begin{aligned} y_1 + 100y_2 &\geq 30 \\ y_1 + 40y_2 &\geq 10 \\ y_1 + 70y_2 &\geq 15. \end{aligned}$$

Without considering slack variables just yet, let us write the augmented matrix of coefficients of the system of inequalities, including the objective function as the last row in the matrix.

$$\left[\begin{array}{cc|c} 1 & 100 & 30 \\ 1 & 40 & 10 \\ 1 & 70 & 15 \\ \hline 100 & 5000 & 0 \end{array}\right]$$

For the sake of comparison, we now write the corresponding augmented matrix for the problem about the farmer, discussed in the last section.

$$\left[\begin{array}{ccc|c} 1 & 1 & 1 & 100 \\ 100 & 40 & 70 & 5000 \\ \hline 30 & 10 & 15 & 0 \end{array}\right]$$

These two matrices are related. In fact, the rows of the first matrix for the minimizing problem are the columns of the matrix for the farmer's maximizing problem. Because of this relationship, these two problems are called *duals*. The solution to this minimizing problem depends on the solution to the farmer's maximizing problem as we will see by Theorem 7.3. ■

For every given maximizing linear programming problem, there is a minimizing problem which is called its **dual problem,** and vice-versa. The given problem is called the **primal problem.** In general, the matrix of the dual problem is formed from the matrix of the primal problem by writing its rows as columns. This process is called **transposing,** and each of the two matrices is called the **transpose** of the other.

Example 2 Find the transpose of matrix A, where

$$A = \begin{bmatrix} 2 & -1 & 5 \\ 6 & 8 & 0 \\ -3 & 7 & -1 \end{bmatrix}.$$

Write the rows of matrix A as the columns of the transpose matrix.

$$\text{transpose of } A = \begin{bmatrix} 2 & 6 & -3 \\ -1 & 8 & 7 \\ 5 & 0 & -1 \end{bmatrix} \quad ■$$

Work Problem 1 at the side.

1. Give the transpose of the following matrix.

$$\begin{bmatrix} 1 & 2 & 4 & 0 \\ 2 & 1 & 7 & 6 \end{bmatrix}$$

Answer: $\begin{bmatrix} 1 & 2 \\ 2 & 1 \\ 4 & 7 \\ 0 & 6 \end{bmatrix}$

Example 3 Transpose to the dual of the following linear programming problem.

$$\begin{aligned} &\text{Maximize } z = 2x_1 + 5x_2 \text{ subject to} \\ &\qquad x_1 + x_2 \le 10 \\ &\qquad 2x_1 + x_2 \le 8 \\ &\text{and } x_1 \ge 0,\ x_2 \ge 0. \end{aligned}$$

We begin by writing the augmented matrix for the primal problem.

$$\left[\begin{array}{cc|c} 1 & 1 & 10 \\ 2 & 1 & 8 \\ \hline 2 & 5 & 0 \end{array}\right]$$

Writing the rows of this augmented matrix as columns, we get the matrix

$$\left[\begin{array}{cc|c} 1 & 2 & 2 \\ 1 & 1 & 5 \\ \hline 10 & 8 & 0 \end{array}\right]$$

of the dual problem. We can now state the dual problem from this second matrix, as follows (using y instead of x).

$$\text{Minimize } w = 10y_1 + 8y_2 \text{ subject to}$$
$$y_1 + 2y_2 \geq 2$$
$$y_1 + y_2 \geq 5$$
$$\text{and } y_1 \geq 0,\ y_2 \geq 0. \quad ■$$

In Example 3, note that the constraints of the given maximizing problem were all $\leq$ inequalities, while those in the dual minimizing problem were given as $\geq$ inequalities. This is generally the case; inequalities are reversed when the dual problem is stated.

Also, in Example 3, we used the variable x in the original maximizing problem, and the variable y in stating the dual minimizing problem. For clarity, we continue to use y for the variables of a minimizing problem and x for the variables of a maximizing problem.

Work Problem 2 at the side.

2. Write the dual of the following linear programming problem.

Minimize $w = 7y_1 + 5y_2 + 8y_3$ subject to

$$3y_1 + 2y_2 + y_3 \geq 10$$
$$y_1 + y_2 + y_3 \geq 8$$
$$4y_1 + 5y_2 \geq 25$$

and $y_1 \geq 0,\ y_2 \geq 0,\ y_3 \geq 0$.

Answer:
Maximize $z = 10x_1 + 8x_2 + 25x_3$ subject to

$$3x_1 + x_2 + 4x_3 \leq 7$$
$$2x_1 + x_2 + 5x_3 \leq 5$$
$$x_1 + x_2 \leq 8$$

and $x_1 \geq 0,\ x_2 \geq 0,\ x_3 \geq 0$.

The following theorem allows us to solve a minimizing problem by solving the dual maximizing problem.

Theorem 7.3 The objective function z of a maximizing linear programming problem takes on a maximum value if and only if the objective function w of the corresponding dual problem takes on a minimum value. The maximum value of z equals the minimum value of w.

Using Theorem 7.3, the minimum value of w in Example 1 is 1500, which was the maximum value we found for the farmer's maximizing problem in Section 7.4. In general to solve a minimizing linear programming problem, the following steps are required.

1. Form the dual maximizing problem.
2. Solve the maximizing problem using the simplex method.
3. The minimum value of the objective function w is given by the maximum value of the objective function z.

4. The optimum solution is given by the last entries in the columns corresponding to the slack variables.

This method is illustrated in the following example.

Example 4 Minimize $w = 3y_1 + 2y_2$ subject to

$$\begin{aligned} y_1 + 3y_2 &\geq 6 \\ 2y_1 + \ y_2 &\geq 3 \end{aligned}$$

and $y_1 \geq 0, y_2 \geq 0$.

Step 1 The matrix of the primal problem is

$$\left[\begin{array}{cc|c} 1 & 3 & 6 \\ 2 & 1 & 3 \\ \hline 3 & 2 & 0 \end{array}\right].$$

By transposing, we get the following matrix for the dual problem.

$$\left[\begin{array}{cc|c} 1 & 2 & 3 \\ 3 & 1 & 2 \\ \hline 6 & 3 & 0 \end{array}\right]$$

From this matrix we can form the dual problem, as follows.

$$\text{Maximize } z = 6x_1 + 3x_2 \text{ subject to}$$
$$\begin{aligned} x_1 + 2x_2 &\leq 3 \\ 3x_1 + \ x_2 &\leq 2 \end{aligned}$$
$$\text{and } x_1 \geq 0, x_2 \geq 0.$$

Step 2 These constraint inequalities and the objective function are rewritten using slack variables to give the system

$$\begin{array}{rcrcrcrcrcl} x_1 & + & 2x_2 & + & x_3 & & & & & = & 3 \\ 3x_1 & + & x_2 & & & + & x_4 & & & = & 2 \\ -6x_1 & - & 3x_2 & & & & & + & z & = & 0 \end{array}$$
$$x_1 \geq 0, x_2 \geq 0, x_3 \geq 0, x_4 \geq 0.$$

The first tableau for this system is given below with the entering and departing variables and the pivot as indicated.

$$\begin{array}{l} \text{Quotients} \\ 3/1 = 3 \\ 2/3 \\ \\ \\ \end{array} \quad \begin{array}{c} \begin{array}{cccc} x_1 & x_2 & x_3 & x_4 \end{array} \\ \left[\begin{array}{cccc|c} 1 & 2 & 1 & 0 & 3 \\ \textcircled{3} & 1 & 0 & 1 & 2 \\ \hline -6 & -3 & 0 & 0 & 0 \end{array}\right] \\ \uparrow \end{array} \begin{array}{l} \\ = x_3 \\ = x_4 \quad \leftarrow \\ = z \\ \\ \end{array}$$

Go through the steps of the simplex method. The final tableau is shown here.

$$\begin{array}{c} \begin{array}{cccc} x_1 & x_2 & x_3 & x_4 \end{array} \\ \left[\begin{array}{cccc|c} 0 & 1 & \frac{3}{5} & -\frac{1}{5} & \frac{7}{5} \\ 1 & 0 & -\frac{1}{5} & \frac{2}{5} & \frac{1}{5} \\ \hline 0 & 0 & \frac{3}{5} & \frac{9}{5} & \frac{27}{5} \end{array}\right] \begin{array}{l} = x_2 \\ = x_1 \\ = z \end{array} \end{array}$$

Step 3 From the final tableau in Step 2, we find that the minimum value of the objective function is 27/5 (the same as the maximum value of the objective function of the dual).

Step 4 The slack variable x_3 was introduced in the first row of the transposed matrix, which corresponds to the first column of the given matrix, the column of coefficients of y_1. Hence, the entry in the x_3 column is identified with y_1. In the same way, x_4 is identified with y_2. Thus, from the last entries in the columns corresponding to the slack variables x_3 and x_4, we find that the optimum solution to the primal problem occurs when $y_1 = \frac{3}{5}$ and $y_2 = \frac{9}{5}$. In summary, the minimum value of $w = 3y_1 + 2y_2$ subject to the given constraints is 27/5 and occurs when $y_1 = 3/5$ and $y_2 = 9/5$. ■

3. Minimize $w = 10y_1 + 8y_2$ subject to

$$y_1 + 2y_2 \geq 2$$
$$y_1 + y_2 \geq 5$$

and $y_1 \geq 0$, $y_2 \geq 0$.

Answer: $y_1 = 0$, $y_2 = 5$, for a minimum of 40

Work Problem 3 at the side.

In this chapter, we certainly have not covered all the possible complications that can arise in using the simplex method. Some of the difficulties (which may have occurred to you) include the following.

1. The minimum quotient may be zero in which case the entering variable retains the value of zero (rather than taking on a nonzero value) in the new basic solution. In this case, there is no improvement in the value of z.
2. There may be two or more equal quotients which are smallest. This leads to a basic solution in which one or more of the basic variables is zero. Each "tie" produces a zero basic variable in the row of the tying quotient.
3. Occasionally, a transformation will cycle—that is, produce a "new" solution which was an earlier solution in the process. These situations are known as degeneracies and special methods are available for handling them.

7.5 EXERCISES

Find the transpose of each of the following matrices. (See Example 2.)

1. $\begin{bmatrix} 1 & 2 & 3 \\ 3 & 2 & 1 \\ 1 & 10 & 0 \end{bmatrix}$

2. $\begin{bmatrix} 2 & 5 & 8 & 6 & 0 \\ 1 & -1 & 0 & 12 & 14 \end{bmatrix}$

3. $\begin{bmatrix} -1 & 4 & 6 & 12 \\ 13 & 25 & 0 & 4 \\ -2 & -1 & 11 & 3 \end{bmatrix}$

4. $\begin{bmatrix} 1 & 11 & 15 \\ 0 & 10 & -6 \\ 4 & 12 & -2 \\ 1 & -1 & 13 \\ 2 & 25 & -1 \end{bmatrix}$

State the dual problem for each of the following. (See Example 3.)

5. Maximize $z = 4x_1 + 3x_2 + 2x_3$ subject to

$$\begin{aligned} x_1 + x_2 + x_3 &\leq 5 \\ x_1 + x_2 &\leq 4 \\ 2x_1 + x_2 + 3x_3 &\leq 15 \end{aligned}$$

and $x_1 \geq 0$, $x_2 \geq 0$, $x_3 \geq 0$.

6. Maximize $z = 8x_1 + 3x_2 + x_3$ subject to

$$\begin{aligned} 7x_1 + 6x_2 + 8x_3 &\leq 18 \\ 4x_1 + 5x_2 + 10x_3 &\leq 20 \end{aligned}$$

and $x_1 \geq 0$, $x_2 \geq 0$, $x_3 \geq 0$.

7. Minimize $w = y_1 + 2y_2 + y_3 + 5y_4$ subject to

$$\begin{aligned} y_1 + y_2 + y_3 + y_4 &\geq 50 \\ 3y_1 + y_2 + 2y_3 + y_4 &\geq 100 \end{aligned}$$

and $y_1 \geq 0$, $y_2 \geq 0$, $y_3 \geq 0$, $y_4 \geq 0$.

8. Minimize $w = y_1 + y_2 + 4y_3$ subject to

$$\begin{aligned} y_1 + 2y_2 + 3y_3 &\geq 115 \\ 2y_1 + y_2 + 8y_3 &\geq 200 \\ y_1 + y_3 &\geq 50 \end{aligned}$$

and $y_1 \geq 0$, $y_2 \geq 0$, $y_3 \geq 0$.

Use the simplex method to solve the following problems. (See Example 4.)

9. Find $y_1 \geq 0$, $y_2 \geq 0$ such that

$$\begin{aligned} 2y_1 + 3y_2 &\geq 6 \\ 2y_1 + y_2 &\geq 7 \end{aligned}$$

and $w = 5y_1 + 2y_2$ is minimized.

10. Find $y_1 \geq 0$, $y_2 \geq 0$ such that

$$\begin{aligned} 3y_1 + y_2 &\geq 12 \\ y_1 + 4y_2 &\geq 16 \end{aligned}$$

and $w = 2y_1 + y_2$ is minimized.

11. Find $y_1 \geq 0$, $y_2 \geq 0$ such that

$$\begin{aligned} 10y_1 + 5y_2 &\geq 100 \\ 20y_1 + 10y_2 &\geq 150 \end{aligned}$$

and $w = 4y_1 + 5y_2$ is minimized.

12. Minimize $w = 3y_1 + 2y_2$ subject to

$$\begin{aligned} 2y_1 + 3y_2 &\geq 60 \\ y_1 + 4y_2 &\geq 40 \end{aligned}$$

and $y_1 \geq 0$, $y_2 \geq 0$.

13. Minimize $w = 2y_1 + y_2 + 3y_3$ subject to

$$\begin{aligned} y_1 + y_2 + y_3 &\geq 100 \\ 2y_1 + y_2 \quad\quad &\geq 50 \end{aligned}$$

and $y_1 \geq 0$, $y_2 \geq 0$, $y_3 \geq 0$.

14. Minimize $w = 3y_1 + 2y_2$ subject to

$$\begin{aligned} y_1 + 2y_2 &\geq 10 \\ y_1 + y_2 &\geq 8 \\ 2y_1 + y_2 &\geq 12 \end{aligned}$$

and $y_1 \geq 0$, $y_2 \geq 0$.

APPLIED PROBLEMS

Use the simplex method to solve each of the following problems.

Management

15. Sam, who is dieting, requires two food supplements, I and II. He can get these supplements from two different products, A and B, as shown in the following table.

		Supplement (grams per serving) I	II
Product	A	3	2
	B	2	4

Sam's physician has recommended that he include at least 15 grams of each supplement in his daily diet. If product A costs 25¢ per serving and product B costs 40¢ per serving, how can he satisfy his requirements most economically? (See Exercise 21, Section 7.1.)

16. Brand X Canners produce canned whole tomatoes and tomato sauce. This season, they ordered and must use 3,000,000 kilograms of tomatoes for these two products. To meet the demands of regular customers, they must produce at least 80,000 kilograms of sauce and 800,000 kilograms of whole tomatoes. The cost per kilogram is $4 to produce canned whole tomatoes and $3.25 to produce tomato sauce. How many kilograms of tomatoes should they use for each product to minimize cost?

17. Mark, who is ill, takes vitamin pills. Each day he must have at least 16 units of vitamin A, 5 units of vitamin B_1, and 20 units of vitamin C. He can choose between pill #1 which costs 10 cents and contains 8 units of A, 1 of B_1, and 2 of C, and pill #2 which costs 20 cents and contains 2 units of A, 1 of B_1, and 7 of C. How many of each pill should he buy in order to minimize his cost? (See Exercise 14, Section 7.1.)

18. A brewery produces regular beer and a lower-carbohydrate "light" beer. Steady customers of the brewery buy 12 units of regular beer and 10 units of light beer. While setting up the brewery to produce the beers, the management decides to produce extra beer, beyond that needed to satisfy the steady customers. The cost per unit of regular beer is $36,000 and the cost per unit of light beer is $48,000. The number of units of light beer should not exceed twice the number of units of regular beer. At least twenty additional units of beer can be sold. How much of each type beer should be made so as to minimize total production costs?

19. The chemistry department at a local college decides to stock at least 800 small test tubes and 500 large test tubes. It wants to buy at least 1500 test tubes to take advantage of a special price. Since the small tubes are broken twice as often as the larger, the department will order at least twice as many small tubes as large. If the small test tubes cost 15¢ each and the large ones 12¢ each, how many of each size should they order to minimize cost?

20. Topgrade Turf lawn seed mixtures contain three types of seeds: bluegrass, rye, and bermuda. The costs per pound of the three types of seed are 20¢, 15¢, and 5¢. In each mixture there must be at least 20% bluegrass seed and the amount of bermuda must be no more than the amount of rye. To fill current orders, the company must make at least 5000 pounds of the mixture. How much of each kind of seed should be used to minimize cost?

Natural Science

21. A biologist must make a nutrient for her algae. The nutrient must contain the three basic elements D, E, and F, and must contain at least 10 kilograms of D, 12 kilograms of E, and 20 kilograms of F. The nutrient is made from three ingredients, I, II, and III. The quantity of D, E, and F in one unit of each of the ingredients is as given in the following chart.

One unit of ingredient	Contains the following elements in kilograms D	E	F	Cost of one unit of ingredient
I	4	3	0	4
II	1	2	4	7
III	10	1	5	5

How many units of each ingredient are required to meet her needs at minimum cost?

CHAPTER 7 TEST

[7.1]

1. Use the given region to find the maximum and minimum values of the objective function $z = 2x + 4y$.

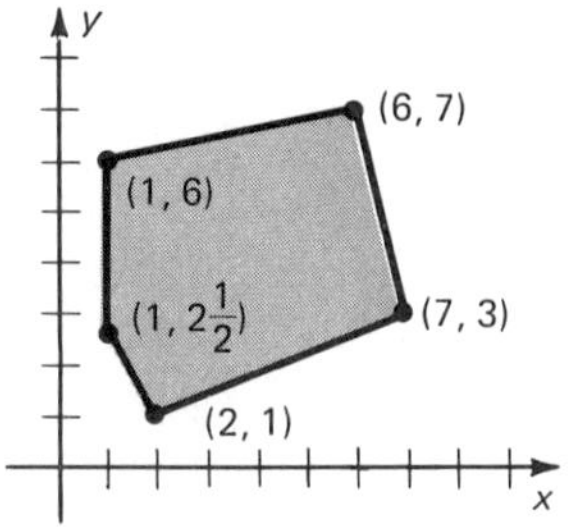

2. Use graphical methods to solve the following problem. Find $x \geq 0$ and $y \geq 0$ such that $3x + 2y \leq 18$, $x + 4y \leq 15$, $x + y \leq 6$, and $5x + 3y$ is maximized.

For each of the following problems, (a) select appropriate variables, (b) write the objective function, (c) write the constraints as inequalities.

3. Roberta Hernandez sells three items, A, B, and C, in her gift shop. Each unit of A costs her \$5 to buy, \$1 to sell, and \$2 to deliver. For each unit of B, the costs are \$3, \$2, and \$1 respectively, and for each unit of C the costs are \$6, \$2, and \$5 respectively. Assume she can sell all she buys. The profit on the three items is \$4, \$4, and \$5, respectively. How many of each should she get to maximize her profit if she can spend \$1200 to buy, \$800 on selling costs, and \$500 on delivery costs?

4. An investor is considering three types of investment: a high risk venture into oil leases with a potential return of 15%, a medium risk investment in bonds with a 9% return, and a relatively safe stock investment with a 5% return. He has \$50,000 to invest. Because of the risk, he will limit his investment in oil leases and bonds to 30% and his investment in oil leases and stock to 50%. How much should he invest in each to maximize his return assuming investment returns are as expected?

[7.2] and [7.3]

For each of the following problems, (a) add slack variables, (b) set up the first simplex tableau, and (c) determine the pivot.

5. Maximize $z = 5x_1 + 3x_2$ subject to

$$\begin{aligned} 2x_1 + 6x_2 &\leq 50 \\ x_1 + 3x_2 &\leq 25 \\ 4x_1 + x_2 &\leq 18 \\ x_1 + x_2 &\leq 12 \end{aligned}$$

and $x_1 \geq 0$, $x_2 \geq 0$.

6. Maximize $z = 2x_1 + 3x_2 + 4x_3$ subject to

$$\begin{aligned} x_1 + x_2 + x_3 &\leq 100 \\ 2x_1 + 3x_2 &\leq 500 \\ x_1 + 2x_3 &\leq 350 \end{aligned}$$

and $x_1 \geq 0$, $x_2 \geq 0$, $x_3 \geq 0$.

[7.4] **7.** Use the simplex method to solve the following linear programming problem, given the first tableau.

$$\begin{array}{ccccc|c} x_1 & x_2 & x_3 & x_4 & x_5 & \\ 1 & 2 & 3 & 1 & 0 & 28 \\ 2 & 4 & 1 & 0 & 1 & 32 \\ \hline -5 & -2 & -3 & 0 & 0 & 0 \end{array}$$

$(x_1, x_2, x_3, x_4, x_5) = ($ $\qquad\qquad$ $)$
The maximum value = _______________.

[7.5] **8.** Find the transpose of each of the following.

(a) $\begin{bmatrix} 2 & 5 & 3 \\ -1 & 4 & 2 \\ 6 & -2 & 5 \end{bmatrix}$

(b) $\begin{bmatrix} 1 & 7 & 12 & 11 \\ -3 & 4 & 8 & 20 \\ 0 & 10 & -6 & 9 \end{bmatrix}$

Write the dual of the following problem.

9. Minimize $6y_1 + 4y_2 + y_3$ subject to

$$\begin{aligned} 2y_1 + 2y_2 + y_3 &\leq 8 \\ y_1 + 3y_2 + y_3 &\leq 10 \\ y_1 + y_3 &\leq 5 \end{aligned}$$

and $y_1 \geq 0$, $y_2 \geq 0$, $y_3 \geq 0$.

10. The following tableau is the final tableau of the dual of a minimizing problem. (a) State the solution to the minimizing problem. (b) State the solution to the dual maximizing problem.

$$\begin{array}{cccccc|c} x_1 & x_2 & x_3 & x_4 & x_5 & x_6 & \\ 1 & 0 & 0 & 3 & 1 & 2 & 12 \\ 0 & 0 & 1 & 4 & 5 & 3 & 5 \\ 0 & 1 & 0 & -2 & 7 & -6 & 8 \\ \hline 0 & 0 & 0 & 5 & 7 & 3 & 172 \end{array}$$

Hint: Both problems have three variables.

CASE 6 AIRLINE FLEET ASSIGNMENT— BOEING COMMERCIAL AIRPLANE COMPANY*

Boeing Commercial Airplane Company has developed a mathematical model to help an airline decide on its fleet's needs. This model helps airlines choose the best mix of small, medium, and large jets for the routes that it serves. To begin, the airline estimates the potential revenue from passengers and freight from one city to all the others that can be reached from that city.

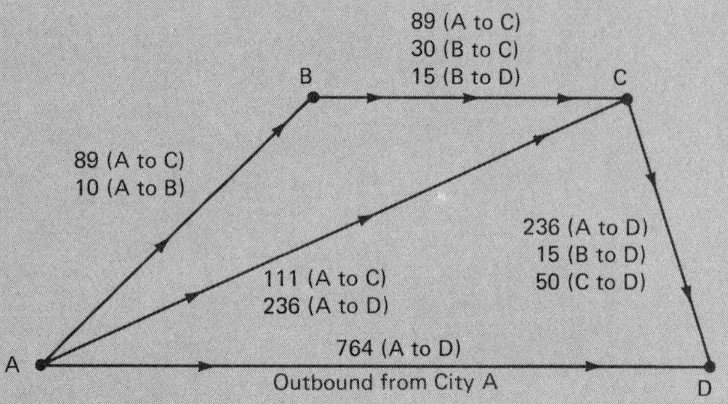

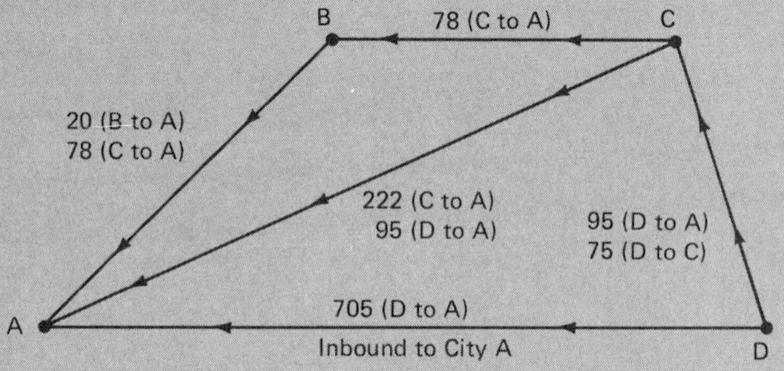

For example, the figure shows that 10 units of revenue ($10,000) is earned between city A and city B. Also, $89,000 is received from traffic that goes from A to C, through B, while $111,000 is received from traffic going from A directly to C. Thus, a total of $89,000 + $111,000 = $200,000 is received from traffic going from A to C. Using the information given in the figure above, a *demand matrix* can be set up. (See Exercise 1 below.)

After the expected demand has been decided upon, the mathematical model for optimum fleet utilization can be discussed. The model includes three types of variables—demand variables, itinerary variables, and fleet variables. A demand variable represents the demand from customers for flights from one city to another, including nonstop flights, through flights

* Based on "Applications of an Airline Fleet Assignment Model" by A. Maimon and R. M. Peterson. Reprinted by permission of Boeing Commercial Airplane Company, Seattle, Washington.

(not nonstop, but no change of planes), and connecting flights (which involve a change of plane). Itinerary variables show how the airplanes owned by the company can be applied to the airline system. Finally, the fleet variables give the number of each type of plane that the airline should have.

The mathematical model is designed so that linear programming, using the simplex method, can be used to maximize profit. Profit (the objective function) is given by

Profit = P = passenger revenue + cargo revenue − flight operating costs − aircraft ownership costs − costs of any new aircraft.

(So many variables are involved here that we will not try to write out the complete mathematical formulation of the profit function.)

The profit function must be maximized subject to many constraints. For example, the number of aircraft services for a city pair must be greater than some minimum marketing requirement. This constraint is of the form

$$\sum_{i=1} \sum_{j=1}^{a} N_{ij}^{pq} \geq ms_{pq}$$

where ms_{pq} is the minimum service requirement for city pair $p - q$, and N_{ij}^{pq} is the number of frequencies of aircraft type j on those routes i that serve the city pair $p - q$.

System fuel usage must not exceed the total amount of fuel available.

$$\sum_{i=1}^{r} \sum_{j=1}^{a} FB_{ij} \cdot N_{ij} \leq FA$$

Here FB_{ij} is the amount of fuel burned by an aircraft of type j on route i. Also, FA represents the total amount of fuel available.

The "load factor" (the fraction of available seats that are occupied) must not exceed some preassigned value.

$$\sum_{i=1}^{a} LF \cdot S_i \cdot N_{ri} - \sum p \geq 0$$

In this constraint, LF is the maximum load factor on a given route ($LF < 1$), S_i is the number of seats on an aircraft of a certain type, and N_{ri} represents the number of aircraft of one type on route r. Also, $\sum p$ represents the total number of passengers on the route.

Other constraints relate to the number of seats available on a route, the frequency of service to a city, and so on. When a given airline wishes to use this model, the values of all necessary variables must be found. (Normally, this is the most time-consuming part of the process.) These values are entered into a large computer, which sets up a simplex tableau, goes through the steps of the simplex method, and produces a conclusion.

The conclusions are quite lengthy, since they must specify the number of flights made by several different types of aircraft between many different pairs of cities. The following table shows the size of the simplex tableau for the mathematical model of this case for several different types of airlines.

Airline type—domestic or international	Passenger or freight or combination	Number of types of aircraft	Number of itineraries	Number of city pairs	Size of simplex tableau rows	Size of simplex tableau columns	Computer running time (seconds)
Domestic	Passenger	7	46	89	353	384	100
International	Combination	8	80	46	590	1283	190
Domestic	Passenger	8	203	287	1373	2008	1800
Domestic	Passenger	3	169	736	1638	3594	2700
International	Combination	7	259	151	2293	4233	1560
International	Combination	3	66	112	649	1139	210
Domestic	Freight	5	44	177	401	864	190
International	Passenger	4	140	204	604	1213	150

The size of these simplex tableaus shows why a large computer is usually necessary for realistic problems when mathematical models involve linear programming.

EXERCISES

1. Complete the following demand matrix using the information in the figure at the beginning of this case.

$$\text{From}\quad \begin{array}{c} \\ A \\ B \\ C \\ D \end{array} \overset{\text{To}}{\begin{array}{cccc} A & B & C & D \end{array}} \atop \left[\begin{array}{cccc} 0 & 10 & 200 & \\ 20 & 0 & & \\ & & 0 & \\ & & 75 & 0 \end{array}\right]$$

Find the total number of entries in the simplex tableaus for each of the following airlines in the table above. (Hint: find the number of rows and columns in each tableau.)

2. the first airline

3. the second airline

4. the third airline

5. the fourth airline

6. the last airline

(In practice, the number of entries is not as important as the fraction of the numbers that are not zero. In a realistic use of this model only 2% of the entries are not zero.)

CASE 7 MERIT PAY—THE UPJOHN COMPANY*

Individuals doing the same job within the management of a company often receive different salaries. These salaries may differ because of length of service, productivity of an individual worker, and so on. However, for each job there is usually an established minimum and maximum salary.

Many companies make annual reviews of the salary of each of their management employees. At these reviews, an employee may receive a general cost of living increase, an increase based on merit, both, or neither. (*The Wall Street Journal* tells us that in a recent year, about 99.97% of all employees of the Department of Agriculture who were eligible for merit increases actually received them.)

In this case, we look at a mathematical model for distributing merit increases in an optimum way. An individual who is due for salary review may be described as shown in the figure below. Here i represents the number of the employee whose salary is being reviewed.

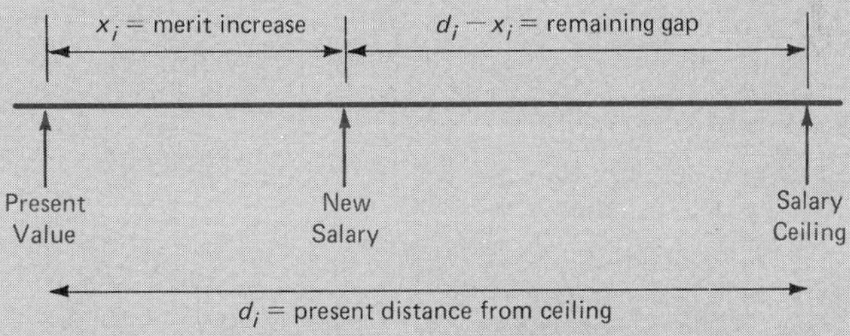

Here the salary ceiling is the maximum salary for the job classification, x_i is the merit increase to be awarded to the individual ($x_i \geq 0$), d_i is the present distance of the current salary from the salary ceiling, and the difference, $d_i - x_i$, is the remaining gap.

We let w_i be a measure of the relative worth of the individual to the company. This is the most difficult variable of the model to actually calculate. One way to evaluate w_i is to give a rating sheet to a number of co-workers and supervisors of employee i. An average rating can be obtained and then divided by the highest rating received by any employee with that same job. This number then gives the worth of employee i relative to all other employees with that job.

The best way to allocate money available for merit pay increases is to minimize

$$\sum_{i=1}^{n} w_i(d_i - x_i).$$

This sum is found by multiplying the relative worth of employee i and the distance of employee i from the salary gap, after any merit increase. Here

* Based on part of a paper by Jack Northam, Head, Mathematical Services Department, The Upjohn Company, Kalamazoo, Michigan.

n represents the total number of employees who have this job. The one constraint here is that the total of all merit increases cannot exceed P, the total amount available for merit increases. That is,

$$\sum_{i=1}^{n} x_i = P.$$

Also, the increases for an employee must not put that employee over the maximum salary. That is, for employee i,

$$x_i \leq d_i.$$

We can simplify the objective function, using rules from algebra.

$$\sum_{i=1}^{n} w_i(d_i - x_i) = \sum_{i=1}^{n} (w_i d_i - w_i x_i)$$

$$= \sum_{i=1}^{n} w_i d_i - \sum_{i-1}^{n} w_i x_i$$

For a given individual, w_i and d_i are constant. Therefore, $w_i d_i$ is some constant, say Z, and

$$\sum_{i=1}^{n} w_i(d_i - x_i) = Z - \sum_{i=1}^{n} w_i x_i.$$

We want to minimize the sum on the left; we can do so by *maximizing* the sum on the right. (Why?) Thus, the original model simplifies to maximizing

$$\sum_{i=1}^{n} w_i x_i,$$

subject to $\sum_{i=1}^{n} x_i \leq P$ and $x_i \leq d_i$

for each i.

EXERCISES

Here are current salary information and job evaluation averages for six employees who have the same job. The salary ceiling is $1700 *per month.*

Employee number	Evaluation average	Current salary
1	570	$1600
2	500	$1550
3	450	$1500
4	600	$1610
5	520	$1530
6	565	$1420

1. Find w_i for each employee by dividing that employee's evaluation average by the highest evaluation average.

2. Use the simplex method to find the merit increase for each employee. Assume that P is 400.

8
Probability

If you buy six bags of french fries from a local hamburger establishment at 36¢ per bag, you know the *exact* price of your total purchase is \$2.16. On the other hand, the manager of the establishment is faced with the problem of ordering potatoes for making the fries. She may have a good estimate of the number of bags of fries that will be sold during the day, but she can't know *exactly*.

A great many problems that come up in applications of mathematics are of the type in which we really have no way of knowing the exact answer. The best we can do is to construct a mathematical model which gives the **probability** of certain outcomes.

We study probability in this chapter and use it to construct mathematical models. We begin in Section 8.1 with some basic definitions.

8.1 BASIC PROPERTIES OF PROBABILITY

In probability, each repetition of an experiment is called a **trial.** The possible results of each trial are called **outcomes.** For example, in the experiment of tossing a coin, there are two possible outcomes of each trial (here a trial is a toss of the coin). One outcome is heads and the other is tails (abbreviated h and t, respectively). If the coin is not loaded to favor one outcome over the other, it is called a "fair" coin. This means that the two possible outcomes, h and t, are equally likely or **equiprobable.** For a fair coin, this "equally likely" assumption is made for each trial.

Since there are *two* equally likely outcomes possible, heads and tails, and just *one* of them is heads, the probability of tossing a coin and getting heads is 1 divided by 2.

$$\text{Probability (heads)} = \frac{1}{2} \qquad \text{or} \qquad P(h) = \frac{1}{2}$$

1. Find $P(t)$.

Answer: 1/2

Work Problem 1 at the side.

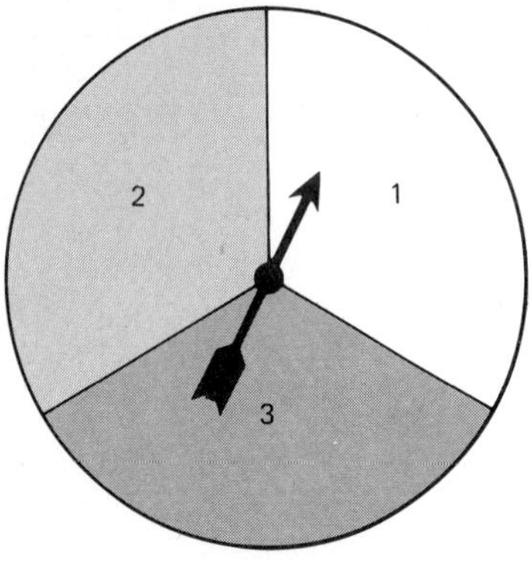

Figure 8.1

Example 1 Suppose we spin the spinner of Figure 8.1.

(a) Find the probability that it will point to 1.

Since there are three equally likely outcomes possible on a single spin, and one of them is getting the number 1, then $P(1) = 1/3$.

(b) Find the probability that it will point to 2.

Since getting the number 2 is one of three possible outcomes, then $P(2) = 1/3$. ■

An ordinary die is a cube whose six faces show the numbers 1, 2, 3, 4, 5, and 6. If the die is not "loaded" to favor certain faces over others, then any one of the six faces is equally likely to come up when the die is rolled.

Example 2 **(a)** If a single fair die is rolled, find the probability of rolling the number 4.

Since one out of six faces is a 4, $P(4) = 1/6$.

(b) Using the same die, find the probability of rolling the number 6.

Since one out of six faces is a 6, $P(6) = 1/6$. ■

Sometimes we are interested in a result that is satisfied by more than one of the possible outcomes. To find the probability that the spinner in Example 1 will point to an odd number, we notice that two of the three possible outcomes are odd numbers, 1 and 3. Therefore,

$$P(\text{odd}) = \frac{2}{3}.$$

The set of all possible outcomes for an experiment is the **sample space** for the experiment. A sample space for the experiment of tossing a coin is made up of the two outcomes, heads and tails. If we use S to represent this sample space, then

$$S = \{h, t\}.$$

2. Write the sample space for the experiment of rolling a single fair die.

Answer: $\{1, 2, 3, 4, 5, 6\}$

Work Problem 2 at the side.

An **event** is a subset of a sample space. If the sample space for tossing a coin is $S = \{h, t\}$, then one event is $E = \{h\}$, which represents the outcome "heads." In this book, we assume that sample spaces and events will be finite sets. The **probability of an event** is defined as

$$\textbf{Probability (event } E\textbf{)} = P(E) = \frac{\textbf{number of elements in } E}{\textbf{number of elements in } S},$$

3. A fair die is rolled. Find the probability of rolling
(a) an odd number;
(b) 3, 4, 5, or 6;
(c) a number greater than 5;
(d) the number 7.

Answer:
(a) 1/2
(b) 2/3
(c) 1/6
(d) 0

Figure 8.2

where S is the sample space for the event. In other words, the probability of an event is the number of favorable outcomes divided by the total number of outcomes. In these definitions, don't forget that all outcomes must be equally likely.

Example 3 A single die is rolled. Given a sample space $\{1, 2, 3, 4, 5, 6\}$, write each of the following events in set notation and give its probability.

(a) E: the number rolled is even.

$$E = \{2, 4, 6\} \qquad P(E) = \frac{3}{6} = \frac{1}{2}$$

(b) F: the number rolled is greater than 4.

$$F = \{5, 6\} \qquad P(F) = \frac{2}{6} = \frac{1}{3}$$

(c) G: the number rolled is less than 10.

$$G = \{1, 2, 3, 4, 5, 6\} \qquad P(G) = \frac{6}{6} = 1$$

(d) H: the number rolled is 8.

$$H = \varnothing \qquad P(H) = 0$$

(e) J: the number rolled is a multiple of 5.

$$J = \{5\} \qquad P(J) = \frac{1}{6} \quad ■$$

Work Problem 3 at the side.

Example 4 If a single playing card is drawn at random from an ordinary 52-card bridge deck (Figure 8.2), find the probability of each of the following events:

(a) Drawing an ace.
There are four aces in the deck, so

$$P(\text{ace}) = \frac{4}{52} = \frac{1}{13}.$$

(b) Drawing a face card.
Since there are 12 face cards,

$$P(\text{face card}) = \frac{12}{52} = \frac{3}{13}.$$

(c) Drawing a spade.
The deck contains 13 spades, so

$$P(\text{spade}) = \frac{13}{52} = \frac{1}{4}.$$

(d) Drawing a spade or a heart.
Besides the 13 spades, the deck contains 13 hearts, so

$$P(\text{spade or heart}) = \frac{26}{52} = \frac{1}{2}. \quad ■$$

4. A single playing card is drawn at random from an ordinary 52-card deck. Find the probability of drawing
(a) a queen;
(b) a diamond;
(c) a red card.

Answer:
(a) 1/13
(b) 1/4
(c) 1/2

Work Problem 4 at the side.

The set of all outcomes in the sample space that do not belong to an event E is called the **complement** of E, written E'. For example, in the experiment of drawing a single card from a well-shuffled deck of 52 cards, let E be the event "the card is an ace." Then E' is the event "the card is not an ace." From the definition of E', we see that for any event E from a sample space S, we have

$$E \cup E' = S \qquad \text{and} \qquad E \cap E' = \varnothing.$$

Example 5 In a particular experiment, $P(E) = 2/7$. Find $P(E')$.
Since $E \cup E' = S$ and E occurs 2 times out of 7, then E' must occur $7 - 2 = 5$ times out of 7, or

$$P(E') = \frac{5}{7}. \quad ■$$

5. Suppose $P(E) = 11/20$. Find $P(E')$.

Answer: 9/20

Work Problem 5 at the side.

Sometimes probability statements are given in terms of **odds,** a comparison of $P(E)$ with $P(E')$. The odds in favor of an event E is defined as the ratio of $P(E)$ to $P(E')$, or the fraction $P(E)/P(E')$.

Example 6 Suppose the weather forecaster says that the probability of rain tomorrow is 1/3. Find the odds in favor of rain tomorrow.

Let E be the event "rain tomorrow." Then E' is the event "no rain tomorrow." Since $P(E) = 1/3$, we have $P(E') = 2/3$. By definition,

$$\text{odds in favor of rain} = \frac{1/3}{2/3} = \frac{1}{2} \qquad \text{written 1 to 2.}$$

On the other hand, the odds that it will *not* rain are

$$\frac{2/3}{1/3} = \frac{2}{1} \quad \text{written 2 to 1.} \quad \blacksquare$$

6. Suppose $P(E) = 9/10$. Find the odds
(a) in favor of E
(b) against E.

Answer:
(a) 9 to 1
(b) 1 to 9

Work Problem 6 at the side.

If we know that the odds in favor of an event are, say, 3 to 5, then the probability of the event is 3/8, while the probability of the complement of the event is 5/8. (Odds of 3 to 5 indicate 3 outcomes in favor of the event out of a total of 8 outcomes.) In general, if the odds favoring event E are m to n, then

$$P(E) = \frac{m}{m+n} \quad \text{and} \quad P(E') = \frac{n}{m+n}.$$

Example 7 The odds in favor of Carla Jones marrying Seymour West are 4 to 5. Find the probability that the couple will marry.

Odds of 4 to 5 show 4 favorable chances out of $4 + 5 = 9$ chances altogether. Thus,

$$P\text{ (marriage)} = \frac{4}{4+5} = \frac{4}{9}. \quad \blacksquare$$

7. If the odds in favor of event E are 1 to 5, find
(a) $P(E)$;
(b) $P(E')$.

Answer:
(a) 1/6
(b) 5/6

Work Problem 7 at the side.

The formulas above enable us to find the probability of an event when all outcomes are equally likely. However, we would also like to be able to assign probabilities to many occurrences which cannot be exactly repeated. For example, we might want to know the probability of rain on the day of the picnic or the probability that our team will win the game. These are known as **subjective** probabilities and must be assigned, if at all, on the basis of personal judgment. A sales manager may use his experience to assign a probability to the success of his current July Sale, but since it can never be exactly like those of past Julys (or future Julys), it is still a subjective assignment of probability. In weather forecasting, the forecaster compares atmospheric conditions with similar conditions in the past, and predicts the weather according to what happened then under such conditions. For example, if in the past, under certain kinds of atmospheric conditions, rain resulted 80 times out of 100, the forecaster will announce an 80% chance of rain. In many cases, no information is available and probability must be assigned on the basis of hunches or expectation. In fact, sometimes we want to assign probabilities to unique occurrences that happen only once.

One difficulty with subjective probability is that different probabilities may be assigned to the same event. Nevertheless, subjective probabilities can be assigned to many occurrences where the objective approach to probability cannot be used. In Chapter 9 we illustrate the use of subjective probability in more detail.

8.1 EXERCISES

A single fair die is rolled. Find the probability (see Examples 2 and 3) of rolling

1. the number 2;
2. an odd number;
3. a number less than 5;
4. a number greater than 2.

A card is drawn from a well-shuffled deck of 52 cards. Find the probability (see Example 4) of drawing

5. a 9;
6. black;
7. a black 9;
8. a heart;
9. the 9 of hearts;
10. a face card.

A single fair die is rolled. Find the odds (see Examples 6 and 7) in favor of rolling

11. the number 5;
12. 3, 4, or 5;
13. 1, 2, 3, 4;
14. some number less than 2.

Write a sample space with equiprobable outcomes for each of the following experiments.

15. A two-headed coin is tossed once.

16. Two coins are tossed.

17. Three coins are tossed.

18. Slips of paper marked with the numbers 1, 2, 3, 4, and 5 are placed in a box. After being mixed, two slips are drawn.

19. An unprepared student takes a three-question true/false quiz in which he guesses the answer to all three questions.

20. A die is rolled and then a coin is tossed.

Write the following events in set notation and give the probability of each event. (See Example 3.)

21. In the experiment of Exercise 16:
(a) both coins show the same face;
(b) at least one coin turns up heads.

22. In Exercise 15:
(a) the result of the toss is heads;
(b) the result of the toss is tails.

23. In Exercise 18:
(a) both slips are marked with even numbers;
(b) both slips are marked with odd numbers;
(c) both slips are marked with the same number;
(d) one slip is marked with an odd number, the other with an even number.

24. In Exercise 19:
(a) the student gets all three answers correct;
(b) he gets all three answers wrong;
(c) he gets exactly two answers correct;
(d) he gets at least one answer correct.

A marble is drawn at random from a box containing 3 yellow, 4 white, and 8 blue marbles. Find the probability of drawing

25. A yellow marble;

26. A blue marble;

27. A white marble.

28. What are the odds in favor of drawing a yellow marble?

29. What are the odds in favor of drawing a blue marble?

30. What are the odds in favor of drawing a white marble?

31. A baseball player with a batting average of .300 comes to bat. What are the odds in favor of his getting a hit?

32. In Exercise 18, what are the odds that the sum of the numbers on the two slips of paper is 5?

33. If the odds that it will rain are 4 to 5, what is the probability of rain?

34. If the odds that a candidate will win an election are 3 to 2, what is the probability that the candidate will lose?

Which of the following are examples of subjective probability?

35. the probability of heads on five consecutive tosses of a coin

36. the probability that a freshman entering a four-year college will graduate with a degree

37. the probability that a person is allergic to penicillin

38. the probability of drawing an ace from a standard deck of 52 playing cards

39. the probability that an individual will get lung cancer from smoking cigarettes

40. A weather forecaster allows a 70% chance of rain tomorrow.

41. A gambler claims that on a roll of a fair die, $P(\text{even}) = \frac{1}{2}$.

42. A surgeon gives a patient a 90% chance of full recovery.

43. A bridge player has a $\frac{1}{4}$ probability of being dealt a diamond.

44. A forest ranger states that the probability of a short fire season this year is only three in ten.

8.2 PROBABILITY OF ALTERNATE EVENTS

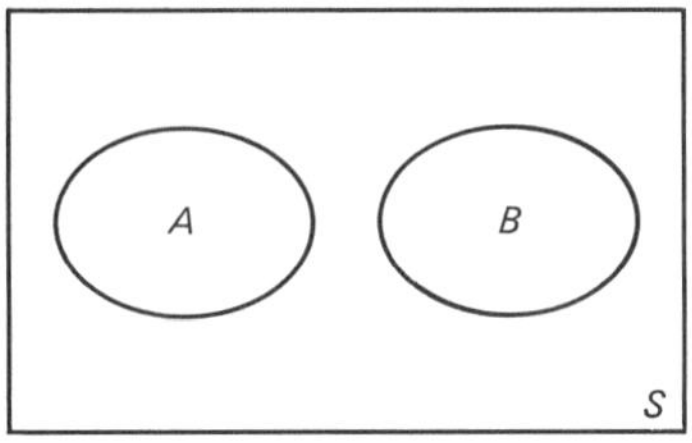

Figure 8.3

Two events which have no outcomes in common are said to be **mutually exclusive events.** Any set E and its complement E' satisfy this definition. Also, for the experiment of rolling a die, the events "rolling a 2" and "rolling a 4" are mutually exclusive. If we use a Venn diagram to illustrate mutually exclusive events, they will be disjoint, as shown in Figure 8.3 where the sample space is the universal set. Three or more events are mutually exclusive if each pair of events is mutually exclusive.

The probability that event E or event F occurs is written as $P(E \text{ or } F)$, or in set symbols, as $P(E \cup F)$. If E and F are mutually exclusive events, we can find $P(E \cup F)$ by using the *addition rule* for mutually exclusive events:

For mutually exclusive events E and F,

$$\mathbf{P(E \cup F) = P(E) + P(F).}$$

This rule can be used in the same way for more than two mutually exclusive events.

Example 1 A single fair die is rolled. Let E represent the event "rolling a 5" and F represent "rolling an even number." Find $P(E \cup F)$.

$$P(E \cup F) = P\text{ (rolling a 5 or an even number)}$$

Since E and F are mutually exclusive events (why?), we have

$$\begin{aligned} P(E \cup F) &= P(E) + P(F) \\ &= P\text{ (rolling a 5)} + P\text{ (rolling an even number)} \\ &= \frac{1}{6} + \frac{3}{6} = \frac{4}{6} = \frac{2}{3}. \quad \blacksquare \end{aligned}$$

Work Problem 1 at the side.

1. A single card is drawn from an ordinary deck (52 cards). Let E represent the event "drawing an ace" and let F represent "drawing a jack, queen, or king." Find $P(E \cup F)$.

Answer: 16/52 or 4/13

For any experiment, events E and E' (the complement of E) are mutually exclusive. Therefore,

$$P(E \cup E') = P(E) + P(E').$$

However, $E \cup E' = S$, the entire sample space, and $P(S) = 1$, so that

$$\mathbf{P(E) + P(E') = 1.}$$

This fact can be restated in a more useful form:

$$\mathbf{P(E) = 1 - P(E')} \qquad \text{and} \qquad \mathbf{P(E') = 1 - P(E).}$$

Example 2 The probability that Sarah will finish her term paper on time is 5/8. Find the probability that she will not finish on time.

Let E represent the event "will finish on time." Then E' represents "will not finish on time." We apply the formula above to get

$$P(E') = 1 - P(E) = 1 - \frac{5}{8} = \frac{3}{8}. \quad ■$$

2. Let $P(K) = 2/3$. Find $P(K')$.

Answer: 1/3

Work Problem 2 at the side.

Example 3 Francisco figures that there is only a slight chance that he can finish his homework in just 1 hour tonight. In fact, he assigns probabilities to the various numbers of hours he may actually need as shown in the table.

Hours	Probability
1	0.05
2	0.10
3	0.20
4	0.40
5	0.10
6	0.15

Find each of the following probabilities.

(a) It will take him fewer than 3 hours.

"Fewer than 3" means 1 or 2. So P (fewer than 3) $= 0.05 + 0.10 = 0.15$

(b) It will take him more than 2 hours.

$$\begin{aligned} P\text{ (more than 2)} &= P(3) + P(4) + P(5) + P(6) \\ &= 0.20 + 0.40 + 0.10 + 0.15 \\ &= 0.85 \end{aligned}$$

Alternatively, P (more than 2) $= 1 - P$ (2 or fewer)

$$= 1 - [P(1) + P(2)]$$
$$= 1 - [.05 + .10]$$
$$= 1 - .15$$
$$= .85$$

(c) It will take him more than 1 hour but no more than 5 hours.

P (more than 1 but no more than 5) $= P(2) + P(3) + P(4) + P(5)$

$$= 0.80 \quad \blacksquare$$

3. The probability that a salesperson will have weekly sales of certain amounts is as shown below.

Sales	Probability
\$500	.25
1000	.40
1500	.20
2000	.10
2500	.05

Find the probability that the salesperson has sales
(a) less than \$1500;
(b) more than \$1000.

Answer:
(a) .65
(b) .35

Work Problem 3 at the side.

Example 4 Two dice are rolled. Find each of the following probabilities.

(a) The sum of points rolled is greater than 3.

There are 36 equiprobable outcomes in the sample space, which can be displayed as follows.

		Die 2					
		1	2	3	4	5	6
Die 1	1	1, 1	1, 2	1, 3	1, 4	1, 5	1, 6
	2	2, 1	2, 2	2, 3	2, 4	2, 5	2, 6
	3	3, 1	3, 2	3, 3	3, 4	3, 5	3, 6
	4	4, 1	4, 2	4, 3	4, 4	4, 5	4, 6
	5	5, 1	5, 2	5, 3	5, 4	5, 5	5, 6
	6	6, 1	6, 2	6, 3	6, 4	6, 5	6, 6

(Although these are actually ordered pairs, we have eliminated the parentheses.) The sample space contains 33 outcomes in which the sum of the points is greater than 3. Thus the probability of a sum greater than 3 is 33/36, or 11/12. A different approach to this problem would be to use the relationship

$$P(E) = 1 - P(E').$$

For the event "the sum is greater than 3," the complement is "the sum is 3 or less," which is satisfied by 3 of the 36 outcomes, and so has probability of 1/12. The required probability, then, is

$$1 - \frac{1}{12} = \frac{11}{12}.$$

(b) The sum of points rolled is exactly 7.

As the sample space above shows, there are 6 ways to get a sum of 7: 1, 6; 2, 5; 3, 4; 4, 3; 5, 2; and 6, 1. These 6 ways are from a total of 36 possible outcomes, so

$$P\text{ (sum of 7)} = \frac{6}{36} = \frac{1}{6}.$$

(c) The sum of points rolled is exactly 11.
From the sample space above,

$$P\text{ (sum of 11)} = \frac{2}{36} = \frac{1}{18}. \quad \blacksquare$$

Work Problem 4 at the side.

4. Two dice are rolled. Find the probability of
(a) a sum of 4 or less;
(b) a sum of exactly 5;
(c) a sum of 7 or more.

Answer:
(a) 1/6
(b) 1/9
(c) 7/12

Example 5 Janet Passman has appointments for three job interviews. Based on her own assessment of her qualifications for these jobs and her knowledge of the job market, she assigns probabilities of being hired as follows: Company X, 2/5; Company Y, 3/10; and Company Z, 1/10. Assume that being hired by Company X, by Company Y, and by Company Z are mutually exclusive events.
(a) What is the probability that she will get none of these jobs?
There are four outcomes in the sample space: she is hired by Company X, she is hired by Company Y, she is hired by Company Z, or she is not hired at all. The probabilities of the four outcomes in the sample space must add up to 1. Therefore, the probability that she is not hired is

$$1 - \left[\frac{2}{5} + \frac{3}{10} + \frac{1}{10}\right] = \frac{2}{10} = \frac{1}{5}.$$

(b) Find the probability that she is hired by Company X or Y.
Since the two events are mutually exclusive, the probability of being hired by either Company X or Y is the sum

$$\frac{2}{5} + \frac{3}{10} = \frac{7}{10}.$$

(c) What is the probability that she is hired by one of the three companies?
The complement of this event is the event that she gets none of the jobs. In part (a), we found the probability of the complement to be 1/5. Therefore, the desired probability is

$$1 - \frac{1}{5} = \frac{4}{5}. \quad \blacksquare$$

Work Problem 5 at the side.

5. In Example 5, find the probability that Janet is
(a) not hired by X or Y;
(b) not hired by Z;
(c) not hired by X or Z.

Answer:
(a) 3/10
(b) 9/10
(c) 1/2

The addition rule given earlier in this section applies only to mutually exclusive events. For two events that are *not* mutually exclusive, use the following property.

For any events E and F,

$$P(E \cup F) = P(E) + P(F) - P(E \cap F).$$

(In words: the probability that E or F happens is the probability that E happens plus the probability that F happens, minus the probability that both happen.)

Example 6 If a single card is drawn from an ordinary deck, find the probability that it will be red or a face card.

Let R and F represent the events "red" and "face card" respectively. Then

$$P(R) = \frac{26}{52}, \qquad P(F) = \frac{12}{52}, \qquad \text{and} \qquad P(R \cap F) = \frac{6}{52}.$$

(There are six red face cards in the deck.) Thus,

$$\begin{aligned} P(R \cup F) &= P(R) + P(F) - P(R \cap F) \\ &= \frac{26}{52} + \frac{12}{52} - \frac{6}{52} \\ &= \frac{32}{52} \\ &= \frac{8}{13}. \quad \blacksquare \end{aligned}$$

Work Problem 6 at the side.

6. A single card is drawn from an ordinary deck. Find the probability that it is black or a 9.

Answer: 7/13

8.2 EXERCISES

Two dice are rolled. Find the probability of rolling each of the following sums. (See Example 4.)

1. 2 **2.** 4 **3.** 5 **4.** 6

5. 8 **6.** 9 **7.** 10 **8.** 13

Find the following probabilities for the experiment of rolling two dice. The sum rolled is

9. 9 or more;

10. less than 7;

11. between 5 and 8;

12. not more than 5;

13. not less than 8;

14. between 3 and 7.

The table below gives a certain golfer's probabilities of scoring in various ranges on a par-70 course.

Range	Probability
below 60	.05
60–64	.08
65–69	.15
70–74	.28
75–79	.22
80–84	.08
85–89	.06
90–94	.04
95–99	.02
100 or more	.02

In a given round, find the probability that the golfer's score will be

15. 90 or higher;

16. below par of 70;

17. in the 70's;

18. in the 90's;

19. not in the 60's;

20. not in the 60's or 70's.

21. Find the odds in favor of the golfer shooting below par.

Mrs. Elliott invites 10 *relatives to a party: her mother,* 2 *uncles,* 3 *brothers, and* 4 *cousins. If the chances of any one guest arriving first are equally likely, find the probability that the first guest to arrive is*

22. an uncle or brother;

23. a brother or cousin;

24. a brother or her mother;

25. an uncle or a cousin.

One card is drawn from an ordinary deck of cards. Find the probability of drawing

26. a 9 or a 10;

27. red or a 3;

28. a 9 or a black 10;

29. a heart or black;

30. less than a 4 (count aces as 1's);

31. a diamond or a 7;

32. a black card or an ace;

33. a heart or a card less than 5.

A student estimates that the probability of his getting an A *in a certain course is* .4; *getting a* B *is* .3; *getting a* C *is* .2; *and getting a* D *is* .1.

34. Assuming that only the grades A, B, C, D, and F are possible, what is the probability that he will fail the course?

35. What is the probability that he will receive a grade of C or better?

36. What is the probability that he will receive at least a B in the course?

37. What is the probability that he will get at most a C in the course?

The numbers 1, 2, 3, 4, and 5 are written on slips of paper and two slips are drawn at random.

38. What is the probability that the sum of the numbers drawn is 10?

39. What is the probability that the numbers drawn are both even?

40. What is the probability that the sum of the numbers drawn is at most 5?

41. What is the probability that at least one of the numbers is even or greater than 3?

APPLIED PROBLEMS

Social Science

Twenty television programs will be shown this evening. Of these programs, 8 are educational, 9 are interesting, and 5 are both educational and interesting. Bruce decides to choose one program at random. Find the probability that the program chosen is

42. not educational;

43. not interesting;

44. educational or interesting.

Natural Science

Color blindness is an inherited characteristic which is sex-linked, so that it is more common in males than in females. If M represents male and C represents red-green color blindness, we use the relative frequencies of the incidence of males and of red-green color blindness as probabilities to get $P(C) = .049$, $P(M \cap C) = .042$, $P(M \cup C) = .534$. *Find the following. (Hint: Use a Venn diagram with two circles labeled M and C.)*

45. $P(C')$ **46.** $P(M)$ **47.** $P(M')$

48. $P(M' \cap C')$ **49.** $P(C \cap M')$ **50.** $P(C \cup M')$

Gregor Mendel, an Austrian monk, was the first to use probability in the study of genetics. In an effort to understand the mechanism of character transmittal from one generation to the next in plants, he counted the number of occurrences of various characteristics. Mendel found that the flower color in certain pea plants obeyed this scheme:

Pure red crossed with pure white produces red.

The red offspring received from its parents genes for both red (R) and white (W) but in this case red is dominant *and white* recessive, *so the offspring exhibits the color red. However, the offspring still carries both genes, and when two such offspring are crossed, several things can happen in the third generation as shown in the table below, which is called a* Punnet square.

		2nd parent R	2nd parent W
1st parent	R	RR	RW
	W	WR	WW

This table shows the possible combinations of genes. Recalling that red is dominant over the recessive white character (one or more red genes leads to a red-flowered offspring), find

51. P (red); **52.** P (white).

Mendel found no dominance in snapdragons, with one red gene and one white gene producing pink-flowered offspring. These second generation pinks, however, still carry one red and one white gene, and when they are crossed, the next generation still yields the Punnet square above. Find

53. P (red); **54.** P (pink); **55.** P (white).

(Mendel verified these probability ratios experimentally with large numbers of observations, and did the same for many character units other than flower color. The importance of his work, published in 1866, was not recognized until 1900.)

In most animals and plants, it is very unusual for the number of main parts of the organism (arms, legs, toes, flower petals, etc.) to vary from generation to generation. Some species, however, have meristic variability, *in which the number of certain body parts varies from generation to generation. One researcher* studied the front feet of certain guinea pigs and produced the following probabilities.*

$$P\text{ (only four toes, all perfect)} = .77$$
$$P\text{ (one imperfect toe and four good ones)} = .13$$
$$P\text{ (exactly five good toes)} = .10$$

Find the probability of each of the following events.

56. no more than four good toes

57. five toes, whether perfect or not

Management

The table below shows the probability that a customer of a department store will make a purchase in the indicated range.

Cost	Probability
below \$2	.07
\$2–\$4.99	.18
\$5–\$9.99	.21
\$10–\$19.99	.16
\$20–\$39.99	.11
\$40–\$69.99	.09
\$70–\$99.99	.07
\$100–\$149.99	.08
\$150 or over	.03

* From "An Analysis of Variability in Guinea Pigs" by J. R. Wright, from *Genetics*, Vol. 19, 1934, pp. 506–536. Reprinted by permission.

Find the probability that a customer makes a purchase which is

58. less than \$5; **59.** \$10 to \$69.99; **60.** \$20 or more;

61. more than \$4.99; **62.** less than \$100; **63.** \$100 or more.

8.3 CONDITIONAL PROBABILITY

The Dean of Students at Lincoln Tech feels that students who come to college with a specific goal do better than those who have no goal. To test this theory, the dean conducted a survey of 100 students; he found the results shown in the table below.

	Graduated	Did not	Totals
Goal	20	10	30
No goal	20	50	70
Totals	40	60	100

Letting A represent the event "will graduate" and B represent the event "entered college with a goal," we can assign the following probabilities.

$$P(A) = \frac{40}{100} = .4 \qquad P(A') = \frac{60}{100} = .6$$

$$P(B) = \frac{30}{100} = .3 \qquad P(B') = \frac{70}{100} = .7$$

Suppose we want to find the probability that a student who entered college with a goal will graduate. From the chart above, of the 30 students who entered with a goal, there are 20 who did graduate. Thus, the desired probability is $20/30 = .67$. Note that this is a larger number than the probability that a student will graduate, .4, because we had additional information which reduced the sample space. In other words, we found the probability that a student would graduate given that he or she entered college with a goal. This is called the **conditional probability** of event A given that B has occurred, written $P(A|B)$.

In the example above, we found

$$P(A|B) = \frac{20}{30}$$

which can be written as

$$P(A|B) = \frac{20/100}{30/100} = \frac{P(A \cap B)}{P(B)},$$

1. The table shows the results of a survey of 1000 new or used car buyers of a certain model car.

	Satisfied	Not satisfied	Totals
New	300	100	400
Used	450	150	600
Totals	750	250	1000

Let S represent the event "satisfied" and N the event "bought car new." Find each of the following.

(a) $P(N|S)$
(b) $P(N'|S)$
(c) $P(S|N')$
(d) $P(S'|N')$

Answer:
(a) 2/5
(b) 3/5
(c) 3/4
(d) 1/4

where $P(A \cap B)$ represents the probability that A and B will both occur. We can then define the conditional probability of an event E given that F has occurred as

$$P(E|F) = \frac{P(E \cap F)}{P(F)}, \qquad P(F) \neq 0$$

Example 1 Use the information above to find each of the following probabilities.

(a) $P(B|A)$
By the definition above,

$$P(B|A) = \frac{P(B \cap A)}{P(A)},$$

In our example, $P(B \cap A) = 20/100$, and $P(A) = 40/100$.

$$P(B|A) = \frac{20/100}{40/100} = \frac{1}{2}.$$

If we know that a student graduated, then there is a probability of 1/2 that the student had a goal.

(b) $$P(A'|B) = \frac{P(A' \cap B)}{P(B)} = \frac{10/100}{30/100} = \frac{1}{3}$$

(c) $$P(B'|A') = \frac{P(B' \cap A')}{P(A')} = \frac{50/100}{60/100} = \frac{5}{6}. \quad ■$$

Work Problem 1 at the side.

Venn diagrams can be used to illustrate conditional probability problems. A Venn diagram for the example above, in which the probabilities are used to indicate the number in the set defined by each region, is given in Figure 8.4. In the diagram, $P(B|A)$ is found by reducing the sample space to set A. Then $P(B|A)$ is the ratio of the number in that part of set B which is also in A to the number in set A—that is, $20/40 = .5$.

The following example shows how a sketch called a tree diagram can be used in answering several kinds of questions about conditional probabilities.

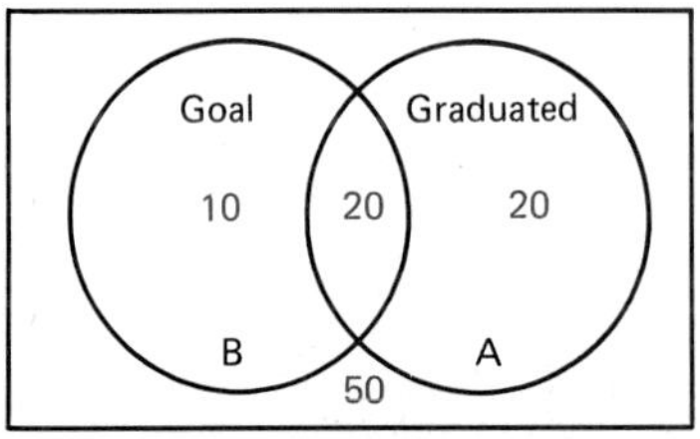

Figure 8.4

Example 2 Your local garage employs two mechanics, A and B, but the service manager never tells you which one worked on your car. All you know is the following set of facts discovered by a newspaper reporter: A does twice as many jobs as B, A does a satisfactory job three out of four times, and B does satisfactorily on only two out of five jobs. Suppose you take your car in for work. Find the probabilities of the following events.

(a) Mechanic A does the work with satisfactory results.

We first construct a **tree diagram** which shows the various possible outcomes to this experiment. (See Figure 8.5.) Since A does twice as many jobs as B, the (unconditional) probabilities of A and B are 2/3 and 1/3 respectively, as shown on the first stage of the tree. The second stage of the tree shows four different conditional probabilities. The job ratings are S(satisfactory) and U (unsatisfactory). For example, along the branch from B to S we have

$$P(\mathrm{S}|\mathrm{B}) = 2/5$$

since mechanic B does a satisfactory job on two out of five attempts.

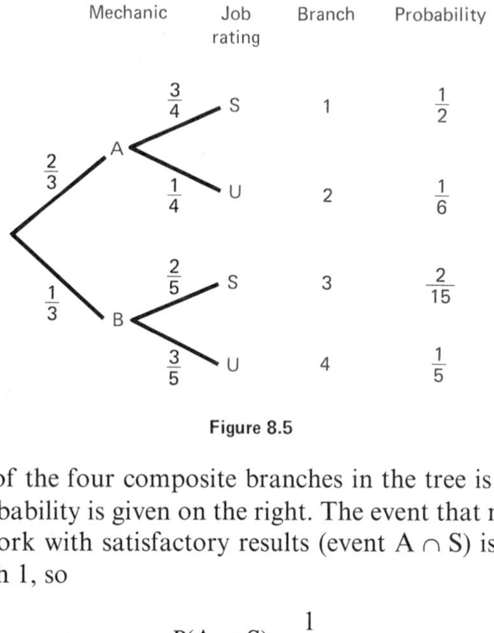

Figure 8.5

Each of the four composite branches in the tree is numbered, and its probability is given on the right. The event that mechanic A does the work with satisfactory results (event $\mathrm{A} \cap \mathrm{S}$) is associated with branch 1, so

$$P(\mathrm{A} \cap \mathrm{S}) = \frac{1}{2}.$$

(b) Mechanic B does the work with satisfactory results.

The event $\mathrm{B} \cap \mathrm{S}$ is shown on branch 3, so

$$P(\mathrm{B} \cap \mathrm{S}) = \frac{2}{15}.$$

(c) The results are satisfactory.

The event S combines branches 1 and 3, so

$$P(\mathrm{S}) = \frac{1}{2} + \frac{2}{15} = \frac{19}{30}.$$

(d) The results are unsatisfactory.
The event U combines branches 2 and 4, so

$$P(\text{U}) = \frac{1}{6} + \frac{1}{5} = \frac{11}{30}.$$

(e) Either A does the work or the results are satisfactory (or both).
Event A combines branches 1 and 2, while event S combines branches 1 and 3. Thus we use branches 1, 2, and 3.

$$P(\text{A} \cup \text{S}) = \frac{1}{2} + \frac{1}{6} + \frac{2}{15} = \frac{4}{5}. \quad ■$$

2. Find each of the following probabilities for Example 2.
(a) A did the work, given that the results are satisfactory.
(b) $P(\text{B}|\text{U})$

Answer:
(a) 15/19
(b) 6/11

Work Problem 2 at the side.

In the definition of conditional probability given earlier, we can multiply both sides of the equation by $P(F)$ to get the following *product rule* of probability.

For any events E and F,

$$P(E \cap F) = P(F) \cdot P(E|F).$$

This equation gives us a method for finding the probability that E and F both occur, as illustrated by the next example.

Example 3 From a box containing 1 red, 3 white, and 2 green marbles, two marbles are drawn one at a time without replacing the first before the second is drawn. Find the probability that one white and one green marble are drawn.

A tree diagram showing the various possible outcomes is given in Figure 8.6. In this diagram, W represents the event "drawing a white marble" and G represents "drawing a green marble." On the first draw, $P(W \text{ on the 1st}) = 3/6 = 1/2$ because three of the six marbles in the box are white. On the second draw, $P(G \text{ on the 2nd}|W \text{ on the 1st}) = 2/5$. One white marble has been removed, leaving 5, of which 2 are green.

Now we want to find the probability of drawing one white marble and one green marble. This event can occur in two ways: drawing a white marble first and then a green one (branch 2 of the tree diagram), or drawing a green marble first and then a white one (branch 4). For branch 2, we have

$$P(W \text{ on 1st}) \cdot P(G \text{ on 2nd}|W \text{ on 1st}) = \frac{1}{2} \cdot \frac{2}{5} = \frac{1}{5}.$$

For branch 4, when the green marble is drawn first, we have

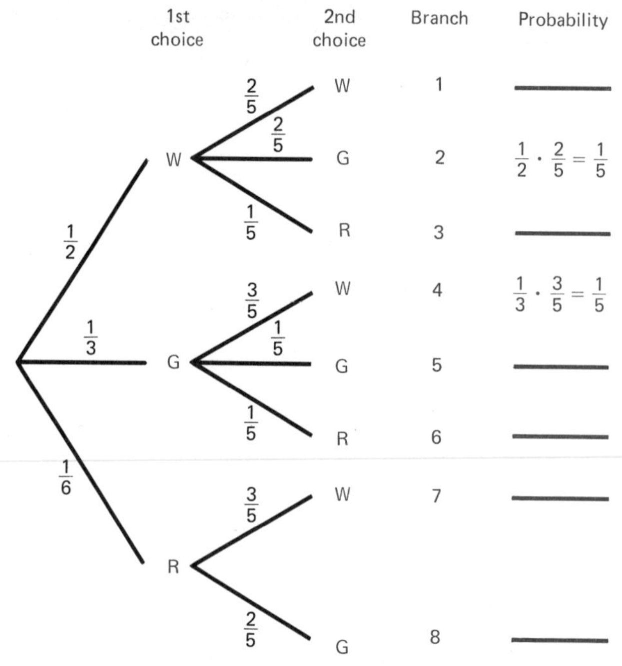

Figure 8.6

$$P(G \text{ on 1st}) \cdot P(W \text{ on 2nd} | G \text{ on 1st}) = \frac{1}{3} \cdot \frac{3}{5} = \frac{1}{5}.$$

The final probability is the sum of these two probabilities.

$$P(\text{one } W, \text{ one } G) = P(W \text{ on 1st}) \cdot P(G \text{ on 2nd} | W \text{ on 1st}) + P(G \text{ on 1st}) \cdot P(W \text{ on 2nd} | G \text{ on 1st}) = \frac{2}{5} \quad ■$$

3. In Example 3 find the probability that both marbles drawn are green.

Answer: 1/15

Work Problem 3 at the side.

Example 4 Two cards are drawn without replacement from an ordinary deck (52 cards). Find the probability that the first card is a heart and the second card is red.

On the first draw, since there are 13 hearts in the 52 cards, the probability of drawing a heart first is $13/52 = 1/4$. On the second draw, since a heart has been drawn already, there are 25 red cards in the remaining 51 cards. Thus, the probability of drawing a red card on the second draw, given that the first is a heart is $25/51$. Therefore,

$$
\begin{aligned}
&P(\text{heart on 1st and red on 2nd}) \\
&\quad = P(\text{heart on 1st}) \cdot P(\text{red on 2nd}|\text{heart on 1st}) \\
&\quad = \frac{1}{4} \cdot \frac{25}{51} = \frac{25}{204} = .1225. \quad \blacksquare
\end{aligned}
$$

4. Find the probability of drawing a heart on the first draw and a black card on the second, if two cards are drawn without replacement.

Answer: 26/204, or 13/102, or .1275

Work Problem 4 at the side.

When $P(F|E)$ equals $P(F)$, the occurrence or nonocurrence of event E has no effect on event F. In this case, events E and F are said to be **independent**, and our earlier product rule becomes

$$P(E \cap F) = P(E) \cdot P(F).$$

In Example 4 if we had replaced the first card before drawing the second, then the probability of drawing a red card on the second draw, given that a heart was drawn on the first, would be the same as the probability of drawing a red card on a single draw, $26/52 = 1/2$. Thus, the two events would be independent and the probability of drawing a heart on the first draw and a red card on the second draw would be

$$\frac{1}{4} \cdot \frac{1}{2} = \frac{1}{8} = .125.$$

Events which are not independent are called **dependent events.** In Example 4 drawing the second card when the first was not replaced is an illustration of a dependent event.

Example 5 Find the probability of getting heads on 5 successive tosses of a coin.

Let H represent the event "heads." For each toss of the coin, $P(H) = 1/2$. Since each toss is independent of the other tosses, we have

$$P(H \cap H \cap H \cap H \cap H) = \frac{1}{2} \cdot \frac{1}{2} \cdot \frac{1}{2} \cdot \frac{1}{2} \cdot \frac{1}{2} = \left(\frac{1}{2}\right)^5 = \frac{1}{32}. \quad \blacksquare$$

5. Find the probability of getting 4 sixes in 4 rolls of a die.

Answer: $(1/6)^4 = 1/1296$

Work Problem 5 at the side.

Example 6 When black-coated mice are crossed with brown-coated mice, a pair of genes, one from each parent, determines the coat color of the offspring. Let b represent the gene for brown and B the gene for black. If a mouse carries either one B gene and one b

gene (Bb or bB) or two B genes (BB), the coat will be black. If the mouse carries two b genes (bb), the coat will be brown. Find the probability that a mouse born to a brown-coated female and a black-coated male who is known to carry the Bb combination will be brown.

To be brown-coated, the offspring must receive one b gene from each parent. The brown-coated parent carries two b genes, so that the probability of getting one b gene from the mother is 1. The probability of getting one b gene from the black-coated father is 1/2. Therefore, since these are independent events, the probability of a brown-coated offspring from these parents is $1 \cdot 1/2 = 1/2$. ■

6. Find the probability that the mice of Example 6 will have a black-coated offspring.

Answer: 1/2

Work Problem 6 at the side.

8.3 EXERCISES

If a single fair die is rolled, find the probabilities of rolling

1. a 2, given that the number rolled was odd;

2. a 4, given that the number rolled was even;

3. an even number, given that the number rolled was a 6.

If two fair dice are rolled (recall the 36-outcome sample space), find the probabilities of rolling

4. a sum of 8, given the sum was greater than 7;

5. a sum of 6, given the roll was a "double" (two identical numbers);

6. a double, given that the sum was 9.

If two cards are drawn without replacement from an ordinary deck (see Example 4), find the probability that

7. the second is a heart, given that the first is a heart;

8. they are both hearts;

9. the second is black, given that the first is a spade;

10. the second is a face card, given that the first is a jack.

If five cards are drawn without replacement from an ordinary deck, find the probability that all the cards are

11. diamonds;

12. diamonds, given that the first and second were diamonds;

13. diamonds, given that the first four were diamonds;

14. clubs, given that the third was a spade;

15. the same suit.

A smooth-talking young man has a 1/3 *probability of talking a policeman out of giving him a speeding ticket. The probability that he is stopped for speeding during a given weekend is* 1/2. *Find the probability that*

16. he will receive no tickets on a given weekend;

17. he will receive no tickets on three consecutive weekends.

Slips of paper marked with the digits 1, 2, 3, 4, *and* 5 *are placed in a box and mixed well. If two slips are drawn (without replacement), find the probability that*

18. the first is even and the second is odd;

19. the first is a 3 and the second a number greater than 3;

20. both are even;

21. both are marked 3.

Two marbles are drawn without replacement from a jar with four black and three white marbles. Find the probability that

22. both are white;

23. both are black;

24. the second is white given that the first is black;

25. the first is black and the second is white;

26. one is black and the other is white.

APPLIED PROBLEMS

Management

The Midtown Bank has found that most customers at the tellers' windows either cash a check or make a deposit. The chart below indicates the transactions for one teller for one day.

	Cash check	No check	Totals
Make deposit	50	20	70
No deposit	30	10	40
Totals	80	30	110

Letting C represent "cashing a check" and D represent "making a deposit," express each of the following probabilities in words and find its value.

27. $P(C|D)$ **28.** $P(D'|C)$ **29.** $P(C'|D')$

30. $P(C'|D)$ **31.** $P[(C \cap D)']$

A pet shop has 10 *puppies,* 6 *of them males. There are* 3 *beagles* (1 *male*),

1 cocker spaniel (male), and 6 poodles. Construct a table similar to the one above and find the probability that one of these puppies, chosen at random, is

32. a beagle;

33. a beagle, given that it is a male;

34. a male, given that it is a beagle;

35. a cocker spaniel, given that it is a female;

36. a poodle, given that it is a male;

37. a female, given that it is a beagle.

A bicycle factory runs two assembly lines, A and B. If 95% of line A's products pass inspection, while only 90% of line B's products pass inspection, and 60% of the factory's bikes come off assembly line B (the rest off A), find the probability that one of the factory's bikes did not pass inspection and came off

38. assembly line A;

39. assembly line B.

Natural Science

40. Both of a certain pea plant's parents had a gene for red and a gene for white flowers. (See the exercises for Section 8.2.) If the offspring has red flowers, find the probability that it combined a gene for red and a gene for white (rather than two for red).

Assume that boy and girl babies are equally likely.

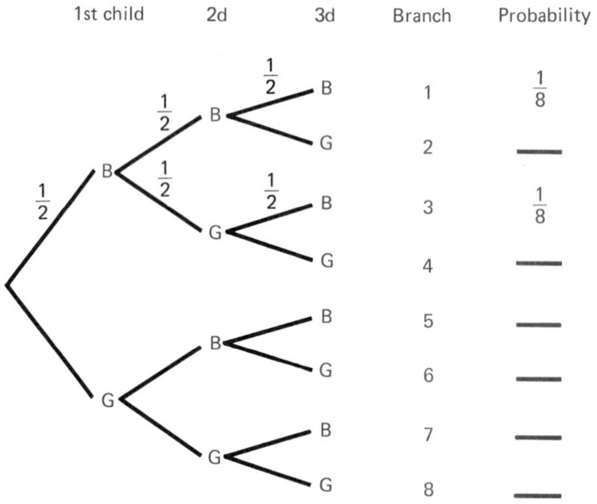

Fill in the remaining probabilities on the tree diagram above and use the information to find the probability that a family with three children has all girls, given that

41. the first is a girl;

42. the third is a girl;

43. the second is a girl;

44. at least two are girls;

45. at least one is a girl.

The following table shows frequencies for red-green color blindness, where M represents male and C represents color-blind.

	M	M'	Totals
C	.042	.007	.049
C'	.485	.466	.951
Totals	.527	.473	1.000

Use this table to find the following probabilities.

46. $P(M)$

47. $P(C)$

48. $P(M \cap C)$

49. $P(M \cup C)$

50. $P(M|C)$

51. $P(C|M)$

52. $P(M'|C)$

53. Are the events C and M described above dependent? [Recall that two events E and F are dependent if $P(E|F) \neq P(E)$.]

54. A scientist wishes to determine if there is any dependence between color blindness (C) and deafness (D). Given the probabilities listed in the table below, what should his findings be? (See Exercises 46–53.)

	D	D'	Totals
C	.0004	.0796	.0800
C'	.0046	.9154	.9200
Totals	.0050	.9950	1.0000

Social Science

The Motor Vehicle Department has found that the probability of a person passing the test for a driver's license on the first try is .75. The probability that an individual who fails on the first test will pass on the second try is .80, and the probability that an individual who fails the first and second tests will pass the third time is .70. Find the probability that an individual

55. fails both the first and second tests;

56. will fail three times in a row;

57. will require at least two tries to pass the test.

Management

According to a booklet put out by Frontier Airlines, 98% of all scheduled Frontier flights actually take place. (The other flights are cancelled due to weather, equipment problems, and so on.) Assume that the event that a given flight takes place is independent of the event that another flight takes place.

58. Elizabeth Thornton plans to visit her company's branch offices; her journey requires three separate flights on Frontier. What is the probability that all these flights will take place?

59. Based on the reasons we gave for a flight to be cancelled, how realistic is the assumption of independence that we made?

8.4 BAYES' FORMULA

We saw in the last section that in general, $P(E|F) \neq P(F|E)$. Many times we know one of these probabilities and wish to calculate the other. To see how they are related, recall the product rule

$$P(E \cap F) = P(E) \cdot P(F|E),$$

which can also be written as $P(F \cap E) = P(F) \cdot P(E|F)$. From the fact that $P(E \cap F) = P(F \cap E)$, we have

$$P(E) \cdot P(F|E) = P(F) \cdot P(E|F),$$

or

$$P(F|E) = \frac{P(F) \cdot P(E|F)}{P(E)}. \qquad \textbf{(1)}$$

Given the two events E and F, if E occurs, then either F also occurs or F' also occurs. The probabilities of $E \cap F$ and $E \cap F'$ can be expressed as follows.

$$P(E \cap F) = P(F) \cdot P(E|F)$$
$$P(E \cap F') = P(F') \cdot P(E|F')$$

We know that $(E \cap F) \cup (E \cap F') = E$ (because F and F' together form the sample space), so that

$$P(E) = P(E \cap F) + P(E \cap F')$$

or

$$P(E) = P(F) \cdot P(E|F) + P(F') \cdot P(E|F').$$

From this, equation (1) above becomes

$$P(F|E) = \frac{P(F) \cdot P(E|F)}{P(F) \cdot P(E|F) + P(F') \cdot P(E|F')}. \qquad \textbf{(2)}$$

Example 1 Experience has shown that the probability of worker error on the production line is .1, the probability that an accident will occur when there is a worker error is .3, and the probability that an accident will occur when there is no worker error is .2. Find the probability that if there is an accident, there is also a worker error.

Let A represent "accident" and E represent "worker error." From the information above,

$$P(E) = .1, \qquad P(A|E) = .3, \qquad \text{and} \qquad P(A|E') = .2.$$

These probabilities are shown in the tree diagram in Figure 8.7.

We need to find $P(E|A)$. By equation (2) above, we have

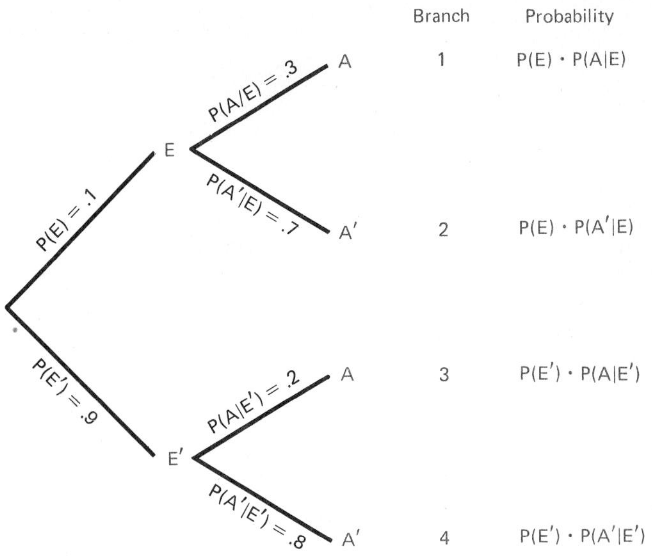

Figure 8.7

$$P(E|A) = \frac{P(E) \cdot P(A|E)}{P(E) \cdot P(A|E) + P(E') \cdot P(A|E')}$$

$$= \frac{(.1)(.3)}{(.1)(.3) + (.9)(.2)} = \frac{1}{7}. \quad \blacksquare$$

1. In Example 1, find $P(E'|A)$.

Answer: 6/7

Work Problem 1 at the side.

Equation (2) can be generalized to more than two possibilities as illustrated in the tree diagram of Figure 8.8 where $F_1, F_2, \ldots, F_n$ are mutually exclusive events whose union is the sample space, and where E is an event which has occurred.

To find the probability that any branch of the tree occurs, we have the following general result, known as **Bayes' formula.** For $1 \leq i \leq n$,

$$P(F_i|E) = \frac{P(F_i)P(E|F_i)}{P(F_1)P(E|F_1) + P(F_2)P(E|F_2) + \cdots + P(F_n) \cdot P(E|F_n)}.$$

Example 2 The grass in front of the office of Robertson and Grossman Accountants is dying. The firm's gardener decides that it either needs fertilizer, that fungus is retarding its growth, or that grub worms are attacking the root system. Upon consulting with the local nursery, she discovers that the probabilities of a "sick"

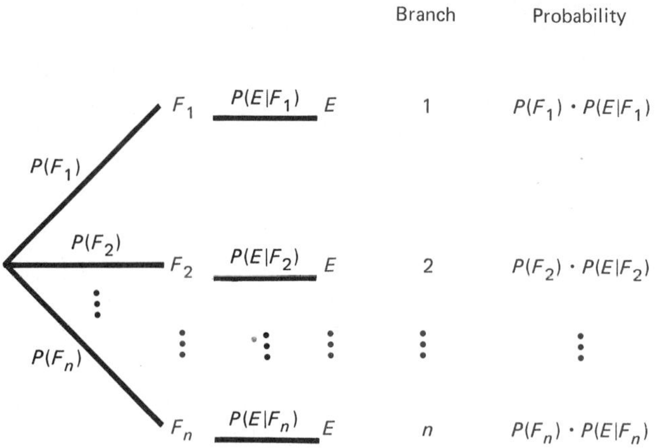

Figure 8.8

lawn given these conditions are .5, .2, and .3 respectively. The nursery also tells her that the probability that the lawn needs fertilizer is .6; fungicide, .1; treatment for grub worms, .3. What is the probability that the lawn has a fungus?

Let E represent "the lawn is dying," F_1 represent "the lawn needs fertilizer," F_2 represent "the lawn has fungus," and F_3 represent "the lawn has grub worms." Then we have the following.

$$\begin{aligned} P(F_1) &= .6 \qquad & P(E|F_1) &= .5 \\ P(F_2) &= .1 \qquad & P(E|F_2) &= .2 \\ P(F_3) &= .3 \qquad & P(E|F_3) &= .3 \end{aligned}$$

To find the probability that the lawn has fungus, $P(F_2|E)$, we use Bayes' formula, as follows.

$$P(F_2|E) = \frac{(.1)(.2)}{(.6)(.5) + (.1)(.2) + (.3)(.3)} = \frac{2}{41} \quad \blacksquare$$

Work Problem 2 at the side.

2. In Example 2, find the probability that the sick lawn is caused by
(a) lack of fertilizer;
(b) worms.

Answer:
(a) 30/41
(b) 9/41

8.4 EXERCISES

For certain events, M and N, $P(M) = .4$, $P(N|M) = .3$, and $P(N|M') = .4$. Find each of the following.

1. $P(M|N)$

2. $P(M'|N)$

For events R_1, R_2, R_3, we have $P(R_1) = .05$, $P(R_2) = .6$, and $P(R_3) = .35$. Also, $P(Q|R_1) = .40$, $P(Q|R_2) = .30$, and $P(Q|R_3) = .60$. Find each of the following.

3. $P(R_1|Q)$ **4.** $P(R_2|Q)$ **5.** $P(R_3|Q)$ **6.** $P(R_1'|Q)$

Suppose you have three jars with the following contents: 2 black balls and 1 white ball in the first; 1 black ball and 2 white balls in the second; 1 black ball and 1 white ball in the third. If the probability of selecting one of the three jars is 1/2, 1/3, and 1/6, respectively, find the probability that if a white ball is drawn, it came from the

7. second jar; **8.** third jar.

APPLIED PROBLEMS

Management

Of all the people applying for a certain job, 70% are qualified, and 30% are not. The personnel manager claims that she approves qualified people 85% of the time; she approves an unqualified person 20% of the time. Find each of the following probabilities.

9. A person is qualified if he or she was approved by the manager.

10. A person is unqualified if he or she was approved by the manager.

A building contractor buys 70% of his cement from supplier A, and 30% from supplier B. A total of 90% of the bags from A arrive undamaged, while 95% of the bags from B come undamaged. Give the probability that a damaged bag is from supplier

11. A; **12.** B.

During "Ronald McPickle Week" at McDonald's, each hamburger is given an extra pickle slice. Anita makes 25% of all the hamburgers; she forgets the extra slice 1 time out of 20. The rest of the hamburgers are made by Wayne, who leaves off the pickle 1 time in 12. Suppose a hamburger is chosen at random and found to have no extra pickle. Find the probability that it was made by

13. Anita; **14.** Wayne.

The probability that a customer of a local department store will be a "slow pay" is .02. The probability that a "slow pay" will make a large down payment when buying a refrigerator is .14. The probability that a person who is not a "slow pay" will make a large down payment when buying a refrigerator is .50. Suppose a customer makes a large down payment on a refrigerator. Find the probability that the customer is

15. a "slow pay;" **16.** not a "slow pay."

Companies A, B, and C produce 15%, 40%, and 45% respectively of the major appliances sold in a certain area. In that area, 1% of the Company A appliances, $1\frac{1}{2}$% of the Company B appliances, and 2% of the Company C appliances need service within the first year. Suppose a defective appliance is chosen at random; find the probability that it was manufactured by Company

17. A; **18.** B.

On a given weekend in the fall, a tire company can buy television advertising time for a college football game, a baseball game, or a professional football game. If the company sponsors the college game, there is a 70% chance of a high rating, a 50% chance if they sponsor a baseball game, and a 60% chance if they sponsor a professional football game. The probability of the company sponsoring these various games is .5, .2, and .3 respectively. Suppose the company does get a high rating; find the probability that it sponsored

19. a college game;

20. a professional football game.

According to your readings in business publications, there is a 50% chance of a booming economy next summer, a 20% chance of a mediocre economy, and a 30% chance of a recession. The probabilities that a particular investment strategy will produce a huge profit under each of these possibilities are .1, .6, and .3 respectively. Suppose it turns out that the strategy does produce huge profits; find the probability that the economy was

21. booming;

22. in recession.

Natural Science

23. The probability that a person with certain symptoms has hepatitis is .8. The blood test used to confirm this diagnosis gives positive results for 90% of those who have the disease and 5% of those without the disease. What is the probability that an individual with the symptoms who reacts positively to the test has hepatitis?

A recent issue of Newsweek described a new test for toxemia, a disease that affects pregnant women. To perform the test, the woman lies on her left side and then rolls over on her back. The test is considered positive if there is a 20 mm rise in her blood pressure within one minute. The article gives the following probabilities, where T represents having toxemia at some time during the pregnancy, and N represents a negative test.

$$P(T'|N) = .90, \quad \text{and} \quad P(T|N') = .75.$$

Assume that $P(N') = .11$, and find each of the following.

24. $P(N|T)$

25. $P(N'|T)$

Social Science

26. In a certain county, the Democrats have 53% of the registered voters, 12% of whom are under 21. The Republicans have 47% of all registered voters, of whom 10% are under 21. If Kay is a registered voter who is under 21, what is the probability that she is a Democrat?

8.5 COUNTING

After making do with your old automobile for several years, you finally decide to replace it with a shiny new one. You drive over to Ned's New Car Emporium to choose the car of your dreams. Once there, you find that you can select from 5 models, each with 14 power options, with a choice of 8 exterior colors and 4 interior colors.

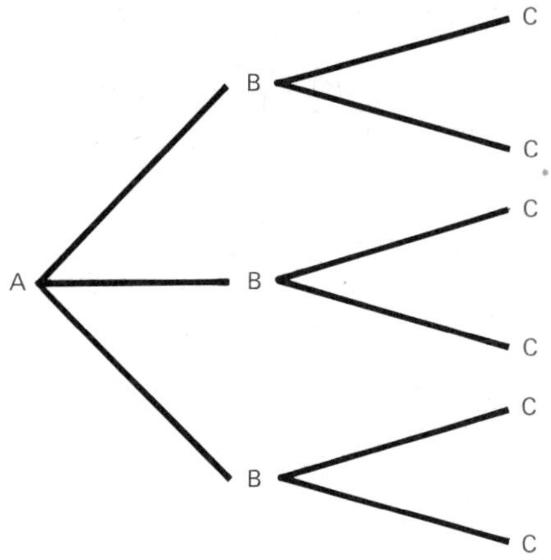

Figure 8.9

How many different new cars are available to you? Problems of this sort are best solved by means of the counting principles which are discussed in this section. These counting methods are very useful in probability.

Let us begin with a simpler example. If there are three roads from town A to town B and two roads from town B to town C, how many ways can we travel from A to C by way of B? For each of the three roads from A there are two different routes leading from B to C; hence there are $3 \cdot 2 = 6$ different ways to make the trip, as shown in Figure 8.9. This example illustrates a general principle of counting, sometimes called the **multiplication axiom:**

> If an event can occur in m ways and a second event can occur in n ways, the first event followed by the second event can occur in mn ways, assuming the second event is in no way influenced by the first.

The multiplication axiom can be extended to any number of events, provided that no one event influences another. Thus we can apply it to our problem of determining how many choices we actually have in selecting a new car at Ned's. Applying the axiom, we find that

$$5 \cdot 14 \cdot 8 \cdot 4 = 2240$$

different new cars are available from this one dealer.

Example 1 A teacher has 5 different books which he wishes to arrange on his desk. How many different arrangements are possible?

There are 5 choices (events) to be made, one for each space which will hold a book. To select a book for the first space, the teacher has 5 choices, for the second space, 4 choices (one book has already been put in the first space), for the third space, 3 choices, and so on. By the multiplication axiom, we see that the number of different arrangements is $5 \cdot 4 \cdot 3 \cdot 2 \cdot 1 = 120$. ■

1. In how many ways can 6 business tycoons line up their golf carts at the country club?

Answer: $6 \cdot 5 \cdot 4 \cdot 3 \cdot 2 \cdot 1 = 720$

Work Problem 1 at the side.

In using the multiplication axiom, we frequently encounter such products as $5 \cdot 4 \cdot 3 \cdot 2 \cdot 1$. For convenience, the symbol $n!$ (read **"*n* factorial"**) is used to denote such products. For any counting number n, we define

$$\boldsymbol{n! = n(n-1)(n-2)(n-3)\cdots(2)(1).}$$

Thus, we may write $5 \cdot 4 \cdot 3 \cdot 2 \cdot 1$ as $5!$. Similarly $3! = 3 \cdot 2 \cdot 1 = 6$. By this definition of $n!$, we see that $n[(n-1)!] = n!$ for all natural numbers $n \geq 2$. It is convenient to have this relation hold also for $n = 1$, so we define

$$\mathbf{0! = 1.}$$

Example 2 Suppose the teacher in Example 1 wishes to place only 3 of the 5 books on his desk. How many arrangements of 3 books are possible?

The teacher again has 5 ways to fill the first space, 4 ways to fill the second space, and 3 ways to fill the third. Since he wants to use only 3 books, there are only 3 spaces to be filled (3 events) instead of 5. Thus, there are $5 \cdot 4 \cdot 3 = 60$ arrangements. ■

2. In how many ways may 3 of 7 items be arranged?

Answer: $7 \cdot 6 \cdot 5 = 210$

Work Problem 2 at the side.

The answer of 60 in Example 2 is called the number of permutations of 5 things taken 3 at a time, written $P(5, 3)$. In other words, by **permutations** we mean the number of arrangements or orderings which can be formed from a given set. Note that we can use factorial notation to express the product $5 \cdot 4 \cdot 3$.

$$5 \cdot 4 \cdot 3 = \frac{5 \cdot 4 \cdot 3 \cdot 2 \cdot 1}{2 \cdot 1} = \frac{5!}{2!} = \frac{5!}{(5-3)!}$$

Generalizing from these examples, we get Theorem 8.1, which is a direct consequence of the multiplication axiom.

Theorem 8.1 If $P(n, r)$ (where $r \leq n$) denotes the number of permutations of n elements taken r at a time, then

$$P(n, r) = \frac{n!}{(n-r)!}.$$

To find $P(n, r)$, either Theorem 8.1 or direct application of the multiplication axiom may be used, as the following example demonstrates.

Example 3 Find the number of permutations of 8 elements taken 3 at a time.

Using the multiplication axiom, we note that there are 3 choices to be made, so that $P(8, 3) = 8 \cdot 7 \cdot 6 = 336$. However, we can instead use the formula given above for $P(n, r)$.

$$P(8, 3) = \frac{8!}{5!} = \frac{8 \cdot 7 \cdot 6 \cdot 5 \cdot 4 \cdot 3 \cdot 2 \cdot 1}{5 \cdot 4 \cdot 3 \cdot 2 \cdot 1} = 8 \cdot 7 \cdot 6 = 336 \quad ■$$

3. Find the number of permutations of
(a) 5 things taken 2 at a time;
(b) 9 things taken 3 at a time.

Answer:
(a) 20
(b) 504

Work Problem 3 at the side.

In Example 2 above, we found that there are 60 ways that a teacher can arrange 3 of 5 different books on his desk. That is, there are 60 permutations of 5 things taken 3 at a time. Suppose now that the teacher does not wish to arrange the books on his desk, but rather wishes to choose, at random, any 3 of the 5 books to give to a book sale to raise money for his school. In how many ways can he do this?

At first glance, we might say 60 again, but this is incorrect. The number 60 counts all possible arrangements of 3 books chosen from 5. However, the following arrangements would all lead to the same set of three books being given to the book sale.

mystery-biography-textbook	biography-textbook-mystery
mystery-textbook-biography	textbook-biography-mystery
biography-mystery-textbook	textbook-mystery-biography

We have here 6 different *arrangements* of 3 books, but only one *set* of 3 books for the book sale. If we choose a subset of items from a larger set, without worrying about order, we are forming a **combination.** The number of combinations of 5 things taken 3 at a time is written $\binom{5}{3}$.

To evaluate $\binom{5}{3}$, we note first that there are $5 \cdot 4 \cdot 3$ *permutations* of 5 things taken 3 at a time. However, we don't care about order, and each subset of 3 items from the set of 5 items can have its elements rearranged in $3 \cdot 2 \cdot 1 = 3!$ ways. Therefore, $\binom{5}{3}$ can be found by dividing the number of permutations by 3!, or

$$\binom{5}{3} = \frac{5 \cdot 4 \cdot 3}{3!} = \frac{5 \cdot 4 \cdot 3}{3 \cdot 2 \cdot 1} = 10.$$

There are 10 ways that the teacher can choose 3 books at random for the book sale. The general formula for the number of combinations of n elements taken r at a time is

$$\binom{n}{r} = \frac{P(n, r)}{r!}.$$

For purposes of calculation, the formula given in Theorem 8.2 is probably the most useful.

Theorem 8.2 If $\binom{n}{r}$ denotes the number of combinations of n elements taken r at a time, then

$$\binom{n}{r} = \frac{n!}{(n - r)!r!}.$$

Table 11 in the Appendix gives the values of $\binom{n}{r}$ for $n \leq 20$.

Example 4 How many committees of 3 people can be formed from a group of 8 people?

A committee is an unordered set, so we want $\binom{8}{3}$. Using Theorem 8.2, we have

$$\binom{8}{3} = \frac{8!}{5!3!} = \frac{8 \cdot 7 \cdot 6 \cdot 5 \cdot 4 \cdot 3 \cdot 2 \cdot 1}{5 \cdot 4 \cdot 3 \cdot 2 \cdot 1 \cdot 3 \cdot 2 \cdot 1} = \frac{8 \cdot 7 \cdot 6}{3 \cdot 2 \cdot 1} = 56. \quad ■$$

Work Problem 4 at the side.

4. Find the following.

(a) $\binom{6}{3}$

(b) $\binom{5}{1}$

(c) $\binom{7}{0}$

Answer:
(a) 20
(b) 5
(c) 1

Example 5 From a group of 30 employees, 3 are to be selected to work on a special project.

(a) In how many different ways can the employees be selected?

Here we wish to know how many 3-element combinations can be formed from a set of 30 elements. (We want combinations, not permutations, since order within the group of 3 is irrelevant.)

$$\binom{30}{3} = \frac{30!}{27!3!} = \frac{30(29)(28)(27) \cdots (3)(2)(1)}{27(26)(25) \cdots (2)(1)(3)(2)(1)}$$
$$= \frac{30(29)(28)}{3(2)(1)}$$
$$= 4060$$

There are 4060 ways to select the project group.

(b) In how many ways can the group of three be selected if it has been decided that a particular man must work on the project?

Since one man has already been selected for the project, the problem is reduced to selecting two more from the remaining 29 employees.

$$\binom{29}{2} = \frac{29!}{27!2!} = 406$$

In this case, the project group can be selected in 406 ways. ■

5. Five orchids from a collection of twenty are to be selected for a flower show.
(a) In how many ways can this be done?
(b) In how many different ways can the group of five be selected if two particular orchids must be included?

Answer:

(a) $\binom{20}{5} = 15{,}504$

(b) $\binom{18}{3} = 816$

Work Problem 5 at the side.

The rest of the examples of this section show how counting methods involving combinations and permutations are used in probability.

Example 6 A manager must select 4 employees for a promotion; 12 employees are eligible.

(a) In how many different ways can the four be chosen?

The manager may select 4 employees from 12 in $\binom{12}{4} = 495$ ways.

(b) Suppose the group of 12 includes 8 women and 4 men. Find the probability that the final group of 4 that is chosen is made up of 2 women and 2 men.

There are $\binom{8}{2}$ ways that 2 women can be chosen from 8 in the group, and $\binom{4}{2}$ ways that the 2 men can be chosen. The probability of choosing 2 women and 2 men is

$$\frac{\text{number of successes}}{\text{total number possible}} = \frac{\binom{8}{2} \times \binom{4}{2}}{\binom{12}{4}} = \frac{28 \times 6}{495} = \frac{56}{165}. \quad ■$$

6. In Example 6, find the probability that the final group of 4 contains
(a) 3 women, 1 man;
(b) all women.

Answer:

(a) $\frac{224}{495}$

(b) $\frac{14}{99}$

Work Problem 6 at the side.

Example 7 What is the probability of being dealt a heart flush in five-card poker? (A heart flush is a five-card hand of all hearts.)

The total number of ways that 5 cards can be dealt from the 52 cards in an ordinary deck is $\binom{52}{5}$. (This number equals 2,598,960.) The number of hands containing 5 hearts is $\binom{13}{5}$, since there are 13 hearts in the deck. The probability of being dealt a heart flush is therefore

$$\frac{\binom{13}{5}}{\binom{52}{5}} = \frac{\dfrac{13 \cdot 12 \cdot 11 \cdot 10 \cdot 9}{5 \cdot 4 \cdot 3 \cdot 2 \cdot 1}}{\dfrac{52 \cdot 51 \cdot 50 \cdot 49 \cdot 48}{5 \cdot 4 \cdot 3 \cdot 2 \cdot 1}} = \frac{33}{66{,}640}. \quad ■$$

7. What is the probability of being dealt *any* flush?

Answer: $4\left(\frac{33}{66,640}\right) = \frac{33}{16,660}$

Work Problem 7 at the side.

Example 8 In five-card poker, what is the probability of being dealt a full house of aces and eights (3 aces and 2 eights)?

There are $\binom{4}{3}$ ways to get 3 aces from the 4 aces in the deck, and $\binom{4}{2}$ ways to get 2 eights. The probability of this hand is

$$\frac{\binom{4}{3} \times \binom{4}{2}}{\binom{52}{5}} = \frac{1}{108,290}. \quad ■$$

8. Find the probability of a full house of eights and aces (3 eights and 2 aces).

Answer: 1/108,290

Work Problem 8 at the side.

The following table outlines the similarities of permutations and combinations as well as their differences.

Permutations	Combinations
Number of ways of selecting r items out of n items	
Repetitions are not allowed	
Order is important	Order is not important
Arrangements of n items taken r at a time	Subsets of n items taken r at a time
$P(n, r) = \frac{n!}{(n-r)!}$	$\binom{n}{r} = \frac{n!}{(n-r)!r!}$

It should be stressed that not all counting problems lend themselves to either of these techniques. Whenever a tree diagram or the product rule can be used directly, then use them.

8.5 EXERCISES

Evaluate the following factorials, permutations, and combinations.

1. $P(4, 2)$
2. $3!$
3. $\binom{8}{3}$
4. $7!$
5. $P(8, 1)$
6. $\binom{8}{1}$
7. $4!$
8. $P(4, 4)$
9. $\binom{12}{5}$
10. $\binom{10}{8}$
11. $P(13, 2)$
12. $P(12, 3)$

13. How many different two-card hands can be dealt from an ordinary deck (52 cards)?

14. How many different thirteen-card bridge hands can be dealt from an ordinary deck?

15. Five cards are marked with the numbers 1, 2, 3, 4, and 5, then shuffled, and two cards are drawn. How many different two-card combinations are possible?

16. Marbles are drawn without replacement from a bag containing 15 marbles.
(a) How many samples of 2 marbles can be drawn?
(b) How many samples of 4 marbles can be drawn?
(c) If the bag contains 3 yellow, 4 white, and 8 blue marbles, how many samples of 2 marbles can be drawn in which both marbles are blue?

Use the multiplication axiom to decide how many 7-digit telephone numbers are possible if the first digit cannot be zero and

17. only odd digits may be used;

18. the telephone number must be a multiple of 10 (that is, it must end in zero);

19. the telephone number must be a multiple of 100;

20. the first three digits are 481;

21. no repetitions are allowed?

22. How many different license numbers consisting of 3 letters followed by 3 digits are possible?

In a club with 8 men and 11 women members, find the probability that a 5-member committee chosen randomly will have

23. all men;

24. all women;

25. 3 men and 2 women.

If 5 cards are drawn at random from an ordinary deck, find the probability of drawing

26. all queens;

27. all face cards;

28. no face card;

29. exactly 2 face cards;

30. 1 heart, 2 diamonds, and 2 clubs.

31. If a baseball coach has 5 good hitters and 4 poor hitters on the bench and chooses 3 players at random, find the probability that he will choose at least 2 good hitters.

32. If your college offers 400 courses, 20 of which are in mathematics, and your counselor arranges your schedule of 4 courses by random selection, find the probability that you will not get a math course.

A bag contains 5 black, 1 red, and 3 yellow jelly beans; you reach in and take 3 at random. Find the probability of getting

33. all black;

34. all red;

35. all yellow;

36. 2 black, 1 red;

37. 2 black, 1 yellow;

38. 2 yellow, 1 black;

39. 2 red, 1 yellow.

40. There are 5 rotten apples in a crate of 25 apples.
(a) How many samples of 3 apples can be drawn from the crate?
(b) How many samples of 3 could be drawn in which all 3 are rotten?
(c) How many samples of 3 could be drawn in which there are two good apples and one rotten one?

APPLIED PROBLEMS

Management

41. How many different types of homes are available if a builder offers a choice of 5 basic plans, 3 roof styles, and 2 exterior finishes?

42. An auto manufacturer produces 7 models, each available in 6 different colors, with 4 different upholstery fabrics, and 5 interior colors. How many varieties of the auto are available?

43. How many different 4-letter radio station call letters can be made
(a) if the first letter must be K or W and no letter may be repeated;
(b) if repeats are allowed (but the first letter is K or W)?
(c) How many of the possible 4-letter call letters with no repeats end in R? (The first letter is still K or W.)

44. A business school gives courses in typing, shorthand, transcription, business English, technical writing, and accounting. In how many ways can a student arrange a schedule if 3 courses are taken?

45. In how many ways can an employer select 2 new employees from a group of 4 applicants?

46. Hal's Hamburger Hamlet sells hamburgers with cheese, relish, lettuce, tomato, mustard, or catsup. How many different kinds of hamburgers can be made using any three of the extras?

47. A group of 7 workers decides to send a delegation of 2 to their supervisor to discuss their grievances.
(a) How many delegations are possible?
(b) If it is decided that a particular employee must be in the delegation, how many different delegations are possible?
(c) If there are 2 women and 5 men in the group, how many delegations would include at least 1 woman?

Natural Science

48. In how many ways can 7 of 10 monkeys be arranged in a row for a genetics experiment?

49. A group of 3 students is to be selected from a group of 12 students to take part in a special class in cell biology.
(a) In how many ways can this be done?
(b) In how many ways can the group which will *not* take part be chosen?

50. In an experiment on plant hardiness, a researcher gathers 6 wheat plants, 3 barley plants, and 2 rye plants. She wishes to select 4 plants at random.
(a) In how many ways can this be done?
(b) In how many ways can this be done if 2 wheat plants must be included?

Social Science

51. In an experiment on social interaction, 6 people will sit in 6 seats in a row. In how many ways can this be done?

52. A couple has narrowed down the choice of a name for their new baby to 3 first names and 5 middle names. How many different first and middle name arrangements are possible?

53. A concert to raise money for an economics prize is to consist of 5 works: 2 overtures, 2 sonatas, and a piano concerto. In how many ways can the program be arranged?

54. How many different license plate numbers can be formed using 3 letters followed by 3 digits if no repeats are allowed?

55. How many license plate numbers (see Exercise 54) are possible if there are no repeats and either numbers or letters can come first?

56. An economics club has 30 members. If a committee of 4 is to be selected, in how many ways can it be done?

57. A city council is composed of 5 liberals and 4 conservatives. A delegation of 3 is to be selected to attend a convention.
(a) How many delegations are possible?
(b) How many delegations could have all liberals?
(c) How many delegations could have 2 liberals and 1 conservative?
(d) If one member of the council serves as mayor, how many delegations which include the mayor are possible?

58. The coach of the Morton Valley Softball Team has 6 good hitters and 8 poor hitters. He chooses three hitters at random.
(a) In how many ways can he choose 2 good hitters and 1 poor hitter?
(b) In how many ways can he choose all good hitters?

8.6 BINOMIAL PROBABILITY DISTRIBUTION

Many probability problems are concerned with experiments in which an event is repeated many times. For example, we might want to find the probability of getting 7 heads in 8 tosses of a coin, or hitting a target 6 times out of 6, or finding 1 defective item in a sample of 15 items. Probability problems of this kind are called **repeated trials** problems, or **Bernoulli processes.** In each case, some outcome is designated a success, and any other outcome is considered a failure. Thus, if the probability of a success in a single trial is p, the probability of failure will be $1 - p$. Repeated trials problems, or **binomial problems,** all have certain things in common.

1. The same experiment is repeated several times.
2. The probability of the desired outcome remains the same for each trial.
3. The repeated trials are independent.

Let us consider the solution of a problem of this type. Suppose we want to find the probability of getting 5 ones on 5 rolls of a die. The probability of getting a one on 1 roll is 1/6, while the probability of any other result is 5/6.

$$P(5 \text{ ones on } 5 \text{ rolls}) = P(1) \cdot P(1) \cdot P(1) \cdot P(1) \cdot P(1) = \left(\frac{1}{6}\right)^5$$

$$\approx .00013$$

Now, let us find the probability of getting a one exactly 4 times in 5 rolls of the die. The desired outcome for this experiment can occur in more than one way as shown below, where S represents getting a success (a one) and F represents getting a failure (any other result).

$$\begin{matrix} S & S & S & S & F \\ S & S & S & F & S \\ S & S & F & S & S \\ S & F & S & S & S \\ F & S & S & S & S \end{matrix}$$

The probability of each of these five outcomes is

$$\left(\frac{1}{6}\right)^4\left(\frac{5}{6}\right).$$

Since the five outcomes represent mutually exclusive alternative events, we add the five probabilities.

$$P\,(4 \text{ ones in } 5 \text{ rolls}) = 5\left(\frac{1}{6}\right)^4\left(\frac{5}{6}\right) = \frac{5^2}{6^5} \approx .0032$$

In the same way, we can compute the probability of rolling a one exactly 3 times in 5 rolls of a die. The probability of 3 successes and 2 failures will be

$$\left(\frac{1}{6}\right)^3\left(\frac{5}{6}\right)^2.$$

Again the desired outcome can occur in more than one way. To find the number of alternative outcomes, we can use combinations. We want to find the number of ways in which we can combine 3 successes and 2 failures—that is, we want $\binom{5}{3}$. Since $\binom{5}{3} = 5!/(3!\,2!) = 10$, we have

$$P\text{ (3 ones in 5 rolls)} = 10\left(\frac{1}{6}\right)^3\left(\frac{5}{6}\right)^2 = \frac{250}{6^5} = .032.$$

In general, for experiments of this kind where the same experiment is repeated with the probability of success in a single trial the same for all trials, we have Theorem 8.3.

Theorem 8.3 If p is the probability of success in a single trial, the probability of x successes and $n - x$ failures in n independent repeated trials of an experiment is

$$\binom{n}{x}p^x(1 - p)^{n-x}.$$

Example 1 The advertising agency which handles the Diet Supercola account thinks that 40% of all consumers prefer this product over its competitors. Suppose a random sample of six people is chosen. Assume that all responses are independent of each other. Find the probability of the following.

(a) Exactly 3 people prefer Diet Supercola.

In this example, we have $p = .4$. The sample is made up of six people, so $n = 6$. We want the probability that exactly 3 people prefer this drink, so $x = 3$.

$$\begin{aligned} P\text{ (exactly 3)} &= \binom{6}{3}(.4)^3(1 - .4)^{6-3} \\ &= 20(.4)^3(.6)^3 \qquad \text{Use Table 11} \\ &= 20(.064)(.216) \\ &= .27648 \end{aligned}$$

(b) None of the people prefer Diet Supercola.

Let $x = 0$.

$$\begin{aligned} P\text{ (exactly 0)} &= \binom{6}{0}(.4)^0(1 - .4)^6 \\ &= 1(1)(.6)^6 \\ &\approx .0467 \quad \blacksquare \end{aligned}$$

1. Eighty percent of all students at a certain school ski. If we select a sample of 5 students at this school, and if their responses are independent, find the probability that exactly
(a) 1 student skis
(b) 4 students ski.

Answer:

(a) $\binom{5}{1}(.8)^1(.2)^4 = .0064$

(b) $\binom{5}{4}(.8)^4(.2)^1 = .4096$

Work Problem 1 at the side.

Example 2 Find the probability of getting 7 heads in 8 tosses of a coin.

The probability of success, getting a head in a single toss, is 1/2. Thus the probability of failure, getting a tail, is 1/2. We have

$$P\text{ (7 heads in 8 tosses)} = \binom{8}{7}\left(\frac{1}{2}\right)^7\left(\frac{1}{2}\right) = 8\left(\frac{1}{2}\right)^8 = .03125. \quad ■$$

Work Problem 2 at the side.

2. Find the probability of getting 2 fours in 8 tosses of a die.

Answer: $\binom{8}{2}\left(\frac{1}{6}\right)^2\left(\frac{5}{6}\right)^6 \approx .2605$

Example 3 Assuming that selection of items for a sample can be treated as independent trials, find the probability of the occurrence of one defective item in a random sample of 15 items from a production line, if the probability that one item is defective is .01.

The probability of success (a defective item) is .01, while the probability of failure (an acceptable item) is .99. Thus,

$$\begin{aligned} P\text{ (one defective in 15 items)} &= \binom{15}{1}(.01)(.99)^{14} \\ &= 15(.01)(.99)^{14} \\ &\approx .130. \quad ■ \end{aligned}$$

Work Problem 3 at the side.

3. Five percent of the clay pots fired in a certain way are defective. Find the probability of getting exactly 2 defective pots in a sample of 12.

Answer: $\binom{12}{2}(.05)^2(.95)^{10} \approx .0988$

Example 4 A new style of shoe is sweeping the country. In one area, 30% of all the shoes sold are of this new style. Assume that these sales are independent events. Find the following probabilities.

(a) Out of 10 customers in a shoe store, at least 8 buy the new shoe style.

For at least 8 people out of 10 to buy the shoe, it must be sold to 8, 9, or 10 people. Thus,

$$\begin{aligned} P\text{ (at least 8)} &= P(8) + P(9) + P(10) \\ &= \binom{10}{8}(.3)^8(.7)^2 + \binom{10}{9}(.3)^9(.7)^1 + \binom{10}{10}(.3)^{10}(.7)^0 \\ &\approx .0014467 + .0001378 + .0000059 \\ &= .0015904. \end{aligned}$$

(b) Out of 10 customers in a shoe store, no more than 7 buy the new shoe style.

"No more than 7" means 0, 1, 2, 3, 4, 5, 6, or 7 people buy the shoe. We could add $P(0)$, $P(1)$, and so on, but it is easier to use the formula $P(E) = 1 - P(E')$. The complement of "no more than

7" is "8 or more." Thus,

$$\begin{aligned} P\,(\text{no more than } 7) &= 1 - P\,(8 \text{ or more}) \\ &= 1 - .0015904 \qquad \text{Answer from part (a)} \\ &= .9984096. \end{aligned}$$ ■

4. In Example 4, find the probability that out of 10 customers
(a) 6 or more buy the shoe,
(b) no more than 5 buy the shoe.

Answer:
(a) .0473490
(b) .9526510

Work Problem 4 at the side.

8.6 EXERCISES

Suppose that a family has 5 children. Also, suppose that the probability of having a girl is 1/2. Find the probability that the family will have

1. exactly 2 girls;
2. exactly 3 girls;
3. no girls;
4. no boys;
5. at least 4 girls;
6. at least 3 boys;
7. no more than 3 boys;
8. no more than 4 girls.

A die is rolled 12 times. Find the probability of rolling

9. exactly 12 ones;
10. exactly 6 ones;
11. exactly 1 one;
12. exactly 2 ones;
13. no more than 3 ones;
14. no more than 1 one.

A coin is tossed 5 times. Find the probability of getting

15. all heads;
16. exactly 3 heads;
17. no more than 3 heads;
18. at least 3 heads.

APPLIED PROBLEMS

Management

A factory tests a random sample of 20 transistors for defectives. The probability that a particular transistor will be defective has been established by past experience to be .05.

19. What is the probability that there are no defectives in the sample?

20. What is the probability that the number of defectives in the sample is at most 2?

A company gives prospective employees a 6-question multiple-choice test. Each question has 5 possible answers, so that there is a 1/5 or 20% chance of answering a question correctly just by guessing. Find the probability of answering

21. exactly 2 questions correctly;
22. no questions correctly;
23. at least 4 correctly;
24. no more than 3 correctly.

25. Five out of the fifty clients of a certain stockbroker will lose their life savings as a result of his advice. Find the probability that in a sample of 3 clients, exactly 1 loses all his money. (Assume independence.)

According to a recent article in a business publication, only 20% of the population of the United States has never had a Big Mac hamburger at McDonalds. Assume independence and find the probability that in a random sample of 10 people

26. exactly 2 never had a Big Mac;
27. exactly 5 never had a Big Mac;
28. 3 or fewer never had a Big Mac;
29. 4 or more *have* had a Big Mac.

Natural Science

A new drug cures 70% of the people taking it. Suppose 20 people take the drug; find the probability that

30. exactly 18 are cured;
31. exactly 17 are cured;
32. at least 17 are cured;
33. at least 18 are cured.

In a 10-question multiple-choice biology test with 5 choices for each question, a student who did not prepare guesses on each item. Find the probability that he answers

34. exactly 6 questions correct;
35. exactly 7 correct;
36. at least 8 correct;
37. less than 8 correct.

Assume that the probability that a person will die within a month after a certain operation is 20%. Find the probability that in 3 such operations

38. all 3 people survive;
39. exactly 1 person survives;
40. at least 2 people survive;
41. no more than 1 person survives.

Six mice from the same litter, all suffering from a vitamin A deficiency are fed a certain dose of carrots. If the probability of recovery under such treatment is .70, find the probability that

42. none recover;
43. exactly 3 of the 6 recover;
44. all recover;
45. no more than 3 recover.

46. In an experiment on the effects of a radiation dose on cells, a beam of radioactive particles is aimed at a group of 10 cells. Find the probability that 8 of the cells will be hit by the beam, if the probability that any single cell will be hit is .6. (Assume independence.)

47. The probability of a mutation of a given gene under a dose of 1 roentgen of radiation is approximately 2.5×10^{-7}. What is the probability that in 10,000 genes, at least 1 mutation occurs?

48. A new drug being tested causes a serious side effect in 5 out of 100 patients. What is the probability that no side effects occur in a sample of 10 patients taking the drug?

Social Science

An economist feels that the probability that a person at a certain income level will buy a new car this year is .2. Find the probability that among 12 such people,

49. exactly 4 buy a new car;

50. exactly 6 buy a new car;

51. no more than 3 buy a new car;

52. at least 3 buy a new car.

8.7 MARKOV CHAINS (OPTIONAL)

In this section we deal with a sequence of trials where each trial depends only on the results of the preceding trial. Such a sequence of trials is called a **Markov chain.** To see how Markov chains work, let's look at an example.

Example 1 Johnson and Company has 40% of the dry cleaning market in a certain town; NorthClean has the other 60% of the market. After taking a course in business mathematics, Johnson's daughter comes in and takes over the business. She is determined to increase the company's share of the market. After an extensive promotional campaign, she finds that 80% of the Johnson customers come back the following week, while 35% of the NorthClean customers switch to Johnson. This information is presented in the following *transition matrix.*

		Following week: Johnson	Following week: NorthClean
First week	Johnson	.8	.2
	NorthClean	.35	.65

For example, the entry of .65 means that 65% of the NorthClean customers remain with that firm, while 35% switch to Johnson. In general, a **transition matrix** is a square matrix having no negative entries and the sum of the entries in each row equal to 1.

When the new promotional campaign began, Johnson had 40% of the market and NorthClean had 60%. This information gives the probability vector [.4 .6]. A **probability vector** is a row vector having no negative entries and the sum of its entries equal to 1.

To find the share of the market that each firm has after one week, multiply the probability vector [.4 .6] and the transition matrix from above.

$$[.4 \quad .6]\begin{bmatrix} .8 & .2 \\ .35 & .65 \end{bmatrix} = [.53 \quad .47]$$

After one week, Johnson will have 53% of the market, and NorthClean will have 47%. To find the market share after two weeks, we would take the vector [.53 .47] and multiply it with the transition matrix given above. ■

1. Find the market share for each firm after two weeks.

Answer: Johnson has 58.85% and NorthClean has 41.15%.

Work Problem 1 at the side.

To get the share after three weeks, multiply the vector [.5885 .4115] and the transition matrix. By continuing this process, we get the results shown in the following table.

Week	Johnson	NorthClean
Start	[.4	.6]
1	[.53	.47]
2	[.5885	.4115]
3	[.6148	.3852]
4	[.62666	.37334]
5	[.631997	.368003]

The probability vector seems to be approaching [.63 .37]. Let us multiply [.63 .37] and the transition matrix from above.

$$[.63 \quad .37]\begin{bmatrix} .8 & .2 \\ .35 & .65 \end{bmatrix} = [.63 \quad .37] \text{ approximately}$$

Here multiplying by [.63 .37] leads to the same probability vector, [.63 .37]. This vector is called the **fixed probability vector** for the given transition matrix. The fixed probability vector gives a long range prediction—the shares of the market will stabilize at 63% for Johnson and 37% for NorthClean.

We can also find the fixed probability vector without doing all the work shown above. Let $v = [v_1 \quad v_2]$ represent the desired vector. Let P represent the transition matrix. By definition, v is the fixed probability vector if $vP = v$. In our example, we have

$$[v_1 \quad v_2]\begin{bmatrix} .8 & .2 \\ .35 & .65 \end{bmatrix} = [v_1 \quad v_2].$$

Since $[v_1 \quad v_2]$ is a probability vector, we have $v_1 + v_2 = 1$. Multiply on the left;

$$[.8v_1 + .35v_2 \quad .2v_1 + .65v_2] = [v_1 \quad v_2]$$

from which we get the following equations (after setting corresponding components equal).

$$.8v_1 + .35v_2 = v_1 \qquad .2v_1 + .65v_2 = v_2$$

Simplify each equation.

$$-.2v_1 + .35v_2 = 0 \qquad .2v_1 - .35v_2 = 0$$

The last two equations are really the same. Finally, we must solve the system

$$-.2v_1 + .35v_2 = 0$$
$$v_1 + v_2 = 1.$$

From the second equation, $v_1 = 1 - v_2$. Substitute for v_1 in the first equation.

$$\begin{aligned} -.2(1 - v_2) + .35v_2 &= 0 \\ -.2 + .2v_2 + .35v_2 &= 0 \\ -.2 + .55v_2 &= 0 \\ .55v_2 &= .2 \\ v_2 &= .364 \end{aligned}$$

Since $v_1 = 1 - v_2$, we have $v_1 = .636$, with a fixed probability vector of [.636 .364].

Example 2 Find the fixed probability vector for the transition matrix

$$P = \begin{bmatrix} \frac{1}{3} & \frac{2}{3} \\ \frac{3}{4} & \frac{1}{4} \end{bmatrix}.$$

We need a vector $[v_1 \quad v_2]$ such that

$$[v_1 \quad v_2]\begin{bmatrix} \frac{1}{3} & \frac{2}{3} \\ \frac{3}{4} & \frac{1}{4} \end{bmatrix} = [v_1 \quad v_2].$$

Multiply on the left, obtaining the two equations

$$\frac{1}{3}v_1 + \frac{3}{4}v_2 = v_1 \qquad \frac{2}{3}v_1 + \frac{1}{4}v_2 = v_2.$$

Simplify each of these equations.

$$-\frac{2}{3}v_1 + \frac{3}{4}v_2 = 0 \qquad \frac{2}{3}v_1 - \frac{3}{4}v_2 = 0$$

By the definition of a probability vector, we know that $v_1 + v_2 = 1$, or $v_1 = 1 - v_2$. Substitute this into either of the two given equations.

$$\begin{aligned} -\frac{2}{3}(1 - v_2) + \frac{3}{4}v_2 &= 0 \\ -\frac{2}{3} + \frac{2}{3}v_2 + \frac{3}{4}v_2 &= 0 \\ -\frac{2}{3} + \frac{17}{12}v_2 &= 0 \\ \frac{17}{12}v_2 &= \frac{2}{3} \\ v_2 &= \frac{8}{17} \end{aligned}$$

Thus, $v_1 = 1 - 8/17 = 9/17$. The fixed probability vector is [9/17 8/17]. ■

2. Find the fixed probability vector for the transition matrix

$$P = \begin{bmatrix} .3 & .7 \\ .5 & .5 \end{bmatrix}.$$

Answer: $[\frac{5}{12} \quad \frac{7}{12}]$

Work Problem 2 at the side.

Example 3 The probability that a complex assembly line works correctly depends on whether or not the line worked correctly the last time it was used. The various probabilities are as given in the following transition matrix.

	Works properly now	Does not
Worked properly before	.9	.1
Did not	.7	.3

Find the long range probability that the assembly line will work properly.

Begin by finding the fixed probability vector $[v_1 \quad v_2]$. We have

$$[v_1 \quad v_2]\begin{bmatrix} .9 & .1 \\ .7 & .3 \end{bmatrix} = [v_1 \quad v_2].$$

From the left side we get the equations

$$.9v_1 + .7v_2 = v_1 \quad \text{and} \quad .1v_1 + .3v_2 = v_2$$

or

$$-.1v_1 + .7v_2 = 0 \quad \text{and} \quad .1v_1 - .7v_2 = 0.$$

Substitute $v_1 = 1 - v_2$ in the second of these equations.

$$-.1(1 - v_2) + .7v_2 = 0$$
$$-.1 + .1v_2 + .7v_2 = 0$$
$$-.1 + .8v_2 = 0$$
$$.8v_2 = .1$$
$$v_2 = \frac{1}{8}$$

Also, $v_1 = 1 - 1/8 = 7/8$. The fixed probability vector is [7/8 1/8]. The company can expect the assembly line to run properly 7/8 of the time, in the long run. ■

3. In Example 3, suppose the company modifies the line so that the transition matrix becomes

$$\begin{bmatrix} .95 & .05 \\ .8 & .2 \end{bmatrix}.$$

Find the long range probability that the assembly line will work properly.

Answer: 16/17

Work Problem 3 at the side.

8.7 EXERCISES

Which of the following could be a probability vector?

1. $[\frac{2}{3} \quad \frac{1}{3}]$ **2.** $[\frac{1}{2} \quad 1]$ **3.** $[0 \quad 1]$

4. $[.1 \quad .1]$ **5.** $[.4 \quad .2 \quad 0]$ **6.** $[\frac{1}{4} \quad \frac{1}{8} \quad \frac{5}{8}]$

Which of the following could be transition matrices, by definition?

7. $\begin{bmatrix} .5 & 0 \\ 0 & .5 \end{bmatrix}$

8. $\begin{bmatrix} \frac{2}{3} & \frac{1}{3} \\ 1 & 0 \end{bmatrix}$

9. $\begin{bmatrix} \frac{1}{4} & \frac{3}{4} \\ \frac{1}{2} & \frac{1}{2} \end{bmatrix}$

10. $\begin{bmatrix} \frac{1}{4} & \frac{3}{4} & 0 \\ 2 & 0 & 1 \\ 1 & \frac{2}{3} & 3 \end{bmatrix}$

11. $\begin{bmatrix} \frac{1}{3} & \frac{1}{3} & \frac{1}{3} \\ 0 & 1 & 0 \\ \frac{1}{2} & 0 & \frac{1}{2} \end{bmatrix}$

12. $\begin{bmatrix} \frac{1}{3} & \frac{1}{2} & 1 \\ 0 & 1 & 0 \\ \frac{1}{2} & 0 & \frac{1}{2} \end{bmatrix}$

13. $\begin{bmatrix} \frac{1}{3} & \frac{1}{2} & 1 \\ \frac{1}{3} & 0 & 0 \\ \frac{1}{3} & \frac{1}{2} & 0 \end{bmatrix}$

14. $\begin{bmatrix} .9 & .1 & 0 \\ .1 & .6 & .3 \\ 0 & .3 & .7 \end{bmatrix}$

APPLIED PROBLEMS

Management

Years ago, about 10% *of all cars sold were small, while* 90% *were large. This has changed drastically over the years;* 20% *of small car owners have switched to large cars, while* 60% *of large car owners have switched to small cars.*

15. Write a transition matrix for this information.

16. Write a probability vector for the initial distribution of cars.

17. What percent of the market will each class of automobile have in 4 years?

18. What is the long-range prediction?

An insurance company classifies its drivers into three groups: G_0 (no accidents), G_1 (one accident), and G_2 (more than one accident). The probability that a driver in G_0 will stay in G_0 after one year is .85, *that he will become a G_1 is* .10, *and that he will become a G_2 is* .05. *A driver in G_1 cannot move to G_0 (this insurance company has a long memory). There is an* .80 *probability that a G_1 driver will stay in G_1 and a* .20 *probability that he will become a G_2. A driver in G_2 must stay in G_2.*

19. Write a transition matrix using this information.

Suppose that the company accepts 50,000 *new policyholders, all of whom are in group G_0. Find the number in each group*

20. after 1 year;

21. after 2 years;

22. after 3 years;

23. after 4 years.

24. Find the fixed probability vector here. Interpret your result.

The difficulty with the mathematical model of Exercises 19–24 is that no "grace period" is provided; there should be a certain probability of moving from G_1 or G_2 back to G_0 (say, after four years with no accidents). A new system with this feature might produce the following transition matrix.

$$\begin{bmatrix} .85 & .10 & .05 \\ .15 & .75 & .10 \\ .10 & .30 & .60 \end{bmatrix}$$

Suppose that when this new policy is adopted, the company has 50,000 policy-holders in group G_0. Find the number of these in each group

25. after 1 year; **26.** after 2 years; **27.** after 3 years.

28. Find the fixed probability vector here. Interpret your result.

Research done by the Gulf Oil Corporation produced the following transition matrix for the probability that a person with one form of home heating would switch to another.*

		Will switch to		
		Oil	Gas	Electric
Now has	Oil	.825	.175	0
	Gas	.060	.919	.021
	Electric	.049	0	.951

The current share of the market held by these three types of heat is given by [.26 .60 .14]. *Find the share of the market held by each type of heat after*

29. 1 year; **30.** 2 years; **31.** 5 years.

32. What is the long-range prediction?

Natural Science

In one state, a land-use survey showed that 35% of the land was used for agricultural purposes, while 10% was urban. Ten years later, of the agricultural land, 15% had become urban and 80% had remained agricultural. (The remainder lay idle.) Of the idle land, 20% had become urbanized and 10% had been converted for agricultural use. Of the urban land, 90% remained urban and 10% was idle. Assume that these trends continue.

33. Write a transition matrix for this information.

34. Write a probability vector for the initial distribution of land.

35. What is the land-use pattern after four years?

36. What is the long-range prediction for land use in this state?

Social Science

In a survey investigating change in housing patterns in one urban area, it was found that 75% of the population lived in single-family dwellings and 25% in multiple housing of some kind. Five years later, in a follow-up survey, of those who had been living in single-family dwellings, 90% still did so, but 10% had moved to multiple-family dwellings. Of those in multiple housing, 95% were still living in that type of housing, while 5% had moved to single-family dwellings. Assume that these trends continue.

---

* Reprinted by permission from Ali Ezzati, "Forecasting Market Shares of Alternative Home Heating Units by Markov Process Using Transition Probabilities Estimated from Aggregate Time Series Data," *Management Science*, Volume 21, #4, December, 1974, Copyright 1974 The Institute of Management Sciences.

37. Write a transition matrix for this information.

38. Write a probability vector for the initial distribution of housing.

39. What percent of the population can be expected in each category five years later?

40. What is the long-range prediction?

At the end of June, in a presidential election year, 40% of the electorate was registered as liberal, 45% as conservative, and 15% as independent. Over a one-month period, the liberals retained 80% of their constituency, while 15% switched to conservative and 5% to independent. The conservatives retained 70%, lost 20% to the liberals, and 10% to the independents. The independents retained 60% and lost 20% each to the conservatives and liberals. Assume that these trends continue.

41. Write a 3×3 transition matrix showing the information of the problem.

42. Write a 1×3 fixed probability vector.

43. Find the percent of each type of voter at the end of July.

44. Find the percent of each type of voter at the end of August.

45. Find the percent of each type of voter at the end of September.

46. Find the long range prediction.

CHAPTER 8 TEST

[8.1–8.3]

When a single card is drawn from an ordinary deck, find the probability that it will be

1. a heart;

2. a red queen;

3. a face card;

4. black or a face card;

5. red, given it is a queen;

6. a jack, given it is a face card;

7. a face card, given it is a king.

Find the odds in favor of a card drawn from an ordinary deck being

8. a club;

9. a black jack;

10. a red face card or a queen.

[8.2]

The square at the side shows the four possible (equally likely) combinations when both parents are carriers of the sickle cell anemia trait. Each carrier parent has normal cells (N) and trait cells (T).

		2nd parent	
		N_2	T_2
1st parent	N_1		N_1T_2
	T_1		

11. Complete the table.

12. If the disease occurs only when two trait cells combine, find the probability that a child born to these parents will have sickle cell anemia.

13. The child will carry the trait but not have the disease if a normal cell combines with a trait cell. Find the probability of this.

14. Find the probability that the child is neither a carrier nor has the disease.

[8.3] *Two fair dice are rolled. Find the probability that the resulting sum will be*

15. 0;

16. odd and greater than 8;

17. 12, given it is greater than 10.

[8.4] **18.** Box A contains 5 red balls and 1 black ball; box B contains 2 red and 3 black. A box is chosen at random, and a ball is selected from it. The probability of choosing box A is 3/8. If the selected ball is black, what is the probability that it came from box A?

19. Find the probability that the ball in Exercise 18 came from box B, given that it is red.

[8.6] *A certain machine is used to manufacture screws. One percent of the screws made by the machine are defective. A random sample of* 20 *screws is selected. Find the probability that this sample contains*

20. exactly 4 defective screws;

21. exactly 3 defective screws;

22. no more than 4 defective screws.

23. *Set up* the probability that the sample has 12 or more defective screws. (Do not evaluate.)

[8.7] *Currently, 35% of all hot dogs sold in one state are made by Dogkins, while 65% are made by Long Dog. Suppose that Dogkins starts a heavy advertising campaign; suppose also, that the campaign produces the following transition matrix.*

		After campaign: Dogkins	After campaign: Long Dog
Before campaign	Dogkins	.8	.2
	Long Dog	.4	.6

24. Find the share of the market for each company
(a) after the campaign; (b) after 3 such campaigns.

25. Predict the long-range share of the market for Dogkins.

CASE 8 DRUG SEARCH PROBABILITIES—SMITH KLINE AND FRENCH LABORATORIES *

In searching for a new drug with commercial possibilities, drug company researchers use the ratio

$$N_S : N_A : N_D : 1.$$

That is, if the company gives preliminary screening to N_S substances, it may find that N_A of them are worthy of further study, with N_D of these surviving into full scale development. Finally, 1 of the substances will result in a marketable drug. Typical numbers used by Smith Kline and French Laboratories in planning research budgets might be 2000:30:8:1.

EXERCISES

1. Suppose a compound has been chosen for preliminary study. Use the ratio 2000:30:8:1 to find the probability that the compound will survive and become a marketable drug.
2. Find the probability that the compound will not lead to a marketable drug.
3. Suppose a such compounds receive preliminary screening. Set up the probability that none of them produces a marketable drug. (Assume independence throughout these exercises.)
4. Use your results from Exercise 3 to set up the probability that at least one of the drugs will prove marketable.
5. Suppose now that N scientists are employed in the preliminary screening, and that each scientist can screen c compounds per year. Set up the probability that no marketable drugs will be discovered in a year.
6. Set up the probability that at least one marketable drug will be discovered.

For the following exercises, evaluate your answer in Exercise 6, which should have been $1 - \left(\frac{1999}{2000}\right)^{Nc}$, *for the following values of N and c. Use a calculator with an* x^y *key, or a computer.*

7. $N = 100$, $c = 6$
8. $N = 25$, $c = 10$

---

* Reprinted by permission from E. B. Pyle III, et. al., "Scientific Manpower Allocation to New Drug Screening Programs," *Management Science*, Volume 19, #12, August, 1973, Copyright 1973 The Institute of Management Sciences.

CASE 9 A CREDIT GRANTING PROCEDURE—FEDERATED DEPARTMENT STORES*

Federated Department Stores owns many large local department stores, such as Bloomingdale's and Abraham and Straus of New York, Bullock's of Los Angeles, I. Magnin of San Francisco, Burdine's of Miami, Sanger-Harris of Dallas, and Foley's of Houston. This example describes a procedure that can be used to determine which customers of a store are to receive credit. Since credit sales currently amount to over 50% of total sales in major department stores, the question of who is to receive credit (and how much he or she is to receive) is of major importance to the management. The total dollar amount of credit purchases allowed is limited by the store's credit manager based on his feeling about the credit reliability of the customer. In fact, the credit manager may be afraid to extend any credit at all to some persons because of the fear of losses due to delinquent payments.

In deciding whether or not to extend consumer credit, the initial piece of information that the credit manager has available is the credit application blank which the customer has completed. An experienced credit granter learns to recognize whether or not the information on a credit application indicates financial responsibility and stability.

There are several problems associated with the granting of consumer credit that a store may encounter. The first of these is the necessity of minimizing delinquent payments and the costs of a bill collection agency, while not overly restricting good customers. Second, there is a shortage of qualified credit analysts available. Third, there is very little consistency from day to day and from store to store in credit granting decisions. Last, there is an inconsistent pendulum effect between the credit manager and the top management. During economic depressions, the credit manager is told to tighten up, and during boom times to loosen up. If there is no quantitative fix on the degree of loosening or tightening, chaos can result.

Mathematical methods helped to resolve all four of the above problems. The factors involved in making a credit decision were listed, and then mathematics was used to help arrive at an optimum solution. Top management had complete control all the way by helping to select the important factors and by participating in decisions at each stage in the development of what is effectively a mathematical screening system.

To develop such a screening system, the analyst at Federated Department Stores began by selecting a random sample of credit applicants who had applied at the store $1\frac{1}{2}$ to 2 years ago. This sample of several thousand contained a number of rejected applicants, a small number of "bad" customers to whom credit was extended but who defaulted, and a large number of "good" customers who paid promptly. Hopefully, this sample was highly representative of incoming credit applicants (the only reason for going back in time is to permit identification of "goods," "bads," and "rejects").

The next step was to begin analyzing the information available on the credit application. Lists were made up for every possible item of information on the credit application blanks. By careful study of all such items, those

---

* This example was supplied by Rodney C. Burress, Operations Research Department, Federated Department Stores.

Chart 1

Credit factor	Points	Credit factor	Points
Living status		Bank accounts	
(a) Rent—furnished	0	(a) Checking only	10
(b) Rent—unfurnished; all other	7	(b) Checking and savings	15
		(c) None or savings only	0
Telephone		Finance company or small-loan company	
(a) One or more	11	(a) Yes	0
(b) None or no answer	0	(b) No	8
Marital status		Sears', Ward's, or Penney's accounts	
(a) Married	6	(a) Yes	10
(b) All other	0	(b) No	0
Number of dependent children		Oil company reference	
(a) None (regardless of marital status)	15	(a) Yes	15
(b) All other	0	(b) No	0
Occupation		Jeweler, furniture store, music store, appliance store, radio and TV store accounts	
(a) Professional	14	(a) Yes	0
(b) All other	0	(b) No	10
Length of time on present job		Fashion, clothing, specialty accounts	
(a) 7 months to 4 years	5	(a) Yes	6
(b) 4 to 10 years	12	(b) No	0
(c) Over 10 years	17		
(d) 6 months or less	0		
Monthly income			
(a) Under $500	0		
(b) All other	16		

which showed promise as discriminators between "good" and "bad" credit customers were selected for further study.

After all the information on the applications was studied, a point scoring system was developed whereby all significant items on the application blank were assigned a numerical value. Responses such as home ownership, checking and savings account references, department store account references, and so on, generally received high point values, while separated marital status, small-loan company references, and so on, typically received low values. An applicant's total point score was arrived at by summing the point values of all those credit application factors which applied to him. Chart 1 shows an example of a credit point scoring system. The magnitude of the total point score is an indicator of credit risk, such that the higher the point score, the lower the risk.

The factors on the application blank (which can be treated as variables) must be analyzed on a computer, using fairly sophisticated mathematical techniques to determine the appropriate point values, as shown in the chart. Multiple linear regression is one of the techniques normally used, with the factors from the application blank treated as the independent variables, and the final disposition of the applicant's account (*good*, *bad*, or *reject*) treated as the dependent variable.

To test the results produced by the mathematics, a random sample of 1000 credit customers' applications was taken, and the point scores for each of the 1000 individuals compiled. It was found that 800 of the 1000 were "good" customers, and 200 were "bad," as shown in Chart 2. The cumulative

Chart 2

Score	Frequency Good	Frequency Bad	Probability of default	Cumulative Good	Cumulative Bad	Net profit
0 to 4	0	5	100.0	800	200	10,000
5 to 9	0	7	100.0	800	195	10,750
10 to 14	0	11	100.0	800	188	11,800
15 to 19	0	13	100.0	800	177	13,450
20 to 24	2	14	88.0	800	164	15,400
25 to 29	4	15	79.0	798	150	17,400
30 to 34	6	16	73.0	794	135	19,450
35 to 39	8	18	69.0	788	119	21,550
40 to 44	9	20	69.0	780	101	23,850
45 to 49	11	20	65.0	771	81	26,400
50 to 54	25	15	37.0	760	61	28,850
55 to 59	35	10	22.0	735	46	29,850
60 to 64	72	8	10.0	700	36	29,600
65 to 69	75	7	8.5	628	28	27,200
70 to 74	84	6	6.7	553	21	24,500
75 to 79	96	5	4.9	469	15	21,200
80 to 84	117	4	3.3	373	10	17,150
85 to 89	124	3	2.5	256	6	11,900
90 to 94	53	2	4.0	132	3	6,150
95 to 99	47	1	2.1	79	1	3,800
100 to 104	23	0	0.0	32	0	1,600
105 up	9	0	0.0	9	0	450
	800	200				

good and bad columns show the total number of the 800 goods and 200 bads that the store will get at any particular minimum cutoff level. For example, if the minimum acceptable point score is 40, the company will get 780 (out of 800) of the goods, and 101 (out of 200) of the bads. With a minimum score of 75, the company will get 469 goods and 15 bads. If we assume a profit of \$50 for each good and a loss of \$150 for each bad, we can calculate the expected profit at each possible minimum score. For example, at a minimum score of 40, we know the company will have 780 goods and 101 bads, for a total profit of

$$780(50) - 101(150) = \$23{,}850,$$

while at a minimum score of 75, the profit would be

$$469(50) - 15(150) = \$21{,}200.$$

The profits are summarized in the right-hand column of the chart. Note that a minimum cutoff of 55 produces the maximum expected profit.

A point scoring system constructed in the manner described in this example can objectively predict the performance of incoming credit applicants, while at the same time permitting optimum cutoff points to be selected. The system does not depend on experienced credit analysts, but can be applied by well-supervised credit clerks who have received a minimum of training. Since an individual will receive the same score no matter which credit clerk takes his application, a high degree of consistency will be achieved. This is important in that it removes the effect of personal biases that credit managers may build up over the years against minority races, single women, and so on. Lastly, top management can have complete quantitative control over the reject rate by knowing in advance the effect on profits of raising or lowering cutoff points.

EXERCISE

In the example in the text, find the maximum expected profit if a "bad" customer creates a loss of only \$100 instead of \$150. (Hint: To save you a lot of arithmetic, we can tell you that the minimum cutoff here is between 35 and 70.)

9

Decision Theory

John F. Kennedy once remarked that he had assumed that as president it would be difficult to choose between distinct, opposite alternatives when a decision needed to be made. Actually, however, he found that such decisions were easy to make; the hard decisions came when he was faced with choices that were not as clear-cut. Most decisions that we are faced with fall in this last category—decisions which must be made under conditions of uncertainty. In this chapter we look at **decision theory,** which provides a systematic way to attack the problem of decision making when not all alternatives are clear and unambiguous.

In order to be able to construct mathematical models of decisions, we define the making of a rational decision to mean the choosing of an alternative which offers the best expected value. Thus, we begin the chapter by studying expected values.

9.1 EXPECTED VALUE

In a recent year, a citizen of the United States could expect to complete 12.8 years of school, to earn $8941 per year, and to be a member of a family of 3.21 persons. What do we mean by "expect?" We know of many people who have completed less than 12.8 years of school, while many others have completed more. Many people earn less than $8941 per year; others earn more. The idea of a family of 3.21 members is a little hard to swallow, to say the least. The numbers above refer to averages, or to imaginary *typical* people. When the term "expectation" is used in this way, it refers to *mathematical expectation*, which, as we will see, is a kind of average.

As an example, let us find the "typical" number of offspring of a species of pheasant. We begin by gathering the data shown in the table.

Number of offspring	0	1	2	3	4
Percent of families having this number of offspring	8%	14%	29%	32%	17%

We might be tempted to find the typical number of offspring by averaging the numbers 0, 1, 2, 3, and 4, which represent the numbers of offspring possible. This won't work, however, since the various numbers of offspring do not occur equally often: for example, 3 offspring are much more common than 0 or 1. We can take the differing frequencies of occurrence into account if we find a **weighted average.** We do this by multiplying each of the possible numbers of offspring by its corresponding percentage, as follows.

$$\begin{aligned}\text{typical number of offspring} &= 0(.08) + 1(.14) + 2(.29) \\ &\quad + 3(.32) + 4(.17) \\ &= 0 + .14 + .58 + .96 + .68 \\ &= 2.36\end{aligned}$$

The typical family of pheasants, based on our data, has 2.36 offspring.

We can use the idea of this example to define **expected value,** or **mathematical expectation,** as follows.

> Suppose an experiment has n outcomes, $x_1, x_2, x_3, \ldots, x_n$. Also, suppose the probabilities that each of these outcomes takes place are respectively $p_1, p_2, p_3, \ldots, p_n$. Then the expected value is
>
> $$x_1p_1 + x_2p_2 + x_3p_3 + \cdots + x_np_n.$$

Example 1 The local church decides to raise money by raffling off a microwave oven worth \$400. A total of 2000 tickets are sold at \$1 each. Find the expected value of winning for a person who buys one ticket in the raffle.

Here the possible outcomes are the possible amounts of net winnings, where net winnings = amount of winning − cost of ticket. The net winnings of the person winning the oven are \$400 (amount of winning) − \$1 (cost of ticket) = \$399. The net winnings for each losing ticket are \$0 − \$1 = −\$1.

The probability of winning is 1 in 2000, or 1/2000, while the probability of losing is 1999/2000. In summary, we have the information shown in the chart.

Outcome (net winning)	\$399	−\$1
Probability	$\frac{1}{2000}$	$\frac{1999}{2000}$

Expected winnings for a person buying one ticket are thus

$$399\left(\frac{1}{2000}\right) + (-1)\left(\frac{1999}{2000}\right) = \frac{399}{2000} - \frac{1999}{2000} = -\frac{1600}{2000} = -.80.$$

On the average, a person buying one ticket in the raffle will lose \$.80, or 80¢.

It is not possible to lose 80¢ in this raffle—you either lose \$1, or you win a \$400 prize. However, if you bought tickets in many such raffles over a long period of time, you would find that you lose 80¢ per ticket, on the average. ■

Work Problem 1 at the side.

Example 2 What is the expected number of girls in a family having three children?

Some families with three children will have 0 girls, others will have 1 girl, and so on We need to find the probabilities associated with 0, 1, 2, or 3 girls in a family of three children. We can find these probabilities if we first write the sample space S of all possible three-child families: $S = \{GGG, GGB, BGG, GBB, BGB, BBG, BBB, GBG\}$. From this sample space, we get the probabilities shown in the following chart.

Outcome (number of girls)	0	1	2	3
Probability	$\frac{1}{8}$	$\frac{3}{8}$	$\frac{3}{8}$	$\frac{1}{8}$

The expected number of girls can now be found by multiplying each outcome (number of girls) and its corresponding probability and finding the sum of these values.

$$\begin{aligned}\text{expected number of girls} &= 0 \cdot \frac{1}{8} + 1 \cdot \frac{3}{8} + 2 \cdot \frac{3}{8} + 3 \cdot \frac{1}{8} \\ &= \frac{3}{8} + \frac{6}{8} + \frac{3}{8} \\ &= \frac{12}{8} = \frac{3}{2} = 1\frac{1}{2}\end{aligned}$$

On the average, a three-child family will have $1\frac{1}{2}$ girls. ■

Work Problem 2 at the side.

Example 3 Each day Donna and Mary toss a coin to see who buys the coffee (40¢ a cup). One tosses and the other calls the outcome. If the person who calls the outcome is correct, the other buys the coffee; otherwise the caller pays. Find Donna's expected winnings.

Let's assume that an honest coin is used, that Mary tosses

1. Suppose you buy one of 1000 tickets at 10¢ each in a lottery where the first prize is \$50. What is your expected winning?

Answer: −5¢

2. Find the expected outcome for throwing one die. Begin by completing this table.

Outcome	1	2	3	4	5	6
Probability	$\frac{1}{6}$					

Answer: All probabilities are 1/6, expected outcome is $3\frac{1}{2}$.

the coin, and that Donna calls the outcome. The possible results and corresponding probabilities are shown below.

	Possible results			
Result of toss	H	H	T	T
Call	H	T	H	T
Caller wins?	Yes	No	No	Yes
Probability	$\frac{1}{4}$	$\frac{1}{4}$	$\frac{1}{4}$	$\frac{1}{4}$

Donna wins a 40¢ cup of coffee whenever the results and calls match, and loses a 40¢ cup when there is no match. Her expected winnings are

$$E = (.40)\left(\frac{1}{4}\right) + (-.40)\left(\frac{1}{4}\right) + (-.40)\left(\frac{1}{4}\right) + (.40)\left(\frac{1}{4}\right) = 0.$$

On the average, over the long run, Donna neither wins nor loses. ■

A game with an expected value of 0 (such as the one of Example 3) is called a **fair game.** Casinos do not offer fair games. If they did, they would win on the average \$0, and have a hard time paying the help! Casino games have expected winnings for the house that vary from 1 1/2¢ per dollar to 60¢ per dollar. Exercises 14–17 below show how to find the expected value for certain games of chance.

The idea of expected value can be very useful in decision making, as shown by the next example.

Example 4 At age 50, you receive a letter from the Mutual of Mauritania Insurance Company. According to the letter, you must tell the company immediately which of the following two options you will choose: take \$20,000 at age 60 (if you are alive, \$0 otherwise) or \$30,000 at age 70 (again, if you are alive, \$0 otherwise). Based on the idea of expected value, which should you choose?

Life insurance companies have constructed elaborate tables showing the probability of a person living a given number of years into the future. By consulting the most recent such table, we find that the probability of living from age 50 to age 60 is .88, while the probability of living from age 50 to age 70 is .64. The expected values of the two options are given below.

First option:

$$(20{,}000)(.88) + (0)(.12) = 17{,}600$$

Second option:

$$(30{,}000)(.64) + (0)(.36) = 19{,}200$$

Based strictly on expected values, we would choose the second option. ■

3. I can take one of two jobs. With job A, there is a 50% chance that I will make \$60,000 per year after 5 years and a 50% chance of making \$30,000. With job B, there is a 30% chance that I will make \$90,000 per year after 5 years and a 70% chance that I will make \$20,000. Based strictly on expected value, which job should I take?

Answer: Job A has an expected salary of \$45,000; job B has an expected salary of \$41,000. Take job A.

Work Problem 3 at the side.

9.1 EXERCISES

Find the expected value for each of the following. Identify any fair games.

1.

Outcome	2	3	4	5
Probability	.1	.4	.3	.2

2.

Outcome	4	6	8	10
Probability	.4	.4	.05	.15

3.

Outcome	9	12	15	18	21
Probability	.14	.22	.36	.18	.10

4.

Outcome	30	32	36	38	44
Probability	.31	.30	.29	.06	.04

5. A raffle offers a first prize of \$100, and two second prizes of \$40 each. One ticket costs \$1, and 500 tickets are sold. Find the expected winnings for a person who buys one ticket.

6. A raffle offers a first prize of \$1000, two second prizes of \$300 each, and twenty prizes of \$10 each. If 10,000 tickets are sold at 50¢ each, find the expected winnings for a person buying one ticket.

Many of the following exercises use the idea of combinations, which we discussed in the last chapter.

7. If 3 marbles are drawn from a bag containing 3 yellow and 4 white marbles, what is the expected number of yellow marbles in the sample?

8. If 5 apples in a barrel of 25 apples are known to be rotten, what is the expected number of rotten apples in a sample of 2 apples?

9. A delegation of 3 is selected from a city council made up of 5 liberals and 4 conservatives.
   (a) What is the expected number of liberals on the committee?
   (b) What is the expected number of conservatives?

10. From a group of 2 women and 5 men, a delegation of 2 is selected. Find the expected number of women in the delegation.

11. In a club with 20 senior and 10 junior members, what is the expected number of junior members on a 3-member committee?

12. If 2 cards are drawn at one time from a deck of 52 cards, what is the expected number of diamonds?

13. Suppose someone offers to pay you \$5 if you draw 2 diamonds in the game of Exercise 12. He says that you should pay 50¢ for the chance to play. Is this a fair game?

Find the expected winnings for each of the following games of chance.

14. In one form of roulette, you bet \$1 on "even." If one of the 18 even numbers comes up, you get your dollar back, plus another one. If one of the 20 noneven (18 odd, 0, and 00) numbers comes up, you lose.

15. In another form of roulette, there are only 19 non-even numbers. (no 00)

16. Numbers is an illegal game where you bet \$1 on any three digit number from 000 to 999. If your number comes up, you get \$500.

17. In Keno, the house has a pot containing 80 balls, each marked with a different number from 1 to 80. You buy a ticket for \$1 and mark one of the 80 numbers on it. The house then selects 20 numbers at random. If your number is among the 20, you get \$3.20 (for a net winning of \$2.20).

18. Use the assumptions of Example 3 to find Mary's expected winnings. If Mary tosses and Donna calls, is it still a fair game?

19. Suppose one day Mary brings a two-headed coin and uses it to toss for the coffee. Since Mary tosses, Donna calls.
    (a) Is this still a fair game?
    (b) What is Donna's expected gain if she calls heads?
    (c) If she calls tails?

APPLIED PROBLEMS

Social Science

20. Find the expected number of girls in a family of four children.

21. Find the expected number of boys in a family of five children.

22. Jack must choose at age 40 to inherit either \$25,000 at age 50 (if he is still alive) or \$30,000 at age 55 (if he is still alive). If the probabilities for a person of age 40 to live to be 50 and 55 are .90 and .85, respectively, which choice gives him the larger expected inheritance?

23. An insurance company has written 100 policies of \$10,000, 500 of \$5000, and 1000 policies of \$1000 on people of age 20. If experience shows that the probability of dying during the twentieth year of life is .001, how much can the company expect to pay out during the year the policies were written?

Management

24. A builder is considering a job which promises a profit of \$30,000 with

a probability of .7 or a loss (due to bad weather, strikes, and such) of $10,000 with a probability of .3. What is her expected profit?

25. Experience has shown that a ski lodge will be full (160 guests) during the Christmas holidays if there is a heavy snow pack in December, while a light snowfall in December means that they will have only 90 guests. What is the expected number of guests if the probability for a heavy snow in December is .40? (Assume that there must either be a light snowfall or a heavy snowfall.)

26. A magazine distributor offers a first prize of $100,000, two second prizes of $40,000 each, and two third prizes of $10,000 each. A total of 2,000,000 entries are received in the contest. Find the expected winnings if you submit one entry to the contest. If it would cost you 25¢ in time, paper, and stamps to enter, would it be worth it?

27. The local Saab dealer gets complaints about his cars, as shown in the following table.

Number of complaints per day	0	1	2	3	4	5	6
Probability	.01	.05	.15	.26	.33	.14	.06

Find the expected number of complaints per day.

28. Levi Strauss and Company* uses expected value to help its salespeople rate their accounts. For each account, a salesperson estimates potential additional volume and the probability of getting it. The product of these gives the expected value of the potential, which is added to the existing volume. The totals are then classified as A, B, or C as follows: below $40,000, class C; between $40,000 and $55,000, class B; above $55,000, class A. Complete the following chart for one of its salespeople.

Account no.	Existing volume	Potential add'l vol.	Probability of getting it	Expected value of potent'l	Existing vol. + Expected value of potent'l	Class
1	$15,000	$10,000	.25	$2,500	$17,500	C
2	40,000	0	—	—	40,000	C
3	20,000	10,000	.20			
4	50,000	10,000	.10			
5	5,000	50,000	.50			
6	0	100,000	.60			
7	30,000	20,000	.80			

29. At the end of play in a major golf tournament, two players, an "old pro" and a "new kid" are tied. Suppose first prize is $80,000 and second prize is $20,000. Find the expected winnings for the old pro if

* This example was supplied by James McDonald, Levi Strauss and Company, San Francisco.

(a) both players are of equal ability,
(b) the new kid will freeze up, giving the old pro a 3/4 chance of winning.

Natural Science

30. In a certain animal species, the probability that a healthy adult female will have no offspring in a given year is .31, while the probability of 1, 2, 3, or 4 offspring are respectively .21, .19, .17, and .12. Find the expected number of offspring.

According to an article in a magazine not known for its accuracy, a male decreases his life expectancy by one year, on the average, for every point that his blood pressure is above 120. *The average life expectancy for a male is* 76 *years. Find the life expectancy for a male whose blood pressure is*

31. 135; **32.** 150; **33.** 115; **34.** 100.

35. Suppose a certain male has a blood pressure of 145. Find his life expectancy. How would you interpret the result to him?

9.2 DECISION MAKING

In the previous section we saw how to use expected values to help make a decision. We extend these ideas in this section, where we consider decision making in the face of uncertainty. Let us begin with an example.

Freezing temperatures are endangering the orange crop in central California. A farmer can protect his crop by burning smudge pots—the heat from the pots keeps the oranges from freezing. However, burning the pots is expensive; the cost is $2000. The farmer knows that if he burns smudge pots he will be able to sell his crop for a net profit (after smudge pot costs are deducted) of $5000, provided that the freeze does develop and wipes out other orange crops in California. If he does nothing he will either lose $1000 in planting costs if it does freeze, or make a profit of $4800 if it does not freeze. (If it does not freeze, there will be a large supply of oranges, and thus his profit will be lower than if there was a small supply.)

What should the farmer do? He should begin by carefully defining the problem. He must begin by deciding on the **states of nature,** the possible alternatives over which he has no control. Here there are two: freezing temperatures, or no freezing temperatures. Next, the farmer should list the things he can control—his actions or **strategies.** The farmer has two possible strategies: use smudge pots or not use smudge pots. The consequences of each action under each state of nature, called **payoffs,** can be summarized in a **payoff matrix,** as shown below. The payoffs in this case represent the profit for each possible combination of events.

		States of nature	
		Freeze	No freeze
Strategies of farmer	Use smudge pots	\$5000	\$2800
	Do not use pots	−\$1000	\$4800

To get the \$2800 entry in the payoff matrix, we took the profit if there is no freeze, \$4800, and subtracted the \$2000 cost of using the pots.

Once the farmer makes the payoff matrix, what then? The farmer might be an optimist (some might call him a gambler); in this case he might assume that the best will happen and go for the biggest number on the matrix (\$5000). To get this profit, he must adopt the strategy "use smudge pots."

On the other hand, the farmer might be a pessimist. As a pessimist, he would want to minimize the worst thing that could happen. If he uses smudge pots, the worst thing that could happen to him would be a profit of \$2800, which will result if there is no freeze. If he does not use smudge pots, he might well face a loss of \$1000. To minimize the worst thing that could happen to him, he once again should adopt the strategy "use smudge pots."

Suppose the farmer decides that he is neither an optimist nor a pessimist, but would like further information before choosing a strategy. For example, he might call the weather forecaster and ask for the probability of a freeze. Suppose the forecaster says that this probability is only .1. What should the farmer do? He should recall our discussion of expected value from the previous section and work out the expected profit for each of his two possible strategies. If the probability of a freeze is .1, then the probability that there is no freeze is .9. Using this information, we get the following expected values.

$$\text{If smudge pots are used:}\quad 5000(.1) + 2800(.9) = 3020$$
$$\text{If no smudge pots are used:}\quad -1000(.1) + 4800(.9) = 4220$$

Here the maximum expected profit, \$4220, is obtained if smudge pots are *not* used.

1. What should the farmer do if the probability of a freeze is .6? What is his expected profit?

Answer: Use smudge pots, \$4120

Work Problem 1 at the side.

As the example and problem have shown, as the farmer's beliefs about the probabilities of a freeze change, so might his choice of strategies.

Example 1 A small Christmas card manufacturer must decide in February about the type of cards she should emphasize in her fall line of cards. She has three possible strategies: emphasize modern cards, emphasize old-fashioned cards, or a mixture of the two. Her success is dependent on the state of the economy in December. If the economy is strong, she will do well with her modern cards, while in a weak economy people long for the old days and buy old-fashioned cards. In an in-between economy, her mixture of lines would do the best. She first prepares a payoff matrix for all three possibilities. The numbers in the matrix represent her profits in thousands of dollars.

		States of nature		
		Weak economy	In-between	Strong economy
Strategies	Modern	\$40	\$85	\$120
	Old-fashioned	\$106	\$46	\$83
	Mixture	\$72	\$90	\$68

(a) If the manufacturer is an optimist, she should aim for the biggest number on the matrix, the \$120 (representing \$120,000 in profit). Her strategy in this case would be to produce modern cards.
(b) A pessimistic manufacturer wants to find the worst of all bad things that can happen. If she produces modern cards, the worst that can happen is a profit of \$40,000. For old-fashioned cards, the worst is a profit of \$46,000, while the worst that can happen from a mixture is a profit of \$68,000. Her strategy here is to use a mixture.
(c) Suppose the manufacturer reads an article in *The Wall Street Journal* that claims that leading experts feel there is a 50% chance of a weak economy at Christmas, a 20% chance of an in-between economy, and a 30% chance of a strong economy. The manufacturer should now use this information to find her expected profit for each possible strategy.

Modern: $40(.50) + 85(.20) + (120)(.30) = 73$
Old-fashioned: $106(.50) + 46(.20) + 83(.30) = 87.1$
Mixture: $72(.50) + 90(.20) + 68(.30) = 74.4$

Here the best strategy is old-fashioned cards; the expected profit is 87.1, or \$87,100. ■

2. Suppose the manufacturer reads another article which gives the following predictions: 35% chance of a weak economy, 25% chance of an in-between economy, and a 40% chance of a strong economy. What is the best strategy now? What is the expected profit?

Answer: modern, \$83,250

Work Problem 2 at the side.

9.2 APPLIED PROBLEMS

Management

1. An investor has \$20,000 to invest in stocks. She has two possible strategies: buy conservative blue-chip stocks or buy highly speculative stocks. There are two states of nature: the market goes up or the market goes

down. The following payoff matrix shows the net amounts she will have under the various circumstances.

$$\begin{array}{lcc} & \text{Market up} & \text{Market down} \\ \text{Buy blue-chip} & \$25{,}000 & \$18{,}000 \\ \text{Buy speculative} & \$30{,}000 & \$11{,}000 \end{array}$$

What should the investor do if she is
(a) an optimist;
(b) a pessimist?
(c) Suppose there is a .7 probability of the market going up. What is the best strategy? What is the expected profit?
(d) What is the best strategy if the probability of a market rise is .2?

2. A developer has $100,000 to invest in land. He has a choice of two parcels (at the same price), one on the highway and one on the coast. With both parcels, his ultimate profit depends on whether he faces light opposition from environmental groups or heavy opposition. He estimates that the payoff matrix is as follows (the numbers represent his profit).

$$\begin{array}{lcc} & \multicolumn{2}{c}{\text{Opposition}} \\ & \text{Light} & \text{Heavy} \\ \text{Highway} & \$70{,}000 & \$30{,}000 \\ \text{Coast} & \$150{,}000 & -\$40{,}000 \end{array}$$

What should the developer do if he is
(a) an optimist;
(b) a pessimist?
(c) Suppose the probability of heavy opposition is .8. What is his best strategy? What is the expected profit?
(d) What is the best strategy if the probability of heavy opposition is only .4?

3. Hillsdale College has sold out all tickets for a jazz concert to be held in the stadium. If it rains, the show will have to be moved to the gym, which has a much smaller seating capacity. The dean must decide in advance whether to set up the seats and the stage in the gym or in the stadium, or both, just in case. The payoff matrix below shows the net profit in each case.

$$\begin{array}{llcc} & & \multicolumn{2}{c}{\text{States of nature}} \\ & & \text{Rain} & \text{No rain} \\ & \text{Set up in stadium} & -\$1550 & \$1500 \\ \text{Strategies} & \text{Set up in gym} & \$1000 & \$1000 \\ & \text{Set up both} & \$750 & \$1400 \end{array}$$

What strategy should the dean choose if she is
(a) an optimist;
(b) a pessimist?
(c) If the weather forecaster predicts rain with a probability of .6, what strategy should she choose to maximize expected profit? What is the maximum expected profit?

4. An analyst must decide what fraction of the items produced by a certain machine are defective. He has already decided that there are three possibilities for the fraction of defective items: .01, .10, and .20. He may recommend two courses of action: repair the machine or make no repairs. The payoff matrix below represents the *costs* to the company in each case.

$$
\begin{array}{llccc}
 & & \multicolumn{3}{c}{\text{States of nature}} \\
 & & .01 & .10 & .20 \\
\text{Strategies} & \begin{array}{l}\text{Repair} \\ \text{No repair}\end{array} & \multicolumn{3}{c}{\begin{bmatrix} \$130 & \$130 & \$130 \\ \$25 & \$200 & \$500 \end{bmatrix}}
\end{array}
$$

What strategy should the analyst recommend if he is

(a) an optimist;

(b) a pessimist?

(c) Suppose the analyst is able to estimate probabilities for the three states of nature as follows.

Fraction of defectives	Probability
.01	.70
.10	.20
.20	.10

Which strategy should he recommend? Find the expected cost to the company if this strategy is chosen.

5. The research department of the Allied Manufacturing Company has developed a new process which it believes will result in an improved product. Management must decide whether or not to go ahead and market the new product. The new product may be better than the old or it may not be better. If the new product is better, and the company decides to market it, sales should increase by \$50,000. If it is not better and they replace the old product with the new product on the market, they will lose \$25,000 to competitors. If they decide not to market the new product they will lose \$40,000 if it is better, and research costs of \$10,000 if it is not.

(a) Prepare a payoff matrix.

(b) If management believes the probability that the new product is better to be .4, find the expected profits under each strategy and determine the best action.

6. A businessman is planning to ship a used machine to his plant in Nigeria. He would like to use it there for the next four years. He must decide whether or not to overhaul the machine before sending it. The cost of overhaul is \$2600. If the machine fails when in operation in Nigeria, it will cost him \$6000 in lost production and repairs. He estimates the probability that it will fail at .3 if he does not overhaul it, and .1 if he does overhaul it. Neglect the possibility that the machine might fail more than once in the four years.

(a) Prepare a payoff matrix.

(b) What should the businessman do to minimize his expected costs?

7. A contractor prepares to bid on a job. If all goes well, his bid should be \$30,000, which will cover his costs plus his usual profit margin of \$4500. However, if a threatened labor strike actually occurs, his bid should be \$40,000 to give him the same profit. If there is a strike and he bids \$30,000, he will lose \$5500. If his bid is too high, he may lose the job entirely, while if it is too low, he may lose money.
   (a) Prepare a payoff matrix.
   (b) If the contractor believes that the probability of a strike is .6, how much should he bid?

Natural Science

8. A community is considering an anti-smoking campaign.* The city council will choose one of three possible strategies: a campaign for everyone over age 10 in the community, a campaign for youths only, or no campaign at all. The two states of nature are a true cause-effect relationship between smoking and cancer and no cause-effect relationship. The costs to the community (including loss of life and productivity) in each case are as shown below.

		States of nature	
		Cause-effect relationship	No cause-effect relationship
Strategies	Campaign for all	\$100,000	\$800,000
	Campaign for youth	\$2,820,000	\$20,000
	No campaign	\$3,100,100	\$0

What action should the city council choose if it is
   (a) optimistic;
   (b) pessimistic?
   (c) If the Director of Public Health estimates that the probability of a true cause-effect relationship is .8, which strategy should the city council choose?

Sometimes the numbers (or payoffs) in a payoff matrix do not represent money (profits or costs, for example), but utility. A ***utility*** *is a number which measures the satisfaction (or lack of it) that results from a certain action. The numbers must be assigned by each individual, depending on how he or she feels about a situation. For example, one person might assign a utility of* +20 *for a week's vacation in San Francisco, with* −6 *being assigned if the vacation were moved to Sacramento. Work the following problems in the same way as those above.*

Social Science

9. A politician must plan her reelection strategy. She can emphasize jobs or she can emphasize the environment. The voters can be concerned about jobs or about the environment. A payoff matrix showing the utility of each possible outcome is shown below.

		Voters	
		Jobs	Environment
Candidate	Jobs	+25	−10
	Environment	−15	+30

---

* This problem is based on an article by B. G. Greenberg in the September 1969 issue of the *Journal of the American Statistical Association.*

The political analysts feel that there is a .35 chance that the voters will emphasize jobs. What strategy should the candidate adopt? What is its expected utility?

Management

10. In an accounting class, the instructor permits the students to bring a calculator or a reference book (but not both) to an examination. The examination itself can emphasize either numerical problems or definitions. In trying to decide which aid to take to an examination, a student first decides on the utilities shown in the following payoff matrix.

		Exam emphasizes	
		Numbers	Definitions
Student chooses	Calculator	+50	0
	Book	+10	+40

(a) What strategy should the student choose if the probability that the examination will emphasize numbers is .6? What is the expected utility in this case?

(b) Suppose the probability that the examination emphasizes numbers is .4. What strategy should be chosen by the student?

9.3 AN APPLICATION—DECISION MAKING IN LIFE INSURANCE*

When a life insurance company receives an application from an agent requesting insurance on the life of an individual, it knows from experience that the applicant will be in one of three possible states of risk, with proportions as shown.

States of risk	Proportions
s_1 = Standard risk	.90
s_2 = Substandard risk (greater risk)	.07
s_3 = Sub-substandard risk (greatest risk)	.03

A particular applicant could be correctly placed if all possible information about the applicant were known. This is not realistic in a practical situation; the company's problem is to obtain the maximum information at the lowest possible cost.

The company can take any of three possible strategies when it receives the application.

* This example was supplied by Donald J. vanKeuren, actuary of Metropolitan Life Insurance Company, and Dave Halmstad, senior actuarial assistant. It is based on a paper by Donald Jones.

Strategies
a_1 = Offer a standard policy
a_2 = Offer a substandard policy (higher rates)
a_3 = Offer a sub-substandard policy (highest rates)

The payoff matrix in Table I below shows the payoffs associated with the possible strategies of the company and the states of the applicant. Here M represents the face value of the policy in thousands of dollars (for a \$20,000 policy we have $M = 20$). For example, if the applicant is substandard (s_2) and the company offers him or her a standard policy (a_1), the company makes a profit of $13M$ (13 times the face value of the policy in thousands). Strategy a_2 would result in a larger profit of $20M$. However, if the prospective customer is a standard risk (s_1) but the company offers a substandard policy (a_2), the company loses \$50 (the cost of preparing a policy) since the customer would reject the policy because it has higher rates than he or she could obtain elsewhere.

Table I

		States of nature		
		s_1	s_2	s_3
Strategies of company	a_1	$20M$	$13M$	$3M$
	a_2	-50	$20M$	$10M$
	a_3	-50	-50	$20M$

Before deciding on the policy to be offered, the company can perform any of three experiments to help it decide.

e_0 = No inspection report (no cost)
e_1 = Regular inspection report (cost: \$5)
e_2 = Special life report (cost: \$20)

On the basis of this report, the company can classify the applicant as follows.

T_1 = Applicant seems to be a standard risk
T_2 = Applicant seems to be a substandard risk
T_3 = Applicant seems to be a sub-substandard risk

We let $P(s|T)$ represent the probability that an applicant is in state s when the report indicates that he or she is in state T. For example, $P(s_1|T_2)$ represents the probability that an applicant is a standard risk (s_1) when the report indicates that he or she is a substandard risk (T_2). These probabilities, shown in Table II, are based on Bayes' formula.

Table II

True state	Regular report $P(s_i\|T_1)$	$P(s_i\|T_2)$	$P(s_i\|T_3)$	Special report $P(s_i\|T_1)$	$P(s_i\|T_2)$	$P(s_i\|T_3)$
s_1	.9695	.8411	.7377	.9984	.2081	.2299
s_2	.0251	.1309	.1148	.0012	.7850	.0268
s_3	.0054	.0280	.1475	.0004	.0069	.7433

From Table II we see that $P(s_2|T_2)$, the probability that an applicant actually is substandard (s_2) if the regular report indicates substandard (T_2) is only .1309, while $P(s_2|T_2)$, using the special report, is .7850.

We now have probabilities and payoffs, and we can use them to find expected values for each possible strategy the company might adopt. There are many possibilities here: the company can use one of three experiments, the experiments can indicate one of three states, the company can offer one of three policies, and the applicant can actually be in one of three states. Figure 9.1 on page 364 shows some of these possibilities in a **decision tree.**

In order to find an optimum strategy for the company, let us consider an example. Suppose the company decides to perform experiment e_2 (special life report) with the report indicating a substandard risk, T_2. Then the expected values E_1, E_2, E_3 for the three possible actions a_1, a_2, a_3, respectively, are as shown below. (Recall: M is a variable, representing the face amount of the policy in thousands.)

For action a_1 (offer standard policy):

$$\begin{aligned} E_1 &= [P(s_1|T_2)](20M) + [P(s_2|T_2)](13M) + [P(s_3|T_2)](3M) \\ &= (.2081)(20M) + (.7850)(13M) + (.0069)(3M) \\ &= 4.162M + 10.205M + .021M \\ &= 14.388M \end{aligned}$$

For action a_2 (offer a substandard policy):

$$E_2 = [P(s_1|T_2)](-50) + [P(s_2|T_2)](20M) + [P(s_3|T_2)](10M)$$

Work Problem 1 at the side.

1. Simplify E_2.

Answer: $15.769M - 10.405$

Work Problem 2 at the side.

2. Find E_3 for action a_3 (offer a sub-substandard policy).

Answer: $.138M - 49.655$

Strategy a_2 is better than a_3 (for any positive M, $15.769M - 10.405 > .138M - 49.655$). Hence, we are reduced to a choice

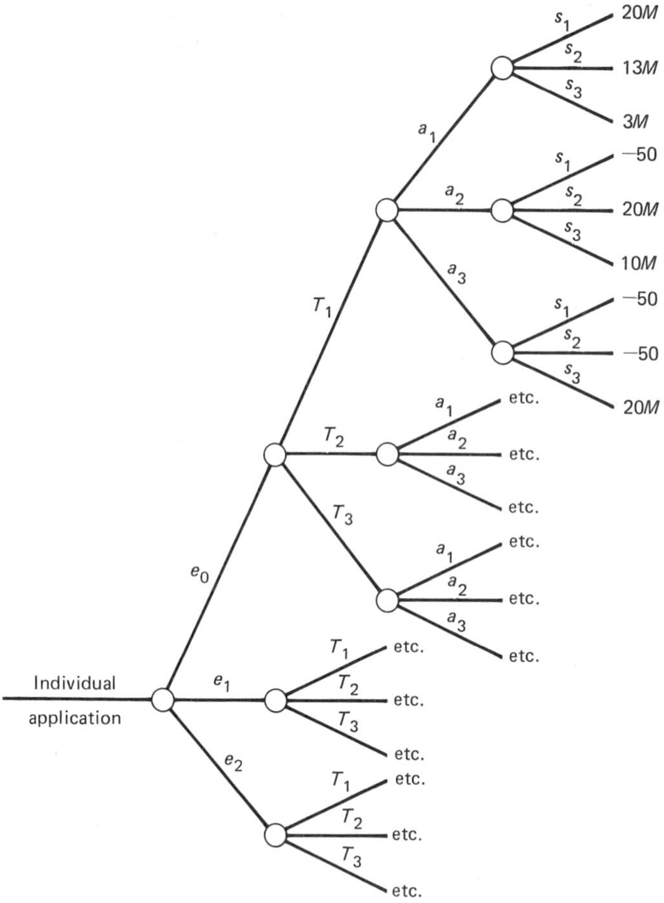

Figure 9.1

between strategies a_1 and a_2. Strategy a_2 is superior if it leads to a higher expected value than a_1. This happens for all values of M such that

$$15.769M - 10.405 > 14.388M$$
$$1.381M > 10.405$$
$$M > 7.53.$$

If the applicant applies for \$7530 or less of insurance, the company should use strategy a_1; otherwise it should use a_2.

Similar analyses can be performed for all possible strategies from the decision tree above to find the best strategy. It turns out that the company will maximize its expected profits if it offers a standard policy to all people applying for less than \$50,000 in life insurance, with a special report form required for all others.

9.3 EXERCISES

1. Find the expected values for each strategy a_1, a_2, and a_3 if the insurance company performs experiment e_1 (a regular report) with the report indicating that the applicant is a substandard risk.

2. Find the expected values for each action if the company performs e_1 with the report indicating that the applicant is a standard risk.

3. Find the expected values for each strategy if e_0 (no report) is selected. (Hint: use the proportions given for the three states s_1, s_2, and s_3 as the probabilities.)

9.4 STRICTLY DETERMINED GAMES

The word *game* in the title of this section may have led you to think of checkers or perhaps some card game. While **game theory** does have some application to these recreational games, it was developed in the 1940's as a means of analyzing competitive situations in business, warfare, and social situations. Game theory deals with how we make decisions when in competition with an aggressive opponent.

A game can be set up with a payoff matrix, such as the one shown here.

$$\begin{array}{cc} & B \\ & \begin{array}{cc} \text{I} & \text{II} \end{array} \\ A \begin{array}{c} 1 \\ 2 \end{array} & \begin{bmatrix} 2 & -1 \\ -3 & 4 \end{bmatrix} \end{array}$$

This game involves the two players A and B, and is called a **two-person game.** Player A can choose either of the two rows, 1 or 2, while player B can choose either column I or column II. A player's choice is called a **strategy,** just as before. The payoff is at the intersection of the row and column selected. As a general agreement, a positive number represents a payoff from B to A; a negative number represents a payoff from A to B.

Example 1 In the payoff matrix shown above, suppose A chooses row 1 and B chooses column II. Who gets what?

Row 1 and column II lead to the number -1. This number represents a payoff of \$1 from A to B. ■

1. Find the payoff in the game above if A chooses row 2 and B chooses column II.

Answer: a payoff of \$4 from B to A

Work Problem 1 at the side.

We assumed above that the numbers in the payoff matrix

represented money. They could just as well represent goods or other property.

In the game above, no money enters the game from the outside; whenever one player wins, the other loses. Such a game is called a **zero-sum game.** The stock market is not a zero-sum game. Stocks can go up or down according to outside forces. Therefore, it is possible that all investors can make or lose money.

We only discuss two-person zero-sum games in the rest of this chapter. Each player can have many different options. In particular, an $m \times n$ matrix game is one in which player A has m strategies (rows) and player B has n strategies (columns).

In the rest of this section we look for the best possible strategies for each player. Let us begin with the 3×3 game defined by the following matrix.

$$\begin{array}{cc} & \quad B \\ & \begin{array}{ccc} \text{I} & \text{II} & \text{III} \end{array} \\ A \begin{array}{c} 1 \\ 2 \\ 3 \end{array} & \begin{bmatrix} -3 & -6 & 10 \\ 3 & 0 & 1 \\ 5 & -4 & -8 \end{bmatrix} \end{array}$$

From B's viewpoint, strategy II is better than strategy I no matter which strategy A selects. This can be seen by comparing the two columns. If A chooses row 1, receiving \$6 from A is better than receiving \$3, in row 2 breaking even is better than paying \$3, and in row 3, getting \$4 from A is better than paying \$5. Therefore, B should never select strategy I. Strategy II is said to *dominate* strategy I, and strategy I (the dominated strategy) can be removed from consideration, producing the following reduced matrix.

$$\begin{array}{cc} & \quad B \\ & \begin{array}{cc} \text{II} & \text{III} \end{array} \\ A \begin{array}{c} 1 \\ 2 \\ 3 \end{array} & \begin{bmatrix} -6 & 10 \\ 0 & 1 \\ -4 & -8 \end{bmatrix} \end{array}$$

Either player may have dominated strategies. In fact, after a dominated strategy for one player is removed, the other player may then have a dominated strategy where there was none before.

In summary,

> a row for A **dominates** another row if every entry in the first row is *larger* than the corresponding entry in the second row. For a column of B to dominate another, each entry must be *smaller*.

Example 2 Find any dominant strategies in the game having the following payoff matrix.

$$A \begin{array}{c} \\ 1 \\ 2 \end{array} \begin{array}{c} B \\ \begin{array}{cccc} \text{I} & \text{II} & \text{III} & \text{IV} \end{array} \\ \begin{bmatrix} -8 & -4 & -6 & -9 \\ -3 & 0 & -9 & 12 \end{bmatrix} \end{array}$$

Here every entry in column III is smaller than the corresponding entry in column II. Thus, column III dominates column II. By removing the dominated column II, the final game is as follows.

$$A \begin{array}{c} \\ 1 \\ 2 \end{array} \begin{array}{c} B \\ \begin{array}{ccc} \text{I} & \text{III} & \text{IV} \end{array} \\ \begin{bmatrix} -8 & -6 & -9 \\ -3 & -9 & 12 \end{bmatrix} \end{array} \quad \blacksquare$$

2. Locate any dominant strategies in the following game.

$$A \begin{array}{c} \\ 1 \\ 2 \\ 3 \end{array} \begin{array}{c} B \\ \begin{array}{cc} \text{I} & \text{II} \end{array} \\ \begin{bmatrix} 3 & -2 \\ 0 & 8 \\ 6 & 4 \end{bmatrix} \end{array}$$

Answer: Row 3 dominates row 1.

Work Problem 2 at the side.

Suppose all dominated strategies have been removed from a game, such as in the one shown here.

$$A \begin{array}{c} \\ 1 \\ 2 \\ 3 \end{array} \begin{array}{c} B \\ \begin{array}{ccc} \text{I} & \text{II} & \text{III} \end{array} \\ \begin{bmatrix} -9 & 11 & -4 \\ 0 & 3 & 5 \\ -1 & -9 & 6 \end{bmatrix} \end{array}$$

Which strategies should the players choose? The goal of game theory is to find **optimum strategies,** those which are the most profitable to the respective players. The payoff which results from choosing the optimum strategy is called the **value** of the game.

In the game above, A could choose row 1, in hopes of getting the payoff of \$11. However, B would quickly discover this, and start playing column I. Thus, B would receive \$9 from A, and A would not be happy. Each player will do best in the long run if they maximize the smallest amount that they can win. For the three rows of A, the smallest amounts that can be won are, respectively, a loss of \$9, a gain of \$0, and a loss of \$9. Thus, to maximize the smallest amount that can be won, A should play row 2.

What about B? The smallest amounts that can be won by B are a gain of \$0 in column I, a loss of \$11 in column II, and a loss of \$6 in column III. Thus, B should maximize the smallest amount that can be won by playing column I.

The payoff here is the number in row 2, column I, or \$0. The

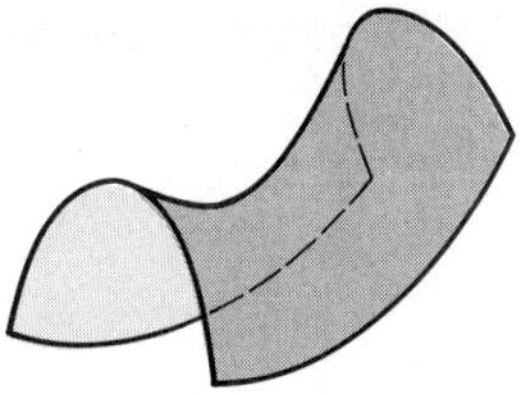

The name *saddle point* comes from a saddle. The seat of the saddle is the maximum from one direction and the minimum from another direction.

value of the game is \$0. Recall that a game having a value of \$0 is called a fair game.

The pair of strategies found above, (2, I), is called the **saddle point** of the game. The saddle point is the smallest number in its row and the largest number in its column. A game with a saddle point is called a **strictly determined game.**

Example 3 Find the saddle point in the following game.

$$\begin{array}{cc} & B \\ & \begin{array}{cc} \text{I} & \text{II} \end{array} \\ A \begin{array}{c} 1 \\ 2 \\ 3 \\ 4 \end{array} & \begin{bmatrix} 2 & 2 \\ 0 & 4 \\ 1 & 6 \\ 3 & 7 \end{bmatrix} \end{array}$$

The number that is both the smallest number in its row and largest number in its column is 3. Thus, 3 is the value of the game, and the saddle point is at strategies (4, I). ■

Work Problem 3 at the side.

Example 4 Locate all saddle points in the game

$$\begin{array}{cc} & B \\ & \begin{array}{cccc} \text{I} & \text{II} & \text{III} & \text{IV} \end{array} \\ A \begin{array}{c} 1 \\ 2 \end{array} & \begin{bmatrix} 4 & 6 & 4 & 12 \\ -8 & -9 & 3 & 2 \end{bmatrix}. \end{array}$$

Here there are two saddle points, (1, I) and (1, III). However, each leads to the same number, 4, for the value of the game. ■

Work Problem 4 at the side.

Example 5 Is the following game strictly determined?

$$\begin{array}{cc} & B \\ & \begin{array}{ccc} \text{I} & \text{II} & \text{III} \end{array} \\ A \begin{array}{c} 1 \\ 2 \end{array} & \begin{bmatrix} 3 & 6 & -2 \\ 8 & -3 & 5 \end{bmatrix} \end{array}$$

A strictly determined game is one with a saddle point. This game has no saddle point, and thus is not strictly determined. In the next section we look at methods for finding optimum strategies for such games. ■

3. Locate all saddle points in the following games.

(a)

$$\begin{array}{cc} & B \\ & \begin{array}{ccc} \text{I} & \text{II} & \text{III} \end{array} \\ A \begin{array}{c} 1 \\ 2 \end{array} & \begin{bmatrix} 3 & -2 & 5 \\ 2 & -1 & 7 \end{bmatrix} \end{array}$$

(b)

$$\begin{array}{cc} & B \\ & \begin{array}{ccc} \text{I} & \text{II} & \text{III} \end{array} \\ A \begin{array}{c} 1 \\ 2 \\ 3 \end{array} & \begin{bmatrix} 0 & -5 & -7 \\ 9 & 7 & 5 \\ 3 & -6 & 3 \end{bmatrix} \end{array}$$

Answer:
(a) (2, II), value is -1
(b) (2, III), value is 5

4. Find the saddle point and the value of the following game. Is the game fair?

$$\begin{array}{cc} & B \\ & \begin{array}{ccc} \text{I} & \text{II} & \text{III} \end{array} \\ A \begin{array}{c} 1 \\ 2 \\ 3 \end{array} & \begin{bmatrix} 1 & -9 & -4 \\ 2 & 3 & -2 \\ 9 & 6 & 0 \end{bmatrix} \end{array}$$

Answer: (3, III), 0, yes

5. Are the following games strictly determined?

(a)

$$A \begin{array}{c} \\ 1 \\ 2 \end{array} \overset{B}{\begin{array}{cc} \text{I} & \text{II} \\ \end{array}} \begin{bmatrix} 2 & 1 \\ 6 & 0 \end{bmatrix}$$

(b)

$$A \begin{array}{c} 1 \\ 2 \end{array} \overset{\begin{array}{ccc} & B & \\ \text{I} & \text{II} & \text{III} \end{array}}{\begin{bmatrix} 6 & -4 & -3 \\ 7 & 2 & -5 \end{bmatrix}}$$

Answer:

(a) yes; saddle point at (1, II)

(b) no; no saddle point

Work Problem 5 at the side.

9.4 EXERCISES

In the following game, decide on the payoff when the indicated strategies are used.

$$A \begin{array}{c} 1 \\ 2 \\ 3 \end{array} \overset{\begin{array}{ccc} & B & \\ \text{I} & \text{II} & \text{III} \end{array}}{\begin{bmatrix} 6 & -4 & 0 \\ 3 & -2 & 6 \\ -1 & 5 & 11 \end{bmatrix}}$$

1. (1, I) **2.** (1, II) **3.** (2, II)

4. (2, III) **5.** (3, I) **6.** (3, II)

7. Does the game have any dominated strategies?

8. Does it have a saddle point?

Remove any dominated strategies in the following games. (From now on, we will save space by deleting the names of the strategies.)

9. $\begin{bmatrix} 0 & -2 & 8 \\ 3 & -1 & -9 \end{bmatrix}$

10. $\begin{bmatrix} 6 & 5 \\ 3 & 8 \\ -1 & -4 \end{bmatrix}$

11. $\begin{bmatrix} 1 & 4 \\ 4 & -1 \\ 3 & 5 \\ -4 & 0 \end{bmatrix}$

12. $\begin{bmatrix} 2 & 3 & 1 & -5 \\ -1 & 5 & 4 & 1 \\ 1 & 0 & 2 & -3 \end{bmatrix}$

13. $\begin{bmatrix} 8 & 12 & -7 \\ -2 & 1 & 4 \end{bmatrix}$

14. $\begin{bmatrix} 6 & 2 \\ -1 & 10 \\ 3 & 5 \end{bmatrix}$

When it exists, find the saddle point and the value of the game for each of the following. Identify any games that are strictly determined. (See Examples 3 and 4.)

15. $\begin{bmatrix} 3 & 5 \\ 2 & -5 \end{bmatrix}$

16. $\begin{bmatrix} 7 & 8 \\ -2 & 15 \end{bmatrix}$

17. $\begin{bmatrix} 3 & -4 & 1 \\ 5 & 3 & -2 \end{bmatrix}$

18. $\begin{bmatrix} -4 & 2 & -3 & -7 \\ 4 & 3 & 5 & -9 \end{bmatrix}$

19. $\begin{bmatrix} -6 & 2 \\ -1 & -10 \\ 3 & 5 \end{bmatrix}$

20. $\begin{bmatrix} 1 & 4 & -3 & 1 & -1 \\ 2 & 5 & 0 & 4 & 10 \\ 1 & -3 & 2 & 5 & 2 \end{bmatrix}$

21. $\begin{bmatrix} 2 & 3 & 1 \\ -1 & 4 & -7 \\ 5 & 2 & 0 \\ 8 & -4 & -1 \end{bmatrix}$

22. $\begin{bmatrix} 3 & 8 & -4 & -9 \\ -1 & -2 & -3 & 0 \\ -2 & 6 & -4 & 5 \end{bmatrix}$

23. $\begin{bmatrix} -6 & 1 & 4 & 2 \\ 9 & 3 & -8 & -7 \end{bmatrix}$

24. $\begin{bmatrix} 6 & -1 \\ 0 & 3 \\ 4 & 0 \end{bmatrix}$

APPLIED PROBLEMS

Social Science

25. When a football team has the ball and is planning its next play, it can choose one of several plays or strategies. The success of the chosen play depends largely on how well the other team "reads" the chosen play. Suppose a team with the ball (team A) can choose from three plays, while the opposition (team B) has four possible strategies. The numbers shown in the following payoff matrix represent yards of gain to team A.

$$\begin{bmatrix} 9 & -3 & -4 & 16 \\ 12 & 9 & 6 & 8 \\ -5 & -2 & 3 & 18 \end{bmatrix}$$

Find the saddle point. Find the value of the game.

26. Two armies, A and B, are involved in a war game. Each army has available three different strategies, with payoffs as shown below. These payoffs represent square kilometers of land with positive numbers representing gains by A.

$$\begin{bmatrix} 3 & -8 & -9 \\ 0 & 6 & -12 \\ -8 & 4 & -10 \end{bmatrix}$$

Find the saddle point and the value of the game.

9.5 MIXED STRATEGIES

As we saw earlier, not every game has a saddle point. However, two-person zero-sum games still have optimum strategies, even if the strategy is not as simple as the ones we saw earlier. In a game with a saddle point, the optimum strategy for player A is to pick the row containing the saddle point. Such a strategy is called a **pure strategy,** since the same row is always chosen.

If there is no saddle point, then it will be necessary for both players to mix their strategies. For example, A will sometimes play row 1, sometimes row 2, and so on. If this were done in some specific pattern, the competitor would soon guess it and play accordingly.

For this reason, it is best to mix strategies according to previously determined probabilities. For example, if a player has only two strategies and has decided to play them with equal probability, the random choice could be made by tossing a fair coin, letting heads represent one strategy and tails the other. This would result in each strategy being used about equally over the long run. However, on a particular play it would not be possible to predetermine the strategy to be used. Some other device, such as a spinner, is necessary for more than two strategies, or when the probabilities are not 1/2.

Let us find the strategies to be used in the following game.

$$\begin{bmatrix} -1 & 2 \\ 1 & 0 \end{bmatrix}$$

There is no saddle point, so both players should mix their strategies. To find the probabilities with which player A should randomly select a strategy, proceed as follows.

Assume that player A chooses strategy 1 with probability x. Then the probability that strategy 2 will be chosen is $1 - x$. (The sum of the two probabilities must be 1.) If player B chooses strategy I, then player A's expectation is

$$\begin{aligned} E_{\text{I}} &= -1 \cdot x + 1 \cdot (1 - x) \\ &= -x + 1 - x \\ E_{\text{I}} &= -2x + 1. \end{aligned}$$

On the other hand, if B selects strategy II, then A's expectation is

$$E_{\text{II}} = 2 \cdot x + 0 \cdot (1 - x) = 2x.$$

Next, draw a graph of each of the straight lines $E_{\text{I}} = -2x + 1$ and $E_{\text{II}} = 2x$. (Think of E_{I} and E_{II} as being y for the purposes of drawing the graph.) The graphs are shown in Figure 9.2.

As before, A wants to maximize the smallest amounts that can be won. On the graph, the smallest amounts that can be won

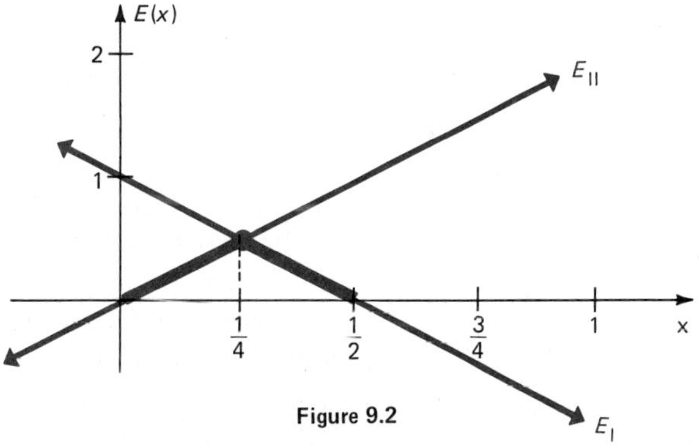

Figure 9.2

1. Find the strategy for player B in the game above. Let y represent the probability with which player B chooses column I. Find the value of the game.

Answer: Choose each column with probability 1/2, value 50¢.

are represented by the points of E_{II} up to the intersection point. To the right of the intersection point, the smallest amounts that can be won are represented by the points of the line E_I. Player A can maximize the smallest amounts that can be won by choosing the point of intersection itself, the peak of the heavily shaded line in Figure 9.2.

To find this point of intersection, we find the simultaneous solution of the two equations. At the point of intersection, we have $E_I = E_{II}$. Substitute $-2x + 1$ for E_I, and $2x$ for E_{II}, giving

$$\begin{aligned} E_I &= E_{II} \\ -2x + 1 &= 2x \\ 1 &= 4x \\ x &= 1/4. \end{aligned}$$

Player A should choose strategy 1 one-fourth of the time, and strategy 2 three-fourths $(1 - 1/4 = 3/4)$ of the time. By doing so, expected winnings will be maximized. To find the maximum winnings (which is also the value of the game), we can substitute 1/4 for x in either E_I or E_{II}. If we choose E_{II} we have

$$E_{II} = 2x = 2\left(\frac{1}{4}\right) = \frac{1}{2}$$

dollar, or 50¢.

Work Problem 1 at the side.

In the game above, we found that player A can maximize expected winnings by playing row 1 with a probability 1/4 and row 2 with a probability 3/4. Such a strategy is called a **mixed strategy.** To actually decide which row to choose on a given play of the game, A could use a spinner, such as the one shown in Figure 9.3.

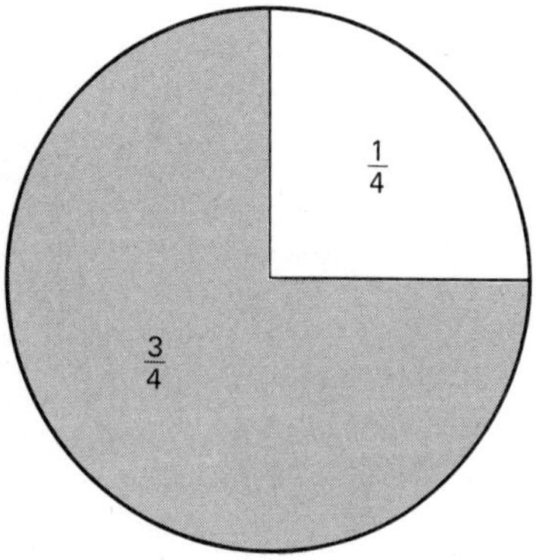

Figure 9.3

We found above that the value of the game for player A is 50¢. In the problem, we found that the value of the game for B is also 50¢. This always happens—even though different mixed strategies may be selected, the value is always the same.

Example 1 Boll weevils threaten the cotton crop near Hattiesburg. Charlie Dawkins owns a small farm; he can protect his crop by spraying with a potent (and expensive) insecticide. In trying to decide what to do, he first sets up a payoff matrix. The numbers represent his profits.

		States of nature	
		Boll weevil attack	No attack
Strategies	Spray	\$14,000	\$7000
	Don't spray	−\$3000	\$8000

Let x represent the probability with which Dawkins chooses strategy I, so $1 - x$ is the probability with which he chooses 2. If nature chooses strategy I (an attack), then Dawkins' expectation is

$$\begin{aligned} E_{\text{I}} &= 14{,}000x - 3000(1 - x) \\ &= 14{,}000x - 3000 + 3000x \\ E_{\text{I}} &= 17{,}000x - 3000. \end{aligned}$$

For nature's strategy II (no attack), Dawkins' expectation is

$$\begin{aligned} E_{\text{II}} &= 7000x + 8000(1 - x) \\ &= 7000x + 8000 - 8000x \\ E_{\text{II}} &= 8000 - 1000x. \end{aligned}$$

To maximize his expected profit, Dawkins should find the value of x for which $E_{\text{I}} = E_{\text{II}}$.

$$\begin{aligned} E_{\text{I}} &= E_{\text{II}} \\ 17{,}000x - 3000 &= 8000 - 1000x \\ 18{,}000x &= 11{,}000 \\ x &= \frac{11}{18} \end{aligned}$$

Thus, $1 - x = 1 - \frac{11}{18} = \frac{7}{18}$.

Dawkins will maximize his expected profit if he chooses strategy 1 with probability 11/18 and strategy 2 with probability 7/18. His expected profit from this mixed strategy can be found by substituting 11/18 for x in either E_{I} or E_{II}. If we choose E_{I} we have

$$\text{expected profit} = 17{,}000\left(\frac{11}{18}\right) - 3000 = \frac{133{,}000}{18} = \$7388.89. \quad ■$$

Work Problem 2 at the side.

2. Find the mixed strategy for player A in the following game. Find the value of the game.

$$\begin{bmatrix} -1 & 2 \\ 3 & 1 \end{bmatrix}$$

Answer: **Choose strategy 1 with probability 2/5, and strategy 2 with probability 3/5; value is 7/5.**

9.5 EXERCISES

Find the optimum strategy for both player A and player B in each of the following games. Find the value of the game. Be sure to look for a saddle point first.

1. $\begin{bmatrix} 5 & 1 \\ 3 & 4 \end{bmatrix}$

2. $\begin{bmatrix} -4 & 5 \\ 3 & -4 \end{bmatrix}$

3. $\begin{bmatrix} -2 & 0 \\ 3 & -4 \end{bmatrix}$

4. $\begin{bmatrix} 6 & 2 \\ -1 & 10 \end{bmatrix}$

5. $\begin{bmatrix} 4 & -3 \\ -1 & 7 \end{bmatrix}$

6. $\begin{bmatrix} 0 & 6 \\ 4 & 0 \end{bmatrix}$

7. $\begin{bmatrix} -2 & \frac{1}{2} \\ 0 & -3 \end{bmatrix}$

8. $\begin{bmatrix} 6 & \frac{3}{4} \\ \frac{2}{3} & -1 \end{bmatrix}$

9. $\begin{bmatrix} \frac{8}{3} & -\frac{1}{2} \\ \frac{3}{4} & -\frac{5}{12} \end{bmatrix}$

10. $\begin{bmatrix} -\frac{1}{2} & \frac{2}{3} \\ \frac{7}{8} & -\frac{3}{4} \end{bmatrix}$

APPLIED PROBLEMS

Management

11. Suppose Allied Manufacturing Company (see Exercise 5, Section 9.2) decides to put its new product on the market with a big television and radio advertising campaign. At the same time, the company finds out that its major competitor, Bates Manufacturing, has also decided to launch a big advertising campaign for a similar product. The payoff matrix below shows the increased sales (in millions) for Allied, as well as the decreased sales for Bates.

		Bates TV	Bates Radio
Allied	TV	1.0	−.7
	Radio	−.5	.5

Find the optimum strategy for Allied Manufacturing and the value of the game.

12. The payoffs in the table below represent the differences between Boeing Aircraft Company's profit and its competitor's profit for two prices (in millions) on commercial jet transports, with positive payoffs being in Boeing's favor. What should Boeing's price strategy be?*

		Competitor's Price Strategy 4.75	Competitor's Price Strategy 4.9
Boeing's Strategy	4.9	−4	2
	4.75	2	0

Natural Science

13. The number of cases of African flu has reached epidemic levels. The disease is known to have two strains with similar symptoms. Doctor De Luca has two medicines available: the first is 60% effective against the first strain and 40% effective against the second. The second medicine is completely effective against the second strain but ineffective against the first. Use the matrix below to decide which medicine she should use and the results she can expect.

---

* Based on "Pricing, Investment, and Games of Strategy," by Georges Brigham in *Management Sciences Models and Techniques*, Volume 1, 1960, by permission of Pergamon Press, Ltd.

		Strain 1	Strain 2
Medicine	1	.6	.4
	2	0	1

Social Science

14. Players A and B play a game in which each shows either one or two fingers at the same time. If there is a match, A wins the amount equal to the total number of fingers shown. If there is no match, B wins the amount of dollars equal to the number of fingers shown.
    (a) Write the payoff matrix.
    (b) Find optimum strategies for A and B and the value of the game.

15. Repeat Exercise 14 if each player may show either 0 or 2 fingers with the same sort of payoffs.

16. In the game of matching coins, two players each flip a coin. If both coins match (both show heads or both show tails), player A wins \$1. If there is no match, player B wins \$1, as in the payoff matrix. Find the optimum strategies for the two players and the value of the game.

$$\begin{bmatrix} 1 & -1 \\ -1 & 1 \end{bmatrix}$$

Management

17. **The Huckster*** Merrill has a concession at Yankee Stadium for the sale of sunglasses and umbrellas. The business places quite a strain on him, the weather being what it is. He has observed that he can sell about 500 umbrellas when it rains, and about 100 when it is sunny; in the latter case he can also sell 1000 sunglasses. Umbrellas cost him 50 cents and sell for \$1; glasses cost 20 cents and sell for 50 cents. He is willing to invest \$250 in the project. Everything that is not sold is considered a total loss.

    He assembles the facts regarding profit in a table.

		Selling during Rain	Selling during Shine
Buying for	Rain	250	−150
	Shine	−150	350

He immediately takes heart, for this is a mixed-strategy game, and he should be able to find a stabilizing strategy which will save him from the vagaries of the weather. Find the best mixed strategy for Merrill.

18. **The Squad Car*** This is a somewhat more harrowing example. A police dispatcher was conveying information and opinion, as fast as she could speak, to Patrol Car 2, cruising on the U.S. Highway: ". . . in a Cadillac; just left Hitch's Tavern on the old Country Road. Direction of flight unknown. Suspect Plesset is seriously wounded but may have an even chance if he finds a good doctor, like Doctor Haydon, soon— even Veterinary Paxson might save him, but his chances would be halved. Plesset shot Officer Flood, who has a large family."

---

*From *The Compleat Strategist* by J. D. Williams. Copyright © 1966 by McGraw-Hill Book Company. Used with their permission. This is an excellent non-technical book on game theory.

Deputy Henderson finally untangled the microphone from the riot gun and his size 14 shoes. He replied: "Roger. We can cut him off if he heads for Haydon's and we have a fifty-fifty chance of cutting him off at the State Highway if he heads for the vet's. We must cut him off because we can't chase him—Deputy Root got this thing stuck in reverse a while ago, and our cruising has been a disgrace to the department ever since."

The headquarter's carrier-wave again hummed in the speaker, but the dispatcher's musical voice was now replaced by the grating tones of Sheriff Lipp. "If you know anything else, don't tell it. He has a hi-fi radio in that Cad. Get him."

Root suddenly was seized by an idea and stopped struggling with the gearshift. "Henderson, we may not need a gun tonight, but we need a pencil: this is just a two-by-two game. The dispatcher gave us all the dope we need." "You gonna use *her* estimates?" "You got better ones? She's got intuition; besides, that's information from headquarters. Now, let's see . . . . Suppose we head for Haydon's. And suppose Plesset does too; then we rack up one good bandit, if you don't trip on that gun again. But if he heads for Paxson, the chances are three out of four that old doc will kill him."

"I don't get it." "Well, it didn't come easy. Remember, Haydon would have an even chance—one-half—of saving him. He'd have half as good a chance with Paxson; and half of one-half is one-quarter. So the chance he dies must be three-quarters—subtracting from one, you know."

"Yeah, it's obvious." "Huh. Now if we head for Paxson's it's tougher to figure. First of all, *he* may go to Haydon's, in which case we have to rely on the doc to kill him, of which the chance is only one-half."

"You ought to subtract that from one." "I did. Now suppose he too heads for Paxson's. Either of two things can happen. One is, we catch him, and the chance is one-half. The other is, we don't catch him—and again the chance is one-half—but there is a three-fourths chance that the doc will have a lethal touch. So the overall probability that he will get by us, but not by the doc, is one-half times three-fourths, or three-eighths. Add to that the one-half chance that he doesn't get by us, and we have seven-eighths."

"I don't like this stuff. He's probably getting away while we're doodling." "Relax. He has to figure it out too, doesn't he? And he's in worse shape than we are. Now let's see what we have."

$$
\begin{array}{lll}
 & & \text{Cad goes to} \\
 & & \begin{array}{cc}\text{Haydon} & \text{Paxson}\end{array} \\
\text{Patrol car goes to} & \begin{array}{l}\text{Haydon}\\ \text{Paxson}\end{array} & \begin{bmatrix} 1 & \frac{3}{4} \\ \frac{1}{2} & \frac{7}{8} \end{bmatrix}
\end{array}
$$

"Fractions aren't so good in this light," Root continues. "Let's multiply everything by eight to clean it up. I hear it doesn't hurt anything."

		Cad	
		Haydon	Paxson
Patrol car	Haydon	8	6
	Paxson	4	7

"It is now clear that this is a very messy business . . ." "I know." "There is no single strategy which we can safely adopt. I shall therefore compute the best mixed strategy."

What mixed strategy should deputies Root and Henderson pursue?

CHAPTER 9 TEST

[9.1]

1. You pay \$6 to play in a game where you will roll a die, with payoffs as follows: \$8 for a 6, \$7 for a 5, \$4 otherwise. What is your mathematical expectation?

2. A developer can buy a piece of property that will produce a profit of \$16,000 with probability .7, or a loss of \$9000 with probability .3. What is her expected profit?

3. The probability that a female Chihuahua will give birth to a litter of 1, 2, 3, or 4 babies is as follows.

Number	1	2	3	4
Probability	.28	.46	.18	.08

Find the expected number of babies in a Chihuahua litter.

[9.2]

In labor-management relations, both labor and management can adopt either a friendly or a hostile attitude. The results are shown in the following payoff matrix. The numbers give the wage gains won by an average worker.

		Management	
		Friendly	Hostile
Labor	Friendly	\$600	\$800
	Hostile	\$400	\$950

4. Suppose the chief negotiator for labor is an optimist. Which strategy should he choose?

5. Which strategy should he choose if he is a pessimist?

6. The chief negotiator for labor feels that there is a 70% chance that the company will be hostile. Which strategy should he adopt? What is the expected payoff?

7. Just before negotiations begin, a new management is installed in the company. There is only a 40% chance that the new management will be hostile. Which strategy should be adopted by labor?

[9.4] *Find any saddle points for the following games. Give the value of the game. Identify any fair games.*

8. $\begin{bmatrix} -2 & 3 \\ -4 & 5 \end{bmatrix}$

9. $\begin{bmatrix} -4 & 0 & 2 & -5 \\ 6 & 9 & 3 & 8 \end{bmatrix}$

10. $\begin{bmatrix} -4 & -1 \\ 6 & 0 \\ 8 & -3 \end{bmatrix}$

Remove any dominated strategies in the following games.

11. $\begin{bmatrix} -11 & 6 & 8 & 9 \\ -10 & -12 & 3 & 2 \end{bmatrix}$

12. $\begin{bmatrix} -1 & 9 & 0 \\ 4 & -10 & 6 \\ 8 & -6 & 7 \end{bmatrix}$

[9.5] *Find the optimum mixed strategies in each of the following games. Find the value of the game.*

13. $\begin{bmatrix} 1 & 0 \\ -2 & 3 \end{bmatrix}$

14. $\begin{bmatrix} 2 & -3 \\ -3 & 5 \end{bmatrix}$

15. $\begin{bmatrix} -3 & 5 \\ 1 & 0 \end{bmatrix}$

16. $\begin{bmatrix} 8 & -3 \\ -6 & 2 \end{bmatrix}$

For each of the following games, remove any dominated strategies, then solve the game. Find the value of the game.

17. $\begin{bmatrix} -4 & 8 & 0 \\ -2 & 9 & -3 \end{bmatrix}$

18. $\begin{bmatrix} 1 & 0 & 3 & -3 \\ 4 & -2 & 4 & -1 \end{bmatrix}$

CASE 10 DECISION THEORY IN THE MILITARY*

The example presented in this case has been reproduced with only minor change from the *Journal of the Operations Research Society of America*, November 1954, pages 365–369. The article is titled "Military Decision and Game Theory," by O. G. Haywood, Jr. This case is presented unedited so that you can get an idea of the type of articles published in the journals.

Military-Decision Doctrine

A military commander may approach decision with either of two philosophies. He may select his course of action on the basis of his estimate of what his enemy *is able to do* to oppose him. Or, he may make his selection on the basis of his estimate of what his enemy *is going to do*. The former is a doctrine of decision based on enemy capabilities; the latter, on enemy intentions.

The doctrine of decision of the armed forces of the United States is a doctrine based on enemy capabilities. A commander is enjoined to select the course of action which offers the greatest promise of success in view of the enemy capabilities. The process of decision, as approved by the Joint Chiefs of Staff and taught in all service schools, is formalized in a five-step analysis called the *Estimate of the Situation*. These steps are illustrated in the following analysis of an actual World War II battle situation.

The Rabaul-Lae Convoy Situation

General Kenney was Commander of the Allied Air Forces in the Southwest Pacific Area. The struggle for New Guinea reached a critical stage in February 1943. Intelligence reports indicated a Japanese troop and supply convoy was assembling at Rabaul (see Figure 1). Lae was expected to be the unloading point. With this general background Kenney proceeded to make his five-step Estimate of the Situation.

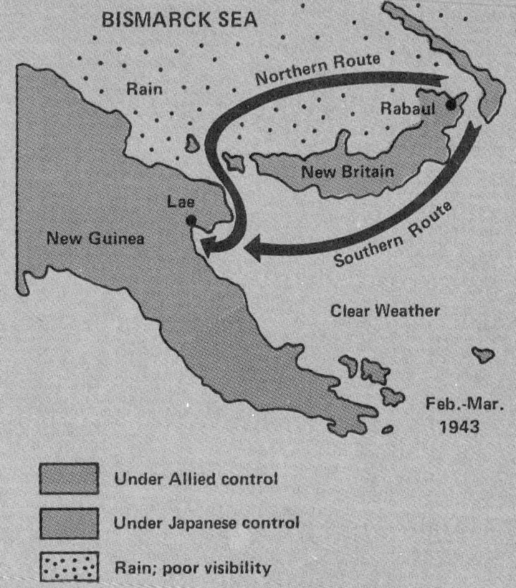

Figure 1 *The Rabaul-Lae Convoy Situation.* The problem is the distribution of reconnaissance to locate a convoy which may sail by either one of two routes.

* From "Military Decision and Game Theory," by O. G. Haywood, Jr. from *Journal of the Operations Research Society of America*, November 1954.

Step 1. The Mission
General MacArthur as Supreme Commander had ordered Kenney to intercept and inflict maximum destruction on the convoy. This then was Kenney's mission.

Step 2. Situation and Courses of Action
The situation as outlined above was generally known. One new critical factor was pointed out by Kenney's staff. Rain and poor visibility were predicted for the area north of New Britain. Visibility south of the island would be good.

The Japanese commander had two choices for routing his convoy from Rabaul to Lae. He could sail north of New Britain, or he could go south of that island. Either route required three days.

Kenney considered two courses of action, as he discusses in his memoirs. He could concentrate most of his reconnaissance aircraft either along the northern route where visibility would be poor, or along the southern route where clear weather was predicted. Mobility being one of the great advantages of air power, his bombing force could strike the convoy on either route once it was spotted.

Step 3. Analysis of the Opposing Courses of Action
With each commander having two alternative courses of action, four possible conflicts could ensue. These conflicts are pictured in Figure 2.

Step 4. Comparison of Available Courses of Action
If Kenney concentrated on the northern route, he ensured one of the two battles of the top row of sketches. However, he alone could not determine which one of these two battles in the top row would result from his decision. Similarly, if Kenney concentrated on the southern route, he ensured one of the battles of the lower row. In the same manner, the Japanese commander could not select a particular battle, but could by his decision assure that the battle would be one of those pictured in the left column or one of those in the right column.

Kenney sought a battle which would provide the maximum opportunity for bombing the convoy. The Japanese commander desired the minimum exposure to bombing. But neither commander could determine the battle which would result from his own decision. Each commander had full and independent freedom to select either one of his alternative strategies. He had to do so with full realization of his opponent's freedom of choice. The particular battle which resulted would be determined by the two independent decisions.

The U.S. doctrine of decision—the doctrine that a commander base his action on his estimate of what the enemy is capable of doing to oppose him—dictated that Kenney select the course of action which offered the greatest promise of success in view of all of the enemy capabilities. If Kenney concentrated his reconnaissance on the northern route, he could expect two days of bombing regardless of his enemy's decision. If Kenney selected his other strategy, he must accept the possibility of a less favorable outcome.

Step 5. The Decision
Kenney concentrated his reconnaissance aircraft on the northern route.

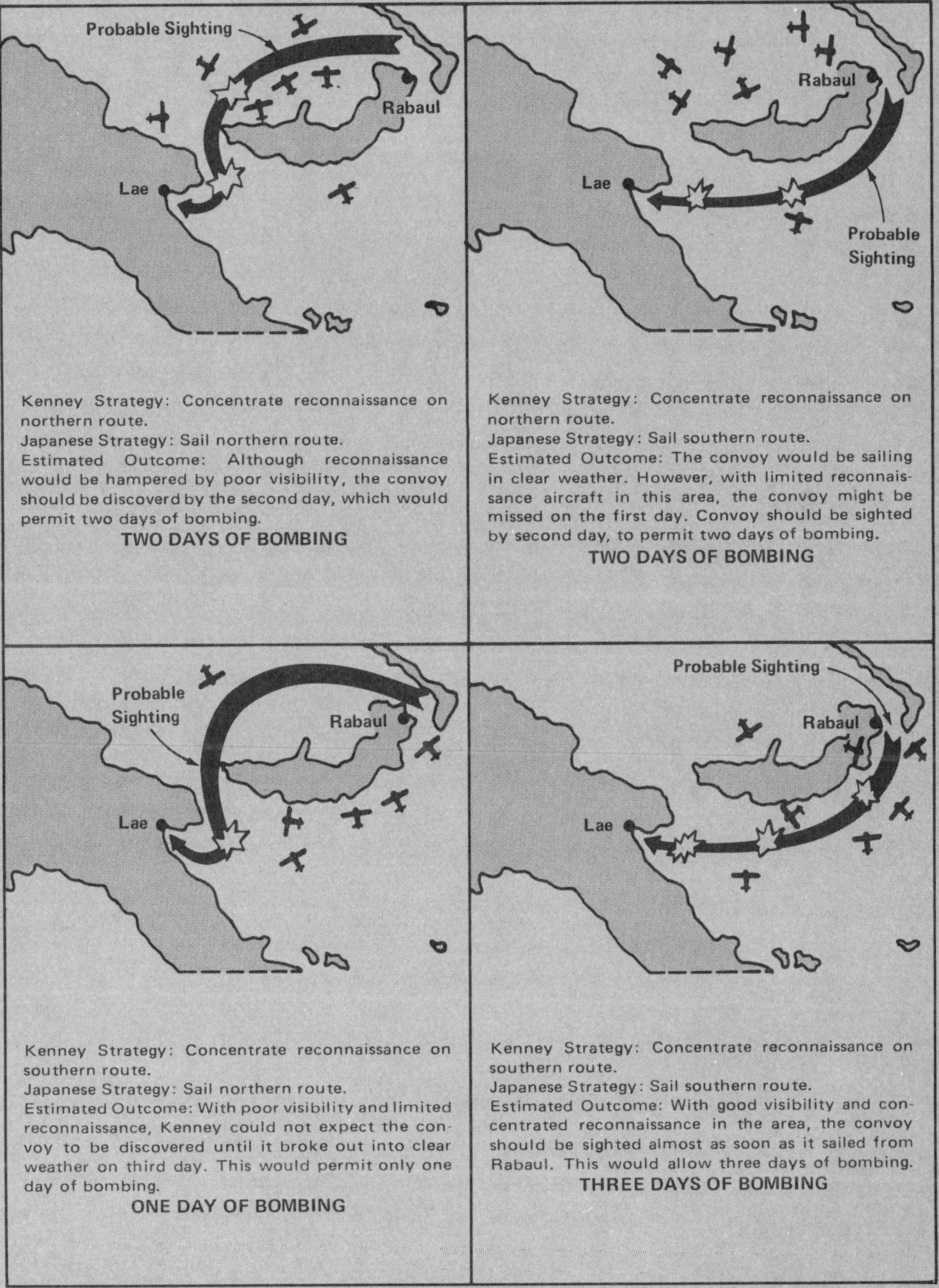
Probable Sighting
Rabaul
Lae
Kenney Strategy: Concentrate reconnaissance on northern route.
Japanese Strategy: Sail northern route.
Estimated Outcome: Although reconnaissance would be hampered by poor visibility, the convoy should be discoverd by the second day, which would permit two days of bombing.
TWO DAYS OF BOMBING
Rabaul
Lae
Probable Sighting
Kenney Strategy: Concentrate reconnaissance on northern route.
Japanese Strategy: Sail southern route.
Estimated Outcome: The convoy would be sailing in clear weather. However, with limited reconnaissance aircraft in this area, the convoy might be missed on the first day. Convoy should be sighted by second day, to permit two days of bombing.
TWO DAYS OF BOMBING
Probable Sighting
Rabaul
Lae
Kenney Strategy: Concentrate reconnaissance on southern route.
Japanese Strategy: Sail northern route.
Estimated Outcome: With poor visibility and limited reconnaissance, Kenney could not expect the convoy to be discovered until it broke out into clear weather on third day. This would permit only one day of bombing.
ONE DAY OF BOMBING
Probable Sighting
Rabaul
Lae
Kenney Strategy: Concentrate reconnaissance on southern route.
Japanese Strategy: Sail southern route.
Estimated Outcome: With good visibility and concentrated reconnaissance in the area, the convoy should be sighted almost as soon as it sailed from Rabaul. This would allow three days of bombing.
THREE DAYS OF BOMBING

Figure 2 *Possible Battles for the Rabaul-Lae Convoy Situation.* Four different engagements of forces may result from the interaction of Kenney's two strategies with the two Japanese strategies. Neither commander alone can determine which particular battle will result.

Discussion

Let us assume that the Japanese commander used a similar philosophy of decision, basing his decision on his enemy's capabilities. Considering the four battles as sketched, the Japanese commander could select either the left or the right column, but could not select the row. If he sailed the northern route, he exposed the convoy to a maximum of two days of bombing. If he sailed the southern route, the convoy might be subjected to three days of bombing. Since he sought minimum exposure to bombing, he should select the northern route.

These two independent choices were the actual decisions which led to the conflict known in history as the Battle of the Bismarck Sea. Kenney concentrated his reconnaissance on the northern route; the Japanese convoy sailed the northern route; the convoy was sighted approximately one day after it sailed; and Allied bombing started shortly thereafter. Although the Battle of the Bismarck Sea ended in a disastrous defeat for the Japanese, we cannot say the Japanese commander erred in his decision. A similar convoy had reached Lae with minor losses two months earlier. The need was critical, and the Japanese were prepared to pay a high price. They did not know that Kenney had modified a number of his aircraft for low-level bombing and had perfected a deadly technique. The U.S. victory was the result of careful planning, thorough training, resolute execution, and tactical surprise of a new weapon—not of error in the Japanese decision.

EXERCISES

1. Use the results in Figure 2 to make a 2×2 game. Let the numbers in the matrix represent days of bombing.

2. Find any saddle point of the game in Exercise 1.

3. Read the rest of the article used as the source for this case and prepare a discussion of the game theory aspects of the Avranches-Gap Situation. (Don't expect this answer to be in the back of the book!)

10 Statistics

Statistics is commonly broken into two broad areas. **Descriptive statistics** deals with the collection and summarization of data. **Inferential statistics** deals with making inferences or conclusions about a population based on data obtained from a limited sample of the population.

Many times we may wish to learn something about the characteristics of a given population, but because the population is very large or mobile, we may not be able to examine all the elements of the population. Our only recourse is to examine a limited sample drawn from the population and to try to infer the characteristics of the population from the sample. For our inferences to be correct, the sample we choose must be representative of the whole population in the long run. To assure that a sample is representative, it must be chosen so that every element of the population is equally likely to be chosen. This is called **random sampling.**

When a hand is dealt from a well-shuffled deck of cards, a random sample has been obtained. In other situations, a random sample might be selected by assigning a number to each element in the population and then using some chance method to select those numbers which would be included in the sample. For example, suppose a sample of 15 members is to be selected from a club with 75 members. For a random sample, one could assign each club member a number. These numbers, each written on a slip of paper, could then be put into a container and thoroughly mixed. A random sample could then be selected by drawing 15 slips of paper from the container, using the numbers on those slips to determine which members would be in the sample. A method similar to this, using birth dates, was used by the Selective Service for the draft. Random sampling typically relies on a chance device.

In recent years, inferential statistics has assumed greater importance partly because of new and more sophisticated methods, but primarily because it is applicable to so many varied fields. The methods of inferential statistics have become increasingly useful in manufacturing, government, agriculture, medicine, the social sciences, and in all kinds of research.

10.1 FREQUENCY DISTRIBUTIONS

Once a sample has been chosen and the data collected, it is helpful to organize the data so that we can draw conclusions from it. One way to do this is to group the data into intervals; equal intervals are usually chosen. For example, suppose the sample resulted in the following data, which represent the scores of 30 students on a mathematics test.

60	90	76	81	59	46
48	61	57	78	86	65
63	54	68	93	71	78
79	67	75	87	76	74
86	57	62	95	80	70

Since the highest score on the list is 95 and the lowest is 46, one convenient way to group the data would be in intervals of size 10 (containing 10 scores), starting with 40–49, and ending with 90–99. This gives an interval for each score in the data and results in six equal intervals of a convenient size. Too many intervals of smaller size would not simplify the data enough, while too few intervals of larger size would conceal information that the data might provide. A rule of thumb is to use from six to fifteen intervals. Within certain bounds, the number of intervals and their size is actually quite arbitrary, and dependent on the judgment of the statistician.

In Table 1 below, the six intervals chosen above are listed in the first column, the number of grades falling into each interval are tallied in the second column, and the tallies are totaled and their frequency is then entered in the third column. We call Table 1 a **grouped frequency distribution.**

Table 1

Interval	Tally	Frequency	
40–49	||	2	
50–59	||||	4	
60–69	~~||||~~ ||	7	
70–79	~~||||~~ ||||	9	
80–89	~~||||~~	5	
90–99	|||	3	
		30	Total

Notice that the frequency distribution in Table 1 clearly shows information about the data that was not noticed before. For example, we see that the interval with the largest number of scores is 70–79 and that 16 scores (more than half) were between 59 and

80. We might also note that the number of scores in each interval increases rather evenly (up to 9) and then decreases at about the same pace.

Example 1 Suppose we ask 30 business executives to give their recommendations as to the number of college units in management that a business major should have. Use the following results to make a grouped frequency distribution for this data. Use intervals 0–4, 5–9, and so on.

3	25	22	16	0	9	14	8	34	21
15	12	9	3	8	15	20	12	28	19
17	16	23	19	12	14	29	13	24	18

We first tally the number of college units falling into each interval. We then find the total of the tallies in each interval, as follows.

Interval	Tally	Frequency
0–4	|||	3
5–9	||||	4
10–14	卌 |	6
15–19	卌 |||	8
20–24	卌	5
25–29	|||	3
30–34	|	1
		30 Total ■

Work Problem 1 at the side.

1. The Internal Revenue Service selected 24 personal tax returns prepared by a certain incompetent tax preparer. The number of errors per return were as follows.

8 12 0 6 10 8 0 14 8 12 14 16
4 14 7 11 9 12 17 5 11 21 22 19

Prepare a grouped frequency distribution for this data. Use intervals 0–4, 5–9, and so on.

Answer:

Interval	Frequency
0–4	3
5–9	7
10–14	9
15–19	3
20–24	2
	24 Total

The information in a grouped frequency distribution can be displayed in a special kind of bar graph called a **histogram.** In a histogram, interval sizes correspond to widths of the bars, and since equal intervals are usually used, the bars in a histogram are all the same width. The heights of the bars are determined by the frequencies. A histogram for the data in Table 1 is shown in Figure 10.1.

A **frequency polygon** is another form of graph which illustrates a grouped frequency distribution. The polygon is formed by joining consecutive midpoints of the tops of the histogram bars with straight line segments. The midpoints of the first and last bars are joined to endpoints on the horizontal axis. A frequency polygon for the data of Table I is illustrated in Figure 10.2.

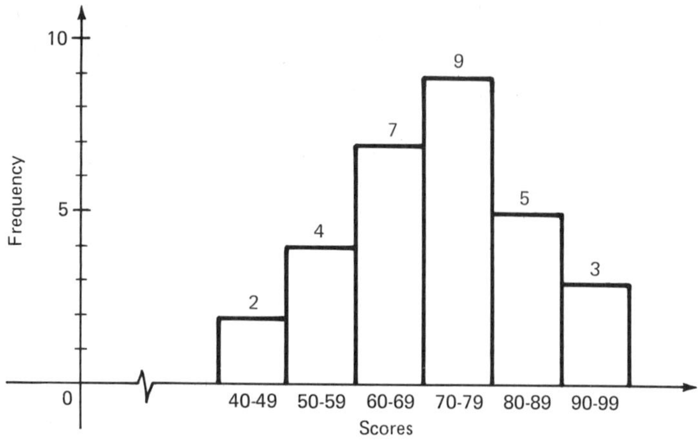

Figure 10.1

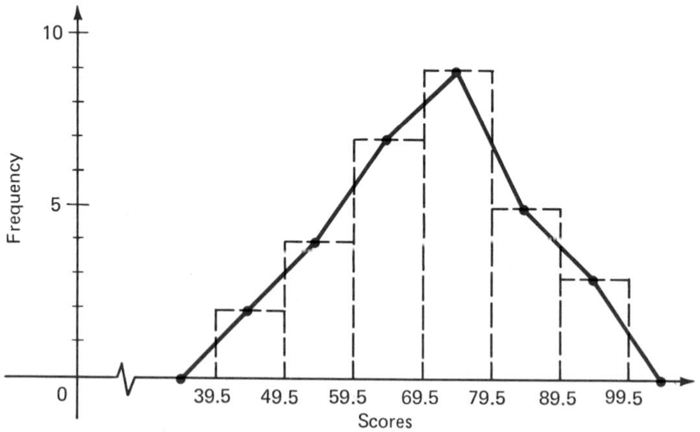

Figure 10.2

Example 2 In Example 1 we ended up with the grouped frequency distribution of college units shown below. (Note that we omitted the column of tallies here, since the column of frequencies tells us the same thing.) Draw a histogram and a frequency polygon for this distribution.

Interval	Frequency
0–4	3
5–9	4
10–14	6
15–19	8
20–24	5
25–29	3
30–34	1

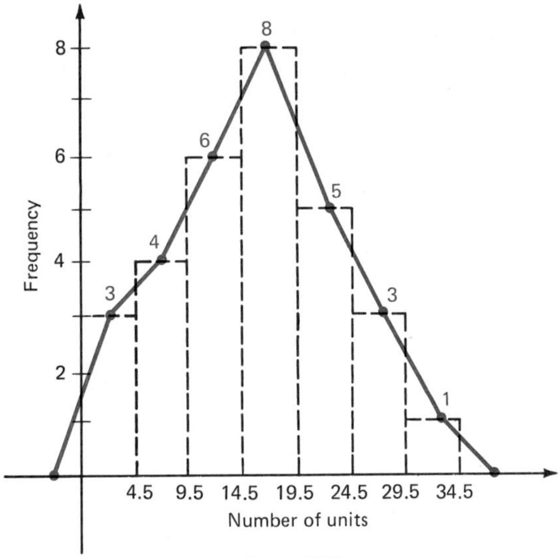

Figure 10.3

First draw a histogram, as shown in black in Figure 10.3. To get a frequency polygon, connect consecutive midpoints of the tops of the bars. The frequency polygon is shown in color in Figure 10.3. ■

Work Problem 2 at the side.

We can obtain further information from Table 1, given at the beginning of this section, by removing the column of tallies and adding a new column, called *cumulative frequency*, as shown in Table 2. The entries in this new column give the cumulative number of scores up to and including a given interval. For example, from Table 2, we see that 13 of the scores were below 70, 22 were below 80, and 30 (all scores) were below 100. We call Table 2 a **cumulative frequency distribution.**

Table 2

Interval	Frequency	Cumulative frequency
40–49	2	2
50–59	4	6
60–69	7	13
70–79	9	22
80–89	5	27
90–99	3	30

2. In Problem 1, we obtained the following grouped frequency distribution.

Interval	Frequency
0–4	3
5–9	7
10–14	9
15–19.	3
20–24	2

Make a histogram and a frequency polygon for this distribution.

Answer:

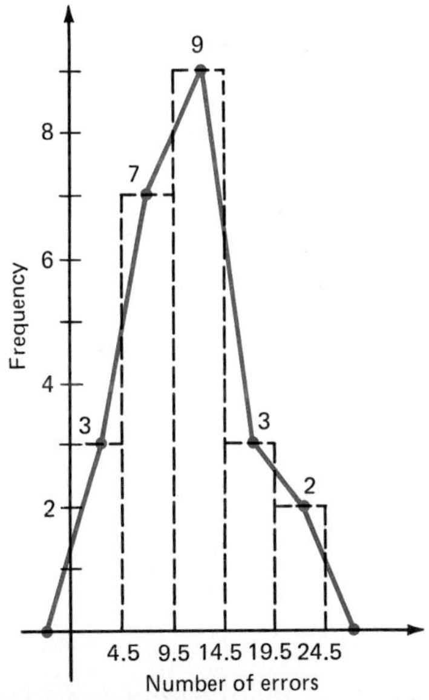

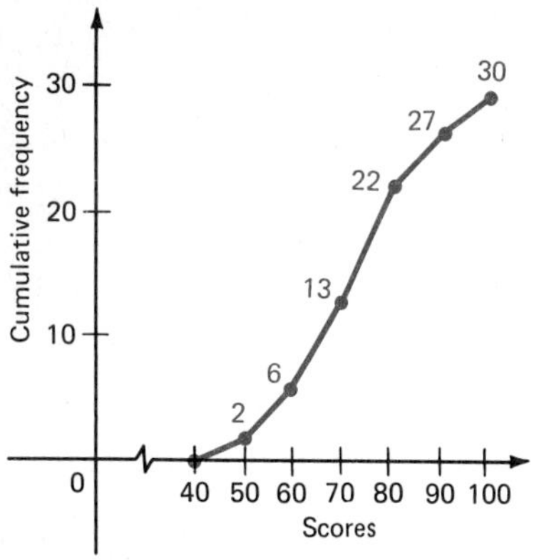

Figure 10.4

A graph which illustrates the cumulative frequencies, called a **cumulative frequency polygon,** is shown in Figure 10.4. Note that the points are not at the midpoints of the intervals, as in a frequency polygon, but at the beginning cf each interval, to correspond with the information that the cumulative frequency provides.

Example 3 Complete a cumulative frequency distribution for the data of Example 2. Draw a cumulative frequency polygon.

Write the data from Example 2. Add a cumulative frequency column. The first number in this column is the same as the first frequency, the second number is the same as the sum of the first two frequencies, and so on.

Interval	Frequency	Cumulative frequency
0–4	3	3
5–9	4	$3 + 4 = 7$
10–14	6	$7 + 6 = 13$
15–19	8	$13 + 8 = 21$
20–24	5	$21 + 5 = 26$
25–29	3	$26 + 3 = 29$
30–34	1	$29 + 1 = 30$

The cumulative frequency polygon is shown in Figure 10.5. ■

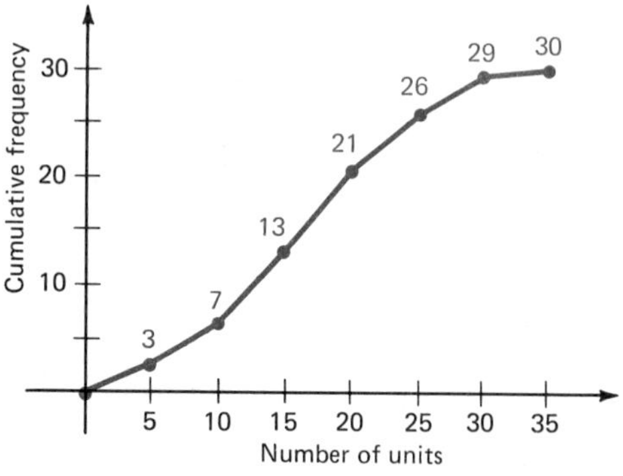

Figure 10.5

Work Problem 3 at the side.

3. Make a cumulative frequency distribution for the data of Problem 2. Draw a cumulative frequency polygon.

Answer:

Interval	Frequency	Cumulative frequency
0–4	3	3
5–9	7	10
10–14	9	19
15–19	3	22
20–24	2	24

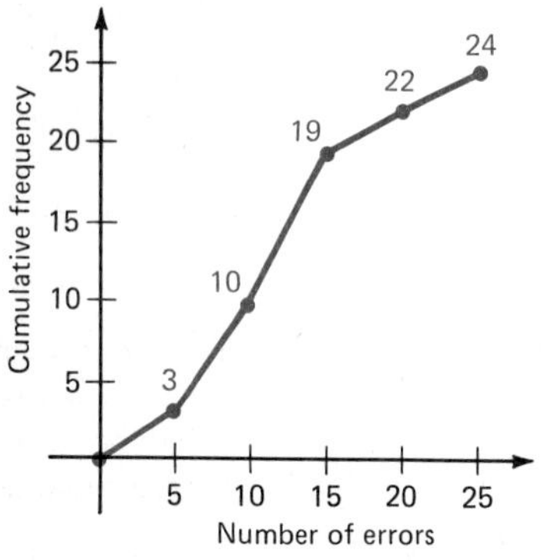

10.1 APPLIED PROBLEMS

For exercises 1–4,
(a) group the data as indicated;

(*b*) *prepare a frequency distribution with a column for intervals, frequencies, and cumulative frequencies;*
(*c*) *construct a histogram;*
(*d*) *construct a frequency polygon;*
(*e*) *construct a cumulative frequency polygon.*

Management

1. The data below gives the number of college units completed by 30 of the EZ Life Company's employees. Group the data into six intervals, starting with 0–24.

74	133	4	127	20	30
103	27	139	118	138	121
149	132	64	141	130	76
42	50	95	56	65	104
4	140	12	88	119	64

General

2. The scores of 80 students on a business law test were as follows. Group the data into seven intervals, starting with 30–39.

79	71	78	87	69	50	63	51
60	46	65	65	56	88	94	56
74	63	87	62	84	76	82	67
59	66	57	81	93	93	54	88
55	69	78	63	63	48	89	81
98	42	91	66	60	70	64	70
61	75	82	65	68	39	77	81
67	62	73	49	51	76	94	54
83	71	94	45	73	95	72	66
71	77	48	51	54	57	69	87

Natural Science

3. The daily high temperatures in Tucson for the month of July for one year were as follows. Group the data using 70–74 as the first interval.

79	84	88	96	102	104	110	108	106	106	
104	99	97	92	94	90	82	74	72	83	
85	92	100	99	101	107	111	102	97	94	92

4. The heights of 40 women (in centimeters) are as follows. Group the data using 140–149 as the first interval.

174	190	172	182	179	186	171	152	174	185
180	170	160	173	163	177	165	157	149	167
169	182	178	158	182	169	181	173	183	176
170	162	159	147	150	192	179	165	148	188

Social Science

5. The frequency with which letters occur in a large sample of any written language does not vary much. Therefore, determining the frequency of each letter in a coded message is usually the first step in deciphering it. The percent frequencies of the letters in the English language are as follows.

Letter	%	Letter	%
E	13	L	3.5
T	9	C, U, M	3
A, O	8	F, P, Y	2
N	7	W, G, B	1.5
I, R	6.5	V	1
S, H	6	K, X, J	0.5
D	4	Q, Z	0.2

Use the introductory paragraph of this exercise as a sample of the English language. Find the percent frequency for each letter in the sample. Compare your results with the frequencies given above.

6. The following message is written in a code in which the frequency of the symbols is the main key to the solution.

```
)?– –8))y* +8506*3×6;4?*7* &×*–6.48()985)?
(8+2: ;48) 81&? (;46*3)y*; 48&(+8(*509+&8 ( ) 8=8
(5* – 8 – 5(81?098; 4& + )&15*50:)6)6*; ?6; 6&*0? – 7
```

   (a) Find the frequency of each symbol.
   (b) By comparing the high-frequency symbols with the high-frequency letters in English, and the low-frequency symbols with the low-frequency letters, try to decipher the message. (Hint: Look for repeated two-symbol combinations and double letters for added clues. Try to identify vowels first.)

7. Get five coins and toss them all at once. Keep track of the number of heads. Repeat this experiment 64 times. Make a histogram of your results.

8. Use the results of Exercise 7 to make a frequency polygon.

9. The frequency distribution at the side gives the theoretical results of tossing 5 coins 64 times. Use this distribution to make a frequency polygon. Compare it to the one you got in Exercise 7.

Number of heads	Frequency
0	2
1	10
2	20
3	20
4	10
5	2

10. Toss six coins a total of 64 times. Keep track of the number of heads. Make a histogram of your results.

11. Make a frequency polygon using the results of Exercise 10.

12. The frequency distribution at the side gives the theoretical results of tossing 6 coins 64 times. Use the distribution to make a frequency polygon. Compare it to the one you made in Exercise 11.

Number of heads	Frequency
0	1
1	6
2	15
3	20
4	15
5	6
6	1

10.2 MEASURES OF CENTRAL TENDENCY

In summarizing data, it is often useful to have a single number, a typical or "average" number, which is representative of the entire collection of data. We customarily refer to such numbers as **measures of central tendency.** Everyone is familiar with some type of average. For example, we can compare our heights and weights to those of the typical or "average" person on weight charts. Students are quite familiar with the "class average" and their own "average" at any time in a given course.

These kinds of averages usually refer to the arithmetic average, or mean. The **arithmetic mean** (the **mean**) of a set of numbers is the sum of the numbers, divided by the total number of numbers. To write the sum of the n numbers $x_1, x_2, x_3, \cdots, x_n$, in a compact way, it is customary to use **summation notation:**

$$x_1 + x_2 + x_3 + \cdots + x_n = \sum_{i=1}^{n} x_i.$$

To save time and space, we often abbreviate $\sum_{i=1}^{n} x_i$ as just $\sum(x)$. The symbol $\bar{x}$ (read x-bar) is used to represent the mean.

$$\bar{x} = \frac{\sum(x)}{n}$$

Example 1 Sales at Lupe's Laundry for the days of last week were

\$86. \$103. \$118. \$117. \$126. \$158. \$149.

Find the mean, $\bar{x}$, of these sales.

We can let $x_1 = 86$, $x_2 = 103$, and so on. ($n = 7$ since there are 7 numbers.) Then

$$\bar{x} = \text{mean} = \frac{\sum(x)}{n} = \frac{86 + 103 + 118 + 117 + 126 + 158 + 149}{7}$$

$$= \frac{857}{7}$$

$$\bar{x} = 122.43.$$

The mean sales at the laundry were \$122.43 per day, rounded to the nearest cent. ■

1. Find the mean of the following list of numbers.

\$25.12	\$42.58
\$76.19	\$32
\$81.11	\$26.41
\$19.76	\$59.32
\$71.18	\$21.03

Answer: \$45.47

Work Problem 1 at the side.

When data have been arranged in a frequency distribution, we can find the mean by the following approach. Suppose we have the following data.

Value	Frequency
84	2
87	4
88	7
93	4
99	3
	20 Total

The value 84 appears twice, 87 four times, and so on. To find the mean, we must first add 84 two times, 87 four times, and so on. We get the same result faster if we multiply 84 and 2, 87 and 4, and so on. Then divide by 20, the total of the frequencies. We have

$$\bar{x} = \frac{(84 \cdot 2) + (87 \cdot 4) + (88 \cdot 7) + (93 \cdot 4) + (99 \cdot 3)}{20}$$

$$= \frac{168 + 348 + 616 + 372 + 297}{20}$$

$$= \frac{1801}{20}$$

$$\bar{x} = 90.05.$$

Example 2 Find the mean for the data shown in the following frequency distribution.

Value	Frequency	Value × Frequency
30	6	$30 \cdot 6 =$ 180
32	9	$32 \cdot 9 =$ 288
33	7	$33 \cdot 7 =$ 231
37	12	$37 \cdot 12 =$ 444
42	6	$42 \cdot 6 =$ 252
	40 Total	1395 Total

An additional column, "Value × Frequency," is included. Adding the products from this column gives us a total of 1395. The total from the frequency column is 40. Thus,

$$\bar{x} = \frac{1395}{40} = 34.875. \quad ■$$

Work Problem 2 at the side.

We can find the mean of grouped data in the same way. For grouped data, we are not given single values, but rather intervals

2. Find $\bar{x}$ for the following frequency distribution.

Value	Frequency
7	2
9	3
11	6
13	4
15	1
17	4

Answer: $\bar{x} = 12.1$

of values. To find the mean, we assume that all these values are located at the midpoint of the interval.

Example 3 Find the mean for the following grouped frequency distribution (Table 1 of the previous section).

Interval	Midpoint, x	Frequency, f	Product, xf
40–49	44.5	2	89
50–59	54.5	4	218
60–69	64.5	7	451.5
70–79	74.5	9	670.5
80–89	84.5	5	422.5
90–99	94.5	3	283.5
		30 Total	2135 Total

An additional column for the midpoint of each interval has been included. To get the numbers in this column, we add the endpoints of each interval and divide by 2. For the interval 40–49, the midpoint is $(40 + 49)/2 = 44.5$.

We then introduce the product column on the right. To get the numbers in this column, we multiply frequencies and corresponding midpoints. Finally, we divide the total of the product column by the total of the frequency column to get

$$\bar{x} = \frac{2135}{30} = 71.2 \text{ (to the nearest tenth).} \quad \blacksquare$$

In general, the formula for the mean of a grouped frequency distribution is

$$\bar{x} = \frac{\sum(xf)}{n},$$

where x represents the midpoints, f the frequencies, and $n = \sum f$ (that is, n is the sum of the numbers in the frequency column).

3. Find the mean of the following grouped frequency distribution.

Interval	Frequency
0–4	6
5–9	4
10–14	7
15–19	3

Answer: 8.75

Work Problem 3 at the side.

Asked by a reporter to give the average height of the players on his team, the Little League coach lined up his 15 players by increasing height. He picked out the boy in the middle and pronounced him to be of average height. This kind of average, called the **median,** is defined as the middle entry in a set of data arranged in either increasing or decreasing order. If there is an even number of entries, the median is defined to be the mean of the two center entries.

Odd number of entries	Even number of entries
8	2
7	3
median = 4	4 ⎤
3	7 ⎦ median $= \frac{4 + 7}{2} = 5.5$
1	9
	12

The procedure for finding the median of a grouped frequency distribution is more complicated. We omit it here because it is more common to find the mean when working with grouped frequency distributions.

Example 4 Find the median for the following lists of numbers.

(a) 11, 12, 17, 20, 23, 28, 29

The median is the middle number, in this case, 20. (Note that the numbers are arranged in numerical order.) In this list, three numbers are smaller than 20 and three are larger.

(b) 15, 13, 7, 11, 19, 30, 39, 5, 10

First arrange the numbers in numerical order, from smallest to largest.

5, 7, 10, 11, 13, 15, 19, 30, 39

The middle number can now be picked out: the median is 13.

(c) 47, 59, 32, 81, 74, 153

Write the numbers in numerical order.

32, 47, 59, 74, 81, 153

There are six numbers here, so there is no single middle number. Therefore, the median is the average of the two middle numbers, or

$$\text{median} = \frac{59 + 74}{2} = \frac{133}{2} = 66\frac{1}{2}. \quad ■$$

Work Problem 4 at the side.

4. Find the median for each of the following lists of numbers.
(a) 12, 15, 17, 19, 35, 42, 58
(b) 28, 68, 7, 15, 47, 59, 13, 74, 32, 25

Answer:
(a) 19
(b) 30

In some situations, the median gives a truer representation or average of the data than the mean. For example, in Donna's office, there are 10 salespersons, 4 secretaries, the sales manager, and Ms. Daly, who owns the business. Their annual salaries are as follows: secretaries, \$6000 each; salespersons, \$10,000 each; manager, \$18,000; and owner, \$50,000. The mean salary is

$$\bar{x} = \frac{(6000)4 + (10{,}000)10 + 18{,}000 + 50{,}000}{16} = \$12{,}000.$$

However, since 14 people earn less than \$12,000 and only 2 earn more, this does not seem very representative. The median salary is found by ranking the salaries by size: \$6000, \$6000, \$6000, \$6000, \$10,000, \$10,000, . . . , \$50,000. Since there are 16 salaries (an even number) in the list, the mean of the 8th and 9th entries will give the value of the median. The 8th and 9th entries are both \$10,000, so the median is \$10,000. In this example, the median gives a truer average than the mean.

Sue's scores on ten class quizzes include one 7, two 8's, six 9's, and one 10. She claims that her average grade on quizzes is 9, because most of her scores are 9's. This kind of "average," found by selecting the most frequent entry, is called the **mode.** Similarly, in a grouped frequency distribution, the interval with the greatest frequency is the **modal class.**

Example 5 Find the mode for each list of numbers.

(a) 57, 38, 55, 55, 80, 87, 98

The number 55 occurs more often than any other, and is the mode. It is not necessary to place the numbers in numerical order when looking for the mode.

(b) 182, 185, 183, 185, 187, 187, 189

Both 185 and 187 occur twice. This list has *two* modes.

(c) 10,708, 11,519, 10,972, 17,546 13,905, 12,182

No number here occurs more than once. This list has no mode. ■

Work Problem 5 at the side.

5. Find the mode for each of the following lists of numbers.

(a) 29, 35, 29, 18, 29, 56, 48

(b) 13, 17, 19, 20, 20, 13, 25, 27, 13, 20

(c) 512, 546, 318, 729, 854, 253

Answer:

(a) 29

(b) 13 and 20

(c) no mode

The mode has the advantages of being easily found and not being influenced by extremes—data which are very large or very small compared to the rest of the data. It is often used in samples where the data are not numerical. For example, if students were asked to select one of six candidates as the "typical student," the one with the most votes would be selected. A major disadvantage to the mode is that there may be more than one, in case of ties, or there may be no mode at all when all entries occur with the same frequency.

The mean is the most commonly used measure of central tendency. Its advantages are that it is easy to compute, it takes all the data into consideration, and it is reliable—that is, repeated samples are likely to give very similar means. A disadvantage of the mean is that it is influenced by extreme values, as illustrated in the salary example above.

The median can be easy to compute and is influenced very little by extremes. Like the mode, the median can be found in situations where the data are not numerical. For example, in a taste test, people are asked to rank five soft drinks from the one they like best to the one they like least. The combined rankings then produce an ordered sample from which the median can be identified. A disadvantage of the median is the need to rank the data in order; this is difficult when the number of items is large.

10.2 EXERCISES

Find the mean for each list of numbers. Round to the nearest tenth. (See Example 1.)

1. 8, 10, 16, 21, 25

2. 44, 41, 25, 36, 67, 51

3. 130, 141, 149, 152, 158, 163, 139, 170

4. 42, 48, 54, 62, 69, 75, 90, 94

5. 21,900, 22,850, 24,930, 29,710, 28,340, 40,000

6. 38,500, 39,720, 42,183, 21,982, 43,250

7. 9.4, 11.3, 10.5, 7.4, 9.1, 8.4, 9.7, 5.2, 1.1, 4.7

8. 30.1, 42.8, 91.6, 51.2, 88.3, 21.9, 43.7, 51.2

9. .06, .04, .05, .08, .03, .14, .18, .29, .07, .01

10. .31, .09, .08, .22, .46, .51, .48, .42, .53, .42

Find the mean for each of the following. Round to the nearest tenth. (See Example 2.)

11.

Value	Frequency
3	4
5	2
9	1
12	3

12.

Value	Frequency
9	3
12	5
15	1
18	1

13.

Value	Frequency
12	4
13	2
15	5
19	3
22	1
23	5

14.

Value	Frequency
25	1
26	2
29	5
30	4
32	3
33	5

15.

Value	Frequency
104	6
112	14
115	21
119	13
123	22
127	6
132	9

16.

Value	Frequency
246	2
291	4
295	3
304	8
307	9
319	2

Find the median for each of the following lists of numbers. Don't forget to first place the numbers in numerical order if necessary. (See Example 4.)

17. 12, 18, 32, 51, 58, 92, 106

18. 596, 604, 612, 683, 719

19. 100, 114, 125, 135, 150, 172

20. 298, 346, 412, 501, 515, 521, 528, 621

21. 32, 58, 97, 21, 49, 38, 72, 46, 53

22. 1072, 1068, 1093, 1042, 1056, 1005, 1009

23. 576, 578, 542, 551, 565, 525, 590, 559

24. 7, 15, 28, 3, 14, 18, 46, 59, 1, 2, 9, 21

25. 28.4, 9.1, 3.4, 27.6, 59.8, 32.1, 47.6, 29.8

26. .6, .4, .9, 1.2, .3, 4.1, 2.2, .4, .7, .1

Find the mode or modes for each of the following lists of numbers. (See Example 5.)

27. 4, 9, 8, 6, 9, 2, 1, 3

28. 21, 32, 46, 32, 49, 32, 49

29. 80, 72, 64, 64, 72, 53, 64

30. 97, 95, 94, 95, 94, 97, 97

31. 74, 68, 68, 68, 75, 75, 74, 74, 70

32. 158, 162, 165, 162, 165, 157, 163

33. 5, 9, 17, 3, 2, 8, 19, 1, 4, 20

34. 12, 15, 17, 18, 21, 29, 32, 74, 80

35. 6.1, 6.8, 6.3, 6.3, 6.9, 6.7, 6.4, 6.1, 6.0

36. 12.75, 18.32, 19.41, 12.75, 18.30, 19.45, 18.33

For grouped data, we give the modal class *(interval containing the most data values.) Give the mean and modal class for each of the following collections of grouped data.*

37.

College units	Frequency
0–24	4
25–49	3
50–74	6
75–99	3
100–124	5
125–149	9

38. (You will need to use Exercise 3, Section 10.1, to obtain the frequencies for this distribution before finding the mean and modal class.)

Temperature	Frequency
70–74	
75–79	
80–84	
85–89	
90–94	
95–99	
100–104	
105–109	
110–114	

39. Use the distribution of Exercise 2, Section 10.1

40. Use the distribution of Exercise 4, Section 10.1

APPLIED PROBLEMS

Management

Many Kentucky Fried Chicken outlets are owned by the company itself, while others are franchised by individual operators. To see which outlets produced the largest annual sales, the company found the average sales for each. (The results were given by Business Week magazine.) Find the mean for each of the following.

41. 6 company-owned outlets had sales of \$240,000, \$320,000, \$300,000, \$340,000, \$250,000, \$350,000.

42. 7 franchised outlets had sales of \$240,000, \$220,000, \$300,000, \$320,000, \$260,000, \$250,000, \$270,000.

Denny's Restaurants did the same for its restaurants. Find each of the following means.

43. 8 company-owned outlets had sales of \$382,520, \$321,710, \$308,512, \$371,519, \$382,710, \$297,413, \$314,725, \$303,603.

44. 5 franchised restaurants had sales of \$217,941, \$223,825, \$234,818, \$239,513, \$238,403.

Social Science

In the following problems, find the average salary for players on the team. The averages come from a recent Harper's magazine article.

45.

Baseball salary	Number of players
$16,000	3
$24,000	2
$32,000	5
$41,000	3
$45,000	4
$59,000	3
$63,000	2
$81,000	1

46.

Hockey salary	Number of players
$15,000	1
$25,000	2
$37,000	4
$51,000	3
$82,000	3
$104,000	3
$108,000	2
$130,000	2

Natural Science

Find the mean, median, and mode for each set of ungrouped data.

47. The weight gains of 10 experimental rats fed on a special diet were $-1, 0, -3, 7, 1, 1, 5, 4, 1, 4$.

48. A sample of 7 measurements of the thickness of a copper wire were .010, .010, .009, .008, .007, .009, .008.

49. The times in minutes that 12 patients spent in a doctor's office were 20, 15, 18, 22, 10, 12, 16, 17, 19, 21, 23, 13.

50. The scores on a 10-point botany quiz were 6, 6, 8, 10, 9, 7, 6, 5, 6, 8, 3.

10.3 MEASURES OF VARIATION

The mean gives us a measure of central tendency of a list of numbers, but the mean tells us nothing about the spread of the numbers in the list. For example, look at the following three samples.

I	3	5	6	3	3
II	4	4	4	4	4
III	10	1	0	0	9

Each of these three samples has a mean of 4, and yet it is clear that they are quite different; the amount of dispersion or variation within the samples is different. In addition to a measure of central tendency, we need another kind of measure, called a **measure of variation,** which describes how much the numbers vary.

Note that the largest number in Sample I is 6, while the smallest is 3, a difference of 3. In Sample II this difference is 0; in Sample III it is 10. The difference between the largest and smallest number in a sample is called the **range,** and is one example of a measure of variation. The range of Sample I is 3, of Sample II is 0, and of Sample III is 10. The range has the advantage of being very easy to compute and gives a rough estimate of the variation among the data in the sample. However, it depends only on the two extremes and tells nothing about how the other data are distributed between the extremes.

Example 1 Find the range for each list of numbers.

(a) 12, 27, 6, 19, 38, 9, 42, 15

The highest number here is 42; the lowest is 6. The range is the difference of these numbers, or

$$42 - 6 = 36.$$

(b) 74, 112, 59, 88, 200, 73, 92, 175

$$\text{range} = 200 - 59 = 141 \quad ■$$

Work Problem 1 at the side.

1. Find the range for the numbers

159, 283, 490, 375, 390, 297.

Answer: 331

One of the most useful measures of variation is the *standard deviation.* Before defining it however, we must find the *deviations from the mean.* To find the **deviations from the mean,** find the mean and then subtract the mean from each number in the list.

Example 2 Find the deviations from the mean for the numbers

32, 41, 47, 53, 57.

Adding these numbers and dividing by 5 gives us a mean of 46. To find the deviations from the mean, subtract 46 from each number in the list.

Number	32	41	47	53	57
Deviations from mean	−14	−5	1	7	11

(To check your work, find the sum of these deviations. It should always equal 0.) ■

Work Problem 2 at the side.

2. Find the deviations from the mean for the numbers

19, 25, 36, 41, 52, 61.

Answer: Mean is 39; deviations are −20, −14, −3, 2, 13, 22.

To find a measure of variation, we might be tempted to find the mean of the deviations. However, this number turns out to always be 0, no matter how much dispersion in the data. (The answer is always 0 because the positive and negative numbers cancel each other out.) We can get around this problem by *squaring* each deviation.

Number	32	41	47	53	57
Deviations from mean	−14	−5	1	7	11
Square of deviation	196	25	1	49	121

Now we are ready to define the **standard deviation:** it is the square root of the mean of the squares of the deviations. Luckily, the standard deviation is harder to say than to find. In our example, the mean of the squares of the deviations is

$$\frac{196 + 25 + 1 + 49 + 121}{5} = \frac{392}{5} = 78.4.$$

The standard deviation is the square root of this number, or $\sqrt{78.4}$. The letter s is used to represent standard deviation. In summary, for the numbers 32, 41, 47, 53, 57, we have

$$s = \sqrt{78.4} = 8.9 \text{ (to the nearest tenth).}$$

The square root of 78.4 can be found in tables or by using a calculator.

In summary, if we have a list of n numbers, we can find the standard deviation of these numbers by using the formula

$$s = \sqrt{\frac{\sum(x - \bar{x})^2}{n}}.^{*}$$

Example 3 Find the standard deviation of the numbers

7, 9, 18, 22, 27, 29, 32, 40.

Step 1 The mean of the numbers is

$$\frac{7 + 9 + 18 + 22 + 27 + 29 + 32 + 40}{8} = 23.$$

Step 2 Find the deviations from the mean.

Number	7	9	18	22	27	29	32	40
Deviation	−16	−14	−5	−1	4	6	9	17

Step 3 Square each deviation.

Square of deviation	256	196	25	1	16	36	81	289

* Dividing by $n - 1$ gives an answer that is theoretically better. Thus, some texts recommend dividing by $n - 1$, but we prefer n for simplicity. In any case, in most practical problems, the difference is very slight.

Step 4 Find the mean of the squares.

$$\frac{256 + 196 + 25 + 1 + 16 + 36 + 81 + 289}{8} = \frac{900}{8} = 112.5$$

Step 5 The standard deviation is the square root of the answer in Step 4. Thus,

$$s = \sqrt{112.5} = 10.6. \blacksquare$$

3. Find the standard deviation of the numbers

19, 25, 36, 41, 52, 61.

Answer: $\sqrt{210.3} = 14.5$

Work Problem 3 at the side.

The square of the standard deviation, s^2, is called the **variance.** In Example 3 above, the variance is 10.6^2, or 112.5.

The definition of standard deviation can be changed algebraically to give the following **shortcut formula for standard deviation.**

$$s = \frac{1}{n}\sqrt{n\sum(x^2) - (\sum x)^2}$$

We call this the shortcut formula since it is not necessary to find the mean to use it.

Example 4 Use the shortcut formula to find the standard deviation, s, for the numbers

7, 9, 18, 22, 27, 29, 32, 40.

To use the shortcut formula, we need $\sum x$ and $\sum x^2$. Arrange your work as shown here.

x	x^2	
7	49	
9	81	
18	324	
22	484	
27	729	
29	841	
32	1024	
40	1600	
184	5132	Totals

Now use the shortcut formula.

$$s = \frac{1}{n}\sqrt{n\sum(x^2) - (\sum x)^2}$$

$$= \frac{1}{8}\sqrt{8 \cdot (5132) - (184)^2} \qquad \text{Let } n = 8, \textstyle\sum(x^2) = 5132, \sum x = 184$$

$$= \frac{1}{8}\sqrt{41{,}056 - 33{,}856}$$

$$= \frac{1}{8}\sqrt{7200}$$

$$= \frac{1}{8}(84.85)$$

$$s = 10.6,$$

the same answer we found in Example 3. ■

4. Use the shortcut method to find the standard deviation, s, for the numbers 19, 25, 36, 41, 52, 61.

Answer: $\sqrt{210.3} = 14.5$

Work Problem 4 at the side.

For data in a grouped frequency distribution, the formula for the standard deviation is

$$s = \sqrt{\frac{\sum(x - \bar{x})^2 f}{n}}.$$

The formula indicates that the product $(x - \bar{x})^2 f$ is to be found for each interval. (Recall, x is the interval midpoint.) Then, these products are summed, and the sum is divided by n, the total frequency. The square root of this result is s, the standard deviation.

Example 5 Find s for the grouped data of Example 3, Section 10.2.

We begin by including columns for $x - \bar{x}$, $(x - \bar{x})^2$, and $(x - \bar{x})^2 \cdot f$. Recall from Section 10.2 that $\bar{x} = 71.2$.

Interval	f	x	$x - \bar{x}$	$(x - \bar{x})^2$	$(x - \bar{x})^2 \cdot f$	
40–49	2	44.5	−26.7	712.89	1425.78	
50–59	4	54.5	−16.7	278.89	1115.56	
60–69	7	64.5	−6.7	44.89	314.23	
70–79	9	74.5	3.3	10.89	98.01	
80–89	5	84.5	13.3	176.89	884.45	
90–99	3	94.5	23.3	542.89	1628.67	
	30				5466.70	Totals

5. Find the standard deviation for the grouped data that follows (Hint: $\bar{x} = 28.5$)

Value	Frequency
20–24	3
25–29	2
30–34	4
35–39	1

Answer: $\sqrt{25.25} = 5.02$

Use the appropriate formula for s.

$$s = \sqrt{\frac{\sum(x - \bar{x})^2 \cdot f}{n}} = \sqrt{\frac{5466.70}{30}} = \sqrt{182.2} = 13.5 \quad ■$$

Work Problem 5 at the side.

10.3 EXERCISES

Find the range and standard deviation for each of the following sets of numbers. (See Examples 1 and 3.)

1. 6, 8, 9, 10, 12

2. 12, 15, 19, 23, 26

3. 7, 6, 12, 14, 18, 15

4. 4, 3, 8, 9, 7, 10, 1

5. 42, 38, 29, 74, 82, 71, 35

6. 122, 132, 141, 158, 162, 169, 180

7. 241, 248, 251, 257, 252, 287

8. 51, 58, 62, 64, 67, 71, 74, 78, 82, 93

9. 3, 7, 4, 12, 15, 18, 19, 27, 24, 11

10. 15, 42, 53, 7, 9, 12, 28, 47, 63, 14

11. 21, 28, 32, 42, 51

12. 76, 78, 92, 104, 111

Find the standard deviation for the following grouped data.

13. (From Exercise 1, Section 10.1)

College units	Frequency
0–24	4
25–49	3
50–74	6
75–99	3
100–124	5
125–149	9

14. (You will need to use Exercise 3, Section 10.1, to obtain the frequencies for this distribution before finding the standard deviation.)

Temperature	Frequency
70–74	
75–79	
80–84	
85–89	
90–94	
95–99	
100–104	
105–109	
110–114	

15. Use the distribution of Exercise 2, Section 10.1.

16. Use the distribution of Exercise 4, Section 10.1.

An application of standard deviations is given by **Chebyshev's theorem.** *(P. L. Chebyshev was a Russian mathematician who lived from 1821 to 1894.) His theorem applies to* any *distribution of numbers. It states:*

> For any distribution of numbers, at least $1 - 1/k^2$ of the numbers lie within k standard deviations of the mean.

Example For any distribution, at least

$$1 - 1/3^2 = 1 - 1/9 = 8/9$$

of the numbers lie within 3 standard deviations of the mean. ■

Find the fraction of all the numbers of a data set lying within the following numbers of standard deviations from the mean.

17. 2 **18.** 4 **19.** 5

In a certain distribution of numbers, the mean is 50 *with a standard deviation of* 6. *Use Chebyshev's theorem to tell what percent of the numbers are*

20. between 38 and 62;

21. between 32 and 68;

22. between 26 and 74;

23. between 20 and 80;

24. less than 38 or more than 62;

25. less than 32 or more than 68;

26. less than 26 or more than 74.

APPLIED PROBLEMS

Management

27. The Forever Power Company conducted tests on the life of its batteries and those of a competitor (Brand X) with the following results for samples of 10 batteries of each brand.

	Hours of use									
Forever Power	20	22	22	25	26	27	27	28	30	35
Brand X	15	18	19	23	25	25	28	30	34	38

Compute the mean and standard deviation for each sample.
(a) Which batteries have a more uniform life in hours?
(b) Which batteries have the highest average life in hours?

28. The weekly wages of the seven employees of Harold's Hardware Store are: \$90, \$95, \$120, \$128, \$150, \$180, and \$500.
(a) Find the mean and standard deviation of this distribution.
(b) How many of the seven employees earn within one standard deviation of the mean? How many earn within two standard deviations of the mean?

The Quaker Oats Company conducted a survey to determine if a proposed premium, to be included in their cereal, was appealing enough to generate new sales. Four cities were used as test markets, where the cereal was distributed with the premium, and four cities as control markets, where the cereal was distributed without the premium. The eight cities were chosen on the basis of their similarity in terms of population, per capita income, and total cereal purchase volume. The results were as follows.*

		Percent change in average market shares per month
Test cities	1	+18
	2	+15
	3	+7
	4	+10
Control cities	1	+1
	2	−8
	3	−5
	4	0

29. Find the mean of the change in market share for the four test cities.

30. Find the mean of the change in market share for the four control cities.

31. Find the standard deviation of the change in market share for the test cities.

32. Find the standard deviation of the change in market share for the control cities.

33. Find the difference between the mean of Exercise 29 and the mean of Exercise 30. This represents the estimate of the percent change in sales due to the premium.

34. The two standard deviations from Exercise 31 and Exercise 32 were used to calculate an "error" of ± 7.95 for the estimate in Exercise 33. With this amount of error what is the smallest and largest estimate of the increase in sales? [Hint: Use the answer to Exercise 33.]

On the basis of the interval estimate of Exercise 34, the company decided to mass produce the premium and distribute it nationally.

* This example was supplied by Jeffery S. Berman, Senior Analyst, Marketing Information, Quaker Oats Company.

10.4 THE NORMAL CURVE

In this section we discuss a theoretical distribution in contrast to the frequency distributions we have been dealing with thus far. A frequency distribution is the result of an actual experiment, while a **theoretical distribution** describes the outcomes of an experimental situation on the basis of probability. Some theoretical distributions are **discrete,** which means that there is only a finite number of results. For example, a distribution of the results of tossing a die would be discrete, since only the numbers, 1, 2, 3, 4, 5, or 6 could result. We look at an example of a discrete distribution in the next section.

On the other hand, the distribution of heights (in inches) of college freshman women includes infinitely many possible measurements, such as 53 in., $58\frac{1}{2}$ in., 66.5 in., 72.33 . . . in., and so on. In such a set of numbers one cannot determine the next larger or next smaller number. Such a distribution is called a **continuous** (in contrast to a discrete) **distribution.** Figure 10.6 illustrates the continuous distribution of heights of college freshman women. Note that the most frequent heights occur near the center of the graph.

Another continuous curve, which approximates the distribution of yearly incomes in the United States, is shown in Figure 10.7. From the graph, it can be seen that the most frequent incomes are grouped near the low end of the graph. This kind of distribution, where the peak is not at the center, is called **skewed.**

The distribution of heights of college freshmen women and the distribution of the lengths of a plant species' leaves are examples of natural distributions which approximate an important theoretical distribution known as the **normal distribution.** The normal distribution has several distinguishing characteristics, including its well-known "bell" shape, called the **normal curve,** which is symmetrical about a vertical center line. Its mean is located in the center and coincides with its median. The area between the graph and the

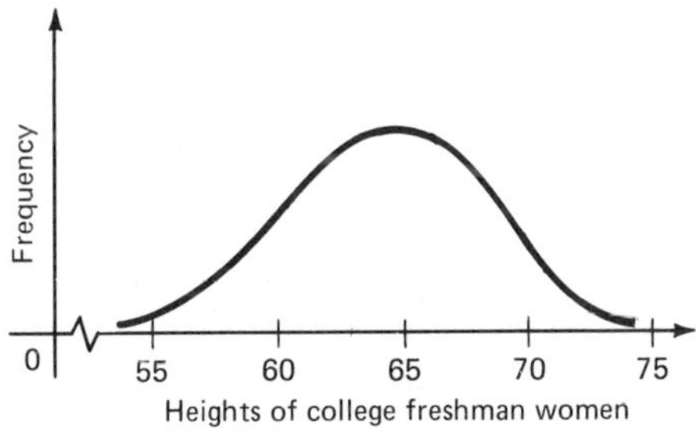

Figure 10.6

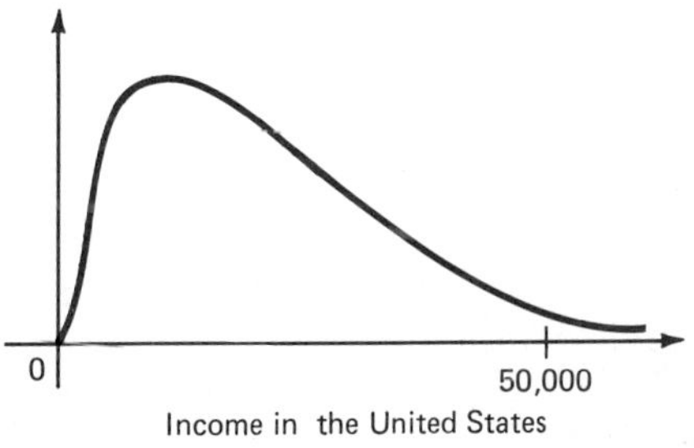

Figure 10.7

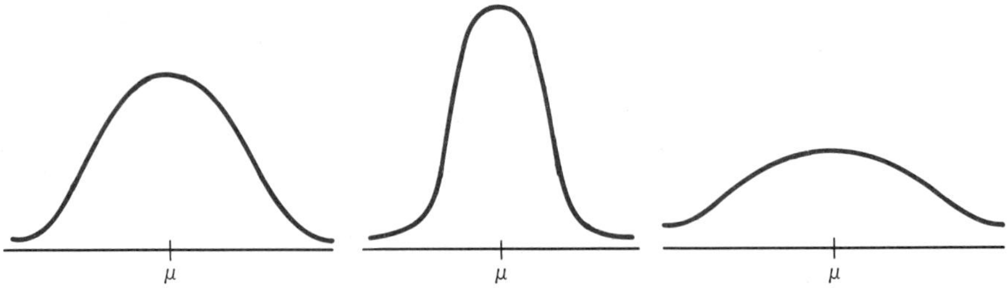

Three normal distributions

Figure 10.8

x-axis always equals 1. Its importance results from the fact that many natural and social phenomena have approximately normal distributions. There are actually many normal distributions, some tall and thin, others short and wide, as shown in Figure 10.8, but the area beneath the curve is always equal to 1. Each normal distribution is determined by its mean and standard deviation.

To distinguish a theoretical distribution from the frequency distributions we studied earlier, we use the Greek letters μ (mu) for the mean of a theoretical distribution and σ (sigma) for the standard deviation of a theoretical distribution.

To simplify work with normal distributions, the normal distribution with $\mu = 0$ and $\sigma = 1$ is called the **standard normal distribution.** Since the total area under a normal curve is equal to 1, parts of the area can be used to determine probabilities. In Figure 10.9, the shaded area under the curve from a to b represents the probability that an observed data value will be between a and b. The areas under portions of the standard normal curve have been calculated, and can be found by consulting Table 12 in the Appendix.

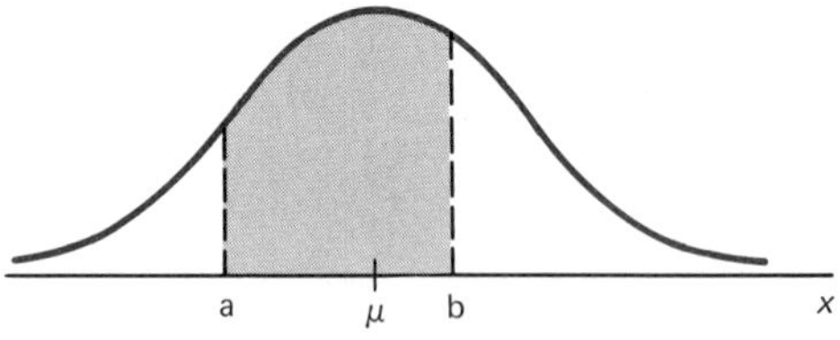

Figure 10.9

Example 1 Use Table 12 to find the fraction of all the scores in a standard normal distribution that lie between the mean and

(a) 1 standard deviation above the mean;

We use z in Table 12 to represent the number of standard deviations from the mean, written as a decimal to the nearest hundredth. Here $z = 1.00$. Find 1.00 in the z column; you should find .3413 next to it. This entry means that .3413 or 34.13% of all values lie between the mean and one standard deviation above the mean.

In other words, the shaded area of Figure 10.10 represents 34.13% of the total area under the normal curve.

(b) 2.43 standard deviations below the mean.

Here, $z = -2.43$. Since we are interested only in area, we look for 2.43 in the z column of Table 12. This should lead you to the entry .4925. A total of .4925 or 49.25% of all values lie between the mean and 2.43 standard deviations below the mean. This region is shaded in Figure 10.11. ■

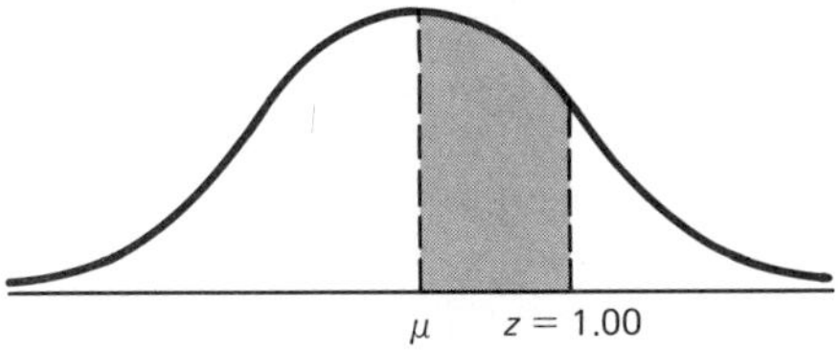

Figure 10.10

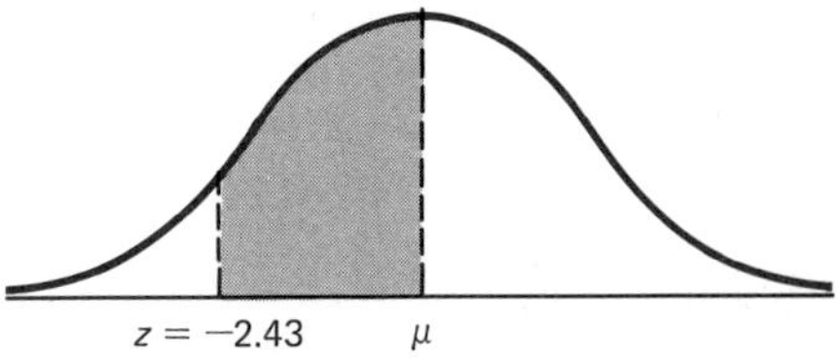

Figure 10.11

1. Find the percent of area between the mean and
(a) 1.51 standard deviations above the mean;
(b) 2.04 standard deviations below the mean.

Answer:
(a) 43.45%
(b) 47.93%

Work Problem 1 at the side.

Example 2 Find the following areas under the standard normal curve.

(a) the area between 1.41 standard deviations *below* the mean and 2.25 standard deviations *above* the mean

First, draw a sketch showing the desired area, as in Figure 10.12. By using Table 12, the area from the mean to 1.41 standard deviations below the mean is .4207. Also, the area from the mean to 2.25 standard deviations above the mean is .4878. As our sketch shows, the total desired area can be found by *adding* these numbers.

$$\begin{array}{r} .4207 \\ +.4878 \\ \hline .9085 \end{array}$$

The total shaded area in Figure 10.12 represents 90.85% of the total area under the normal curve.

(b) the area between .58 standard deviations above the mean and 1.94 standard deviations above the mean

Figure 10.13 shows the desired area. The area between the mean and .58 standard deviations above the mean is .2190. The area between the mean and 1.94 standard deviations above the mean is .4738. As our sketch shows, the desired area is found by *subtracting* the two areas.

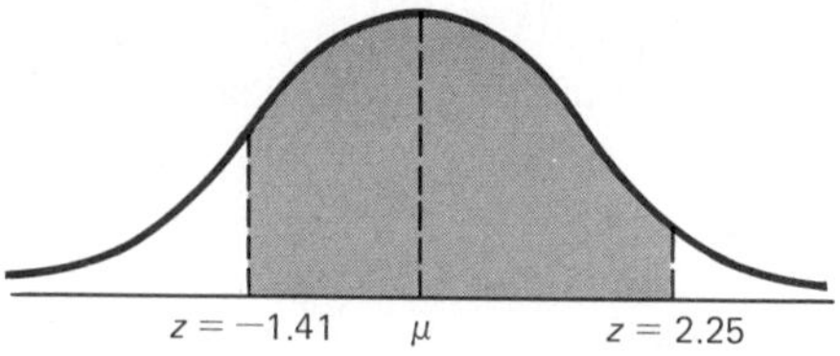

Figure 10.12

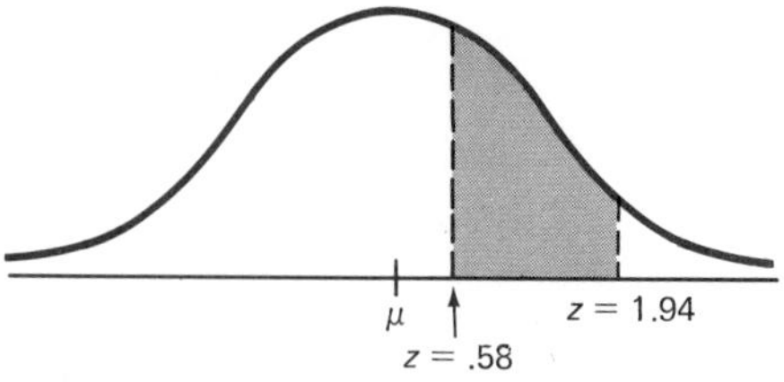

Figure 10.13

$$\begin{array}{r} .4738 \\ -.2190 \\ \hline .2548 \end{array}$$

The shaded area of Figure 10.13 represents 25.48% of the total area under the normal curve.

(c) the area to the right of 2.09 standard deviations above the mean

The total area under a normal curve is 1. Thus, the total area to the right of the mean is 1/2, or .5000. From Table 12, the area from the mean to 2.09 standard deviations above the mean is .4817. The area we want, to the right of 2.09 standard deviations, is found by subtracting .4817 from .5000.

$$\begin{array}{r} .5000 \\ -.4817 \\ \hline .0183 \end{array}$$

A total of 1.83% of the total area is to the right of 2.09 standard deviations above the mean. Figure 10.14 shows the desired area. ■

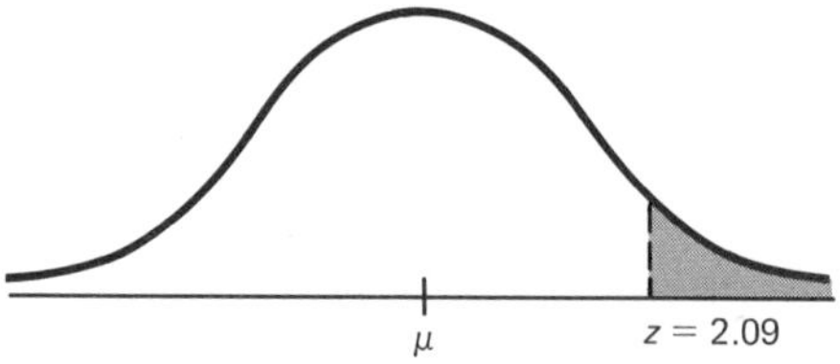

Figure 10.14

Work Problem 2 at the side.

2. Find the following normal curve areas.

(a) between .31 standard deviations below the mean and 1.01 standard deviations above the mean

(b) between .38 and 1.98 standard deviations below the mean

(c) to the right of 1.49 standard deviations above the mean

Answer:

(a) 46.55%

(b) 32.82%

(c) 6.81%

We can use the standard normal distribution to find probabilities for other normal distributions by converting given x values to comparable values, called ***z*-scores,** in the standard normal distribution. An x value from a normal distribution is converted to a z-score for the standard normal distribution by means of the formula

$$z = \frac{x - \mu}{\sigma},$$

where μ and σ are the mean and the standard deviation respectively of the given distribution.

Example 3 Suppose that the average salesperson for Dixie Office Supplies drives $\mu = 1200$ miles per month in a company car, with standard deviation $\sigma = 150$ miles. Assume that the number of miles is closely approximated by a normal curve. Find the percent of all drivers traveling

(a) between 1200 and 1600 miles per month;

We first need to find the number of standard deviations that 1600 miles is above the mean. Use the z-score formula given above.

$$\begin{aligned} z &= \frac{x - \mu}{\sigma} \\ &= \frac{1600 - 1200}{150} \qquad \text{Let } x = 1600,\ \mu = 1200,\ \sigma = 150 \\ &= \frac{400}{150} \\ z &= 2.67 \end{aligned}$$

(Since $\mu = 1200$, the z-score for 1200 is 0.) From Table 12, .4962, or 49.62% of all drivers travel between 1200 and 1600 miles per month. See Figure 10.15.

(b) between 1000 and 1500 miles per month.

As shown in Figure 10.16, we need to find the z-score for both $x = 1000$ and $x = 1500$.

For 1000,

$$z = \frac{1000 - 1200}{150} = \frac{-200}{150} = -1.33.$$

For 1500,

$$z = \frac{1500 - 1200}{150} = \frac{300}{150} = 2.00.$$

From the table, $z = 1.33$ leads to an area of .4082, while $z = 2.00$

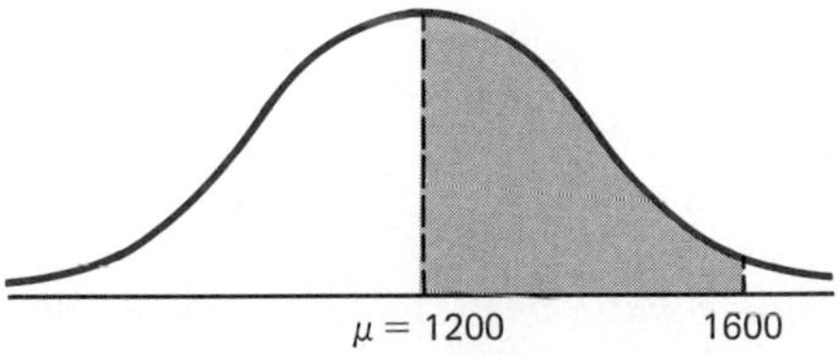

Figure 10.15

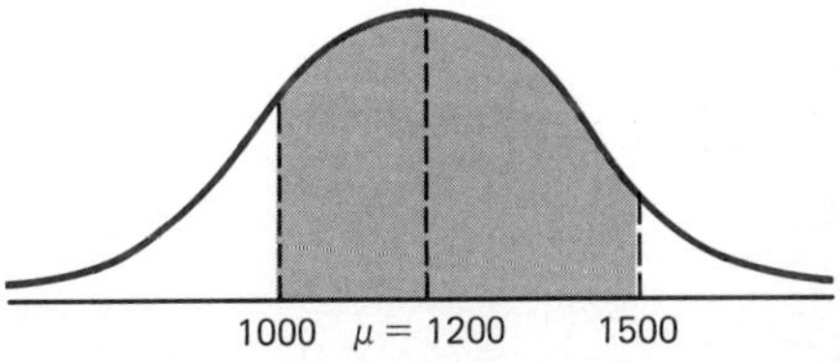

Figure 10.16

gives .4773. Thus, a total of $.4082 + .4773 = .8855$, or 88.55%, of all drivers travel between 1000 and 1500 miles per month. ■

3. The heights of female sophomore college students at one school have $\mu = 172$ centimeters, with $\sigma = 10$ centimeters. Find the percent of the female students who are
(a) between 172 cm and 185 cm tall;
(b) between 160 cm and 180 cm tall;
(c) less than 165 cm tall.

Answer:
(a) 40.32%
(b) 67.30%
(c) 24.20%

Work Problem 3 at the side.

10.4 EXERCISES

Find the percent of area under a normal curve between the mean and the following number of standard deviations from the mean. (See Example 2.)

1. 2.50 **2.** 1.68 **3.** 0.45 **4.** 0.81

5. -1.71 **6.** -2.04 **7.** 3.11 **8.** 2.80

Find the percent of the total area under the normal curve between the following z-scores.

9. $z = 1.41$ and $z = 2.83$ **10.** $z = 0.64$ and $z = 2.11$

11. $z = -2.48$ and $z = -0.05$ **12.** $z = -1.74$ and $z = -1.02$

13. $z = -3.11$ and $z = 1.44$ **14.** $z = -2.94$ and $z = -0.43$

15. $z = -0.42$ and $z = 0.42$ **16.** $z = -1.98$ and $z = 1.98$

Find a z-score satisfying each of the following conditions. (Hint: Use Table 12 backwards.)

17. 5% of the total area is to the right of z.

18. 1% of the total area is to the left of z.

19. 15% of the total area is to the left of z.

20. 25% of the total area is to the right of z.

APPLIED PROBLEMS

Management

A certain type of light bulb has an average life of 500 hours, with a standard deviation of 100 hours. The length of life of the bulb can be closely approximated by a normal curve. An amusement park buys and installs 10,000 such bulbs. Find the total number that can be expected to last

21. at least 500 hours; **22.** less than 500 hours;

23. between 500 and 650 hours; **24.** between 300 and 500 hours;

25. between 650 and 780 hours; **26.** between 290 and 540 hours;

27. less than 740 hours; **28.** more than 300 hours;

29. more than 790 hours; **30.** less than 410 hours.

A box of oatmeal must contain 16 ounces. The machine that fills the oatmeal boxes is set so that, on the average, a box contains 16.5 ounces. The boxes filled by the machine have weights that can be closely approximated by a normal curve. What fraction of the boxes filled by the machine are underweight if the standard deviation is

31. .5 ounce; **32.** .3 ounce; **33.** .2 ounce; **34.** .1 ounce?

Natural Science

The chickens at Colonel Thompson's Ranch have a mean weight of 1850 grams with a standard deviation of 150 grams. The weights of the chickens are closely approximated by a normal curve. Find the percent of all chickens having weights

35. more than 1700 grams;

36. less than 1800 grams;

37. between 1750 grams and 1900 grams;

38. between 1600 grams and 2000 grams;

39. less than 1550 grams;

40. more than 2100 grams.

In nutrition, the Recommended Daily Allowance of vitamins is a number set by the government as a guide to an individual's daily vitamin intake. Actually, vitamin needs vary drastically from person to person, but the needs are very closely approximated by a normal curve. To calculate the Recommended Daily Allowance, the government first finds the average need for vitamins among people in the population, and the standard deviation. The Recommended Daily Allowance is then defined as the mean plus 2.5 times the standard deviation.

41. What percentage of the population will receive adequate amounts of vitamins under this plan?

Find the recommended daily allowance for the following vitamins.

42. mean = 1800 units, standard deviation = 140 units

43. mean = 159 units, standard deviation = 12 units

44. mean = 1200 units, standard deviation = 92 units

Assume the following distributions are all normal, and use the areas under the normal curve given in Table 12 to answer the questions.

Management

45. A machine produces bolts with an average diameter of .25 inches and a standard deviation of .02 inches. What is the probability that a bolt will be produced with a diameter greater than .3 inches?

46. The mean monthly income of the trainees of an engineering firm is \$600 with a standard deviation of \$100. Find the probability that an individual trainee earns less than \$500 per month.

47. A machine which fills quart milk cartons is set up to average 32.2 ounces per carton, with a standard deviation of 1.2 ounces. What is the probability that a filled carton will contain less than 32 ounces of milk?

Social Science

48. The average contribution to the campaign of Polly Potter, a candidate for city council, was \$50 with a standard deviation of \$15. How many of the 200 people who contributed to Polly's campaign gave between \$30 and \$100?

49. At the Discount Market, the average weekly grocery bill is \$32.25 with a standard deviation of \$9.50. What are the largest and smallest amounts spent by the middle 50% of this market's customers?

Natural Science

50. The mean clotting time of blood is 7.45 seconds with a standard deviation of 3.6 seconds. What is the probability that an individual's blood clotting time will be less than 7 seconds or greater than 8 seconds?

51. The average size of the fish in Lake Amotan is 12.3 inches with a standard deviation of 4.1 inches. Find the probability of catching a fish there which is longer than 18 inches.

52. To be graded extra large, an egg must weigh at least 2.2 ounces. If the average weight for an egg is 1.5 ounces with a standard deviation of .4 ounces, how many of five dozen eggs would you expect to grade extra large?

Social Science

A teacher gives a test to a large group of students. The results are closely approximated by a normal curve. The mean is 74, with a standard deviation of 6. The teacher wishes to give A's to the top 8% of the students and F's to the bottom 8%. A grade of B is given to the next 15%, with D's given similarly. All other students get C's. Find the bottom cutoff (rounded to the nearest whole number) for the following grades. (Hint: use the table in the Appendix backwards.)

53. A   54. B   55. C   56. D

10.5 THE BINOMIAL DISTRIBUTION

A **binomial distribution** is an example of a discrete distribution which has only two possible outcomes (such as *heads* or *tails*, or *success* or *failure*). Repetitions of the experiment are independent of each other, and the probability of a desired result never changes. We can obtain examples of binomial distributions by tossing a coin, rolling a 5 on a die, or selecting a defective radio from a batch produced by a factory.

As an example, let us consider an experiment in which a die is tossed 5 times. Each time that the toss results in a 1 or a 2, the toss is considered a success; otherwise it is a failure. How many successes are possible in such an experiment? Since each trial can result in either success or failure, it is possible to have from 0 to 5 successes. We now wish to consider the probability of each of these six outcomes. From Theorem 8.3 for independent repeated trials, we see that the required probabilities are given by

$$P(x) = \binom{n}{x} p^x(1 - p)^{n-x}$$

where n is the number of trials, x is the number of successes, and p is the probability of success in a single trial. In this example, $n = 5$ and $p = 1/3$. The results are tabulated at the side. In a binomial distribution, the probabilities can always be calculated by the repeated trials formula.

x	$P(x)$
0	$\binom{5}{0}\left(\frac{1}{3}\right)^0\left(\frac{2}{3}\right)^5 = \frac{32}{243}$
1	$\binom{5}{1}\left(\frac{1}{3}\right)^1\left(\frac{2}{3}\right)^4 = \frac{80}{243}$
2	$\binom{5}{2}\left(\frac{1}{3}\right)^2\left(\frac{2}{3}\right)^3 = \frac{80}{243}$
3	$\binom{5}{3}\left(\frac{1}{3}\right)^3\left(\frac{2}{3}\right)^2 = \frac{40}{243}$
4	$\binom{5}{4}\left(\frac{1}{3}\right)^4\left(\frac{2}{3}\right)^1 = \frac{10}{243}$
5	$\binom{5}{5}\left(\frac{1}{3}\right)^5\left(\frac{2}{3}\right)^0 = \frac{1}{243}$

By definition, the mean μ of a theoretical distribution is given by the expected value of x. Expected value is found by finding the products of x and $P(x)$, as follows.

$$\mu = 0\left(\frac{32}{243}\right) + 1\left(\frac{80}{243}\right) + 2\left(\frac{80}{243}\right) + 3\left(\frac{40}{243}\right) + 4\left(\frac{10}{243}\right) + 5\left(\frac{1}{243}\right)$$
$$= \frac{405}{243} = 1\frac{2}{3}$$

For a binomial distribution, which is a special kind of theoretical distribution, the method for finding the mean reduces to the formula

$$\boldsymbol{\mu = np},$$

where n is the number of trials and p is the probability of success for a single trial. Using this simplified formula, the computation of the mean in the example above is

$$\mu = np = 5(1/3) = 1\frac{2}{3},$$

which agrees with the result we obtained using the expected value.

Like the mean, the variance, σ^2, of a theoretical probability distribution is an expected value—the expected value of the squared deviations from the mean, $(x - \mu)^2$. To find the variance for the example given above, we first find the quantities $(x - \mu)^2$. (Recall, the mean μ is 5/3.)

x	$P(x)$	$x - \mu$	$(x - \mu)^2$
0	$\frac{32}{243}$	$-\frac{5}{3}$	$\frac{25}{9}$
1	$\frac{80}{243}$	$-\frac{2}{3}$	$\frac{4}{9}$
2	$\frac{80}{243}$	$\frac{1}{3}$	$\frac{1}{9}$
3	$\frac{40}{243}$	$\frac{4}{3}$	$\frac{16}{9}$
4	$\frac{10}{243}$	$\frac{7}{3}$	$\frac{49}{9}$
5	$\frac{1}{243}$	$\frac{10}{3}$	$\frac{100}{9}$

We now find σ^2 by adding the products $[(x - \mu)^2][P(x)]$.

$$\sigma^2 = \frac{25}{9}\left(\frac{32}{243}\right) + \frac{4}{9}\left(\frac{80}{243}\right) + \frac{1}{9}\left(\frac{80}{243}\right) + \frac{16}{9}\left(\frac{40}{243}\right) + \frac{49}{9}\left(\frac{10}{243}\right) + \frac{100}{9}\left(\frac{1}{243}\right)$$
$$= \frac{10}{9} = 1\frac{1}{9}$$

If we wish to find the standard deviation σ, we evaluate $\sqrt{10/9}$ to get $\frac{1}{3}\sqrt{10}$, or approximately 1.05.

As one might expect, the formula for the variance, like that for the mean, is greatly simplified for a binomial distribution. The complicated procedure for finding σ^2 above reduces to the formula

$$\boldsymbol{\sigma^2 = np(1 - p), \qquad \text{or} \qquad \sigma = \sqrt{np(1 - p)}.}$$

By substituting the appropriate values for n and p from the example into this new formula, we have

$$\sigma^2 = 5(1/3)(2/3) = 10/9 = 1\frac{1}{9},$$

which agrees with our previous result.

Summarizing, we see that a theoretical distribution is one in which probabilities are used in place of frequencies. A binomial distribution is a theoretical probability distribution where the experiment is a series of n independent repeated trials in which we observe the number of successes, x, and where the probability of a success in a single trial is always p. The probability that exactly x successes will occur is found by the formula $\binom{n}{x}p^x(1-p)^{n-x}$. The mean and variance of a binomial distribution are respectively $\mu = np$ and $\sigma^2 = np(1-p)$. The standard deviation is $\sigma = \sqrt{np(1-p)}$.

Example 1 The probability that a plate picked at random from the assembly line in a china factory will be defective is .01. A sample of three is to be selected. Write the distribution for the number of defective plates in the sample, and give its mean and standard deviation.

Since three plates will be selected, the possible number of defective plates ranges from 0 to 3. Here, n (the number of trials) is 3; p (the probability of selecting a defective on a single trial) is .01. The distribution and the probability of each outcome are shown below.

x	$P(x)$
0	$\binom{3}{0}(.01)^0(.99)^3 = .970$
1	$\binom{3}{1}(.01)(.99)^2 = .029$
2	$\binom{3}{2}(.01)^2(.99) = .0003$
3	$\binom{3}{3}(.01)^3(.99)^0 = .000001$

The mean of the distribution is

$$\mu = np = 3(.01) = .03.$$

The standard deviation is

$$\sigma = \sqrt{np(1-p)} = \sqrt{3(.01)(.99)} = \sqrt{.0297} = .17. \quad \blacksquare$$

Work Problem 1 at the side.

The methods above are too complicated for most practical problems. For this reason, practical problems are usually worked with the *normal curve approximation to the binomial distribution.*

1. The probability that a can of Budweiser from a certain plant is defective is .005. A sample of 4 cans is selected at random. Write a distribution for the number of defective cans in the sample, and give its mean and standard deviation.

Answer: possible number of defective cans: 0, 1, 2, 3, or 4; $\mu = .02$, $\sigma = .14$

x	$P(x)$
0	.961
1	.020
2	.015
3	.00002
4	.00000000

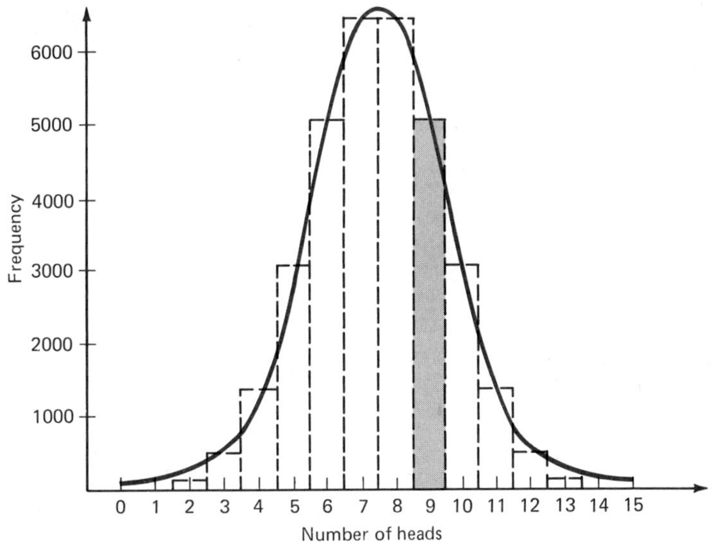

Figure 10.17

To see how the normal curve is used, look at the bar graph and normal curve in Figure 10.17. This bar graph shows the expected results if one coin is tossed 15 times, with the experiment repeated 32,768 times.

Suppose we need to know the fraction of the time that we would get exactly 9 heads in the 15 tosses. We could work this out using the methods above. After performing a huge amount of arithmetic, we would get .153. This answer is the same fraction we would get by dividing the area of the shaded bar in Figure 10.17 by the total area of all 16 bars in the graph. (Note: Some of the bars at the extreme left and right ends of the graph are too short to show up.)

The area of the shaded bar is also approximately equal to the area under the normal curve from $x = 8.5$ to $x = 9.5$. The normal curve runs higher than the top of the bar in the left half, but lower in the right half.

To find the area under the normal curve from $x = 8.5$ to $x = 9.5$, we need to find z-scores, as we did in the last section. To do so, we find the mean and the standard deviation for the distribution by using the formulas

$$\mu = np \qquad \text{and} \qquad \sigma = \sqrt{np(1-p)}.$$

Using $n = 15$ and $p = 1/2$ in the formulas, we have

$$\begin{aligned} \mu &= 15 \cdot 1/2 & \sigma &= \sqrt{15 \cdot 1/2 \cdot (1 - 1/2)} \\ &= 15/2 & &= \sqrt{15 \cdot 1/2 \cdot 1/2} \\ \mu &= 7.5 & &= \sqrt{3.75} \\ & & \sigma &= 1.94 \end{aligned}$$

Using these results we can find z-scores for $x = 8.5$ and $x = 9.5$.

For $x = 8.5$

$$z = \frac{8.5 - 7.5}{1.94} = \frac{1.00}{1.94}$$

$$z = .52$$

For $x = 9.5$

$$z = \frac{9.5 - 7.5}{1.94} = \frac{2.00}{1.94}$$

$$z = 1.03$$

From Table 12, $z = .52$ gives an area of .1985, while $z = 1.03$ gives .3485. To find the result we need, we subtract these two numbers.

$$.3485 - .1985 = .1500$$

This answer (.1500) is not far from the exact answer, .153, that we found above.

2. Find the probability of getting exactly the following number of heads in 15 tosses of a coin.
(a) 7
(b) 10

Answer:
(a) .1985
(b) .0909

Work Problem 2 at the side.

Example 2 About 6% of the bolts produced by a certain machine are defective.
(a) Find the probability that in a sample of 100 bolts, 3 or fewer are defective.

Notice that this problem satisfies the conditions of the definition of a binomial distribution. For this reason, we can use the normal curve approximation. We must first find the mean and the standard deviation. Here $n = 100$ and $p = 6\% = .06$. Thus

$$\mu = 100(.06) \qquad \sigma = \sqrt{100(.06)(1 - .06)}$$

$$\mu = 6. \qquad = \sqrt{100(.06)(.94)}$$

$$= \sqrt{5.64}$$

$$\sigma = 2.37$$

As the graph of Figure 10.18 shows, we need to find the area to the left of $x = 3.5$ (since we want 3 or fewer defective bolts). We have

$$z = \frac{3.5 - 6}{2.37} = \frac{-2.5}{2.37} = -1.05.$$

From the table, $z = -1.05$ leads to an area of .3531. Finally, we get the result we need by subtracting .3531 from .5000 to get .1469. The probability that we will get 3 or fewer defective bolts in a set of 100 bolts is .1469, or 14.69%.
(b) Find the probability of getting exactly 11 defective bolts in a sample of 100 bolts.

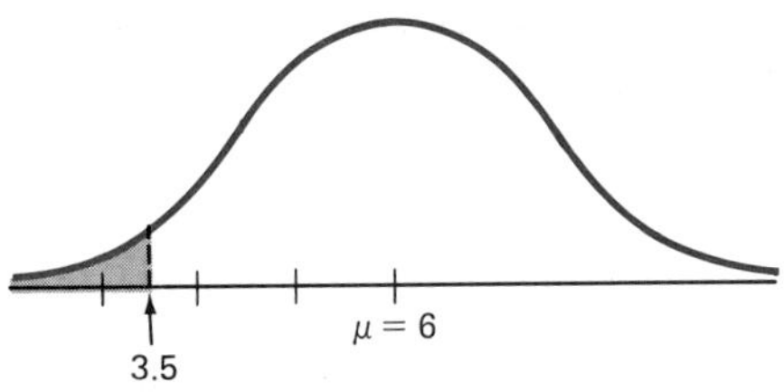

Figure 10.18

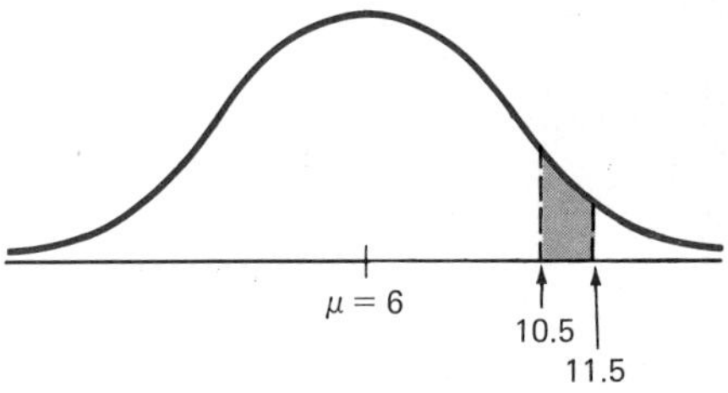

Figure 10.19

As Figure 10.19 shows, we need the area between $x = 10.5$ and $x = 11.5$.

$$\text{If } x = 10.5, \text{ then } z = \frac{10.5 - 6}{2.37} = 1.90.$$

$$\text{If } x = 11.5, \text{ then } z = \frac{11.5 - 6}{2.37} = 2.32.$$

Look in Table 12; $z = 1.90$ gives an area of .4713, while $z = 2.32$ yields .4898. The final answer is the difference of these numbers, or

$$.4898 - .4713 = .0185.$$

There is about a 1.85% chance of having exactly 11 defective bolts. ■

Work Problem 3 at the side.

3. About 9% of the transistors produced by a certain factory are defective. Find the probability that in a sample of 200, we will find the following numbers of defective transistors. (Hint: $\sigma = 4.05$)
(a) exactly 11
(b) 16 or fewer
(c) more than 14

Answer:
(a) .0226
(b) .3557
(c) .8051

The normal curve approximation to the binomial distribution is usually quite accurate, especially for practical problems. As a rule of thumb, the normal curve approximation can be used as long as both np and $n(1 - p)$ are at least 5.

10.5 EXERCISES

In Exercises 1–6, several binomial experiments are described. For each one, answer the questions presented by
(a) giving the distribution;
(b) finding the mean;
(c) finding the standard deviation.

1. A die is rolled six times and the number of 1's that come up is tallied. Write the distribution of 1's that can be expected to occur.

2. A 6-item multiple choice test has four possible answers for each item. If a student selects all his answers randomly, how many can he expect to answer correctly? With what standard deviation?

3. To maintain quality control on the production line, the Bright Lite Company randomly selects three light bulbs each day for testing. Experience has shown a defective rate of .02. Write the distribution for the number of defectives in the daily samples.

4. In a taste test, each member of a panel of four is given two glasses of Super Cola, one made using the old formula and one with the new formula, and asked to identify the new formula. Assuming the panelists operate independently, write the distribution of the number of successful identifications, if each judge actually guesses.

5. The probability that a radish seed will germinate is .7. Joe's mother gives him four seeds to plant. Write the distribution for the number of seeds which he can expect to germinate.

6. Five patients in Ward 8 of Memorial Hospital have a disease with a known mortality rate of .1. Write the distribution of the number who can be expected to survive.

Work the following problems involving binomial experiments.

7. The probability that an infant will die in the first year of life is about .025. In a group of 500 babies, what are the mean and standard deviation of the number of babies who can be expected to die in their first year of life?

8. The probability that a particular kind of mouse will have a brown coat is 1/4. In a litter of 8, assuming independence, how many could be expected to have a brown coat? With what standard deviation?

9. A certain drug is effective 80% of the time. Give the mean and standard deviation of the number of patients using the drug who recover, out of a group of 64 patients.

10. The probability that a newborn infant will be a girl is .49. If 50 infants are born on Susan B. Anthony's birthday, how many can be expected to be girls? With what standard deviation?

For the remaining exercises, use the normal curve approximation to the binomial distribution.
Suppose 16 *coins are tossed. Find the probability of getting exactly*

11. 8 heads; **12.** 7 heads; **13.** 10 tails; **14.** 12 tails.

Suppose 1000 *coins are tossed. Find the probability of getting each of the following. (Hint:* $\sqrt{250} = 15.8$*)*

15. exactly 500 heads

16. exactly 510 heads

17. 480 heads or more

18. less than 470 tails

19. less than 518 heads

20. more than 550 tails

A die is tossed 120 *times. Find the probability of getting each of the following. (Hint:* $\sigma = 4.08$*)*

21. exactly 20 fives

22. exactly 24 sixes

23. exactly 17 threes

24. exactly 22 twos

25. more than 18 threes

26. fewer than 22 sixes

Two percent of the hamburgers sold at Tom's Burger Queen are defective. Tom sold 10,000 *burgers last week. Find the probability that among these burgers*

27. fewer than 170 were defective;

28. more than 222 were defective.

A new drug cures 80% *of the patients to whom it is administered. It is given to* 25 *patients. Find the probability that among these patients*

29. exactly 20 are cured;

30. exactly 23 are cured;

31. all are cured;

32. no one is cured;

33. 12 or fewer are cured;

34. between 17 and 23 are cured.

10.6 CURVE FITTING—THE LEAST SQUARES METHOD

To produce a mathematical model, we frequently must work with data; we want to obtain an equation which describes the data. The simplest equation is a linear equation, so we usually start by trying to fit a straight line through the data points. Figure 10.20 shows several data points, and a straight line which fits through the points fairly well.

How do we decide on the "best" possible line? The line that has proven to be "best" in many different applications is the one

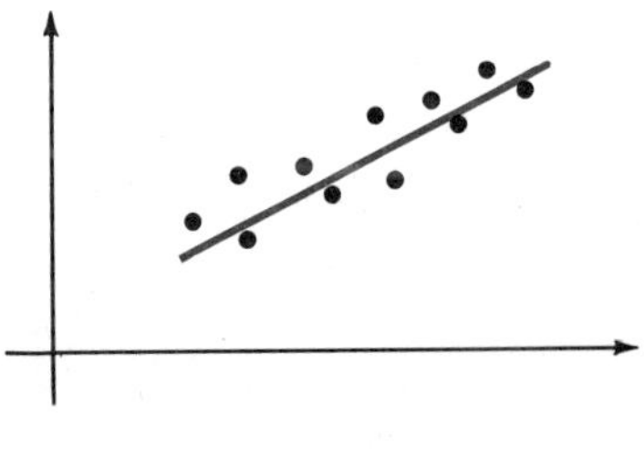

Figure 10.20

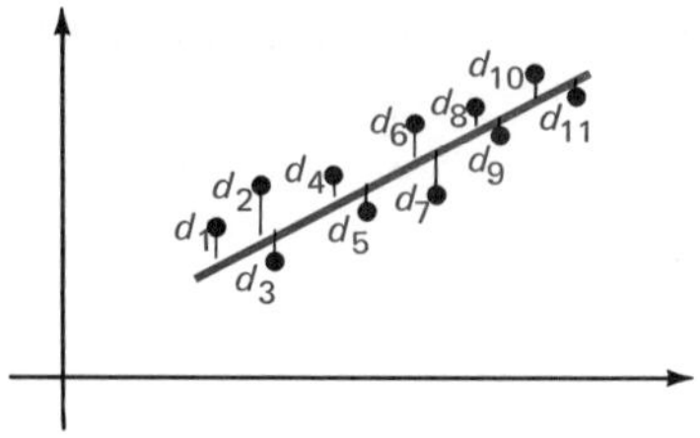

Figure 10.21

in which the sum of the squares of the distances from the data points to the line is as small as possible. We call such a line the **least squares line.** In Figure 10.21, the distances from the data points to the line are represented by d_1, d_2, d_3, and so on. Then the least squares line is found by minimizing the sum $(d_1)^2 + (d_2)^2 + (d_3)^2 + \cdots + (d_n)^2$.

The method of finding the equation of the least squares line requires calculus, and will not be given here; we only give the results. We use y' instead of y to distinguish these predicted values from the y-values of the given pairs. The **least squares line** $y' = mx + b$ which provides the best fit to the data points (x_1, y_1), (x_2, y_2), . . . , (x_n, y_n) has slope

$$m = \frac{n(\sum xy) - (\sum x)(\sum y)}{n(\sum x^2) - (\sum x)^2}$$

and y-intercept

$$b = \frac{\sum y - m(\sum x)}{n}.$$

Example 1 A college registrar wants to see if college grade-point averages and high school grade-point averages can be closely approximated by a linear function. To begin, the registrar chooses 10 students at random and finds their college and high school grade-point averages, with the results shown in the following table. Find the least squares line which relates these averages.

Student	1	2	3	4	5	6	7	8	9	10
High school GPA. x	2.5	2.8	2.9	3.0	3.2	3.3	3.3	3.4	3.6	3.9
College GPA. y	2.0	2.5	2.5	2.8	3.0	3.0	3.5	3.3	3.4	3.6

A graph of the ten pairs of averages is shown in Figure 10.22. This graph is called a **scatter diagram.** In order to use the formulas above to get m and b, we need $\sum x$, $\sum y$, $\sum xy$, and $\sum x^2$. Find these sums as shown below. (The column headed y^2 will be used later.)

x	y	xy	x^2	y^2
2.5	2.0	5.00	6.25	4.00
2.8	2.5	7.00	7.84	6.25
2.9	2.5	7.25	8.41	6.25
3.0	2.8	8.40	9.00	7.84
3.2	3.0	9.60	10.24	9.00
3.3	3.0	9.90	10.89	9.00
3.3	3.5	11.55	10.89	12.25
3.4	3.3	11.22	11.56	10.89
3.6	3.4	12.24	12.96	11.56
3.9	3.6	14.04	15.21	12.96
31.9	29.6	96.20	103.25	90.00

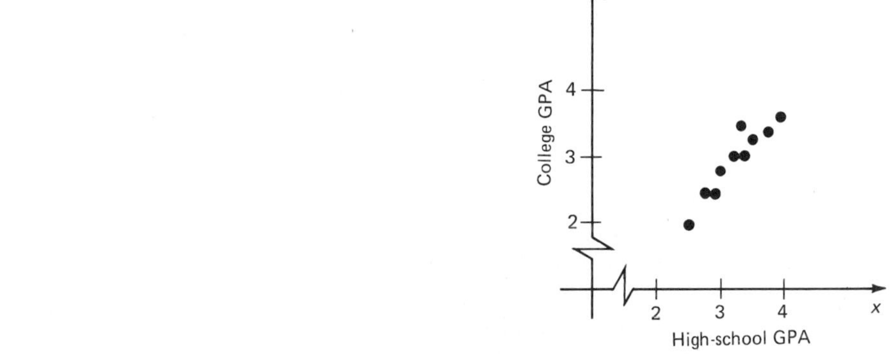

Figure 10.22

We now compute m and b as shown below.

$$m = \frac{10(96.2) - (31.9)(29.6)}{10(103.25) - (31.9)^2} = 1.19$$

$$b = \frac{29.6 - 1.19(31.9)}{10} = -.84$$

The least squares line is $y' = mx + b$, or

$$y' = 1.19x - .84.$$

This equation can be used to predict y from a given value of x. See Example 2. ■

1. Find the least squares line for the points (3, 4), (5, 11), (7, 20), and (8, 23).

Answer: $y' = 3.90x - 7.92$

Work Problem 1 at the side.

Example 2 In Example 1, we ended up with the least squares line $y' = 1.19x - .84$, when x represents high school grade-point averages and y' represents predicted college averages. Use this equation to answer the following question.

Lupe Renoso had a grade-point average of 3.5 in high school. Predict her college average to the nearest tenth.

Let $x = 3.5$ in our equation $y' = 1.19x - .84$.

$$\begin{aligned} y' &= 1.19(3.5) - .84 \\ &= 4.165 - .84 \\ y' &= 3.325 \\ &\approx 3.3 \end{aligned}$$

We would predict that Renoso will have a college grade-point average of about 3.3. (In more advanced courses, you will see methods for deciding how much faith to place in this prediction.) ■

2. In Problem 1 we found that $y' = 3.90x - 7.92$ is the equation of the line that best fits through the given data points. Use this equation to predict y when x is 6.

Answer: 15.48

Work Problem 2 at the side.

Once we find an equation for the line of best fit (the least squares line), we might well ask, "Just how good is this line for prediction purposes?" If the points already observed fit the line quite closely, then we can expect future pairs of scores to do so. If the points are widely scattered about even the "best fitting" line, then predictions are not likely to be accurate.

In order to have a quantitative basis for confidence in our predictions, we need a measure of the "goodness of fit" of our original data to the prediction line. One such measure is called the **coefficient of correlation,** denoted r. We can calculate r by using the following formula.

$$r = \frac{n(\sum xy) - (\sum x)(\sum y)}{\sqrt{n(\sum x^2) - (\sum x)^2} \cdot \sqrt{n(\sum y^2) - (\sum y)^2}}$$

The coefficient of correlation, r, is always equal to or between 1 and -1. Values of exactly 1 or -1 indicate that the data points lie *exactly* on the least squares line. If $r = 1$, the least squares line has a positive slope; $r = -1$ gives a negative slope. If $r = 0$, there is no linear correlation between the data points. (However, some other nonlinear function might provide an excellent fit for the data.) Typical values of r are shown in Figure 10.23.

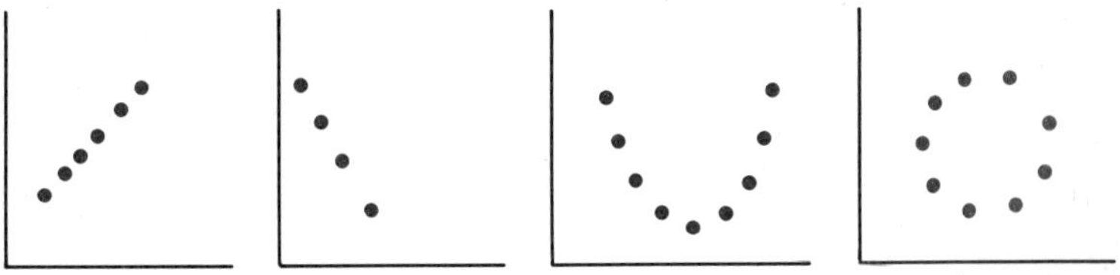

Figure 10.23

Example 3 Find r for the data of Example 1.

In Example 1, we found that $\sum x = 31.9$, $\sum y = 29.6$, $\sum xy = 96.20$, $\sum x^2 = 103.25$, and $\sum y^2 = 90.00$. To find the coefficient of correlation between the college grade-point averages and the high school grade-point averages given above, we substitute these values into the formula for r.

$$r = \frac{n(\sum xy) - (\sum x)(\sum y)}{\sqrt{n(\sum x^2) - (\sum x)^2} \cdot \sqrt{n(\sum y^2) - (\sum y)^2}}$$

$$= \frac{10(96.2) - (31.9)(29.6)}{\sqrt{10(103.25) - (31.9)^2} \cdot \sqrt{10(90) - (29.6)^2}}$$

$$r = \frac{17.76}{\sqrt{14.89} \cdot \sqrt{23.84}} = .94$$

As discussed above, the fact that the coefficient of correlation is close to 1 indicates that there is a close relationship between the two grade-point averages. ■

Work Problem 3 at the side.

Examples of curve fitting and correlation are shown in Figure 10.24.

3. In Problem 1 of this section, you should have found that $n = 4$, $\sum x = 23$, $\sum y = 58$, $\sum xy = 391$, $\sum x^2 = 147$, and $\sum y^2 = 1066$. Use these values to find r.

Answer: $r = .998$

The linear relation of human weight to climatic temperature (after D. F. Roberts); $r = -.60$

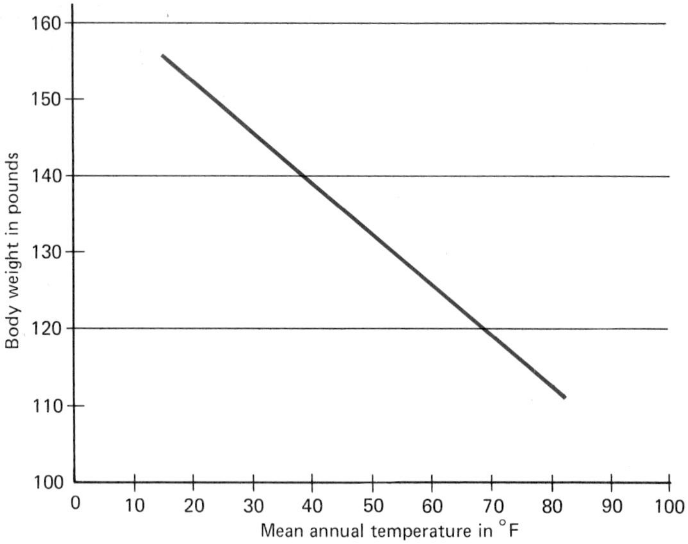

Figure 10.24 (a)

Logarithmic relation between land and rainfall in aboriginal Australia; $r = .8$

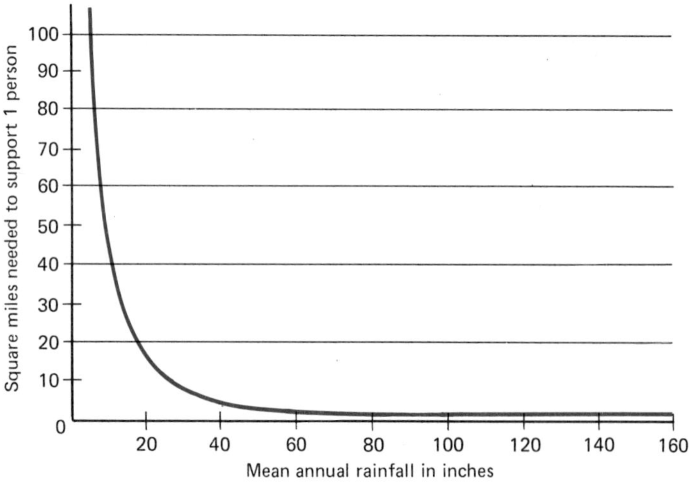

Figure 10.24 (b)

10.6 APPLIED PROBLEMS

Natural Science

1. In a study to determine the linear relationship between the size (in decimeters) of an ear of corn (y) and the amount (in tons per acre) of fertilizer used (x), the following data were collected.

$$n = 10 \qquad \sum xy = 75$$
$$\sum x = 30 \qquad \sum x^2 = 100$$
$$\sum y = 24 \qquad \sum y^2 = 80$$

(a) Find an equation for the least squares line.
(b) Find the coefficient of correlation.
(c) If 3 tons per acre of fertilizer are used, what length (in decimeters) would the equation in (a) predict for an ear of corn?

2. In an experiment to determine the linear relationship between temperatures on the Celsius scale (y) and on the Fahrenheit scale (x), a student got the following results.

$$n = 5 \qquad \sum xy = 28{,}050$$
$$\sum x = 376 \qquad \sum x^2 = 62{,}522$$
$$\sum y = 120 \qquad \sum y^2 = 13{,}450$$

(a) Find an equation for the least squares line.
(b) Find the reading on the Celsius scale that corresponds to a reading of 120° Fahrenheit, using the equation of part (a).
(c) Find the coefficient of correlation.

3. A sample of 10 adult men gave the following data on their heights and weights.

Height (inches)	(x)	62	62	63	65	66	67	68	68	70	72
Weight (pounds)	(y)	120	140	130	150	142	130	135	175	149	168

(a) Find the equation of the least squares line.
Using the results of (a), predict the weight of a man whose height is
(b) 60 inches;
(c) 70 inches.
(d) Compute the coefficient of correlation.

4. (For those who have studied common logarithms) Sometimes the scatter diagram of the data does not have a linear pattern. This is particularly true in biological applications. In these applications, however, often the scatter diagram of the *logarithms* of the data has a linear pattern. A least squares line then can be used to predict the logarithm of any desired value from which the value itself can be found. Suppose that a certain kind of bacterium grows in number as shown in Table A. The actual number of bacteria present at each time period is replaced with the common logarithm of that number (Table B).

Table A

Time in hours	Number of bacteria
0	1000
1	1649
2	2718
3	4482
4	7389
5	12182

Table B

Time x	Log y
0	3.0000
1	3.2172
2	3.4343
3	3.6515
4	3.8686
5	4.0857

We can now find a least squares line which will predict y, given x.

(a) Plot the original pairs of numbers. The pattern should be nonlinear.

(b) Plot the log values against the time values. The pattern should be almost linear.

(c) Find the equation of the least squares line. (First round off the log values to the nearest hundredth.)

(d) Predict the log value for a time of 7 hours. Find the number whose logarithm is your answer. This will be the predicted number of bacteria.

Management

It is sometimes possible to get a better prediction for a variable by considering its relationship with more than one other variable. For example, one should be able to predict college GPAs more precisely if both high school GPAs and scores on the ACT are considered. To do this, we alter the equation used to find a least squares line by adding a term for the new variable as follows. If y represents college GPAs, x_1 high school GPAs, and x_2 ACT scores, then y', the predicted GPAs, is given by

$$y' = ax_1 + bx_2 + c.$$

This equation represents a ***least squares plane.*** *The equations for the constants a, b, and c are more complicated than those given in the text for m and b, so that calculating a least squares equation for three variables is more likely to require the aid of a computer.*

5. Alcoa* used a least squares line with two independent variables, x_1 and x_2, to predict the effect on revenue of the price of aluminum forged truck wheels, as follows.

x_1 = the average price per wheel

x_2 = large truck production in thousands

y = sales of aluminum forged truck wheels in thousands

Using data for the past eleven years, the company found the equation of the least squares line to be

$$y' = 49.2755 - 1.1924x_1 + 0.1631x_2,$$

for which the correlation coefficient was .902. The following figures were then forecast for truck production.

1976	1977	1978	1979	1980	1981
160.0	165.0	170.0	175.0	180.0	185.0

Three possible price levels per wheel were considered: \$42, \$45, and \$48.

(a) Use the least squares plane equation given above to find the estimated sales of wheels (y) for 1978 at each of the three price levels.

(b) Repeat part (a) for 1981.

(c) For which price level, on the basis of the 1978 and 1981 figures, are total estimated sales greatest?

* This example supplied by John H. Van Denender, Public Relations Department, Aluminum Company of America.

(By comparing total estimated sales for the years 1976 through 1981 at each of the three price levels, the company found that the selling price of \$42 per wheel would generate the greatest sales volume over the six-year period.)

6. Records show that the annual sales of the EZ Life Company in 5-year periods for the last 20 years were as follows.

Year (x)	Sales (in millions) (y)
1955	1.0
1960	1.3
1965	1.7
1970	1.9
1975	2.1

The company wishes to estimate sales from these records for the next few years. Code the years so that 1955 = 0, 1956 = 1, and so on.
(a) Plot the 5 points on a graph.
(b) Find the equation of the least squares line, and graph it on the graph of part (a).
(c) Predict the company's sales for 1979 and 1980.
(d) Compute the correlation coefficient.

7. Sales, in thousands of dollars, of a certain company are shown here.

Year (x)	0	1	2	3	4	5
Sales (y)	48	59	66	75	80	90

Find the equation of the least squares line. Compare your answer with the answer to Exercise 47, Section 3.3. Find the coefficient of correlation.

8. McDonald's* wished to find the relationship between annual store sales and pre-tax profit in order to estimate increases in profit due to increased sales volume. The data shown below were obtained from a sample of McDonald's stores throughout the country.

Annual store sales in \$1000 (x)	250	300	375	425	450	475	500	575	600	650
Pre-tax percent profit (y)	9.3	10.8	14.8	14.0	14.2	15.3	15.9	19.1	19.2	21.0

(a) Plot the ten pairs of (x, y) values on a graph.
(b) Find the equation of the least squares line which gives the desired relationship between the variables.
(c) Graph the line of part (b) on the same graph as part (a).

---

* This example supplied by Harvey Lubelchek, Operations Research Manager, McDonald's System, Inc.

(d) Using the equation of part (b), predict the pre-tax percent profit for annual sales of \$700,000. (Remember to convert the sales figure to 1000's.)
(e) Find the coefficient of correlation.

9. The following data, furnished by a major brewery, which asked that its name not be given, were used to determine if there is a relationship between repair costs and barrels of beer produced. The data in thousands are given for a 10-month period.

Month	Barrels of beer X	Repairs Y
Jan	369	299
Feb	379	280
Mar	482	393
April	493	388
May	496	385
June	567	423
July	521	374
Aug	482	357
Sept	391	106
Oct	417	332

(a) Find the equation of the least squares line.
(b) Find the coefficient of correlation.
(c) If 500,000 barrels of beer are produced, what will the equation from part (a) give as the predicted repair costs?

10. Eastman Kodak Company* used the method of least squares to find an equation which would enable them to predict personal consumption expenditures (PCE) by finding a relationship between PCE and disposable income (DI), which was easier to forecast than PCE.

The data used were the annual rate of PCE for each of the years 1971 through 1977 (the latest years for which figures were available) compared with the annual rate of DI for the same period. From the data pairs, they were able to determine that a linear relationship of the form $y = mx + b$ could be used. A portion of their data is given here.

Year	DI(x) (\$ billion)	PCE(y) (\$ billion)
1971	743	668
1972	801	733
1973	902	810
1974	983	890
1975	1081	980
1976	1206	1094
1977	1316	1194

* This example was based on material furnished by Carol J. Irwin, Corporate Information Department, Eastman Kodak Company.

From the data, they found that $m = .91$ and $b = -5.87$, which gave the least squares equation

$$y' = .91x - 5.87. \qquad (^\dagger)$$

(a) Use equation (†) together with the disposable income given for the years 1972 and 1977 to find the predicted personal consumption expenditure. Compare your result with the actual amount given in the table.

(b) Use equation (†) to predict the 1980 PCE if the forecast for DI is 1650.

Social Science

11. The admission test scores of 8 students were compared with their grade-point averages after one year of college. The results are shown below.

Admission test score (x)	19	20	22	24	25	26	27	29
Grade-point average (y)	2.2	2.4	2.7	2.6	3.0	3.5	3.4	3.8

(a) Plot the 8 points on a graph.

(b) Find the equation of the least squares line and graph it on the same graph used in (a).

(c) Using the results of part (b), predict the grade-point average of a student with an admission test score of 28.

(d) Compute the coefficient of correlation.

CHAPTER 10 TEST

[10.1]

The following numbers give the sales in dollars for lunch hour at the local Wendy's hamburger store for the last twenty Fridays.

480	451	501	478	512	473	509	515	458	566
516	535	492	558	488	547	461	475	492	471

1. Write a frequency distribution for these numbers. Use intervals 450–474, 475–499, and so on.

2. Use the results of Exercise 1 to draw a histogram.

[10.2]

Find the mean for each of the following.

3. 41, 60, 67, 68, 72, 74, 78, 83, 90, 97

4.

Interval	Frequency
10–19	6
20–29	12
30–39	14
40–49	10
50–59	8

Find the median and the mode (or modes) for each of the following.

5. 32, 35, 36, 44, 46, 46, 59

6. 38, 36, 42, 44, 38, 36, 48, 35

7. Find the modal class for the distribution of Exercise 4 above.

[10.3] *Find the range and standard deviation for each of the following lists of numbers.*

8. 14, 17, 18, 19, 32

9. 26, 43, 51, 29, 37, 56, 29, 82, 74, 93

10. Find the standard deviation for the distribution of Exercise 4 above.

[10.4] *On standard IQ tests, the mean is* 100, *with a standard deviation of* 15. *The results are very close to fitting a normal curve. Suppose an IQ test is given to a very large group of people. Find the percentage of those people whose IQ score is*

11. more than 130;

12. less than 85;

13. between 85 and 115.

Find the following areas under the normal curve.

14. between $z = 0$ and $z = 1.27$

15. between $z = -1.88$ and $z = 2.10$

16. to the left of $z = -.41$

The average resident of a certain Eastern suburb spends 42 *minutes per day commuting, with a standard deviation of* 12 *minutes. Find the percent of all residents of this suburb who commute*

17. at least 50 minutes per day;

18. no more than 35 minutes per day;

19. between 32 and 40 minutes per day;

20. between 38 and 60 minutes per day.

[10.5] *About* 6% *of the frankfurters produced by a certain machine are overstuffed, and thus defective. Find the probability that in a sample of* 500 *frankfurters,*

21. 25 or fewer are overstuffed; (Hint: $\sigma = 5.3$)

22. exactly 30 are overstuffed.

[10.6] *The following data show the connection between blood sugar levels, x, and cholesterol levels, y, for* 8 *different patients.*

Patient	1	2	3	4	5	6	7	8
Blood sugar level, x	130	138	142	159	165	200	210	250
Cholesterol level, y	170	160	173	181	201	192	240	290

For this data, $\sum x = 1394$, $\sum y = 1607$, $\sum xy = 291{,}990$, $\sum x^2 = 255{,}214$, *and* $\sum y^2 = 336{,}155$.

23. Find the equation of the least squares line, $y' = mx + b$.

24. Predict the cholesterol level for a person whose blood sugar level is 190.

25. Find r.

CASE 11 INVENTORY SHORTAGES OR OVERAGES—THE COLEMAN COMPANY, INC.*

A shipping department may experience overages and shortages on shipments to customers. Since customers complain if their order is short, shortages are usually caught and corrected. However, overages are seldom reported by customers.

The data in the following table represent the results of 15 shipments with errors from a sample of 200 shipments.

A correlation between "dollar value of the shipment" and "absolute value of the amount over or under" (the error) can be calculated as follows. Let X represent the dollar value of the shipment and Y represent the error. Then we have the following sums: $\sum X = 607{,}300$, $\sum Y = 52$, $\sum XY = 1{,}322{,}400$, $\sum X^2 = 42{,}886{,}300{,}000$, $\sum Y^2 = 276$. There are 15 pairs, so n is 15. Substituting these values into the formula for the correlation coefficient gives

$$r = \frac{n\sum(XY) - (\sum X)(\sum Y)}{\sqrt{[n(\sum X^2) - (\sum X)^2][n(\sum Y^2) - (\sum Y)^2]}}$$

$$r = \frac{15(1{,}323{,}400) - (607{,}300)(52)}{\sqrt{[15(42{,}886{,}300{,}000) - (607{,}300)^2][15(276) - (52)^2]}}$$

$$r = -.592.$$

Dollar value of shipment	Size of pieces (cu. ft.)	Driver count of shipment at delivery	Shipping count	Absolute value of amount over or short	Average seniority of loading team in months	Average value of pieces missing
$120,000	12.1	145	144	1	25	$550
17,500	5.0	203	200	3	30	180
28,000	6.2	152	150	2	52	290
15,000	3.8	295	300	5	7	90
4500	2.7	205	200	5	9	50
97,000	8.4	152	150	2	37	450
10,500	7.5	122	125	3	30	400
27,000	2.9	104	100	4	6	250
36,000	3.1	177	175	2	48	275
18,000	1.3	1011	1000	11	5	100
14,700	2.6	796	800	4	10	110
88,900	5.4	501	500	1	42	350
63,200	6.9	249	250	1	35	300
55,000	10.3	102	100	2	24	325
12,000	2.2	306	300	6	8	75

* Based on an example provided by Larry D. Landrith, Corporate Personnel Manager, The Coleman Company, Inc., Wichita, Kansas.

For 15 pairs of numbers, a negative correlation less than or equal to $-.514$ is significant. Thus the value of the shipment and the absolute value of the error are related. The negative number indicates that as the value of the shipment increases, the errors decrease. This may be because there are more precautions taken for valuable shipments.

Since the two quantities "dollar value of the shipment" and "absolute value of amount over or under" are related, it is of interest to find an equation of the least squares line. Using the sums given above and $n = 15$, we have

$$m = \frac{n\sum(XY) - (\sum X)(\sum Y)}{n(\sum X^2) - (\sum X)^2}$$

$$m = \frac{15(1{,}323{,}400) - (607{,}300)(52)}{15(42{,}886{,}300{,}000) - (607{,}300)^2},$$

$$m = -.0000428,$$

also,

$$b = \frac{\sum Y - m(\sum X)}{n}$$

$$b = \frac{52 - (-.0000428)(607{,}300)}{15}$$

$$b = 5.20.$$

The equation of the least squares line is

$$y' = mx + b$$

$$y' = -.0000428x + 5.20.$$

Note that the line is almost horizontal because the slope is very close to 0.

The equation of the least squares line can be used to predict the size of the error when the value of the shipment is known. For example, if a shipment has a value of \$10,000, then the absolute value of the error is

$$y' = (-.0000428)(10000) + 5.20$$

$$y' = 4.77,$$

or about 5 items over or under.

EXERCISES

Find the coefficient of correlation r for each of the following. If $r \geq .514$ or $r \leq -.514$, find the equation of the least squares line.

1. size of pieces (X) and absolute value of the number over or short (Y)
2. average seniority of loading team (X) and absolute value of the number over or short (Y)
3. average value of pieces missing (X) and absolute value of the number over or short (Y)
4. If the value of r in Exercise 1 is significant, predict Y for $X = 4.1$.
5. If the value of r in Exercise 2 is significant, predict Y for $X = 21$.
6. If the value of r in Exercise 3 is significant, predict Y for $X = \$100$.

11 Exponential and Logarithmic Functions

Exponential functions are probably the single most important type of function used in practical applications. Exponential functions, and the closely related logarithmic functions, are used mostly to describe growth and decay, which are vital concepts in the fields of management, social science, and biology. Thus, in this chapter we see exponential functions used to describe growth of populations, increases in sales, or growth of money with time, for example. Other exponential functions describe decay—the decay of a radioactive sample, or the decay of sales in the absence of advertising. In the first two sections of this chapter, we discuss exponential functions. Logarithmic functions are discussed in the next two sections, and an appendix on common logarithms is included at the end of the chapter.

11.1 EXPONENTIAL FUNCTIONS

In Chapter 1 we saw how to evaluate 2^x for rational values of x. For example,

$$\begin{aligned} 2^3 &= 8 \\ 2^{-1} &= \frac{1}{2} \\ 2^{1/2} &= \sqrt{2} \\ 2^{3/4} &= \sqrt[4]{2^3} = \sqrt[4]{8}. \end{aligned}$$

Now we want to consider how to evaluate 2^x for irrational values of x. For example, what do we mean by $2^{\sqrt{2}}$? We know that $\sqrt{2} \approx 1.414$, and we know that we can evaluate 2^1, $2^{1.4}$, $2^{1.41}$, and so on. (It is possible to evaluate $2^{1.4}$ and $2^{1.41}$ by using logarithms or a good calculator.) Thus, we can approximate the value of $2^{\sqrt{2}}$ by replacing $\sqrt{2}$ with a decimal approximation of $\sqrt{2}$ to as many decimal places as we require. By replacing $\sqrt{2}$ with 1.414, we get $2^{\sqrt{2}} \approx 2.665$. All irrational numbers can be approximated by rational numbers as closely as necessary in the same way. Therefore, we assume that 2^x can be evaluated for *any* value of x, rational or irrational, so that $y = 2^x$ represents a function which has the real numbers as domain. In general, for $a > 0$,

$$f(x) = a^x$$

is an **exponential function** with the real numbers as domain.

To graph the exponential function $y = 2^x$, we can make a table of values of x and y, as shown beside Figure 11.1. If we then plot these points and draw a smooth curve through them, we get the graph shown in Figure 11.1. The graph approaches the negative x-axis but will never actually touch it, since y cannot be 0. (Why?) Thus the x-axis is a horizontal asymptote. This graph is typical of the graphs of exponential functions of the form $y = a^x$, where $a > 1$.

x	y
−3	1/8
−2	1/4
−1	1/2
0	1
1	2
2	4
3	8
4	16

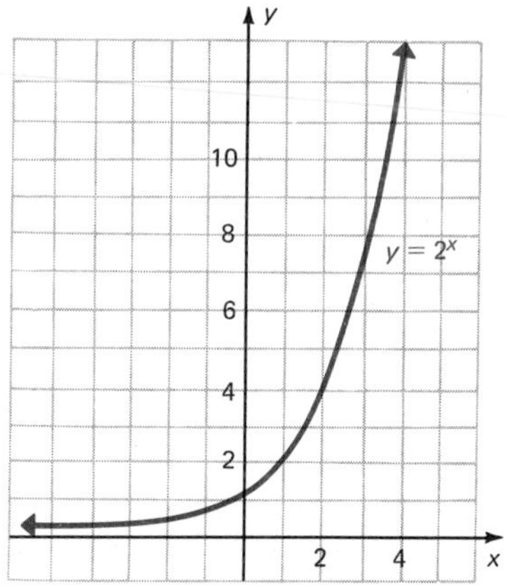

Figure 11.1

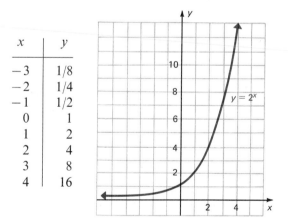

Example 1 Graph $y = 2^{-x}$.

By the properties of exponents, we know that

$$2^{-x} = \frac{1}{2^x} = \left(\frac{1}{2}\right)^x.$$

Once again we construct a table of values and draw a smooth curve through the resulting points (see Figure 11.2). This graph is typical of the graphs of exponential functions of the form $y = a^x$ where $0 < a < 1$. ■

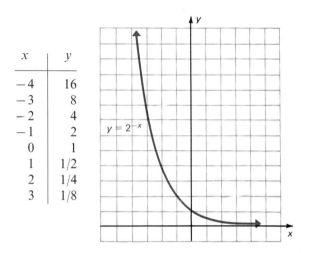

x	y
-4	16
-3	8
-2	4
-1	2
0	1
1	1/2
2	1/4
3	1/8

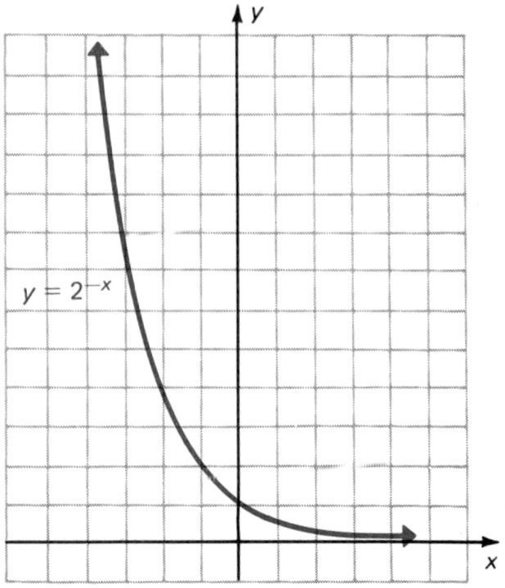

Figure 11.2

1. Graph $y = \left(\frac{1}{3}\right)^x$.

Answer:

x	y
-2	9
-1	3
0	1
1	1/3

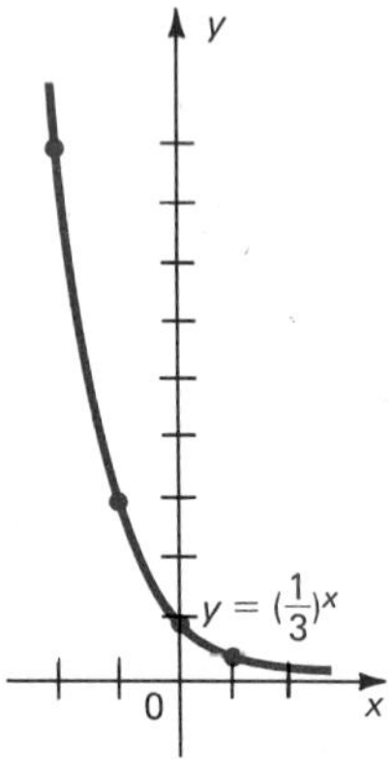

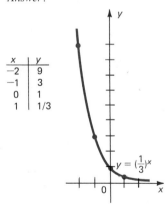

Work Problem 1 at the side.

In general, for $a > 0$, the graph of

$$f(x) = a^{kx}$$

depends on k. If k is positive, then $f(x)$ increases as x increases. If k is negative, then $f(x)$ decreases as x increases. The y-intercept is always (0, 1) and the graph always has the x-axis as an asymptote.

When discussing exponential functions, notice that we restrict a to $a > 0$, that is, we do not consider negative or zero bases. If we should try to graph $y = (-4)^x$, the domain would not include $x = 1/2$ or $x = 1/4$, for example. (Why?) The resulting graph would be at best a series of separate points having little practical use. What does the graph look like if $a = 1$?

Example 2 Graph $y = 2^{-x^2}$.

If we plot several values of x and y, we get the graph shown in Figure 11.3. Such graphs are important in probability theory. The normal curve of probability (see Section 10.4) has an equation similar to the one in this example. ■

x	y
-2	$1/16$
-1	$1/2$
0	1
1	$1/2$
2	$1/16$

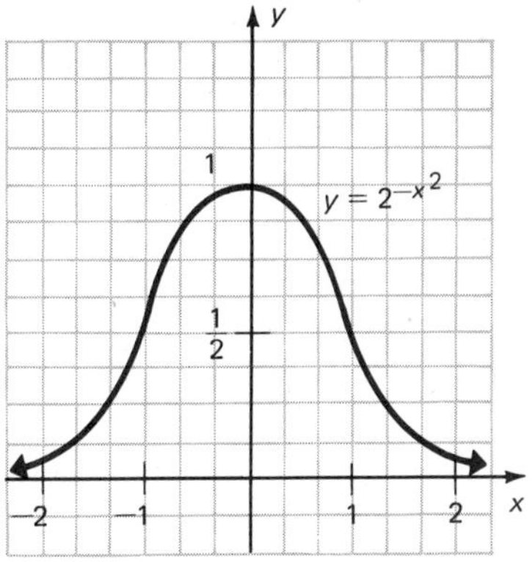

Figure 11.3

An equation with the variable in the exponent can often be solved using the following property.

If $a \neq 1$ and $a^x = a^y$, then $x = y$.

(Note that both bases must be the same.) We exclude $a = 1$ since $1^2 = 1^3$, for example, even though $2 \neq 3$. To solve $2^{3x} = 2^7$ using this property, we have

$$2^{3x} = 2^7$$
$$3x = 7$$
$$x = \frac{7}{3}.$$

Example 3 Solve $9^x = 27$.

First rewrite both sides of the equation so the bases are the same. Since $9 = 3^2$ and $27 = 3^3$,

$$(3^2)^x = 3^3$$
$$3^{2x} = 3^3$$
$$2x = 3$$
$$x = \frac{3}{2}. \quad \blacksquare$$

2. Solve $8^{2x} = 4$.

Answer: $x = \frac{1}{3}$

Work Problem 2 at the side.

11.1 EXERCISES

Graph each of the following exponential functions. (See Examples 1 and 2.)

1. $y = 3^x$ **2.** $y = 4^x$ **3.** $y = 3^{-x}$ **4.** $y = 4^{-x}$

5. $y = \left(\frac{1}{4}\right)^x$ **6.** $y = \left(\frac{1}{3}\right)^x$ **7.** $y = \left(\frac{1}{3}\right)^{-x}$ **8.** $y = \left(\frac{1}{4}\right)^{-x}$

9. $y = 3^{2x}$ **10.** $y = 4^{x/2}$ **11.** $y = 2^{-x/2}$ **12.** $y = 3^{-2x}$

13. $y = 2^{x^2}$ **14.** $y = 3^{x^2}$ **15.** $y = 3^{-x^2}$ **16.** $y = 2^{-x^2}$

17. $y = 2^{x+1}$ **18.** $y = 3^{x+1}$ **19.** $y = 2^{1-x}$ **20.** $y = 3^{1-x}$

Solve each of the following equations. (See Example 3.)

21. $5^x = 25$ **22.** $3^x = \frac{1}{9}$ **23.** $2^x = \frac{1}{8}$

24. $4^x = 64$ **25.** $4^x = 8$ **26.** $25^x = 125$

27. $16^x = 64$ **28.** $\left(\frac{1}{8}\right)^x = 8$ **29.** $\left(\frac{3}{4}\right)^x = \frac{16}{9}$

30. $5^{-2x} = \frac{1}{25}$ **31.** $3^{x-1} = 9$ **32.** $16^{-x+1} = 8$

33. $25^{-2x} = 3125$ **34.** $16^{x+2} = 64^{2x-1}$ **35.** $81^{-2x} = 3^{x-1}$

36. $7^{-x} = 49^{x+3}$

APPLIED PROBLEMS

Management

If \$1 is deposited into an account paying 6% compounded annually, then the account will contain

$$y = (1 + .06)^t = (1.06)^t$$

dollars after t years.

37. Use a calculator to help you complete the following table.

t	0	1	2	3	4	5	6	7	8	9	10
y	1					1.34					1.79

38. Graph $y = (1.06)^t$

If money loses value at the rate of 8% per year, the value of \$1 in t years is given by

$$y = (1 - .08)^t = (.92)^t.$$

39. Use a calculator to help you complete the following table.

t	0	1	2	3	4	5	6	7	8	9	10
y	1					.66					.43

40. Graph $y = (.92)^t$.

Use the results of Exercise 39 to answer the following questions.

41. Suppose a house costs \$50,000 today. Estimate the cost of a similar house in 10 years. (Hint: Solve the equation $.43x = 50{,}000$.)

42. Find the cost in 8 years of a textbook costing \$14 today.

Natural Science

Under certain conditions, the number of individuals of a species that is newly introduced into an area can double every year. That is, if t represents the number of years since the species was introduced into the area, and y represents the number of individuals, then

$$y = 6 \cdot 2^t$$

if 6 animals were introduced into the area originally.

43. Complete the following table.

t	0	1	2	3	4	5	6	7	8	9	10
y	6					192					6144

44. Graph $y = 6 \cdot 2^t$.

Escherichia coli *is a strain of bacteria that occurs naturally in many different situations. Under certain conditions, the number of these bacteria present in a colony is given by*

$$E(t) = E_0 \cdot 2^{t/30},$$

where $E(t)$ is the number of bacteria present t minutes after the beginning of an experiment, and E_0 is the number present when $t = 0$. Let $E_0 = 1{,}000{,}000$, and use a calculator with an x^y key to find the number of bacteria at the following times.

45. $t = 5$ **46.** $t = 10$ **47.** $t = 15$ **48.** $t = 20$

49. $t = 30$ **50.** $t = 60$ **51.** $t = 90$ **52.** $t = 120$

11.2 APPLICATIONS OF EXPONENTIAL FUNCTIONS

As mentioned before, exponential functions have many practical applications. One important application of exponential functions, compound interest, was discussed at length in Chapter 4. Several other applications are discussed in this section. Most of the applications we discuss involve the irrational number e. It is shown in more advanced courses that, to seven decimal places of accuracy,

$$e = 2.7182818.$$

It can be shown (see Section 12.7) that in many situations involving growth or decay of a population, the amount or number present at time t can be closely approximated by a function of the form

$$y = y_0 e^{kt},$$

where y_0 is the amount or number present at time $t = 0$ and k is a constant. (In other words, once the numbers y_0 and k have been determined, population is a function of time.)

For example, suppose the population of a southeastern city, $P(t)$, is given by

$$P(t) = 10{,}000e^{.04t},$$

where t represents time measured in years. The population at time $t = 0$ is given by

$$\begin{aligned} P(0) &= 10{,}000e^{.04(0)} \\ &= 10{,}000e^{0} \\ &= 10{,}000 \cdot 1 \qquad \text{(Recall: } e^0 = 1.) \\ P(0) &= 10{,}000. \end{aligned}$$

Thus, the population of the city is 10,000 at time $t = 0$. The number present initially is often expressed using a subscript 0. For example, here we could express the fact that the population is 10,000 at a time $t = 0$ by saying $P_0 = P(0) = 10{,}000$. The population of the city in year $t = 5$ is given by

$$\begin{aligned} P(5) &= 10{,}000e^{.04(5)} \\ &= 10{,}000e^{.2}. \end{aligned}$$

The number $e^{.2}$ can be found in Table 3. From this table, $e^{.2} \approx 1.22140$, so that

$$P(5) \approx 10{,}000(1.22140) = 12{,}214.$$

Thus, in 5 years the population of the city will be about 12,200.

Example 1 Suppose the amount, y, of a certain radioactive substance present at time t is given by

$$y = 1000e^{-.1t},$$

where t is measured in days and y is measured in grams.

(a) How much of the substance will be present at time $t = 0$? When $t = 0$, we have

$$\begin{aligned} y &= 1000e^{-.1(0)} \\ &= 1000 \cdot 1 \\ &= 1000. \end{aligned}$$

Thus, $y_0 = 1000$ grams.

(b) How much will be present at time $t = 5$?

$$\begin{aligned} y &= 1000e^{-.1(5)} \\ &= 1000e^{-.5} \\ &\approx 1000(.60653) \qquad \text{(using Table 3)} \\ &= 606.53 \text{ grams.} \quad ■ \end{aligned}$$

1. Suppose the number of bacteria in a culture at time t is

$$y = 500e^{.4t},$$

where t is measured in hours.

(a) How many bacteria are present at $t = 0$?

(b) How many bacteria are present at $t = 10$?

Answer:

(a) 500

(b) about 27,500

Work Problem 1 at the side.

Example 2 Sales of a new product often grow rapidly at first and then begin to level off with time. For example, suppose the sales, $S(x)$, in some appropriate unit, of a new model typewriter are approximated by the function

$$S(x) = 1000 - 800e^{-x},$$

where x represents the number of years the typewriter has been on the market. Calculate S_0, $S(1)$, $S(2)$, and $S(4)$. Graph $S(x)$.

Verify that $S_0 = S(0) = 1000 - 800 \cdot 1 = 200$. Using Table 3, we have

$$\begin{aligned} S(1) &= 1000 - 800e^{-1} \\ &\approx 1000 - 800(.36787) \\ &= 1000 - 294 \\ &= 706. \end{aligned}$$

In the same way, verify that $S(2) = 892$ and $S(4) = 985$. By plotting several such points we get the graph shown in Figure 11.4. Note that the line $y = 1000$ is an asymptote for the graph. This shows

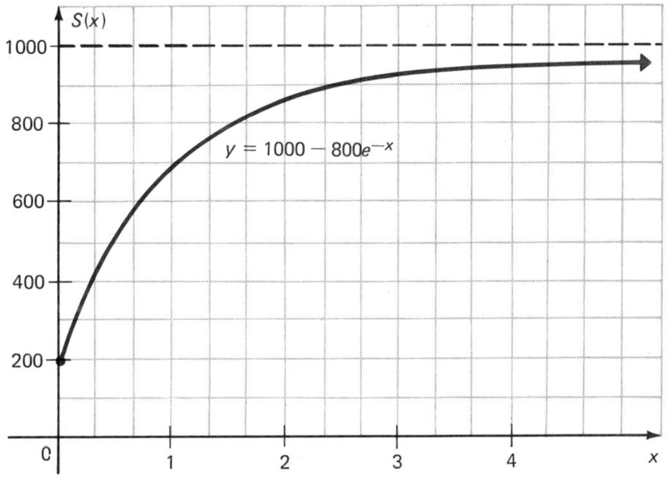

Figure 11.4

that sales will tend to level off with time and gradually approach a level of 1000 units. ■

Example 3 Assembly line operations tend to have a high turnover of employees. Thus, the companies involved must spend much time and effort in training new workers. It has been found that workers new to the operation of a certain machine (or a certain task on the assembly line) will produce items according to the function

$$P(x) = 25 - 25e^{-.3x}$$

where $P(x)$ items are produced on day x. Using this function, how many items will be produced by a new worker on day 8?

We must evaluate $P(8)$.

$$\begin{aligned} P(8) &= 25 - 25e^{-.3(8)} \\ &= 25 - 25e^{-2.4} \\ &\approx 25 - 25(.09071) \\ &= 25 - 2.3 \\ &= 22.7. \end{aligned}$$

Thus, on the eighth day, the worker can be expected to produce about 23 items. By plotting several such points, we can obtain the graph of $P(x)$, as shown in Figure 11.5. ■

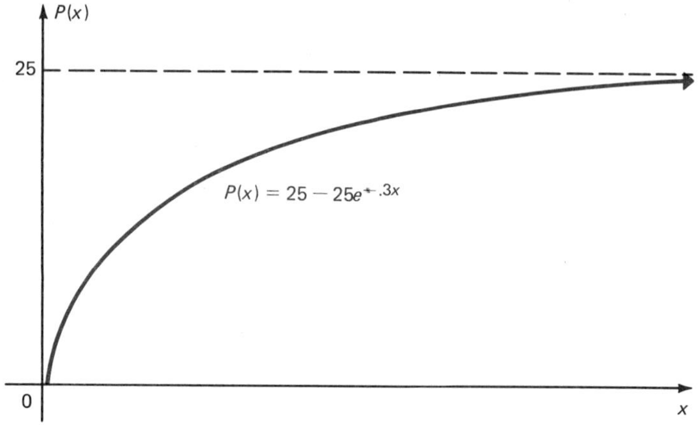

Figure 11.5

Graphs such as the one in Figure 11.5 are called **learning curves.** According to such a graph, a new worker tends to learn quickly at first, and then learning tapers off and approaches some upper limit. This is characteristic of the learning of certain types of skills involving the repetitive performance of the same task.

2. Suppose the value of the assets of a certain company at time t are given by

$$y = 100{,}000 - 75{,}000e^{-.2t}$$

where t is measured in years. Find y for the following values of t.

(a) $t = 0$
(b) $t = 5$
(c) $t = 10$
(d) $t = 25$
(e) Graph the function.

Answer:
(a) 25,000
(b) 72,410
(c) 89,850
(d) 99,495
(e)

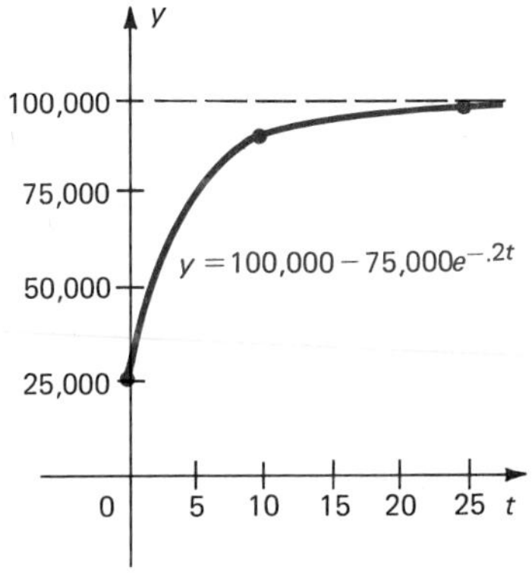

Work Problem 2 at the side.

Example 4 Under certain conditions the total number of facts of a certain kind that are remembered is approximated by

$$N(t) = y_0\left(\frac{1+e}{1+e^{t+1}}\right)$$

where $N(t)$ is the number of facts remembered at time t, measured in days, and y_0 is the number of facts remembered initially. Graph the function.

Here we can plot some points as an aid in graphing the function. For example, if $t = 0$, we have

$$\begin{aligned} N(0) &= y_0\left(\frac{1+e}{1+e^1}\right) \\ &= y_0(1) \\ &= y_0. \end{aligned}$$

If $t = 1$, we have

$$\begin{aligned} N(1) &= y_0\left(\frac{1+e}{1+e^2}\right) \\ &\approx y_0\left(\frac{3.718}{8.389}\right) \\ &= .44y_0. \end{aligned}$$

Plotting several such points gives the graph of Figure 11.6. ■

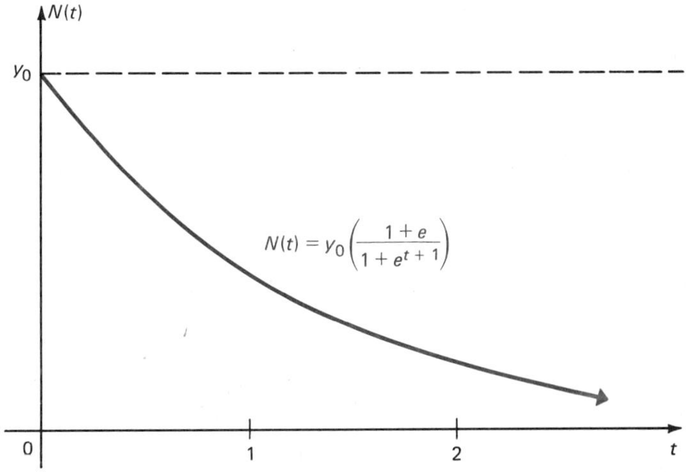

Figure 11.6

Graphs such as the one in Figure 11.6 are called **forgetting curves.**

11.2 APPLIED PROBLEMS

Social Science

Suppose the population of a city, P(t), is given by

$$P(t) = 1{,}000{,}000e^{.02t},$$

where t represents time measured in years. Find each of the following values.

1. P_0 **2.** $P(2)$ **3.** $P(4)$ **4.** $P(10)$

Natural Science

A population of lice L(t) is given by

$$L(t) = 100e^{.1t},$$

where t is measured in months. Find each of the following values.

5. L_0 **6.** $L(1)$ **7.** $L(6)$ **8.** $L(12)$

Suppose the quantity, Q(t), measured in grams, of a radioactive substance present at time t is given by

$$Q(t) = 500e^{-.05t},$$

where t is measured in days. Find the quantity present at each of the following times.

9. $t = 0$ **10.** $t = 4$ **11.** $t = 8$ **12.** $t = 20$

The amount of a chemical in grams that will dissolve in a solution is given by

$$C(t) = 10e^{.02t},$$

where t is the temperature of the solution. Find each of the following.

13. $C(0^\circ)$ **14.** $C(10^\circ)$ **15.** $C(30^\circ)$ **16.** $C(100^\circ)$

Let the number of bacteria, B(t), present in a certain culture be given by

$$B(t) = 25{,}000e^{.2t},$$

where t is time measured in hours, and t = 0 corresponds to noon. Find the number of bacteria present at each of the following times.

17. noon **18.** 1 p.m. **19.** 2 p.m. **20.** 5 p.m.

When a bactericide is introduced into a certain culture, the number of bacteria present, D(t), is given by

$$D(t) = 50{,}000e^{-.01t},$$

where t is time measured in hours. Find the number of bacteria present at each of the following times.

21. $t = 0$ **22.** $t = 5$ **23.** $t = 20$ **24.** $t = 50$

The number of fish in a small pond is given by

$$F(t) = 27 - 15e^{-.8t},$$

where t is in years. Find each of the following.

25. F_0 **26.** $F(1)$ **27.** $F(5)$ **28.** $F(10)$

Management

In Chapter 4, we discussed compound interest. It can be shown that P dollars compounded continuously (every instant) at i% annually would amount to

$$A = Pe^{ni}$$

at the end of n years. How much would \$20,000 amount to at 8% compounded continuously for the following number of years?

29. 1 year **30.** 5 years **31.** 10 years

32. Find the present value of \$1000 compounded continuously for 10 years at 5%. (Hint: Use $A = Pe^{ni}$ and solve for P.)

Sales of a new model of can opener are approximated by

$$S(x) = 5000 - 4000e^{-x},$$

where x represents the number of years the can opener has been on the market, and S(x) represents the sales in thousands. Find each of the following.

33. $S(0)$ **34.** $S(1)$ **35.** $S(2)$

36. $S(5)$ **37.** $S(10)$ **38.** Graph $S(x)$.

The number of words per minute a typist can type is given by

$$W(x) = 60 - 30e^{-.5t},$$

where t is time in months after the beginning of a typing class. Find each of the following.

39. W_0 **40.** $W(1)$ **41.** $W(4)$ **42.** $W(6)$

Assume that a person new to an assembly line will produce

$$P(x) = 500 - 500e^{-x}$$

items per day, where x is time measured in days. Find each of the following.

43. $P(0)$ **44.** $P(1)$ **45.** $P(2)$

46. $P(5)$ **47.** $P(10)$ **48.** Graph $P(x)$.

Experiments have shown that sales of a product, under relatively stable market conditions, but in the absence of promotional activities such as advertising, tend to decline at a constant yearly rate. This rate of sales decline varies considerably from product to product, but seems to remain the same for any particular product. The sales decline can often be expressed by a function of the form

$$S(t) = S_0e^{-at},$$

where S(t) is the rate of sales at time t measured in years, S_0 *is the rate of sales at time* $t = 0$*, and a is the sales decay constant.*

49. Suppose the sales decay constant for a particular product is $a = 0.10$.

Let $S_0 = 50{,}000$. Find
(a) $S(1)$;
(b) $S(3)$.

50. Suppose $S_0 = 80{,}000$ and $a = .05$. Find
(a) $S(2)$;
(b) $S(10)$.

Social Science

A sociologist has shown that the fraction $y(t)$ of people who have heard a rumor after t days is approximated by

$$y(t) = \frac{y_0 e^{kt}}{1 - y_0(1 - e^{kt})},$$

where y_0 is the fraction of people who have heard the rumor at time $t = 0$, and k is a constant. A graph of $y(t)$ for a particular value of k is shown in the figure.

51. If $k = 0.1$ and $y_0 = .05$, find $y(10)$.

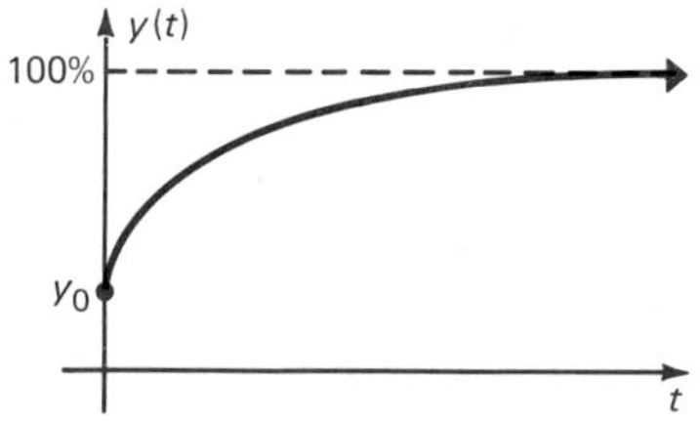

52. If $k = .2$ and $y_0 = 0.10$, find $y(5)$.

The higher a student's grade-point average, the fewer applications he has to send to medical schools (other things being equal). Using information given in a guidebook for prospective medical school students, we constructed the following mathematical model of the number of applications that a student should send out:

$$y = 540e^{-1.3x},$$

where y is the number of applications that should be sent out by a person whose grade-point average is x. Here $2.0 \le x \le 4.0$. Use a calculator with an x^y key to find the number of applications that should be sent out by students having a grade-point average of

53. 2.0; **54.** 2.5; **55.** 3.0;

56. 3.5; **57.** 3.9; **58.** 4.0.

11.3 LOGARITHMIC FUNCTIONS

In the previous section we discussed exponential functions. In this section, we discuss **logarithmic functions,** which can be obtained from exponential functions using the following definition. For $a > 0$,

$$y = a^x \text{ means that } x = \log_a y.$$

Read $x = \log_a y$ as "x is the logarithm of y to the base a." For example, since $16 = 2^4$, we can write $4 = \log_2 16$. Also, since $10^3 = 1000$, we can write $\log_{10} 1000 = 3$.

Example 1 In this example we show several pairs of equivalent statements. The same statement is written in both exponential and logarithmic form.

	Exponential form	*Logarithmic form*
(a)	$3^2 = 9$	$\log_3 9 = 2$
(b)	$\left(\frac{1}{5}\right)^{-2} = 25$	$\log_{1/5} 25 = -2$
(c)	$10^5 = 100{,}000$	$\log_{10} 100{,}000 = 5$
(d)	$4^{-3} = 1/64$	$\log_4 1/64 = -3$
(e)	$2^{-4} = 1/16$	$\log_2 1/16 = -4$
(f)	$e^0 = 1$	$\log_e 1 = 0$ ■

Work Problems 1 and 2 at the side.

1. Write the logarithmic form of
(a) $5^3 = 125$;
(b) $3^{-4} = 1/81$;
(c) $8^{2/3} = 4$.

Answer:
(a) $\log_5 125 = 3$
(b) $\log_3(1/81) = -4$
(c) $\log_8 4 = 2/3$

2. Write the exponential form of
(a) $\log_{16} 4 = 1/2$;
(b) $\log_3(1/9) = -2$;
(c) $\log_{16} 8 = 3/4$.

Answer:
(a) $16^{1/2} = 4$
(b) $3^{-2} = 1/9$
(c) $16^{3/4} = 8$

The exponential function $f(x) = 2^x$ and the logarithmic function $g(x) = \log_2 x$ are closely related. Note that $f(3) = 2^3 = 8$, while $g(8) = \log_2 8 = 3$. Hence, $f(3) = 8$ and $g(8) = 3$. Similarly, $f(5) = 32$ and $g(32) = 5$. In fact, for any number m, if $f(m) = n$, then $g(n) = m$. Functions related in this way are called **inverses** of each other. Do you see that every logarithmic function is the inverse of some exponential function? A general discussion of inverse functions would carry us too far afield; such a discussion can be found in books listed in the bibliography.

If we rewrite both $y = \log_2 x$ and $y = \log_{1/2} x$ in exponential form, we can calculate a series of points for each equation. By plotting the points and connecting them with a smooth curve, we get the graphs shown in Figure 11.7. Note that these graphs are the graph of functions. The graph of Figure 11.7(a) is typical of the graphs of logarithm functions of base $a > 1$, while the graph of Figure 11.7(b) is typical of the graphs of logarithm functions of

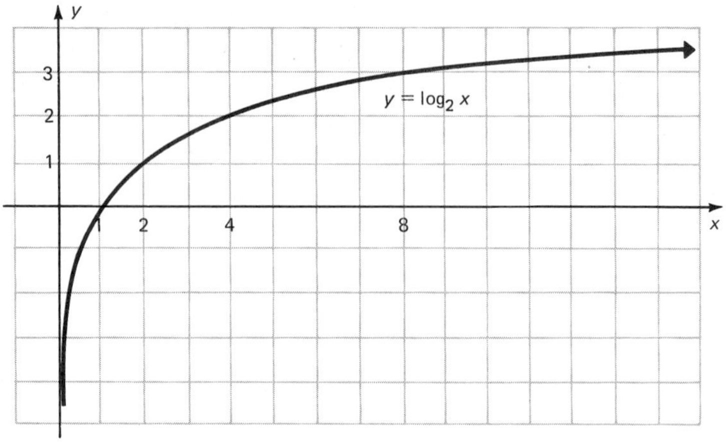

Figure 11.7 (a)

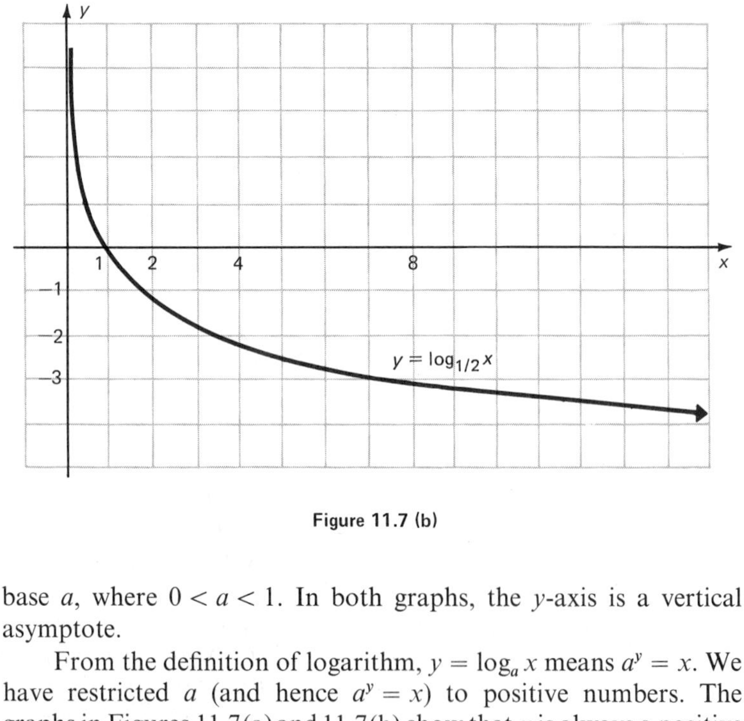

Figure 11.7 (b)

base a, where $0 < a < 1$. In both graphs, the y-axis is a vertical asymptote.

From the definition of logarithm, $y = \log_a x$ means $a^y = x$. We have restricted a (and hence $a^y = x$) to positive numbers. The graphs in Figures 11.7(a) and 11.7(b) show that x is always a positive number.

The usefulness of logarithmic functions arises from the properties of logarithms given in the following theorem.

Theorem 11.1 Let x and y be any positive real numbers and r any real number. Let a be a positive real number, $a \neq 1$. Then

(a) $\log_a xy = \log_a x + \log_a y;$

(b) $\log_a \frac{x}{y} = \log_a x - \log_a y;$

(c) $\log_a x^r = r \log_a x;$

(d) $\log_a a = 1;$

(e) $\log_a 1 = 0.$

Example 2 Using the results of Theorem 11.1, if $\log_6 7 \approx 1.09$ and $\log_6 5 \approx .9$, we have
(a) $\log_6 35 = \log_6 7 \cdot 5 = \log_6 7 + \log_6 5 \approx 1.09 + .9 = 1.99;$
(b) $\log_6 5/7 = \log_6 5 - \log_6 7 \approx -.19;$
(c) $\log_6 5^3 = 3 \log_6 5 \approx 3(.9) = 2.7;$
(d) $\log_6 6 = 1;$
(e) $\log_6 1 = 0.$ ■

3. Use Theorem 11.1 to rewrite each of the following, given $\log_3 7 \approx 1.77$ and $\log_3 5 \approx 1.46$.
(a) $\log_3 35$
(b) $\log_3 7/5$
(c) $\log_3 25$
(d) $\log_3 3$
(e) $\log_3 1$

Answer:
(a) 3.23
(b) .31
(c) 2.92
(d) 1
(e) 0

Work Problem 3 at the side.

Historically, one of the main applications of logarithms has been as an aid to numerical calculation. Using the properties above and tables of logarithms, many numerical problems can be simplified. Since our number system has base 10, logarithms to base 10 are most convenient for numerical calculations. Such logarithms are called **common logarithms.** For simplicity, $\log_{10} x$ is abbreviated $\log x$. Using this notation, we have, for example,

$$\log 1000 = 3,$$
$$\log 100 = 2,$$
$$\log 1 = 0,$$
$$\log .01 = -2.$$

The use of common logarithms to simplify numerical calculations is discussed in most intermediate algebra or college algebra texts. Common logarithms have few applications other than numerical calculations. In most other practical applications of logarithms, the number e is used as base. (Recall: To seven decimal places, $e = 2.7182818$.) Logarithms to base e are called **natural logarithms,** written in the form $\ln x$. While common logarithms may seem more "natural" than logarithms to base e, there are several good reasons for using natural logarithms instead. The most important reason is discussed in Section 13.2.

A table of natural logarithms is given in Table 4. From this table we find, for example, that

$$\ln 55 = 4.0073$$
$$\ln 1.9 = 0.6419$$
$$\ln 0.4 = -.9163.$$

As Figure 11.7(a) shows, if $0 < x < 1$, then $\log_a x$ is negative.

Example 3 Use Table 4 to find the following logarithms.

(a) $\ln 81$

Table 4 does not give $\ln 81$. However, we can use the properties of logarithms given in Theorem 11.1 to write

$$\begin{aligned} \ln 81 &= \ln(8.1 \times 10) \\ &= \ln 8.1 + \ln 10 \\ &\approx 2.0919 + 2.3026 \\ &= 4.3945. \end{aligned}$$

(b) $\ln 36$

Since 36 is not listed in Table 4, we write

$$\begin{aligned} \ln 36 &= \ln 6^2 \\ &= 2 \ln 6 \\ &\approx 2(1.7918) \\ &= 3.5836. \quad ■ \end{aligned}$$

4. Use Table 4 and the properties of logarithms to find the following.
(a) $\ln 15$
(b) $\ln 28$
(c) $\ln 270$

Answer:
(a) 2.7081
(b) 3.3322
(c) 5.5984

Work Problem 4 at the side.

Table 4 gives the natural logarithms of many numbers, but often it is necessary to find natural logarithms of numbers that cannot be found with this table. To do this, the following theorem may be used to convert common logarithms to natural logarithms.

Not required

Theorem 11.2 For any positive number x,

$$\ln x \approx 2.3026 \log x,$$

where $\log x$ is the common logarithm of x.*

Example 4 Use Theorem 11.2 above and the common logarithm table in the Appendix to find the following logarithms.

(a) $\ln 83$

$$\begin{aligned} \ln 83 &\approx 2.3026 \log 83 \\ &\approx 2.3026(1.9191) \\ &= 4.4189. \end{aligned}$$

(b) $\ln 6000$

$$\begin{aligned} \ln 6000 &\approx 2.3026(3.7782) \\ &= 8.6997 \quad ■ \end{aligned}$$

* $2.3026 \approx 1/\log e$.

5. Use Table 4, Theorem 11.1 or 11.2 to find each of the following.
(a) ln 2.8
(b) ln 160
(c) ln .126
(d) ln 753

Answer:
(a) 1.0296
(b) 5.0752
(c) −2.0715
(d) 6.6241

Work Problem 5 at the side.

11.3 EXERCISES

Write each of the following using logarithms.

1. $2^3 = 8$ **2.** $5^2 = 25$ **3.** $3^4 = 81$

4. $6^3 = 216$ **5.** $\left(\frac{1}{3}\right)^{-2} = 9$ **6.** $\left(\frac{3}{4}\right)^{-2} = \frac{16}{9}$

Write each of the following using exponents.

7. $\log_2 8 = 3$ **8.** $\log_3 27 = 3$ **9.** $\log_{10} 100 = 2$

10. $\log_2(1/8) = -3$ **11.** $\log 10{,}000 = 4$ **12.** $\ln e^5 = 5$

Evaluate each of the following.

13. $\log_{10} 10{,}000$ **14.** $\log_9 81$ **15.** $\log_4 64$

16. $\log_2(1/4)$ **17.** $\log .01$ **18.** $\log .00001$

19. Complete the following table of values for $y = \log_3 x$.

x	1/27					9	
y	−3	−2	−1	0	1	2	3

Graph the function using the same scale on both axes.

20. Complete the following table of values for $y = 3^x$.

x	−3	−2	−1	0	1	2	3
y		1/9					27

Graph the function on the same axes you used in Exercise 19. Compare the two graphs. How are they related?

Given $\ln 2 \approx .7$ *and* $\ln 3 \approx 1.1$, *evaluate the following.*

21. $\ln 6$ **22.** $\ln 12$ **23.** $\ln 3/2$ **24.** $\ln 2/3$

25. $\ln 3^4$ **26.** $\ln 2^5$ **27.** $\ln 2^{.1}$ **28.** $\ln 3^{.2}$

29. $\ln 3^{-.2}$ **30.** $\ln 2^{-.1}$ **31.** $\ln 18$ **32.** $\ln 24$

33. $\ln e$ **34.** $\ln 1/e$ **35.** $\ln 2e$ **36.** $\ln 3e$

Find each of the following natural logarithms.

37. ln 20 **38.** ln 35 **39.** ln 60 **40.** ln 50

41. ln 800	**42.** ln 920	**43.** ln 532	**44.** ln 255
45. ln 768	**46.** ln 324	**47.** ln 58,500	**48.** ln 12,400

11.4 APPLICATIONS OF LOGARITHMIC FUNCTIONS

In this section we discuss several applications of logarithmic functions. Some of these applications depend on the following theorem.

Theorem 11.3 Let a and b be positive numbers.
(a) If $a = b$, then $\ln a = \ln b$.
(b) If $\ln a = \ln b$, then $a = b$.

Theorem 11.3 can be used to solve exponential equations as in the following example.

Example 1 Solve $3^x = 5$.

Since 3 and 5 cannot easily be written with the same base, we cannot use the methods of Section 11.1 to solve this equation. We use Theorem 11.3(a) as follows.

$$\begin{aligned} 3^x &= 5 \\ \ln 3^x &= \ln 5 \\ x \ln 3 &= \ln 5 \qquad \text{(Theorem 11.1(c))} \\ x &= \frac{\ln 5}{\ln 3} \end{aligned}$$

From Table 4, we find $\ln 5 \approx 1.6094$ and $\ln 3 \approx 1.0986$. Thus,

$$x \approx \frac{1.6094}{1.0986} \approx 1.5. \quad \blacksquare$$

1. Solve $2^x = 7$.

Answer: About 2.8

Work Problem 1 at the side.

Example 2 Suppose that $A(t)$, the amount of a certain radioactive substance present at time t, is given by

$$A(t) = 1000e^{-.1t},$$

where t is measured in days and $A(t)$ in grams. At time $t = 0$, we have

$$A(0) = 1000e^{-.1(0)}$$
$$= 1000$$

grams of the substance present. Also,

$$A(5) = 1000e^{-.1(5)}$$
$$= 1000e^{-.5}$$
$$\approx 1000(.60653)$$
$$= 606.53,$$

so that about 607 grams are still present after 5 days. Now let us find the half-life of the substance. (The **half-life** of a radioactive substance is the time it takes for exactly half the sample to decay.)

To find the half-life, we must find a value of t such that $A(t) = \frac{1}{2}(1000) = 500$ grams. That is, we must solve the equation

$$500 = 1000e^{-.1t}.$$

To solve this equation, first divide through by 1000, obtaining

$$\frac{1}{2} = e^{-.1t}.$$

Now use Theorem 11.3, and take the natural logarithms of both sides. This gives

$$\ln \frac{1}{2} = \ln e^{-.1t}.$$

Using one of the properties of logarithms, we have

$$\ln \frac{1}{2} = (-.1t)(\ln e),$$

and since $\ln e = 1$, we have

$$\ln \frac{1}{2} = -.1t,$$

or

$$t = \frac{\ln \frac{1}{2}}{-.1}.$$

Since $\ln \frac{1}{2} = \ln .5$, we can use Table 4 to write

$$t \approx \frac{-.6931}{-.1} \approx 6.9.$$

Thus, it will take about 6.9 days for half the sample to decay. ■

Work Problems 2 and 3 at the side.

2. The amount of a substance present at time t (in hours) is given by

$$A(t) = 530e^{-.2t},$$

and $A(t)$ is measured in grams. How much of the substance will remain after 5 hours? That is, find $A(5)$.

Answer: about 195 grams

3. Find the half-life of the substance.

Answer: about 3.5 hours

Example 3 Carbon 14 is a radioactive isotope of carbon which has a half-life of about 5600 years. The earth's atmosphere contains much carbon, mostly in the form of carbon dioxide gas, with small traces of carbon 14. Most atmospheric carbon is the nonradioactive isotope, carbon 12. The ratio of carbon 14 to carbon 12 is virtually constant in the atmosphere. However, as a plant absorbs carbon dioxide from the air in the process of photosynthesis, the carbon 12 stays in the plant while the carbon 14 is converted into nitrogen. Thus, in a plant, the ratio of carbon 14 to carbon 12 is smaller than it is in the atmosphere. Even when the plant is eaten by an animal, this ratio will continue to decrease. Based on these facts, a method of dating objects called **carbon-14 dating** has been developed.

(a) Suppose an Egyptian mummy has been discovered in which the ratio of carbon 14 to carbon 12 is only about half the ratio found in the atmosphere. How long ago did the individual who became the mummy die?

To solve this, we need to know that in 5600 years, half the carbon 14 in a specimen will decay. Thus, the individual who became the mummy died about 5600 years ago.

(b) Let R be the (nearly constant) ratio of carbon 14 to carbon 12 found in the atmosphere, and let r be the ratio as found in an observed specimen. It can then be shown that the relationship between R and r is given by

$$R = r \cdot e^{(t \ln 2)/5600},$$

where t is the age of the specimen in years. Verify the formula for $t = 0$.

To do this, substitute 0 for t in the formula. This gives

$$\begin{aligned} R &= r \cdot e^{(0 \cdot \ln 2)/5600} \\ &= r \cdot e^0 \\ &= r \cdot 1 \\ &= r. \end{aligned}$$

This result is correct; when $t = 0$, the specimen has just died, so that R and r should be the same.

(c) Suppose a specimen is found in which $r = \frac{2}{3}R$. Estimate the age of the specimen.

Here we use the relationship given in (b) above, and then substitute $\frac{2}{3}R$ for r.

$$\begin{aligned} R &= r \cdot e^{(t \ln 2)/5600} \\ &= \frac{2}{3}R \cdot e^{(t \ln 2)/5600} \end{aligned}$$

Dividing through by R, and multiplying through by 3/2, we have

$$\frac{3}{2} = e^{(t \ln 2)/5600}.$$

Taking natural logarithms of both sides, we have

$$\ln \frac{3}{2} = \ln e^{(t \ln 2)/5600}.$$

Using properties of logarithms, we have

$$\ln \frac{3}{2} = \left(\frac{t \ln 2}{5600}\right) \ln e.$$

Since $\ln e = 1$, we have

$$\ln \frac{3}{2} = \frac{t \ln 2}{5600}.$$

To solve this equation for t, the age of the specimen, multiply both sides by $5600/\ln 2$. This gives

$$\frac{5600 \ln 3/2}{\ln 2} = t.$$

Using Table 4, we have $\ln \frac{3}{2} = \ln 1.5 \approx .4055$ and $\ln 2 \approx .6931$. Thus,

$$t \approx \frac{5600(.4055)}{.6931} = 3280,$$

so that the specimen is about 3280 years old. ■

4. What is the age of a specimen in which $r = \frac{1}{3}R$?

Answer: about 8876 years

Work Problem 4 at the side.

11.4 EXERCISES

Solve each of the following equations by using Theorem 11.3. (See Example 1.)

1. $5^{2x} = 6$

2. $3^{-x} = 8$

3. $e^{-.3x} = 5$

4. $e^{.1x} = 40$

5. $e^{.02x} = 10$

6. $e^{-.01x} = 12$

7. $e^{x+2} = 4$

8. $e^{1-x} = 7$

9. $e^{2x/5} = 8$

10. $e^{x/2} = 15$

APPLIED PROBLEMS

Natural Science

The amount of a certain radioactive specimen present at time t (measured in seconds) is given by

$$A(t) = 5000e^{-.02t},$$

where $A(t)$ is the amount measured in grams. Find each of the following.

11. $A(0)$ **12.** $A(5)$

13. $A(20)$ **14.** the half-life of the specimen

The number of bacteria in a certain culture, $B(t)$, is approximated by

$$B(t) = 250{,}000e^{-.04t},$$

where t is time measured in hours. Find each of the following.

15. B_0 **16.** $B(5)$ **17.** $B(20)$

18. the time it will take until only 125,000 bacteria are present

19. the time it will take until only 25,000 bacteria are present

Using Example 3, find the age of a specimen for each of the following.

20. $r = .8R$ **21.** $r = .4R$ **22.** $r = .1R$ **23.** $r = .01R$

A large cloud of radioactive debris from a nuclear explosion has floated over the Pacific Northwest, contaminating much of the hay supply. Consequently, farmers in the area are concerned that the cows who eat this hay will give contaminated milk. (The tolerance level for radioactive iodine in milk is 0.) The percent of the initial amount of radioactive iodine still present in the hay after t days is approximated by $P(t)$, which is given by the mathematical model

$$P(t) = 100e^{-.1t}$$

where t is time measured in days.

24. Find the percent remaining after 4 days.

25. Find the percent remaining after 10 days.

26. Some scientists feel that the hay is safe after the percent of radioactive iodine has declined to 10% of the original amount. Solve the equation $10 = 100e^{-.1t}$ to find the number of days before the hay may be used.

27. Other scientists believe that the hay is not safe until the level of radioactive iodine has declined to only 1% of the original level. Find the number of days that this would take.

Social Science

The number of years, $N(r)$, since two independently evolving languages split off from a common ancestral language is approximated by

$$N(r) = -5000 \ln r,$$

where r is the proportion of the words from the ancestral language that are common to both languages now. Find each of the following.

28. $N(.9)$ **29.** $N(.5)$ **30.** $N(.3)$

31. How many years have elapsed since the split if 70% of the words of the ancestral language are common to both languages today?

32. If two languages split off from a common ancestral language about 1000 years ago, find r.

Management

Use the formula $A = Pe^{ni}$ for continuous compounding of interest for Exercises 33 and 34.

33. Find the number of years it will take for \$1000 compounded continuously at 5% to double.

34. Find the interest rate at which \$1000 compounded continuously for 10 years will double.

CHAPTER 11 TEST

[11.1] and [11.4]

Solve each of the following equations.

1. $3^x = 81$ **2.** $2^{3x} = \frac{1}{8}$ **3.** $\left(\frac{9}{16}\right)^x = \frac{3}{4}$ **4.** $2^{5x} = 7$

[11.1] and [11.3]

Graph each of the following functions.

5. $y = 5^x$ **6.** $y = 5^{-x}$ **7.** $y = (\frac{1}{5})^x$ **8.** $y = \log_5 x$

Write each of the following using logarithms.

9. $2^6 = 64$ **10.** $3^{1/2} = \sqrt{3}$ **11.** $1000^{-1} = .001$

Write each of the following using exponents.

12. $\log_2 32 = 5$ **13.** $\log_{10} 100 = 2$ **14.** $\log_{27} 3 = \frac{1}{3}$

Evaluate each of the following.

15. $\log_7 49$ **16.** $\log_{25} 5$

Given $\log_5 3 \approx .68$ and $\log_5 2 \approx .43$, evaluate each of the following.

17. $\log_5 18$ **18.** $\log_5 3/4$ **19.** $\log_5 \sqrt{8}$

Find each of the following natural logarithms.

20. $\ln 6.2$ **21.** $\ln 700$ **22.** $\ln 483$

Solve each of the following problems.

[11.2]

23. Sales of a new product are approximated by $S(x) = 100{,}000e^{.2x}$, where x represents years since the product was introduced. Find each of the following.
(a) S_0 (b) $S(2)$ (c) $S(5)$

[11.4]

24. At $A(t)$ represent the amount of a certain radioactive substance present at time t (in days), where $A(t) = 5000e^{-.2t}$. Find each of the following.
(a) $A(0)$ (b) $A(2)$ (c) the half-life

25. A new worker produces $P(x) = 100 - 100e^{-.8x}$ items per day, after x days on the job. Find each of the following.
(a) $P(0)$ (b) $P(1)$ (c) $P(5)$
(d) How many items per day would you expect an experienced worker to produce?

CASE 12 THE VAN MEEGEREN ART FORGERIES*

After the liberation of Belgium at the end of World War II, officials began a search for Nazi collaborators. One person arrested as a collaborator was a minor painter, H. A. Van Meegeren; he was charged with selling a valuable painting by the Dutch artist Vermeer (1632–1675) to the Nazi Hermann Goering. He defended himself from the very serious charge of collaboration by claiming that the painting was a fake—he had forged it himself.

He also claimed that the beautiful and famous painting "Disciples at Emmaus," as well as several other supposed Vermeers, was his own work. To prove this, he did another "Vermeer" in his prison cell. An international panel of experts was assembled, which pronounced as forgeries all the "Vermeers" in question.

Many people would not accept the verdict of this panel for the painting "Disciples at Emmaus"; it was felt to be too beautiful to be the work of a minor talent such as Van Meegeren. In fact, the painting was declared genuine by a noted art scholar and sold for $170,000. The question of the authenticity of this painting continued to trouble art historians, who began to insist on conclusive proof one way or the other. This proof was given by a group of scientists at Carnegie-Mellon University, using the idea of radioactive decay.

The dating of objects is based on radioactivity; the atoms of certain radioactive elements are unstable, and within a given time period a fixed fraction of such atoms will spontaneously disintegrate, forming atoms of a new element. If t_0 represents some initial time, N_0 represents the number of atoms present at time t_0, and N represents the number present at some later time t, then it can be shown (using physics and calculus) that

$$t - t_0 = \frac{1}{\lambda} \cdot \ln \frac{N_0}{N}$$

where λ is a "decay constant" that depends on the radioactive substance under consideration.

If t_0 is the time that the substance was formed or made, then $t - t_0$ is the age of the item. Thus, the age of an item is given by

$$\frac{1}{\lambda} \cdot \ln \frac{N_0}{N}.$$

The decay constant λ can be readily found, as can N, the number of atoms present now. The problem is N_0—we can't find a value for this variable. However, it is possible to get reasonable ranges for the values of N_0. This is done by studying the white lead in the painting. This pigment has been used by artists for over 2000 years. It contains a small amount of the radioactive substance lead-210 and an even smaller amount of radium-226.

Radium-226 disintegrates through a series of intermediate steps to pro-

---

* From "The Van Meegeren Art Forgeries" by Martin Braun from *Applied Mathematical Sciences*, Vol. 15. Copyright © 1975. Published by Springer-Verlag New York, Inc. Reprinted by permission.

duce lead-210. The lead-210, in turn, decays to form polonium-210. This last process, lead-210 to polonium-210, has a half-life of 22 years. That is, in 22 years half the initial quantity of lead-210 will decay to polonium-210.

When lead ore is processed to form white lead, most of the radium is removed with other waste products. Thus, most of the supply of lead-210 is cut off, with the remainder beginning to decay very rapidly. This process continues until the lead-210 in the white lead is once more in equilibrium with the small amount of radium then present. Let $y(t)$ be the amount of lead-210 per gram of white lead present at time of manufacture of the pigment, t_0. Let r represent the number of disintegrations of radium-226 per minute per gram of white lead. (Actually, r is a function of time, but the half-life of radium-226 is so long in comparison to the time interval in question that we assume it to be a constant.) If λ is the decay constant for lead-210, then it can be shown that

$$y(t) = \frac{r}{\lambda}[1 - e^{-\lambda(t - t_0)}] + y_0 e^{-\lambda(t - t_0)}. \qquad \textbf{(1)}$$

All variables in this result can be evaluated except y_0. To get around this problem, we use the fact that the original amount of lead-210 was in radioactive equilibrium with the larger amount of radium-226 in the ore from which the metal was extracted. We therefore take samples of different ores and compute the rate of disintegration of radium-226. The results are as shown in the table.

Location of ore	Disintegrations of radium-226 per minute per gram of white lead
Oklahoma	4.5
S.E. Missouri	.7
Idaho	.18
Idaho	2.2
Washington	140
British Columbia	.4

The numbers in the table vary from .18 to 140—quite a range. Since the number of disintegrations is proportional to the amount of lead-210 present originally, we must conclude that y_0 also varies over a tremendous range. Thus, equation (1) cannot be used to obtain even a crude estimate of the age of a painting. However, we want to distinguish only between a modern forgery and a genuine painting that would be 300 years old.

To do this, we observe that if the painting is very old compared to the 22 year half-life of lead-210, then the amount of radioactivity from the lead-210 will almost equal the amount of radioactivity from the radium-226. On the other hand, if the painting is modern, then the amount of radioactivity from the lead-210 will be much greater than the amount from the radium-226.

We want to know if the painting is modern or 300 years old. To find out let $t - t_0 = 300$ in equation (1), getting

$$\lambda y_0 = \lambda \cdot y(t) \cdot e^{300\lambda} - r(e^{300\lambda} - 1) \qquad \textbf{(2)}$$

after some rearrangement of terms. If the painting is modern, then λy_0 should be a very large number; λy_0 represents the number of disintegrations of the lead-210 per minute per gram of white lead at the time of manufacture. By studying samples of white lead, we can conclude that λy_0 should never be anywhere near as large as 30,000. Thus, we use equation (2) to calculate λy_0; if our result is greater than 30,000 we conclude that the painting is a modern forgery. The details of this calculation are left for the exercises.

EXERCISES

1. To calculate λ, use the formula

$$\lambda = \frac{\ln 2}{\text{half-life}}.$$

Find λ for lead-210, whose half-life is 22 years.

2. For the painting "Disciples at Emmaus," the current rate of disintegration of the lead-210 was measured and found to be $\lambda \cdot y(t) = 8.5$. Also, r was found to be .8. Use this information and equation (2) to calculate λy_0. Based on your results, what do you conclude about the age of the painting?

The table below lists several other possible forgeries. Decide which of them must be modern forgeries.

Title	$\lambda \cdot y(t)$	r
3. "Washing of Feet"	12.6	.26
4. "Lace Maker"	1.5	.4
5. "Laughing Girl"	5.2	6
6. "Woman Reading Music"	10.3	.3

12 The Derivative

The management of a firm is willing to increase its advertising budget as long as its profit also increases. The expenditure on advertising that produces maximum profit is at the point where the rate of growth of profit slows to 0. However, if an increase in expenditure on advertising causes the rate of growth of the profit to become negative, then too much is spent on advertising.

In general, the task of finding the optimum level of expenditure is not at all easy. In this chapter we will see how one branch of calculus, called differential calculus, is used to optimize such variables. In addition, we will see how differential calculus is used to describe rates of change in business, social science, and biology.

The basic idea of differential calculus is the *derivative*, which we define in Section 4 of this chapter. The definition of a derivative requires the idea of a *limit*, which we discuss in the first section of this chapter.

12.1 LIMITS

Suppose we let the values of x in the domain of some function f get closer and closer to a fixed number, a. What will happen to the values of $f(x)$? Do they also approach a fixed number? Figure 12.1 shows the graph of a function f. The arrowheads show that as x gets closer and closer to the number 5 (on the x-axis), the values of $f(x)$ get closer and closer to the number 4 (on the y-axis). In this case, we say that the **limit** of $f(x)$ as x approaches 5 is the number 4, written as

$$\lim_{x \to 5} f(x) = 4.$$

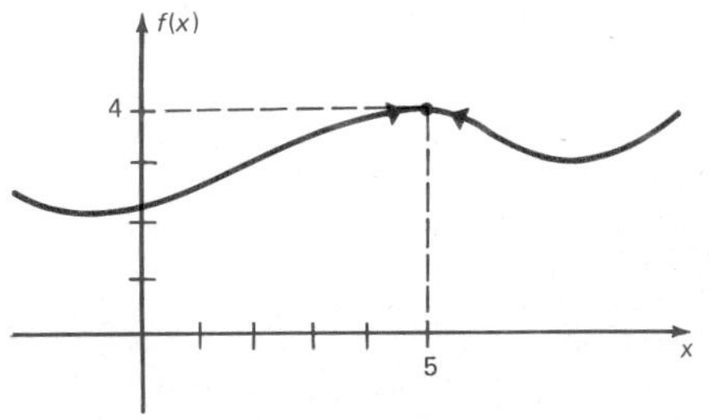

Figure 12.1

Example 1 The graph in Figure 12.2 shows a function g. The open circle at $(3, -1)$ shows that the function is not defined at $x = 3$. (That is, $g(3)$ does not exist.) What about

$$\lim_{x \to 3} g(x)?$$

By studying the graph, we see that as x gets closer and closer to 3, $g(x)$ gets closer and closer to the number -1, so that

$$\lim_{x \to 3} g(x) = -1,$$

even though $g(3)$ does not exist. It is important to note that the phrase "x gets closer and closer to 3" does not require that x ever actually *reach* 3. ■

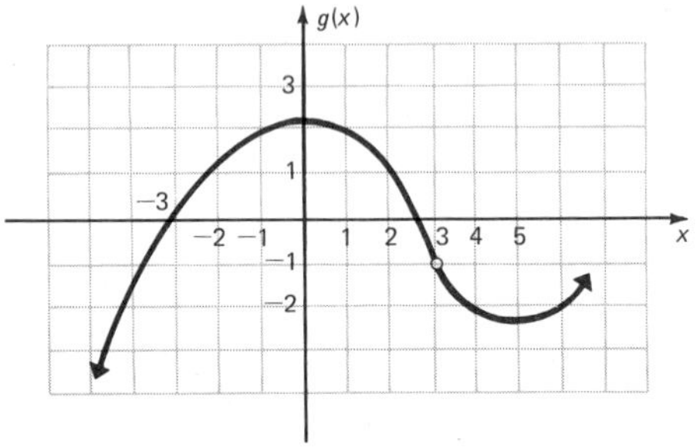

Figure 12.2

1. Find $\lim_{x \to 2} f(x)$ for the following

(a)

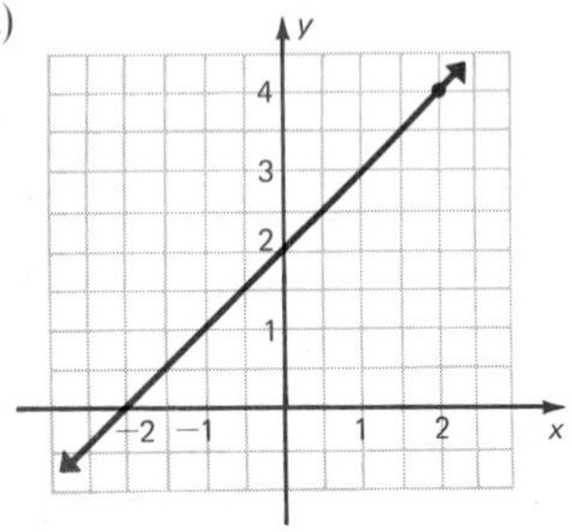

(b)

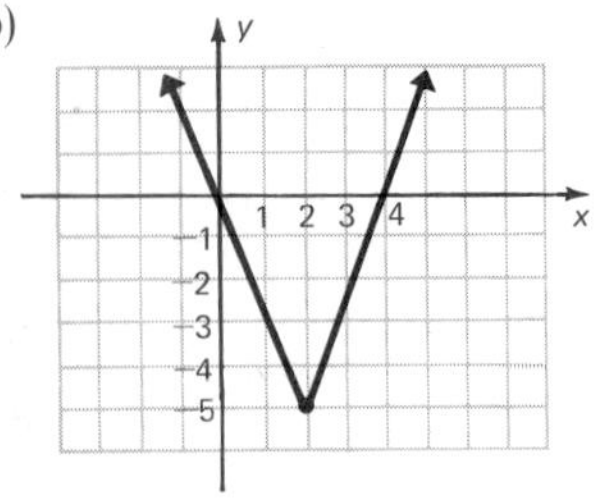

Answer:

(a) 4

(b) -5

Work Problem 1 at the side.

Example 2 The graph of function h in Figure 12.3 shows that as x approaches 4 from the left, the values of $h(x)$ approach 3. However,

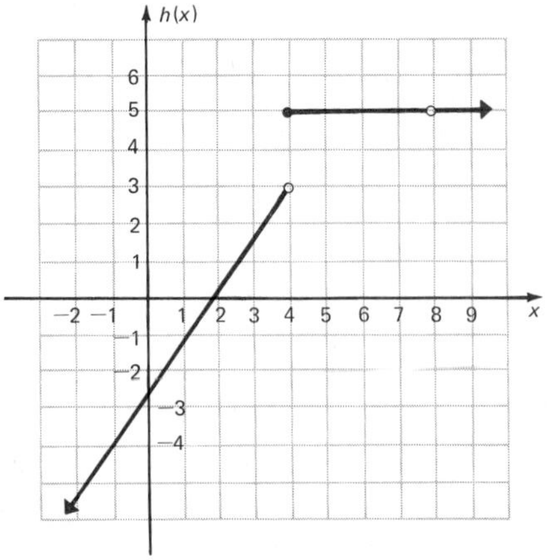

Figure 12.3

as x gets closer and closer to 4 from the right, the values of $h(x)$ approach 5. Since the values of $h(x)$ approach two different numbers depending on whether x approaches 4 from the left or the right,

$$\lim_{x \to 4} h(x) \text{ does not exist.} \quad \blacksquare$$

2. Use the graph of Figure 12.3 to find the following.

(a) $\lim_{x \to 6} h(x)$

(b) $\lim_{x \to -1} h(x)$

(c) $\lim_{x \to 8} h(x)$

Answer:

(a) 5

(b) -4

(c) 5

Work Problem 2 at the side.

Example 3 Graph $f(x) = \dfrac{1}{x}$ and use the graph to find the following.

(a) $\lim_{x \to 4} f(x)$

The graph is shown in Figure 12.4. As x gets closer and closer to 4, the values of $f(x) = \dfrac{1}{x}$ get closer and closer to 1/4, so that

$$\lim_{x \to 4} f(x) = \frac{1}{4}.$$

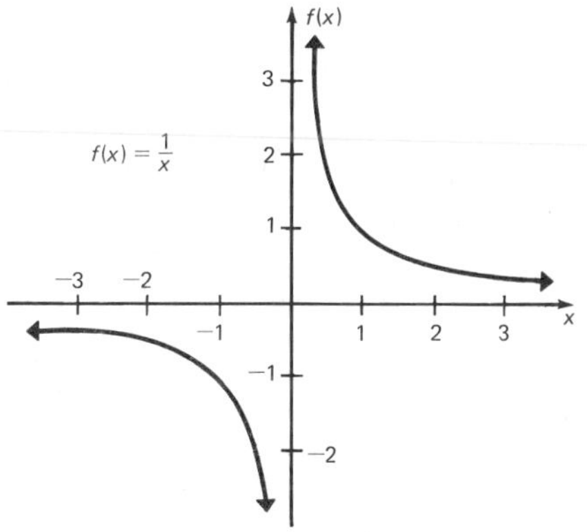

Figure 12.4

(b) $\lim_{x \to 0} f(x)$

As x gets close to 0 from the left, the values of $f(x)$ get more and more negative. As x approaches 0 from the right, the values get larger and larger. Thus, the values of $f(x)$ approach no one fixed number, and

$$\lim_{x \to 0} f(x) \text{ does not exist.} \quad \blacksquare$$

3. Use Figure 12.4 to find the following.
(a) $\lim_{x \to 1} f(x)$
(b) $\lim_{x \to -3} f(x)$
(c) $\lim_{x \to 2} f(x)$

Answer:
(a) 1
(b) $-1/3$
(c) $1/2$

Work Problem 3 at the side.

In the examples above, we found limits by drawing a graph. There are other methods for finding limits; for example, we can find several values of the function as x approaches a given number. The next few examples show how this works.

Example 4 Find $\lim_{x \to 1} \dfrac{x^2 - 4}{x - 2}$.

We want to find the limit of the function

$$f(x) = \frac{x^2 - 4}{x - 2}$$

as x approaches 1. We can do this by choosing several values of x on either side of 1 and close to 1, and evaluating $f(x)$ for these values. (Don't use 1 itself.) The results for several different values of x are shown in the following table.

Closer and closer to 1 ↓

x	0.8	0.9	0.99	0.9999	1.0000001	1.0001	1.001	1.01	1.05	1.1
$f(x)$	2.8	2.9	2.99	2.9999	3.0000001	3.0001	3.001	3.01	3.05	3.1

↑ Closer and closer to 3

From the table we see that the values of $f(x)$ get closer and closer to 3 as the values of x get closer and closer to 1. Thus the limit of $f(x)$ as x approaches 1 is 3, or

$$\lim_{x \to 1} \frac{x^2 - 4}{x - 2} = 3. \quad ■$$

4. Find $\lim_{x \to 4} \dfrac{x^2 - 4}{x - 2}$ by completing the following table. A calculator will be helpful.

x	$f(x)$
3.8	
3.9	
3.99	
3.999	
4.001	
4.1	
4.2	

Answer: 5.8; 5.9; 5.99; 5.999; 6.001; 6.1; 6.2; limit is 6

Work Problem 4 at the side.

Example 5 Find $\lim_{x \to 2} \dfrac{x^2 - 4}{x - 2}$.

The value $x = 2$ is not in the domain of the function, since 2 makes the denominator equal 0. Even though $f(2)$ does not exist, we can still try to find the limit. (Recall Example 1 above; the limit existed, even though the function was not defined.) First, make a table of values.

Closer and closer to 2

x	1.8	1.9	1.99	1.9999	↓	2.0000001	2.00001	2.001	2.05	2.1
$f(x)$	3.8	3.9	3.99	3.9999	↑	4.0000001	4.00001	4.001	4.05	4.1

Closer and closer to 4

From this table we see that the values of $f(x)$ get closer and closer to 4 as x gets closer and closer to 2. Therefore,

$$\lim_{x\to 2} \frac{x^2-4}{x-2} = 4. \quad \blacksquare$$

5. Complete the following table of values to find $\lim\limits_{x\to 3} \dfrac{x^2-9}{x-3}$.

x	$f(x)$
2.9	
2.99	
2.999	
3.001	
3.01	
3.1	

Answer: 5.9; 5.99; 5.999; 6.001; 6.01; 6.1; limit is 6

Work Problem 5 at the side.

In the last two examples above, you may have noticed that

$$\frac{x^2-4}{x-2} = \frac{(x+2)(x-2)}{x-2} = x+2.$$

The values in the tables could have been found more easily by evaluating $x+2$ instead of $(x^2-4)/(x-2)$; the limits could also have been found in this way. For example,

$$\lim_{x\to 1} \frac{x^2-4}{x-2} = \lim_{x\to 1}(x+2).$$

As x gets closer and closer to 1, the expression $x+2$ will get closer and closer to $1+2=3$. Thus,

$$\lim_{x\to 1} \frac{x^2-4}{x-2} = \lim_{x\to 1}(x+2) = 3.$$

Example 6 Let $f(x) = \dfrac{x^2-x-2}{x+1}$. Find the following limits.

(a) $\lim\limits_{x\to 0} f(x)$

First, simplify $\dfrac{x^2-x-2}{x+1}$ by factoring.

$$\frac{x^2-x-2}{x+1} = \frac{(x+1)(x-2)}{x+1} = x-2$$

Now we can find the limit.

$$\lim_{x\to 0} f(x) = \lim_{x\to 0} \frac{x^2-x-2}{x+1} = \lim_{x\to 0}(x-2)$$

As x approaches 0, $x - 2$ will approach $0 - 2$, or -2.

$$\lim_{x \to 0} f(x) = \lim_{x \to 0}(x - 2) = -2$$

(b) $\lim_{x \to 5} f(x) = \lim_{x \to 5}(x - 2) = 3$

As x approaches 5, $x - 2$ approaches $5 - 2 = 3$.

(c) $\lim_{x \to -1} f(x) = \lim_{x \to -1} (x - 2) = -3$ ■

6. Let $f(x) = \dfrac{2x^2 - 3x - 2}{x - 2}$.
Find the following.
(a) $\lim_{x \to 3} f(x)$
(b) $\lim_{x \to 0} f(x)$
(c) $\lim_{x \to 2} f(x)$

Answer:
(a) 7
(b) 1
(c) 5

Work Problem 6 at the side.

Example 7 Find $\lim_{x \to 3}(x^2 + 2x + 5)$.

There are no denominators here that might become 0, so we can find this limit by reasoning as follows: As x approaches 3, x^2 will approach 3^2, $2x$ will approach $2 \cdot 3$, and 5 will not change. Therefore,

$$\lim_{x \to 3} (x^2 + 2x + 5) = 3^2 + 2 \cdot 3 + 5 = 9 + 6 + 5 = 20. \quad ■$$

7. Find the following.
(a) $\lim_{x \to 4}(3x - 9)$
(b) $\lim_{x \to -1} (2x^2 - 4x + 1)$
(c) $\lim_{x \to 0}(9x^3 - 8x^2 + 6x + 4)$

Answer:
(a) 3
(b) 7
(c) 4

Work Problem 7 at the side.

Example 8 Find $\lim_{x \to 9} \sqrt{2x - 2}$.

As x approaches 9, $\sqrt{2x - 2}$ approaches $\sqrt{2 \cdot 9 - 2}$. We have

$$\lim_{x \to 9} \sqrt{2x - 2} = \sqrt{2 \cdot 9 - 2} = \sqrt{18 - 2} = \sqrt{16} = 4. \quad ■$$

8. Find the following.
(a) $\lim_{x \to 2} \sqrt{3x + 3}$
(b) $\lim_{x \to -4} \sqrt{8 - 7x}$
(c) $\lim_{x \to 2} \sqrt{x^2 + 3x + 4}$

Answer:
(a) 3
(b) 6
(c) $\sqrt{14}$

Work Problem 8 at the side.

Example 9 Find $\lim_{x \to 4} \dfrac{\sqrt{x} - 2}{x - 4}$.

As x approaches 4, we have

$$\lim_{x \to 4} \frac{\sqrt{x} - 2}{x - 4} = \frac{\sqrt{4} - 2}{4 - 4} = \frac{2 - 2}{4 - 4} = \frac{0}{0}.$$

The expression 0/0 is not a real number; in fact 0/0 is called an **indeterminate form.** Even though we got 0/0, the limit might still exist. We can look for the limit by making a table of values or by using a little algebra. To use algebra, multiply numerator and denominator of $(\sqrt{x}-2)/(x-4)$ by $\sqrt{x}+2$. This gives

$$\frac{\sqrt{x}-2}{x-4}\cdot\frac{\sqrt{x}+2}{\sqrt{x}+2}=\frac{\sqrt{x}\cdot\sqrt{x}-2\sqrt{x}+2\sqrt{x}-4}{(x-4)(\sqrt{x}+2)}$$

$$=\frac{x-4}{(x-4)(\sqrt{x}+2)} \qquad \sqrt{x}\cdot\sqrt{x}=x$$

$$=\frac{1}{\sqrt{x}+2}.$$

We now have

$$\lim_{x\to 4}\frac{\sqrt{x}-2}{x-4}=\lim_{x\to 4}\frac{1}{\sqrt{x}+2}=\frac{1}{\sqrt{4}+2}=\frac{1}{2+2}=\frac{1}{4}. \blacksquare$$

Whenever the indeterminate form 0/0 is obtained, the limit may still exist. You must do further work to find out.

9. Find the following.

(a) $\lim_{x\to 1}\frac{\sqrt{x}-1}{x-1}$

(b) $\lim_{x\to 25}\frac{\sqrt{x}-5}{x-25}$

Answer:
(a) 1/2
(b) 1/10

Work Problem 9 at the side.

Example 10 The graph of Figure 12.5 shows the profit from producing x units of a certain item in a given day. Notice the break in the graph at $x = 25$. This break occurs because $x = 25$ items is the maximum that can be produced by one shift of workers. If more than 25 items are needed, the night shift must be called to work. This increases the fixed costs and causes a drop in profits unless many more than 25 items are needed. (By inspecting the graph, we see that 35 items must be needed before it pays to bring on the second shift.) From the graph,

$$\lim_{x\to 25} p(x) \text{ does not exist.} \quad \blacksquare$$

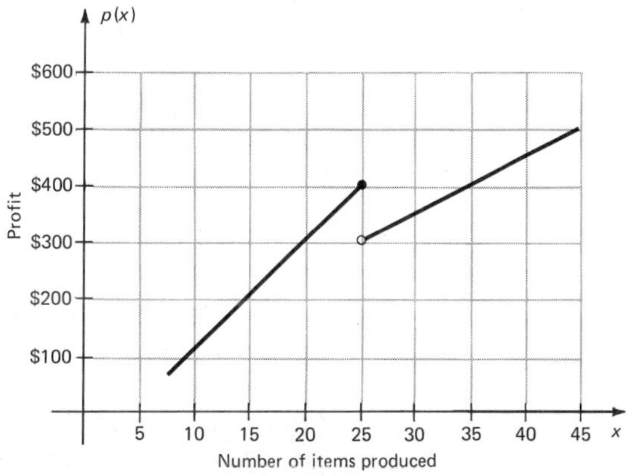

Figure 12.5

10. Use the graph of Figure 12.5 to find the following.

(a) $\lim_{x \to 20} p(x)$

(b) $\lim_{x \to 30} p(x)$

Answer:

(a) \$300

(b) \$340

Work Problem 10 at the side.

Limits to Infinity We sometimes need to find the limit of a function as values of x get larger and larger without bound. This can be done by using **limits to infinity** as shown in the following example.

Example 11 Find the limit of $f(x) = \dfrac{6x+1}{2x-5}$ as x increases without bound.

This limit is written

$$\lim_{x \to \infty} \frac{6x+1}{2x-5}.$$

To find the limit, we can complete a table of values.

x	1000	100,000	1,000,000	1,000,000,000
$f(x)$	3.00802	3.00008	3.000008	3.000000008

From this table, $\lim_{x \to \infty} \dfrac{6x+1}{2x-5} = 3.$ ■

11. Make a table of values and find the following limits.

(a) $\lim_{x \to \infty} \dfrac{15x+9}{3x-6}$

(b) $\lim_{x \to \infty} \dfrac{1}{x}$

(c) $\lim_{x \to \infty} \dfrac{1}{x^2}$

Answer:

(a) 5

(b) 0

(c) 0

Work Problem 11 at the side.

The results of parts (b) and (c) of Problem 11 can be generalized as follows:

$$\lim_{x \to \infty} \frac{1}{x^n} = 0 \qquad \text{for any positive integer } n.$$

This fact can be used to find limits to infinity, as shown in the next example.

Example 12 Find the following limits.

(a) $\lim_{x \to \infty} \dfrac{8x+6}{3x-1}$

The highest power of x here is x^1, so we divide numerator and denominator by x^1.

$$\lim_{x \to \infty} \frac{8x+6}{3x-1} = \lim_{x \to \infty} \frac{\frac{8x}{x} + \frac{6}{x}}{\frac{3x}{x} - \frac{1}{x}} = \lim_{x \to \infty} \frac{8 + \frac{6}{x}}{3 - \frac{1}{x}} = \frac{8+0}{3-0} = \frac{8}{3}$$

(b) $\lim_{x\to\infty} \dfrac{4x^2 - 6x + 3}{2x^2 - x + 4}$

Divide numerator and denominator by x^2, the highest power of x.

$$\lim_{x\to\infty} \frac{4x^2 - 6x + 3}{2x^2 - x + 4} = \lim_{x\to\infty} \frac{4 - \frac{6}{x} + \frac{3}{x^2}}{2 - \frac{1}{x} + \frac{4}{x^2}} = \frac{4 - 0 + 0}{2 - 0 + 0} = \frac{4}{2} = 2$$

(c) $$\lim_{x\to\infty} \frac{3x + 2}{4x^2 - 1} = \lim_{x\to\infty} \frac{\frac{3}{x} + \frac{2}{x^2}}{4 - \frac{1}{x^2}} = \frac{0 + 0}{4 - 0} = \frac{0}{4} = 0$$

(d) $$\lim_{x\to\infty} \frac{9x^2 - 1}{3x + 5} = \lim_{x\to\infty} \frac{9 - \frac{1}{x^2}}{\frac{3}{x} + \frac{5}{x^2}} = \frac{9 - 0}{0 + 0} = \frac{9}{0}$$

Since 9/0 is not a real number, this limit does not exist. ■

Work Problem 12 at the side.

12. Find the following limits.

(a) $\lim_{x\to\infty} \dfrac{4x^2 + 6x}{9x^2 - 3}$

(b) $\lim_{x\to\infty} \dfrac{3x^2 + 4x + 5}{8x^3 + 2x + 1}$

(c) $\lim_{x\to\infty} \dfrac{4x^2 + 9x + 1}{2x + 3}$

Answer:
(a) 4/9
(b) 0
(c) does not exist

12.1 EXERCISES

Use the following graphs to find the indicated limits if they exist. (See Examples 1–3.)

1. $\lim_{x\to 3} f(x)$

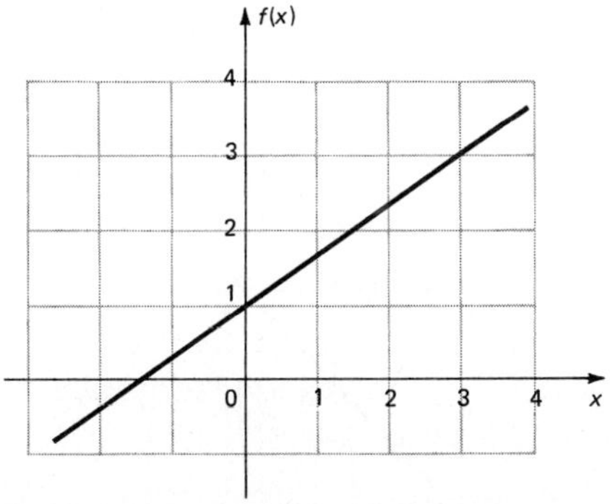

2. $\lim_{x\to 2} F(x)$

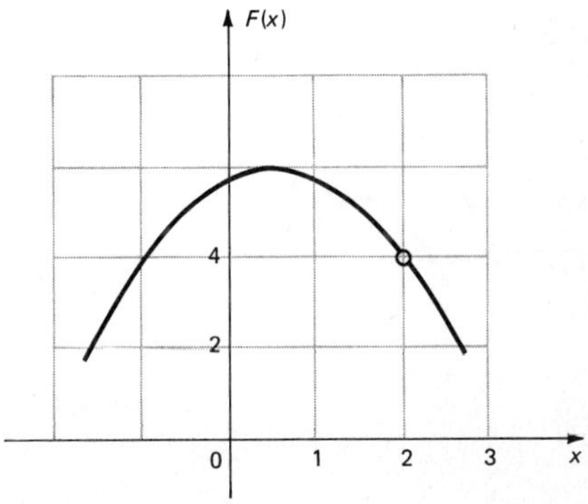

3. $\lim_{x \to -2} f(x)$

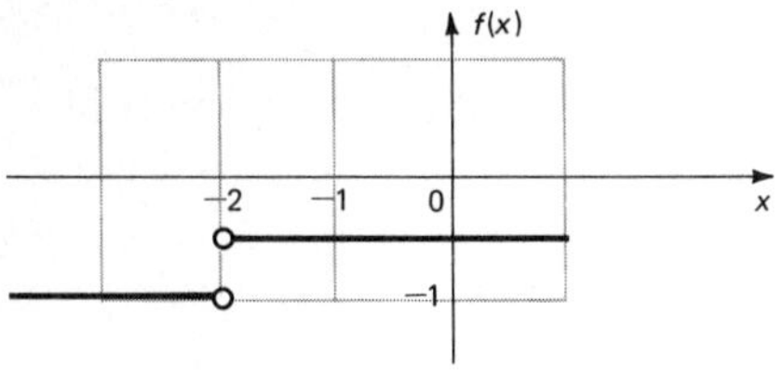

4. $\lim_{x \to 3} g(x)$

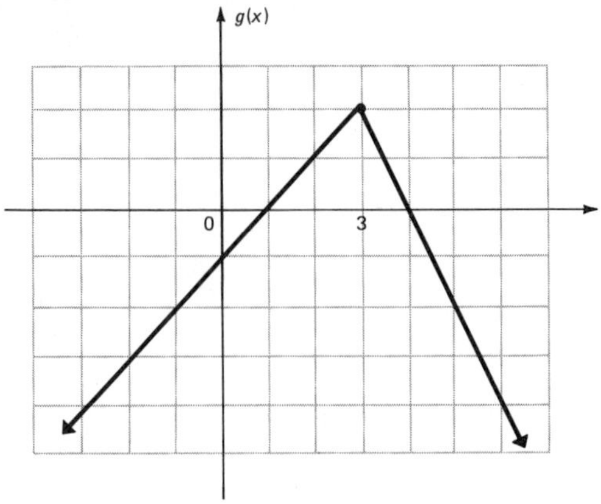

5. $\lim_{x \to 0} f(x)$

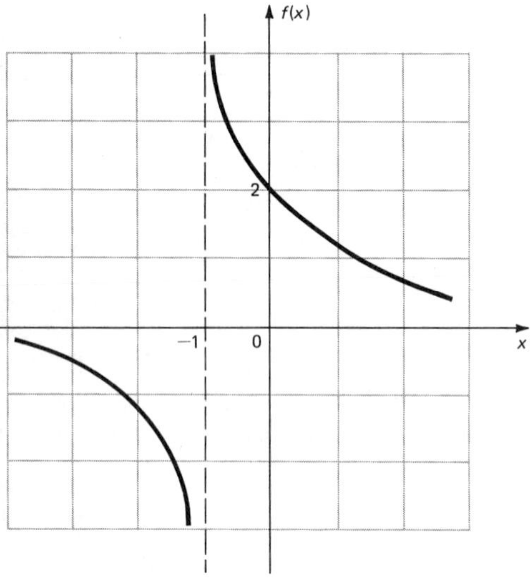

6. $\lim_{x \to 1} h(x)$

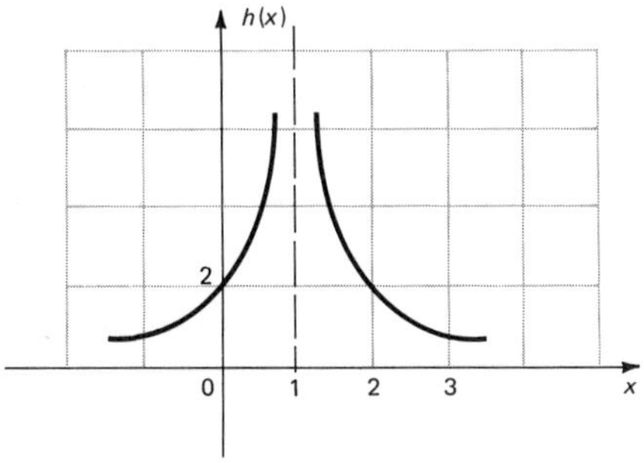

7. $\lim_{x \to 0} f(x)$

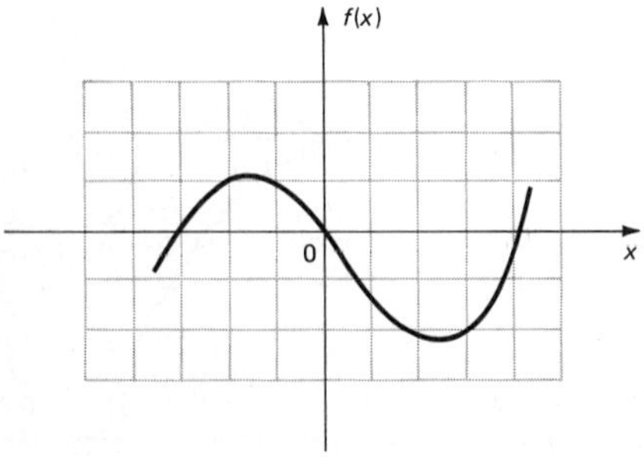

8. $\lim_{x \to 1} g(x)$

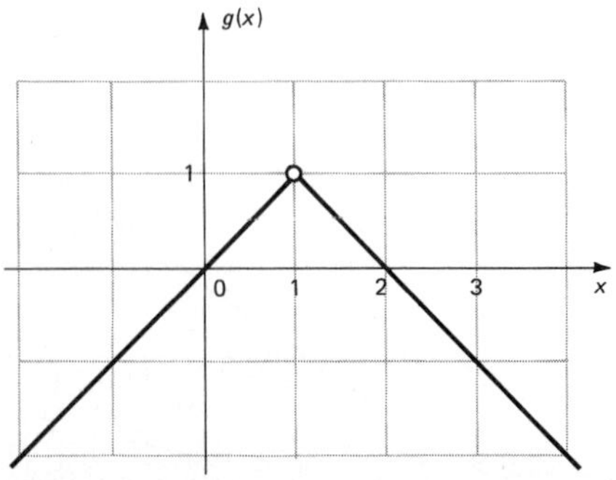

Complete the following table and use the results to find $\lim_{x \to a} f(x)$. *You will need a calculator for Exercises 15–18 and a* $\sqrt{x}$ *key for Exercises 17 and 18. (See Examples 4 and 5.)*

9. $f(x) = 4 + 5x$, find $\lim_{x \to -3} f(x)$.

x	-3.1	-3.01	-3.001	-3.0001	-2.9999	-2.999	-2.99	-2.9
$f(x)$								

10. $g(x) = 2x - 5$, find $\lim_{x \to 3} g(x)$.

x	2.9	2.99	2.999	2.9999	3.0001	3.001	3.01	3.1
$g(x)$								

11. $h(x) = x^2 + 2x + 1$, find $\lim_{x \to 2} h(x)$.

x	1.9	1.99	1.999	2.001	2.01	2.1
$h(x)$			8.994001	9.006001		

12. $f(x) = 2x^2 - 4x + 3$, find $\lim_{x \to 1} f(x)$.

x	.9	.99	.999	1.001	1.01	1.1
$f(x)$			1.000002	1.000002		

13. $g(x) = 2/x$, find $\lim_{x \to 0} g(x)$.

x	$-.1$	$-.01$	$-.001$	.001	.01	.1
$g(x)$						

14. $h(x) = -5/x$, find $\lim_{x \to 0} h(x)$.

x	$-.1$	$-.01$	$-.001$	.001	.01	.1
$h(x)$						

15. $k(x) = \dfrac{x^3 - 2x - 4}{x - 2}$, find $\lim_{x \to 2} k(x)$.

x	1.9	1.99	1.999	2.001	2.01	2.1
$k(x)$						

16. $f(x) = \dfrac{2x^3 + 3x^2 - 4x - 5}{x + 1}$, find $\lim_{x \to -1} f(x)$.

x	-1.1	-1.01	-1.001	$-.999$	$-.99$	$-.9$
$f(x)$						

17. $h(x) = \dfrac{\sqrt{x} + 2}{x - 1}$, find $\lim_{x \to 1} h(x)$.

x	.9	.99	.999	1.001	1.01	1.1
$h(x)$						

18. $f(x) = \dfrac{\sqrt{x} - 3}{x - 3}$, find $\lim_{x \to 3} f(x)$.

x	2.999	2.99	2.9	3.001	3.01	3.1
$f(x)$						

Use any method to find the following limits that exist. (See Examples 6–9.)

19. $\lim_{x \to 4} 3x$

20. $\lim_{x \to -3} -4x$

21. $\lim_{x \to -3} (4x^2 + 2x - 1)$

22. $\lim_{x \to 2} (2x^2 - 3x + 5)$

23. $\lim_{x \to 2} (2x^3 + 5x^2 + 2x + 1)$

24. $\lim_{x \to -1} (4x^3 - x^2 + 3x - 1)$

25. $\lim_{x \to 3} \dfrac{5x - 6}{2x + 1}$

26. $\lim_{x \to -2} \dfrac{2x + 1}{3x - 4}$

27. $\lim_{x \to 1} \dfrac{2x^2 - 6x + 3}{3x^2 - 4x + 2}$

28. $\lim_{x \to 2} \dfrac{-4x^2 + 6x - 8}{3x^2 + 7x - 2}$

29. $\lim_{x \to 3} \dfrac{x^2 - 9}{x - 3}$

30. $\lim_{x \to -2} \dfrac{x^2 - 4}{x + 2}$

31. $\lim_{x \to -2} \dfrac{x^2 - x - 6}{x + 2}$

32. $\lim_{x \to 5} \dfrac{x^2 - 3x - 10}{x - 5}$

33. $\lim_{x \to 0} \dfrac{x^2 - x}{x}$

34. $\lim_{x \to 0} \dfrac{3x^2 - 4x}{x}$

35. $\lim_{x \to 3} \sqrt{x^2 - 4}$

36. $\lim_{x \to 3} \sqrt{x^2 - 5}$

37. $\lim_{x \to 25} \dfrac{\sqrt{x} - 5}{x - 25}$

38. $\lim_{x \to 36} \dfrac{\sqrt{x} - 6}{x - 36}$

The next graph shows the profit from the daily production of x thousand kilograms of an industrial chemical. Use the graph to find the following limits. (See Example 10.)

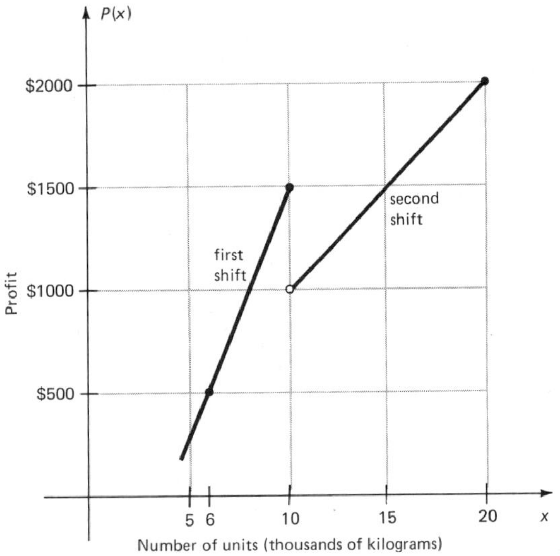

39. $\lim_{x \to 6} P(x)$ **40.** $\lim_{x \to 10} P(x)$ **41.** $\lim_{x \to 15} P(x)$

42. Use the graph to estimate the number of units of the chemical that must be produced before the second shift is profitable.

Find each of the following limits that exist. (*See Example 12.*)

43. $\lim_{x \to \infty} \dfrac{3x}{5x - 1}$

44. $\lim_{x \to \infty} \dfrac{5x}{3x - 1}$

45. $\lim_{x \to \infty} \dfrac{2x + 3}{4x - 7}$

46. $\lim_{x \to \infty} \dfrac{8x + 2}{2x - 5}$

47. $\lim_{x \to \infty} \dfrac{x^2 + 2x}{2x^2 - 2x + 1}$

48. $\lim_{x \to \infty} \dfrac{x^2 + 2x - 5}{3x^2 + 2}$

49. $\lim_{x \to \infty} \dfrac{3x^3 + 2x - 1}{2x^4 - 3x^3 - 2}$

50. $\lim_{x \to \infty} \dfrac{2x^2 + 11x - 10}{5x^3 + 3x^2 + 2x}$

51. $\lim_{x \to \infty} \dfrac{2x^4 + 3x + 4}{x^2 - 4x + 1}$

52. $\lim_{x \to \infty} \dfrac{2x^2 - 1}{3x^4 + 2}$

APPLIED PROBLEMS

Social Science

Members of a legislature must often vote repeatedly on the same issue. As time goes on, the members may change their vote. A formula for the chance*

* See John W. Bishir and Donald W. Drewes, *Mathematics in the Behavioral and Social Sciences* (New York: Harcourt Brace Jovanovich, 1970), p. 538.

that a legislator will vote yes *on the* nth *roll call vote on the same issue is*

$$p_n = \frac{1}{2} + \left(p_0 - \frac{1}{2}\right)(1 - 2p)^n,$$

where n is the number of roll calls taken, p_n is the chance that the member will vote yes *on the* nth *roll call vote, p is the chance that the legislator will change his or her position on successive roll calls, and p_0 is the chance that the member favors the issue at the beginning (p_n, p, and p_0 are always between* 0 *and* 1, *inclusive). Suppose that $p_0 = .7$ and $p = .2$. Use a calculator to find p_n for*

53. $n = 2$; **54.** $n = 4$; **55.** $n = 8$.

56. Find $\lim_{n \to \infty} p_n$.

12.2 CONTINUITY

In the previous section, we saw a graph showing the profit from producing x units of a certain item. This graph is repeated in Figure 12.6. The graph is smooth and unbroken except for a sudden drop at the point $x = 25$. A break in a graph such as this is called a **discontinuity.** The function is **discontinuous** at the given point.

1. Find any discontinuities for the following functions.

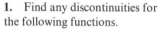

(a)

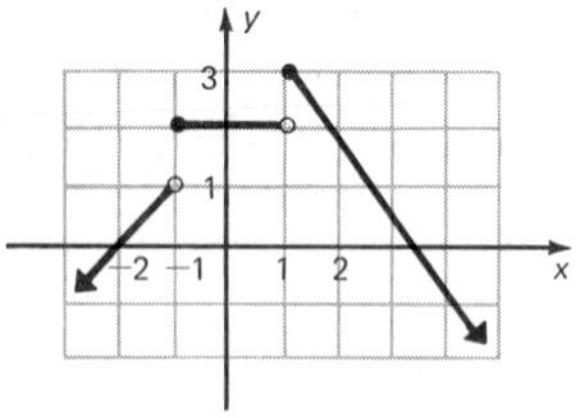

(b)

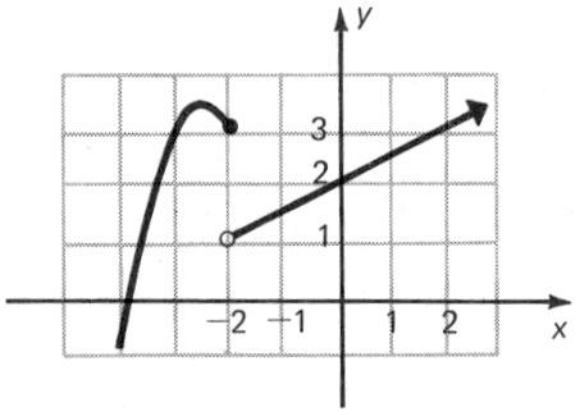

(c)

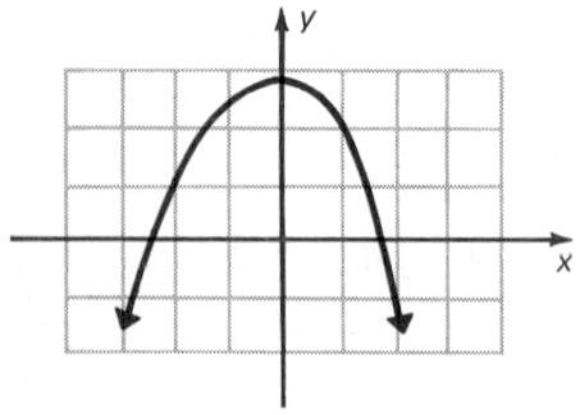

Answer:
(a) $-1, 1$
(b) -2
(c) discontinuous nowhere

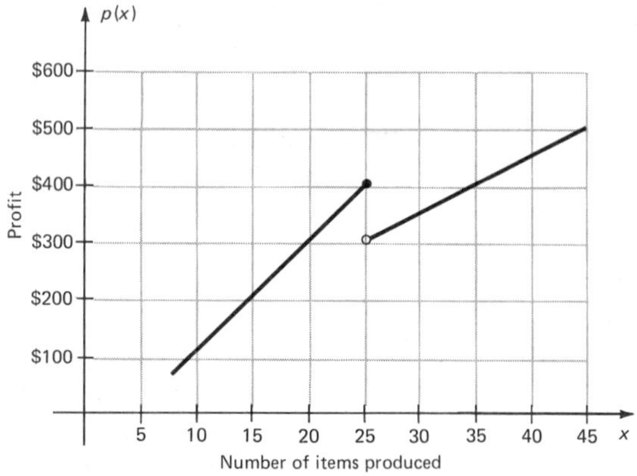

Figure 12.6

Work Problem 1 at the side.

Example 1 A trailer rental firm charges a flat $4 to rent a hitch. The trailer itself is rented for $11 per day or fraction of a day. Let $C(x)$ represent the cost of renting a hitch and trailer for x days.
(a) Graph $C(x)$.

2. A chain-saw rental firm charges a flat \$6 sharpening fee, plus \$5 per day or fraction of a day. Let $C(x)$ represent the cost of renting a saw for x days.
(a) Graph $C(x)$.
(b) Find any places where the function is discontinuous.

Answer:
(a)

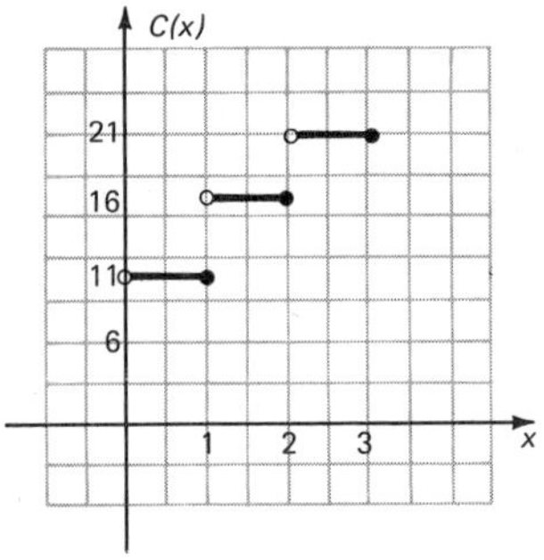

(b) at $x = 1, 2, 3, 4, 5, 6, \ldots$

3. Which of the following functions are continuous?
(a)

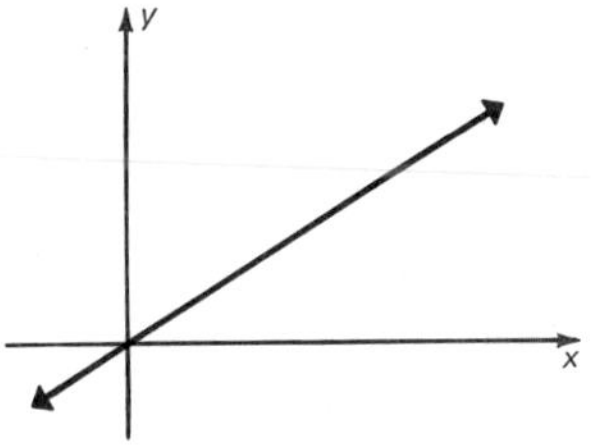

(b)

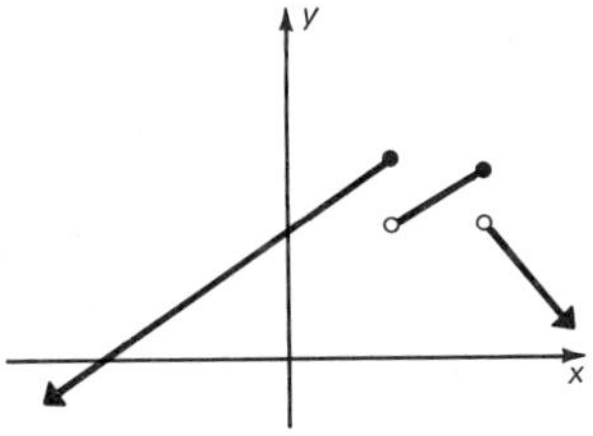

(c)

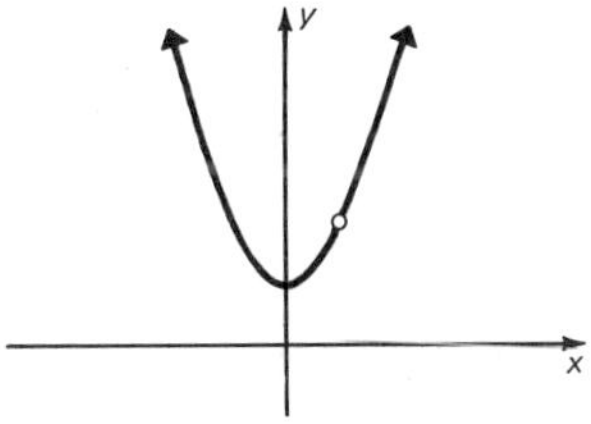

Answer:
(a) continuous
(b) not continuous
(c) not continuous

To rent a trailer for one day, we pay \$4 for the hitch and \$11 for the trailer. The total rental is thus \$4 + \$11 = \$15. In fact, if $0 < x \leq 1$, then $C(x) = 15$. If we rent the trailer for more than one day, but not more than two, we pay $4 + 2 \cdot 11 = 26$ dollars. Thus, if $1 < x \leq 2$, then $C(x) = 26$. Also, if $2 < x \leq 3$, then $C(x) = 37$. These results lead to the graph of Figure 12.7.

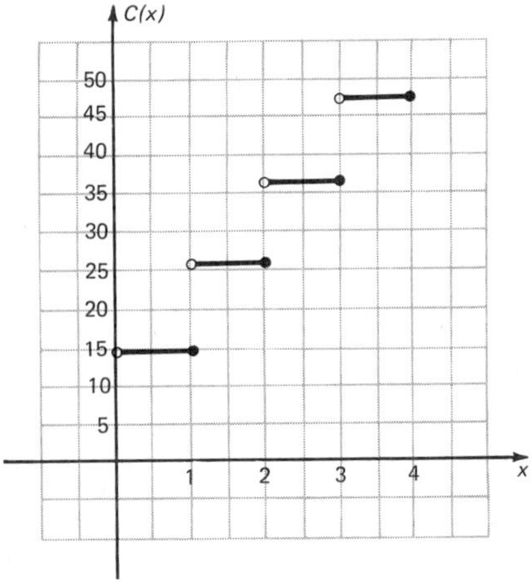

Figure 12.7

(b) Find any places where the function $C(x)$ is discontinuous.

The graph of Figure 12.7 shows that the function is discontinuous at $x = 1, 2, 3, 4, 5, 6, \ldots$. ■

Work Problem 2 at the side.

If a function has a discontinuity, then it is not possible to draw a graph of the function without lifting the pencil from the paper. On the other hand, if a function *can* be drawn without lifting the pencil from the paper, the function is *continuous*. Many useful functions are continuous. In particular, the linear, quadratic, and polynomial functions of Chapter 3 are continuous; the rational functions are usually not.

Work Problem 3 at the side.

The graph of Figure 12.8 is that of a continuous function—the graph can be drawn without lifting the pencil from the paper. Let us see why this function is continuous by studying a sample point

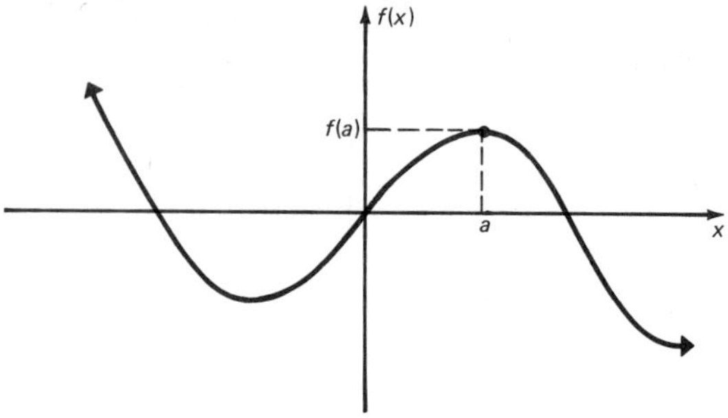

Figure 12.8

such as the point $x = a$. First of all, $f(a)$ is defined. That is, there is a point on the graph corresponding to $x = a$.

Second, notice that $\lim_{x \to a} f(x)$ exists. As x approaches a from either side of a, the values of $f(x)$ approach one particular number. The number that the values of $f(x)$ approach is in fact $f(a)$.

In general, the function f is **continuous at $x = a$** if all of the following conditions hold:

(a) $f(a)$ exists

(b) $\lim_{x \to a} f(x)$ exists

(c) $\lim_{x \to a} f(x) = f(a)$

Example 2 Tell why the following functions are discontinuous at the indicated points.

(a)

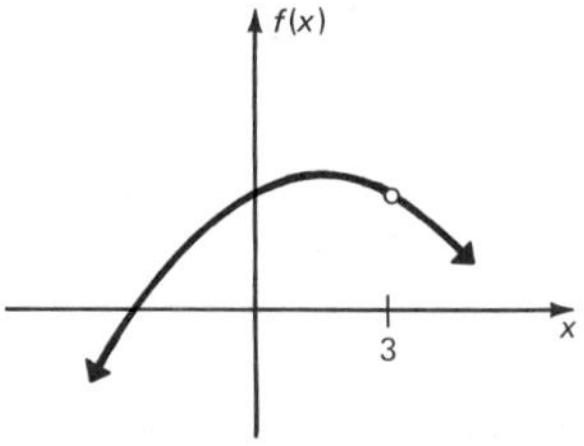

The open circle on the graph shows that there is no value of the function for $x = 3$. Thus, $f(3)$ does not exist, and part (a) of the definition fails.

4. Tell why the following functions are discontinuous at the indicated points.

(a)

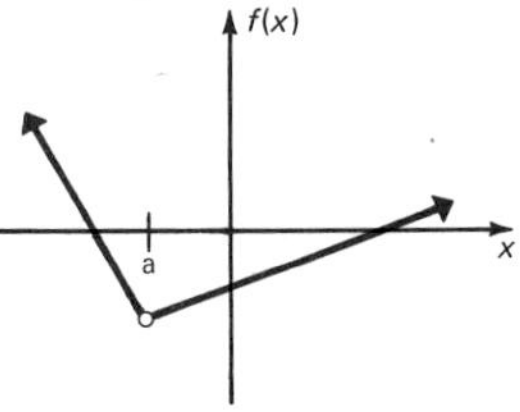

(b)

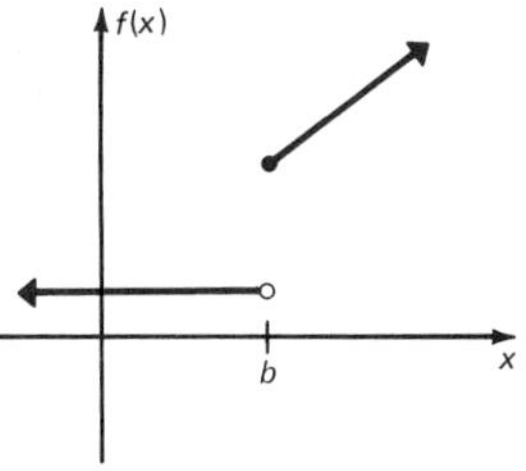

(c)

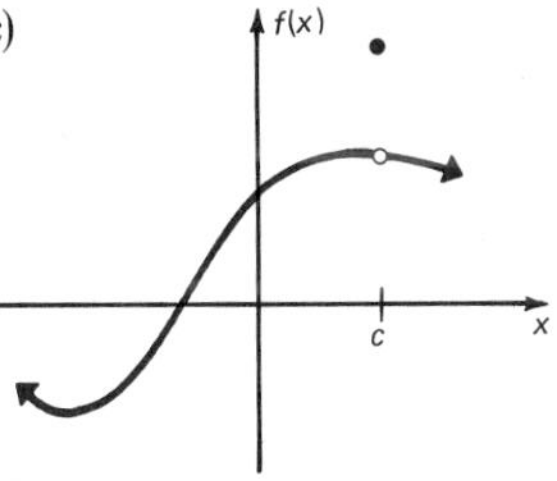

Answer:

(a) $f(a)$ does not exist

(b) $\lim_{x \to b} f(x)$ does not exist

(c) $\lim_{x \to c} f(x) \neq f(c)$

(b)

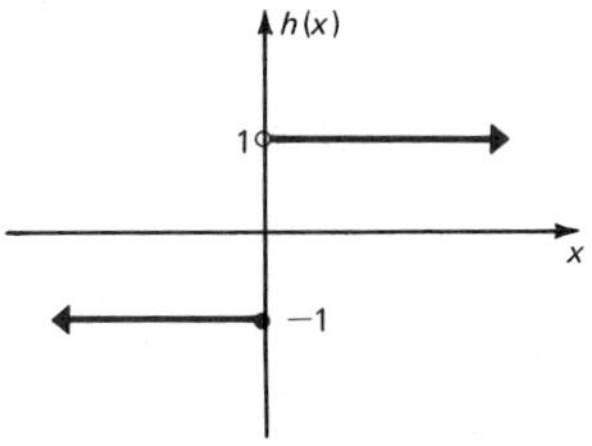

The graph shows that $f(0)$ exists. As x approaches 0 from the left, $h(x)$ is -1. However, as x approaches 0 from the right, $h(x)$ is 1. Thus, $\lim_{x \to 0} h(x)$ does not exist and part (b) of the definition fails.

(c)

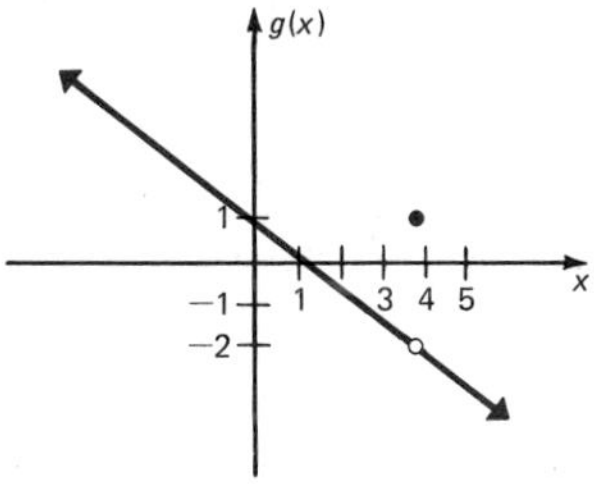

Here, the heavy dot above 4 shows that $g(4)$ exists. In fact, $g(4) = 1$. However, check that

$$\lim_{x \to 4} g(x) = -2.$$

Thus, $\lim_{x \to 4} g(x) \neq g(4)$, and part (c) of the definition fails. ■

Work Problem 4 at the side.

When discussing the continuity of a function, it is often helpful to use **interval notation.** This notation is defined below.

Interval	Name	Description	Interval notation
	Open interval	$-2 < x < 3$	$(-2, 3)$
	Closed interval	$-2 \le x \le 3$	$[-2, 3]$
	Open interval	$x < 3$	$(-\infty, 3)$
	Open interval	$x > -5$	$(-5, \infty)$

The symbol ∞ (read "infinity") does not represent a number; ∞ is used for convenience in interval notation.

5. Write each of the following using interval notation.

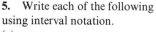

(a) 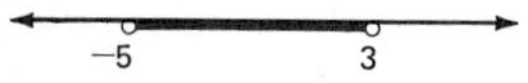

(b) 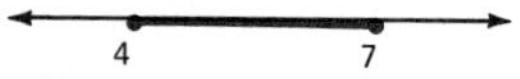

(c) 

(d) 

Answer:
(a) $(-5, 3)$
(b) $[4, 7]$
(c) $(-\infty, -1)$
(d) $(10, \infty)$

Work Problem 5 at the side.

If a function is continuous at each point of an open interval, it is said to be **continuous on the open interval.**

Example 3 Find all open intervals where the function of Figure 12.9 is continuous.

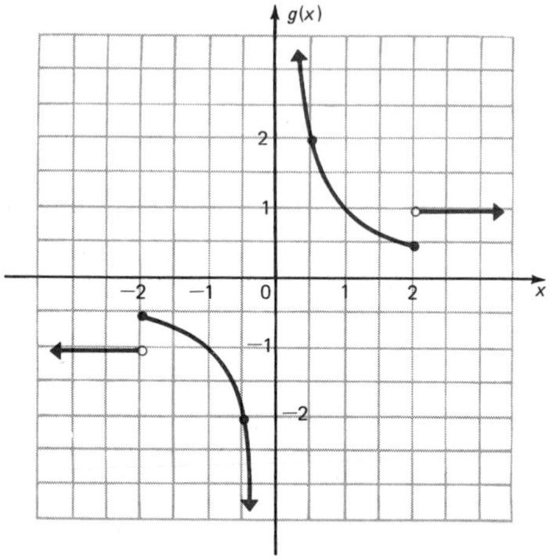

Figure 12.9

As the graph shows, the function is discontinuous at $x = -2, 0,$ and 2. Also from the graph, the function is continuous on the open intervals $(-\infty, -2)$, $(-2, 0)$, $(0, 2)$, and $(2, \infty)$. ■

6. Find all open intervals where the following functions are continuous.

(a)

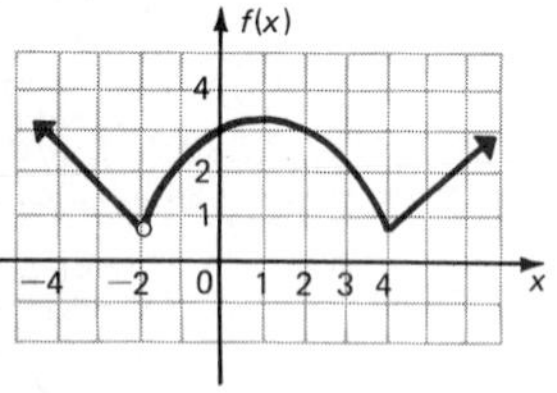

(b)

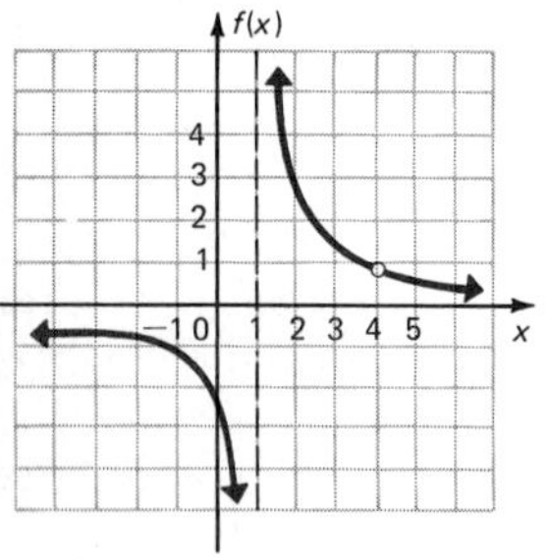

Answer:
(a) $(-\infty, -2)$, $(-2, \infty)$
(b) $(-\infty, 1)$, $(1, 4)$, $(4, \infty)$

Work Problem 6 at the side.

12.2 EXERCISES

Find all points of discontinuity for the following functions. (See Example 1.)

1.

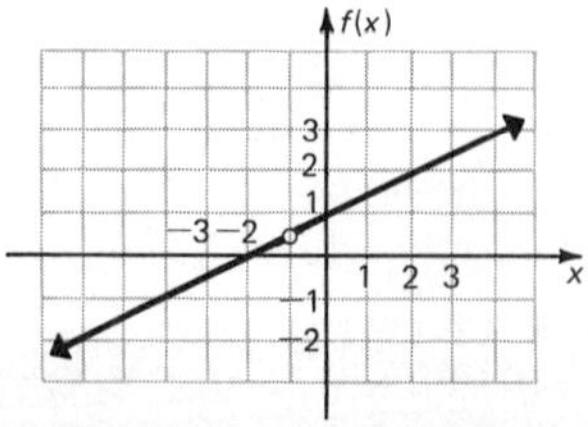

2.

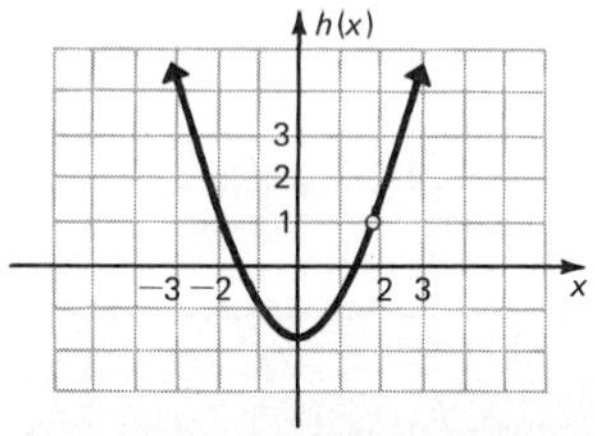

3.

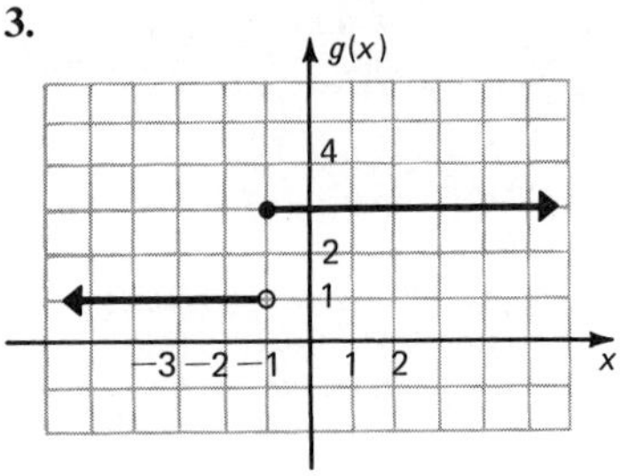

4.

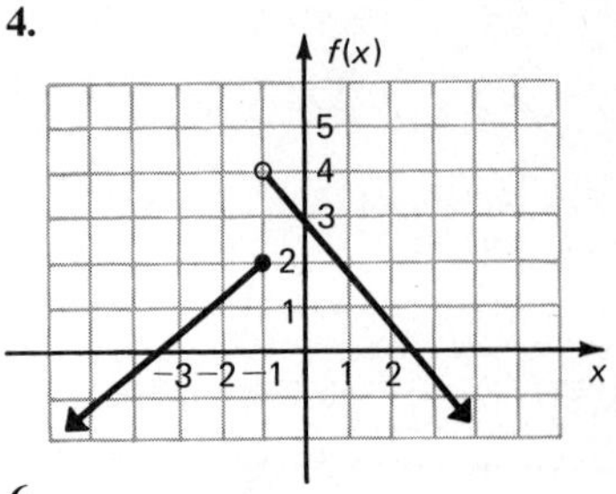

5.

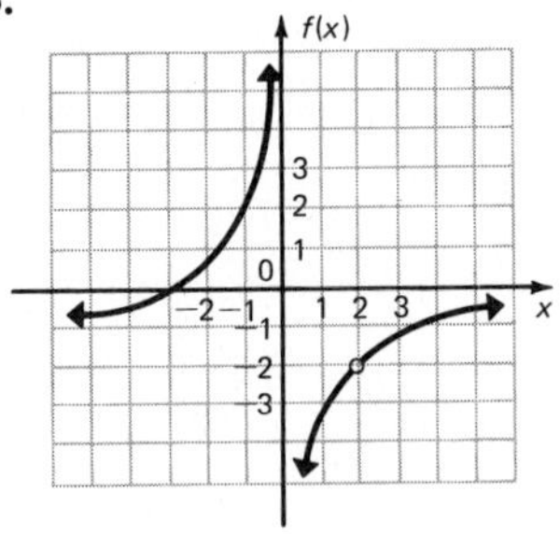

6.

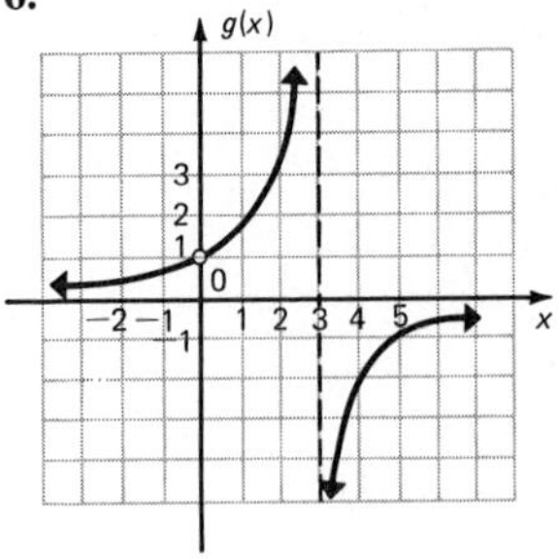

7.

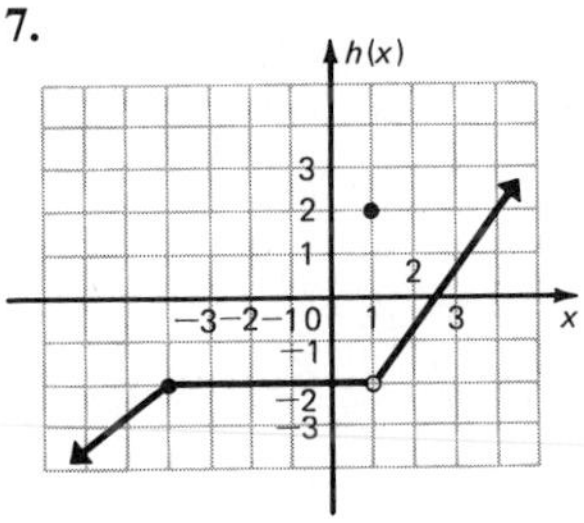

8.

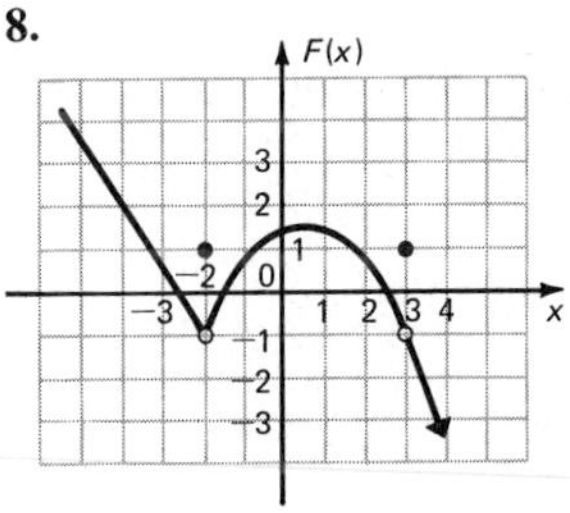

Are the following functions continuous at the given points? (*See Example 2.*)

9. $f(x) = \dfrac{2}{x-3}$; $x = 0,\ x = 3$

10. $f(x) = \dfrac{6}{x}$; $x = 0,\ x = -1$

11. $g(x) = \dfrac{1}{x(x-2)}$; $x = 0,\ x = 2,\ x = 4$

12. $h(x) = \dfrac{-2}{3x(x+5)}$; $x = 0,\ x = 3,\ x = -5$

13. $h(x) = \dfrac{1+x}{(x-3)(x+1)}$; $x = 0,\ x = 3,\ x = -1$

14. $g(x) = \dfrac{-2x}{(2x+1)(3x+6)}$; $x = 0,\ x = -1/2,\ x = -2$

15. $k(x) = \dfrac{5+x}{2+x}$; $x = 0,\ x = -2,\ x = -5$

16. $f(x) = \dfrac{4-x}{x-9}$; $x = 0,\ x = 4,\ x = 9$

17. $g(x) = \dfrac{x^2 - 4}{x - 2}$; $x = 0, x = 2, x = -2$

18. $h(x) = \dfrac{x^2 - 25}{x + 5}$; $x = 0, x = 5, x = -5$

19. $p(x) = x^2 - 4x + 11$; $x = 0, x = 2, x = -1$

20. $q(x) = -3x^3 + 2x^2 - 4x + 1$; $x = -2, x = 3, x = 1$

Management

A company charges \$1.20 per pound for a certain fertilizer on all orders not over 100 pounds, and \$1 per pound for orders over 100 pounds. Let $F(x)$ represent the cost for buying x pounds of the fertilizer.

21. Find $F(80)$.

22. Find $F(150)$.

23. Graph $y = F(x)$.

24. Where is F discontinuous?

The cost to transport a mobile home depends on the distance, x, in miles that the home is moved. Let $C(x)$ represent the cost to move a mobile home x miles. One firm charges as follows.

Cost per mile	Distance in miles
\$2	if $0 < x \leq 150$
\$1.50	if $150 < x \leq 400$
\$1.25	if $400 < x$

25. Find $C(130)$.

26. Find $C(210)$.

27. Find $C(350)$.

28. Find $C(500)$.

29. Graph $y = C(x)$.

30. Where is C discontinuous?

31. There are certain skills (such as music) where learning is rapid at first and then levels off. All of a sudden, an insight may be had, causing learning to speed up. A typical graph of such learning is shown in the figure. Where is the function discontinuous?

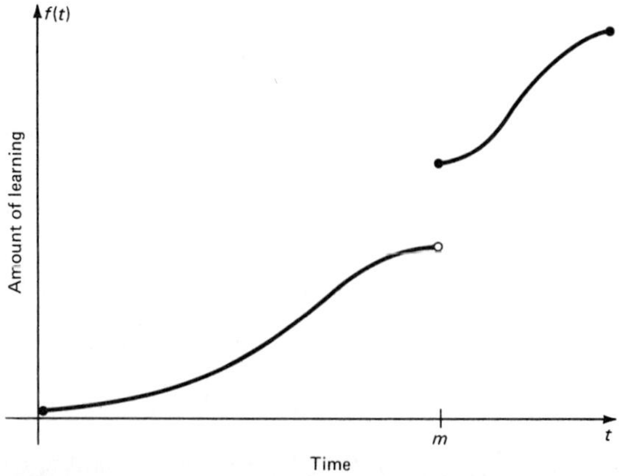

Write each of the following in interval notation.

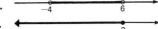

32. 

33. 

34. 

35. 

36. $\{x \mid -11 \le x \le 9\}$

37. $\{x \mid -8 \le x \le 0\}$

38. $\{x \mid -6 < x < -2\}$

39. $\{x \mid 12 < x < 20\}$

40. $\{x \mid x > -4\}$

41. $\{x \mid x < 3\}$

42. $\{x \mid x < 0\}$

43. $\{x \mid x > -10\}$

Identify all open intervals where the following functions are continuous. (*See Example 3.*)

44. 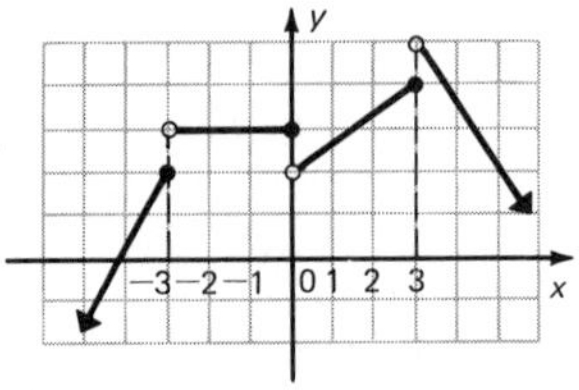

45. 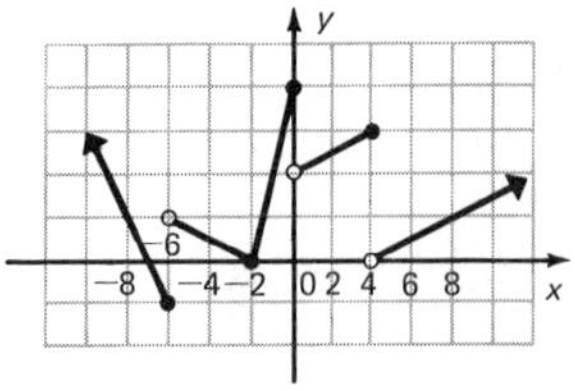

46. 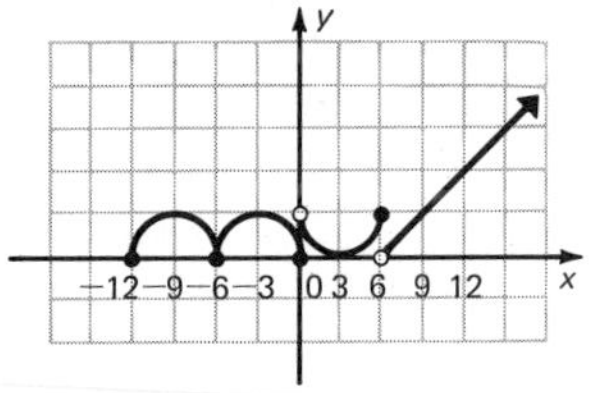

47. 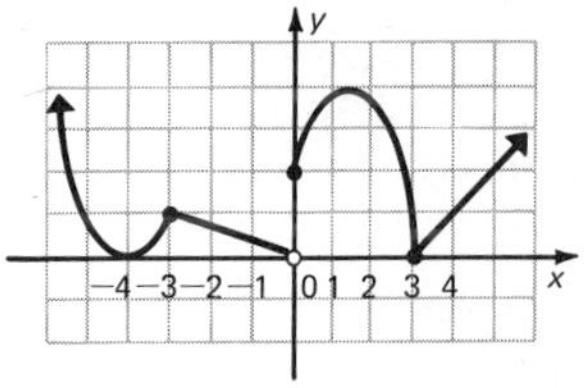

12.3 RATES OF CHANGE

One of the main applications of calculus is telling how one variable changes in relation to another. A person in business wants to know how profit changes with respect to advertising, while a person in medicine wants to know how a patient's reaction to a drug changes with respect to the dose.

For example, the graph of Figure 12.10 shows the profit $P(x)$ for a company when x thousand dollars is spent on advertising. Notice that as advertising expenditures increase, profit also increases, but at a slower rate. When $x = 1$, profit is \$2000, when $x = 3$, profit is \$3000, and so on. The **average rate of change** of profit with respect to advertising is defined as the change in profit divided by the change in advertising costs.

$$\text{average rate of change} = \frac{\text{change in profit}}{\text{change in advertising costs}}$$

$$= \frac{3000 - 2000}{3 - 1} = \frac{1000}{2} = 500$$

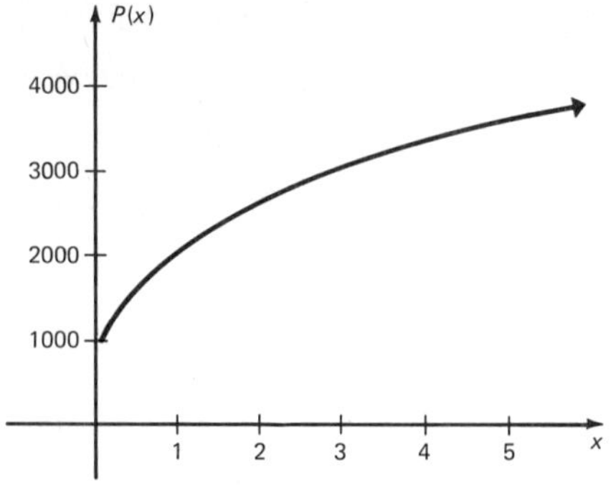

Figure 12.10

In the interval from $x = 1$ to $x = 3$, an increase of \$1000 in advertising produces an average increase of \$500 in profit.

1. Use Figure 12.10 and find the average rate of change in profit if advertising increases from
(a) $x = 1$ to $x = 4$;
(b) $x = 2$ to $x = 4$;
(c) $x = 3$ to $x = 4$.

Answer:
(a) about \$430
(b) about \$350
(c) about \$300

Work Problem 1 at the side.

Example 1 Figure 12.11 shows a graph of the population of rodents, y, in thousands, in a given location when average midday temperatures are x degrees Celsius.

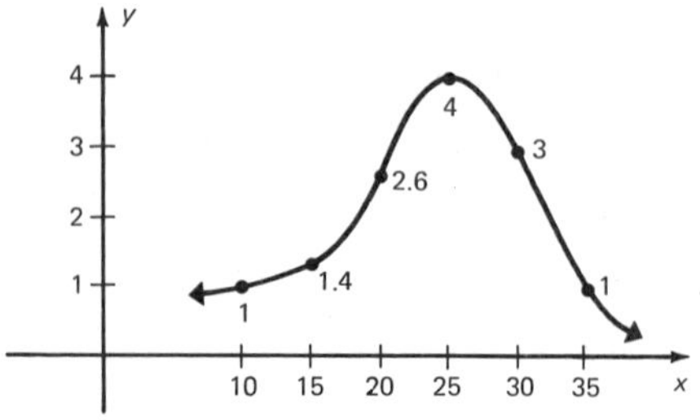

Figure 12.11

(a) Find the average rate of change in the rodent population when the temperature changes from 15° to 25°.

From the graph, there are 1.4 thousand rodents at a temperature of 15° and 4 thousand at a temperature of 25°. The average rate of change of the rodent population with respect to the temperature is given by the quotient of the change in population and the change in temperature, or

$$\frac{4 - 1.4}{25 - 15} = \frac{2.6}{10} = .26.$$

In the interval from $x = 15$ to $x = 25$, an increase of 1° in temperature produces an average of .26 thousands (or 260) additional rodents.

(b) Find the average rate of change in the rodent population when the temperature changes from 25° to 30°.

We have

$$\frac{3 - 4}{30 - 25} = \frac{-1}{5} = -.20.$$

In the interval from $x = 25$ to $x = 30$, an increase of 1° in the temperature produces an average *decrease* of .2 thousand (or 200) rodents.

(c) Find the average rate of change in the rodent population when the temperature changes from 10° to 35°.

Read the necessary values from the graph of Figure 12.11.

$$\frac{1 - 1}{35 - 10} = \frac{0}{25} = 0$$

In the interval from $x = 10$ to $x = 35$, an increase in the temperature of 1° produces an average change in the rodent population of 0. ■

2. Use Figure 12.11 and find the average rate of change of the population if temperature changes from

(a) $x = 15$ to $x = 20$;
(b) $x = 20$ to $x = 25$;
(c) $x = 25$ to $x = 35$;
(d) $x = 30$ to $x = 35$.

Answer:
(a) .24
(b) .28
(c) $-.3$
(d) $-.4$

Work Problem 2 at the side.

As Example 1(c) showed, we can get answers that aren't very helpful if we find the average rate of change of a function over a large interval. Our results are more accurate and usable if we find the average rate of change over a fairly narrow interval. In fact, the most useful result of all comes if we take the limit as the width of the interval approaches 0.

To see how this works, let us suppose that the profit, $P(x)$, for a certain firm is related to the volume of production according to the mathematical model

$$P(x) = 16x - x^2,$$

where x represents the number of units produced. We will assume that the company can produce any nonnegative number of units. A graph of the profit function is shown in Figure 12.12. An inspection of the graph shows that an increase in production from 1 unit to 2 units will increase profit more than an increase in production from 6 units to 7 units. An increase from 7 units to 8 units produces very little increase in profit, while an increase in production from 8 units to 9 units actually produces a decline in total profit.

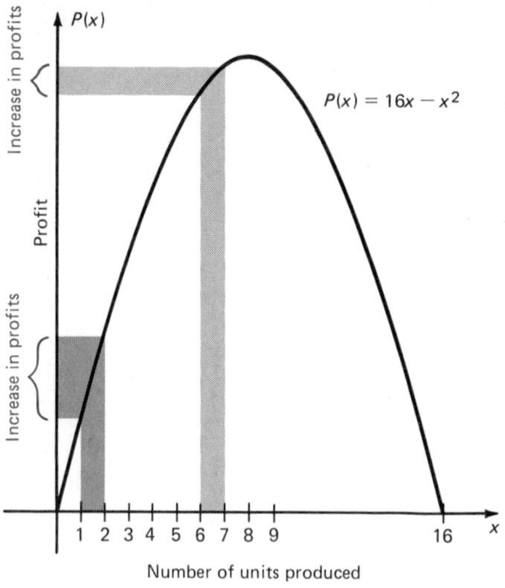

Figure 12.12

Using the profit function $P(x) = 16x - x^2$, we find that the profit from producing 1 unit is

$$P(1) = 16(1) - 1^2 = 15.$$

The profit from producing 2 units is

$$P(2) = 16(2) - 2^2 = 28.$$

This 1-unit increase of production (from 1 unit to 2 units) increased profit by $28 - 15 = 13$ dollars.

3. (a) Find the profit from producing 4 units.
(b) Find the increase in profit if production increases from 1 unit to 4 units.

Answer:
(a) 48
(b) 33

Work Problem 3 at the side.

As Problem 3 shows, if production increases from 1 unit to 4 units, profit increases by \$33. The average rate of increase in profit for this 3-unit increase in production is found by dividing the change in profits by the change in production.

$$\text{average rate of increase} = \frac{P(4) - P(1)}{4 - 1} = \frac{48 - 15}{3} = \frac{33}{3} = 11$$

We have seen that an increase in production from 1 unit to 4 units leads to an average rate of increase in profit of 11, while an increase from 1 unit to 2 units leads to an average rate of increase in profit of 13. In general, if we let production increase from 1 unit to $1 + h$ units, where h is a small positive number, what happens to the average rate of increase in profit? To find out, we make a table which shows the average rate of increase of profit for some selected values of h.

h	1	0.1	0.01	0.001	0.0001
$1 + h$	2	1.1	1.01	1.001	1.0001
$P(1 + h)$	28	16.39	15.1399	15.013999	15.00139999
$P(1)$	15	15	15	15	15
Increase in profit	13	1.39	.1399	.013999	.00139999
Average rate of increase in profit	13	13.9	13.99	13.999	13.9999

We found the numbers in the bottom row of the table by dividing the increase in profit (row 5) by the increase in production (row 1). In symbols, the numbers in the bottom row are found by evaluating the quotient

$$\frac{P(1 + h) - P(1)}{h}$$

for various values of h.

From the table it seems that as h approaches 0 (or, as production gets closer and closer to 1), the average rate of increase in profit approaches the limit 14. This limit,

$$\lim_{h \to 0} \frac{P(1 + h) - P(1)}{h},$$

is called the **instantaneous rate of change,** or just the **rate of change,** of the profit at a production level of $x = 1$.

Think of the rate of change in profit varying all the time as the level of production changes. However, at the exact instant when production is 1 unit, the rate of change in profit is 14 dollars per unit.

4. Make a table similar to the one in the text and find the instantaneous rate of change of profit at $x = 4$.

Answer: 8

Work Problem 4 at the side.

Velocity One additional application of the idea of an instantaneous rate of change is given by velocity. Suppose a particle is moving along a straight line, with its distance from some fixed point given by a function $s(t)$, where t represents time. The quotient

$$\frac{s(t + h) - s(t)}{(t + h) - t} = \frac{s(t + h) - s(t)}{h}$$

represents the average rate of change of the distance; this is the **average velocity** of the particle. We can get the **instantaneous velocity at time t** (often called just the **velocity** at time t) by taking the limit as h approaches 0. That is, if $v(t)$ represents the velocity at time t,

$$v(t) = \lim_{h \to 0} \frac{s(t + h) - s(t)}{h}.$$

Example 2 The position of a red blood cell in the capillaries is given by

$$s(t) = 1.2t + 5,$$

where $s(t)$ gives the position of a cell in millimeters from some initial point and t is time in seconds. Find the velocity of this red blood cell.

We must evaluate the limit given above. To find $s(t + h)$, replace t in $s(t) = 1.2t + 5$ with $t + h$.

$$s(t + h) = 1.2(t + h) + 5$$

Now use the definition of velocity.

$$\begin{aligned} v(t) &= \lim_{h\to 0} \frac{s(t + h) - s(t)}{h} \\ &= \lim_{h\to 0} \frac{1.2(t + h) + 5 - (1.2t + 5)}{h} \\ &= \lim_{h\to 0} \frac{1.2t + 1.2h + 5 - 1.2t - 5}{h} \\ &= \lim_{h\to 0} \frac{1.2h}{h} \\ &= 1.2 \end{aligned}$$

The velocity of the red blood cell is a constant 1.2 millimeters per second. ■

5. Find the velocity for each of the following particles.
(a) $s(t) = 6.4t - 4$
(b) $s(t) = -2t + 12$
(c) $s(t) = 12t$

Answer:
(a) 6.4
(b) −2
(c) 12

Work Problem 5 at the side.

12.3 APPLIED PROBLEMS

Management

The graph shows the total sales in thousands of dollars from the distribution of x thousand catalogs. Find the average rate of change of sales with respect to the number of catalogs distributed for the following changes in x. (See Example 1.)

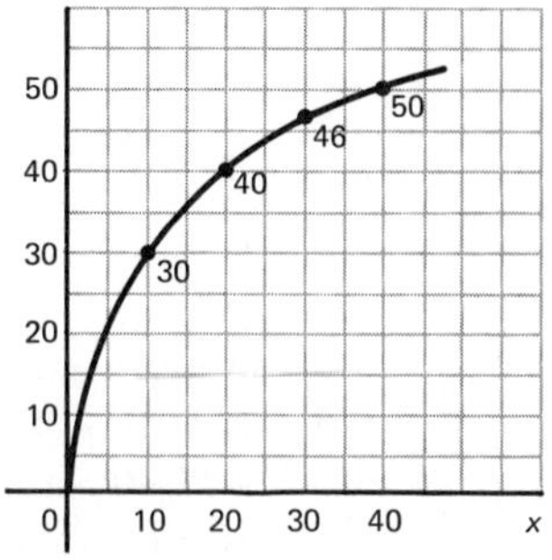

1. 10 to 20
2. 10 to 40
3. 20 to 30
4. 30 to 40

Natural Science

The graph shows the population in millions of bacteria t minutes after a bactericide is introduced into the culture. Find the average rate of change of population with respect to time for the following time intervals.

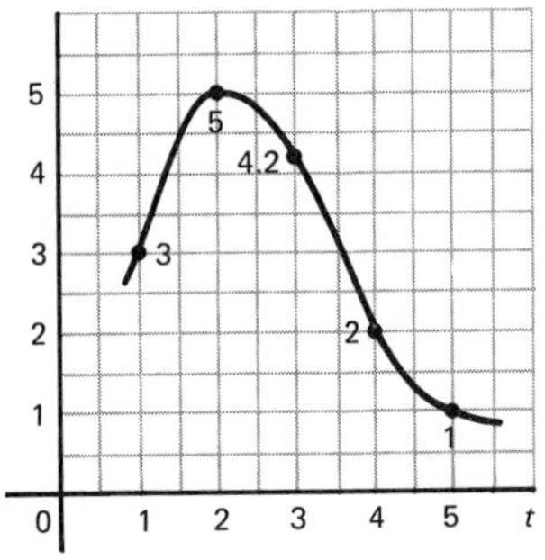

5. 1 to 2

6. 1 to 3

7. 2 to 3

8. 2 to 5

9. 3 to 4

10. 4 to 5

Management

The graph shows annual sales (in units) of a typical product. Sales increase slowly at first to some peak, hold steady for a while, and then decline as the product goes out of style. (See also the exercises of Section 13.3.) Find the average annual rate of change in sales for the following changes in years.

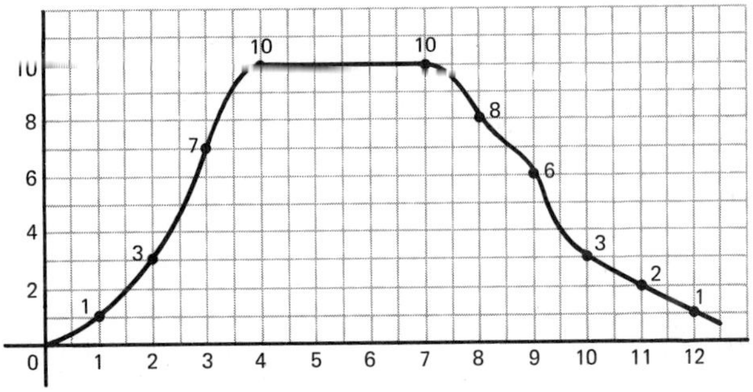

11. 1 to 3

12. 2 to 4

13. 3 to 6

14. 5 to 7

15. 7 to 9

16. 8 to 11

17. 9 to 10

18. 10 to 12

The next graph is from Business Week *magazine.* It shows the annual profits for Pan American World Airways for several past years. Find the average rate of change of profits for the following intervals.*

* Reprinted from the September 5, issue of *Business Week* by special permission,

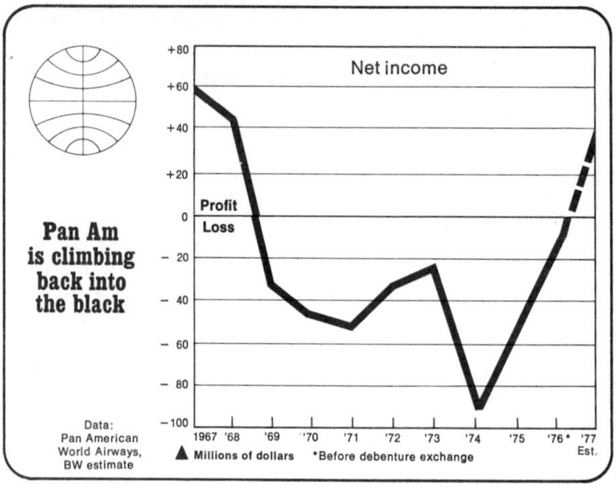

19. 1970 to 1972

20. 1970 to 1974

21. 1974 to 1976

22. 1972 to 1974

23. 1967 to 1974

24. 1967 to 1976

Find the average rate of change for the following functions.

25. $y = x^2 + 2x$ between $x = 0$ and $x = 3$

26. $y = -4x^2 - 6$ between $x = 2$ and $x = 5$

27. $y = 2x^3 - 4x^2 + 6x$ between $x = -1$ and $x = 1$

28. $y = -3x^3 + 2x^2 - 4x + 1$ between $x = 0$ and $x = 1$

29. $y = \sqrt{x}$ between $x = 1$ and $x = 4$

30. $y = \sqrt{3x - 2}$ between $x = 1$ and $x = 2$

31. $y = 6x - 2$ between $x = -2$ and $x = 1$

32. $y = -4x + 5$ between $x = -1$ and $x = 3$

33. $y = \dfrac{1}{x - 1}$ between $x = -2$ and $x = 0$

34. $y = \dfrac{-5}{2x - 3}$ between $x = 2$ and $x = 4$

Suppose that the total cost in dollars to produce x items is given by

$$C(x) = 2x^2 - 5x + 6.$$

Find the average rate of change of cost as x increases from

35. 2 to 4;

36. 2 to 3.

37. Use $C(x) = 2x^2 - 5x + 6$ to complete the following table.

h	1	0.1	0.01	0.001	0.0001
$2+h$	3	2.1	2.01	2.001	2.0001
$C(2+h)$	9	4.32	4.0302	4.003002	4.00030002
$C(2+h)-C(2)$	5	.32	.0302	.003002	.00030002
$\frac{C(2+h)-C(2)}{h}$	5	3.2	____	____	____

38. Use the bottom row of the chart to find $\lim_{h\to 0} \frac{C(2+h)-C(2)}{h}$.

39. In Exercises 37 and 38, what is the instantaneous rate of change of cost with respect to the number of items produced when $x = 2$? (This number, called the **marginal cost** at $x = 2$, will be discussed in more detail later; it is the approximate cost of producing the third item.)

40. Redo the chart of Exercise 37. This time, change the second line to $4 + h$, the third line to $C(4 + h)$, and so on. Find

$$\lim_{h\to 0} \frac{C(4+h)-C(4)}{h}.$$

Natural Science

The distance of a particle from some fixed point is given by

$$s(t) = t^2 + 5t + 2,$$

where t is time measured in seconds. Find the average velocity of the particle from:

41. 4 seconds to 6 seconds;

42. 4 seconds to 5 seconds.

43. Complete the following table.

h	1	0.1	0.01	0.001	0.0001
$4+h$	5	4.1	4.01	4.001	4.0001
$s(4+h)$	52	39.31	38.1301	38.013001	38.0013000
$s(4)$	38	38	38	38	38
$s(4+h)-s(4)$	14	1.31	.1301	.013001	.0013000
$\frac{s(4+h)-s(4)}{h}$	14	13.1	____	____	____

44. Find $\lim_{h\to 0} \frac{s(4+h)-s(4)}{h}$, and give the instantaneous velocity of the particle when $x = 4$. (*See Example 2.*)

Find each of the following for $s(t) = t^2 + 5t + 2$. (See Example 2.)

45. $\lim_{h\to 0} \frac{s(6+h)-s(6)}{h}$

46. $\lim_{h\to 0} \frac{s(1+h)-s(1)}{h}$

12.4 DEFINITION OF THE DERIVATIVE

The idea of an instantaneous rate of change, discussed in the previous section, is very useful in applications. The methods that we used in that section to find this rate were cumbersome—if we wanted to change a value of x, we had to start all over. To get around these problems, we use the idea of a *derivative*. The derivative is one of the two main ideas of calculus (the other, the integral, is discussed in Chapter 14).

We define the derivative using the idea of a *tangent line* to a curve. Perhaps the easiest tangent line to define is the one for a circle: a tangent line to a circle at a point on the circle is a line which touches the circle in only one point. See Figure 12.13. Tangent lines for more complicated curves are not as easy to define. In Figure 12.14 we would probably agree that the lines at P_1, P_3, and P_4 are tangent lines to the curve, while the ones at P_2 and P_5 are not.

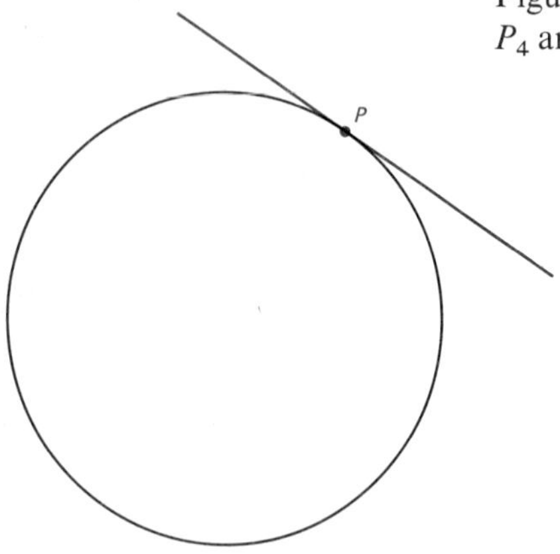

Figure 12.13

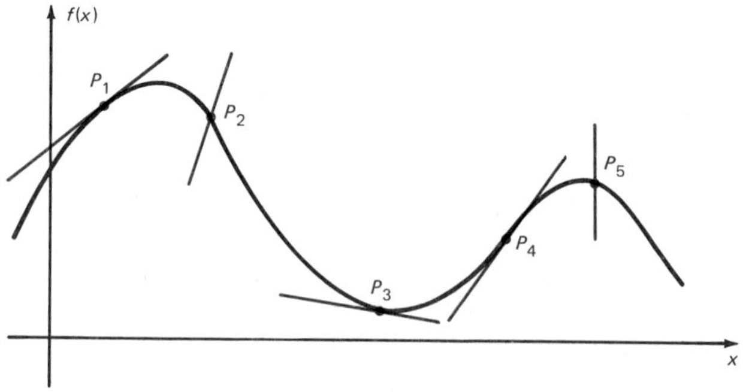

Figure 12.14

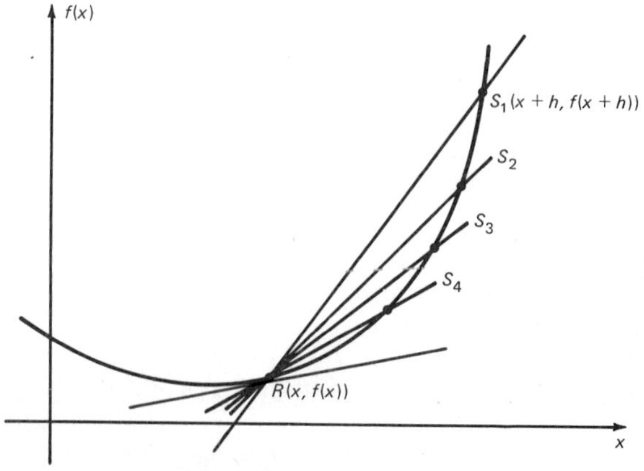

Figure 12.15

We define the **tangent line** for a point on a curve by finding the slope of the tangent line. The slope and the point then determine the tangent line. To find the slope of a tangent line, let us begin with the curve $y = f(x)$ shown in Figure 12.15. The tangent line corresponding to the point x is shown in color. The tangent line goes through point R on the curve, with the coordinates of R given by $(x, f(x))$.

Figure 12.15 also shows a number of **secant lines** (lines cutting the graph in at least two points). Since the secant lines cut the graph in two points, we can find their slope using the definition of slope from Chapter 3.

For example, let point S_1 have x-coordinate given by $x + h$ (h represents "a little change in x"). If $x + h$ represents the x-coordinate of point S_1, then its y-coordinate is $f(x + h)$; its coordinates are thus $(x + h, f(x + h))$. The slope of the secant line through R and S_1 is

$$\text{slope} = \frac{y_2 - y_1}{x_2 - x_1} = \frac{f(x+h) - f(x)}{(x+h) - x} = \frac{f(x+h) - f(x)}{h}.$$

Suppose we now let h approach 0. As h approaches 0, $x + h$ will approach $x + 0$, or x. This forces point S_1 to slide down the graph in Figure 12.15, passing, in turn, through S_2, S_3, S_4, and so on. As h gets extremely small, the secant line will get extremely close to the desired tangent line. In fact,

the **slope of the tangent line** for the curve $y = f(x)$ at the point $(x, f(x))$ is defined as

$$\lim_{h \to 0} \frac{f(x+h) - f(x)}{h}$$

provided that this limit exists.

The *slope of the tangent line* at a given point on its graph is called the *derivative* of the function at that point.

The **derivative** for the function $y = f(x)$ is written $f'(x)$ and is defined as

$$f'(x) = \lim_{h \to 0} \frac{f(x+h) - f(x)}{h}$$

provided that this limit exists.

Example 1 Let $f(x) = x^2$.

(a) Find the derivative.

The derivative is given by $f'(x)$, where

$$f'(x) = \lim_{h \to 0} \frac{f(x+h) - f(x)}{h}.$$

We can find this limit through a sequence of steps.

1. Find $f(x+h)$.
   Replace x with $x+h$ in the equation for $f(x)$.

$$\begin{aligned} f(x) &= x^2 \\ f(x+h) &= (x+h)^2 \\ &= x^2 + 2xh + h^2 \end{aligned}$$

2. Find $f(x+h) - f(x)$.
   Since $f(x) = x^2$, we have

$$\begin{aligned} f(x+h) - f(x) &= x^2 + 2xh + h^2 - x^2 \\ &= 2xh + h^2. \end{aligned}$$

3. Form and simplify the *difference quotient* $\dfrac{f(x+h) - f(x)}{h}$.

$$\frac{f(x+h) - f(x)}{h} = \frac{2xh + h^2}{h} = \frac{h(2x+h)}{h} = 2x + h$$

4. Take the limit as h approaches 0.

$$f'(x) = \lim_{h \to 0} \frac{f(x+h) - f(x)}{h} = \lim_{h \to 0}(2x + h) = 2x + 0 = 2x$$

Thus, for $f(x) = x^2$, we have $f'(x) = 2x$.

(b) Calculate and interpret $f'(3)$.

$$f'(3) = 2(3) = 6$$

The number 6 is the slope of the tangent line to the graph of $y = x^2$ at the point where $x = 3$. See Figure 12.16. ■

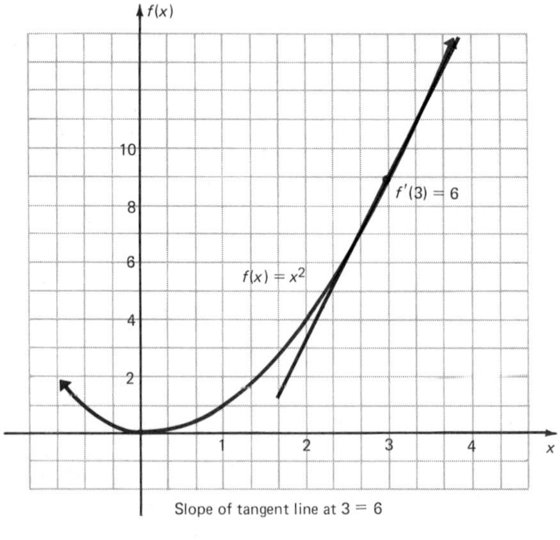

Figure 12.16

1. Let $f(x) = -2x^2 + 7$. Find the following.

(a) $f(x+h)$
(b) $f(x+h) - f(x)$
(c) $\dfrac{f(x+h) - f(x)}{h}$
(d) $f'(x)$
(e) $f'(4)$
(f) $f'(0)$

Answer:
(a) $-2x^2 - 4xh - 2h^2 + 7$
(b) $-4xh - 2h^2$
(c) $-4x - 2h$
(d) $-4x$
(e) -16
(f) 0

Work Problem 1 at the side.

Example 2 Let $f(x) = 2x^3 + 4x$. Find $f'(x)$.
Go through the four steps used above.

1. Find $f(x+h)$.

$$f(x+h) = 2(x+h)^3 + 4(x+h) \qquad \text{Replace } x \text{ with } x+h$$

Now we need some algebra. Multiply $x+h$, $x+h$, and $x+h$ to get

$$(x+h)^3 = x^3 + 3x^2h + 3xh^2 + h^3.$$

Thus,

$$\begin{aligned} f(x+h) &= 2(x^3 + 3x^2h + 3xh^2 + h^3) + 4(x+h) \\ &= 2x^3 + 6x^2h + 6xh^2 + 2h^3 + 4x + 4h. \end{aligned}$$

2. $$\begin{aligned} f(x+h) - f(x) &= 2x^3 + 6x^2h + 6xh^2 + 2h^3 + 4x + 4h - 2x^3 - 4x \\ &= 6x^2h + 6xh^2 + 2h^3 + 4h \end{aligned}$$

3. $$\begin{aligned} \frac{f(x+h) - f(x)}{h} &= \frac{6x^2h + 6xh^2 + 2h^3 + 4h}{h} \\ &= \frac{h(6x^2 + 6xh + 2h^2 + 4)}{h} \\ &= 6x^2 + 6xh + 2h^2 + 4 \end{aligned}$$

4. $$\begin{aligned} f'(x) = \lim_{h\to 0} \frac{f(x+h) - f(x)}{h} &= \lim_{h\to 0}(6x^2 + 6xh + 2h^2 + 4) \\ &= 6x^2 + 6x(0) + 2(0)^2 + 4 \\ &= 6x^2 + 4 \end{aligned}$$

If $f(x) = 2x^3 + 4x$, then $f'(x) = 6x^2 + 4$. ■

2. Let $f(x) = -3x^3 + 2x$. Find the following.

(a) $f(x+h)$
(b) $f(x+h) - f(x)$
(c) $\dfrac{f(x+h) - f(x)}{h}$
(d) $f'(x)$
(e) $f'(-1)$

Answer:
(a) $-3x^3 - 9x^2h - 9xh^2 - 3h^3 + 2x + 2h$
(b) $-9x^2h - 9xh^2 - 3h^3 + 2h$
(c) $-9x^2 - 9xh - 3h^2 + 2$
(d) $-9x^2 + 2$
(e) -7

Work Problem 2 at the side.

Example 3 Let $f(x) = \dfrac{4}{x}$. Find $f'(x)$.

1. $$f(x+h) = \frac{4}{x+h}$$

2. $$\begin{aligned} f(x+h) - f(x) &= \frac{4}{x+h} - \frac{4}{x} \\ &= \frac{4x - 4(x+h)}{x(x+h)} && \text{Find a common denominator} \\ &= \frac{4x - 4x - 4h}{x(x+h)} && \text{Simplify the numerator} \\ &= \frac{-4h}{x(x+h)} \end{aligned}$$

$$3.\ \frac{f(x+h)-f(x)}{h} = \frac{\dfrac{-4h}{x(x+h)}}{h}$$

$$= \frac{-4h}{x(x+h)} \cdot \frac{1}{h} \qquad \text{Invert and multiply}$$

$$= \frac{-4}{x(x+h)}$$

$$4.\ f'(x) = \lim_{h \to 0} \frac{f(x+h)-f(x)}{h} = \lim_{h \to 0} \frac{-4}{x(x+h)}$$

$$= \frac{-4}{x(x+0)}$$

$$= \frac{-4}{x(x)}$$

$$= \frac{-4}{x^2} \quad ■$$

3. Let $f(x) = -5/x$. Find the following.

(a) $f(x+h)$

(b) $f(x+h) - f(x)$

(c) $\dfrac{f(x+h)-f(x)}{h}$

(d) $f'(x)$

(e) $f'(-1)$

Answer:

(a) $\dfrac{-5}{x+h}$

(b) $\dfrac{5h}{x(x+h)}$

(c) $\dfrac{5}{x(x+h)}$

(d) $\dfrac{5}{x^2}$

(e) 5

Work Problem 3 at the side.

Let us summarize the four steps used when finding $f'(x)$.

1. Find $f(x+h)$.

2. Find $f(x+h) - f(x)$.

3. Find and simplify $\dfrac{f(x+h)-f(x)}{h}$.

4. $f'(x) = \lim_{h \to 0} \dfrac{f(x+h)-f(x)}{h}$ if this limit exists.

Not all functions have derivatives at all points in their domains. Figure 12.17 shows a graph of the function $f(x) = |x|$. We cannot

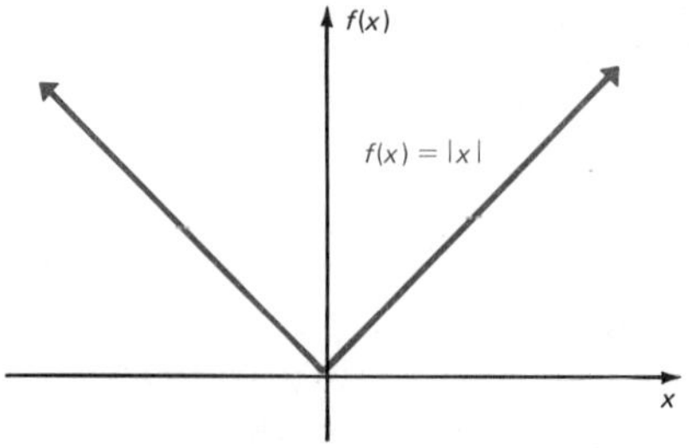

Figure 12.17

really draw a single unique tangent at the point (0, 0) on the graph—the graph has a sharp "corner" and more than one "tangent" could be drawn there. Therefore, the function $f(x) = |x|$ does not have a derivative at (0, 0). The function of Figure 12.18 does not have a derivative at point x_1, x_2, or x_3. A function which is not continuous at a point cannot have a derivative there.

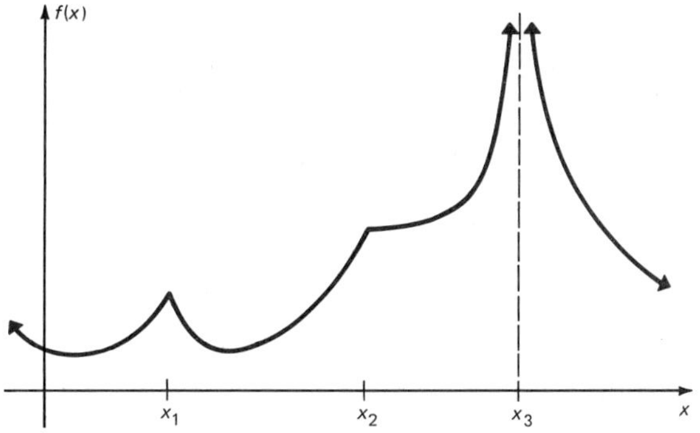

Figure 12.18

Often we are given a graph and we want to find the point on the graph where the slope of the tangent line equals a given number. The next example shows how this is done.

Example 4 Find all points on the graph of $f(x) = x^2 + 6x + 5$ where the slope of the tangent line is 0.

The slope of the tangent line is given by the derivative, $f'(x)$. Go through the four steps given above and verify that

$$f'(x) = 2x + 6.$$

We want to find all values of x where the slope of the tangent is 0. In other words, we want to find all values of x so that $f'(x) = 0$.

$$\begin{aligned} f'(x) &= 0 \\ 2x + 6 &= 0 \\ 2x &= -6 \\ x &= -3 \end{aligned}$$

When $x = -3$, the slope of the tangent is 0. Recall from Chapter 3 that a horizontal line has a slope of 0. Thus, the tangent to $f(x) = x^2 + 6x + 5$ at $x = -3$ is horizontal. This is shown in Figure 12.19 When $x = -3$, $y = -4$, so the slope of the tangent line is 0 at the point $(-3, -4)$. ■

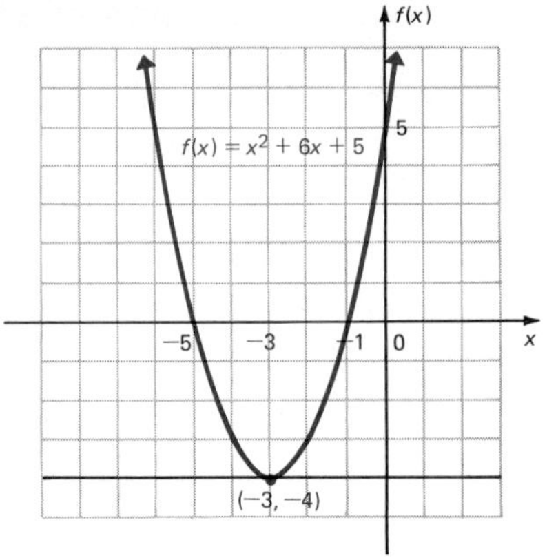

Figure 12.19

4. Find all points on the graph of $f(x) = -4x^2 + 16x + 5$ where the slope of the tangent line is 0.

Answer: (2, 21)

Work Problem 4 at the side.

The derivative was defined as the limit

$$\lim_{h \to 0} \frac{f(x+h) - f(x)}{h}.$$

This is exactly the same limit that we used in the previous section to define the instantaneous rate of change of a function. Therefore,

the *instantaneous rate of change* of a function is given by the *derivative* of the function.

From now on, we will use "rate of change" to mean "instantaneous rate of change."

Example 5 The cost in dollars to manufacture x electric pencil sharpeners is given by $C(x) = x^3 - 4x^2 + 3x$. Find the rate of change of cost when 5 pencil sharpeners are made.

The rate of change of cost is given by the derivative of the cost function. If we went through the four steps above, we would find that

$$C'(x) = 3x^2 - 8x + 3.$$

When $x = 5$,

$$C'(5) = 3(5)^2 - 8(5) + 3 = 38.$$

Thus, at the point where exactly 5 pencil sharpeners are made, the rate of change of cost is \$38. ■

5. The revenue from the sale of x can openers is given by $R(x) = 2x^3 - 4x^2 + x$. Find the rate of change of revenue when
(a) $x = 2$;
(b) $x = 8$.
(Hint: $R'(x) = 6x^2 - 8x + 1$.)

Answer:
(a) 9
(b) 321

Work Problem 5 at the side.

12.4 EXERCISES

Find $f'(x)$ for each of the following. Then find $f'(2)$, $f'(0)$, and $f'(-3)$. (See Examples 1–3.)

1. $f(x) = 2x^2$

2. $f(x) = -3x^2$

3. $f(x) = -4x^2 + 11x$

4. $f(x) = 6x^2 - 4x$

5. $f(x) = -9x$

6. $f(x) = 4x$

7. $f(x) = 8x + 6$

8. $f(x) = -9x - 5$

9. $f(x) = x^3 + 3x$

10. $f(x) = 2x^3 - 14x$

11. $f(x) = -\frac{2}{x}$

12. $f(x) = \frac{6}{x}$

13. $f(x) = \frac{4}{x - 1}$

14. $f(x) = \frac{3}{x + 2}$

15. $f(x) = \sqrt{x}$ (Hint: In Step 3, multiply numerator and denominator by $\sqrt{x + h} + \sqrt{x}$.)

16. $f(x) = -3\sqrt{x}$

Find all points where the following functions do not have derivatives.

17.

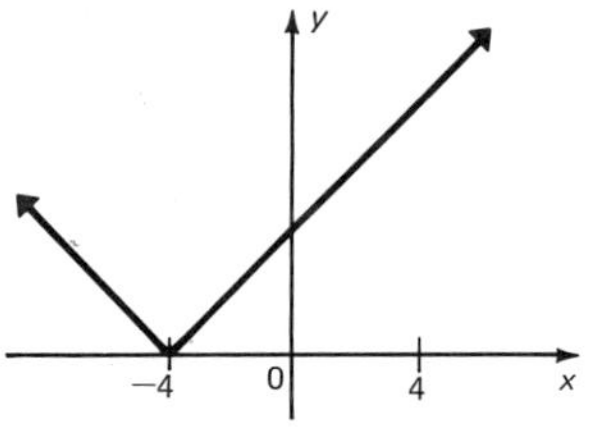

18.

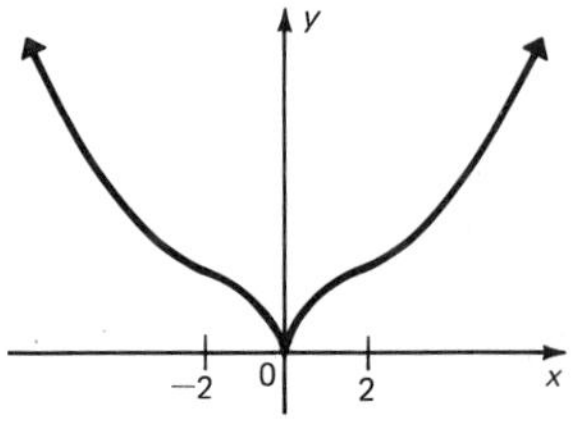

19.

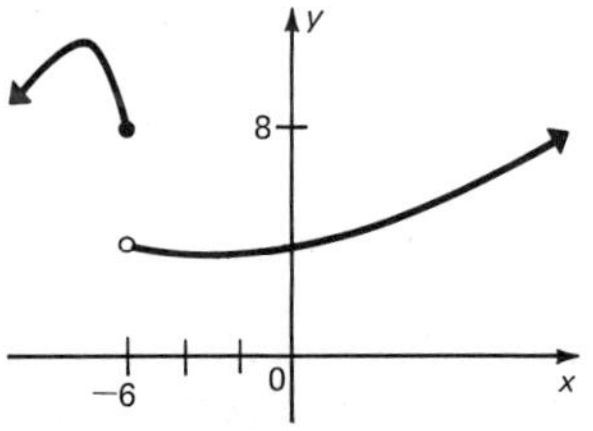

20.

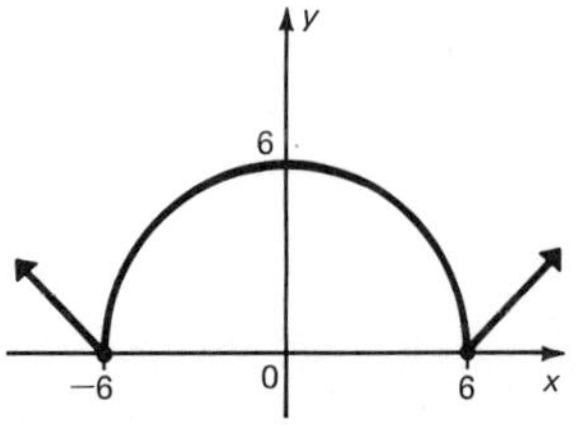

21.

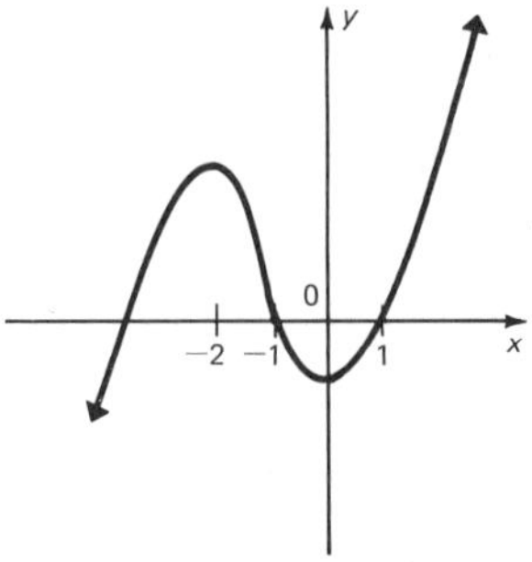

22.

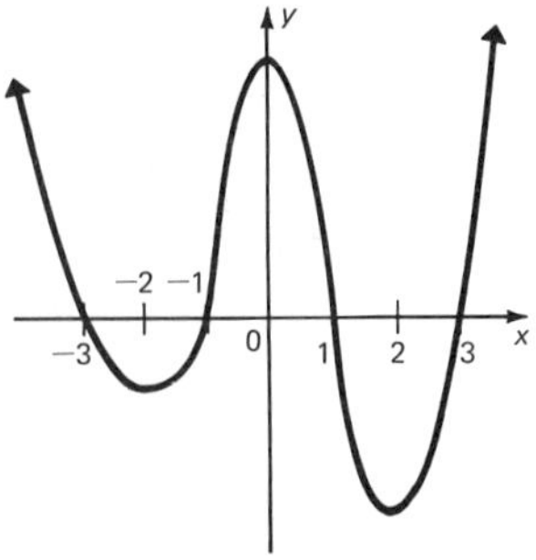

Find all points (if there are any) where the following functions have tangent lines whose slope is 0. In Exercises 27–30, it is necessary to factor. (See Example 4.)

23. $f(x) = 9x^2$

24. $f(x) = -2x^2$

25. $f(x) = 8x^2 + 32x$

26. $f(x) = 2x^2 + 6x$

27. $f(x) = 2x^3 - 3x^2 - 12x$

28. $f(x) = 2x^3 - 6x^2 + 6x$

29. $f(x) = 4x^3 + 24x^2$

30. $f(x) = 4x^3 - 18x^2$

APPLIED PROBLEMS

Management

Suppose the demand for a certain item is given by

$$D(x) = -2x^2 + 4x + 6,$$

where x represents the price of the item. Find the rate of change of demand with respect to price for the following values of x. (See Example 5.)

31. $x = 3$

32. $x = 6$

Suppose the profit from the expenditure of x thousand dollars on advertising is given by

$$P(x) = 1000 + 32x - 2x^2.$$

Find the rate of change of profit with respect to the expenditure on advertising for the following amounts. In each case, decide if the firm should increase the expenditure.

33. $x = 8$

34. $x = 6$

35. $x = 12$

36. $x = 20$

Natural Science

A biologist estimates that when a bactericide is introduced into a culture of bacteria, the number of bacteria present is given by

$$B(t) = 1000 + 50t - 5t^2,$$

where $B(t)$ is the number of bacteria, in millions, present at time t, measured in hours. Find the rate of change of the number of bacteria with respect to time for the following values of t.

37. $t = 2$

38. $t = 3$

39. $t = 4$

40. $t = 5$

41. $t = 6$

42. When does the population of bacteria start to decline?

12.5 SOME SHORTCUTS FOR FINDING DERIVATIVES

The 4-step method for finding derivatives is hard to use—it is too easy to make an algebraic error. In the remainder of this chapter, we look at some shortcut ways of finding derivatives.

In the previous section we used $f'(x)$ to represent the derivative of the function $y = f(x)$. Alternate notations for the derivative include

$$y' \quad \text{or} \quad \frac{dy}{dx} \quad \text{or} \quad \frac{d}{dx} f(x).$$

We will use these notations interchangeably.

We begin our shortcuts with the formula for the derivative of a constant; it is the simplest derivative of all.

Theorem 12.1 *Constant rule* If $y = k$, where k is any real number, then

$$\mathbf{y' = 0.}$$

(The derivative of a constant is 0.)

Example 1 If $y = 9$, then $y' = 0$. ■

1. Find the derivatives of the following.
(a) $y = -4$
(b) $f(x) = \pi$
(c) $y = 0$

Answer:
(a) 0
(b) 0
(c) 0

Work Problem 1 at the side.

We can use the definition of derivative from the previous section to show that the constant rule is correct. Let $f(x) = k$, a real number. By the definition of derivative,

$$f'(x) = \lim_{h \to 0} \frac{f(x+h) - f(x)}{h}.$$

Since $f(x) = k$, we also have $f(x + h) = k$. (To see this, let $f(x) = k + 0x$.) Then

$$f'(x) = \lim_{h \to 0} \frac{k - k}{h} = \lim_{h \to 0} \frac{0}{h} = 0.$$

Functions of the form $y = x^n$, where n is a real number, occur often in applications. To find a general rule for the derivative of $y = x^n$, we work out the derivative for various special cases. Use the definition of derivative to check the entries in the table at the side. The results in the table suggest the following rule.

Function	Derivative
$y = x^2$	$y' = 2x^1$
$y = x^3$	$y' = 3x^2$
$y = x^4$	$y' = 4x^3$
$y = x^5$	$y' = 5x^4$
$y = x^{-1}$	$y' = -1 \cdot x^{-2}$

Theorem 12.2 *Power rule* If $y = x^n$, then

$$y' = n \cdot x^{n-1}.$$

(The derivative of $y = x^n$ is found by decreasing the exponent on x by 1 and multiplying the result by the exponent n.)

The actual proof of the power rule requires algebra which is more advanced than we have studied.

Example 2 Find the derivatives of the following functions.

(a) $y = x^6$

Multiply x^{6-1} by 6.

$$y' = 6 \cdot x^{6-1} = 6x^5$$

(b) $y = x^{15}$

$$y' = 15 \cdot x^{15-1} = 15x^{14}$$

(c) If $y = x^1$, then $y' = 1 \cdot x^{1-1} = 1 \cdot x^0 = 1 \cdot 1 = 1$. (Recall that $x^0 = 1$ if $x \neq 0$.)

(d) If $y = x^{-3}$, then $y' = -3 \cdot x^{-3-1} = -3 \cdot x^{-4} = -3/x^4$.

(e) If $y = x^{4/3}$, then $y' = \frac{4}{3}x^{1/3}$.

(f) $y = \sqrt{x}$

Replace $\sqrt{x}$ by the equivalent expression $x^{1/2}$. Then

$$y' = \frac{1}{2} \cdot x^{(1/2)-1} = \frac{1}{2}x^{-1/2} = \frac{1}{2x^{1/2}} = \frac{1}{2\sqrt{x}}. \quad \blacksquare$$

2. Find the derivatives of the following.

(a) $y = x^4$

(b) $y = x^{17}$

(c) $y = x^{-2}$

(d) $y = x^{-5}$

(e) $y = x^{3/2}$

Answer:

(a) $y' = 4x^3$

(b) $y' = 17x^{16}$

(c) $y' = -2/x^3$

(d) $y' = -5/x^6$

(e) $y' = \frac{3}{2}x^{1/2}$

Work Problem 2 at the side.

The next rule tells us how to handle the derivative of the product of a constant and a function.

Theorem 12.3 *Constant times a function* Let k be a real number. Then the derivative of $y = k \cdot f(x)$ is

$$y' = k \cdot f'(x).$$

(The derivative of a constant times a function is the constant times the derivative of the function.)

Example 3 Find the derivatives of the following functions.

(a) $y = 8x^4$

Since the derivative of $f(x) = x^4$ is $f'(x) = 4x^3$, we have

$$y' = 8(4x^3) = 32x^3.$$

(b) $y = -\frac{3}{4}x^{12}$

$$y' = -\frac{3}{4}(12x^{11}) = -9x^{11}$$

(c) If $y = 15x$, then $y' = 15(1) = 15$.

(d) If $y = 10x^{3/2}$, then $y' = 10(\frac{3}{2}x^{1/2}) = 15x^{1/2}$.

(e) If $y = 1000x^{.4}$, then $y' = 1000(.4x^{-.6}) = 400x^{-.6}$.

(f) $y = \dfrac{6}{x}$

Replace $\dfrac{6}{x}$ with $6 \cdot \dfrac{1}{x}$, or $6x^{-1}$. Then

$$y' = 6(-1x^{-2}) = -6x^{-2} = \frac{-6}{x^2}. \quad \blacksquare$$

3. Find the derivatives of the following.
(a) $y = 12x^3$
(b) $f(x) = 30x^7$
(c) $y = -35x$
(d) $y = 10x^{.2}$
(e) $y = 5\sqrt{x}$
(f) $y = -10/x$

Answer:
(a) $36x^2$
(b) $210x^6$
(c) -35
(d) $2x^{-.8}$
(e) $\frac{5}{2}x^{-1/2}$ or $\dfrac{5}{2\sqrt{x}}$
(f) $10x^{-2}$ or $\dfrac{10}{x^2}$

Work Problem 3 at the side.

We have one final rule in this section:

Theorem 12.4 *Sum or difference rule* If $y = f(x) + g(x)$, then

$$y' = f'(x) + g'(x);$$

if $y = f(x) - g(x)$, then

$$y' = f'(x) - g'(x).$$

(The derivative of a sum or difference of two functions is the sum or difference of the derivatives of the functions.)

Example 4 Find the derivatives of the following functions.

(a) $y = 6x^3 + 9x^2$

The derivative of $f(x) = 6x^3$ is $f'(x) = 18x^2$, and the derivative of $g(x) = 9x^2$ is $g'(x) = 18x$. Thus, if

$$y = 6x^3 + 9x^2, \qquad \text{then} \qquad y' = 18x^2 + 18x.$$

(b) If $y=12x^2+11x-8$, then $y'=24x+11-0=24x+11$.

(c) If $y = 8x^4 - 6\sqrt{x} + \dfrac{5}{x}$, we find y' by first rewriting y as $y = 8x^4 - 6x^{1/2} + 5x^{-1}$. Then $y' = 32x^3 - 3x^{-1/2} - 5x^{-2}$. If desired, this result can be written as $y' = 32x^3 - 3/\sqrt{x} - 5/x^2$. ■

4. Find the derivatives of the following.
(a) $y = -4x^5 - 8x + 6$
(b) $y = 32x^5 - 100x^2 + 12x$
(c) $y = 8x^{3/2} + 2x^{1/2}$
(d) $f(x) = -\sqrt{x} + 6/x$

Answer:
(a) $y' = -20x^4 - 8$
(b) $y' = 160x^4 - 200x + 12$
(c) $y' = 12x^{1/2} + x^{-1/2}$ or $12x^{1/2} + 1/x^{1/2}$
(d) $f'(x) = -1/[2\sqrt{x}] - 6/x^2$

Work Problem 4 at the side.

As we saw in the previous section, the derivative of a function can be used to find the rate of change of the function.

Example 5 A tumor has the approximate shape of a cone. The radius of the tumor is fixed by the bone structure at 2 centimeters, but the tumor is growing along the height of the cone. The formula for the volume of a cone is $V = \frac{1}{3}\pi r^2 h$, where r is the radius of the base and h is the height of the cone. Find the rate of change in the volume of the tumor with respect to the height.

To emphasize that we need the rate of change of the volume with respect to the height, we use the symbol dV/dh for the derivative. In our example, r is fixed at 2 cm. Thus

$$V = \frac{1}{3}\pi r^2 h \quad \text{becomes} \quad V = \frac{1}{3}\pi 2^2 \cdot h \quad \text{or} \quad V = \frac{4}{3}\pi h.$$

Since $4\pi/3$ is a constant,

$$\frac{dV}{dh} = \frac{4\pi}{3} \approx 4.2 \text{ cu cm per cm.}$$

For each additional centimeter that the tumor grows in height, its volume will increase approximately 4.2 cubic centimeters. ■

5. A balloon is spherical. The formula for the volume of a sphere is $V = \frac{4}{3}\pi r^3$, where r is the radius of the sphere. Find the following.
(a) dV/dr
(b) the rate of change of the volume when $r = 3$ inches

Answer:
(a) $4\pi r^2$
(b) 36π cubic inches per inch

Work Problem 5 at the side.

Example 6 *Marginal cost* Suppose a function $y = C(x)$ gives the total cost to manufacture x units of an item. The derivative of $y = C(x)$ gives the rate of change of cost with respect to the number of units produced. This rate of change of cost is called the **marginal cost**; think of the marginal cost as the approximate cost of producing one more item, after x items have already been produced. Suppose that the total cost in hundreds of dollars to produce x thousand barrels of beer is given by

$$C(x) = 4x^2 + 100x + 500.$$

Find the marginal cost for the following values of x.

(a) $x = 5$

To find the marginal cost, we need the derivative of the total cost function.

$$C'(x) = 8x + 100$$

When $x = 5$,

$$C'(5) = 8(5) + 100 = 140.$$

After 5 thousand barrels of beer have been produced, the cost to produce 1 thousand more barrels will be approximately 140 hundred dollars, or $14,000.

(b) $x = 30$

After 30 thousand barrels have been produced, the cost to produce 1 thousand more barrels will be approximately

$$C'(30) = 8(30) + 100 = 340,$$

or $34,000.

Management must be careful to keep track of marginal costs. If the marginal cost of producing an extra unit exceeds the revenue received from selling it, then the company will lose money on that unit. ■

Work Problem 6 at the side.

6. The cost to produce x units of wheat is given by $C(x) = 5000 + 20x + 10\sqrt{x}$. Find the marginal cost when
(a) $x = 9$;
(b) $x = 16$;
(c) $x = 25$.

Answer:
(a) 65/3
(b) 85/4
(c) 21

The idea of marginal cost also applies to revenue or profit.

Work Problem 7 at the side.

7. The revenue from the sale of x units of ice cream is given by $R(x)$, where $R(x) = 800x + 10x^2$. Find the marginal revenue when
(a) $x = 10$;
(b) $x = 25$;
(c) $x = 32$.

Answer:
(a) 1000
(b) 1300
(c) 1440

12.5 EXERCISES

Find the derivatives of the following functions. (See Examples 1–4.)

1. $y = 5x^2$

2. $y = -8x^2$

3. $y = -6x^5$

4. $y = 12x^6$

5. $y = 15x^2 + 7$

6. $y = -4x^2 + 15$

7. $y = 9x^2 - 8x + 4$

8. $y = 10x^2 + 4x - 9$

9. $y = 3x^2 - 4x + 15$

10. $y = 2x^2 + 16x - 8$

11. $y = -4x^3 + 2x^2 - 6$

12. $y = 9x^3 + 8x + 5$

13. $y = 10x^3 - 9x^2 + 6x$

14. $y = 3x^3 - x^2 - 12x$

15. $y = x^4 - 5x^3 + 9x^2 + 5$

16. $y = 3x^4 + 11x^3 + 2x^2 - 4x$

17. $y = 6x^{1.5}$

18. $y = -2x^{2.5}$

19. $y = -15x^{3.2}$

20. $y = 18x^{1.6}$

21. $y = -18x^{3/2}$

22. $y = 24x^{5/2}$

23. $y = 32x^{1/2}$

24. $y = -10x^{1/2}$

25. $y = 50x^{3/2} - 6x^{1/2}$

26. $y = 32x^{3/2} + 12x^{1/2}$

27. $y = 8\sqrt{x} + 6x$

28. $y = -100\sqrt{x} - 11x$

29. $y = 6x^{-5}$

30. $y = -2x^{-4}$

31. $y = -4x^{-2} + 3x^{-1}$

32. $y = 8x^{-4} - 9x^{-2}$

33. $y = 10x^{-2} + 3x^{-4} - 6x$

34. $y = x^{-5} - x^{-2} + 5x^{-1}$

35. $y = \frac{6}{x} - \frac{8}{x^2}$

36. $y = \frac{4}{x} + \frac{2}{x^3}$

37. $y = 12x^{-1/2}$

38. $y = -30x^{-1/2}$

39. $y = -10x^{-1/2} + 8x^{-3/2}$

40. $y = x^{-1/2} - 14x^{-3/2}$

For the following functions, find all points where the tangent line is horizontal. (As we saw in the last section, we do this by finding points where the derivative is 0.)

41. $y = 6x^2 - 20x + 5$

42. $y = 9x^2 - 18x - 4$

43. $y = x^3 - \frac{5}{2}x^2 + 2x - 1$

44. $y = \frac{4}{3}x^3 - 13x^2 + 12x - 6$

45. $y = \frac{2}{3}x^3 - \frac{3}{2}x^2 - 5x + 8$

46. $y = x^3 - 2x^2 - 15x + 6$

47. $y = x^4 - 16x$

48. $y = x^4 - 256x$

APPLIED PROBLEMS

Management

49. If the price of a product is given by

$$P(x) = \frac{-1000}{x} + 1000,$$

where x represents the demand for the product, find the rate of change of price when the demand is $x = 10$. Is the price increasing or decreasing at that point?

Often sales of a new product grow rapidly at first and then level off with time. This is the case with the sales represented by the function

$$S(t) = 100 - 100t^{-1},$$

where t represents time in years. Find the rate of change of sales for the following values of t.

50. $t = 1$

51. $t = 10$

Suppose $P(x) = 100/x$ represents the percent of cheap shoes made by a company that are still wearable after x days of wearing. Find the percent of shoes wearable after the following number of days.

52. 1 day

53. 100 days

54. Find and interpret $P'(100)$.

Natural Science

Insulin affects the glucose, or blood-sugar, level of some diabetics according to the function

$$G(x) = -.2x^2 + 450,$$

where $G(x)$ is the blood-sugar level one hour after x units of insulin are injected. (This mathematical model is only approximate and valid only for values of x less than about 40.)

55. Find $G(0)$.

56. Find $G(25)$.

Find dG/dx for the following values of x. Interpret your answer. (*See Example* 5.)

57. $x = 10$

58. $x = 25$

A short length of a blood vessel has the approximate shape of a cylinder. The volume of a cylinder is given by $V = \pi r^2 h$. Suppose we set up an experimental device to measure the volume of blood in a blood vessel of fixed length 80 mm.

59. Find a formula for dV/dr.

Suppose now a drug which causes a blood vessel to expand is administered. Evaluate dV/dr for the following values of r.

60. 4 mm

61. 6 mm

62. 8 mm

Management

The profit in dollars from the sale of x expensive tape recorders is given by

$$P(x) = x^3 - 5x^2 + 7x + 10.$$

Find the marginal profit for the following values of x. (*See Example* 6.)

63. $x = 4$

64. $x = 8$

65. $x = 10$

66. $x = 12$

The total cost to produce x handcrafted weather vanes is given by

$$C(x) = 100 + 8x - x^2 + 4x^3.$$

Find the marginal cost for the following values of x.

67. $x = 0$

68. $x = 4$

69. $x = 6$

70. $x = 8$

The cost to manufacture x units of tacos is given by

$$T(x) = 500 + 2x + 4x^{1/2}.$$

Find the marginal cost for the following values of x.

71. $x = 1$

72. $x = 4$

73. $x = 9$

74. $x = 16$

Natural Science

We saw in Section 3 of this chapter that the velocity of a particle is given by

$$\lim_{h \to 0} \frac{s(t+h) - s(t)}{h},$$

where s(t) gives the position of the particle at time t. This limit is actually the derivative of s(t), so the velocity of a particle is given by s′(t). If v(t) represents velocity at time t, then v(t) = s′(t). For each of the following position functions, find (a) a formula for v(t); (b) the velocity when t = 0, t = 5, and t = 10.

75. $s(t) = 6t + 5$

76. $s(t) = 9 - 2t$

77. $s(t) = 11t^2 + 4t + 2$

78. $s(t) = 25t^2 - 9t + 8$

79. $s(t) = 4t^3 + 8t^2$

80. $s(t) = -2t^3 + 4t^2 - 1$

12.6 DERIVATIVES OF PRODUCTS AND QUOTIENTS

To find the derivative of a sum of two functions, we find the sum of the derivatives. What about products? Is the derivative of a product equal to the product of the derivatives? Let's try an example.

$$\text{Let } g(x) = 2x + 3 \quad \text{and} \quad h(x) = 3x^2,$$
$$\text{so that } g'(x) = 2 \quad \text{and} \quad h'(x) = 6x.$$

There are two ways we might try to find the derivative of the product $g(x) \cdot h(x)$:

1. Find the product of $g(x)$ and $h(x)$ and take the derivative of this result.

$$f(x) = g(x) \cdot h(x) = (2x + 3)(3x^2) = 6x^3 + 9x^2$$

The derivative of this result is

$$f'(x) = 18x^2 + 18x.$$

2. Find the derivatives of $g(x)$ and $h(x)$ and take the product of the results.

$$g'(x) = 2$$
$$h'(x) = 6x$$
$$f'(x) = g'(x) \cdot h'(x) = 2(6x) = 12x$$

The two results are not the same. Since the method on the left uses only algebra and proven results, we can accept it as correct. The method on the right has given a wrong answer. Thus, in general, the derivative of a product *does not* equal the product of the derivatives. This fact is unfortunate, since the rule that must actually be used is a little more complicated than the other rules we have seen.

Theorem 12.5 *Product rule* If $f(x) = g(x) \cdot k(x)$, and if both g' and k' exist, then

$$f'(x) = g(x) \cdot k'(x) + k(x) \cdot g'(x).$$

(The derivative of a product of two functions is the first function times the derivative of the second, plus the second function times the derivative of the first.)

The proof of the product rule is included in Exercise 59 below.

Example 1 Let $f(x) = (2x + 3)(3x^2)$. Use the product rule to find $f'(x)$.

Here f is given as the product of $g(x) = 2x + 3$ and $k(x) = 3x^2$. Use the product rule and the fact that $g'(x) = 2$ and $k'(x) = 6x$.

$$\begin{aligned} f'(x) &= g(x) \cdot k'(x) + k(x) \cdot g'(x) \\ &= (2x + 3)(6x) + (3x^2)(2) \\ &= 12x^2 + 18x + 6x^2 \qquad \text{Use algebra to simplify} \\ f'(x) &= 18x^2 + 18x \end{aligned}$$

This is the same result we found before. ■

1. Use the product rule to find the derivatives of the following.
(a) $f(x) = (5x^2 + 6)(3x)$
(b) $g(x) = (8x)(4x^2 + 5x)$

Answer:
(a) $45x^2 + 18$
(b) $96x^2 + 80x$

Work Problem 1 at the side.

Example 2 Let $f(x) = (x^2 - 4x)(3x + 5)$. Use the product rule to find $f'(x)$.

Let $g(x) = x^2 - 4x$, so that $g'(x) = 2x - 4$. Let $k(x) = 3x + 5$, with $k'(x) = 3$. Then

$$\begin{aligned} f'(x) &= g(x) \cdot k'(x) + k(x) \cdot g'(x) \\ &= (x^2 - 4x)(3) + (3x + 5)(2x - 4) \\ &= 3x^2 - 12x + 6x^2 - 2x - 20 \\ f'(x) &= 9x^2 - 14x - 20. \end{aligned}$$ ■

2. Find the derivatives of the following.
(a) $f(x) = (x^2 - 3)(x + 5)$
(b) $g(x) = (x + 4)(5x^2 + 6)$

Answer:
(a) $3x^2 + 10x - 3$
(b) $15x^2 + 40x + 6$

Work Problem 2 at the side.

You may have noticed that many of the problems above could have been worked by multiplying out the original functions. The product rule would not then be needed. This is correct; however, in the next section we shall see products of functions where this cannot be done—the product rule is essential.

What about *quotients* of functions? To find the derivative of the quotient of two functions, we must use the result of the next theorem.

Theorem 12.6 *Quotient rule* If $f(x) = \frac{g(x)}{k(x)}$, where g and k are functions whose derivatives exist, and if $k(x) \neq 0$, then

$$f'(x) = \frac{k(x) \cdot g'(x) - g(x) \cdot k'(x)}{[k(x)]^2}.$$

(The derivative of a quotient is the denominator times the derivative of the numerator, minus the numerator times the derivative of the denominator, all over the denominator squared.)

The proof of the quotient rule is included as Exercise 60 below.

Example 3 Let $f(x) = \frac{2x - 1}{4x + 3}$. Find $f'(x)$.

Here we can let $g(x) = 2x - 1$, with $g'(x) = 2$. Also, $k(x) = 4x + 3$, and $k'(x) = 4$. Then

$$f'(x) = \frac{k(x) \cdot g'(x) - g(x) \cdot k'(x)}{[k(x)]^2}$$

$$= \frac{(4x + 3)(2) - (2x - 1)(4)}{(4x + 3)^2}$$

$$= \frac{8x + 6 - 8x + 4}{(4x + 3)^2}$$

$$f'(x) = \frac{10}{(4x + 3)^2}. \quad ■$$

3. Find the derivatives of the following.

(a) $f(x) = \frac{3x + 7}{5x + 8}$

(b) $g(x) = \frac{2x + 11}{5x - 1}$

Answer:

(a) $\frac{-11}{(5x + 8)^2}$

(b) $\frac{-57}{(5x - 1)^2}$

Work Problem 3 at the side.

Example 4 Let $F(x) = \frac{6x^2 - 4x}{8x + 1}$. Find $F'(x)$.

Use the quotient rule.

$$F'(x) = \frac{(8x + 1)(12x - 4) - (6x^2 - 4x)(8)}{(8x + 1)^2}$$

$$= \frac{96x^2 - 20x - 4 - 48x^2 + 32x}{(8x + 1)^2}$$

$$F'(x) = \frac{48x^2 + 12x - 4}{(8x + 1)^2} \quad ■$$

4. Find the derivatives of the following.

(a) $f(x) = \dfrac{9x + 2}{x^2 + 6}$

(b) $F(x) = \dfrac{4x^2 - 8x}{3x - 5}$

(c) $h(x) = \dfrac{x^3 - x^2}{x + 1}$

Answer:

(a) $\dfrac{-9x^2 - 4x + 54}{(x^2 + 6)^2}$

(b) $\dfrac{12x^2 - 40x + 40}{(3x - 5)^2}$

(c) $\dfrac{2x^3 + 2x^2 - 2x}{(x + 1)^2}$

5. The total revenue in thousands of dollars from the sale of x dozen CB radios is given by

$$R(x) = 32x^2 + 7x + 80.$$

(a) Find the average revenue.
(b) Find the derivative of the average revenue.

Answer:

(a) $\dfrac{32x^2 + 7x + 80}{x}$

(b) $\dfrac{32x^2 - 80}{x^2}$

Work Problem 4 at the side.

Suppose $y = C(x)$ gives the total cost to manufacture x items. Then the **average cost per item** can be found by dividing the total cost by the number of items, or

$$\text{average cost} = \frac{C(x)}{x}.$$

A company naturally would be interested in making this average cost as small as possible. We will see in the next chapter that this can be done by taking the derivative of $C(x)/x$. This derivative can be found by using the quotient rule, as the next example shows.

Example 5 The total cost in thousands of dollars to manufacture x electrical generators is given by $C(x)$, where

$$C(x) = -x^3 + 15x^2 + 1000.$$

(a) Find the average cost per generator.

The average cost is given by the total cost divided by the number of items, or

$$\frac{C(x)}{x} = \frac{-x^3 + 15x^2 + 1000}{x}.$$

(b) Find the derivative of the average cost.

We can use the quotient rule to find the derivative of the average cost.

$$\frac{d}{dx}\left[\frac{C(x)}{x}\right] = \frac{x(-3x^2 + 30x) - (-x^3 + 15x^2 + 1000)(1)}{x^2}$$

$$= \frac{-3x^3 + 30x^2 + x^3 - 15x^2 - 1000}{x^2}$$

$$= \frac{-2x^3 + 15x^2 - 1000}{x^2} \quad ■$$

Work Problem 5 at the side.

12.6 EXERCISES

Use the product rule to find the derivative of each of the following functions. (See Examples 1 and 2.) In Exercises 13–16, use the fact that $p^2 = p \cdot p$.

1. $y = 3x(2x - 5)$

2. $y = -4x(8x + 1)$

3. $y = (2x - 5)(x + 4)$

4. $y = (3x + 7)(x - 1)$

5. $y = (8x - 2)(3x + 9)$

6. $y = (4x + 1)(7x + 12)$

7. $y = (3x^2 + 2)(2x - 1)$

8. $y = (5x^2 - 1)(4x + 3)$

9. $y = (x^2 + x)(3x - 5)$

10. $y = (2x^2 - 6x)(x + 2)$

11. $y = (9x^2 + 7x)(x^2 - 1)$

12. $y = (2x^2 - 4x)(5x^2 + 4)$

13. $y = (2x - 5)^2$

14. $y = (7x - 6)^2$

15. $y = (x^2 - 1)^2$

16. $y = (3x^2 + 2)^2$

17. $y = (x + 1)(\sqrt{x} + 2)$

18. $y = (2x - 3)(\sqrt{x} - 1)$

19. $y = (5\sqrt{x} - 1)(2\sqrt{x} + 1)$

20. $y = (-3\sqrt{x} + 6)(4\sqrt{x} - 2)$

Use the quotient rule to find the derivatives of each of the following functions. (See Examples 3 and 4.)

21. $y = \frac{x + 1}{2x - 1}$

22. $y = \frac{3x - 5}{x - 4}$

23. $y = \frac{7x + 1}{3x + 8}$

24. $y = \frac{6x - 11}{8x + 1}$

25. $y = \frac{2}{3x - 5}$

26. $y = \frac{-4}{2x - 11}$

27. $y = \frac{5 - 3x}{4 + x}$

28. $y = \frac{9 - 7x}{1 - x}$

29. $y = \frac{x^2 + x}{x - 1}$

30. $y = \frac{x^2 - 4x}{x + 3}$

31. $y = \frac{x - 2}{x^2 + 1}$

32. $y = \frac{4x + 11}{x^2 - 3}$

33. $y = \frac{3x^2 + x}{2x^2 - 1}$

34. $y = \frac{-x^2 + 6x}{4x^2 + 1}$

35. $y = \frac{x^2 - 4x + 2}{x + 3}$

36. $y = \frac{x^2 + 7x - 2}{x - 2}$

37. $y = \frac{\sqrt{x}}{x - 1}$

38. $y = \frac{\sqrt{x}}{2x + 3}$

39. $y = \frac{5x + 6}{\sqrt{x}}$

40. $y = \frac{9x - 8}{\sqrt{x}}$

APPLIED PROBLEMS

Management

The total cost to produce x units of perfume is given by

$$C(x) = 9x^2 - 4x + 8.$$

Find the average cost per unit to produce (See Example 5.)

41. 10 units; (Find $C(10)/10$.)

42. 20 units;

43. x units.

44. Find the derivative of the average cost function.

The total profit from selling x units of mathematics textbooks is given by

$$P(x) = 10x^2 - 5x - 18.$$

Find the average profit from selling

45. 8 units; **46.** 15 units; **47.** x units.

48. Find the derivative of the average profit function.

Suppose you are the manager of a trucking firm, and one of your drivers reports that, according to her calculations, her truck burns fuel at the rate of

$$G(x) = \frac{1}{200}\left(\frac{800}{x} + x\right)$$

gallons per mile when traveling at x miles per hour on a smooth, dry road. (If a driver did report this to you, it would probably be a good idea to make the driver your mathematical modeler.)

49. If the driver tells you that she wants to travel 20 miles per hour, what should you tell her? (Hint: Take the derivative of G and evaluate it for $x = 20$. Then interpret your results.)

50. If the driver wants to go 40 miles per hour, what should you say? (Hint: Find $G'(40)$. In Example 3 of Section 13.4 we will find the speed that produces the lowest possible cost.)

Natural Science

Assume that the total number of bacteria in millions present at a certain time t is given by

$$N(t) = (t - 10)^2(2t) + 50.$$

51. Find $N'(t)$.

At what rate is the population of bacteria changing

52. when $t = 8$; **53.** when $t = 11$?

54. Answer 52 is negative, while answer 53 is positive. What does this mean in terms of the population of bacteria?

When a certain drug is introduced into a muscle, the muscle responds by contracting. The amount of contraction, s, in millimeters, is related to the concentration of the drug, x, in milliliters, by

$$s(x) = \frac{x}{m + nx},$$

where m and n are constants.

55. Find $s'(x)$.

56. Evaluate $s'(x)$ when $x = 50$, $m = 10$, and $n = 3$.

Social Science

According to the psychologist L. L. Thurstone, the number of facts of a certain type that are remembered after t hours is given by

$$f(t) = \frac{kt}{t + a},$$

where k and a are constants. Find $f'(t)$ if $a = 50$, $k = 100$, and

57. $t = 10$;

58. $t = 20$.

59. Prove the product rule. (Hint: Write $g(x+h) \cdot k(x+h) - g(x) \cdot k(x)$ as $g(x+h) \cdot k(x+h) - g(x+h) \cdot k(x) + g(x+h) \cdot k(x) - g(x) \cdot k(x)$.)

60. Prove the quotient rule.

12.7 THE CHAIN RULE

We can find the derivative of the function $y = (3 + 6x)^2$ by writing the function as $y = (3 + 6x)(3 + 6x)$ and using the product rule, as follows.

$$\begin{aligned} y' = \frac{dy}{dx} &= (3 + 6x)(6) + (3 + 6x)(6) \\ &= 18 + 36x + 18 + 36x \\ &= 36 + 72x \end{aligned}$$

We could find the derivative of $y = (3 + 6x)^3$ or $y = (3 + 6x)^4$ in a similar (but much longer) way. Luckily, there is a shortcut for finding derivatives of functions raised to a power.

Theorem 12.7 *Generalized power rule* Let u be some function of x. Let $y = u^n$. Then

$$y' = \frac{dy}{dx} = n \cdot u^{n-1} \cdot u'.$$

(The derivative of $y = u^n$ is found by decreasing the exponent on u by 1 and multiplying the result by the exponent n and the derivative of u.)

Example 1 Use the generalized power rule to find the derivative of $y = (3 + 6x)^2$.

Here we let $u = 3 + 6x$. Thus, $u' = 6$. By the generalized power rule, we have

$$\frac{dy}{dx} = n \cdot u^{n-1} \cdot u'$$

$$= 2(3 + 6x)^{2-1} \cdot 6 \qquad \text{Let } n = 2,\ u = 3 + 6x, \text{ and } u' = 6$$

$$= 12(3 + 6x)$$

$$= 36 + 72x.$$

This is the same answer we found above. ■

Example 2 Let $y = (x^2 + 5)^4$. Find dy/dx.
We have $u = x^2 + 5$ and $u' = 2x$. Thus

$$\frac{dy}{dx} = 4(x^2 + 5)^3(2x) \qquad \text{Let } n = 4,\ u = x^2 + 5, \text{ and } u' = 2x$$

$$= 8x(x^2 + 5)^3. \ ■$$

Work Problem 1 at the side.

1. Find dy/dx for the following functions.

(a) $y = (2x + 5)^6$
(b) $y = (4x^2 - 7)^3$

Answer:
(a) $12(2x + 5)^5$
(b) $24x(4x^2 - 7)^2$

Example 3 Find dy/dx if $y = 5(8x + 6)^4$.
We have

$$\frac{dy}{dx} = 5(4)(8x + 6)^3(8) = 160(8x + 6)^3. \ ■$$

Example 4 Find dy/dx if $y = \sqrt{9x + 2}$.
Rewrite $y = \sqrt{9x + 2}$ as $y = (9x + 2)^{1/2}$. Then

$$\frac{dy}{dx} = \frac{1}{2}(9x + 2)^{-1/2}(9)$$

$$= \frac{9}{2}(9x + 2)^{-1/2}.$$

This answer can be written as

$$\frac{dy}{dx} = \frac{9}{2(9x + 2)^{1/2}} \qquad (9x + 2)^{-1/2} = \frac{1}{(9x + 2)^{1/2}}$$

or

$$\frac{dy}{dx} = \frac{9}{2\sqrt{9x + 2}}. \ ■$$

Work Problem 2 at the side.

2. Find dy/dx for the following functions.

(a) $y = 12(x^2 + 6)^5$
(b) $y = -4\sqrt{3x - 8}$
(c) $8(4x^2 + 2)^{3/2}$

Answer:
(a) $120x(x^2 + 6)^4$
(b) $-6(3x - 8)^{-1/2}$ or $\dfrac{-6}{\sqrt{3x - 8}}$
(c) $96x(4x^2 + 2)^{1/2}$

Example 5 Suppose the revenue, $R(x)$, produced by selling x units of agar-agar to a biologist is given by

$$R(x) = 4\sqrt{3x+1} + \frac{4}{x}.$$

Assume that $x > 0$. Find the marginal revenue when $x = 5$.

As we saw earlier, the marginal revenue is given by the derivative of the revenue function. By the generalized power rule, we have

$$R'(x) = \frac{6}{\sqrt{3x+1}} - \frac{4}{x^2}.$$

When $x = 5$, we have

$$\begin{aligned} R'(5) &= \frac{6}{\sqrt{3(5)+1}} - \frac{4}{5^2} \\ &= \frac{6}{\sqrt{16}} - \frac{4}{25} \\ &= \frac{6}{4} - \frac{4}{25} \\ &= 1.34. \end{aligned}$$

After 5 units have been sold, the sale of one more unit will increase revenue by about \$1.34. ■

Work Problem 3 at the side.

3. Use the function of Example 5 and find the marginal revenue when
(a) $x = 8$;
(b) $x = 16$;
(c) $x = 20$.

Answer:
(a) 1.14
(b) .84
(c) .76

Sometimes we need to use both the generalized power rule and the product or quotient rule, as the next example shows.

Example 6 Find the derivative of $y = 4x(3x+5)^5$.

By the product rule and the generalized power rule, we have

$$\begin{aligned} \frac{dy}{dx} &= 4x[5(3x+5)^4 \cdot 3] + (3x+5)^5(4) \\ &= 60x(3x+5)^4 + 4(3x+5)^5 \\ &= 4(3x+5)^4[15x + (3x+5)^1] \qquad \text{Factor out the greatest common factor.} \\ &= 4(3x+5)^4(18x+5). \end{aligned}$$ ■

Work Problem 4 at the side.

4. Find the derivatives of the following.
(a) $y = 6x(x+2)^2$
(b) $y = -9x(2x^2+1)^3$
(c) $y = 5x^2(3x-1)^4$

Answer:
(a) $6(x+2)(3x+2)$
(b) $-9(2x^2+1)^2(14x^2+1)$
(c) $10x(3x-1)^3(9x-1)$

Example 7 Find the derivative of $y = \dfrac{(3x+2)^7}{x-1}$.

Use the quotient rule and the generalized power rule.

$$\frac{dy}{dx} = \frac{(x-1)[7(3x+2)^6 \cdot 3] - (3x+2)^7(1)}{(x-1)^2}$$
$$= \frac{21(x-1)(3x+2)^6 - (3x+2)^7}{(x-1)^2}$$
$$= \frac{(3x+2)^6[21(x-1) - (3x+2)]}{(x-1)^2}$$
$$= \frac{(3x+2)^6[21x - 21 - 3x - 2]}{(x-1)^2}$$
$$\frac{dy}{dx} = \frac{(3x+2)^6(18x-23)}{(x-1)^2} \quad ■$$

Work Problem 5 at the side.

5. Find the derivatives of the following.

(a) $y = \frac{(2x+1)^3}{3x}$

(b) $y = \frac{(x-6)^5}{3x-5}$

Answer:

(a) $\frac{(2x+1)^2(4x-1)}{3x^2}$

(b) $\frac{(x-6)^4(12x-7)}{(3x-5)^2}$

Some functions, especially the ones we saw in Chapter 11, have derivatives that cannot be found by the generalized power rule. For these functions we need the chain rule.

Theorem 12.8 *Chain rule* Let u be some function of x. Let y be a function of u. Let all indicated derivatives exist. Then

omit

$$\frac{dy}{dx} = \frac{dy}{du} \cdot \frac{du}{dx}.$$

Note that the generalized power rule is just a special case of the chain rule.

Example 8 Use the chain rule to find the derivatives of each of the following functions.

(a) $y = (9x+2)^5$

Let $u = 9x + 2$. Then $du/dx = 9$. The expression $9x + 2$ is raised to the fifth power, so that $y = u^5$, and $dy/du = 5u^4$. By the chain rule,

$$\frac{dy}{dx} = \frac{dy}{du} \cdot \frac{du}{dx}$$
$$= 5u^4(9)$$
$$= 5(9x+2)^4(9) \qquad \text{since } u = 9x + 2$$
$$\frac{dy}{dx} = 45(9x+2)^4.$$

This is the same result we would get using the generalized power rule.

(b) $y = 25\sqrt{15x^2 + 1}$

Here $y = 25(15x^2 + 1)^{1/2}$. Let $u = 15x^2 + 1$, and let $y = 25u^{1/2}$. Then $dy/du = 25/(2u^{1/2})$, and $du/dx = 30x$. Finally,

$$\frac{dy}{dx} = \frac{25}{2u^{1/2}}(30x)$$

$$= \frac{25(30x)}{2(15x^2 + 1)^{1/2}} \qquad \text{since } u = 15x^2 + 1$$

$$= \frac{375x}{\sqrt{15x^2 + 1}}. \quad \blacksquare$$

Work Problem 6 at the side.

6. Use the chain rule to find the derivatives of the following.

(a) $y = 10(2x^2 + 1)^4$

(b) $y = -8\sqrt{14x + 1}$

Answer:

(a) $160x(2x^2 + 1)^3$

(b) $\dfrac{-56}{\sqrt{14x + 1}}$

12.7 EXERCISES

Find the derivatives of the following functions. (See Examples 1–4.)

1. $y = (2x + 9)^2$

2. $y = (8x - 3)^2$

3. $y = 6(5x - 1)^3$

4. $y = -8(3x + 2)^3$

5. $y = -2(12x^2 + 4)^3$

6. $y = 5(3x^2 - 5)^3$

7. $y = 9(x^2 + 5x)^4$

8. $y = -3(x^2 - 5x)^4$

9. $y = 12(2x + 5)^{3/2}$

10. $y = 45(3x - 8)^{3/2}$

11. $y = -7(4x^2 + 9x)^{3/2}$

12. $y = 11(5x^2 + 6x)^{3/2}$

13. $y = 8\sqrt{4x + 7}$

14. $y = -3\sqrt{7x - 1}$

15. $y = -2\sqrt{x^2 + 4x}$

16. $y = 4\sqrt{2x^2 + 3}$

Use the product or quotient rule to find the following derivatives. (See Examples 6 and 7.)

17. $y = 4x(2x + 3)^2$

18. $y = -6x(5x - 1)^2$

19. $y = (x + 2)(x - 1)^2$

20. $y = (3x + 1)^2(x + 4)$

21. $y = 5(x + 3)^2(x - 1)^2$

22. $y = -9(x + 4)^2(2x - 3)^2$

23. $y = (x + 1)^2\sqrt{x}$

24. $y = (3x + 5)^2\sqrt{x}$

25. $y = \dfrac{1}{(x - 4)^2}$

26. $y = \dfrac{-5}{(2x + 1)^2}$

27. $y = \dfrac{(x + 3)^2}{x - 1}$

28. $y = \dfrac{(x - 6)^2}{x + 4}$

29. $y = \dfrac{x^2 + 4x}{(x + 2)^2}$

30. $y = \dfrac{3x^2 - x}{(x - 1)^2}$

APPLIED PROBLEMS

Management

Assume that the total revenue from the sale of x television sets is given by

$$R(x) = 1000\left(1 - \frac{x}{500}\right)^2.$$

Find the marginal revenue for the following values of x. (See Example 5.)

31. $x = 100$ **32.** $x = 150$

33. $x = 200$ **34.** $x = 400$

35. Find the average revenue from the sale of x sets.

36. Find the derivative of the average revenue.

Natural Science

The total number of bacteria in millions present in a culture is given by

$$N(t) = 2t(5t + 9)^{1/2} + 12,$$

where t represents time in hours after the beginning of an experiment. Find the rate of change of the population of bacteria with respect to time when

37. $t = 0$; **38.** $t = 7/5$;

39. $t = 8$; **40.** $t = 11$.

As a check on a person's use of calcium, a small amount of radioactive calcium is injected into the person's bloodstream with measurements of the calcium remaining in the bloodstream made each day for several days. Suppose the amount of the calcium remaining in the bloodstream in milligrams per cubic centimeter t days after the initial injection is approximated by

$$C(t) = \frac{1}{2}(2t + 1)^{-1/2}.$$

Find the rate of change of C with respect to time when

41. $t = 0$; **42.** $t = 4$;

43. $t = 6$ (use a calculator); **44.** $t = 7.5$.

The strength of a person's reaction to a certain drug is given by

$$R(Q) = Q\left(C - \frac{Q}{3}\right)^{1/2},$$

where Q represents the quantity of the drug given to the patient and C is a constant.

45. The derivative $R'(Q)$ is called the *sensitivity* to the drug. Find $R'(Q)$.

46. Find $R'(Q)$ if $Q = 87$ and $C = 59$.

CHAPTER 12 TEST

[12.1] *Find each of the following limits that exist.*

1. $\lim_{x\to 6} \dfrac{2x+5}{x-3}$

2. $\lim_{x\to 4} \dfrac{x^2-16}{x-4}$

3. $\lim_{x\to -4} \dfrac{2x^2+3x-20}{x+4}$

4. $\lim_{x\to 9} \dfrac{\sqrt{x}-3}{x-9}$

5. $\lim_{x\to \infty} \dfrac{2x+5}{4x-9}$

6. $\lim_{x\to \infty} \dfrac{x^2+6x+8}{x^3+2x+1}$

[12.2] *Find all points of discontinuity for the following functions.*

7.

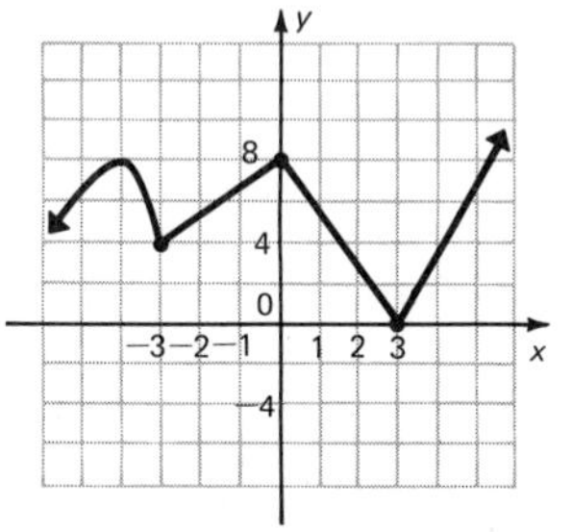

8.

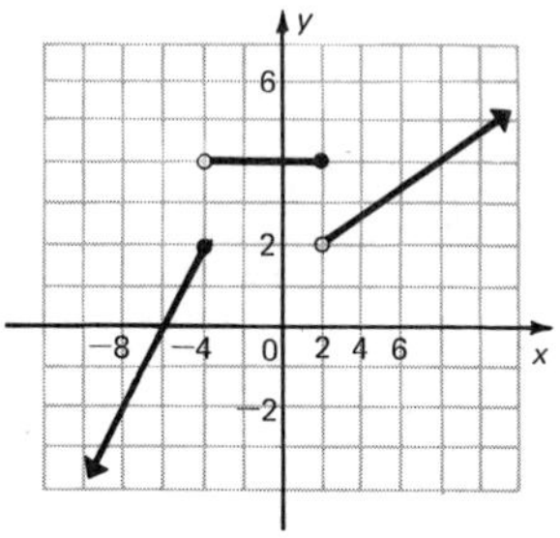

Write each of the following sets in interval notation.

9. $\{x \mid -4 \le x \le 2\}$

10. $\{x \mid x > 4\}$

[12.3] *Find the average rate of change for the following functions.*

11. $y = 6x^2 + 2$, from $x = 1$ to $x = 4$

12. $y = -2x^3 - x^2 + 5$, from $x = -2$ to $x = 6$

13. $y = \dfrac{-6}{3x-5}$, from $x = 4$ to $x = 9$

[12.4] *Use the definition of the derivative to find the derivative of each of the following functions.*

14. $y = 4x + 3$

15. $y = 5x^2 + 6x$

[12.5] *Suppose that the profit in cents from selling x pounds of potatoes is given by*

$$P(x) = 15x + 25x^2.$$

Find the marginal profit when

16. $x = 6$;

17. $x = 20$;

18. $x = 30$.

The sales of a company are related to its expenditure on research by

$$S(x) = 1000 + 50\sqrt{x} + 10x,$$

where $S(x)$ gives sales in millions when x thousand dollars are spent on research. Find dS/dx when

19. $x = 9$; **20.** $x = 16$; **21.** $x = 25$.

For each of the following position functions, find (a) a formula for $v(t)$; (b) the velocity when $t = 0$, $t = 9$, and $t = 20$.

22. $s(t) = 60t + 5$

23. $s(t) = 8t^3 - 4t^2$

Find the derivative of each of the following functions.

24. $y = x^3 - 4x^2$

25. $y = 6x^{7/4}$

[12.6] **26.** $y = (3x + 5)(6x - 1)$

27. $y = 4(x + 2)\sqrt{x}$

28. $y = \dfrac{-8}{2x + 1}$

29. $y = \dfrac{x^2 - x + 1}{x - 1}$

[12.7] **30.** $y = 6(x^2 + 2)^5$

31. $y = \sqrt{6x - 11}$

32. $y = 9x(x + 1)^5$

33. $y = \dfrac{2x}{(x + 5)^2}$

13 Applications of the Derivative

While calculus has been successfully applied to mathematical models in physical science for the past three hundred years, it has been only in the last thirty years or so that calculus has been successfully applied to problems in management, social, and life sciences. One important application of calculus in these fields is finding maximum or minimum values for a function. In this chapter we discuss the basic ideas for finding maximums and minimums and look at practical examples.

13.1 RELATIVE MAXIMUMS AND MINIMUMS

Recall from Section 12.2 that an **open interval** is a set of points such as $\{x|a < x < b\}$ or $\{x|x > a\}$. The first of these open intervals would be written (a, b) and the second would be written (a, ∞). A **closed interval,** $\{x|a \leq x \leq b\}$, is written $[a, b]$. We will use this notation when we look for maximum and minimum values for a function.

A function can have two kinds of maximums: a maximum can be the largest value the function ever takes, or it can be the largest value of the function in a given interval. The largest value a function ever takes, called the *absolute maximum*, is defined in the next section. The largest value of a function in a given interval, called a *relative maximum*, is defined as follows.

A function f has a **relative maximum** at $x = a$ if $f(x) < f(a)$ for all values of x (except a) in some open interval containing a.

A **relative minimum** is defined in a very similar way. Think of a relative maximum as a peak (not necessarily the highest peak) and a relative minimum as a valley on the graph of the function. An *absolute minimum* is defined in the next section.

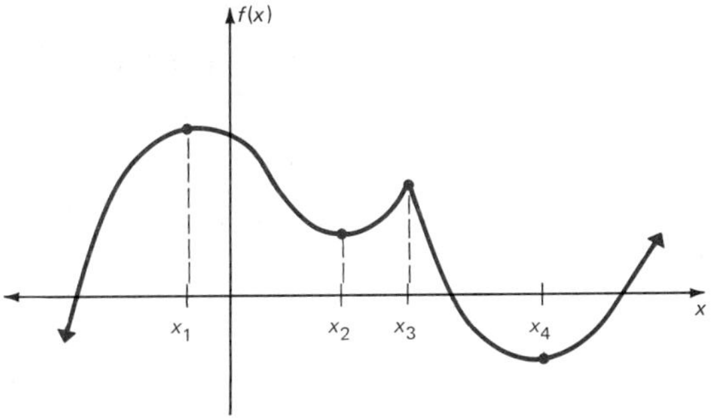

Figure 13.1

Example 1 Identify the x-values of all points where the graph of Figure 13.1 has relative maximums or relative minimums.

The graph has relative maximums at x_1 and x_3 and relative minimums at x_2 and x_4. ■

Work Problem 1 at the side.

How do we use the equation of a function to find the relative maximums or minimums for the function? We do this by using the fact that the derivative of a function can be used to find the slope of a line tangent to the graph of a function.

Figure 13.2 shows the graph of a function; this function has a relative maximum at x_2 and relative minimums at x_1 and x_3. The tangents to the curve at the points x_1, x_2, and x_3 have been drawn; all three of these tangents are horizontal lines and thus have a slope of 0. (Recall that m is used to represent slope.) Therefore, one way to find relative maximums and relative minimums for a function is to find all points where the derivative (the slope of the tangent) is 0. These points *might* lead to relative maximums or minimums.

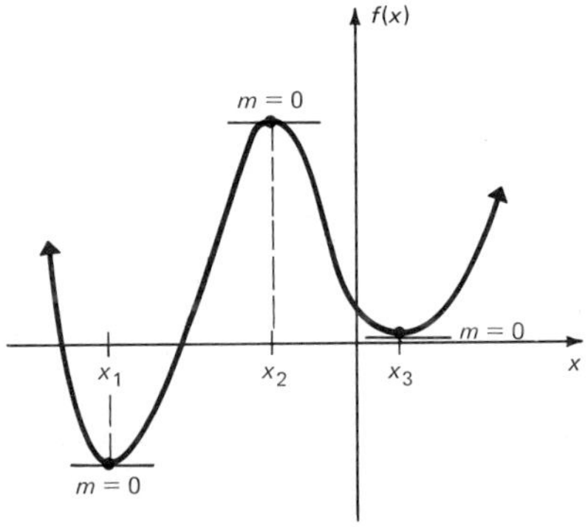

Figure 13.2

1. Identify the x-values of all points where the graphs here have relative maximums or relative minimums.

(a)

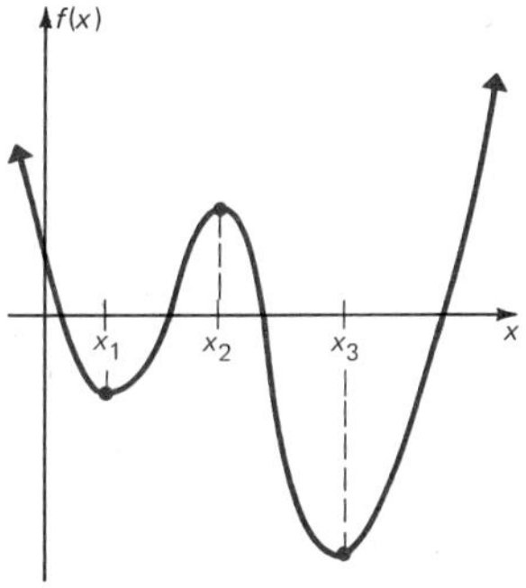

(b)

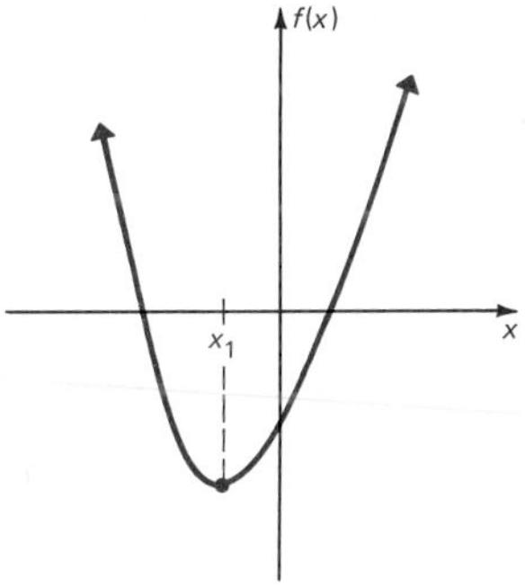

(c)

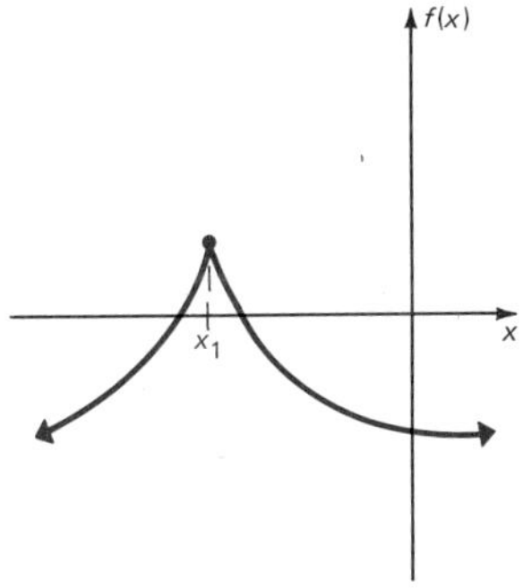

Answer:

(a) relative maximum at x_2, relative minimums at x_1 and x_3

(b) no relative maximum, relative minimum at x_1

(c) relative maximum at x_1, no relative minimum

Unfortunately, the process of finding relative maximums or minimums is not quite as simple as just finding values where the derivative is 0. It is possible for a function to have a relative maximum or minimum even though the derivative does not exist. Two examples of such functions are shown in Figure 13.3.

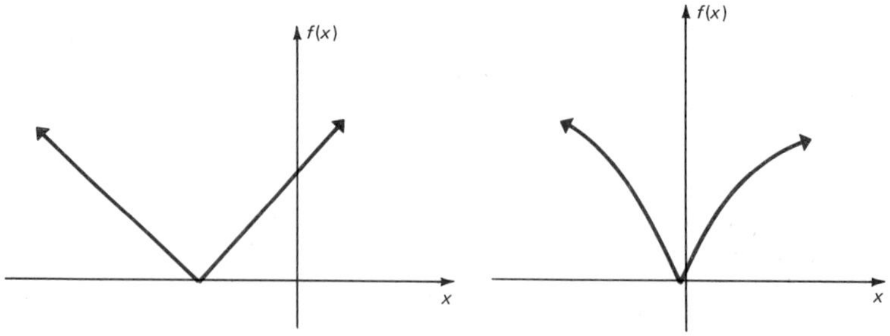

Figure 13.3

Because of this, when looking for relative maximums or minimums, we must check all values that make the derivative equal 0 and all values where the derivative does not exist but the function does. Any of these values might lead to relative maximums or minimums. The points where a derivative equals 0 or does not exist (but the function does exist) are called the **critical points** of the function. The following theorem summarizes these facts.

Theorem 13.1 If a function has any relative maximums or relative minimums, they occur only at critical points of the function.

Be very careful not to get the theorem backward. It does *not* say that a function has either a relative maximum or relative

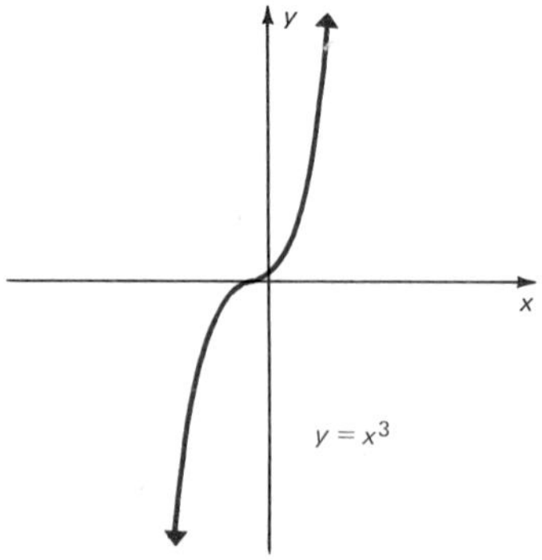

Figure 13.4

minimum at every critical point of the function. For example, Figure 13.4 shows the graph of $y = x^3$. This function has a critical point at $x = 0$, but neither a relative maximum nor a relative minimum at $x = 0$ (or anywhere else, for that matter).

Example 2 Find any relative maximums or minimums for the following functions.

(a) $f(x) = 3x^2 - 8x + 2$

Find the x-values of the critical points. One kind of critical point is found when the derivative equals 0. Here $f'(x) = 6x - 8$. Place this derivative equal to 0.

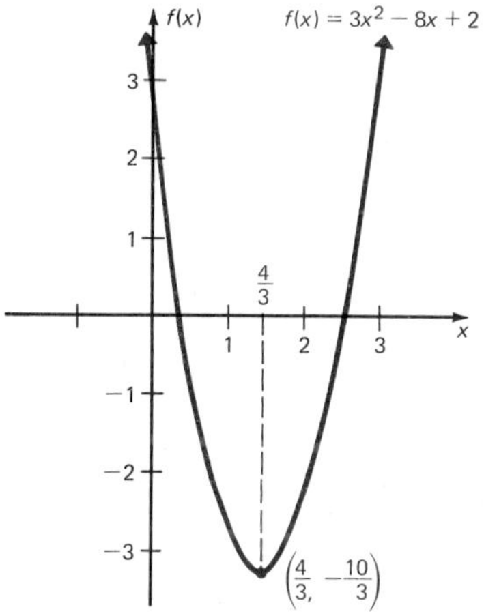

Figure 13.5

$$6x - 8 = 0$$
$$6x = 8$$
$$x = \frac{8}{6} = \frac{4}{3}$$

There are no points where the derivative does not exist, so that the only critical point is at $x = 4/3$. A sketch of the graph of $f(x) = 3x^2 - 8x + 2$ in Figure 13.5 shows that $x = 4/3$ leads to a relative minimum value for the function. The value of the relative minimum is

$$f\left(\frac{4}{3}\right) = 3\left(\frac{4}{3}\right)^2 - 8\left(\frac{4}{3}\right) + 2 = -\frac{10}{3}.$$

(b) $f(x) = 4x^3 + 3x^2 - 18x + 6$

Find the x-values of the critical points by finding the derivative and setting it equal to 0. Here $f'(x) = 12x^2 + 6x - 18$.

$$12x^2 + 6x - 18 = 0$$
$$6(2x^2 + x - 3) = 0 \qquad \text{Factor out the greatest common factor}$$
$$6(2x + 3)(x - 1) = 0 \qquad \text{Factor } 2x^2 + x - 3$$
$$2x + 3 = 0 \quad \text{or} \quad x - 1 = 0 \qquad \text{Place each factor equal to 0}$$
$$2x = -3 \qquad x = 1$$
$$x = -\frac{3}{2}$$

There are no points where the derivative does not exist, so that the only critical points are at $x = -3/2$ and $x = 1$. Figure 13.6 shows a sketch of the graph of $f(x) = 4x^3 + 3x^2 - 18x + 6$. As the graph shows, $-3/2$ leads to a relative maximum, while 1 gives a relative minimum. The relative maximum is $f(-3/2) = 105/4$, and the relative minimum is $f(1) = -5$. ■

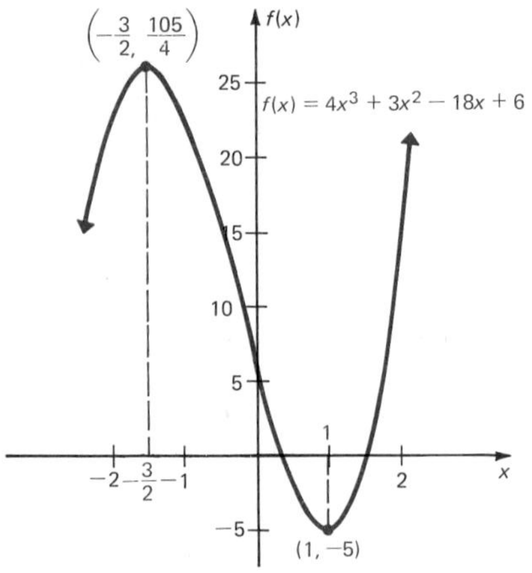

Figure 13.6

2. Find all relative maximums or minimums for the following functions. Sketch the graph to decide if a critical point gives a maximum, a minimum, or neither.
(a) $f(x) = 6x^2 + 24x + 20$
(b) $f(x) = \frac{1}{3}x^3 - x^2 - 15x + 6$

Answer:
(a) relative minimum at $x = -2$, relative minimum is $f(-2) = -4$
(b) relative maximum at $x = -3$, relative maximum is $f(-3) = 33$; relative minimum at $x = 5$, minimum is $f(5) = -157/3$

Work Problem 2 at the side.

In these examples we found the critical points by using algebra, but it was necessary to draw a graph of the function to see if the various critical points led to relative maximums or relative minimums. A better way to decide whether or not a given critical point leads to a relative maximum, a relative minimum, or neither will be developed in the rest of this section.

If we choose any two values x_1 and x_2 in an interval with $x_1 < x_2$ and find that $f(x_1) < f(x_2)$, we say function f is *increasing* on the interval. A function is increasing if its graph goes *up* as we move from left to right along the x-axis. Examples of increasing functions are shown in Figure 13.7.

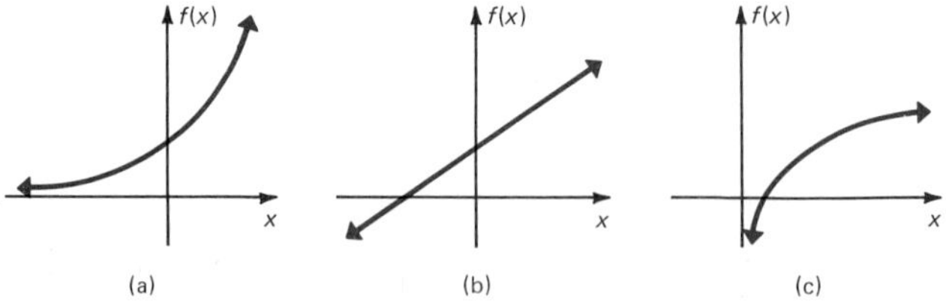

Figure 13.7

If we can choose any two values x_1 and x_2 in an interval with $x_1 < x_2$ and find that $f(x_1) > f(x_2)$, we say function f is *decreasing* on the interval. A function is decreasing if its graph goes *down* as we move from left to right along the x-axis. Examples of decreasing functions are shown in Figure 13.8.

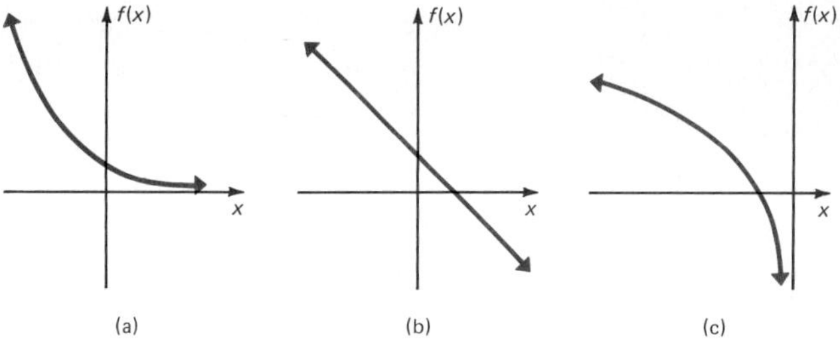

Figure 13.8

Example 3 Find the open intervals where the function of Figure 13.9 is increasing or decreasing.

The function is increasing on the open intervals $(-\infty, -4)$ and $(0, 6)$. It is decreasing on $(-4, 0)$ and $(6, \infty)$. ■

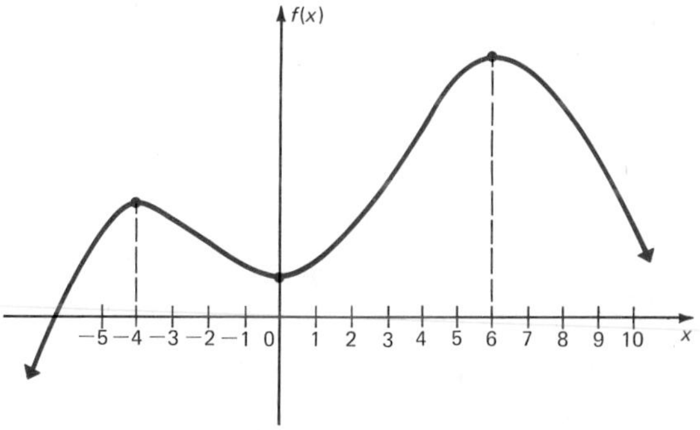

Figure 13.9

Work Problem 3 at the side.

3. Find the open intervals where the function whose graph is shown here is increasing or decreasing.

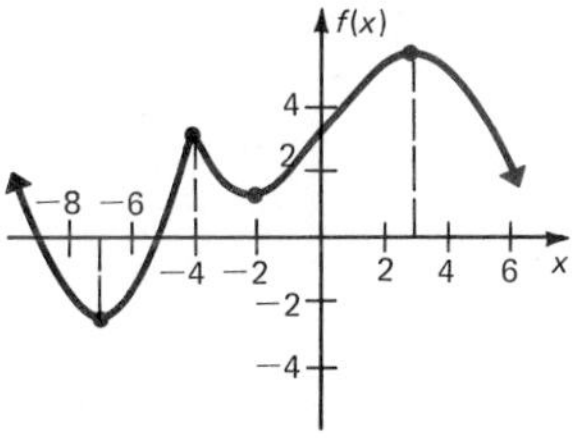

Answer: increasing on $(-7, -4)$ and $(-2, 3)$, decreasing on $(-\infty, -7)$, $(-4, -2)$, and $(3, \infty)$

We can use the derivative of a function to tell whether the function is increasing or decreasing. We know that the derivative gives the slope of the tangent to a curve at any point on the curve. If a function is increasing, the tangent line at any point on the graph will have *positive* slope, while if the function is decreasing, the slopes of the tangents will be *negative*. That is,

A function f is **increasing on an interval** if $f'(x) > 0$ for every value of x in the interval;

the function is **decreasing on an interval** if $f'(x) < 0$ for every x in the interval.

Example 4 Find the open intervals where the following functions are increasing or decreasing.

(a) $f(x) = 2x^2 + 6x - 5$

The derivative here is $f'(x) = 4x + 6$. This derivative equals 0 when $x = -3/2$. Do you see that for any value of x to the *left* of $-3/2$, the derivative is negative? Likewise, for any value of x to the *right* of $-3/2$, $f'(x) > 0$. Thus, the function is decreasing on the interval $(-\infty, -3/2)$ and increasing on $(-3/2, \infty)$. See Figure 13.10.

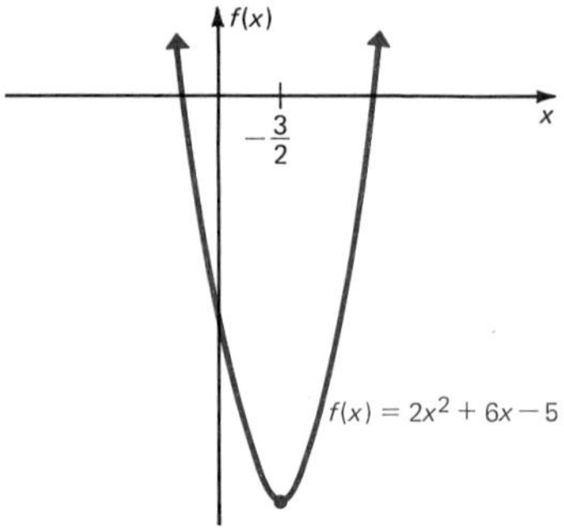

Figure 13.10

(b) $f(x) = x^3 + 3x^2 - 9x + 4$

The derivative is $f'(x) = 3x^2 + 6x - 9$. Place the derivative equal to 0 and solve the resulting equation.

$$3x^2 + 6x - 9 = 0$$

$$3(x^2 + 2x - 3) = 0 \qquad \text{Factor}$$

$$3(x + 3)(x - 1) = 0 \qquad \text{Factor } x^2 + 2x - 3$$

$$x + 3 = 0 \quad \text{or} \quad x - 1 = 0 \qquad \text{Place each factor equal to 0}$$

$$x = -3 \quad \text{or} \quad x = 1$$

The derivative is 0 at $x = -3$ and $x = 1$. To see where the function is increasing or decreasing, make a table showing the intervals where the derivative is positive or negative.

Interval	$x + 3$	$x - 1$	$3(x + 3)(x - 1)$
$(-\infty, -3)$	$-$	$-$	$+$
$(-3, 1)$	$+$	$-$	$-$
$(1, \infty)$	$+$	$+$	$+$

The function $f(x) = x^3 + 3x^2 - 9x + 4$ is increasing on the intervals $(-\infty, -3)$ and $(1, \infty)$ and decreasing on $(-3, 1)$. See the graph of f in Figure 13.11. ■

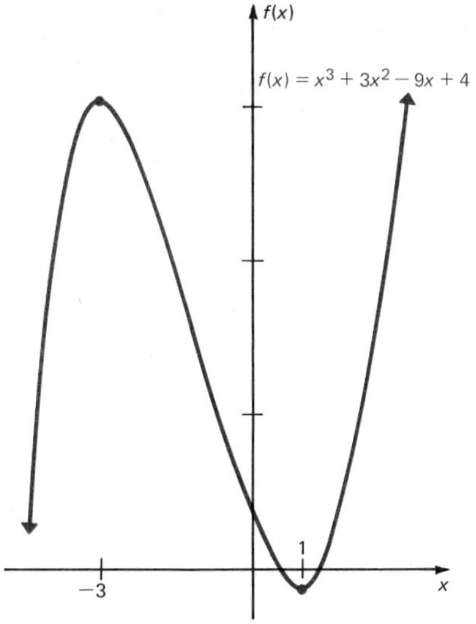

Figure 13.11

4. Find the open intervals where the following functions are increasing or decreasing.

(a) $f(x) = -3x^2 + 18x + 4$

(b) $f(x) = \frac{1}{3}x^3 + 3x^2 + 5x - 8$

Answer:

(a) increasing on $(-\infty, 3)$, decreasing on $(3, \infty)$

(b) increasing on $(-\infty, -5)$ and $(-1, \infty)$, decreasing on $(-5, -1)$

Work Problem 4 at the side.

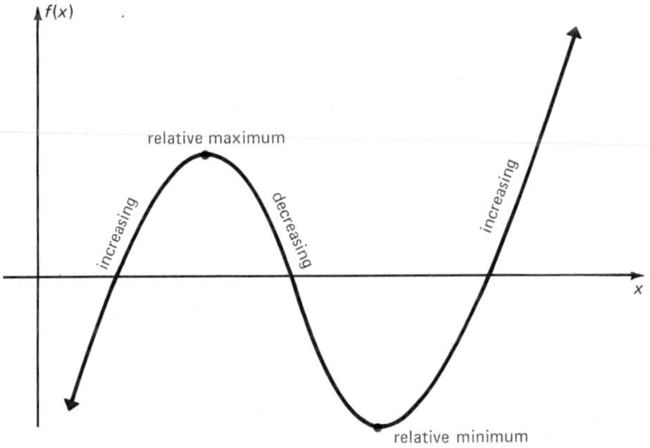

Figure 13.12

Now we can find a method for identifying relative maximums and relative minimums. As shown in Figure 13.12 a function is increasing just to the left of a relative maximum and decreasing just to the right. A function is decreasing just to the left of a relative minimum and increasing just to the right.

Putting these facts together with our methods for identifying intervals where a function is increasing or decreasing, we have the following first derivative test for locating relative maximums and relative minimums.

First derivative test Let a be a critical point for a function f. Suppose that f is continuous at a.

(a) There is a relative maximum at a if the derivative is positive in an interval just to the left of a and negative in an interval just to the right of a.

(b) There is a relative minimum at a if the derivative is negative in an interval just to the left of a and positive in an interval just to the right of a.

This test is called the *first* derivative test to set it apart from the second derivative test, which we discuss later.

Example 5 Find all relative maximums or relative minimums for the following functions.

(a) $f(x) = -x^2 + 8x + 3$

Here $f'(x) = -2x + 8$. This derivative is 0 when $x = 4$. To decide if 4 is a relative maximum or a relative minimum, note that $f'(x) = -2x + 8$ is positive if $x < 4$ and negative if $x > 4$. By part (a) of the first derivative test, this means that f has a relative maximum at $x = 4$. The value of this relative maximum is $f(4) = 19$.

(b) $f(x) = 2x^3 - 3x^2 - 72x + 15$

The derivative is $f'(x) = 6x^2 - 6x - 72$. Place this derivative equal to 0.

$$6x^2 - 6x - 72 = 0$$
$$6(x^2 - x - 12) = 0$$
$$6(x - 4)(x + 3) = 0$$
$$x - 4 = 0 \quad \text{or} \quad x + 3 = 0$$
$$x = 4 \quad \text{or} \quad x = -3$$

Make a table.

Interval	$x - 4$	$x + 3$	$6(x - 4)(x + 3)$
$(-\infty, -3)$	$-$	$-$	$+$
$(-3, 4)$	$-$	$+$	$-$
$(4, \infty)$	$+$	$+$	$+$

The derivative is positive to the left of -3 and negative to the right; thus the function has a relative maximum when $x = -3$. The value of this relative maximum is $f(-3) = 150$.

The derivative is negative to the left of 4 and positive to the right. By part (b) of the first derivative test, this means that f has a relative minimum when $x = 4$. The value of this relative minimum is $f(4) = -193$. ■

5. Find all relative maximums or relative minimums for the following functions.
(a) $f(x) = 5x^2 - 20x + 3$
(b) $f(x) = x^3 + 4x^2 - 3x + 5$

Answer:
(a) relative minimum at 2, minimum is $f(2) = -17$
(b) relative maximum at -3, maximum is $f(-3) = 23$; relative minimum at $1/3$, minimum is $f(1/3) = 121/27$

Work Problem 5 at the side.

We can now summarize the steps in **finding a relative maximum or relative minimum for a function f.**

1. Find the derivative, $f'(x)$.
2. Find all critical points a (those points where $f'(a) = 0$ or $f'(a)$ does not exist but $f(a)$ does exist).
3. If $f'(x) > 0$ to the left of a and if $f'(x) < 0$ to the right of a, then a leads to a relative maximum. The value of the relative maximum is $f(a)$.
4. If $f'(x) < 0$ to the left of a, and if $f'(x) > 0$ to the right of a, then a leads to a relative minimum. The value of the relative minimum is $f(a)$.

13.1 EXERCISES

Find the location and value of all relative maximums and relative minimums for the following functions. Identify all open intervals where the function is increasing or decreasing. (See Examples 1 and 3.)

1.
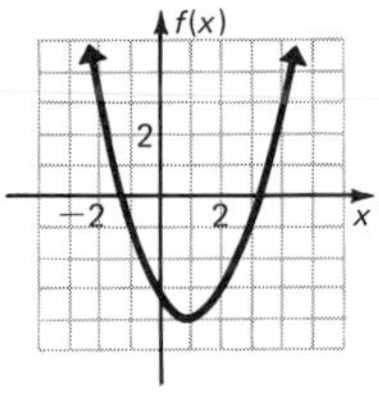

2.
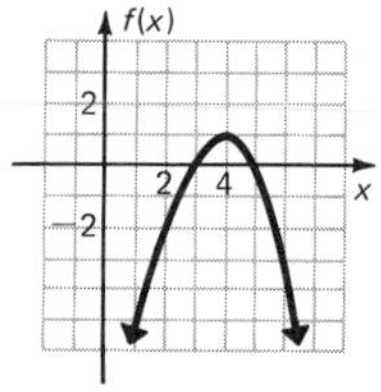

3.
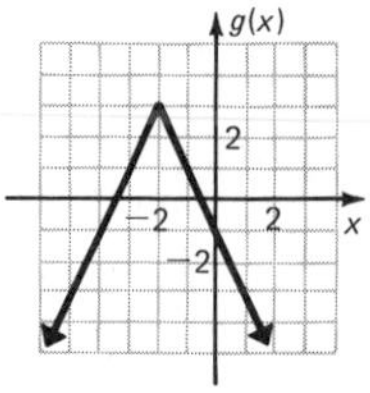

4.
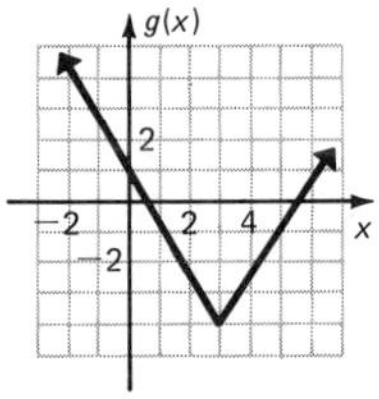

5.
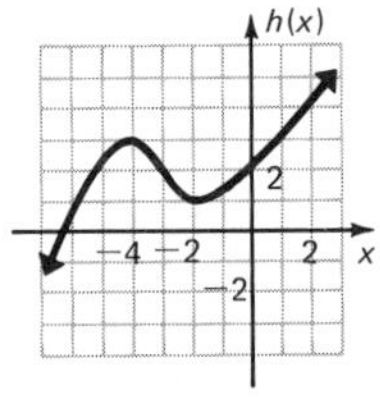

6.
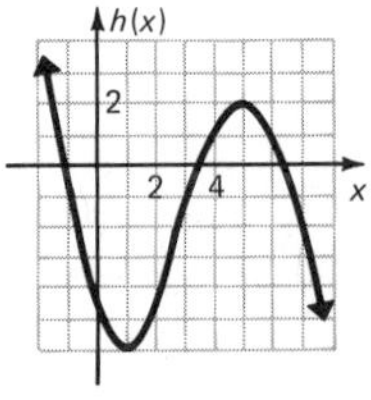

7.
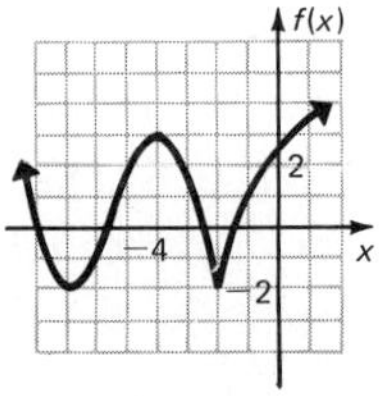

8.
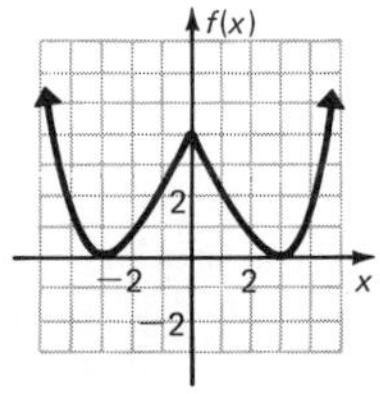

9.
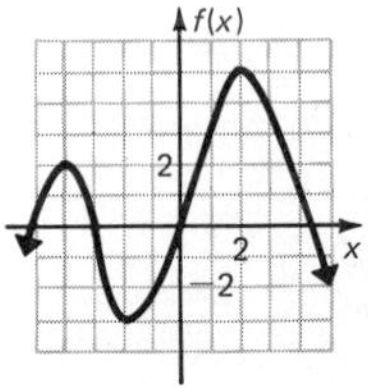

Find the x-values of all points where the following functions have any relative maximums or relative minimums. Find the value of any relative maximums or relative minimums. (See Examples 2 and 5.)

10. $f(x) = x^2 - 4x + 6$

11. $f(x) = x^2 + 12x - 8$

12. $f(x) = x^2 - 8x + 12$

13. $f(x) = x^2 - 14x + 3$

14. $f(x) = -x^2 + 4x - 5$

15. $f(x) = -x^2 - 8x + 6$

16. $f(x) = 3 - 4x - 2x^2$

17. $f(x) = 8 - 6x - x^2$

18. $f(x) = x^3 + 3x^2 - 24x + 2$

19. $f(x) = x^3 + 6x^2 + 9x - 8$

20. $f(x) = x^3 + x^2 - 5x + 1$

21. $f(x) = x^3 + 6x^2 + 12x - 5$

22. $f(x) = 2x^3 + 15x^2 + 36x - 4$

23. $f(x) = 2x^3 - 21x^2 + 60x + 5$

24. $f(x) = -\frac{2}{3}x^3 - \frac{1}{2}x^2 + 3x - 4$

25. $f(x) = -\frac{4}{3}x^3 - \frac{21}{2}x^2 - 5x + 8$

26. $f(x) = x^4 - 8x^2 + 9$

27. $f(x) = x^4 - 18x^2 - 4$

Find the vertex of each of the following parabolas. (Hint: The vertex of a parabola is the highest or lowest point on the parabola.)

28. $y = x^2 - 8x + 4$

29. $y = x^2 + 12x - 6$

30. $y = 2x^2 - 8x + 3$

31. $y = 3x^2 - 12x + 2$

32. $y = -2x^2 + 8x - 1$

33. $y = -x^2 - 2x + 1$

34. $y = 2x^2 - 5x + 2$

35. $y = 3x^2 - 8x + 4$

Find the open intervals where the following functions are increasing or decreasing. (See Examples 3 and 4.)

36. $f(x) = x^2 - 9x + 4$

37. $f(x) = x^2 + 12x - 6$

38. $f(x) = 3 + 4x - 2x^2$

39. $f(x) = -9 + 8x - 3x^2$

40. $f(x) = 2x^3 - 3x^2 - 12x + 2$

41. $f(x) = 2x^3 - 3x^2 - 72x - 4$

42. $f(x) = 4x^3 - 9x^2 - 30x + 6$

43. $f(x) = 4x^3 - 15x^2 - 72x + 5$

44. $f(x) = 6x - 9$

45. $f(x) = -3x + 6$

APPLIED PROBLEMS

General

46. A professor has found that the number of biology students attending class is approximated by

$$S(x) = -x^2 + 20x + 80,$$

where x is the number of hours that the student union is open daily. Find the number of hours that the union should be open so that the number of students attending class is a maximum. Find the maximum number of such students.

47. The number of people visiting Timberline Ski Lodge on Washington's Birthday is approximated by

$$W(x) = -x^2 + 60x + 180,$$

where x is the total snowfall in inches for the previous week. Find the snowfall that will produce the maximum number of visitors. Find the maximum number of visitors.

Management

48. The total profit in thousands of dollars from the sale of x hundred thousand automobile tires is approximated by

$$P(x) = -x^3 + 9x^2 + 120x - 400.$$

(Assume $x \geq 5$.) Find the number of hundred thousands of tires that must be sold to maximize profit. Find the maximum profit.

Natural Science

49. The number of salmon swimming upstream to spawn is approximated by

$$S(x) = -x^3 + 3x^2 + 360x + 5000,$$

where x represents the temperature of the water in degrees Celsius. (This function is valid only if $6 \leq x \leq 20$.) Find the water temperature that produces the maximum number of salmon swimming upstream.

Management

50. The profit in hundreds of dollars from the sale of x hundred pounds of a certain type of chicken feed is given by

$$P(x) = \tfrac{1}{3}x^3 - 12x^2 + 23x + 500.$$

Find all open intervals (with $x > 0$) where profit is increasing.

51. The total profit in dollars from the sale of x hundred citizen's band radios is approximated by

$$P(x) = x^3 - 11x^2 + 35x + 150.$$

Find all intervals (with $x > 0$) where profit is increasing.

52. A company has found through experience that increasing its advertising also increases its sales, up to a point. The company believes that the mathematical model connecting profit in dollars, $P(x)$, and expenditures on advertising in dollars, x, is

$$P(x) = 80 + 108x - x^3.$$

(a) Find the expenditure on advertising that leads to maximum profit.
(b) Find the maximum profit.

53. The total profit in dollars from the sale of x units of a certain prescription drug is given by

$$P(x) = -x^3 + 3x^2 + 72x + 1280.$$

(a) Find the number of units that should be sold in order to maximize the total profit.
(b) What is the maximum profit?

54. The total profit in dollars from the sale of x buckets of Extra Crispy fried chicken is approximated by

$$C(x) = -x^3 + 6x^2 + 288x + 500.$$

(a) Find the number of units that should be sold in order to maximize profits.
(b) Find the maximum profit.

Natural Science

55. The microbe concentration, $B(x)$, in appropriate units, of Lake Tom depends approximately on the oxygen concentration, x, again in appropriate units, according to the function

$$B(x) = x^3 - 7x^2 - 160x + 1800.$$

(a) Find the oxygen concentration that will lead to the minimum microbe concentration.
(b) What is the minimum concentration?

13.2 ABSOLUTE MAXIMUMS AND MINIMUMS

As we saw in the last section, a function might well have several relative maximums or relative minimums. However, a function never has more than one absolute maximum or absolute minimum. A function f has an **absolute maximum** at $x = a$ if $f(x) \leq f(a)$ for all x in the domain of the function. There is an **absolute minimum** at $x = a$ whenever $f(x) \geq f(a)$ for all x in the domain of the function.

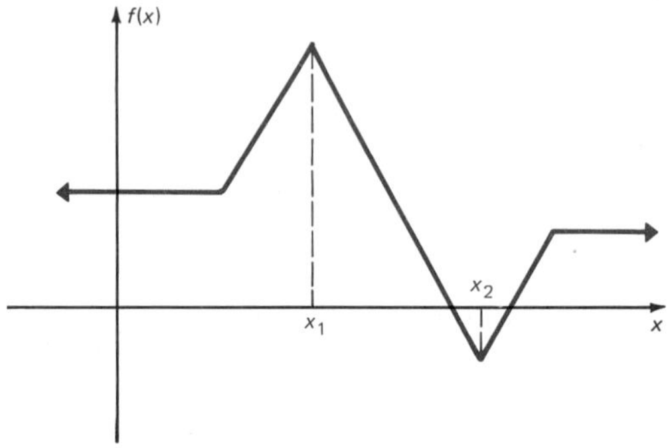

Figure 13.13

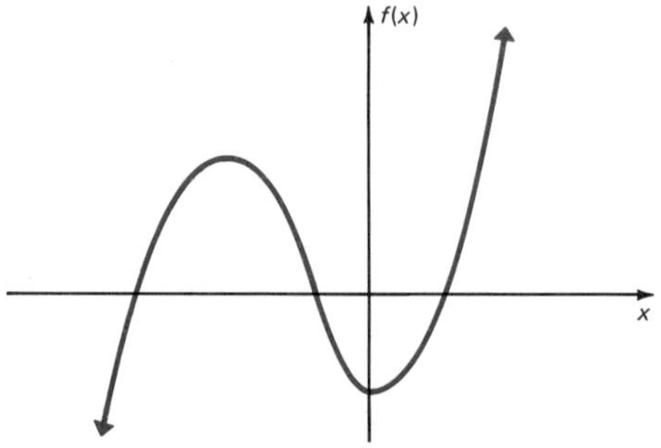

Figure 13.14

The function graphed in Figure 13.13 has an absolute maximum at x_1 and an absolute minimum at x_2. The function graphed in Figure 13.14 has neither an absolute maximum nor an absolute minimum.

A relative maximum may also be an absolute maximum, but an absolute maximum need not be a relative maximum. The function graphed in Figure 13.15 has an absolute maximum which is also a

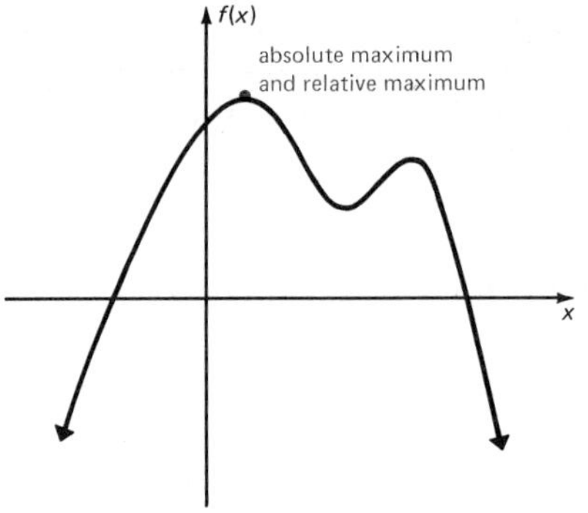

Figure 13.15

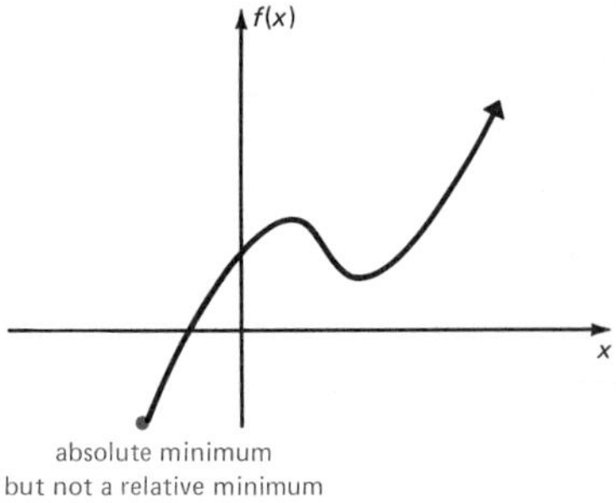

Figure 13.16

relative maximum; the function graphed in Figure 13.16 has an absolute minimum at an endpoint of the domain of the function. This absolute minimum is not a relative minimum because a relative minimum at a point where $x = a$ requires that $f(a)$ be less than all $f(x)$ in some interval containing a. Here, any open interval containing a includes points which are not in the domain of the function.

In more advanced courses it can be shown that a function defined on a closed interval and continuous on the interval will always have both an absolute maximum and an absolute minimum on that interval. These will be found either at a relative maximum or a relative minimum, or at the endpoints of the closed interval on which the function is defined.

Example 1 Find the absolute maximum and absolute minimum of the function

$$f(x) = x^2 - 4x + 7$$

defined on the closed interval $[-3, 4]$.

We first look for critical points of the function in $(-3, 4)$. Here $f'(x) = 2x - 4$, and

$$2x - 4 = 0 \qquad \text{if} \qquad x = 2.$$

This x-value is in $(-3, 4)$. By the first derivative test, check that 2 leads to a relative minimum; the minimum is $f(2) = 3$.

Now we need to check the endpoints of the interval on which the function is defined.

Endpoint	Value of function
-3	$f(-3) = 28$
4	$f(4) = 7$

The relative minimum of 3 occurred at $x = 2$. Since 3 is the smallest of the three numbers 3, 7, and 28, we say that 3 is the absolute minimum. The largest of the three numbers 3, 7, and 28 is 28, which occurred at $x = -3$. Thus, we say that 28 is the absolute maximum.

On the given domain the function never has a value smaller than 3 or larger than 28. A graph of the function $f(x) = x^2 - 4x + 7$ defined on the closed interval $[-3, 4]$ is shown in Figure 13.17. ■

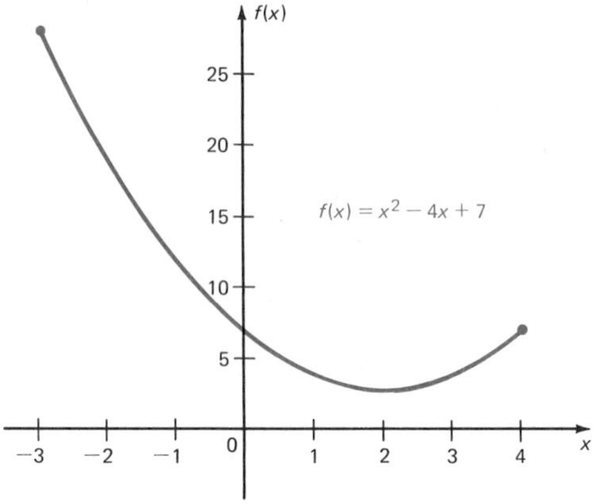

Figure 13.17

1. Find the absolute maximum and absolute minimum values of
(a) $f(x) = -x^2 + 4x - 8$ on $[-4, 4]$;
(b) $y = 2x^2 - 6x + 6$ on $[0, 5]$.

Answer:
(a) absolute maximum of -4 at $x = 2$, absolute minimum of -40 at $x = -4$
(b) absolute maximum of 26 at $x = 5$, absolute minimum of $3/2$ at $x = 3/2$

Work Problem 1 at the side.

Example 2 Find the absolute maximum and absolute minimum of the function

$$f(x) = 2x^3 - x^2 - 20x + 40$$

defined on the interval $[-2, 5]$.

First look for relative maximums or relative minimums in the interval $(-2, 5)$. Here $f'(x) = 6x^2 - 2x - 20$. Now place this derivative equal to 0.

$$6x^2 - 2x - 20 = 0$$
$$2(3x^2 - x - 10) = 0$$
$$2(3x + 5)(x - 2) = 0$$
$$3x + 5 = 0 \quad \text{or} \quad x - 2 = 0$$
$$3x = -5 \qquad\qquad x = 2$$
$$x = -\frac{5}{3}$$

Both x-values are in the interval $(-2, 5)$. By the first derivative test, $x = -5/3$ leads to a relative maximum and $x = 2$ leads to a relative minimum. Now evaluate the function at $-5/3$, 2, and the endpoints of its domain, -2 and 5.

x-value	Value of function	
-2	60	
$-\frac{5}{3}$	$\frac{1655}{27} \approx 61.296$	
2	12	← absolute minimum
5	165	← absolute maximum

The absolute maximum of 165 occurs when $x = 5$, and the absolute minimum of 12 occurs when $x = 2$. ■

2. Find the absolute maximum and absolute minimum for the function

$$f(x) = \tfrac{1}{3}x^3 + \tfrac{1}{2}x^2 - 2x + 8$$

defined on $[-3, 4]$.

Answer: absolute maximum of 88/3 at $x = 4$, absolute minimum of 41/6 at $x = 1$

Work Problem 2 at the side.

Example 3 A company has found that its profit from the sale of x units of plastic turtles is given by

$$P(x) = x^4 - 8x^2 + 100.$$

Production bottlenecks limit the number of units that can be made to no more than 5. Find the maximum possible profit that the firm can make.

Because of the restriction, the profit function is defined only for the domain $[0, 5]$. We first look for critical points of the function in (0, 5). Here $f'(x) = 4x^3 - 16x$. Now place this derivative equal to 0.

$$4x^3 - 16x = 0$$

$$4x(x^2 - 4) = 0 \qquad \text{Factor out the common factor}$$

$$4x(x + 2)(x - 2) = 0$$

$$x = 0 \quad \text{or} \quad x + 2 = 0 \quad \text{or} \quad x - 2 = 0$$

$$x = 0 \quad \text{or} \quad x = -2 \quad \text{or} \quad x = 2$$

Since $x = 0$ and $x = -2$ are not in the interval (0, 5), we disregard them. Now evaluate the function at 2 and at the endpoints of the domain, 0 and 5.

x-value	Value of function	
0	100	
2	84	
5	525	← absolute maximum

Maximum profit of $525 occurs when 5 units are made. ■

3. For a certain firm, the cost to produce x units of an item is given by

$$C(x) = x^3 - 15x^2 + 48x + 50.$$

Because of production problems, the function is defined only for the domain $[1, 6]$. Find the maximum and minimum costs that the firm will face.

Answer: maximum of 94 when $x = 2$, minimum of 14 when $x = 6$

Work Problem 3 at the side.

We can now summarize the steps involved in finding absolute maximums and absolute minimums of a continuous function f defined on a closed interval $[a, b]$.

1. Find $f'(x)$ and locate all critical points in the interval (a, b).
2. Compute the values of f for all critical points from Step 1 and for the endpoints of the interval, a and b.
3. The largest value found in Step 2 is the absolute maximum and the smallest value is the absolute minimum.

13.2 EXERCISES

Find the location of all absolute maximums and absolute minimums (if any) for the following functions.

1.
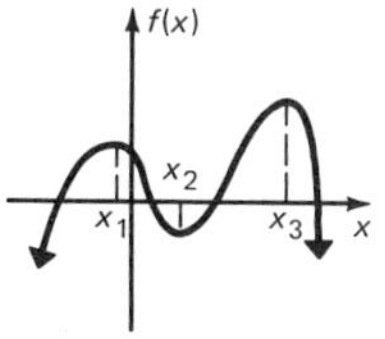

2.
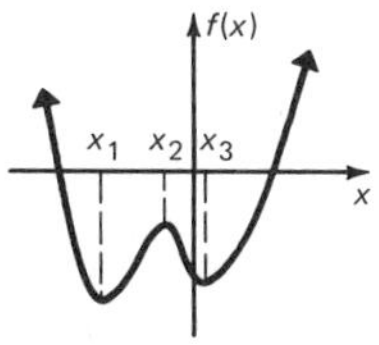

3.
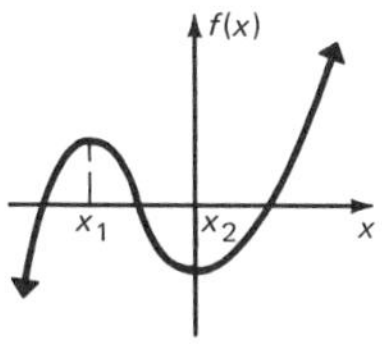

4.
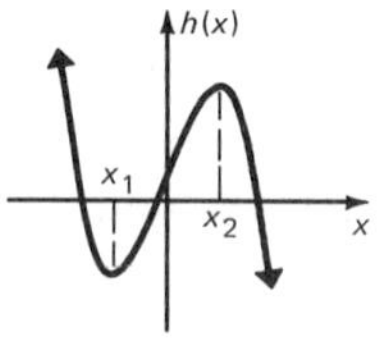

5.
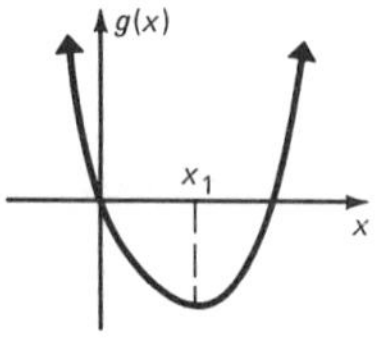

6.
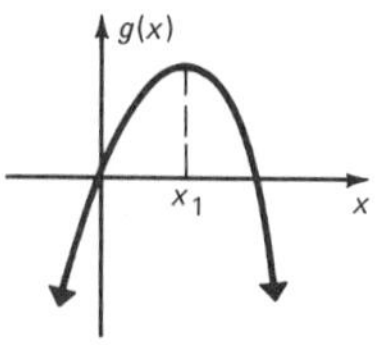

7.
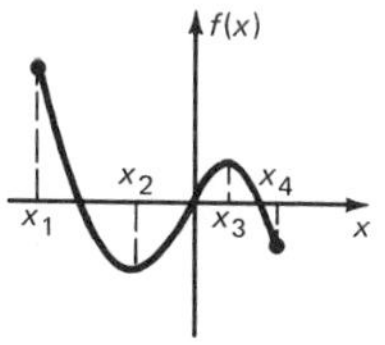

8.
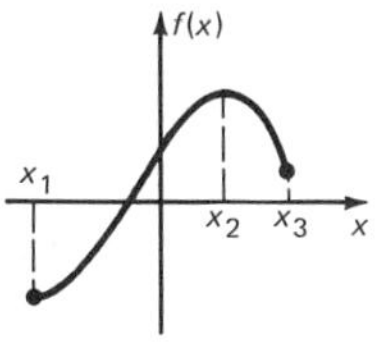

9.
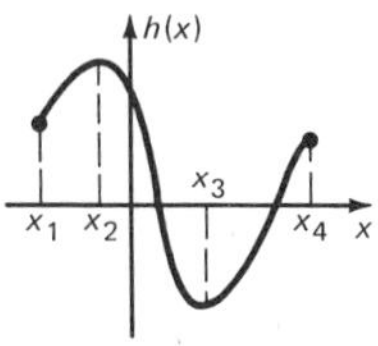

Find the location of all absolute maximums and absolute minimums for the following functions having domains as specified. A calculator will be helpful for many of these problems. (See Examples 1 and 2.)

10. $f(x) = x^2 - 4x + 1, \quad [-6, 5]$

11. $f(x) = x^2 + 6x + 2, \quad [-4, 0]$

12. $f(x) = 3x^2 - 8x - 6, \quad [-2, 4]$

13. $f(x) = 10x^2 + 15x - 8, \quad [-3, 0]$

14. $f(x) = 9 - 6x - 3x^2, \quad [-4, 3]$

15. $f(x) = 5 - 8x - 4x^2, \quad [-5, 1]$

16. $f(x) = x^3 - 6x^2 + 9x - 8, \quad [0, 5]$

17. $f(x) = x^3 - 3x^2 - 24x + 5, \quad [-3, 6]$

18. $f(x) = \frac{1}{3}x^3 + \frac{3}{2}x^2 - 4x + 1, \quad [-5, 2]$

19. $f(x) = \frac{1}{3}x^3 - \frac{1}{2}x^2 - 6x + 3, \quad [-4, 4]$

20. $f(x) = x^3 + 5x - 4, \quad [-3, 2]$

21. $f(x) = x^3 + x + 3, \quad [-1, 4]$

22. $f(x) = x^4 - 18x^2 + 1, \quad [-4, 4]$

23. $f(x) = x^4 - 32x^2 - 7, \quad [-5, 6]$

24. From information given in a recent business publication we constructed the mathematical model

$$M(x) = -\frac{1}{45}x^2 + 2x - 20, \qquad 30 \le x \le 65$$

to represent the miles per gallon used by a certain car at a speed of x miles per hour. Find the absolute maximum miles per gallon and the absolute minimum. See Example 3.

25. For a certain compact car,

$$M(x) = -.018x^2 + 1.24x + 6.2, \qquad 30 \le x \le 60$$

represents the miles per gallon obtained at a speed of x miles per hour. Find the absolute maximum miles per gallon and the absolute minimum.

A piece of wire of length 12 *feet is cut into two pieces. One piece is made into a circle and the other piece is made into a square. (See the figure.)*

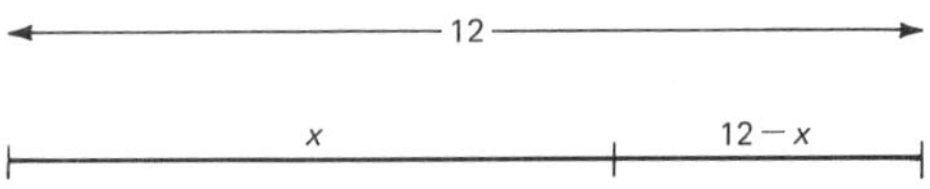

Let the piece of length x be formed into a circle;

radius of circle $= \frac{x}{2\pi}$; area of circle $= \pi\left(\frac{x}{2\pi}\right)^2$;

side of square $= \frac{12-x}{4}$; area of square $= \left(\frac{12-x}{4}\right)^2$.

26. Where should the cut be made in order to make the sum of the areas enclosed by both figures minimum? (Hint: Use 3.14 as an approximation for π. Have a calculator handy.)

27. Where should the cut be made in order to make the sum of the areas maximum? (Hint: Remember to use the endpoints of a domain when looking for absolute maximums and minimums.)

The following functions have derivatives which cannot easily be set equal to 0. To find the maximum and minimum of each function on the given domain, use a calculator to evaluate the function at intervals of .1. Then estimate the absolute maximum and absolute minimum values of the function.

28. $f(x) = x^3 - 3x^2 - 3x + 4$, $[2, 3]$

29. $f(x) = x^3 - 9x^2 + 18x - 7$, $[0, 1.5]$

30. $f(x) = x^3 - 3x^2 - 12x + 4$, $[-1.5, 0]$ and $[3, 4]$

31. $f(x) = x^3 + 6x^2 - 6x + 3$, $[-5, -4]$ and $[0, 1]$

13.3 THE SECOND DERIVATIVE TEST AND CURVE SKETCHING

The derivative of a function gives us a new function for which we can again find a derivative. We call the derivative of the derivative of a function the **second derivative** of the function. If we take the derivative of the second derivative of a function f, we get the *third derivative* of f. In the same way, we could get the *fourth derivative* and other higher derivatives. Thus, if $f(x) = x^4 + 2x^3 + 3x^2 - 5x + 7$, we have

$$f'(x) = 4x^3 + 6x^2 + 6x - 5, \qquad \text{(the first derivative of } f\text{)}$$
$$f''(x) = 12x^2 + 12x + 6, \qquad \text{(the second derivative of } f\text{)}$$
$$f'''(x) = 24x + 12, \qquad \text{(the third derivative of } f\text{)}$$
$$f^{(4)}(x) = 24, \qquad \text{(the fourth derivative of } f\text{)}$$
$$f^{(5)}(x) = 0. \qquad \text{(the fifth derivative of } f\text{)}$$

The second derivative of the function $y = f(x)$ can also be written as

$$y'' \quad \text{or} \quad \frac{d^2y}{dx^2}.$$

The other higher derivatives have corresponding notations.

Example 1 Find the second derivative of the following functions.

(a) $y = 8x^3 - 9x^2 + 6x + 4$

Here $y' = 24x^2 - 18x + 6$. The second derivative is the derivative of y', or

$$y'' = 48x - 18.$$

(b) $y = \dfrac{4x + 2}{3x - 1}$

Use the quotient rule to find y'.

$$y' = \frac{(3x-1)(4)-(4x+2)(3)}{(3x-1)^2} = \frac{12x-4-12x-6}{(3x-1)^2} = \frac{-10}{(3x-1)^2}$$

We again use the quotient rule to find y''.

$$y'' = \frac{(3x-1)^2(0)-(-10)(2)(3x-1)(3)}{[(3x-1)^2]^2}$$

$$= \frac{60(3x-1)}{(3x-1)^4}$$

$$y'' = \frac{60}{(3x-1)^3} \quad ■$$

1. Find the second derivatives of the following.

(a) $y = -9x^3 + 8x^2 + 11x - 6$

(b) $y = -2x^4 + 6x^2$

(c) $y = \dfrac{x+2}{5x-1}$

Answer:

(a) $y'' = -54x + 16$

(b) $y'' = -24x^2 + 12$

(c) $y'' = \dfrac{110}{(5x-1)^3}$

Work Problem 1 at the side.

Example 2 Let $f(x) = x^3 + 6x^2 - 9x + 8$. Find the following.

(a) $f''(0)$

Here $f'(x) = 3x^2 + 12x - 9$, so that $f''(x) = 6x + 12$. Then

$$f''(0) = 6(0) + 12 = 12.$$

(b) $f''(-3) = 6(-3) + 12 = -6$ ■

2. Let $f(x) = 4x^3 - 12x^2 + x - 1$. Find

(a) $f''(0)$;

(b) $f''(4)$;

(c) $f''(-2)$.

Answer:

(a) -24

(b) 72

(c) -72

Work Problem 2 at the side.

We defined increasing and decreasing functions in the first section of this chapter. We saw that a function is increasing on an interval if its derivative is positive on the interval. If the *derivative* is increasing on an interval, the function is said to be concave upward on the interval. If the *derivative* is decreasing, the function is concave downward. The derivative is increasing if its derivative (the second derivative) is positive and decreasing if its derivative is negative. Therefore,

a function f is **concave upward** on an interval if $f''(x) > 0$ for every value of x in the interval;

a function f is **concave downward** on an interval if $f''(x) < 0$ for every value of x in the interval.

The function itself can be either increasing or decreasing, and either concave upward or concave downward on an interval. Examples of various combinations are shown in Figure 13.18.

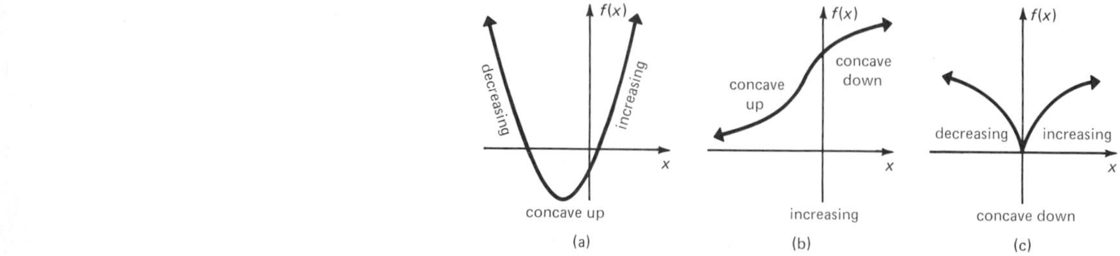

Figure 13.18

Example 3 Find the intervals where $f(x) = x^3 - 3x^2 + 5x - 4$ is concave upward or downward.

Here $f'(x) = 3x^2 - 6x + 5$ and $f''(x) = 6x - 6$. The second derivative is 0 when $x = 1$. Do you see that if we choose any value of x to the left of 1, $f''(x)$ will be negative? Thus, f is concave downward on the interval $(-\infty, 1)$. Likewise, if we choose any value of x to the right of 1, $f''(x)$ will be positive. Therefore, f is concave upward on the interval $(1, \infty)$. A graph of the function f is shown in Figure 13.19. ■

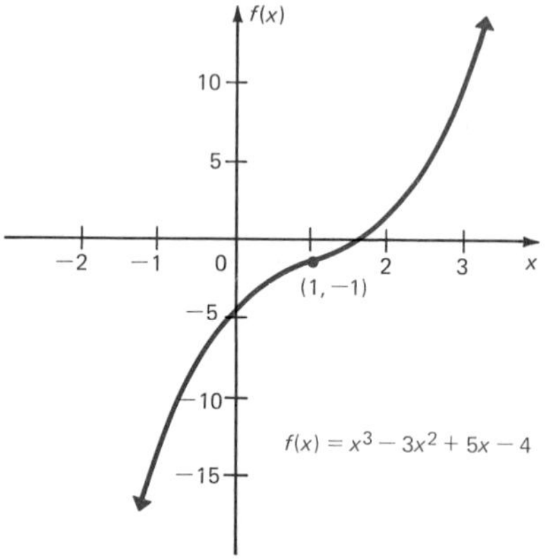

Figure 13.19

In Example 3, the function changed from being concave downward to being concave upward at the point where $x = 1$. Such a point is called a **point of inflection.** In Example 3, the inflection point is $(1, f(1))$ or $(1, -1)$ and occurs at the value of x where $f''(x) = 0$. An inflection point may also occur where $f''(x)$ does not exist.

3. Find the intervals where the following functions are concave upward or concave downward. Identify any inflection points.
(a) $f(x) = 6x^3 - 24x^2 + 9x - 3$
(b) $f(x) = 2x^2 - 4x + 8$

Answer:
(a) concave downward on $(-\infty, 4/3)$, concave upward on $(4/3, \infty)$; point of inflection is $(4/3, -175/9)$
(b) $f''(x) = 4$, which is positive; function is always concave upward, no inflection point

Work Problem 3 at the side.

Example 4 When a new skill is being learned, it is common for learning to be slow at first, then increase, and finally level off at a fairly high degree of skill. A typical function showing such growth of learning is graphed in Figure 13.20. There is a point of inflection at the point $(a, f(a))$. ■

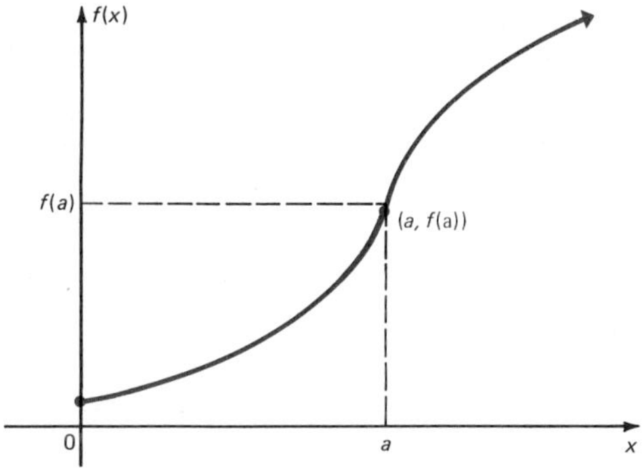

Figure 13.20

4. Find the intervals where the function of Figure 13.20 is concave upward and concave downward.

Answer: concave upward on $(0, a)$ and concave downward on (a, ∞)

Work Problem 4 at the side.

One of the main applications of second derivatives is helping us tell whether a given critical value leads to a relative maximum or a relative minimum. To see how this is done, glance at the graph of Figure 13.21. This graph shows a relative maximum at $x = a$. We could locate an open interval containing a so that the function is concave downward on this interval.

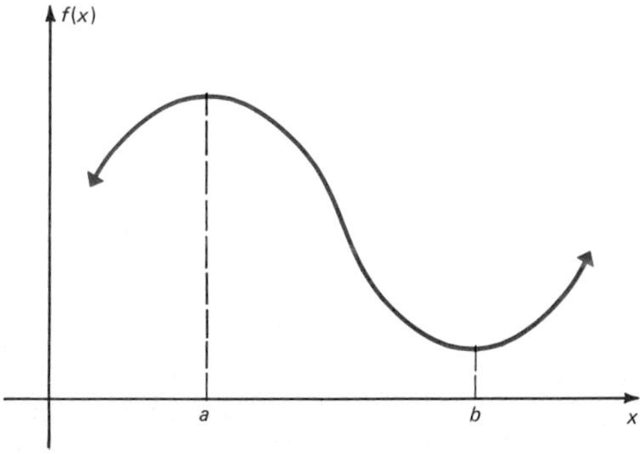

Figure 13.21

Also, since the graph shows a relative minimum at $x = b$, we could find an open interval containing b so that the function is concave upward on the interval. Based on these ideas, we have the second derivative test for deciding about critical points:

Second derivative test Let a be a critical point for the function f. Suppose f' exists for every point in an open interval containing a.
(a) If $f''(a) > 0$, then $f(a)$ is a relative minimum.
(b) If $f''(a) < 0$, then $f(a)$ is a relative maximum.
(c) If $f''(a) = 0$, then the test gives no information.

Example 5 Find all relative maximums and relative minimums for the function

$$f(x) = 4x^3 + 7x^2 - 10x + 8.$$

First, find the critical points. Here $f'(x) = 12x^2 + 14x - 10$. Place this derivative equal to 0.

$$12x^2 + 14x - 10 = 0$$
$$2(6x^2 + 7x - 5) = 0$$
$$2(3x + 5)(2x - 1) = 0$$
$$3x + 5 = 0 \quad \text{or} \quad 2x - 1 = 0$$
$$3x = -5 \qquad 2x = 1$$
$$x = -\frac{5}{3} \qquad x = \frac{1}{2}$$

Now we can use the second derivative test. The second derivative is $f''(x) = 24x + 14$. Since the critical points are $-5/3$ and $1/2$, we have

$$f''\left(-\frac{5}{3}\right) = 24\left(-\frac{5}{3}\right) + 14 = -40 + 14 = -26 < 0,$$

so that $-5/3$ leads to a relative maximum of $f(-\frac{5}{3}) = \frac{691}{27}$.

$$f''\left(\frac{1}{2}\right) = 24\left(\frac{1}{2}\right) + 14 = 12 + 14 = 26 > 0,$$

so $1/2$ gives a relative minimum of $f(\frac{1}{2}) = \frac{21}{4}$. ■

Work Problem 5 at the side.

The second derivative gives the rate of change of the first derivative. In the rest of this section, we look at examples of this idea.

5. Find all relative maximums and relative minimums for the following functions. Use the second derivative test.
(a) $f(x) = 6x^2 + 12x + 1$
(b) $f(x) = x^3 - 3x^2 - 9x + 8$

Answer:
(a) relative minimum of -5 at $x = -1$
(b) relative maximum of 13 at $x = -1$, relative minimum of -19 at $x = 3$

Example 6 In the last chapter we saw that the velocity of a particle is given by the derivative of the position function for the particle. That is, if $y = s(t)$ describes the position of the particle at time t, then $v(t) = s'(t)$ gives the velocity at time t.

The rate of change of the velocity is called the **acceleration.** The acceleration is thus the derivative of the velocity; if $a(t)$ represents the acceleration at time t, then

$$a(t) = \frac{d}{dt}v(t) = s''(t).$$

For example, if a particle is moving along a line with

$$s(t) = t^3 + 2t^2 - 7t + 9,$$

then $\quad v(t) = s'(t) = 3t^2 + 4t - 7,$

and $\quad a(t) = v'(t) = s''(t) = 6t + 4.$

The acceleration is positive if $6t + 4 > 0$ (or $t > -2/3$), and negative if $t < -2/3$. ■

6. If $s(t) = 4t^3 + 3t^2 - 8t + 1$, find the acceleration of a particle at
(a) $t = 0$;
(b) $t = 4$;
(c) $t = 10$.

Answer:
(a) 6
(b) 102
(c) 246

Work Problem 6 at the side.

Example 7 Suppose the revenue from the sale of x units of a certain product is

$$R(x) = 6x^3 + 8x^2 + 9.$$

Then $R'(x) = 18x^2 + 16x$ is called the *velocity of the revenue*, and $R''(x) = 36x + 16$ is called the *acceleration of the revenue*. ■

7. In Example 7, find
(a) $R'(3)$;
(b) $R'(10)$;
(c) $R''(3)$;
(d) $R''(10)$.

Answer:
(a) 210
(b) 1960
(c) 124
(d) 376

Work Problem 7 at the side.

All the methods of this section, together with previous methods such as point plotting and finding asymptotes, permit us to sketch quickly the graphs of many functions. This process is called **curve sketching.**

Example 8 Sketch the graph of $f(x) = x^3 - 2x^2 - 4x + 3$.

To find all intervals for which the function is increasing or decreasing, we find $f'(x)$.

$$\begin{aligned} f'(x) &= 3x^2 - 4x - 4 \\ &= (3x + 2)(x - 2) \end{aligned}$$

Verify that $f'(x) = 0$ when $x = -2/3$ or when $x = 2$. We can use the second derivative test to check for maximums or minimums. Here

$$f''(x) = 6x - 4,$$

which is positive when $x = 2$ and negative when $x = -2/3$. Thus, $f(2)$ is a relative minimum and $f(-2/3)$ is a relative maximum.

The first derivative is positive on the intervals $(-\infty, -2/3)$ and $(2, \infty)$, so the function is increasing on these intervals. The first derivative is negative on the interval $(-2/3, 2)$, which shows that the function is decreasing on this interval.

The second derivative, $f''(x) = 6x - 4$, is 0 when $x = 2/3$. Using this fact, verify that the second derivative is negative for all values of x in the interval $(-\infty, 2/3)$ and positive for all values of x in the interval $(2/3, \infty)$. By this result, the graph is concave downward on $(-\infty, 2/3)$ and concave upward on $(2/3, \infty)$. A summary of this information is given in the chart below. Using these facts, and plotting a few points, we get the graph shown in Figure 13.22.

Interval	f'	f''	
$(-\infty, -2/3)$	+		increasing
$(-2/3, 2)$	−		decreasing
$(2, \infty)$	+		increasing
$(-\infty, 2/3)$		−	concave downward
$(2/3, \infty)$		+	concave upward ■

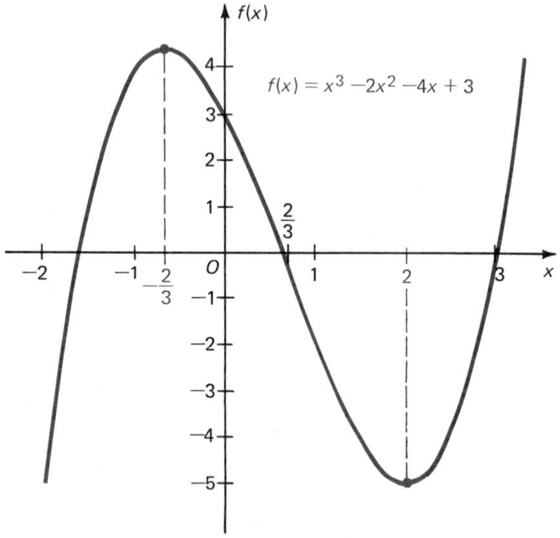

Figure 13.22

In the example above, we found that the function f was concave upward on the interval $(2/3, \infty)$ and concave downward on $(-\infty, 2/3)$. At the point where $x = 2/3$ there is a point of inflection. As shown above, if $(x_0, f(x_0))$ is a point of inflection, then $f''(x_0) = 0$. However, just because $f''(x_0) = 0$, we have no assurance that a point of inflection has been located. For example, if $f(x) = (x - 1)^4$, then $f''(x) = 12(x - 1)^2$, which is 0 at $x = 1$. However, the graph of $f(x) = (x - 1)^4$ is always concave upward and thus has no point of inflection.

The steps involved in sketching the graph of a function are as summarized below.

1. Find f' and f''.

2. Find all values of x such that $f'(x) = 0$. Check these points to identify maximums or minimums.

3. Find any intervals for which $f'(x) < 0$ or $f'(x) > 0$.

4. Find any values of x for which $f''(x) = 0$.

5. Find any intervals for which $f''(x) < 0$ or $f''(x) > 0$.

6. Locate as many points of the graph as needed and sketch the graph, using the following facts.

$f'(x) > 0$, f is increasing	$f''(x) > 0$, f is concave upward
$f'(x) < 0$, f is decreasing	$f''(x) < 0$, f is concave downward

Work Problem 8 at the side.

8. Sketch the graphs.
(a) $f(x) = x^3 - 3x^2$
(b) $f(x) = x + 1/x$

Answer:
(a)

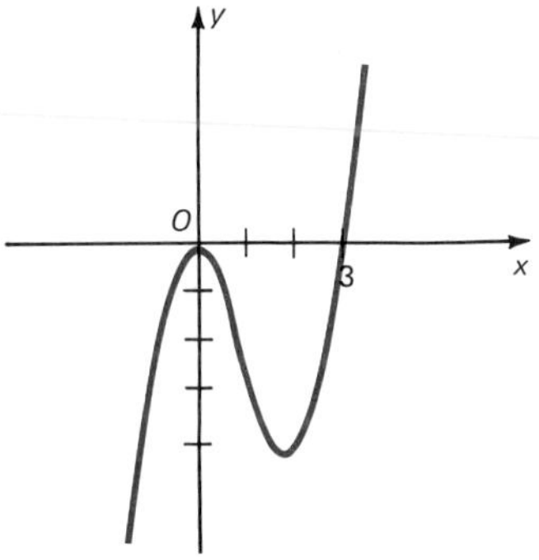

(b)

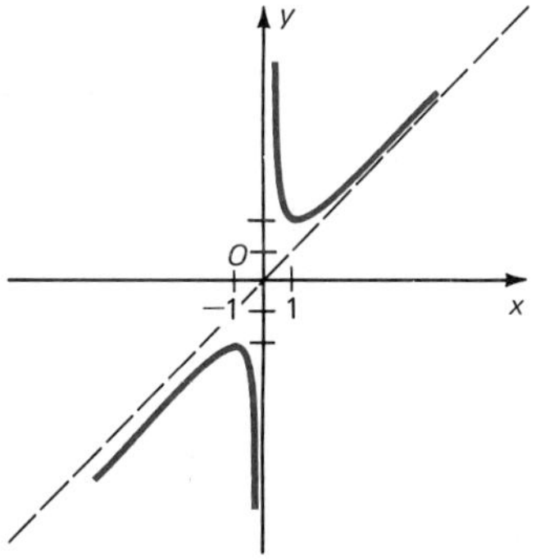

13.3 EXERCISES

For each of the following functions, find f''. Then find $f''(0)$, $f''(2)$, and $f''(-3)$. (See Examples 1 and 2.)

1. $f(x) = x^3 + 4x^2 + 2$

2. $f(x) = 3x^3 - 4x + 5$

3. $f(x) = -x^4 + 2x^3 - x^2$

4. $f(x) = 3x^4 - 5x^3 + 2x^2$

5. $f(x) = 8x^2 + 6x + 5$

6. $f(x) = 3x^2 - 4x + 8$

7. $f(x) = (x - 2)^3$

8. $f(x) = (x + 4)^3$

9. $f(x) = \dfrac{x + 1}{x - 1}$

10. $f(x) = \dfrac{2x + 1}{x - 2}$

For each of the following functions, find $f'''(x)$ and $f^{(4)}(x)$.

11. $f(x) = 2x^4 - 3x^3 + x^2$

12. $f(x) = -x^4 + 2x^2 + 8$

13. $f(x) = 3x^5 - x^4 + 2x^3 - 7x$

14. $f(x) = 4x^5 + 6x^4 - x^2 + 2$

15. $f(x) = \dfrac{x+1}{x}$

16. $f(x) = \dfrac{x-1}{x+2}$

17. $f(x) = \dfrac{x}{2x+1}$

18. $f(x) = \dfrac{3x}{x-2}$

Find any critical points for the following functions and then use the second derivative test to find out if the critical points lead to relative maximums or relative minimums. If $f''(a) = 0$ for a critical point $x = a$, then the second derivative test gives no information. In this case, use the first derivative test instead. Sketch the graph of each function. (See Examples 5 and 8.)

19. $f(x) = x^2 - 12x + 36$

20. $f(x) = -x^2 - 10x - 25$

21. $f(x) = 12 - 8x + 4x^2$

22. $f(x) = 6 + 4x + x^2$

23. $f(x) = 2x^3 - 4x^2 + 2$

24. $f(x) = 3x^3 - 3x^2 + 1$

25. $f(x) = -2x^3 - 9x^2 + 60x - 8$

26. $f(x) = -2x^3 - 9x^2 + 108x - 10$

27. $f(x) = x^3 - \frac{15}{2}x^2 - 18x - 1$

28. $f(x) = 2x^3 + \frac{7}{2}x^2 - 5x + 3$

29. $f(x) = x^3$

30. $f(x) = (x-1)^4$

31. $f(x) = x^4 - 8x^2$

32. $f(x) = x^4 - 18x^2 + 5$

33. $f(x) = 2x + \dfrac{8}{x}$

34. $f(x) = x + \dfrac{1}{x}$

35. $f(x) = \dfrac{x^2+4}{x}$

36. $f(x) = \dfrac{x^2+25}{x}$

37. $f(x) = \dfrac{x}{1+x}$

38. $f(x) = \dfrac{x-1}{x+1}$

Find all intervals where the following functions are concave upward or concave downward. Find the location of any points of inflection. (See Example 3.)

39. $f(x) = x^2 - 4x + 3$

40. $f(x) = x^2 + 10x - 9$

41. $f(x) = 8 - 6x - x^2$

42. $f(x) = -3 + 8x - x^2$

43. $f(x) = 2x^3 - 3x^2 - 12x + 1$

44. $f(x) = x^3 + 3x^2 - 45x - 3$

45. $f(x) = -x^3 - 12x^2 - 45x + 2$

46. $f(x) = -2x^3 + 9x^2 + 168x - 3$

47. $f(x) = -2/(x+1)$

48. $f(x) = 3/(x-5)$

APPLIED PROBLEMS

Management

The figure shows the product life cycle *graph, with typical products marked on it.**

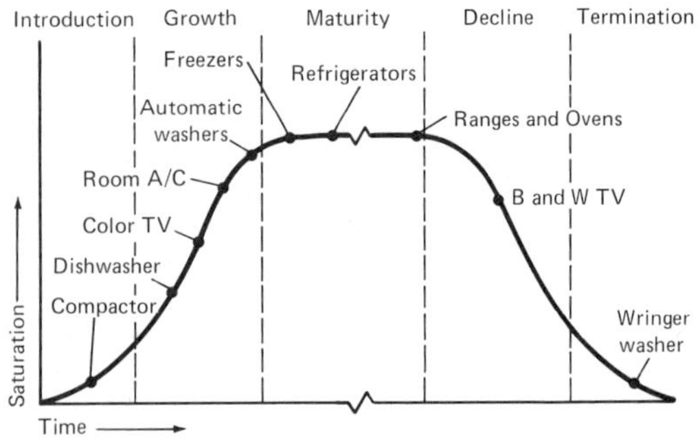

49. Where would you place home-video tape recorders on this graph?

50. Where would you place light bulbs?

51. Which products are closest to the point of inflection on the left of the graph? What does the point of inflection mean here?

52. Which products are closest to the point of inflection on the right of the graph? What does the point of inflection mean here?

Natural Science

53. The number of mosquitos, $M(x)$, in millions, in a certain area depends on the April rainfall, x, measured in inches, approximately as follows:

$$M(x) = 50 - 32x + 14x^2 - x^3.$$

Find the amounts of rainfall that will produce the maximum and the minimum number of mosquitos.

Management

54. Because of raw material shortages, it is increasingly expensive to produce fine cigars. In fact, the profit in thousands of dollars from producing x hundred thousand cigars is approximated by

$$P(x) = x^3 - 23x^2 + 120x + 60.$$

Find the level of production that will lead to maximum profit and to minimum profit.

Natural Science

55. An autocatalytic chemical reaction is one in which the product being formed causes the rate of formation to increase. The rate of a certain autocatalytic reaction is given by

$$V(x) = 12x(100 - x),$$

* John E. Smallwood. "The Product Life Cycle: A Key to Strategic Marketing Planning," *MSU Business Topics* (Winter 1973). Reprinted by permission of the publisher. Division of Research, Graduate School of Business Administration. Michigan State University.

where x is the quantity of the product present and 100 represents the quantity of chemical present initially. For what value of x is the rate of the reaction a maximum?

56. The percent of concentration of a certain drug in the bloodstream x hours after the drug is administered is given by

$$K(x) = \frac{3x}{x^2 + 4}.$$

For example, after one hour the concentration is given by $K(1) = 3(1)/(1^2 + 4) = \frac{3}{5}\% = .6\% = .006$.
(a) Find the time at which concentration is a maximum.
(b) Find the maximum concentration.

57. The percent of concentration of another drug in the bloodstream x hours after the drug is administered is given by

$$K(x) = \frac{4x}{3x^2 + 27}.$$

(a) Find the time at which the concentration is a maximum.
(b) Find the maximum concentration.

58. Assume that the number of bacteria, in millions, present in a certain culture at time t is given by

$$R(t) = t^2(t - 18) + 96t + 1000.$$

(a) At what time before 8 hours will the population be maximized?
(b) Find the maximum population.

Each of the following functions gives the displacement at time t of a particle moving along a line. Find the velocity and acceleration functions. Then find the velocity and acceleration at $t = 0$ and $t = 4$. Assume that time is measured in seconds and distance is measured in centimeters. Velocity will be in centimeters per second (cm/sec) *and acceleration in centimeters per second per second* (cm/sec^2). (*See Example 6.*)

59. $s(t) = -3t^2 - 6t + 2$

60. $s(t) = 8t^2 + 4t$

61. $s(t) = 3t^3 - 4t^2 + 8t - 9$

62. $s(t) = -5t^3 - 8t^2 + 6t - 3$

63. $s(t) = \dfrac{1}{t + 3}$

64. $s(t) = \dfrac{-2}{3t + 4}$

When an object is dropped straight down, the distance in feet that it travels in t seconds is given by

$$s(t) = -16t^2.$$

Find the velocity of an object

65. after 3 seconds;

66. after 5 seconds;

67. after 8 seconds.

68. Find the acceleration. (The answer here is a constant, the acceleration due to the influence of gravity alone.)

Management

The revenue, in thousands of dollars, of a small firm is given by

$$R(x) = 2x^3 - 8x^2 + 4x,$$

where x is the number of units of products that the firm sells. Find the following. (See Example 7.)

69. velocity of revenue when $x = 8$

70. when $x = 12$

71. acceleration of revenue when $x = 4$

72. when $x = 8$

13.4 APPLICATIONS OF MAXIMUMS AND MINIMUMS

To apply our methods of finding maximums and minimums to realistic problems, we first need a mathematical model of the real-life situation. For example, suppose we find that the number of units that can be produced on a production line can be approximated by the mathematical model

$$T(x) = 8x^{1/2} + 2x^{3/2} + 50,$$

where x is the number of employees on the line. Once this mathematical model has been established, we can use it to produce information about the production line. For example, we could use the derivative of $T(x)$ to estimate the marginal production resulting from the addition of an extra worker to the line.

However, in writing the mathematical model itself, we must be aware of restrictions on the values of the variables involved. For example, since x represents the number of employees on a production line, x must certainly be restricted to the positive integers, or perhaps to a few common fractional values (we can conceive of half-time employees, but probably not 1/32-time employees). Certainly, in this example we cannot have $x = \sqrt{2}$ or $x = 3 + \sqrt{7}$.

On the other hand, if we wish to apply the tools of calculus to obtain a maximum value or minimum value for some function, it is necessary that the function be defined and be meaningful at every real-number point in some interval. Because of this, the answer obtained by using a mathematical model of a practical problem might be a number that is not feasible in the setting of the problem.

Usually, the requirement that we use a continuous function instead of a function which can take on only certain selected values is of theoretical interest only. In most cases, calculus gives results which *are* acceptable in a given situation. And if the methods of calculus should be used on a function f and lead to the conclusion that $80\sqrt{2}$ units should be produced in order to get the lowest possible cost, it is usually only necessary to calculate $f(80\sqrt{2})$, and then compare this result to various values of $f(x)$, where x is an acceptable number. The lowest of these values of $f(x)$ then gives minimum cost. In most cases, the result obtained will be very close to the theoretical minimum.

In the rest of this section we give several examples showing applications of calculus to maximum and minimum problems.

Example 1 For a chicken farmer to sell x roosters in a small farming community, the price per rooster must be given by

$$p(x) = 4 - \frac{x}{25}.$$

(a) Find an expression for the total revenue from the sale of x roosters. Let $T(x)$ represent this revenue.

The total revenue is given by the product of the number of roosters sold, x, and the price per rooster, $4 - x/25$. Thus,

$$T(x) = x\left(4 - \frac{x}{25}\right) = 4x - \frac{x^2}{25}.$$

(b) Find the value of x that leads to maximum revenue. Here

$$T'(x) = 4 - \frac{2x}{25}.$$

Place this derivative equal to 0.

$$\begin{aligned} 4 - \frac{2x}{25} &= 0 \\ -\frac{2x}{25} &= -4 \\ 2x &= 100 \\ x &= 50 \end{aligned}$$

We must decide if $x = 50$ leads to maximum revenue or minimum revenue. Use the second derivative test. Since $T'(x) = 4 - 2x/25$, we have $T''(x) = -2/25$, which is negative. Thus, $x = 50$ leads to the maximum revenue, as desired.

(c) Find the maximum revenue.

$$T(50) = 4(50) - \frac{50^2}{25} = 200 - 100 = 100 \text{ dollars.}$$

The function $T(x) = 4x - x^2/25$ is valid only for x in the interval $[0, 100]$, since $p(x)$ is meaningful only for these values. Check the endpoints of the interval and verify that both 0 and 100 lead to a *minimum* revenue of 0. ■

1. A total of x earthmovers will be sold if the price, in thousands of dollars, is given by

$$p(x) = 32 - \frac{x}{8}.$$

Find

(a) an expression for the total revenue $R(x)$;

(b) the value of x that leads to maximum revenue;

(c) the maximum revenue.

Answer:

(a) $R(x) = x(32 - x/8)$
$= 32x - x^2/8$

(b) $x = 128$

(c) \$2048 thousand, or \$2,048,000

Work Problem 1 at the side.

Example 2 When Eastside University charges \$600 for a class that includes a tour of the home where Calvin Coolidge was born, it attracts 1000 students. For each \$20-decrease in the charge, an additional 100 students will attend the class.

(a) Find an expression for the total revenue if there are x \$20-decreases in the price.

The price charged by the university will be

$$\text{price per student} = 600 - 20x,$$

and the number of students taking the class will be

$$\text{number of students} = 1000 + 100x.$$

The total revenue, $R(x)$, is given by the product of the price and the number of students, or

$$\begin{aligned} R(x) &= (600 - 20x)(1000 + 100x) \\ &= 600{,}000 + 40{,}000x - 2000x^2. \end{aligned}$$

(b) Find the value of x that maximizes revenue.

Here $R'(x) = 40{,}000 - 4000x$. Place this derivative equal to 0.

$$\begin{aligned} 40{,}000 - 4000x &= 0 \\ -4000x &= -40{,}000 \\ x &= 10 \end{aligned}$$

Since $R''(x) = -4000$, $x = 10$ leads to maximum revenue.

(c) Find the maximum revenue.

$$R(10) = 600{,}000 + 40{,}000(10) - 2000(10)^2 = 800{,}000$$

dollars. There will be $1000 + 100(10) = 2000$ students in the class; each student will pay $600 - 20(10) = 400$ dollars. ■

2. An investor has built a series of self-storage units near a group of apartment houses. She must now decide on the monthly rental. From past experience, she feels that 200 units will be rented for \$15 per month, with 5 additional rentals for each \$.25-reduction in the rental price. Let x be the number of \$.25-reductions in the price and find

(a) an expression for the number of units rented;
(b) an expression for the price per unit;
(c) an expression for the total revenue;
(d) the value of x leading to maximum revenue;
(e) the maximum revenue.

Answer:
(a) $200 + 5x$
(b) $15 - .25x$
(c) $(200 + 5x)(15 - .25x)$
$= 3000 + 25x - 1.25x^2$
(d) $x = 10$
(e) \$3125

Work Problem 2 at the side.

Example 3 A truck burns fuel at the rate of

$$G(x) = \frac{1}{200}\left(\frac{800}{x} + x\right)$$

gallons per mile when traveling x miles per hour on a straight, level road. If fuel costs \$1 per gallon, find the speed that will produce the minimum total cost for a 1000-mile trip. Find the minimum total cost.

The total cost of the trip, in dollars, is the product of the number of gallons per mile, the number of miles, and the cost per gallon. If we let $C(x)$ represent this cost, then

$$\begin{aligned} C(x) &= \left[\frac{1}{200}\left(\frac{800}{x} + x\right)\right](1000)(1) \\ &= \frac{1000}{200}\left(\frac{800}{x} + x\right) \\ &= 5\left(\frac{800}{x} + x\right) \\ &= \frac{4000}{x} + 5x. \end{aligned}$$

Find the value of x that will minimize cost.

Here

$$C'(x) = \frac{-4000}{x^2} + 5.$$

Place this derivative equal to 0.

$$\begin{aligned} -\frac{4000}{x^2} + 5 &= 0 \\ -\frac{4000}{x^2} &= -5 \\ 4000 &= 5x^2 \\ 800 &= x^2 \\ x &\approx \pm 28.3 \text{ mph} \qquad \text{Take the square root of both sides} \end{aligned}$$

We reject -28.3 as a speed, leaving $x = 28.3$ as the only critical value. To see if we get a minimum when $x = 28.3$, find $C''(x)$.

$$C''(x) = \frac{8000}{x^3}$$

Since $C''(28.3) > 0$, we see that 28.3 leads to a minimum. (We need not actually calculate $C''(28.3)$; we just check and see that $C''(28.3) > 0$.)

The minimum total cost is $C(28.3)$, or

$$C(28.3) = \frac{4000}{28.3} + 5(28.3) \approx 141.34 + 141.50 = 282.84$$

dollars. ■

3. A diesel generator burns fuel at the rate of

$$G(x) = \frac{1}{48}\left(\frac{300}{x} + 2x\right)$$

gallons per hour when producing x thousand kilowatt hours of electricity. Suppose that fuel costs \$.75 a gallon and find the value of x that leads to minimum total cost if the generator is operated for 32 hours. Find the minimum cost.

Answer: $x = \sqrt{150} \approx 12.2$, minimum cost is \$24.50

Work Problem 3 at the side.

Example 4 An open box is to be made by cutting squares from each corner of a 12 inch by 12 inch piece of metal and then folding up the sides. What size square should be cut from each corner in order to produce the box of maximum volume?

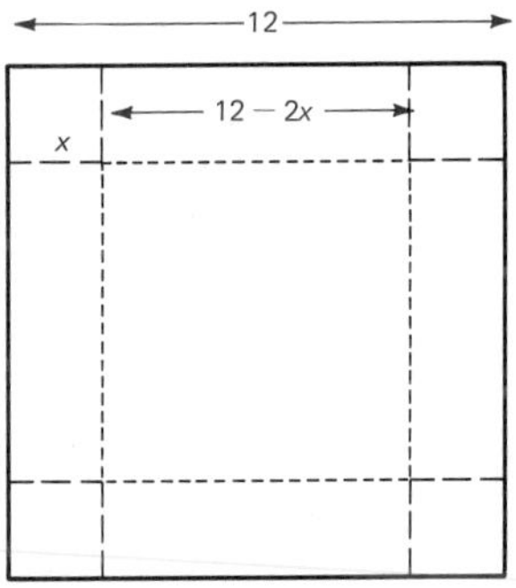

Figure 13.23

Let x represent the length of a side of the square that is cut from each corner, as shown in Figure 13.23. The width of the box that is to be built will be $12 - 2x$, while the length is also $12 - 2x$. The volume of the box is given by the product of the length, width, and height. In our example, the volume, $V(x)$, depends on x:

$$\begin{aligned} V(x) &= x(12 - 2x)(12 - 2x) \\ &= 144x - 48x^2 + 4x^3. \end{aligned}$$

Find the value of x that will maximize volume.

Here $V'(x) = 144 - 96x + 12x^2$. Place this derivative equal to 0.

$$\begin{aligned} 12x^2 - 96x + 144 &= 0 \\ 12(x^2 - 8x + 12) &= 0 \\ 12(x - 2)(x - 6) &= 0 \end{aligned}$$

$$x - 2 = 0 \quad \text{or} \quad x - 6 = 0$$

$$x = 2 \quad \text{or} \quad x = 6$$

Since $V''(x) = -96 + 24x$, we have $V''(2) = -96 + 24(2) = -48$, which is negative. Thus, $x = 2$ leads to maximum volume; the maximum volume is

$$V(2) = 144(2) - 48(2)^2 + 4(2)^3 = 128$$

cubic inches. The function $V(x)$ is defined only on the interval $[0, 6]$. Both the endpoints of this domain, $x = 0$ and $x = 6$, lead to *minimum* volumes of 0. ■

4. An open box is to be made by cutting squares from each corner of a 20 cm by 32 cm piece of metal and folding up the sides. Let x represent the length of the side of the square to be cut out. Find
(a) an expression for the volume of the box, $V(x)$;
(b) $V'(x)$;
(c) the value of x that leads to maximum volume; (Hint: the solutions of the equation $V'(x) = 0$ are 4 and 40/3.)
(d) the maximum volume.

Answer:
(a) $V(x) = 640x - 104x^2 + 4x^3$
(b) $V'(x) = 640 - 208x + 12x^2$
(c) $x = 4$
(d) $V(4) = 1152$ cubic centimeters

Work Problem 4 at the side.

13.4 EXERCISES

Exercises 1–8 involve maximizing and minimizing products and sums of numbers. Work all these problems using the steps shown in Exercise 1.

1. We want to find two numbers x and y such that $x + y = 100$ and the product $P = xy$ is as large as possible.
   (a) We know $x + y = 100$. Solve this equation for y.
   (b) Substitute this result for y into $P = xy$.
   (c) Find P'. Solve the equation $P' = 0$.
   (d) What are the two numbers?
   (e) What is the maximum value of P?
2. Find two numbers whose sum is 250 and whose product is as large as possible. What is the maximum product?
3. Find two numbers whose sum is 200 such that the sum of the squares of the two numbers is minimized.
4. Find two numbers whose sum is 30 such that the sum of the squares of the numbers is minimized.
5. Find numbers x and y such that $x + y = 150$ and x^2y is maximized.
6. Find numbers x and y such that $x + y = 45$ and xy^2 is maximized.
7. Find numbers x and y such that $x - y = 10$ and xy is minimized.
8. Find numbers x and y such that $x - y = 3$ and xy is minimized.

APPLIED PROBLEMS

Management

9. If the price charged for a candy bar is $p(x)$ cents, then x thousand candy bars will be sold in a certain city, where

$$p(x) = 100 - \frac{x}{10}.$$

(a) Find an expression for the total revenue from the sale of x thousand candy bars. (Hint: Find the product of $p(x)$, x, and 1000.) See Examples 1 and 2.
(b) Find the value of x that leads to maximum revenue.
(c) Find the maximum revenue.

10. The sale of cassette tapes of "lesser" performers is very sensitive to price. If a tape manufacturer charges $p(x)$ dollars per tape, where

$$p(x) = 6 - \frac{x}{8},$$

then x thousand tapes will be sold.
(a) Find an expression for the total revenue from the sale of x thousand tapes. (Hint: Find the product of $p(x)$, x, and 1000.)
(b) Find the value of x that leads to maximum revenue.
(c) Find the maximum revenue.

General

11. A truck burns fuel at the rate of $G(x)$ gallons per mile, where

$$G(x) = \frac{1}{32}\left(\frac{64}{x} + \frac{x}{50}\right),$$

while traveling x miles per hour.
(a) If fuel costs \$.80 per gallon, find the speed that will produce minimum total cost for a 400-mile trip. See Example 3.
(b) Find the minimum total cost.

12. A rock and roll band travels from engagement to engagement in a large bus. This bus burns fuel at the rate of $G(x)$ gallons per mile, where

$$G(x) = \frac{1}{50}\left(\frac{200}{x} + \frac{x}{15}\right),$$

while traveling x miles per hour.
(a) If fuel costs \$1 per gallon, find the speed that will produce minimum total cost for a 250-mile trip.
(b) Find the minimum total cost.

13. A farmer has 1200 meters of fencing. He wants to enclose a rectangular field bordering a river, with no fencing needed along the river. (See the sketch.) Let x represent the width of the field.

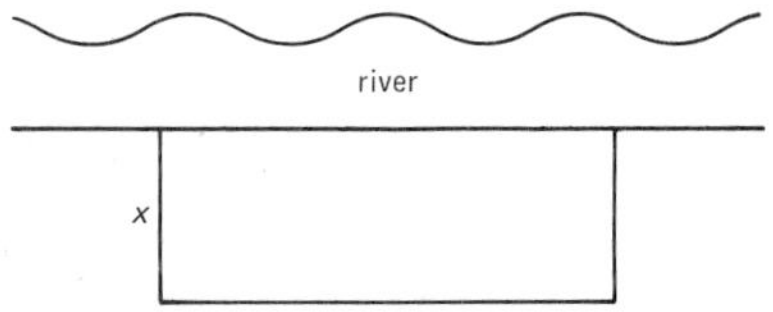

(a) Write an expression for the length of the field.
(b) Find the area of the field (area = length × width).
(c) Find the value of x leading to the maximum area.
(d) Find the maximum area.

14. Find the dimensions of the rectangular field of maximum area that can be made from 200 meters of fencing material. (This fence has four sides.)

15. An ecologist is conducting a research project on breeding pheasants in captivity. She first must construct suitable pens. She wants a rectangular area with two additional fences down the center, as shown in the sketch. Find the maximum area she can enclose with 3600 meters of fencing.

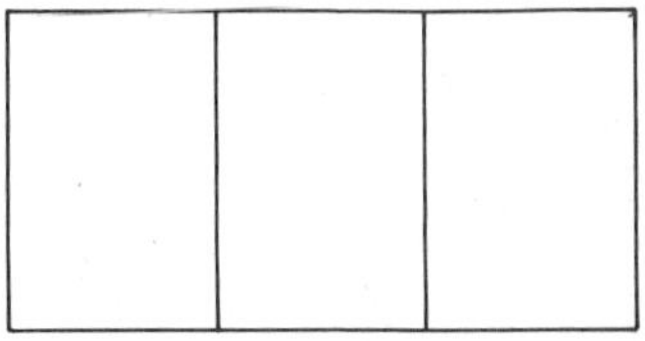

16. A rectangular field is to be enclosed with a fence. One side of the field is against an existing fence, so that no fence is needed on that side. If material for the fence costs \$2 per foot for the two ends and \$4 per foot for the side parallel to the existing fence, find the dimensions of the field of largest area that can be enclosed for \$1000.

17. A rectangular field is to be enclosed on all four sides with a fence. Fencing material costs \$3 per foot for two opposite sides, and \$6 per foot for the other two sides. Find the maximum area that can be enclosed for \$2400.

Management

18. The manager of an 80-unit apartment complex is trying to decide on the rent to charge. It is known from experience that at a rent of \$200, all the units will be full. However, on the average, one additional unit will remain vacant for each \$10 increase in rent.
   (a) Let x represent the number of \$10 increases. Find the amount of rent per apartment. (*See Example 2.*)
   (b) Find the number of apartments rented.
   (c) Find the total revenue from all rented apartments.
   (d) What value of x leads to maximum revenue?
   (e) What is the maximum revenue?

19. The manager of a peach orchard is trying to decide when to arrange for picking the peaches. If they are picked now, the average yield per tree will be 100 pounds, which can be sold for 40¢ per pound. Past experience shows that the yield per tree will increase about 5 pounds per week, while the price will decrease about 2¢ per pound per week.
   (a) Let x represent the number of weeks that the manager should wait. Find the income per pound.
   (b) Find the number of pounds per tree.
   (c) Find the total revenue from a tree.
   (d) When should the peaches be picked in order to produce maximum revenue?
   (e) What is the maximum revenue?

20. A local group of scouts has been collecting old aluminum beer cans for recycling. The group has already collected 12,000 pounds of cans, for

which they could currently receive $4 per hundred pounds. The group can continue to collect cans at the rate of 400 pounds per day. However, a glut in the old-beer-can market has caused the recycling company to announce that it will lower its price, starting immediately, by $.10 per hundred pounds per day. The scouts can make only one trip to the recycling center. Find the best time for that single trip. What total income will be received?

21. In planning a small restaurant, it is estimated that a profit of $5 per seat will be made if the number of seats is between 60 and 80, inclusive. On the other hand, the profit on each seat will decrease by 5¢ for each seat above 80.
(a) Find the number of seats that will produce the maximum profit.
(b) What is the maximum profit?

22. The local hamburger fan club is arranging a charter flight to the island of Hawaii to see the southernmost McDonald's in the United States. The cost of the trip is $425 each for 75 passengers, with a refund of $5 per passenger for each passenger in excess of 75.
(a) Find the number of passengers that will maximize the revenue received from the flight.
(b) Find the maximum revenue.

23. A television manufacturing firm needs to design an open-topped box with a square base. The box must hold 32 cubic inches. Find the dimensions of the box that can be built with the minimum amount of materials.

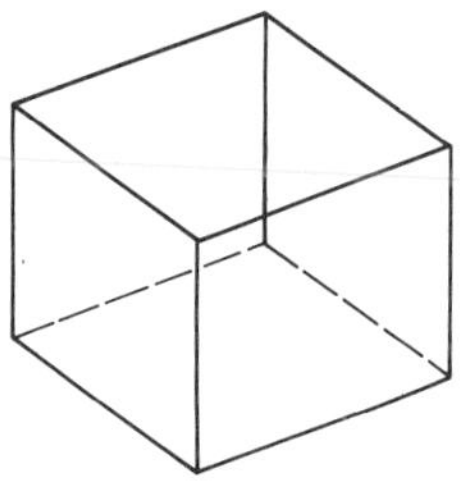

24. A closed box with a square base is to have a volume of 16,000 cubic centimeters. The material for the top and bottom of the box costs $3 per square centimeter, while the material for the sides costs $1.50 per square centimeter. Find the dimensions of the box that will lead to minimum total cost. What is the minimum total cost?

25. A company wishes to manufacture a box with a volume of 36 cubic feet which is open on top and which is twice as long as it is wide. Find the dimensions of the box produced from the minimum amount of material.

26. A mathematics book is to contain 36 square inches of printed matter per page, with margins of 1 inch along the sides, and $1\frac{1}{2}$ inches along the top and bottom. Find the dimensions of the page that will lead to the minimum amount of paper being used for a page.

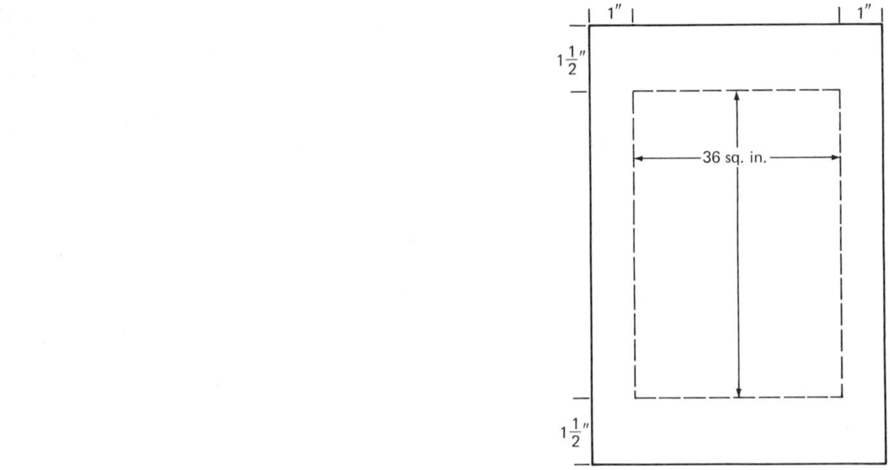

27. In Example 3, we found the speed in miles per hour that minimized cost when we considered only the cost of the fuel. Rework the problem taking into account the driver's salary of \$8 per hour. (Hint: If the trip is 1000 miles at x miles per hour, the driver will be paid for $1000/x$ hours.)

28. Decide what you would do if your assistant brought you the following contract for your signature:

    Your firm offers to deliver 300 tables to a dealer, at \$90 per table, and to reduce the price per table on the entire order by 25¢ for each additional table over 300.

    Find the dollar total involved in the largest possible transaction between the manufacturer and the dealer; find the smallest possible dollar amount.

29. A company wishes to run a utility cable from point A on the shore (see the figure) to an installation at point B on the island. The island is 6 miles from the shore. It costs \$400 per mile to run the cable on land and \$500 per mile underwater. Assume that the cable starts at A and runs along the shoreline, then angles and runs underwater to the island. Find the point at which the line should begin to angle in order to yield the minimum total cost. (Hint: The length of the line underwater is $\sqrt{x^2 + 36}$.)

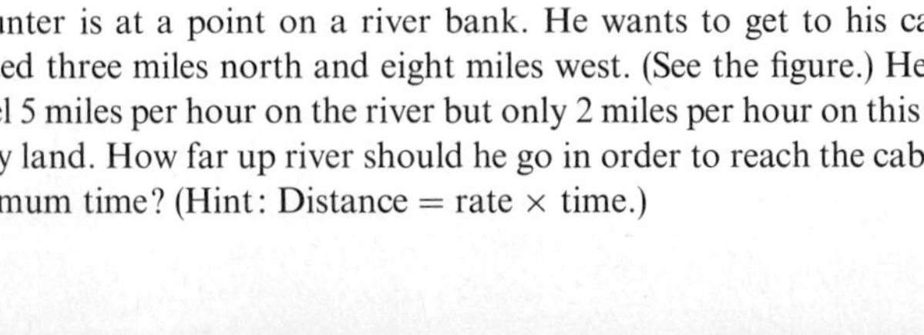

General

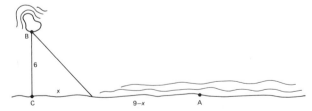

30. A hunter is at a point on a river bank. He wants to get to his cabin, located three miles north and eight miles west. (See the figure.) He can travel 5 miles per hour on the river but only 2 miles per hour on this very rocky land. How far up river should he go in order to reach the cabin in minimum time? (Hint: Distance = rate × time.)

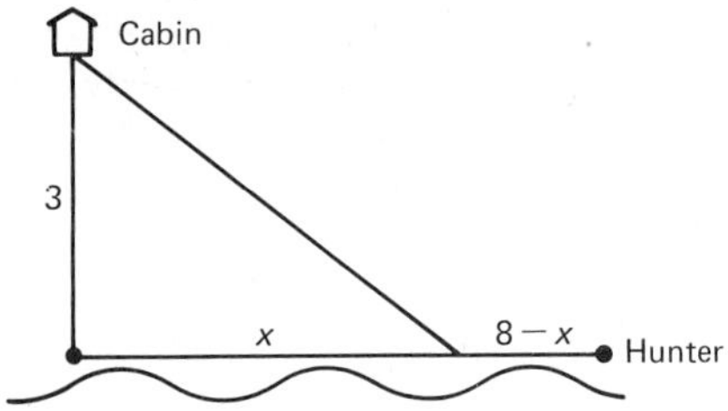

13.5 ECONOMIC LOT SIZE (OPTIONAL)

Suppose that a company manufactures a certain constant number of units of a product per year. Assume that the product can be manufactured in a number of batches of equal size during the year. The company could manufacture the item only once per year to minimize setup costs, but they would then have much higher warehouse costs. On the other hand, they might make many small batches, but this would increase setup costs. We can use calculus to find the number of batches per year that should be manufactured in order to minimize total cost.

Figure 13.24 shows several of the possibilities for a product having an annual demand of 12,000 units. The top graph shows

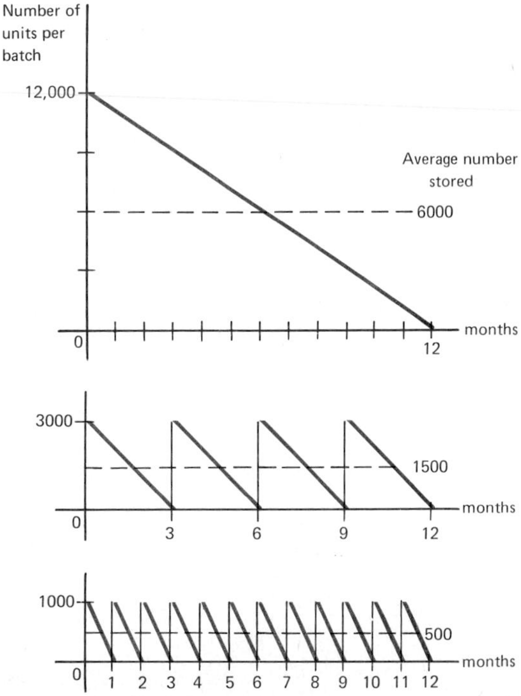

Figure 13.24

the results if only one batch of the product is made annually: an average of 6000 items will be held in a warehouse. If four batches (of 3000 each) are made at equal time intervals during a year, the average number of units in the warehouse falls to only 1500. If twelve batches are made, an average of 500 items will be in the warehouse.

In this section, we use the following variables:

$x =$ number of batches to be manufactured annually

$k =$ cost of storing one unit of the product for one year

$f =$ fixed setup cost to manufacture the product

$g =$ variable cost of manufacturing a single unit of the product

$M =$ total number of units produced annually.

The company has two types of costs associated with the production of its product: a cost associated with manufacturing the item, and a cost associated with storing the finished product.

Let us first consider manufacturing costs. During a year the company will produce x batches of the product. The company will thus produce M/x units of the product per batch. Each batch has a fixed cost f and a variable cost g per unit, so that the manufacturing cost per batch is

$$f + g\left(\frac{M}{x}\right).$$

Since there are x batches per year, the total annual manufacturing cost is

$$\left[f + g\left(\frac{M}{x}\right)\right]x. \tag{1}$$

Now we must find the storage cost. Since each batch consists of M/x units, and since we have assumed demand is constant, it is common to assume an average inventory of

$$\frac{1}{2}\left(\frac{M}{x}\right) = \frac{M}{2x}$$

units per year. It costs k to store one unit of the product for a year, making a total storage cost of

$$k\left(\frac{M}{2x}\right) = \frac{kM}{2x}. \tag{2}$$

The total production cost is the sum of the manufacturing and storage costs, or the sum of expressions (1) and (2). If $T(x)$ is the total cost of producing x batches, we have

$$T(x) = \left[f + g\left(\frac{M}{x}\right)\right]x + \frac{kM}{2x}.$$

Now find the value of x that will minimize $T(x)$. Remember that f, g, k, and M are constants.

Here

$$T'(x) = f - \frac{kM}{2}x^{-2}.$$

Place this derivative equal to 0.

$$f - \frac{kM}{2}x^{-2} = 0$$

$$f = \frac{kM}{2x^2}$$

$$2fx^2 = kM$$

$$x^2 = \frac{kM}{2f}$$

$$x = \sqrt{\frac{kM}{2f}} \qquad \textbf{(3)}$$

Thus, $\sqrt{kM/2f}$ is the annual number of batches which will give minimum total production cost.

Example 1 A paint company has a steady annual demand for 24,500 cans of automobile primer. The cost accountant for the company says that it costs \$2 to hold one can of paint for one year and \$500 to set up the plant for the production of the primer. Find the number of batches of primer that should be produced for the minimum total production cost.

Use equation (3) above.

$$x = \sqrt{\frac{kM}{2f}}$$

$$x = \sqrt{\frac{2(24{,}500)}{2(500)}} \qquad \text{Let } k = 2,\ M = 24{,}500,\ f = 500$$

$$x = \sqrt{49} = 7$$

Seven batches of primer per year will lead to minimum production costs. ■

Work Problem 1 at the side.

1. A manufacturer of business forms has an annual demand for 30,720 units of form letters to people delinquent in their payments of installment debt. It costs \$5 per year to store one unit of the letters and \$1200 to set up the machines to produce them. Find the number of batches that should be made annually to minimize total cost.

Answer: 8 batches

The analysis above also applies to reordering an item that is used at a constant rate throughout the year, as the next example shows.

Example 2 A large pharmacy has an annual need for 200 units of a certain antibiotic. It costs $10 to store one unit for one year. The fixed cost of placing an order (clerical time, mailing, and so on) amounts to $40. Find the number of orders that should be placed annually.

Here $k = 10$, $M = 200$, and $f = 40$. We have

$$x = \sqrt{\frac{10(200)}{2(40)}} = \sqrt{25} = 5.$$

Order the drug 5 times a year. ■

2. An office uses 576 cases of copy-machine paper during the year. It costs $3 per year to store one case. Each reorder costs $24. Find the number of orders that should be placed annually.

Answer: 6

Work Problem 2 at the side.

13.5 APPLIED PROBLEMS

Management

1. Find the approximate number of batches that should be produced annually if 100,000 units are to be manufactured. It costs $1 to store a unit for one year and it costs $500 to set up the factory to produce each batch. (See Example 1.)

2. How many units per batch will be manufactured in Exercise 1?

3. A market has a steady annual demand for 16,800 cases of sugar. It costs $3 to store one case for one year. The market pays $7 for each order that is placed. Find the number of orders for sugar that should be placed each year. (See Example 2.)

4. Find the number of cases per order in Exercise 3.

5. The publisher of a best-selling book has an annual demand for 100,000 copies. It costs 50¢ to store one copy for one year. Setup of the press for a new printing costs $1000. Find the number of batches that should be printed annually.

6. A restaurant has an annual demand for 810 bottles of a California wine. It costs $1 to store one bottle for one year, and it costs $5 to place a reorder. Find the number of orders that should be placed annually.

7. Use the second derivative to show that the value of x obtained in the text [equation (3)] really leads to the minimum cost.

8. Why do you think that the variable cost g does not appear in the answer for x [equation (3)]?

13.6 IMPLICIT DIFFERENTIATION (OPTIONAL)*

In almost every example so far, the functions that we have used have been written in the form $y = f(x)$, in which y is expressed as a function of x. Where $y = f(x)$, the variable y is said to be expressed as an **explicit function** of x. For example,

$$y = 3x - 2,$$
$$y = x^2 + x + 6,$$

and

$$y = -x^3 + 2$$

are all explicit functions of x. We can express $4xy - 3x = 6$ as an explicit function of x by solving for y. This gives

$$4xy - 3x = 6$$
$$4xy = 3x + 6$$
$$y = \frac{3x + 6}{4x}.$$

Work Problem 1 at the side.

1. Write as explicit functions of x.
(a) $2y + 4x = 7$
(b) $xy = 10$
(c) $5x^2y - 3y = 5$

Answer:

(a) $y = \frac{-4x + 7}{2}$

(b) $y = \frac{10}{x}$

(c) $y = \frac{5}{5x^2 - 3}$

On the other hand, some functions are expressed by equations which cannot be solved readily for y, and some by equations which cannot be solved for y at all. For example, it would be possible (but tedious) to use the quadratic formula to solve for y in the equation $y^2 + 2yx + 4x^2 = 0$. On the other hand, it would not be possible to solve for y in the equation $y^5 + 8y^3x + 6y^2x^2 + 2yx^3 + 6 = 0$. Functions such as these last two are said to be expressed as **implicit functions.**

Even though a function is written implicitly, it is still often possible to calculate dy/dx, the derivative of the function. In doing so, we assume that there exists some function f (which we may or may not be able to find) such that $y = f(x)$ and dy/dx exists. (We will use dy/dx here rather than $f'(x)$ for convenience.)

For example, to find dy/dx for the function $3xy + 4y^2 = 10$, first take the derivative of both sides of the equation. The first term $3xy$ can be written as the product $(3x)(y)$. To use the product rule, we need

$$\frac{d(3x)}{dx} = 3 \quad \text{and} \quad \frac{d(y)}{dx} = \frac{dy}{dx}.$$

* The material in this section is used only in Section 13.8 to obtain the derivative of e^x.

By the product rule, the derivative of the first term is

$$3x \cdot \frac{dy}{dx} + y \cdot 3 = 3x\frac{dy}{dx} + 3y.$$

To differentiate the second term, $4y^2$, we use the chain rule, since y is some function of x.

$$\frac{d(4y^2)}{dx} = 4 \cdot 2y^1 \cdot \frac{dy}{dx} = 8y\frac{dy}{dx}$$

On the other side of the equation, the derivative of 10 is 0.

Taking the indicated derivatives term by term, we have

$$3x\frac{dy}{dx} + 3y + 8y\frac{dy}{dx} = 0.$$

To complete the process, solve this result for dy/dx.

$$(3x + 8y)\frac{dy}{dx} = -3y$$

$$\frac{dy}{dx} = \frac{-3y}{3x + 8y}$$

Finding the derivative by this procedure is called **implicit differentiation.**

Example 1 Find dy/dx for $x^2 + 2xy^2 + 3x^2y = 0$.

We have

$$\frac{d(x^2 + 2xy^2 + 3x^2y)}{dx} = \frac{d(0)}{dx}.$$

Now we treat $2xy^2$ and $3x^2y$ as products and differentiate term by term to get

$$2x + 2x\left(2y\frac{dy}{dx}\right) + 2y^2 + 3x^2\left(\frac{dy}{dx}\right) + 6xy = 0$$

or

$$2x + 4xy\left(\frac{dy}{dx}\right) + 2y^2 + 3x^2\left(\frac{dy}{dx}\right) + 6xy = 0,$$

from which

$$(4xy + 3x^2)\frac{dy}{dx} = -2x - 2y^2 - 6xy,$$

or, finally,

$$\frac{dy}{dx} = \frac{-2x - 2y^2 - 6xy}{4xy + 3x^2}. \quad ■$$

Work Problem 2 at the side.

2. Find $\frac{dy}{dx}$ for the following.

(a) $5xy + 6y = 12$

(b) $9x^2y^2 + 2xy = 1$

Answer:

(a) $\frac{dy}{dx} = \frac{-5y}{5x + 6}$

(b) $\frac{dy}{dx} = \frac{-9xy^2 - y}{9x^2y + x}$

Example 2 Find dy/dx for $x + \sqrt{xy} = y^2$.

We have

$$\frac{d(x + \sqrt{xy})}{dx} = \frac{d(y^2)}{dx}.$$

Since $\sqrt{xy} = \sqrt{x} \cdot \sqrt{y} = x^{1/2} \cdot y^{1/2}$, we have, upon taking derivatives,

$$1 + x^{1/2}\left(\frac{1}{2}y^{-1/2} \cdot \frac{dy}{dx}\right) + y^{1/2}\left(\frac{1}{2}x^{-1/2}\right) = 2y\frac{dy}{dx}$$

$$1 + \frac{x^{1/2}}{2y^{1/2}} \cdot \frac{dy}{dx} + \frac{y^{1/2}}{2x^{1/2}} = 2y\frac{dy}{dx}.$$

Multiply both sides by $2x^{1/2} \cdot y^{1/2}$.

$$2x^{1/2} \cdot y^{1/2} + x\frac{dy}{dx} + y = 4x^{1/2} \cdot y^{3/2} \cdot \frac{dy}{dx}$$

Upon combining terms we have

$$2x^{1/2} \cdot y^{1/2} + y = (4x^{1/2} \cdot y^{3/2} - x)\frac{dy}{dx},$$

or

$$\frac{dy}{dx} = \frac{2x^{1/2} \cdot y^{1/2} + y}{4x^{1/2} \cdot y^{3/2} - x}. \quad \blacksquare$$

3. Find $\frac{dy}{dx}$ for the following.

(a) $\sqrt{x^2y} + y^2 = 1$

(b) $y/\sqrt{x} = x + y^4$

(c) $\sqrt{y} + \sqrt{x} = 4$

Answer:

(a) $\frac{dy}{dx} = \frac{-2y}{x + 4y^{3/2}}$

(b) $\frac{dy}{dx} = \frac{2x^{3/2} + y}{2x - 8x^{3/2}y^3}$

(c) $\frac{dy}{dx} = \frac{-y^{1/2}}{x^{1/2}}$

Work Problem 3 at the side.

Example 3 Find an equation of the tangent line to the curve $x^2 + 4y^2 = 17$ at the point (1, 2).

To find the equation of the tangent line, we must first find the slope of the tangent to the curve at the point (1, 2). This can be done by finding dy/dx. Using implicit differentiation, we have

$$2x + 8y\frac{dy}{dx} = 0$$

$$\frac{dy}{dx} = \frac{-2x}{8y}$$

$$= \frac{-x}{4y}.$$

The slope of the tangent line to the curve at the point (1, 2) is thus

$$m = \frac{-x}{4y} = \frac{-1}{4(2)} = \frac{-1}{8}.$$

The equation of the tangent line is then found by using the point-slope form of the equation of a line.

$$y - y_1 = m(x - x_1)$$
$$y - 2 = \frac{-1}{8}(x - 1)$$
$$8y - 16 = -x + 1$$
$$8y + x = 17 \quad ■$$

Work Problem 4 at the side.

4. Find the equation of the tangent line to the given curve at the given point.
(a) $y^2 + x^2 = 25$ at $(-3, 4)$
(b) $x^2y^3 = 27$ at $(1, 3)$
(c) $y^3 - x = 121$ at $(4, 5)$

Answer:
(a) $4y = 3x + 25$
(b) $2x + y = 5$
(c) $75y = x + 371$

Example 4 A biologist has placed a 50-foot ladder against a large building. The base of the ladder is resting on an oil spill, and slips (to the right in Figure 13.25) at the rate of 3 feet per minute. Find the rate of change of the height of the top of the ladder above the ground at the instant when the base of the ladder is 30 feet from the base of the building.

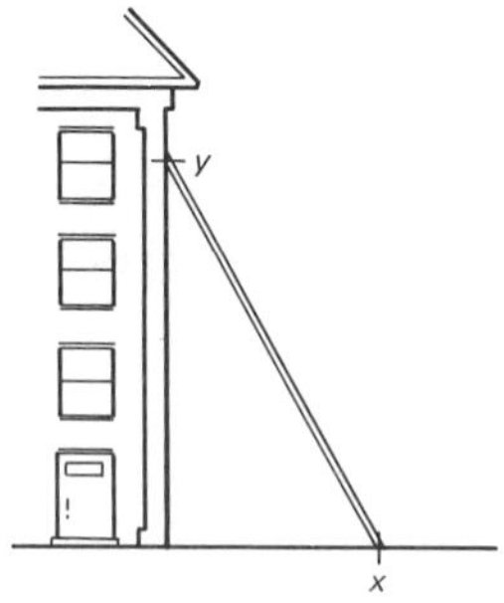

Figure 13.25

Let y be the height of the top of the ladder above the ground, and let x be the distance of the base of the ladder from the base of the building. By the Pythagorean theorem we have

$$50^2 = x^2 + y^2. \quad \textbf{(1)}$$

Both x and y are functions of time, t, measured from the moment that the ladder starts slipping. If we now take the derivative of both sides of equation (1) with respect to time, we get

$$\frac{d(50^2)}{dt} = \frac{d(x^2 + y^2)}{dt},$$

or

$$0 = 2x\frac{dx}{dt} + 2y\frac{dy}{dt}. \qquad \textbf{(2)}$$

In the statement of the problem, we are told that the base is sliding at the rate of 3 feet per minute, so that

$$\frac{dx}{dt} = 3.$$

We are told that the base of the ladder is 30 feet from the base of the building. This fact can be used to find y. We know that $50^2 = x^2 + y^2$. Thus,

$$\begin{aligned} 50^2 &= 30^2 + y^2 \\ 2500 &= 900 + y^2 \\ 1600 &= y^2 \\ y &= 40 \text{ feet.} \end{aligned}$$

We now know that $y = 40$, $x = 30$, and $dx/dt = 3$. Substituting these values into equation (2), we have

$$\begin{aligned} 0 &= 2x\frac{dx}{dt} + 2y\frac{dy}{dt} \\ 0 &= 2(30)(3) + 2(40)\frac{dy}{dt} \\ 0 &= 180 + 80\frac{dy}{dt} \\ 80\frac{dy}{dt} &= -180 \\ \frac{dy}{dt} &= \frac{-180}{80} = \frac{-9}{4} = -2.25 \text{ feet per minute.} \end{aligned}$$

Thus, at the instant when the base of the ladder is 30 feet from the base of the building, the top of the ladder is sliding down the building at the rate of 2.25 feet per minute. ■

5. In Example 4, find the rate at which the ladder is moving down at the instant that the top of the ladder is
(a) 30 feet above the ground;
(b) 20 feet above the ground.

Answer:
(a) 4 feet per minute
(b) 6.9 feet per minute

Work Problem 5 at the side.

13.6 EXERCISES

Find dy/dx by implicit differentiation for the following. (See Examples 1 and 2.)

1. $4x^2 + 3y^2 = 6$

2. $2x^2 - 5y^2 = 4$

3. $2xy + y^2 = 8$

4. $-3xy - 4y^2 = 2$

5. $y^2 = 4x + 1$

6. $y^2 - 2x = 6$

7. $6xy^2 - 8y + 1 = 0$

8. $-4y^2x^2 - 3x + 2 = 0$

9. $x^2 + 2xy = 6$

10. $2x^2 - 3xy = 10$

11. $6x^2 + 8xy + y^2 = 6$

12. $8x^2 = 6y^2 + 2xy$

13. $x^3 = y^2 + 4$

14. $x^3 - 6y^2 = 10$

15. $x^2y = 4$

16. $-2x^2y = 3$

17. $x^2y^2 = 6$

18. $5 - x^2y^2 = 0$

19. $x^2y + y^3 = 4$

20. $2xy^2 + 2y^3 + 5x = 0$

21. $\sqrt{x} + \sqrt{y} = 4$

22. $2\sqrt{x} - \sqrt{y} = 1$

23. $\sqrt{xy} + y = 1$

24. $\sqrt{2xy} - 1 = 3y^2$

Find the equation of the tangent line at the given point on each of the following curves. (See Example 3.)

25. $x^2 + y^2 = 25;\quad (-3, 4)$

26. $x^2 + y^2 = 100;\quad (8, -6)$

27. $x^2y^2 = 1;\quad (-1, 1)$

28. $x^2y^3 = 8;\quad (-1, 2)$

29. $x^2 + \sqrt{y} = 7;\quad (2, 9)$

30. $2y^2 - \sqrt{x} = 4;\quad (16, 2)$

31. $y + \dfrac{\sqrt{x}}{y} = 3;\quad (4, 2)$

32. $x + \dfrac{\sqrt{y}}{3x} = 2;\quad (1, 9)$

APPLIED PROBLEMS

For Exercises 33–40, see Example 4.

General **33.** A 25-foot ladder is placed against a building. The base of the ladder is slipping away from the building at the rate of 4 feet per minute. Find the rate at which the top of the ladder is sliding down the building at the instant when the bottom of the ladder is 7 feet from the base of the building.

34. One car leaves a given point and travels north at 30 miles per hour. Another car leaves the same point at the same time and travels west at 40 miles per hour. At what rate is the distance between the two cars changing at the instant when the cars have traveled 2 hours?

35. A rock is thrown into a still pond. The circular ripples move outward from the point of impact of the rock so that the radius of the area of ripples increases at the rate of 2 feet per minute. Find the rate at which the area is changing at the instant the radius is 3/2 feet.

36. A spherical snowball is placed in the sun. The sun melts the snowball so that its radius decreases 1/4 inch per hour. Find the rate of change of the volume with respect to time at the instant the radius is 4 inches.

Use the ideas of similar triangles from geometry to work the following problems.

37. A sand storage tank used by the highway department for winter storms is leaking. As the sand leaks out, it forms a conical pile. The radius of the base of the pile increases at the rate of 1 inch per minute. The height of the pile is always twice the radius of the base. Find the rate at which the volume of the pile is increasing at the instant the radius of the base is 5 inches. (Hint: The volume of a cone is given by $V = \frac{1}{3}\pi r^2 h$.)

38. A man 6 feet tall is walking away from a lamp post at the rate of 50 feet per minute. When the man is 8 feet from the lamp post, his shadow is 10 feet long. Find the rate at which the length of the shadow is increasing when he is 25 feet from the lamp post.

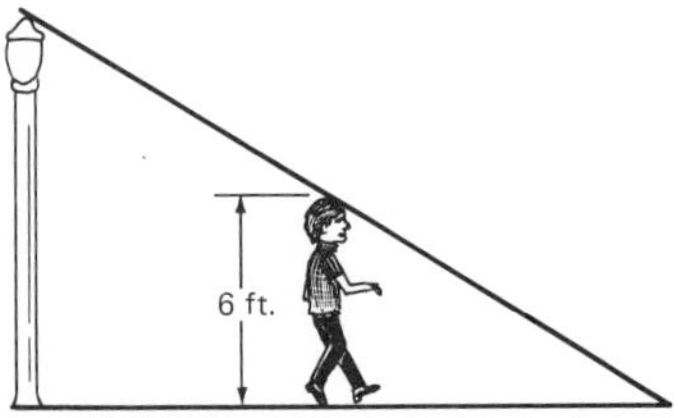

13.7 APPROXIMATION BY DIFFERENTIALS

In this section we look at differentials, which are useful in finding approximate values of functions. To define a differential, look first at Figure 13.26.

Recall that Δx represents a change in x. In particular, Δx is often used to represent a small change in x. In Figure 13.26, PM is the line tangent to the graph of the function $y = f(x)$ at the point P. Line PM then has slope $f'(x)$. Let $PR = \Delta x = dx$ and $MR = dy$. Then by the definition of slope, the slope of line PM is

$$\frac{\text{the change in } y}{\text{the change in } x} = \frac{dy}{dx}.$$

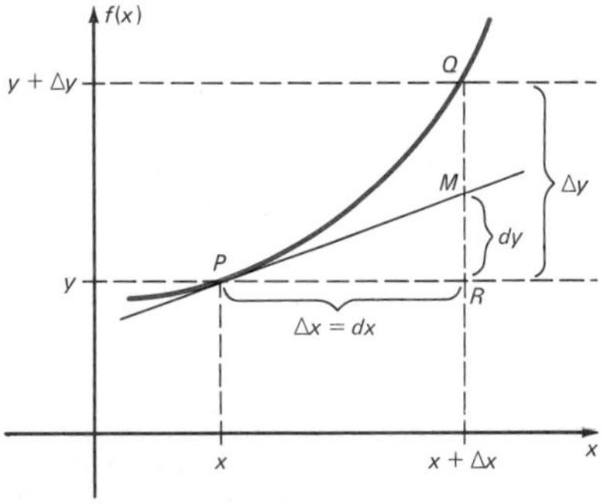

Figure 13.26

Since both $f'(x)$ and dy/dx represent the slope of line PM, we have

$$f'(x) = \frac{dy}{dx}.$$

Until now we have treated the expression dy/dx as a single quantity which represented the derivative of y with respect to x. Now, by defining dy and dx as we have, we are also able to treat dy/dx as a fraction which represents the quotient of two numbers. This allows us to rewrite $f'(x) = dy/dx$ as

$$\boldsymbol{dy = f'(x)\,dx.}$$

We call dy the **differential of y** and dx the **differential of x.**

In Figure 13.25 we see that if we let Δx get smaller and smaller, the value of dy gets closer and closer to Δy, so that for small values of Δx (but $\Delta x \neq 0$),

$$dy \approx \Delta y.$$

Example 1 Find dy for the following functions.

(a) $y = 6x^2$

The derivative is $y' = 12x$, so that

$$dy = 12x \cdot dx.$$

(b) If $y = 8x^{-3/4}$, then $dy = -6x^{-7/4}\,dx.$ ■

Work Problem 1 at the side.

1. Find dy for the following.

(a) $y = 9x^{2/3}$

(b) $y = \dfrac{2+x}{2-x}$

(c) $y = \sqrt{8x-7}$

Answer:

(a) $dy = 6x^{-1/3}\,dx$

(b) $dy = \dfrac{4}{(2-x)^2}dx$

(c) $dy = 4(8x-7)^{-1/2}\,dx$

As shown in Figure 13.25, if a change of Δx is made in the value of x, then the corresponding change in y is Δy, where

$$\Delta y = f(x + \Delta x) - f(x).$$

For small values of Δx, we know that $\Delta y \approx dy$, so that

$$dy \approx f(x + \Delta x) - f(x),$$

or

$$f(x) + dy \approx f(x + \Delta x). \qquad \textbf{(1)}$$

This result can be used to approximate numbers as shown by the next examples.

Example 2 Approximate $\sqrt{38}$.

Since we want to find $\sqrt{38}$, the function we should use is $y = f(x) = \sqrt{x} = x^{1/2}$. The nearest perfect square to 38 is 36. Therefore, in equation (1), we let $x = 36$ and $\Delta x = 2$. Reversed, the equation becomes

$$\sqrt{36+2} = \sqrt{38} \approx \sqrt{36} + dy,$$

where $dy = f'(x) \cdot dx$. Since $f(x) = x^{1/2}$, we have $f'(x) = \frac{1}{2}x^{-1/2}$, and

$$dy = \frac{1}{2}x^{-1/2}\,dx.$$

In this example, $x = 36$ and $dx = \Delta x = 2$;

$$dy = \frac{1}{2}(36)^{-1/2}(2) = \frac{1}{6}.$$

Finally,

$$\sqrt{38} \approx \sqrt{36} + dy = 6 + \tfrac{1}{6} = 6\tfrac{1}{6} \approx 6.167.$$

From a square root table or calculator, $\sqrt{38} \approx 6.164$, so that our approximation is fairly close. ■

Work Problem 2 at the side.

2. Approximate the following.
(a) $\sqrt{10}$
(b) $\sqrt{8}$
(c) $\sqrt[3]{9}$

Answer:
(a) $3\frac{1}{6}$
(b) $2\frac{5}{6}$
(c) $2\frac{1}{12}$

Example 3 A tumor is in the shape of a cone of height 3 centimeters. Find the approximate increase in the volume of the tumor if the radius of the base increases from 1 cm to 1.1 cm.

The volume of a cone is given by $V = \frac{1}{3}\pi r^2 h$. In our example, $h = 3$, so that

$$V = \frac{1}{3}\pi r^2 \cdot 3 = \pi r^2.$$

The increase in volume is approximated by dV, where

$$dV = 2\pi r \cdot dr.$$

In our example, $r = 1$, and $dr = 1.1 - 1 = .1$.

$$dV = 2\pi(1)(.1) \approx 2(3.14)(1)(.1) = .628$$

The approximate increase in volume is .628 cubic centimeters. ■

Work Problem 3 at the side.

3. The volume of a sphere is given by $V = \frac{4}{3}\pi r^3$. Find the approximate change in the volume of a snowball if the radius decreases from 8 cm to 7 cm.

Answer: volume decreases approximately 804 cubic centimeters

13.7 EXERCISES

Find dy for the following functions. (See Example 1.)

1. $y = 6x^2$

2. $y = -8x^4$

3. $y = 2\sqrt{x}$

4. $y = 8\sqrt{2x-1}$

5. $y = 7x^2 - 9x + 6$

6. $y = -3x^3 + 2x^2$

7. $y = \dfrac{8x - 2}{x - 3}$

8. $y = \dfrac{-4x + 7}{3x - 1}$

For the following functions, find $f(x)$, $f(x + \Delta x)$, and $f'(x)$ for the given values of x and Δx. (A calculator will be helpful for finding $f(x + \Delta x)$.) Then
(a) find Δy by using the result $\Delta y = f(x + \Delta x) - f(x)$;
(b) find dy by the result $dy = f'(x) \cdot \Delta x$.

9. $f(x) = 6x - 8; \quad x = 4, \Delta x = .1$

10. $f(x) = -5x + 7; \quad x = 3, \Delta x = .2$

11. $f(x) = 2x^2; \quad x = -2, \Delta x = .2$

12. $f(x) = x^2 - 1; \quad x = 3, \Delta x = .1$

13. $f(x) = x^3 - 2x^2 + 3; \quad x = 1, \Delta x = -.1$

14. $f(x) = 2x^3 + x^2 - 4x; \quad x = 2, \Delta x = -.2$

15. $f(x) = \sqrt{3x}; \quad x = 1, \Delta x = .15$

16. $f(x) = \sqrt{4x - 1}; \quad x = 5, \Delta x = .08$

17. $f(x) = \dfrac{2x - 5}{x + 1}; \quad x = 2, \Delta x = -.03$

18. $f(x) = \dfrac{6x - 3}{2x + 1}; \quad x = 3, \Delta x = -.04$

Find approximations for the following numbers. (See Example 2.)

19. $\sqrt{10}$ 20. $\sqrt{26}$ 21. $\sqrt{15}$ 22. $\sqrt{63}$

23. $\sqrt{123}$ 24. $\sqrt{146}$ 25. $\sqrt[3]{7}$ 26. $\sqrt[3]{25}$

27. $\sqrt[3]{65}$ 28. $\sqrt[3]{127}$ 29. $\sqrt[4]{17}$ 30. $\sqrt[4]{83}$

31. $\sqrt{4.02}$ 32. $\sqrt{16.08}$ 33. $\sqrt[3]{7.9}$ 34. $\sqrt[3]{124.75}$

APPLIED PROBLEMS

For Exercises 35–38, see Example 3.

General

35. A spherical beachball is being inflated. Find the approximate change in volume if the radius increases from 4 cm to 4.2 cm.

36. A spherical snowball is melting; find the approximate change in volume if the radius decreases from 3 cm to 2.8 cm.

37. An oil slick is in the shape of a circle. Find the approximate increase in the area of the slick if its radius increases from 1.2 miles to 1.4 miles.

Natural Science

38. The shape of a colony of bacteria on a Petri dish is circular. Find the approximate increase in its area if the radius increases from 20 mm to 22 mm.

13.8 DERIVATIVES OF EXPONENTIAL AND LOGARITHMIC FUNCTIONS

We approach our discussion of exponential and logarithmic functions for this chapter by discussing methods of finding derivatives of these functions.

We begin by finding the derivative of the function $y = \ln x$, where we assume that $x > 0$. This derivative can be found by going back to the definition of the derivative given in Chapter 12: if $y = f(x)$, then

$$y' = f'(x) = \lim_{h \to 0} \frac{f(x+h) - f(x)}{h}$$

provided this limit exists.

Here $y = \ln x$, so that

$$y' = f'(x) = \lim_{h \to 0} \frac{\ln(x+h) - \ln x}{h}.$$

This limit can be found by using various properties of logarithms.

$$\begin{aligned} y' = \lim_{h \to 0} \frac{\ln(x+h) - \ln x}{h} &= \lim_{h \to 0} \frac{\ln\left(\dfrac{x+h}{x}\right)}{h} \\ &= \lim_{h \to 0} \frac{1}{h} \ln\left(\frac{x+h}{x}\right) \\ &= \lim_{h \to 0} \ln\left(\frac{x+h}{x}\right)^{1/h} \\ &= \lim_{h \to 0} \ln\left(1 + \frac{h}{x}\right)^{1/h} \end{aligned}$$

Now make a substitution: let $m = x/h$, so that $h/x = 1/m$. As $h \to 0$, the fact that $h/x = 1/m$ means that $m \to \infty$. So

$$\begin{aligned} \lim_{h \to 0} \ln\left(1 + \frac{h}{x}\right)^{1/h} &= \lim_{m \to \infty} \ln\left(1 + \frac{1}{m}\right)^{m/x} \\ &= \lim_{m \to \infty} \ln\left[\left(1 + \frac{1}{m}\right)^{m}\right]^{1/x} \\ &= \lim_{m \to \infty} \frac{1}{x} \cdot \ln\left(1 + \frac{1}{m}\right)^{m}. \end{aligned}$$

Earlier, we defined e as $\lim_{m \to \infty} \left(1 + \frac{1}{m}\right)^m$, so

$$\lim_{m \to \infty} \frac{1}{x} \cdot \ln\left(1 + \frac{1}{m}\right)^m = \frac{1}{x} \cdot \ln e = \frac{1}{x} \cdot 1 = \frac{1}{x}.$$

That is, if $y = \ln x$, then $y' = \frac{1}{x}$. From the chain rule, we get a more general statement. If $y = \ln g(x)$, then $y' = \frac{1}{g(x)} \cdot g'(x) = \frac{g'(x)}{g(x)}$. The next theorem summarizes the results. Absolute value bars have been used because the domain of $\ln x$ includes only positive values of x.

Theorem 13.2 **(a)** If $y = \ln|x|$, then $y' = \frac{1}{x}$.

(b) If $y = \ln|g(x)|$, then $y' = \frac{g'(x)}{g(x)}$.

Example 1 Find the derivative of each of the following functions.

(a) $y = \ln|5x|$

Let $g(x) = 5x$, so that $g'(x) = 5$. From Theorem 13.2(b),

$$y' = \frac{g'(x)}{g(x)} = \frac{5}{5x} = \frac{1}{x}.$$

(b) $y = \ln|3x^2 - 4x|$

$$y' = \frac{6x - 4}{3x^2 - 4x} \quad ■$$

Work Problem 1 at the side.

1. Find y' for the following.
(a) $y = \ln|7 + x|$
(b) $y = \ln|4x^2|$
(c) $y = \ln|8x^3 - 3x|$

Answer:

(a) $y' = \frac{1}{7 + x}$

(b) $y' = \frac{2}{x}$

(c) $y' = \frac{24x^2 - 3}{8x^3 - 3x}$

Now we need to find the derivative of the function $y = e^x$. To do so, first take the natural logarithm of both sides.

$$y = e^x$$
$$\ln y = \ln e^x$$

By properties of logarithms,

$$\ln y = x \cdot \ln e$$
$$\ln y = x \qquad \ln e = 1$$

Use implicit differentiation to take the derivative of each side.

$$\frac{1}{y} \cdot \frac{dy}{dx} = 1$$

or

$$\frac{dy}{dx} = y.$$

Since $y = e^x$, we end up with

$$\frac{dy}{dx} = e^x, \qquad \text{or} \qquad y' = e^x.$$

This result shows one of the main reasons for the widespread use of e as a base—the function $y = e^x$ is its own derivative. Theorem 13.3 summarizes this result and the more general result using the chain rule.

Theorem 13.3 **(a)** If $y = e^x$, then $y' = e^x$.

(b) If $y = e^{f(x)}$, then $y' = f'(x) \cdot e^{f(x)}$.

Example 2 Find derivatives of the following.

(a) $y = 4e^{5x}$

Let $f(x) = 5x$, with $f'(x) = 5$. Then

$$y' = 4 \cdot 5e^{5x} = 20e^{5x}.$$

(b) $y = 3e^{-4x}$

$$y' = 3(-4)e^{-4x} = -12e^{-4x}$$

(c) $y = 10e^{3x^2}$

$$y' = 10(6x)e^{3x^2} = 60xe^{3x^2} \quad ■$$

2. Find y' for the following.
(a) $y = 3e^{12x}$
(b) $y = -6e^{-10x+1}$
(c) $y = e^{-x^2}$

Answer:
(a) $y' = 36e^{12x}$
(b) $y' = 60e^{-10x+1}$
(c) $y' = -2xe^{-x^2}$

Work Problem 2 at the side.

Example 3 Let $y = e^x \cdot \ln|x|$. Find y'.

Use the product rule.

$$y' = e^x \cdot \frac{1}{x} + \ln|x| \cdot e^x$$

$$= e^x\left(\frac{1}{x} + \ln|x|\right) \quad ■$$

Example 4 Let $y = \dfrac{100{,}000}{1 + 100e^{-0.3x}}$. Find y'.

Use the quotient rule.

$$y' = \frac{(1 + 100e^{-0.3x})(0) - 100{,}000(-30e^{-0.3x})}{(1 + 100e^{-0.3x})^2}$$

$$= \frac{3{,}000{,}000e^{-0.3x}}{(1 + 100e^{-0.3x})^2} \quad ■$$

3. Find y' for the following.
(a) $y = x^2 \cdot e^x$

(b) $y = \dfrac{e^x}{1 + x}$

(c) $y = \dfrac{\ln|x|}{2 - 3x}$

Answer:
(a) $y' = x(x + 2)e^x$

(b) $y' = \dfrac{xe^x}{(1 + x)^2}$

(c) $y' = \dfrac{2 - 3x + 3x \cdot \ln|x|}{x(2 - 3x)^2}$

Work Problem 3 at the side.

Example 5 Often a population, or the sales of a certain product, will start growing slowly, then grow more rapidly, and then gradually level off. Such growth can often be approximated by a function of the form

$$f(x) = \frac{b}{1 + ae^{kx}}$$

for appropriate constants a, b, and k. For example, suppose that population planners predict that the population of a certain city will be approximated for the next few years by the function

$$P(x) = \frac{100{,}000}{1 + 100e^{-0.3x}},$$

where x is time in years. Find the rate of change of the population at time $x = 4$.

Notice that the derivative of $P(x)$ was given in the previous example. Using this we have

$$P'(4) = \frac{3{,}000{,}000e^{-0.3(4)}}{[1 + 100e^{-0.3(4)}]^2}$$

$$= \frac{3{,}000{,}000e^{-1.2}}{(1 + 100e^{-1.2})^2}.$$

From Table 4, we see that $e^{-1.2} \approx 0.301$. Using this value, we have

$$P'(4) \approx \frac{3{,}000{,}000(0.301)}{[1 + 100(0.301)]^2}$$

$$= \frac{903{,}000}{(1 + 30.1)^2}$$

$$= \frac{903{,}000}{967}$$

$$= 934.$$

Thus the rate of change of the population at time $x = 4$ is an increase of about 934 per year.

4. In Example 5, find the following.
(a) $P'(0)$
(b) $P'(2)$

Answer:
(a) 294
(b) 527

Work Problem 4 at the side.

Example 6 Find all relative maximums or relative minimums for the function $y = (2x + 1)e^{-x}$.

First take the derivative, using the product rule.

$$\begin{aligned} y' &= (2x+1)(-e^{-x}) + e^{-x}(2) \\ &= -2x \cdot e^{-x} - e^{-x} + 2e^{-x} \\ &= -2x \cdot e^{-x} + e^{-x} \\ y' &= e^{-x}(-2x+1) \end{aligned}$$

Place the derivative equal to 0.

$$e^{-x}(-2x+1) = 0$$

Since e^{-x} is never 0, this derivative can equal 0 only when $-2x + 1 = 0$, or when $x = 1/2$.

Use the second derivative test to see that $x = 1/2$ leads to a relative maximum of $2/e^{1/2}$ or about 1.2. ■

5. Find any relative maximums or relative minimums for the following.
(a) $y = 3e^x$
(b) $y = x^2e^{5x}$

Answer:
(a) none
(b) relative minimum of 0 at $x = 0$; relative maximum of about .02 at $x = -2/5$

Work Problem 5 at the side.

13.8 EXERCISES

Find derivatives of the following. (See Examples 1–4.)

1. $y = e^{4x}$

2. $y = e^{-2x}$

3. $y = -6e^{-2x}$

4. $y = 8e^{4x}$

5. $y = -8e^{2x}$

6. $y = 0.2e^{5x}$

7. $y = -16e^{x+1}$

8. $y = -4e^{-0.1x}$

9. $y = e^{x^2}$

10. $y = e^{-x^2}$

11. $y = 3e^{2x^2}$

12. $y = -5e^{4x^3}$

13. $y = 4e^{2x^2-4}$

14. $y = -3e^{3x^2+5}$

15. $y = xe^x$

16. $y = x^2e^{-2x}$

17. $y = (x-3)^2e^{2x}$

18. $y = (3x^2 - 4x)e^{-3x}$

19. $y = \ln|3 - x|$

20. $y = \ln|1 + x^2|$

21. $y = \ln|2x^2 - 7x|$

22. $y = \ln|-8x^2 + 6x|$

23. $y = \ln\sqrt{x+5}$

24. $y = \ln\sqrt{2x+1}$

25. $y = \dfrac{x^2}{e^x}$

26. $y = \dfrac{e^x}{2x+1}$

27. $y = x^2 \cdot \ln|x|$

28. $y = \dfrac{\ln|x|}{x^3}$

29. $y = \dfrac{\ln|x|}{4x + 7}$

30. $y = \dfrac{-2\ln|x|}{3x - 1}$

31. $y = \dfrac{e^x}{\ln|x|}$

32. $y = \dfrac{e^x - 1}{\ln|x|}$

33. $y = \dfrac{e^x + e^{-x}}{x}$

34. $y = \dfrac{e^x - e^{-x}}{x}$

35. $y = \dfrac{5000}{1 + 10e^{0.4x}}$

36. $y = \dfrac{600}{1 - 50e^{0.2x}}$

37. $y = \dfrac{10{,}000}{9 + 4e^{-0.2x}}$

38. $y = \dfrac{500}{12 + 5e^{-0.5x}}$

39. $y = \ln|(\ln|x|)|$

40. $y = \ln 4 \cdot (\ln|3x|)$

APPLIED PROBLEMS

For Exercises 41–45, see Example 5.

Natural Science

41. Suppose that the population of a certain collection of rare Brazilian ants is given by

$$P(t) = 1000e^{0.2t},$$

where t represents the time in days. Find the rate of change of the population when $t = 2$; when $t = 8$. (Hint: Use Table 3 of the Appendix.)

Management

42. Often, sales of a new product grow rapidly and then level off with time. Such a situation can be represented by an equation of the form

$$S(t) = 100 - 90e^{-0.3t},$$

where t represents time in years and $S(t)$ represents sales. Find the rate of change of sales when $t = 1$; when $t = 10$.

43. Suppose $P(x) = e^{-0.02x}$ represents the proportion of shoes manufactured by a given company that are still wearable after x days of use. Find the proportion of shoes wearable after
(a) 1 day;
(b) 10 days;
(c) 100 days.
(d) Calculate and interpret $P'(100)$.

Natural Science

44. Consider an experiment in which equal numbers of male and female insects of a certain species are permitted to intermingle. Assume that

$$M(t) = (e^{0.1t} + 1)\ln\sqrt{t}$$

represents the number of matings observed among the insects in an hour, where t is the temperature in degrees Celsius. (Note: The formula is

an approximation at best and holds only for specific temperature intervals.) Find
(a) $M(15)$;
(b) $M(25)$.
(c) Find the rate of change of $M(t)$ when $t = 15$.

45. The concentration of pollutants, in grams per liter, in the east fork of the Big Weasel River is approximated by

$$P(x) = 0.04e^{-4x},$$

where x is the number of miles downstream from a paper mill that the measurement is taken. Find
(a) $P(0.5)$;
(b) $P(1)$;
(c) $P(2)$.
Find the rate of change of the concentration with respect to distance at
(d) $x = 0.5$;
(e) $x = 1$;
(f) $x = 2$.

Find all relative maximums or relative minimums for the following functions. (See Example 6.)

46. $y = xe^{-x}$

47. $y = -xe^{x}$

48. $y = (x + 4)e^{-2x}$

49. $y = x^2e^{-x}$

50. $y = -x^2e^{x}$

51. $y = e^{x} + e^{-x}$

52. $y = x - \ln|x|$

When a drug is injected into a muscle, the bloodstream concentration increases to some maximum and then declines. According to one mathematical model, the concentration of the drug in the bloodstream, in milliliters per cubic centimeter, is given by

$$C(t) = \frac{k}{b - a}(e^{-at} - e^{-bt}),$$

when a, b, and k are constants. In these exercises, we shall find a value of t (time) that leads to maximum drug concentration.

53. Find $C'(t)$, remembering that $k/(b - a)$ is a constant.

54. Put $C'(t) = 0$, getting

$$\frac{k}{b - a}(-ae^{-at} + be^{-bt}) = 0.$$

Multiply both sides by $(b - a)/k$, then put ae^{-at} on one side of the equals sign and be^{-bt} on the other side.

55. Divide both sides by ae^{-bt}.

56. On one side, you should have e^{-at}/e^{-bt}. Use properties of exponents to show that this equals e^{bt-at}, or $e^{(b-a)t}$.

57. Take the natural logarithm of both sides; you should get

$$\ln e^{(b-a)t} = \ln \frac{b}{a}.$$

By a property of logarithms, we rewrite this as $(b-a)t \cdot \ln e = \ln \frac{b}{a}$. Use the fact that $\ln e = 1$ to simplify this equation.

58. Solve for t.

59. Find t if $b = 50$ and $a = 10$.

60. Find t if $b = 8$ and $a = 4$.

CHAPTER 13 TEST

[13.1] *Find the location of all relative maximums and relative minimums for each of the following functions. Find the value of any of these maximums or minimums.*

1. $f(x) = -x^2 + 4x - 8$

2. $f(x) = -2x^3 - \frac{1}{2}x^2 + x - 3$

3. $f(x) = \dfrac{x-1}{2x+1}$

Find the open intervals where the following functions are increasing or decreasing.

4. $y = x^2 - 5x + 3$

5. $y = -x^3 - 5x^2 + 8x - 6$

[13.2] *Find the location of all absolute maximums and absolute minimums for the following functions defined on the given intervals.*

6. $f(x) = -x^2 + 5x + 1, \quad [1, 4]$

7. $f(x) = x^3 + 2x^2 - 15x + 3, \quad [-4, 2]$

[13.2] and [13.3] *Each of the following functions gives the displacement at time t of a particle moving along a line. Find the velocity and acceleration functions. Then find the velocity and acceleration at time $t = 0$ and at time $t = 3$.*

8. $s(t) = 9t^2 - 7t + 8$

9. $s(t) = -2t^3 + 4t^2 - 6t - 1$

[13.3] *Find $f''(x)$ for each of the following. Then find $f''(0)$ and $f''(-5)$.*

10. $f(x) = 6x^3 - 9x^2 + 2x - 3$

11. $f(x) = \dfrac{5x-1}{2x+3}$

Find $f'''(x)$ and $f^{(4)}(x)$ for each of the following functions.

12. $y = 9x^3 - 8x^2 + 6x - 5$

13. $y = \frac{-2}{x - 4}$

[13.4] *The total profit in dollars from the sale of x hundred boxes of candy is given by*

$$P(x) = -x^3 + 10x^2 - 12x + 106.$$

14. Find the number of boxes of candy that should be sold in order to produce maximum profit.

15. Find the maximum profit.

16. The city park department is planning an enclosed play area in a new park. One side of the area will be against an existing building, with no fence needed there. Find the dimensions of the rectangular field of maximum area that can be made with 900 meters of fence.

[13.5] **17.** A very large camera store sells 320,000 rolls of film annually. It costs 10¢ to store one roll for one year and $10 to place a reorder. Find the number of orders that should be placed annually.

Find dy/dx for the following functions.

[13.6] **18.** $x^2y^3 + 4xy = 2$

19. $\frac{x}{y} - 4y = 3x$

20. Find the equation of the tangent to the graph of $\sqrt{2x} - 4yx = -22$ at the point (2, 3).

[13.7] *Find dy for the following functions.*

21. $y = 8x^3 - 2x^2$

22. $y = 4(x^2 - 1)^3$

23. Use differentials to approximate $\sqrt{166}$.

[13.8] *Find derivatives of the following functions.*

24. $y = e^{6x^2}$

25. $y = \ln|3x - 2|$

26. $y = \ln\sqrt{3 - x}$

27. $y = \frac{e^{2x}}{5x}$

28. $y = e^x \cdot \ln x^2$

29. $y = \frac{6x + 4}{\ln|x|}$

30. Find the location of any relative maximums or relative minimums for $y = (5x + 3)e^{-2x}$.

CASE 13 A TOTAL COST MODEL FOR A TRAINING PROGRAM*

In this case, we set up a mathematical model for determining the total costs in setting up a training program. Then we use calculus to find the time between training programs that produces the minimum total cost. The model assumes that the demand for trainees is constant and that the fixed cost of training a batch of trainees is known. Also, it is assumed that people who are trained, but for whom no job is readily available, will be paid a fixed amount per month while waiting for a job to open up.

The model uses the following variables:

D = demand for trainees per month

N = number of trainees per batch

C_1 = fixed cost of training a batch of trainees

C_2 = variable cost of training per trainee per month

C_3 = salary paid monthly to a trainee who has not yet been given a job after training

m = time interval in months between successive batches of trainees

t = length of training program in months

$Z(m)$ = total monthly cost of program

The total cost of training a batch of trainees is given by $C_1 + NtC_2$. However, $N = mD$, so that the total cost per batch is $C_1 + mDtC_2$.

After training, personnel are given jobs at the rate of D per month. Thus, $N - D$ of the trainees will not get a job the first month, $N - 2D$ will not get a job the second month, and so on. The $N - D$ trainees who do not get a job the first month produce total costs of $(N - D)C_3$, those not getting jobs during the second month produce costs of $(N - 2D)C_3$, and so on. Since $N = mD$, the costs during the first month can be written as

$$(N - D)C_3 = (mD - D)C_3 = (m - 1)DC_3,$$

while the costs during the second month are $(m - 2)DC_3$, and so on. The total cost for keeping the trainees without a job is thus

$$(m - 1)DC_3 + (m - 2)DC_3 + (m - 3)DC_3 + \cdots + 2DC_3 + DC_3,$$

which can be factored to give

$$DC_3[(m - 1) + (m - 2) + (m - 3) + \cdots + 2 + 1].$$

The expression in brackets is the sum of the terms of an arithmetic sequence, discussed in most algebra texts. Using formulas for arithmetic sequences, the

---

* Based on "A Total Cost Model for a Training Program" by P. L. Goyal and S. K. Goyal, Department of Mathematics and Computer Science, The Polytechnic of Wales. Used with permission of the authors.

expression in brackets can be shown to equal $m(m-1)/2$, so that we have

$$DC_3\left[\frac{m(m-1)}{2}\right] \tag{1}$$

as the total cost for keeping jobless trainees.

The total cost per batch is the sum of the training cost per batch, $C_1 + mDtC_2$, and the cost of keeping trainees without a proper job, given by (1). Since we assume that a batch of trainees is trained every m months, the total cost per month, $Z(m)$, is given by

$$Z(m) = \frac{C_1 + mDtC_2}{m} + \frac{DC_3\left[\dfrac{m(m-1)}{2}\right]}{m} = \frac{C_1}{m} + DtC_2 + DC_3\left(\frac{m-1}{2}\right).$$

EXERCISES

1. Find $Z'(m)$.

2. Solve the equation $Z'(m) = 0$.

   As a practical matter, it is usually required that m be a whole number. If m does not come out to be a whole number in Exercise 2, then m^+ and m^-, the two whole numbers closest to m, must be chosen. Calculate both $Z(m^+)$ and $Z(m^-)$; the smaller of the two provides the optimum value of Z.

3. Suppose a company finds that its demand for trainees is 3 per month, that a training program requires 12 months, that the fixed cost of training a batch of trainees is \$15,000, that the variable cost per trainee per month is \$100, and that trainees are paid \$900 per month after training but before going to work. Use your result from Exercise 2 and find m.

4. Since m is not a whole number, find m^+ and m^-.

5. Calculate $Z(m^+)$ and $Z(m^-)$.

6. What is the optimum time interval between successive batches of trainees? How many trainees should be in a batch?

14 Integration

Calculus is divided into two broad areas—differential calculus, which we discussed in Chapters 12 and 13, and integral calculus, which we consider in this chapter. Like the derivative of a function, the definite integral of a function is a special limit with many applications. Geometrically, the definite integral is related to the area under a curve. In Section 14.3 we shall see how differential and integral calculus are related by the fundamental theorem of calculus.

14.1 THE ANTIDERIVATIVE

In Chapter 12 we saw that the marginal cost to produce x items was given by the derivative $f'(x)$ of the cost function $f(x)$. Now suppose we know the marginal cost function and wish to find the cost function. For example, suppose the marginal cost at a level of production of x units is

$$f'(x) = 2x - 10.$$

Can we find the cost function? In other words, can we find a function whose derivative is $2x - 10$?

By trial and error, with a little thought on how the derivative of $f(x)$ is found, we have

$$f(x) = x^2 - 10x,$$

which has $2x - 10$ as its derivative. Is $x^2 - 10x$ the only function with derivative $2x - 10$? No, there are many; for example,

$$f(x) = x^2 - 10x + 2,$$

$$f(x) = x^2 - 10x + 5,$$

$$f(x) = x^2 - 10x - 8,$$

and so on. (Verify that each of these functions has $2x - 10$ as its derivative.) In fact, if we add any real number to $x^2 - 10x$ we have a function whose derivative is $2x - 10$. To express this fact, we write

$$f(x) = x^2 - 10x + C,$$

where C represents any constant. The function $f(x)$ is called an **antiderivative** of $f'(x)$. It can be shown that every antiderivative of $f'(x)$ is of the form $f(x) = x^2 - 10x + C$.

The trial and error method of finding an antiderivative used in the example above is not very satisfactory. We need some rules for finding antiderivatives. First, recall that to take the derivative of x^n we reduce the exponent on x by 1 and multiply by n:

$$\frac{d(x^n)}{dx} = nx^{n-1}.$$

Thus, to antidifferentiate, that is, to undo what was done, we should increase the exponent by 1 and divide by the new exponent $n + 1$. In general, we have the following rule.

Theorem 14.1 If $n \neq -1$, then an antiderivative of $f(x) = x^n$ is

$$F(x) = \frac{1}{n+1}x^{n+1} + C.$$

Note that we are labeling the antiderivative of a function $f(x)$ as $F(x)$. We can check that $F(x)$ is the antiderivative of $f(x)$ by showing that the derivative of $F(x)$ is $f(x)$, or

$$\frac{d}{dx}F(x) = f(x).$$

Here we would have

$$\frac{d}{dx}F(x) = \frac{d}{dx}\left(\frac{1}{n+1}x^{n+1} + C\right) = \frac{n+1}{n+1}x^n + 0 = x^n.$$

Why must we exclude $n = -1$?

Example 1 Find an antiderivative of each of the following.

(a) $f(x) = x^3$

By Theorem 14.1 an antiderivative of $f(x) = x^3$ is

$$F(x) = \frac{1}{3+1}x^{3+1} + C = \frac{1}{4}x^4 + C = \frac{x^4}{4} + C.$$

(b) $f(x) = \sqrt{x}$

First write $\sqrt{x}$ as $x^{1/2}$. Then use Theorem 14.1 to write

$$F(x) = \frac{1}{1/2+1}x^{3/2} + C = \frac{2}{3}x^{3/2} + C.$$

To check this result, show that the derivative of $F(x)$ is $f(x)$.

(c) $f(x) = 1$

To use Theorem 14.1 here, write 1 as x^0, so that the antiderivative of $f(x) = 1 = 1 \cdot x^0$ is given by

$$F(x) = \frac{1}{1}x^1 + C = x + C. \quad \blacksquare$$

1. Find an antiderivative of each of the following.
(a) $f(x) = x^5$
(b) $f(x) = \sqrt[3]{x}$
(c) $f(x) = 5$

Answer:

(a) $\frac{x^6}{6} + C$

(b) $\frac{3x^{4/3}}{4} + C$

(c) $5x + C$

Work Problem 1 at the side

In Chapter 12 we saw that the derivative of the product of a constant and a function is the product of the constant and the derivative of the function. We could expect a similar rule to apply to antidifferentiation and it does. Also, since we differentiate term by term, it seems reasonable to antidifferentiate term by term. The next two theorems state these properties of antiderivatives.

Theorem 14.2 If $F(x)$ is an antiderivative of $f(x)$ and if k is any constant, then the antiderivative of $k \cdot f(x)$ is $k \cdot F(x) + C$, where C is a constant.

Example 2 Find an antiderivative of $f(x) = 2x^3$.

By Theorems 14.1 and 14.2, we have the antiderivative

$$F(x) = 2\left(\frac{1}{4}x^4\right) + C = \frac{x^4}{2} + C. \quad \blacksquare$$

2. Find an antiderivative of each of the following.
(a) $f(x) = -6x^4$
(b) $f(x) = 9x^{2/3}$
(c) $f(x) = 8/x^3$

Answer:

(a) $\frac{-6x^5}{5} + C$

(b) $\frac{27x^{5/3}}{5} + C$

(c) $\frac{-4}{x^2} + C$

Work Problem 2 at the side.

Theorem 14.3 If $F(x)$ is an antiderivative of $f(x)$ and $G(x)$ is an antiderivative of $g(x)$, then an antiderivative of $f(x) \pm g(x)$ is $F(x) \pm G(x) + C$.

Example 3 Find an antiderivative of $f(x) = 3x^2 - 4x + 5$.

Using all the theorems of this section, we have

$$F(x) = 3\left(\frac{1}{3}x^3\right) - 4\left(\frac{1}{2}x^2\right) + 5x + C$$
$$= x^3 - 2x^2 + 5x + C,$$

as an antiderivative of $f(x) = 3x^2 - 4x + 5$. $\blacksquare$

3. Find an antiderivative of each of the following.
(a) $f(x) = 5x^4 - 3x^2 + 6$
(b) $f(x) = -2x^3 + 6x^2 - 3$
(c) $f(x) = 3\sqrt{x} + 2/x^2$

Answer:

(a) $x^5 - x^3 + 6x + C$

(b) $\frac{-x^4}{2} + 2x^3 - 3x + C$

(c) $2x^{3/2} - \frac{2}{x} + C$

Work Problem 3 at the side.

In Chapter 13, we saw that the derivative of $f(x) = e^x$ was $f'(x) = e^x$. Thus, we have the following theorem.

Theorem 14.4 An antiderivative of $f(x) = e^x$ is

$$\mathbf{F(x) = e^x + C.}$$

Example 4 Find an antiderivative of $f(x) = 5e^x$.
By Theorems 14.2 and 14.4,

$$F(x) = 5e^x + C. \quad \blacksquare$$

The restriction $n \neq -1$ in Theorem 14.1 was necessary because $n = -1$ makes the denominator of the antiderivative 0. Recall from Chapter 13 that the derivative of $f(x) = \ln|x|$, where $x \neq 0$, is $f'(x) = 1/x = x^{-1}$. From that, we have the next theorem.

Theorem 14.5 An antiderivative of $f(x) = \frac{1}{x} = x^{-1}$ is

$$\mathbf{F(x) = \ln|x| + C.}$$

The domain of the logarithmic function is the set of positive numbers. However, in the function $1/x$, x can be any nonzero real number. Thus, we must use absolute value in the antiderivative.

Example 5 Find an antiderivative of $f(x) = \frac{4}{x}$.

We use Theorems 14.2 and 14.5.

$$F(x) = 4\ln|x| + C. \quad \blacksquare$$

4. Find an antiderivative of each of the following.
(a) $f(x) = -5e^x$
(b) $f(x) = -6/x$
(c) $f(x) = 3e^x - 2x^{-1}$

Answer:
(a) $-5e^x + C$
(b) $-6\ln|x| + C$
(c) $3e^x - 2\ln|x| + C$

Work Problem 4 at the side.

We know that a marginal function is the derivative of some function $F(x)$. The antiderivative can be used to find $F(x)$ when the marginal function $F'(x)$ is known.

Example 6 Suppose a company has found that the marginal cost at a level of production of x thousand books is given by

$$C'(x) = \frac{50}{\sqrt{x}}$$

and the fixed cost (the cost to produce 0 books) is \$25,000. Find the cost function $C(x)$.

Writing $50/\sqrt{x}$ as $50x^{-1/2}$, we find that an antiderivative of $C'(x)$ is

$$C(x) = 50(2x^{1/2}) + k = 100x^{1/2} + k,$$

where k is a constant. We can find k by using the fact that $C(x)$ is 25,000 when $x = 0$.

$$\begin{aligned} C(x) &= 100x^{1/2} + k \\ 25{,}000 &= 100 \cdot 0 + k \\ k &= 25{,}000 \end{aligned}$$

Therefore, the cost function is $C(x) = 100x^{1/2} + 25{,}000$. ■

5. The marginal cost at a level of production of x items is

$$C'(x) = 2x^3 + 6x - 5.$$

The fixed cost is \$800. Find the cost function $C(x)$.

Answer:
$C(x) = \frac{1}{2}x^4 + 3x^2 - 5x + 800$

Work Problem 5 at the side.

Example 7 If the function $s(t)$ gives the position of a particle at time t, then the velocity of the particle, $v(t)$, and its acceleration, $a(t)$, are given by

$$\begin{aligned} v(t) &= s'(t) \\ a(t) &= v'(t) = s''(t). \end{aligned}$$

(a) Suppose the velocity of an object is $v(t) = 6t^2 - 8t$, with a position of -5 when time is 0. Find $s(t)$.

Since $v(t) = s'(t)$, we find $s(t)$ by identifying the antiderivative of $v(t)$:

$$v(t) = 6t^2 - 8t$$

so that

$$s(t) = 2t^3 - 4t^2 + C$$

for some constant C. We are told that $s = -5$ when $t = 0$.

$$\begin{aligned} s(t) &= 2t^3 - 4t^2 + C \\ -5 &= 2(0)^3 - 4(0)^2 + C \\ -5 &= C \end{aligned}$$

Finally,

$$s(t) = 2t^3 - 4t^2 - 5.$$

(b) Many experiments have shown that when an object is dropped, its acceleration (ignoring air resistance) is constant. This constant has been found to be approximately 32 feet per second every second, or

$$a(t) = -32.$$

The negative sign shows that the object is falling. Suppose an object is dropped so that its initial velocity is -20 feet per second. Find $v(t)$ by taking the antiderivative of $a(t)$:

$$v(t) = -32t + k.$$

When $t = 0$, $v(t) = -20$:

$$-20 = -32(0) + k$$
$$-20 = k$$

and

$$v(t) = -32t - 20.$$

Also, suppose the object was dropped from the top of the 1100-foot tall Sears Tower in Chicago. Then $s(0) = 1100$. Take the antiderivative of $v(t)$ to find $s(t)$.

$$s(t) = -16t^2 - 20t + C$$

We are told that $s(t) = 1100$ when $t = 0$. By substituting these values into our equation for $s(t)$, we end up with

$$s(t) = -16t^2 - 20t + 1100$$

as the distance of the object from the top of the building after t seconds. ■

Work Problem 6 at the side.

6. In Example 7, find the distance of the object from the top of the building after the following times.
(a) 1 second
(b) 4 seconds
(c) 6 seconds
(d) Use the quadratic formula to find the approximate time when the object will hit the ground.

Answer:
(a) 1064 feet
(b) 764 feet
(c) 404 feet
(d) about 7.7 seconds

In later sections, we will discuss applications of antiderivatives in more detail.

14.1 EXERCISES

Find an antiderivative of each of the following. (See Examples 1–5.)

1. $f(x) = 5x^2$

2. $f(x) = 6x^3$

3. $f(x) = 6$

4. $f(x) = 2$

5. $f(x) = 2x + 3$

6. $f(x) = 3x - 5$

7. $f(x) = x^2 - 4x + 5$

8. $f(x) = 5x^2 - 6x + 3$

9. $f(x) = 4x^3 + 3x^2 + 2x - 6$

10. $f(x) = 12x^3 + 6x^2 - 8x + 5$

11. $f(x) = \sqrt{x}$

12. $f(x) = x^{1/3}$

13. $f(x) = x^{1/2} + x^{3/2}$

14. $f(x) = 4\sqrt{x} - 3x^{3/2}$

15. $f(x) = 10x^{3/2} - 14x^{5/2}$

16. $f(x) = 56x^{5/2} + 18x^{7/2}$

17. $f(x) = 9x^{1/2} + 4$

18. $f(x) = -18x^{1/2} - 6$

19. $f(x) = 1/x^2$

20. $f(x) = 4/x^3$

21. $f(x) = 1/x^3 - 1/\sqrt{x}$

22. $f(x) = \sqrt{x} + 1/x^2$

23. $f(x) = -4e^x$

24. $f(x) = -2e^x$

25. $f(x) = 30e^x + 5x^2$

26. $f(x) = 25e^x - 10x^3$

27. $f(x) = 2x^{-1}$

28. $f(x) = -4x^{-1}$

29. $f(x) = \dfrac{3}{x} + e^x$

30. $f(x) = \dfrac{-2}{x} - 3e^x$

Find the cost function for each of the following marginal cost functions. (*See Example 6.*)

31. $C'(x) = 4x - 5$, fixed cost is \$8

32. $C'(x) = 2x + 3x^2$, fixed cost is \$15

33. $C'(x) = 0.2x^2$, fixed cost is \$10

34. $C'(x) = x^2 + 3x$, fixed cost is \$5

35. $C'(x) = \sqrt{x}$, 16 units cost \$40

36. $C'(x) = x^{2/3} + 2$, 8 units cost \$58

37. $C'(x) = x^2 - 2x + 3$, 3 units cost \$15

38. $C'(x) = x + 1/x^2$, 2 units cost \$5.50

APPLIED PROBLEMS

Management

39. The marginal profit of Henrietta's Hamburgers is

$$P'(x) = -2x + 20,$$

where x is the sales volume in thousands of hamburgers. Henrietta knows that her profit is -50 dollars when she sells no hamburgers. What is her profit function?

40. Suppose the marginal profit on x hundred items is

$$P'(x) = 4 - 6x + 3x^2,$$

and the profit on 0 items is -40 dollars. Find the profit function.

General

41. The slope of the tangent line to a curve is given by

$$f'(x) = 6x^2 - 4x + 3.$$

If the point (0, 1) is on the curve, find the equation of the curve.

42. Find the equation of a curve whose tangent line has a slope of

$$f'(x) = x^{2/3},$$

if the point (1, 3/5) is on the curve.

Natural Science

43. For a particular object, $a(t) = t^2 + 1$ and $v(0) = 6$. Find $v(t)$. See Example 7.

44. Suppose $v(t) = 6t^2 - 2/t^2$ and $s(1) = 8$. Find $s(t)$.

45. Repeat the analysis of Example 7(b). Assume that an object is dropped from a small plane flying at 6400 feet, and assume that $v(0)$ is 0. Find $s(t)$. How long will it take the object to hit the ground?

46. According to Fick's Law, the diffusion of a solute across a cell membrane is given by

$$c'(t) = \frac{kA}{V}(C - c(t)), \qquad \textbf{(1)}$$

where A is the area of the cell membrane, V is the volume of the cell, $c(t)$ is the concentration inside the cell at time t, C is the concentration outside the cell, and k is a constant. If c_0 represents the concentration of the solute inside the cell when $t = 0$, then it can be shown that

$$c(t) = (c_0 - C)e^{-kAt/V} + C. \qquad \textbf{(2)}$$

(a) Use this last result to find $c'(t)$.
(b) Substitute back into equation (1) to show that (2) is indeed the correct antiderivative of (1).

14.2 AREA AND THE DEFINITE INTEGRAL

In this section, we consider a method for finding the area of a region bounded on one side by a curve. For example, Figure 14.1 shows the region between the lines $x = 1$, $x = 3$, the x-axis, and the graph of the function defined by

$$f(x) = x^2 + 1.$$

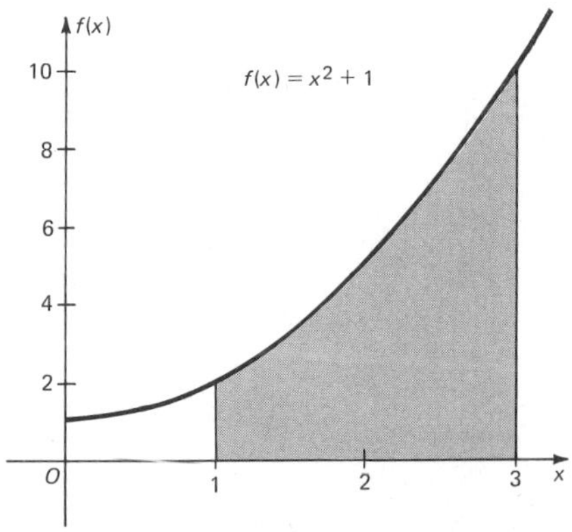

Figure 14.1

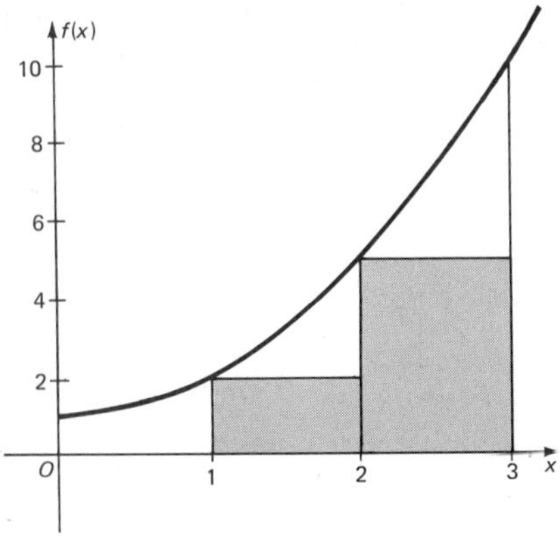

Figure 14.2

To get an approximation of this area, we could use two rectangles as shown in Figure 14.2. We use $f(1) = 2$ as the height of the rectangle on the left and $f(2) = 5$ as the height of the rectangle on the right. The width of each rectangle is 1; thus the total area of the two rectangles is

$$1 \cdot f(1) + 1 \cdot f(2) = 2 + 5 = 7 \text{ square units.}$$

As shown in Figure 14.2 this approximation is smaller than the actual area. To improve the accuracy of the approximation, we can divide the interval from 1 to 3 into four equal parts of width 1/2, as shown in Figure 14.3. The total area of the four rectangles is

$$\begin{aligned}\frac{1}{2} \cdot f(1) + \frac{1}{2} \cdot f\left(1\frac{1}{2}\right) + \frac{1}{2} \cdot f(2) + \frac{1}{2} \cdot f\left(2\frac{1}{2}\right) &\\ = \frac{1}{2}(2) + \frac{1}{2}\left(\frac{13}{4}\right) + \frac{1}{2}(5) + \frac{1}{2}\left(\frac{29}{4}\right) &\\ = 1 + \frac{13}{8} + \frac{5}{2} + \frac{29}{8} &\\ = 8.75 \text{ square units.} &\end{aligned}$$

This approximation is better, but still less than the actual area we seek. To improve the approximation, we could divide the interval from 1 to 3 into 8 parts with equal widths of $\frac{1}{4}$. (See Figure 14.4.) The total area of the rectangles formed would then be

$$\begin{aligned}&\frac{1}{4} \cdot f(1) + \frac{1}{4} \cdot f\left(1\frac{1}{4}\right) + \frac{1}{4} \cdot f\left(1\frac{1}{2}\right) + \frac{1}{4} \cdot f\left(1\frac{3}{4}\right)\\ &\quad + \frac{1}{4} \cdot f(2) + \frac{1}{4} \cdot f\left(2\frac{1}{4}\right) + \frac{1}{4} \cdot f\left(2\frac{1}{2}\right) + \frac{1}{4} \cdot f\left(2\frac{3}{4}\right)\\ &\qquad = 9.69 \text{ square units.}\end{aligned}$$

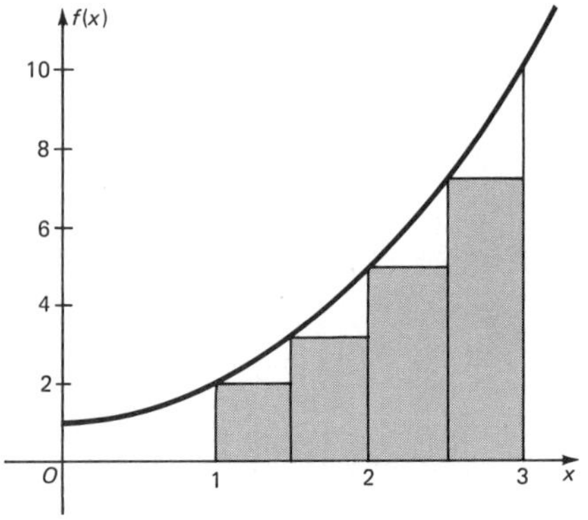

Figure 14.3

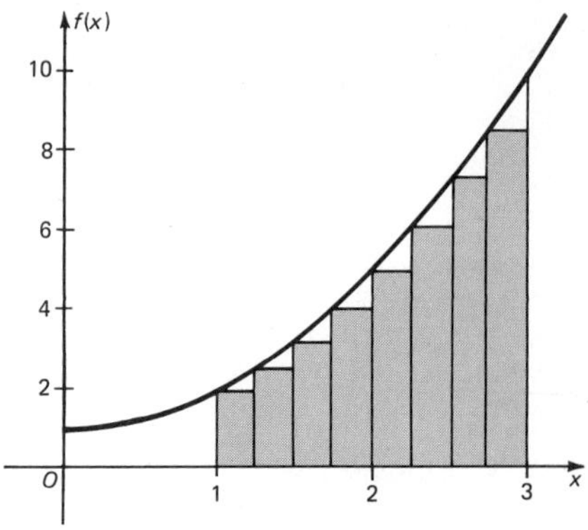

Figure 14.4

The process we have been using here, approximating the area under a curve by using more and more rectangles to get a better and better approximation, can be generalized. To do this, we divide the interval from 1 to 3 into n equal parts. Each of these n intervals will have width

$$\frac{3-1}{n} = \frac{2}{n}.$$

The heights and areas of the n rectangles are shown in the following chart.

Rectangle	Height	Area
1	$f(1)$	$\frac{2}{n} \cdot f(1)$
2	$f\left(1 + \frac{2}{n}\right)$	$\frac{2}{n} \cdot f\left(1 + \frac{2}{n}\right)$
3	$f\left(1 + 2 \cdot \frac{2}{n}\right)$	$\frac{2}{n} \cdot f\left(1 + 2 \cdot \frac{2}{n}\right)$
4	$f\left(1 + 3 \cdot \frac{2}{n}\right)$	$\frac{2}{n} \cdot f\left(1 + 3 \cdot \frac{2}{n}\right)$
$\vdots$	$\vdots$	$\vdots$
n	$f\left(1 + (n-1)\frac{2}{n}\right)$	$\frac{2}{n} \cdot f\left(1 + (n-1)\frac{2}{n}\right)$

The total of the areas listed in the last column approximates the area under the curve, above the x-axis, and between the lines $x = 1$

and $x = 3$. As n becomes larger and larger the approximation is better and better.

The simplification of the expression for the total area in the rectangles is not easy, and fortunately it is not necessary for us to go through all the steps. It is sufficient to say that the sum of the areas of the n rectangles simplifies to

$$\frac{32}{3} - \frac{8}{n} + \frac{4}{3n^2}.$$

Since n is the number of rectangles, this result can be used to evaluate the sum of the areas of any number of rectangles. For example, if $n = 20$, the total area of the rectangles is

$$\frac{32}{3} - \frac{8}{20} + \frac{4}{3(20)^2} = 10.27 \text{ square units,}$$

a result very close to the actual area under the curve which we are trying to find.

1. Find the value of the sum of the areas of 100 rectangles, using the expression given above.

Answer: 10.59 square units

Work Problem 1 at the side.

If we find the limit of the expression

$$\frac{32}{3} - \frac{8}{n} + \frac{4}{3n^2}$$

as the number of rectangles is increased without bound, that is, as $n \to \infty$, we have

$$\lim_{n \to \infty} \left(\frac{32}{3} - \frac{8}{n} + \frac{4}{3n^2} \right) = \frac{32}{3} = 10\frac{2}{3}.$$

This limit is defined to be the area of the region under the graph of $f(x) = x^2 + 1$ and above the x-axis, between the lines $x = 1$ and $x = 3$. This number is called the **definite integral** of $f(x) = x^2 + 1$ from $x = 1$ to $x = 3$, and is written

$$\int_1^3 (x^2 + 1)\, dx = \frac{32}{3}.$$

The symbol $\int$ is called an **integral sign,** 3 is called the **upper limit** of integration, and 1 the **lower limit** of integration.

We can generalize from this example. Figure 14.5 shows the area bounded by the curve $y = f(x)$, the x-axis, and the vertical lines $x = a$ and $x = b$. This area is written

$$\int_a^b f(x)\, dx.$$

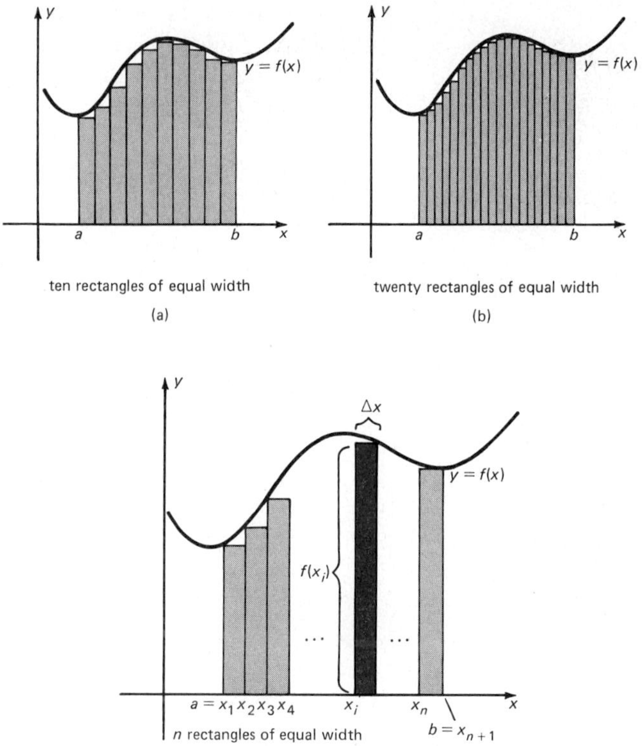

ten rectangles of equal width

(a)

twenty rectangles of equal width

(b)

n rectangles of equal width

(c)

Figure 14.5

To approximate this area, we can divide the area under the curve first into ten rectangles (Figure 14.5(a)) and then into 20 rectangles (Figure 14.5(b)). The sum of the areas of the rectangles gives our approximation to the area under the curve.

To get a number which we can define as the *exact* area, we begin by dividing the interval from a to b into n pieces of equal width and use each of these n pieces as the base of a rectangle. (See Figure 14.5(c).) The sum of the areas of the rectangles gives an approximation to the desired area.

The left endpoints of the n intervals are labeled $x_1, x_2, x_3, \ldots, x_{n+1}$, where $a = x_1$ and $b = x_{n+1}$. In the graph of Figure 14.5(c), the symbol Δx is used to represent the width of each of the intervals. The rectangle in the center is an arbitrary rectangle called the ith rectangle. Its area is given by the product of its length and width. Its width is Δx, and its height is $f(x_i)$:

$$\text{area of } i\text{th rectangle} = f(x_i) \cdot \Delta x.$$

The total area under the curve can be approximated by adding the areas of all n of the rectangles. To represent addition, or summation, the Greek letter sigma, $\sum$, is used. With this symbol, the approximation to the total area becomes

$$\text{area of all } n \text{ rectangles} = \sum_{i=1}^{n} f(x_i) \cdot \Delta x.^*$$

The number that is defined to be the *exact area* is the limit of this sum (if the limit exists) as the number of rectangles increases without bound, or

$$\int_a^b f(x)\,dx = \lim_{n\to\infty} \sum_{i=1}^{n} f(x_i) \cdot \Delta x.$$

The symbol on the left, as we have seen, is the **definite integral** of the function $f(x)$ between the **limits of integration** a and b.

Example 1 Find the following sums.

(a) $\sum_{i=1}^{5} (6i + 2)$

Replace i, in turn, with 1, 2, 3, 4, and 5.

$$\begin{aligned}\sum_{i=1}^{5} (6i+2) &= (6 \cdot 1 + 2) + (6 \cdot 2 + 2) + (6 \cdot 3 + 2) \\ &\quad + (6 \cdot 4 + 2) + (6 \cdot 5 + 2) \\ &= 8 + 14 + 20 + 26 + 32 \\ &= 100\end{aligned}$$

(b) $$\begin{aligned}\sum_{i=1}^{4} 3^i &= 3^1 + 3^2 + 3^3 + 3^4 \\ &= 3 + 9 + 27 + 81 \\ &= 120 \quad ■\end{aligned}$$

Work Problem 2 at the side.

In the next section we shall see how the antiderivative is used in finding the definite integral and thus the area under a curve.

2. Find each sum.

(a) $\sum_{i=1}^{6} (3i - 5)$

(b) $\sum_{i=1}^{8} (-11i + 50)$

(c) $\sum_{i=1}^{4} (2i - 1)(3i + 2)$

Answer:
(a) 33
(b) 4
(c) 182

---

* A computer program for approximating area in this way is given in Margaret L. Lial, *Study Guide with Computer Problems* (Glenview, Ill.: Scott, Foresman, 1979), pp. 75–77.

14.2 EXERCISES

In the following exercises, first approximate the area under the given curve and above the x-axis by using two rectangles. Let the height of the rectangle be given by the value of the function at the left side of the rectangle. Then repeat the process and approximate the area using four rectangles.

1. $f(x) = 3x + 2$ from $x = 1$ to $x = 5$

2. $f(x) = -2x + 1$ from $x = -4$ to $x = 0$

3. $f(x) = x + 5$ from $x = 2$ to $x = 4$

4. $f(x) = 3 + x$ from $x = 1$ to $x = 3$

5. $f(x) = x^2$ from $x = 1$ to $x = 5$

6. $f(x) = x^2$ from $x = 0$ to $x = 4$

7. $f(x) = x^2 + 2$ from $x = -2$ to $x = 2$

8. $f(x) = -x^2 + 4$ from $x = -2$ to $x = 2$

9. $f(x) = e^x - 1$ from $x = 0$ to $x = 4$

10. $f(x) = e^x + 1$ from $x = -2$ to $x = 2$

11. $f(x) = \dfrac{1}{x}$ from $x = 1$ to $x = 5$

12. $f(x) = \dfrac{2}{x}$ from $x = 1$ to $x = 9$

13. Let $f(x) = x^2$. If the region above the x-axis between $x = 1$ and $x = 4$ is divided into n rectangles, as in the text, and the sum of the areas of the n rectangles evaluated, the result simplifies to

$$21 - \frac{45}{2n} + \frac{9}{2n^2}.$$

(a) Approximate the area of the region described above using four rectangles.
(b) Approximate the area of the region using eight rectangles.
(c) Find $\int_1^4 x^2\,dx$. This definite integral gives the exact area.

14. Let $f(x) = 4 - \frac{1}{4}x^2$. If the region from $x = 0$ to $x = 3$ is divided into n rectangles, the sum of the areas of the n rectangles is

$$\frac{39}{4} - \frac{27}{8n} - \frac{9}{8n^2}. \qquad \textbf{(*)}$$

Find $\int_0^3 (4 - \frac{1}{4}x^2)\,dx$ by taking the limit of (*) as $n \to \infty$. This gives the exact area.

In the following exercises, estimate the area under the curve by counting squares.

15. The graph *on the following page* shows the rate of sales of new cars in a recent year. Estimate the total sales during that year.

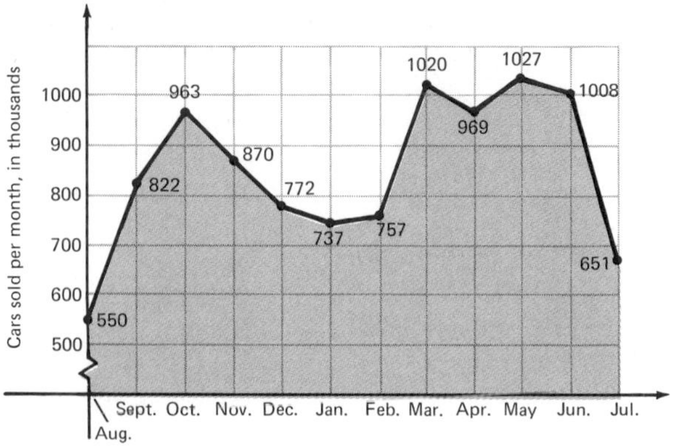

16. The graph *below* shows the rate of usage of electrical energy (in millions of kilowatt hours) in a certain city on a very hot day. Estimate the total usage of electricity on that day.

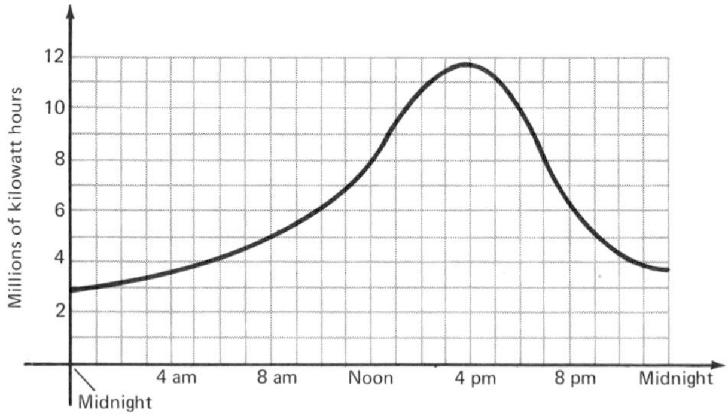

17. The graph *below* shows the approximate concentration of alcohol in a person's bloodstream t hours after drinking 2 ounces of alcohol. Estimate the total amount of alcohol in the bloodstream by estimating the area under the curve.

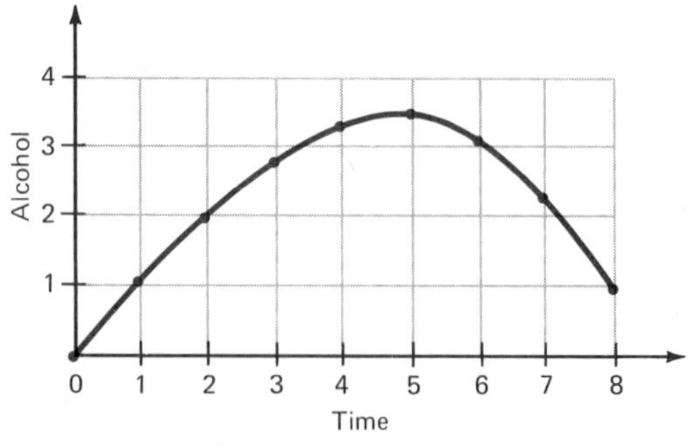

18. The graph *on the following page* shows the rate of inhalation of oxygen by a person riding a bicycle very rapidly for 10 minutes. Estimate the total volume of oxygen inhaled in the first 20 minutes after the beginning of the ride.

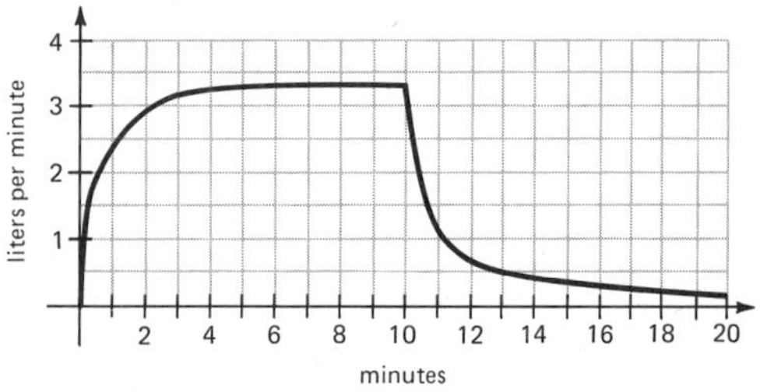

The next two graphs are from Road and Track *magazine.* The curve shows the velocity at time t, in seconds, when the car accelerates from a dead stop. To find the total distance traveled by the car in reaching* 100 *miles per hour, we must estimate the definite integral*

$$\int_0^t v(t)\,dt,$$

where t represents the number of seconds it takes for the car to reach 100 mph.

Use the graphs below to estimate this distance. Each square on the graphs represents 10 *miles on the side and* 5 *seconds across the bottom. Estimate the total number of squares in the desired area and multiply this number by* $5 \times 10 = 50$. *To adjust for the different units, divide this result by* 3600 (*the number of seconds in an hour*). *You then have the number of miles that the car traveled in reaching* 100 mph. *Finally, multiply by* 5280 *feet per mile to convert the answers to feet.*

19. Estimate the distance traveled by the Porsche 928, using the graph below.

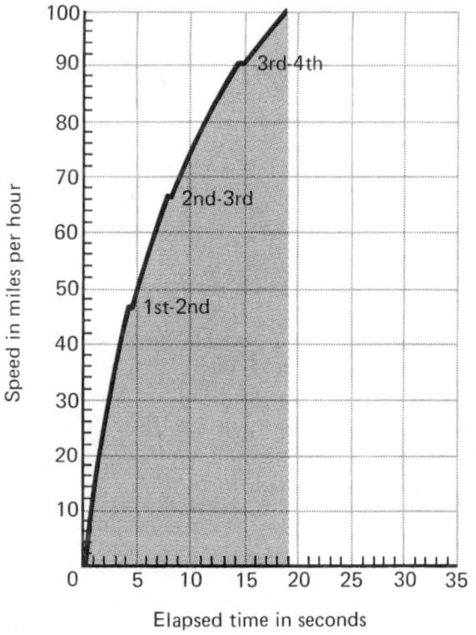

* Two graphs showing acceleration of the Porsche 928 and BMW 733i, from *Road and Track*, April and May 1978.

20. Estimate the distance traveled by the BMW 733i using the graph *below*.

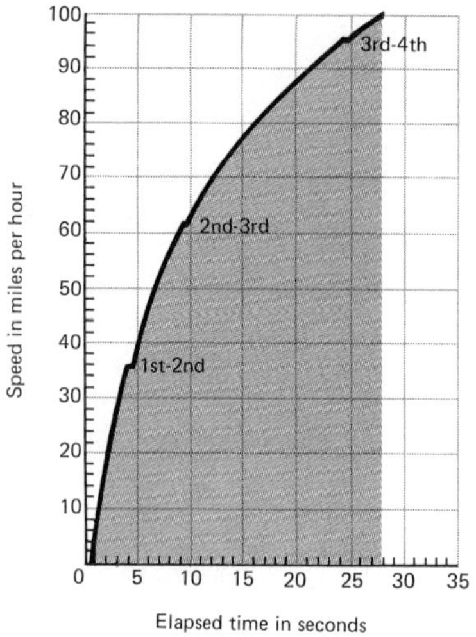

Evaluate the following sums. (See Example 1.)

21. $\sum_{i=1}^{4} 3i$

22. $\sum_{i=1}^{6} -5i$

23. $\sum_{i=1}^{5} (2i + 7)$

24. $\sum_{i=1}^{10} (5i - 8)$

25. $\sum_{i=1}^{7} (2i + 1)(3i - 9)$

26. $\sum_{i=1}^{8} (4i - 9)(5i - 12)$

27. Let $x_1 = -5$, $x_2 = 8$, $x_3 = 7$, and $x_4 = 10$. Find $\sum_{i=1}^{4} x_i$.

28. Let $x_1 = 10$, $x_2 = 15$, $x_3 = -8$, $x_4 = -12$, and $x_5 = 0$. Find $\sum_{i=1}^{5} x_i$.

14.3 THE FUNDAMENTAL THEOREM OF CALCULUS

In the previous section we found that

$$\int_1^3 (x^2 + 1)\,dx = \frac{32}{3}.$$

This definite integral was found by a long calculation using rectangles and a limit. In this section we develop a simpler procedure for finding this number. An antiderivative of $f(x) = x^2 + 1$ is $F(x) = \frac{1}{3}x^3 + x + C$. If we calculate $F(3) - F(1)$ we get

$$\begin{aligned} F(3) - F(1) &= \left(\frac{27}{3} + 3 + C\right) - \left(\frac{1}{3} + 1 + C\right) \\ &= \frac{32}{3}, \end{aligned}$$

which is the value of the definite integral

$$\int_1^3 (x^2 + 1)\,dx,$$

which we found in the last section. This amazing coincidence is expressed as the fundamental theorem of calculus.

Theorem 14.6 *Fundamental theorem of calculus* Let $f(x)$ be the derivative of $F(x)$ and let $f(x)$ be continuous in the interval $[a, b]$. Then

$$\int_a^b f(x)\,dx = F(b) - F(a).$$

To represent $F(b) - F(a)$ we often use

$$F(x)\Big]_a^b.$$

The fundamental theorem of calculus is really quite remarkable. It deserves its name which sets it apart as the most important theorem of calculus. It is the key connection between the differential calculus and the integral calculus, which originally were developed separately without knowledge of this connection between them.

Because of this relationship between the definite integral and the antiderivative given in the fundamental theorem, it is customary to call an antiderivative an indefinite integral (or sometimes just an integral). Indefinite integrals are indicated with the $\int$ symbol, but with no limits of integration. For example, $F(x) = x^3 - 4x^2 + 5x$ is an antiderivative of $f(x) = 3x^2 - 8x + 5$. This fact can be expressed by writing

$$\int (3x^2 - 8x + 5)\,dx = x^3 - 4x^2 + 5x + C.$$

This process of finding an antiderivative is called **integration;** we integrate $3x^2 - 8x + 5$ to obtain $x^3 - 4x^2 + 5x + C$.

Example 1 First find $\int 4x^3\,dx$ and then find $\int_1^2 4x^3\,dx$.

By the theorems in Section 14.1, we have

$$\int 4x^3\,dx = x^4 + C.$$

Thus, by the fundamental theorem, we have

$$\int_1^2 4x^3\,dx = x^4 + C\Big]_1^2$$
$$= (16 + C) - (1 + C)$$
$$= 15. \quad \blacksquare$$

The number 15 represents the area under the curve $f(x) = 4x^3$ that is above the x-axis and between the lines $x = 1$ and $x = 2$. Since the constant C does not appear in the final answer, it can be omitted when evaluating the definite integral. It must be included, however, when finding an indefinite integral.

1. Find the following.

(a) $\int (2x^3 + 4x)\,dx$

(b) $\int_{-1}^{2} (2x^3 + 4x)\,dx$

Answer:

(a) $\frac{x^4}{2} + 2x^2 + C$

(b) $\frac{27}{2}$

Work Problem 1 at the side.

Theorems 14.2 and 14.3 given in Section 14.1 also apply to the definite integral and so are restated here.

Theorem 14.7 For any real constant k,

$$\int_a^b k \cdot f(x)\,dx = k \cdot \int_a^b f(x)\,dx.$$

Theorem 14.8 If all indicated definite integrals exist, then

$$\int_a^b [f(x) \pm g(x)]\,dx = \int_a^b f(x)\,dx \pm \int_a^b g(x)\,dx.$$

2. Find the following.

(a) $\int_1^4 5x^4\,dx$

(b) $\int_0^6 -8x^3\,dx$

(c) $\int_{-2}^{2} x^2\,dx$

Answer:

(a) 1023

(b) -2592

(c) 16/3

Example 2 Evaluate

$$\int_1^4 2x^3\,dx.$$

Using Theorems 14.1 and 14.7 and the fundamental theorem of calculus, we have

$$\int_1^4 2x^3\,dx = 2\int_1^4 x^3\,dx = 2\left(\frac{x^4}{4}\right)\Big]_1^4 = 2\left(\frac{256}{4} - \frac{1}{4}\right) = \frac{255}{2}. \quad \blacksquare$$

Work Problem 2 at the side.

Example 3 Evaluate

$$\int_1^4 (\sqrt{x} + 2x^{3/2})\,dx.$$

By using the theorems of this section and Section 14.1 we have

$$\begin{aligned}
\int_1^4 (\sqrt{x} + 2x^{3/2})\,dx &= \int_1^4 x^{1/2}\,dx + \int_1^4 2x^{3/2}\,dx \\
&= \int_1^4 x^{1/2}\,dx + 2\int_1^4 x^{3/2}\,dx \\
&= \frac{x^{3/2}}{3/2}\Bigg]_1^4 + 2\left(\frac{x^{5/2}}{5/2}\Bigg]_1^4\right) \\
&= \frac{2}{3}\left(x^{3/2}\Big]_1^4\right) + \frac{2\cdot 2}{5}\left(x^{5/2}\Big]_1^4\right) \\
&= \frac{2}{3}(4^{3/2} - 1^{3/2}) + \frac{4}{5}(4^{5/2} - 1^{5/2}) \\
&= \frac{2}{3}(8 - 1) + \frac{4}{5}(32 - 1) \\
&= \frac{14}{3} + \frac{124}{5} \\
&= \frac{442}{15}. \quad \blacksquare
\end{aligned}$$

Work Problem 3 at the side.

Example 4

$$\begin{aligned}
\int_1^2 \frac{dx}{x} &= \ln|x|\Big]_1^2 \\
&= \ln|2| - \ln|1| \\
&= \ln 2 - \ln 1 \\
&= .6931 - 0 \\
&= .6931 \quad \blacksquare
\end{aligned}$$

Work Problem 4 at the side.

Example 5 Find the area of the region between the x-axis and the graph of $f(x) = \frac{1}{2}x^2 + 3x$ from $x = 1$ to $x = 2$.

The area will be given by the definite integral,

$$\int_1^2 \left(\frac{1}{2}x^2 + 3x\right)dx.$$

3. Find the following.

(a) $\int_1^8 x^{2/3}\,dx$

(b) $\int_4^9 (x^{1/2} - 4x^{3/2})\,dx$

Answer:

(a) $\frac{93}{5}$

(b) $\frac{-4874}{15}$

4. Find the following.

(a) $\int_0^4 e^x\,dx$

(b) $\int_3^5 \frac{dx}{x}$

(c) $\int_1^6 e^x\,dx$

(d) $\int_2^8 \frac{4}{x}\,dx$

Answer:

(a) 53.59815

(b) .5108

(c) 400.71051

(d) 5.5452

Evaluating this integral, we have

$$\int_1^2 \left(\frac{1}{2}x^2 + 3x\right) dx = \frac{x^3}{6} + \frac{3x^2}{2}\Bigg]_1^2$$
$$= \left(\frac{8}{6} + 6\right) - \left(\frac{1}{6} + \frac{3}{2}\right)$$
$$= \frac{17}{3}.$$

Thus, the required area is 17/3 square units. ■

5. Find the area of the region between the x-axis and the graph of $f(x) = x^2 + 4$ from $x = 0$ to $x = 3$.

Answer:
21 square units

Work Problem 5 at the side.

14.3 EXERCISES

Evaluate the following definite integrals. (*See Examples 1–4.*)

1. $\int_1^6 x\,dx$
2. $\int_1^5 x\,dx$
3. $\int_0^4 (4x + 3)\,dx$
4. $\int_5^8 (6x - 5)\,dx$
5. $\int_4^8 6\,dx$
6. $\int_3^9 2\,dx$
7. $\int_8^{10} (2x + 3)\,dx$
8. $\int_5^{10} (4x - 7)\,dx$
9. $\int_{-1}^1 (x + 6)\,dx$
10. $\int_1^4 (8x + 3)\,dx$
11. $\int_1^5 3x^2\,dx$
12. $\int_0^3 -9x^2\,dx$
13. $\int_1^3 (6x^2 - 4x + 5)\,dx$
14. $\int_1^4 (15x^2 - 6x + 3)\,dx$
15. $\int_1^4 (3x^2 - 4x + 1)\,dx$
16. $\int_0^3 (3x^2 - 2x + 5)\,dx$
17. $\int_4^9 (10x^{3/2} - 6x^{1/2})\,dx$
18. $\int_4^9 (5x^{3/2} + 3x^{1/2})\,dx$
19. $\int_1^4 \sqrt{x}\,dx$
20. $\int_4^9 \sqrt{x}\,dx$
21. $\int_2^5 \frac{3}{x}\,dx$
22. $\int_1^4 \frac{2}{x}\,dx$
23. $\int_1^2 e^x\,dx$
24. $\int_0^5 e^x\,dx$
25. $\int_1^3 (e^x + x)\,dx$
26. $\int_0^4 (x^2 - e^x)\,dx$
27. $\int_1^4 (x^{-2} + x^{-1})\,dx$
28. $\int_2^5 (3x^{-1} - 2x^{-2})\,dx$

29. Let $f(x) = (x^2 + 2x)^2$. Verify that $f'(x) = 2(x^2 + 2x)(2x + 2) = 4(x + 1)(x^2 + 2x)$, and then find

$$\int_2^3 4(x + 1)(x^2 + 2x)\,dx.$$

30. Let $f(x) = \sqrt{x^2 - 4}$. Verify that $f'(x) = x/\sqrt{x^2 - 4}$, and then find

$$\int_3^4 \frac{x\,dx}{\sqrt{x^2 - 4}}.$$

For Exercises 31–36, see Example 5.

31. Find the area of the region enclosed by the x-axis, the graph of $f(x) = x^2 + 4x + 4$, $x = -1$, and $x = 2$.

32. Find the area of the region enclosed by the x-axis, $y = x$, $x = 0$, and $x = 4$.

33. Find the area of the region enclosed by the x-axis, $y = x^3$, $x = 0$, and $x = 3$.

34. Find the area of the region enclosed by the x-axis, $y = x^2 + x$, $x = 1$, and $x = 4$.

35. Find the area of the region enclosed by the x-axis, $y = \frac{1}{x}$, $x = 1$, and $x = 2$.

36. Find the area of the region enclosed by the x-axis, $y = \frac{1}{x}$, $x = 3$, and $x = 4$.

14.4 SOME APPLICATIONS OF THE DEFINITE INTEGRAL

The definite integral can be used to express *total value over a period of time* as in the following example.

Suppose a leasing company wants to decide on the yearly lease fee for a certain new typewriter. The company expects to lease the typewriter for 5 years and it expects the rate of maintenance, in dollars, to be approximated by

$$M(t) = 10 + 2t + t^2,$$

where t is the number of years the typewriter has been used. Figure 14.6 shows the graph of $M(t)$. The total maintenance charge for the 5-year period will be given by the shaded area of the figure and can be found using the definite integral as follows.

$$\int_0^5 (10 + 2t + t^2)\,dt = 10t + t^2 + \frac{t^3}{3}\Big]_0^5$$

$$= 50 + 25 + \frac{125}{3} - 0$$

$$\approx 116.67$$

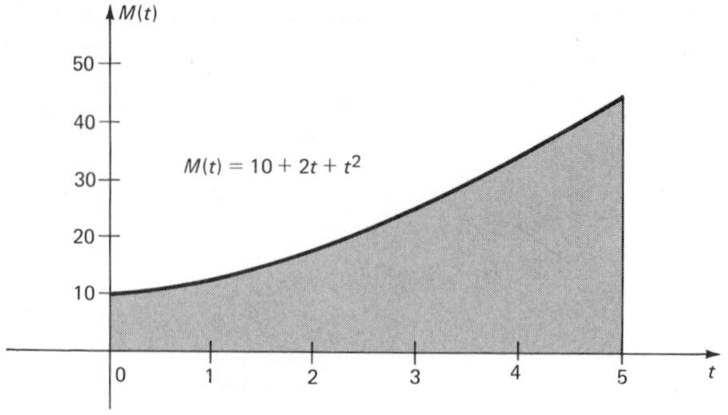

Figure 14.6

The company can expect the total maintenance charge for 5 years to be about \$117. Hence, the company should add about

$$\frac{\$117}{5} = \$23.40$$

to its annual lease price to pay for maintenance.

Work Problem 1 at the side.

1. Find the total maintenance charge for a lease of
(a) 1 year;
(b) 2 years.

Answer:
(a) \$11.33
(b) \$26.67

Example 1 Elizabeth, who runs a factory that makes signs, has been shown a new machine to staple the signs to the handles. She estimates that the rate of savings from the machine will be approximated by

$$S(x) = 3 + 2x,$$

where x represents the number of years the stapler has been used. If the machine costs \$70, would it pay for itself in 5 years?

We need to find the area under the savings curve shown in Figure 14.7 between the lines $x = 0$ and $x = 5$ and the x-axis. Using definite integrals, we have

$$\int_0^5 (3 + 2x)\,dx = 3x + x^2 \Big]_0^5 = 40.$$

The total savings in five years is \$40, so the machine will not pay for itself in this time period. ■

2. Find the amount of savings over the first 6 years for the machine in Example 1.

Answer: \$54

Work Problem 2 at the side.

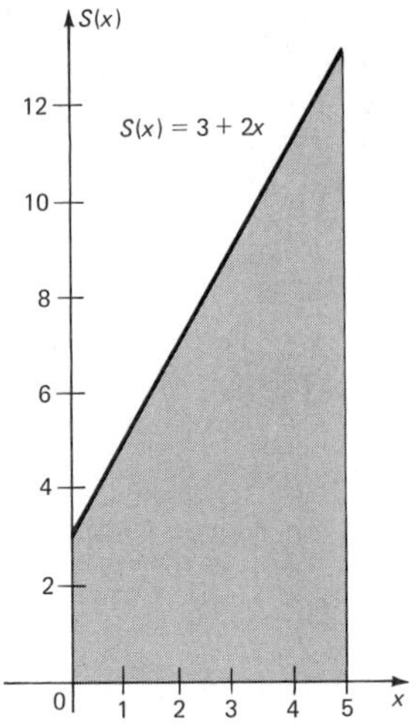

Figure 14.7

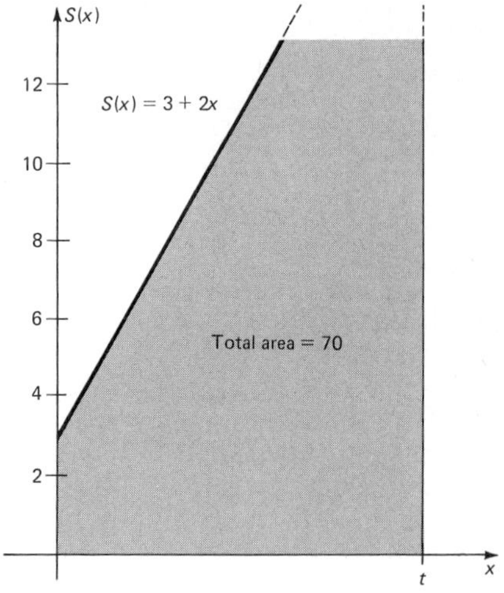

Figure 14.8

To find the number of years it will take for the machine of Example 1 to pay for itself, note that, since the machine costs a total of \$70, it will pay for itself when the area under the savings curve of Figure 14.8 equals 70, or at a time t such that

$$\int_0^t (3 + 2x)\,dx = 70.$$

Evaluating the definite integral we have

$$3x + x^2 \Big]_0^t = (3t + t^2) - (3\cdot 0 + 0^2) = 3t + t^2.$$

Since we want the total savings to equal 70,

$$3t + t^2 = 70.$$

Solve this quadratic equation to verify that the machine will pay for itself in seven years.

Work Problem 3 at the side.

3. Find the number of years in which the machine in Example 1 will pay for itself if the rate of savings is $S(x) = 5 + 4x$.

Answer: 4.8 years

Example 2 Elizabeth, of the previous example, believes that the rate of sales of her signs is given by

$$T(x) = 15 + 10x,$$

where x is the number of years she has been in business. When can she expect to sell her 1000th sign?

Let t be the time at which Elizabeth will sell her 1000th sign. The area under the sales curve between $x = 0$ and $x = t$ gives the total sales during that period. Since we want the total sales to be 1000, we have

$$\int_0^t (15 + 10x)\,dx = 1000.$$

Since

$$\int_0^t (15 + 10x)\,dx = 15x + 5x^2\Big]_0^t = 15t + 5t^2,$$

we have

$$15t + 5t^2 = 1000$$
$$5t^2 + 15t - 1000 = 0.$$

If we divide both sides of this equation by 5, we have

$$t^2 + 3t - 200 = 0,$$

from which we find, using the quadratic formula,

$$t = 12.72 \text{ years.}$$

Hence, Elizabeth can expect to sell her 1000th sign about twelve years and nine months after she goes into business. ■

4. In Example 2, when will Elizabeth sell
(a) the 500th sign;
(b) the 1200th sign?

Answer:
(a) 8.61 years
(b) 14.06 years

Work Problem 4 at the side.

Example 3 The rate of reaction to a given dose of a drug at time t hours after administration is given by

$$R(t) = \frac{1}{t},$$

where $R(t)$ is measured in appropriate units. Evaluate the total reaction to the given dose of the drug from $t = 1$ to $t = 24$ hours.

The total reaction will be given by

$$\begin{aligned}\int_1^{24} \frac{1}{t}\,dt &= \ln|t|\Big]_1^{24}\\ &= \ln 24 - \ln 1\\ &= \ln(4)(6) - 0\\ &= \ln 4 + \ln 6\\ &\approx 1.3863 + 1.7918\\ &= 3.1781. \quad ■\end{aligned}$$

5. The rate of reaction to a given dose of a drug at time t measured in hours is

$$R(t) = \frac{9}{t^2}.$$

(a) Find the total reaction from time $t = 1$ to $t = 12$.
(b) Find the total reaction on the second day.

Answer:
(a) 8.25
(b) 9/48, or .1875

Work Problem 5 at the side.

Example 4 How much is added to the total reaction in Example 3 during the next hour?

We want to find the total reaction from $t = 24$ to $t = 25$.

$$\int_{24}^{25} \frac{1}{t}\,dt = \ln|t|\Big]_{24}^{25}$$
$$= \ln 25 - \ln 24$$
$$\approx 3.2189 - 3.1781$$
$$= .0408$$

We found ln 25 in the table, with ln 24 from Example 3. There is very little change in total reaction after 24 hours. ■

6. In Example 4, find the reaction to the drug from time $t = 6$ to $t = 10$.

Answer: .5108

Work Problem 6 at the side.

14.4 APPLIED PROBLEMS

Management

1. A car-leasing firm must decide how much to charge for maintenance on the cars it leases. After careful study, it is decided that the rate of maintenance on a new car is approximated by

$$M(x) = 60(1 + x^2),$$

where x is the number of years the car has been in use. What total maintenance charge can the company expect for a two-year lease? What amount should it add to the monthly lease payments to pay for maintenance?

2. Using the function of Exercise 1, find the maintenance charge the company can expect during the third year. Find the total charge during the first three years. What monthly charge should be added to take care of a three-year lease which includes maintenance?

3. A company is considering a new manufacturing process. It knows that the rate of savings from the process will be about

$$S(t) = 2t + 30,$$

where t is the number of years of use of the process. Find the total savings during the first year. Find the total savings during the first six years. See Example 1.

4. Assume that the new process in Exercise 3 costs \$1000. About when will it pay for itself?

5. A company is introducing a new product. Production is expected to grow slowly because of difficulties in the start-up process. It is expected that the rate of production will be approximated by

$$P(x) = 600x^{3/2},$$

where x is the number of years since the introduction of the product. Will the company be able to supply 10,000 units during the first four years? See Example 2.

6. Pollution from a factory is entering a lake. The rate of concentration of the pollutant at time t is given by

$$P(t) = 140t^{5/2},$$

where t is the number of years since the factory started introducing pollutants into the lake. Ecologists estimate that the lake can accept a total level of pollution of 4850 units before all the fish life in the lake ends. Can the factory operate for 4 years without killing all the fish in the lake?

7. De Win Enterprises has found that its expenditure rate per day on a certain type of job is given by

$$E(x) = 4x + 2,$$

where x is the number of days since the start of the job. Find the total expenditure if the job takes 10 days.

8. In Exercise 7, how much will be spent on the job from the 10th to the 20th day?

9. In Exercise 7, if the company wants to spend no more than \$5000 on the job, in how many days must they complete it?

10. De Win Enterprises also knows that the rate of income per day for the same job is

$$I(x) = 100 - x,$$

where x is the number of days since the job was started. Find the total income for the first 10 days.

11. In Exercise 10, find the income from the 10th to the 20th day.

12. In Exercise 10, how many days must the job last for the total income to be at least \$5000?

13. After a new firm starts in business, it finds that its rate of profits (in hundreds of dollars) after t years of operation is given by

$$P(t) = 6t^2 + 4t + 5.$$

Find the total profits in the first three years.

14. Find the profit in Exercise 13 in the fourth year of operation.

15. An oil tanker is leaking oil at the rate of $20t + 50$ barrels per hour, where t is time in hours after the tanker hits a hidden rock. Find the total number of barrels that the ship will leak on the first day.

16. Find the number of barrels that the ship of Exercise 15 will leak on the second day.

Natural Science

17. After long study, tree scientists conclude that a eucalyptus tree will grow at the rate of $.2 + 4t^{-4}$ feet per year, when t is time in years. Find the number of feet that the tree will grow in the second year.

18. Find the number of feet the tree in Exercise 17 will grow in the third year.

Management

19. A worker new to a job will improve his efficiency with time so that it takes him fewer hours to produce an item with each day on the job up to a certain point. Suppose the rate of change of the number of hours it takes a worker in a certain factory to produce the xth item is given by

$$H(x) = 20 - 2x.$$

(a) What is the total number of hours required to produce the first 5 items?
(b) What is the total number of hours required to produce the first 10 items?

Natural Science

20. The rate at which a substance grows is given by

$$R(x) = 200e^x,$$

where x is the time in days. What is the total accumulated growth after 2.5 days?

21. For a certain drug, the rate of reaction in appropriate units is given by

$$R(t) = \frac{5}{t} + \frac{2}{t^2},$$

where t is measured in hours after the drug is administered. See Example 3. Find the total reaction to the drug
(a) from $t = 1$ to $t = 12$;
(b) from $t = 12$ to $t = 24$.

22. For another drug, the rate of reaction in appropriate units is

$$R(t) = \frac{4}{t^2} + 1,$$

where t is measured in hours after the drug is administered. Find the total reaction to the drug
(a) from $t = \frac{1}{4}$ to $t = 1$;
(b) from $t = 12$ to $t = 24$.

14.5 THE AREA BETWEEN TWO CURVES

In Section 14.3 we saw that the definite integral $\int_a^b f(x)\,dx$ can be used to find the area below the graph of the function $y = f(x)$, above the x-axis, and between the lines $x = a$ and $x = b$. In this section we extend this idea to find other areas.

Example 1 Find the area between the x-axis and the graph of $f(x) = x^2 - 4$ from $x = 0$ to $x = 4$.

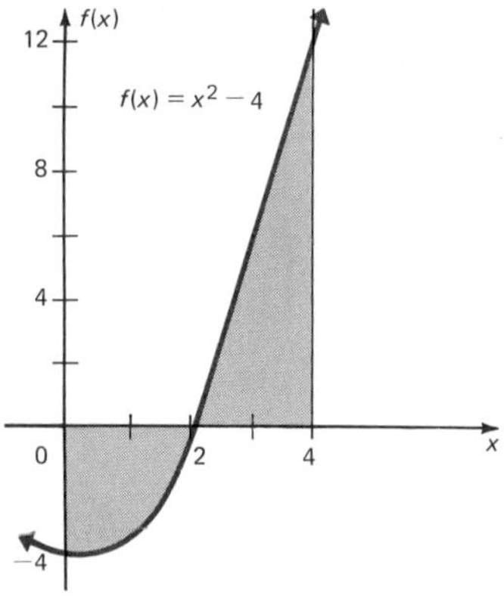

Figure 14.9

As shown in Figure 14.9, part of the area lies above the x-axis and part lies below the x-axis. If we were to use the definite integral to evaluate the area below the x-axis, the result would be a negative number, since the function values there are all negative. Since area is a nonnegative quantity, the correct value of that part of the area is given by the negative of the appropriate definite integral. In cases like this, the two parts of the area must be calculated separately and the two positive numbers then added to give the total area.

To find the area between the x-axis and the graph of $f(x) = x^2 - 4$ from $x = 0$ to $x = 4$, we must first find the point where the graph crosses the x-axis. This is done by solving the equation

$$x^2 - 4 = 0.$$

The solutions of this equation are 2 and -2. We are only interested in the values of x in the interval $[0, 4]$ so we discard the solution -2. Thus, the total area will be given by

$$\begin{aligned} -\int_0^2 (x^2 - 4)\,dx &+ \int_2^4 (x^2 - 4)\,dx \\ &= -\left(\frac{1}{3}x^3 - 4x\right)\Big]_0^2 + \left(\frac{1}{3}x^3 - 4x\right)\Big]_2^4 \\ &= -\left(\frac{8}{3} - 8\right) + \left(\frac{64}{3} - 16\right) - \left(\frac{8}{3} - 8\right) \\ &= 16 \text{ square units. } \blacksquare \end{aligned}$$

Work Problem 1 at the side.

1. Find the area between the x-axis and the graph of $f(x) = x - 4$ from $x = 1$ to $x = 5$.

Answer: 5

Example 2 Find the area between the curves $f(x) = x^{1/2}$ and $g(x) = x^3$ from $x = 0$ to $x = 1$.

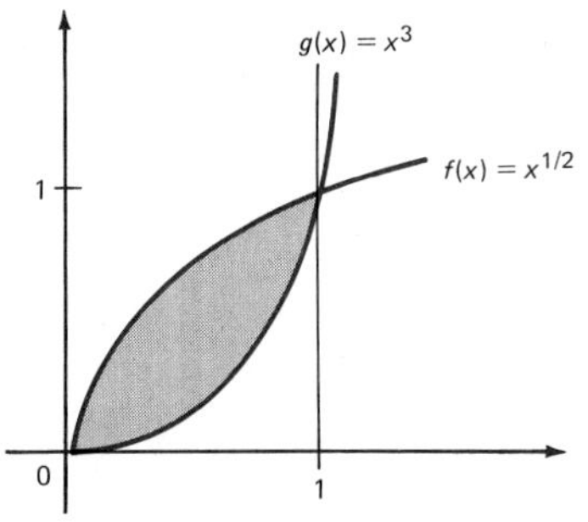

Figure 14.10

To find the area bounded by these two curves and the lines $x = 0$ and $x = 1$, we use definite integrals. The area between $f(x) = x^{1/2}$ and the x-axis from $x = 0$ and $x = 1$ is

$$\int_0^1 x^{1/2}\,dx,$$

while the area between $g(x) = x^3$ and the x-axis from $x = 0$ to $x = 1$ is

$$\int_0^1 x^3\,dx.$$

As shown in Figure 14.10, the area between these two curves is given by the difference between these two integrals, or

$$\int_0^1 x^{1/2}\,dx - \int_0^1 x^3\,dx,$$

which, by Theorem 14.6, can be written as

$$\int_0^1 (x^{1/2} - x^3)\,dx.$$

Using the fundamental theorem of calculus, we have

$$\begin{aligned}\int_0^1 (x^{1/2} - x^3)\,dx &= \frac{x^{3/2}}{3/2} - \frac{x^4}{4}\Big]_0^1 \\ &= \frac{2}{3}x^{3/2} - \frac{x^4}{4}\Big]_0^1 \\ &= \frac{2}{3}\cdot 1 - \frac{1}{4} \\ &= \frac{5}{12} \text{ square units.} \quad \blacksquare\end{aligned}$$

2. Find the area between $y = x$ and $y = x^2$ from $x = 0$ to $x = 1$.

Answer: 1/6 square units

Work Problem 2 at the side.

The difference between two integrals can be used to find the area between the graphs of two functions even if one graph lies below the x-axis or if both graphs lie below the x-axis. In general, if $f(x) \le g(x)$ for all values of x in the interval $[a, b]$, then the area between the two graphs is

$$\int_a^b [g(x) - f(x)]\,dx.$$

3. Find the area between $f(x) = -x^2 + 4$ and $g(x) = x^2 - 4$.

Answer: 64/3 square units

Work Problem 3 at the side.

Example 3 A company is considering introducing a new manufacturing process in one of its plants. The new process provides substantial savings, with the savings declining with time x according to the rate of savings function

$$S(x) = 100 - x^2.$$

At the same time, the cost of operating the new process increases with time x, according to the rate of cost function

$$C(x) = x^2 + \frac{14}{3}x.$$

(a) For how many years will the company realize savings?

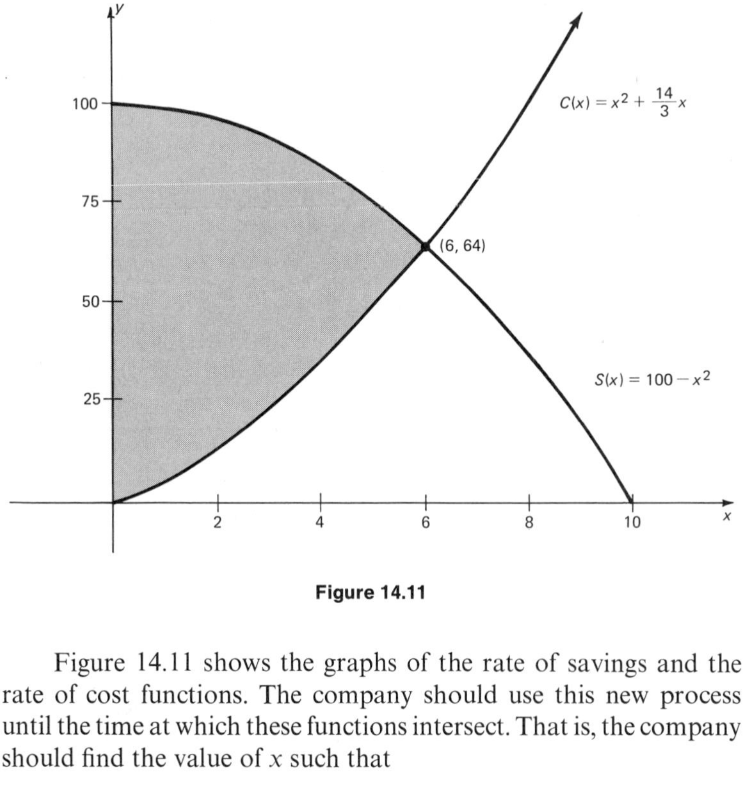

Figure 14.11

Figure 14.11 shows the graphs of the rate of savings and the rate of cost functions. The company should use this new process until the time at which these functions intersect. That is, the company should find the value of x such that

$$C(x) = S(x),$$

or such that

$$100 - x^2 = x^2 + \frac{14}{3}x.$$

Solving this equation, we find that the only valid solution is $x = 6$. Thus, the company should use the new process for 6 years.
(b) What will the total savings be during this period?

The total savings over the 6-year period are given by the area between the rate of cost and the rate of savings curves and the lines $x = 0$ and $x = 6$, which we can evaluate with definite integrals as follows:

$$\begin{aligned}\text{total savings} &= \int_0^6 (100 - x^2)\,dx - \int_0^6 \left(x^2 + \frac{14}{3}x\right)dx\\ &= \int_0^6 \left[(100 - x^2) - \left(x^2 + \frac{14}{3}x\right)\right]dx\\ &= \int_0^6 \left(100 - \frac{14}{3}x - 2x^2\right)dx\\ &= 100x - \frac{7}{3}x^2 - \frac{2}{3}x^3\Big]_0^6\\ &= 100(6) - \frac{7}{3}(36) - \frac{2}{3}(216)\\ &= 372.\end{aligned}$$

Thus, the company will save a total of \$372 over the 6-year period. ■

4. In Example 3, find the total savings if pollution-control regulations permit the new process for only 4 years.

Answer: \$320

Work Problem 4 at the side.

Example 4 A farmer has been using a new fertilizer that gives him a better yield, but because it exhausts the soil of other nutrients he must use other fertilizers in greater and greater amounts, so that his costs increase each year. The new fertilizer produces a rate of increase in revenue (in hundreds of dollars) given by

$$R(t) = -0.4t^2 + 8t + 10,$$

where t is measured in years. The rate of yearly costs due to use of the fertilizer is given by

$$C(t) = 2t + 5.$$

How long can the farmer profitably use the fertilizer? What will be his increase in revenue over this period?

The farmer should use the new fertilizer until the additional costs equal the increase in revenue. Thus, we need to solve the equation $R(t) = C(t)$ as follows.

$$\begin{aligned}-0.4t^2 + 8t + 10 &= 2t + 5\\ -4t^2 + 80t + 100 &= 20t + 50\\ -4t^2 + 60t + 50 &= 0\\ t &= 15.8\end{aligned}$$

The new fertilizer will be profitable for about 15.8 years.

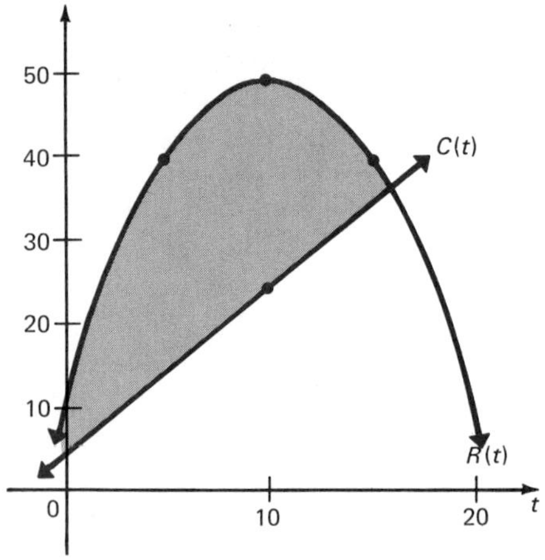

Figure 14.12

To find the total amount of additional revenue over the 15.8-year period, we must find the area between the graphs of the revenue and the cost functions, as shown in Figure 14.12. We have

$$\begin{aligned}\text{total savings} &= \int_0^{15.8} [R(t) - C(t)]\,dt \\ &= \int_0^{15.8} [(-0.4t^2 + 8t + 10) - (2t + 5)]\,dt \\ &= \int_0^{15.8} (-0.4t^2 + 6t + 5)dt \\ &= \frac{-0.4t^3}{3} + \frac{6t^2}{2} + 5t\Big]_0^{15.8} \\ &= 302.01.\end{aligned}$$

The total savings will amount to about \$30,000 over the 15.8-year period.

It is probably not realistic to say that the farmer will need to use the new process for 15.8 years—he will probably have to use it for 15 years or for 16 years. In this case, when the mathematical model produces results that are not in the domain of the function, it will be necessary to find the total savings after 15 years and after 16 years and then select the best result. ■

5. In Example 4, suppose $C(x) = 3t + 5$.
(a) How long should the farmer use the fertilizer?
(b) What is his increase in revenue over that period?

Answer:
(a) about 13.4 years
(b) about \$19,500

Work Problem 5 at the side.

Example 5 Suppose the price, in cents, for a certain product is

$$p(x) = 900 - 20x - x^2,$$

when the demand for the product is x units. Also, suppose the function

$$p(x) = x^2 + 10x$$

gives the price, in cents, when the supply is x units. The graphs of both functions are shown in Figure 14.13 along with the equilibrium point at which supply and demand are equal. To find the equilibrium supply or demand, x^*, we solve the equation

$$\begin{aligned} 900 - 20x^* - (x^*)^2 &= (x^*)^2 + 10x^* \\ 0 &= 2(x^*)^2 + 30x^* - 900 \\ 0 &= (x^*)^2 + 15x^* - 450. \end{aligned}$$

The only positive solution of the equation is $x^* = 15$.

At the equilibrium point where supply = demand = 15 units, the price is

$$p(15) = 900 - 20(15) - 15^2 = 375,$$

or \$3.75.

As the demand graph shows, there are consumers who are willing to pay more than the equilibrium price, so they benefit from the equilibrium price. Some benefit a lot, some less, and so on. Their total benefit is called the *consumer's surplus* and is represented by an area as shown in Figure 14.13. The consumer's surplus is

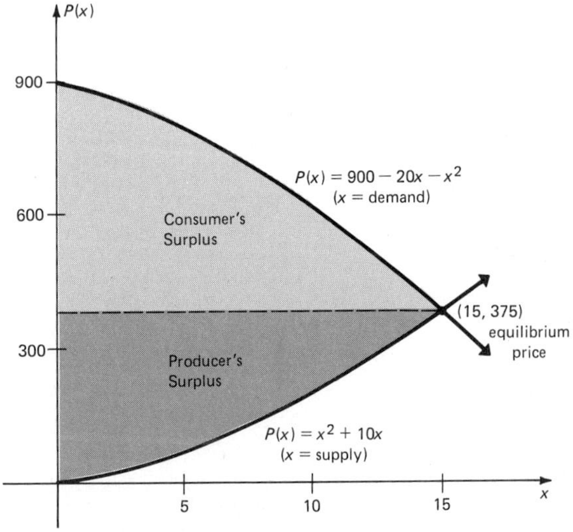

Figure 14.13

evaluated by the definite integral

$$\int_0^{15} (900 - 20x - x^2)\,dx - (15)(375)$$

$$= \left(900x - 10x^2 - \frac{x^3}{3}\Big]_0^{15}\right) - 5625$$

$$= \left(900(15) - 10(15)^2 - \frac{(15)^3}{3}\right) - 5625$$

$$= 4500.$$

Here, the consumer's surplus is 4500 cents, or \$45.00.

On the other hand, some suppliers would have offered the product at a price below the equilibrium price, so they too gain from the equilibrium price. The total of the supplier's gains is called the *producer's surplus* and is represented by the area shown in Figure 14.13. Producer's surplus is given by the definite integral

$$(15)(375) - \int_0^{15} (x^2 + 10x)\,dx = 5625 - \left(\frac{x^3}{3} + 5x^2\right)\Big]_0^{15}$$

$$= 5625 - \left(\frac{15^3}{3} + 5(15)^2\right)$$

$$= 3375.$$

The producer's surplus is \$33.75. ■

6. Given the demand function $p(x) = 12 - .07x$ and the supply function $p(x) = .05x$, where $p(x)$ is the price in dollars, find
(a) the equilibrium point;
(b) the consumer's surplus;
(c) the producer's surplus.

Answer:
(a) $x = 100$
(b) \$350
(c) \$250

Work Problem 6 at the side.

14.5 EXERCISES

Find the areas between the following curves. (See Examples 1 and 2.)

1. $x = -2, \quad x = 1, \quad f(x) = x^2 + 4, \quad y = 0$
2. $x = 1, \quad x = 2, \quad f(x) = x^3, \quad y = 0$
3. $x = -3, \quad x = 1, \quad f(x) = x + 1, \quad y = 0$
4. $x = -2, \quad x = 0, \quad f(x) = 1 - x^2, \quad y = 0$
5. $x = 0, \quad x = 5, \quad f(x) = \frac{7}{5}x, \quad g(x) = \frac{3}{5}x + 10$
6. $x = -1, \quad x = 1, \quad f(x) = x, \quad S(x) = x^2 - 3$
7. $x = 0, \quad x = 2, \quad f(x) = 150 - x^2, \quad S(x) = x^2 + \frac{11}{4}x$
8. $x = 0, \quad x = 8, \quad f(x) = 150 - x^2, \quad g(x) = x^2 + \frac{11}{4}x$
9. $x = 0, \quad x = 2, \quad f(x) = x^2, \quad g(x) = \frac{1}{2}x$
10. $x = 1, \quad x = 4, \quad f(x) = x^2, \quad g(x) = x^3$

APPLIED PROBLEMS

Management

11. Suppose a company wants to introduce a new machine which will produce a rate of annual savings given by

$$S(x) = 150 - x^2,$$

where x is the number of years of operation of the machine, while producing a rate of annual costs of

$$C(x) = x^2 + \frac{11}{4}x.$$

See Examples 3 and 4.

(a) For how many years will it be profitable to use this new machine?

(b) What are the total savings during the first year of the use of the machine?

(c) What are the total savings over the entire period of use of the machine?

12. A new smog-control device will reduce the output of oxides of sulfur from automobile exhausts. It is estimated that the rate of savings to the community from the use of this device will be approximated by

$$S(x) = -x^2 + 4x + 8,$$

where $S(x)$ is the rate of savings in millions of dollars after x years of use of the device. The new device cuts down on the production of oxides of sulfur, but it causes an increase in the production of oxides of nitrogen. The rate of additional costs in millions to the community after x years is approximated by

$$C(x) = \frac{3}{25}x^2.$$

(a) For how many years will it pay to use the new device?

(b) What will be the total savings over this period of time?

In the exercises for Section 13.4, De Win Enterprises had an expenditure rate (in hundreds of dollars) of $E(x) = 4x + 2$, and an income rate (in hundreds of dollars) of $I(x) = 100 - x$ on a particular job, where x was the number of days from the start of the job. Their profit on that job will equal total income less total expenditure. Profit will be maximized if the job ends at the optimum time, which is the point where the two curves meet. Find the following.

13. The optimum number of days for the job to last.

14. The total income for the optimum number of days.

15. The total expenditure for the optimum number of days.

16. The maximum profit for the job.

For Exercises 17–22, refer to Example 5.

17. Find the producer's surplus if the price of some item is given by

$$p(x) = x^2 + 2x + 50,$$

where x is the supply, assuming supply and demand are in equilibrium at $x^* = 20$.

18. Suppose a price for a certain commodity is given by

$$p(x) = 100 + 3x + x^2,$$

where x is the supply. Suppose that supply and demand are in equilibrium at $x^* = 3$. Find the producer's surplus.

19. Find the consumer's surplus if the price for an item is given by

$$p(x) = 50 - x^2$$

at a demand of x, assuming supply and demand are in equilibrium at $x^* = 5$.

20. Find the consumer's surplus if the price for an item is given by

$$p(x) = -(x + 4)^2 + 66,$$

where x is the demand and supply and demand are in equilibrium at $x^* = 3$.

21. Suppose the price of a certain item is given by

$$p(x) = \frac{7}{5}x,$$

when the supply is x items, and the price is given by

$$p(x) = -\frac{3}{5}x + 10,$$

when the demand is for x items.
(a) Graph the supply and demand curves.
(b) Find the point at which supply and demand are in equilibrium.
(c) Find the consumer's surplus.
(d) Find the producer's surplus.

22. Repeat the four steps in Exercise 21 for the supply function

$$p(x) = x^2 + \frac{11}{4}x$$

and the demand function

$$p(x) = 150 - x^2.$$

Suppose that all the people in a country are ranked according to their incomes, starting at the bottom. Let x represent the fraction of the community making the lowest income ($0 \le x \le 1$); $x = .4$, therefore, represents the lower 40% of all income producers. Let $I(x)$ represent the proportion of the total income earned *by the lowest x of all people. Thus, I(.4) represents the fraction of total income earned by the lowest 40% of the population. Suppose*

$$I(x) = .9x^2 + .1x.$$

Find and interpret the following.

23. $I(.1)$ **24.** $I(.4)$ **25.** $I(.6)$ **26.** $I(.9)$

If income were distributed uniformly, we would have $I(x) = x$. The area between the curves $I(x) = x$ and the particular function $I(x)$ for a given country is called the coefficient of inequality *for that country.*

27. Graph $I(x) = x$ and $I(x) = .9x^2 + .1x$ for $0 \leq x \leq 1$ on the same axes.

28. Find the area between the curves.

14.6 INTEGRATION BY SUBSTITUTION

In this section we discuss a technique for integrating functions which is related to the chain rule for derivatives. This technique depends on the idea of a differential, which was discussed in Chapter 13. Recall that

$$dy = f'(x) \cdot dx.$$

For example, if $y = 6x^4$, then $dy = 24x^3\, dx$, and so on.

1. Find dy for the following.
(a) $y = 9x$
(b) $y = 5x^3 + 2x^2$
(c) $y = e^{-2x}$

Answer:
(a) $dy = 9dx$
(b) $dy = (15x^2 + 4x)\,dx$
(c) $dy = -2e^{-2x}\,dx$

Work Problem 1 at the side.

We can use the idea of a differential to help us find the indefinite integral

$$\int 2x(x^2 - 1)^4\, dx.$$

Substitution is used to change the function to one which can be integrated using the methods of Section 1 of this chapter. Let $u = x^2 - 1$. By the definition of the differential, if $u = x^2 - 1$, then $du = 2x\, dx$. Now substitute into the integral above.

$$\int 2x(x^2 - 1)^4\, dx = \int (x^2 - 1)^4 (2x\, dx) = \int u^4\, du$$

This last integral can now be found.

$$\int u^4\, du = \frac{1}{5}u^5 + C$$

Finally, substitute $x^2 - 1$ for u.

$$\int 2x(x^2 - 1)^4\, dx = \frac{1}{5}(x^2 - 1)^5 + C$$

2. Find the following.

(a) $\int 8x(4x^2 - 9)^3\,dx$

(b) $\int 12x^3(3x^4 + 8)^5\,dx$

Answer:

(a) $\frac{1}{4}(4x^2 - 9)^4 + C$

(b) $\frac{1}{6}(3x^4 + 8)^6 + C$

Work Problem 2 at the side.

Example 1 Find $\int 6x(3x^2 + 4)^4\,dx$.

Trial and error must be used to decide on the expression to set equal to u. Try $u = 3x^2 + 4$, so that $du = 6x\,dx$. Then substitute.

$$\int 6x(3x^2 + 4)^4\,dx = \int (3x^2 + 4)^4(6x\,dx) = \int u^4\,du$$

Since

$$\int u^4\,du = \frac{u^5}{5} + C,$$

and since $u = 3x^2 + 4$, we then have

$$\int 6x(3x^2 + 4)^4\,dx = \frac{u^5}{5} + C = \frac{(3x^2 + 4)^5}{5} + C.$$

We can verify this by using the chain rule to find the derivative of $(3x^2 + 4)^5/5 + C$. ■

3. Find the following.

(a) $\int 8x(4x^2 - 1)^5\,dx$

(b) $\int 18x^2(6x^3 - 5)^{3/2}\,dx$

Answer:

(a) $\dfrac{(4x^2 - 1)^6}{6} + C$

(b) $\dfrac{2(6x^3 - 5)^{5/2}}{5} + C$

Work Problem 3 at the side.

Example 2 Find $\int x^2\sqrt{x^3 + 1}\,dx$.

If we let $u = x^3 + 1$, then $du = 3x^2\,dx$. The integral does not contain the constant 3, which is needed for du. To take care of this, multiply by 3 inside the integral sign and $\frac{1}{3}$ outside. (This is permissible with constants, but *not* with variables.)

$$\int x^2\sqrt{x^3 + 1}\,dx = \frac{1}{3}\cdot\int 3x^2\sqrt{x^3 + 1}\,dx$$

Now substitute u for $x^3 + 1$ and du for $3x^2\,dx$.

$$\begin{aligned}\frac{1}{3}\int 3x^2\sqrt{x^3 + 1}\,dx &= \frac{1}{3}\int\sqrt{u}\,du\\ &= \frac{1}{3}\int u^{1/2}\,du\\ &= \frac{1}{3}\cdot\frac{u^{3/2}}{3/2} + C\\ &= \frac{2}{9}u^{3/2} + C\end{aligned}$$

Since $u = x^3 + 1$,

$$\int x^2\sqrt{x^3 + 1}\,dx = \frac{2}{9}(x^3 + 1)^{3/2} + C. \quad ■$$

4. Find the following.

(a) $\int x(5x^2 + 6)^4\,dx$

(b) $\int x^2(x^3 - 2)^5\,dx$

(c) $\int x\sqrt{x^2 + 16}\,dx$

Answer:

(a) $\frac{1}{50}(5x^2 + 6)^5 + C$

(b) $\frac{1}{18}(x^3 - 2)^6 + C$

(c) $\frac{1}{3}(x^2 + 16)^{3/2} + C$

Work Problem 4 at the side.

The substitution method given in the preceding examples will not always work. For example, we might try to find

$$\int x^3\sqrt{x^3+1}\,dx$$

by substituting $u = x^3 + 1$, so that $du = 3x^2\,dx$. There is no constant which can be inserted inside the integral sign to give $3x^2$. This integral, and a great many others, cannot be evaluated by the methods of this course. In fact, a large class of functions do not have antiderivatives at all.

5. Find the following.

(a) $\int x(6x+2)^{1/2}\,dx$

(b) $\int 5x^2(8x^4+2)^3\,dx$

Answer: Neither can be found by the substitution method shown above.

Work Problem 5 at the side.

Example 3 Find the value of

$$\int_0^4 (x^2+5x)^{3/2}(2x+5)\,dx.$$

To evaluate this definite integral, we first find

$$\int (x^2+5x)^{3/2}(2x+5)\,dx.$$

Let $u = (x^2 + 5x)$, so that $du = (2x + 5)\,dx$. This gives

$$\int (x^2+5x)^{3/2}(2x+5)\,dx = \int u^{3/2}\,du$$

$$= \frac{2}{5}u^{5/2} + C.$$

We cannot use the limits of integration given in the original problem, 0 and 4, since they refer to x and this antiderivative contains u. Thus, we must substitute $x^2 + 5x$ for u, obtaining

$$\frac{2}{5}u^{5/2} + C = \frac{2}{5}(x^2+5x)^{5/2} + C.$$

This antiderivative can now be used to evaluate the given definite integral. Using the fundamental theorem of calculus, we have

$$\int_0^4 (x^2+5x)^{3/2}(2x+5)\,dx = \frac{2}{5}(x^2+5x)^{5/2}\Big]_0^4$$

$$= \frac{2}{5}(4^2+5\cdot 4)^{5/2} - \frac{2}{5}(0^2+5\cdot 0)^{5/2}$$

$$= \frac{2}{5}(36)^{5/2} - 0$$

$$= \frac{2}{5}(6)^5. \quad ■$$

Recall that it is not necessary to consider the constant C when evaluating a definite integral.

6. Find the following.

(a) $\int_0^4 \sqrt{x^2 + 4x}\,(2x + 4)\,dx$

(b) $\int_5^8 (x^2 - 5x + 1)^{1/2}(2x - 5)\,dx$

Answer:

(a) $\frac{2}{3}(32)^{3/2}$

(b) $\frac{248}{3}$

Work Problem 6 at the side.

Example 4 Find $\int e^{4x}\,dx$.

Let $u = 4x$, so that $du = 4\,dx$. Place 4 inside the integral sign and $\frac{1}{4}$ outside.

$$\int e^{4x}\,dx = \frac{1}{4}\int 4e^{4x}\,dx$$

Now substitute.

$$\frac{1}{4}\int 4e^{4x}\,dx = \frac{1}{4}\int e^{u}\,du$$
$$= \frac{1}{4}e^{u} + C$$
$$= \frac{1}{4}e^{4x} + C \quad ■$$

Based on results similar to this one, we can state a more general form of Theorem 14.4 as follows.

Theorem 14.4A For any real number $a \neq 0$, we have

$$\int e^{ax}\,dx = \frac{1}{a}e^{ax} + C.$$

7. Find the following.

(a) $\int 6e^{3x}\,dx$

(b) $\int -8e^{-x/2}\,dx$

(c) $\int e^{-.01x}\,dx$

Answer:

(a) $2e^{3x} + C$

(b) $16e^{-x/2} + C$

(c) $-e^{-.01x}/.01 + C$ or $-100e^{-.01x} + C$

Work Problem 7 at the side.

The following example shows further how substitution can be used in integrals involving exponentials.

Example 5 Evaluate $\int xe^{x^2}\,dx$.

If we let $u = x^2$, then $du = 2x\,dx$.

$$\int xe^{x^2}\,dx = \frac{1}{2}\int 2xe^{x^2}\,dx$$
$$= \frac{1}{2}\int e^{u}\,du$$
$$= \frac{1}{2}e^{u} + C$$
$$= \frac{1}{2}e^{x^2} + C \quad ■$$

8. Find the following.

(a) $\int 8xe^{3x^2}\,dx$

(b) $\int x^2 e^{x^3}\,dx$

Answer:

(a) $\frac{4}{3}e^{3x^2} + C$

(b) $\frac{1}{3}e^{x^3} + C$

Work Problem 8 at the side.

Recall that the antiderivative of $f(x) = 1/x$ is $\ln|x|$. The next example shows how substitution can be used to extend the use of this antiderivative.

Example 6 Find $\int \frac{(2x-3)\,dx}{x^2-3x}$.

Let $u = x^2 - 3x$, so that $du = (2x-3)\,dx$. Then,

$$\begin{aligned}\int \frac{(2x-3)\,dx}{x^2-3x} &= \int \frac{du}{u} \\ &= \ln|u| + C \\ &= \ln|x^2-3x| + C. \quad \blacksquare\end{aligned}$$

9. Find the following.

(a) $\int \frac{4\,dx}{x-3}$

(b) $\int \frac{(2x-9)\,dx}{x^2-9x}$

(c) $\int \frac{(3x^2+8)\,dx}{x^3+8x+5}$

Answer:

(a) $4\ln|x-3| + C$

(b) $\ln|x^2-9x| + C$

(c) $\ln|x^3+8x+5| + C$

Work Problem 9 at the side.

14.6 EXERCISES

Find the following indefinite integrals. (See Examples 1, 2, and 4–6.)

1. $\int 4(2x+3)^4\,dx$

2. $\int (-4x+1)^3\,dx$

3. $\int \frac{4}{(x-2)^3}\,dx$

4. $\int \frac{-3}{(x+1)^4}\,dx$

5. $\int \sqrt{x-4}\,dx$

6. $\int (\sqrt{x+5})^3\,dx$

7. $\int 2x(x^2+1)^3\,dx$

8. $\int 3x^2(x^3-4)^3\,dx$

9. $\int (\sqrt{x^2+12x})(2x+12)\,dx$

10. $\int (\sqrt{x^2-6x})(2x-6)\,dx$

11. $\int e^{2x}\,dx$

12. $\int 3e^{-2x}\,dx$

13. $\int (-4e^{2x})\,dx$

14. $\int 5e^{-0.3x}\,dx$

15. $\int 3x^2 e^{2x^3}\,dx$

16. $\int xe^{-x^2}\,dx$

17. $\int \frac{-8}{1+x}\,dx$

18. $\int \frac{9}{2+x}\,dx$

19. $\int \frac{dx}{2x+1}$

20. $\int \frac{dx}{5x-2}$

21. $\int \frac{2x+4}{x^2+4x}\,dx$

22. $\int \frac{x}{3x^2-2}\,dx$

23. $\int \frac{6x\,dx}{(3x^2+2)^4}$

24. $\int \frac{4x\,dx}{(2x^2-5)^3}$

25. $\int \frac{x-1}{(2x^2-4x)^2}dx$

26. $\int \frac{2x+1}{(x^2+x)^3}dx$

27. $\int \left(\frac{1}{x}+x\right)\left(1-\frac{1}{x^2}\right)dx$

28. $\int \left(\frac{2}{x}-x\right)\left(\frac{-2}{x^2}-1\right)dx$

Find the following definite integrals. (See Example 3.)

29. $\int_1^4 \sqrt{x}\,dx$

30. $\int_4^9 2x^{1/2}\,dx$

31. $\int_0^1 2x(x^2+1)^3\,dx$

32. $\int_0^1 3x^2(x^3-4)^3\,dx$

33. $\int_0^4 (\sqrt{x^2+12x})(2x+12)\,dx$

34. $\int_0^8 (\sqrt{x^2-6x})(2x-6)\,dx$

35. $\int_{-1}^{e-2} \frac{1}{x+2}dx$ (Hint: $\ln e = 1$.)

36. $\int_{-2}^{0} \frac{-4}{x+3}dx$

37. $\int_0^1 e^{2x}\,dx$

38. $\int_1^2 e^{-x}\,dx$

39. $\int_2^4 e^{-0.2x}\,dx$

40. $\int_1^4 e^{-0.3x}\,dx$

APPLIED PROBLEMS

Management

When a new technological development comes along, the cost to manufacture the first few units can be very high, but the cost can decline as more experience in manufacturing is obtained. (This effect is called the learning curve.*) Suppose that the rate of change of the cost to manufacture item number x is given by C′(x) dollars, where*

$$C'(x) = 5e^{x/5}.$$

41. Find the total cost of manufacturing the first 20 items by finding $\int_0^{20} C'(x)\,dx$.

42. Find the average cost of each of the first 20 items.

(Most of the numbers in this exercise come from an article in The Wall Street Journal.*) Suppose that the rate of consumption of a natural resource is*

$$C(t) = k \cdot e^{rt},$$

where t is time in years, r is a constant representing the rate of growth in use, and k is the consumption in the year when $t = 0$. In 1977, Texaco sold 1.2 billion barrels of oil. Assume that oil sales increase at the rate of 4% per year, so that $r = .04$.

43. Write $C(t)$ for Texaco, letting $t = 0$ represent 1977.

44. Set up a definite integral for the amount of oil that Texaco will sell in the ten years from 1977 to 1987.

45. Evaluate the definite integral of Exercise 44.

46. Texaco has about 20 billion barrels of oil in reserve. To find the number of years that this amount will last, we must solve the equation

$$\int_0^t 1.2e^{.04t}\,dt = 20.$$

On the left, show that we get $30e^{.04t} - 30$.

47. We now have $30e^{.04t} - 30 = 20$, or $30e^{.04t} = 50$, and, finally, $e^{.04t} = 50/30 = 1.67$. Look down the e^x column of Table 3 until you find the number closest to 1.67. What is the corresponding value of x?

48. You should now have $e^{.04t} = e^{.50}$, or $.04t = .50$. Find t.

49. Rework the last three problems assuming a rate of increase of only 2% per year.

50. Repeat Exercise 49 for 1% per year.

14.7 TABLES OF INTEGRALS

It is a fairly straightforward task to find the derivative of most of the useful functions. The chain rule, together with other formulas of Chapter 12, can be used to find derivatives of almost all functions that are useful in practical applications. However, this is not true for integration. There are many useful functions whose integrals cannot be found by the methods we have discussed. For example, $f(x) = e^{-x^2}$, used to obtain the normal curve of statistics, is one of many such functions. Some calculus courses spend a considerable amount of time on techniques of integration. These techniques often depend on trigonometry and complicated algebraic substitution and manipulation.

Another less time-consuming approach is possible. This is to list all commonly needed integrals in a table of integrals. This table can then be referred to as needed. One such table is included inside the back cover of this text. The remainder of this section will be devoted to examples showing how to use the table in this book.

Example 1 Find $\int \frac{1}{\sqrt{x^2 + 16}}\,dx$.

By inspecting the table, we see that if $a = 4$, this antiderivative is the same as entry 5 of the table. Entry 5 of the table is

$$\int \frac{1}{\sqrt{x^2 + a^2}}\,dx = \ln\left|\frac{x + \sqrt{x^2 + a^2}}{a}\right| + C.$$

By substituting 4 for a in this entry, we get

$$\int \frac{1}{\sqrt{x^2+16}}\,dx = \ln\left|\frac{x+\sqrt{x^2+16}}{4}\right| + C.$$

This result could be verified by taking the derivative of the right-hand side of this last equation. ■

Work Problem 1 at the side.

Example 2 Find $\int \frac{8}{16-x^2}\,dx$.

We can convert this antiderivative into the one given in entry 7 of the table by writing the 8 in front of the integral sign (permissible only with constants) and by letting $a = 4$. Doing this gives

$$8\int \frac{1}{16-x^2}\,dx = 8\left[\frac{1}{2\cdot 4}\ln\left(\frac{4+x}{4-x}\right)\right] + C$$
$$= \ln\left(\frac{4+x}{4-x}\right) + C.$$

In entry 7 of the table, the condition $x^2 < a^2$ is given. Here $a = 4$. Hence, the result given above is valid only for $x^2 < 16$, so that the final answer should be written as

$$\int \frac{8}{16-x^2}\,dx = \ln\left(\frac{4+x}{4-x}\right) + C, \qquad \text{for } x^2 < 16.$$

Because of the condition $x^2 < 16$, the expression in parentheses is always positive, so that absolute value bars are not needed. ■

Work Problem 2 at the side.

Example 3 Evaluate $\int \frac{12x}{2x+1}\,dx$.

For $a = 2$ and $b = 1$, this antiderivative can be rewritten to match entry 11 of this table. Thus

$$\int \frac{12x}{2x+1}\,dx = 12\int \frac{x}{2x+1}\,dx$$
$$= 12\left[\frac{x}{2} - \frac{1}{4}\ln|2x+1|\right] + C$$
$$= 6x - 3\ln|2x+1| + C. \ ■$$

1. Find the following.

(a) $\int \frac{4}{\sqrt{x^2+100}}\,dx$

(b) $\int \frac{-9}{\sqrt{x^2-4}}\,dx$

(c) $\int -6\ln|x|\,dx$

Answer:

(a) $4\ln\left|\frac{x+\sqrt{x^2+100}}{10}\right| + C$

(b) $-9\ln\left|\frac{x+\sqrt{x^2-4}}{2}\right| + C$

(c) $-6x(\ln|x| - 1) + C$

2. *Find the following.*

(a) $\int \frac{1}{x^2-4}\,dx$

(b) $\int \frac{-6}{x\sqrt{25-x^2}}\,dx$

(c) $\int \frac{5}{x\sqrt{36+x^2}}\,dx$

Answer:

(a) $\frac{1}{4}\ln\left(\frac{x-2}{x+2}\right) + C, \quad x^2 > 4$

(b) $\frac{6}{5}\ln\left(\frac{5+\sqrt{25-x^2}}{x}\right) + C,$

$0 < x < 5$

(c) $-\frac{5}{6}\ln\left|\frac{6+\sqrt{36+x^2}}{x}\right| + C$

3. Find the following.

(a) $\int \frac{3x}{5x-2}\,dx$

(b) $\int \frac{-6x}{(9x+1)^2}\,dx$

(c) $\int \frac{2}{x(2x-3)}\,dx$

Answer:

(a) $\frac{3x}{5} + \frac{6}{25}\ln|5x-2| + C$

(b) $\frac{-2}{27(9x+1)} - \frac{2}{27}\ln|9x+1| + C$

(c) $\frac{-2}{3}\ln\left|\frac{x}{2x-3}\right| + C$

Work Problem 3 at the side.

Example 4 Find $\int \sqrt{9x^2+1}\,dx$.

This antiderivative seems most similar to entry 15 of the table. However, entry 15 requires that the coefficient of the x^2 term be 1. We can satisfy that requirement here by factoring out the 9.

$$\begin{aligned}\int \sqrt{9x^2+1}\,dx &= \int \sqrt{9\left(x^2+\frac{1}{9}\right)}\,dx\\ &= \int 3\sqrt{x^2+\frac{1}{9}}\,dx\\ &= 3\int \sqrt{x^2+\frac{1}{9}}\,dx\end{aligned}$$

Now, using entry 15 with $a = 1/3$, we have

$$\begin{aligned}\int \sqrt{9x^2+1}\,dx &= 3\left[\frac{x}{2}\sqrt{x^2+\frac{1}{9}} + \frac{(\frac{1}{3})^2}{2}\cdot \ln\left|x+\sqrt{x^2+\frac{1}{9}}\right|\right] + C\\ &= \frac{3x}{2}\sqrt{x^2+\frac{1}{9}} + \frac{1}{6}\ln\left|x+\sqrt{x^2+\frac{1}{9}}\right| + C. \quad \blacksquare\end{aligned}$$

4. Find the following.

(a) $\int \frac{3}{16x^2-1}\,dx$

(b) $\int \frac{-1}{100x^2-1}\,dx$

Answer:

(a) $\frac{3}{8}\ln\left|\frac{x-1/4}{x+1/4}\right| + C, \quad x^2 > 1/16$

(b) $\frac{1}{20}\ln\left|\frac{1/10+x}{1/10-x}\right| + C$

$= -\frac{1}{20}\ln\left|\frac{x-1/10}{x+1/10}\right| + C,$

$x^2 > \frac{1}{100}$

Work Problem 4 at the side.

Example 5 Find $\int x^2e^x\,dx$.

Using entry 17 with $n = 2$ and $a = 1$, we have

$$\int x^2e^x\,dx = \frac{x^2e^x}{1} - \frac{2}{1}\int xe^x\,dx + C. \tag{1}$$

We must now use entry 17 again, this time to find $\int xe^x\,dx$. Here $n = 1$ and $a = 1$. Thus,

$$\begin{aligned}\int xe^x\,dx &= \frac{xe^x}{1} - \frac{1}{1}\int x^0e^x\,dx + K\\ &= xe^x - \int e^x\,dx + K \qquad \text{Recall: } x^0 = 1\\ &= xe^x - e^x + K.\end{aligned}$$

Substituting this result back into equation (1) gives

$$\begin{aligned}\int x^2e^x\,dx &= x^2e^x - 2(xe^x - e^x + K) + C\\ &= x^2e^x - 2xe^x + 2e^x + M\\ &= e^x(x^2 - 2x + 2) + M,\end{aligned}$$

where M is an arbitrary constant. ■

5. Find the following.

(a) $\int x^4 \ln|x|\,dx$

(b) $\int xe^{2x}\,dx$

Answer:

(a) $x^5\left(\frac{\ln|x|}{5} - \frac{1}{25}\right) + C$

(b) $\frac{xe^{2x}}{2} - \frac{1}{4}e^{2x} + C$

Work Problem 5 at the side.

14.7 EXERCISES

Find the following antiderivatives, using the table inside the back cover. (See Examples 1–5.)

1. $\int \ln|4x|\,dx$

2. $\int \ln\left|\frac{3}{5}x\right|\,dx$

3. $\int \frac{-4}{\sqrt{x^2+36}}\,dx$

4. $\int \frac{9}{\sqrt{x^2+9}}\,dx$

5. $\int \frac{6}{x^2-9}\,dx$

6. $\int \frac{-12}{x^2-16}\,dx$

7. $\int \frac{-4}{x\sqrt{9-x^2}}\,dx$

8. $\int \frac{3}{x\sqrt{121-x^2}}\,dx$

9. $\int \frac{-2x}{3x+1}\,dx$

10. $\int \frac{6x}{4x-5}\,dx$

11. $\int \frac{2}{3x(3x-5)}\,dx$

12. $\int \frac{-4}{3x(2x+7)}\,dx$

13. $\int \frac{4}{4x^2-1}\,dx$

14. $\int \frac{-6}{9x^2-1}\,dx$

15. $\int \frac{3}{x\sqrt{1-9x^2}}\,dx$

16. $\int \frac{-2}{x\sqrt{1-16x^2}}\,dx$

17. $\int x^4 \ln|x|\,dx$

18. $\int 4x^2 \ln|x|\,dx$

19. $\int \frac{\ln|x|}{x^2}\,dx$

20. $\int \frac{-2\ln|x|}{x^3}\,dx$

21. $\int xe^{-2x}\,dx$

22. $\int xe^{3x}\,dx$

23. $\int 2x^2e^{-2x}\,dx$

24. $\int -3x^2e^{-4x}\,dx$

25. $\int x^3e^x\,dx$

26. $\int x^3e^{2x}\,dx$

CHAPTER 14 TEST

[14.1] *Find an antiderivative of each of the following.*

1. $f(x) = 2x^3$

2. $f(x) = x^2 + 5x$

3. $f(x) = (\sqrt{x})^3$

4. $f(x) = \frac{1}{x^4}$

5. $f(x) = 3e^x$

6. $f(x) = \frac{5}{x}$

[14.2] 7. Approximate the area under the graph of $f(x) = 2x + 3$ and above the x-axis from $x = 0$ to $x = 4$ using four rectangles. Let the height of each rectangle be given by the value of the function at the left side of the rectangle.

[14.3] 8. Find the exact area under the graph of $f(x) = 2x + 3$ and above the x-axis from $x = 0$ to $x = 4$.

Evaluate the following definite integrals.

9. $\int_{-1}^{2} (3x^2 + 5)\,dx$

10. $\int_{1}^{3} x^{-1}\,dx$

11. $\int_{0}^{4} -2e^x\,dx$

[14.4] 12. The rate of change of sales of a new brand of tomato soup, in thousands, is given by

$$S(x) = \sqrt{x} + 2,$$

where x is the time in months that the new product has been on the market. Find the total sales after 9 months.

13. The rate of change of a population of prairie dogs, in terms of the number of coyotes, x, which prey on them, is given by

$$P(x) = 25 - .1x.$$

Find the total number of prairie dogs as the coyote population grows from 100 to 200.

[14.5] 14. Find the area between the graphs of $f(x) = 5 - x^2$ and $g(x) = x^2 - 3$.

15. A company has installed new machinery which will produce a savings rate (in thousands of dollars) of

$$S(x) = 225 - x^2,$$

where x is the number of years the machinery is to be used. The rate of additional costs to the company due to the new machinery is expected to be

$$C(x) = x^2 + 25x + 150.$$

For how many years should the company use the new machinery? What will the net savings in thousands of dollars be over this period?

[14.6] *Use substitution to find the following.*

16. $\int x\sqrt{5x^2 + 6}\,dx$

17. $\int (x^2 - 5x)^4(2x - 5)\,dx$

18. $\int e^{-4x}\,dx$

19. $\int \frac{12(2x + 9)}{x^2 + 9x + 1}\,dx$

20. $\int_{2}^{3} (3x^2 - 6x)^2(6x - 6)\,dx$

[14.7] *Use the table of integrals to find the following.*

21. $\int \frac{1}{\sqrt{x^2 - 64}} dx$

22. $\int \frac{5}{x\sqrt{25 + x^2}} dx$

23. $\int \frac{-2}{x(3x + 4)} dx$

24. $\int \sqrt{x^2 + 49}\, dx$

25. $\int 3x^2 \ln|x|\, dx$

15 Further Techniques and Applications of Integration

In this chapter we discuss methods of integrating more complicated functions as well as numerical methods of integration, which are often used with practical data or when a function cannot be integrated at all. An application of integration to probability is also included. The chapter ends by using methods of integration to solve differential equations.

15.1 INTEGRATION BY PARTS

In this section we introduce a technique of integration which often makes it possible to reduce a complicated integral to a simpler integral. If u and v are both differentiable functions, then uv is also differentiable and, by the product rule for derivatives,

$$\frac{d(uv)}{dx} = u\frac{dv}{dx} + v\frac{du}{dx}.$$

We can rewrite this expression, using differentials, as

$$d(uv) = u\,dv + v\,du.$$

Now, integrating both sides of this last equation, we have

$$\begin{aligned}\int d(uv) &= \int u\,dv + \int v\,du,\\ uv &= \int u\,dv + \int v\,du.\end{aligned}$$

By rearranging terms, we obtain the formula:

$$\boldsymbol{\int u\,dv = uv - \int v\,du.}$$

The technique using this formula to find integrals is called **integrating by parts** and is illustrated in the following examples.

Example 1 Find $\int x\sqrt{1-x}\,dx$.

To use integration by parts, first write the expression $x\sqrt{1-x}\,dx$ as a product of functions u and dv in such a way that v can be found. Let us select $u = x$ and $dv = \sqrt{1-x}\,dx$. Then $du = dx$. We find v by integrating dv. Therefore,

$$v = \int \sqrt{1-x}\,dx = -\frac{2}{3}(1-x)^{3/2} + C.$$

(Verify this.) For simplicity, we ignore the constant C and just add it at the end. Thus, $v = -\frac{2}{3}(1-x)^{3/2}$. Now substitute this into the formula for integration by parts as follows.

$$\int u\,dv = uv - \int v\,du,$$

$$\int x\sqrt{1-x}\,dx = x\left[-\frac{2}{3}(1-x)^{3/2}\right] - \int\left[-\frac{2}{3}(1-x)^{3/2}\right]dx$$

$$= -\frac{2}{3}x(1-x)^{3/2} - \frac{2}{3}\int -(1-x)^{3/2}\,dx$$

$$= -\frac{2}{3}x(1-x)^{3/2} - \frac{2}{3}\left[\frac{2}{5}(1-x)^{5/2}\right]$$

$$\int x\sqrt{1-x}\,dx = -\frac{2}{3}x(1-x)^{3/2} - \frac{4}{15}(1-x)^{5/2} + C$$

We added a constant C in the last step. ■

1. Find the following.

(a) $\int x(x+1)^{-2}\,dx$

(b) $\int x\sqrt{x+1}\,dx$

Answer:

(a) $-x(x+1)^{-1} + \ln|x+1| + C$

(b) $\frac{2}{3}x(x+1)^{3/2} - \frac{4}{15}(x+1)^{5/2} + C$

Work Problem 1 at the side.

Example 2 Find $\int xe^{5x}\,dx$.

Let $dv = e^{5x}\,dx$ and let $u = x$. Then $v = \frac{1}{5}e^{5x}$ and $du = dx$. Substitute into the formula for integration by parts.

$$\int u\,dv = uv - \int v\,du$$

$$\int xe^{5x}\,dx = x\left(\frac{1}{5}e^{5x}\right) - \int \frac{1}{5}e^{5x}\,dx$$

$$= \frac{1}{5}xe^{5x} - \frac{1}{25}e^{5x} + C \quad ■$$

2. Find the following.

(a) $\int 8xe^{-2x}\,dx$

(b) $\int -3(x+1)e^{2x}\,dx$

Answer:

(a) $-4xe^{-2x} - 2e^{-2x} + C$

(b) $-\frac{3}{2}(x+1)e^{2x} + \frac{3}{4}e^{2x} + C$

Work Problem 2 at the side.

Sometimes it is necessary to use the technique of integrating by parts more than once as in the following example.

Example 3 Find $\int 2x^2e^x\,dx$.

Since $\int e^x\,dx$ can be found, let us choose $dv = e^x\,dx$, so that $v = e^x$. Then $u = 2x^2$ and $du = 4x\,dx$. Now we substitute into the formula for integrating by parts.

$$\int u\,dv = uv - \int v\,du$$

$$\int 2x^2 e^x\,dx = 2x^2 e^x - \int e^x(4x\,dx) \qquad \textbf{(1)}$$

We must find $\int e^x(4x\,dx)$ by parts. Again, we choose $dv = e^x\,dx$, which gives $v = e^x$, $u = 4x$, and $du = 4\,dx$. Substituting again into the formula for integrating by parts, we have the following.

$$\int e^x(4x\,dx) = 4xe^x - \int e^x(4\,dx)$$
$$= 4xe^x - 4e^x$$

Now we must substitute back into our first result, equation (1), to get the final answer.

$$\int 2x^2 e^x\,dx = 2x^2 e^x - (4xe^x - 4e^x)$$
$$= 2x^2 e^x - 4xe^x + 4e^x + C,$$

where a constant C was added at the last step. ■

3. Find $\int 3x^2 e^{-5x}\,dx$.

Answer:
$-\frac{3}{5}x^2 e^{-5x} - \frac{6}{25}xe^{-5x} - \frac{6}{125}e^{-5x} + C$

Work Problem 3 at the side.

The method of integration by parts requires choosing the factor dv so that $\int dv$ can be found. If this is not possible, or if the remaining factor, which becomes u, does not have a differential du such that $v\,du$ can be integrated, the technique cannot be used. For example, to integrate

$$\int \frac{1}{4 - x^2}\,dx$$

we might choose $dv = dx$ and $u = (4 - x^2)^{-1}$. Then $v = x$ and $du = 2x\,dx/(4 - x^2)^2$. Then we have

$$\int \frac{1}{4 - x^2}\,dx = \frac{x}{4 - x^2} - \int \frac{2x^2\,dx}{(4 - x^2)^2},$$

where the integral on the right is more complicated than the original integral. A second use of integration by parts on the new integral would make matters even worse. Since we cannot choose $dv = (4 - x^2)^{-1}$ because we cannot integrate it, integration by parts is not suitable for this problem. In fact, there are many functions whose integrals cannot be found by any of the methods we have described. Many of these can be found by more advanced methods and are available in tables of integrals, discussed in Section 14.7. However, there are functions which cannot be integrated at all, for example, $f(x) = e^{-x^2}$, a function which is important in statistical theory.

15.1 EXERCISES

Use integration by parts to find the following. (See Examples 1–3.)

1. $\int x(x + 1)^5\, dx$

2. $\int x\sqrt{5x - 1}\, dx$

3. $\int x^3(1 + x^2)^{1/4}\, dx$

4. $\int 2x^5(1 - 3x^3)^{1/2}\, dx$

5. $\int \frac{x}{\sqrt{x - 1}}\, dx$

6. $\int \frac{4x}{\sqrt{8 - x}}\, dx$

7. $\int \frac{2x}{(x + 5)^6}\, dx$

8. $\int \frac{3x^3}{(x^2 - 5)^5}\, dx$

9. $\int \frac{x^3\, dx}{\sqrt{3 - x^2}}$

10. $\int \frac{x^7\, dx}{(x^4 + 3)^{2/3}}$

11. $\int xe^x\, dx$

12. $\int (x + 1)e^x\, dx$

13. $\int (5x - 9)e^{-3x}\, dx$

14. $\int (6x + 3)e^{-2x}\, dx$

15. $\int (x^2 + 1)e^x\, dx$

16. $\int 3x^2e^{-x}\, dx$

17. $\int (1 - x^2)e^{2x}\, dx$

18. $\int x^2e^{2x}\, dx$

19. Use integration by parts to find $\int \ln|x|\, dx$. Let $u = \ln|x|$, so that $du = \frac{1}{x}$.

20. Use integration by parts to find $\int \ln|5x - 1|dx$. $\left(\text{Hint: Write } \frac{5x}{5x - 1} \text{ as } 1 + \frac{1}{5x - 1}.\right)$

Use the results of Exercise 19 and integration by parts to find the following.

21. $\int x \ln|x|\, dx$

22. $\int x^2 \ln|x|\, dx$

23. $\int (2x - 1)\ln|3x|\, dx$

24. $\int (8x + 7)\ln|5x|\, dx$

25. Find the area between $y = (x - 2)e^x$ and the x-axis from $x = 2$ to $x = 4$.

26. Find the area between $y = xe^x$ and the x-axis from $x = 0$ to $x = 1$.

15.2 NUMERICAL INTEGRATION

As mentioned in Section 15.1 some integrals cannot be evaluated by any technique or found in any table. Since $\int_a^b f(x)\,dx$ represents an area, any approximation of that area also approximates the definite integral. Many methods of approximating definite integrals by areas, made feasible by the availability of pocket calculators and the high-speed computer, are in use today. These methods are referred to as **numerical integration.** We shall discuss two methods of numerical integration, *the trapezoidal rule* and *Simpson's rule*.*

As an example, let us consider

$$\int_1^5 \frac{1}{x}\,dx.$$

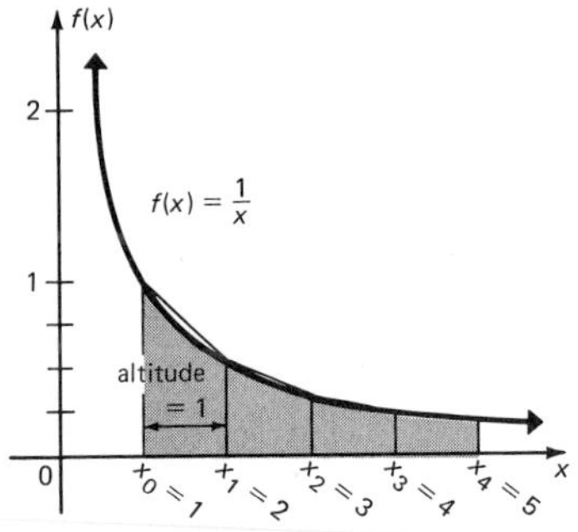

Figure 15.1

The shaded region in Figure 15.1 shows the area under the graph of $f(x) = 1/x$ above the x-axis and between the lines $x = 1$ and $x = 5$. Since $\int (1/x)\,dx = \ln|x| + C$, we have

$$\begin{aligned}\int_1^5 \frac{1}{x}\,dx &= \ln|x|\Big]_1^5\\ &= \ln 5 - \ln 1\\ &= \ln 5 - 0\\ &= \ln 5,\end{aligned}$$

a number which can be found in Table 5. Using the table, we find that $\ln 5 \approx 1.6094$. How was this number in the table obtained? One way to find this number is to approximate the area under the curve in Figure 15.1. This can be done by the trapezoidal rule. To get a first approximation to $\ln 5$ by the trapezoidal rule, we might first find the sum of the areas of the four trapezoids indicated in Figure 15.1. The area of a trapezoid is one-half the product of the sum of the bases and the altitude. Each of the trapezoids in Fig. 15.1 has altitude 1. Thus,

$$\begin{aligned}\ln 5 = \int_1^5 \frac{1}{x}\,dx &\approx \frac{1}{2}\left(\frac{1}{1}+\frac{1}{2}\right)(1)+\frac{1}{2}\left(\frac{1}{2}+\frac{1}{3}\right)(1)+\frac{1}{2}\left(\frac{1}{3}+\frac{1}{4}\right)(1)+\frac{1}{2}\left(\frac{1}{4}+\frac{1}{5}\right)(1)\\ &= \frac{1}{2}\left(\frac{3}{2}+\frac{5}{6}+\frac{7}{12}+\frac{9}{20}\right)\\ &\approx 1.68.\end{aligned}$$

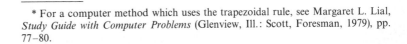

* For a computer method which uses the trapezoidal rule, see Margaret L. Lial, *Study Guide with Computer Problems* (Glenview, Ill.: Scott, Foresman, 1979), pp. 77–80.

To get a better approximation, we would divide the interval $1 \le x \le 5$ into more subintervals. The larger the number of subintervals, the better the approximation will be. In general, suppose f is a continuous function on an interval $a \le x \le b$. Divide the interval from a to b into n equal subintervals by the points $a = x_0$, $x_1, x_2, \ldots, x_n = b$. Use the subintervals to make trapezoids which approximate the area under the curve (as in Figure 15.1). The sum of the areas of the trapezoids is given by

$$\frac{1}{2}\Big[f(x_0) + f(x_1)\Big]\left(\frac{b-a}{n}\right) + \frac{1}{2}\Big[f(x_1) + f(x_2)\Big]\left(\frac{b-a}{n}\right) + \cdots + \frac{1}{2}\Big[f(x_{n-1}) + f(x_n)\Big]\left(\frac{b-a}{n}\right)$$

$$= \left(\frac{b-a}{n}\right)\left[\frac{1}{2}f(x_0) + \frac{1}{2}f(x_1) + \frac{1}{2}f(x_1) + \frac{1}{2}f(x_2) + \frac{1}{2}f(x_2) + \cdots + \frac{1}{2}f(x_{n-1}) + \frac{1}{2}f(x_n)\right]$$

$$= \left(\frac{b-a}{n}\right)\left[\frac{1}{2}f(x_0) + f(x_1) + f(x_2) + \cdots + f(x_{n-1}) + \frac{1}{2}f(x_n)\right].$$

A general statement of the trapezoidal rule follows.

Trapezoidal rule Let f be a continuous function in $a \le x \le b$. Let $a \le x \le b$ be divided into n equal subintervals by the points $a = x_0, x_1, x_2, \ldots, x_n = b$. Then

$$\int_a^b f(x)\,dx \approx \left(\frac{b-a}{n}\right)\left[\frac{1}{2}f(x_0) + f(x_1) + \cdots + f(x_{n-1}) + \frac{1}{2}f(x_n)\right].$$

1. (a) Approximate $\int_2^6 (1 + x)\,dx$ by using the trapezoidal rule with $n = 4$.
(b) Evaluate the definite integral and compare your answers.

Answer:
(a) 20
(b) 20

Work Problem 1 at the side.

Another numerical method, called Simpson's rule, approximates consecutive portions of the curve with portions of parabolas rather than with line segments as in the trapezoidal rule. As shown in Figure 15.2 a parabola is fitted through points A, B, and C, another through C, D, and E, and so on. (It is necessary to have an even number of intervals for this process to come out right.) Then the sum of the areas under these parabolas will approximate the area under the graph of the function.

We do not derive Simpson's rule here.

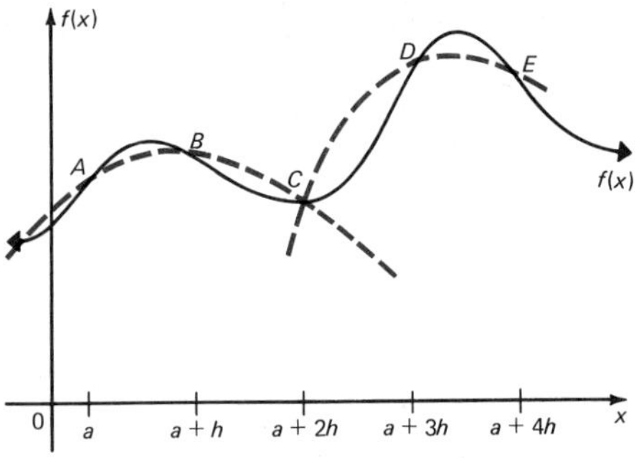

Figure 15.2

Simpson's rule Let f be a continuous function on $a \le x \le b$. Let $a \le x \le b$ be divided into an even number n of equal subintervals by the points $a = x_0, x_1, x_2, \ldots, x_n = b$. Then,

$$\int_a^b f(x)\,dx \approx \frac{b-a}{3n}[f(x_0) + 4f(x_1) + 2f(x_2) + 4f(x_3) + \cdots + 2f(x_{n-2}) + 4f(x_{n-1}) + f(x_n)].$$

Note that it is necessary for n to be an even number. For example, let us use the same function, $1/x$, and divide the interval $1 \le x \le 5$ into 4 subintervals in the same way as in the example of the trapezoidal rule. By Simpson's rule,

$$\int_1^5 \frac{1}{x}\,dx \approx \frac{4}{12}\left[\frac{1}{1} + 4\left(\frac{1}{2}\right) + 2\left(\frac{1}{3}\right) + 4\left(\frac{1}{4}\right) + \frac{1}{5}\right]$$
$$= \frac{1}{3}\left(1 + 2 + \frac{2}{3} + 1 + \frac{1}{5}\right)$$
$$\approx 1.62.$$

Simpson's rule usually gives a better approximation than the trapezoidal rule. However, as n is increased, the two approximations would differ by less and less.

Work Problem 2 at the side.

2. (a) Approximate

$$\int_2^6 (x^2 + 1)\,dx$$

using Simpson's rule with $n = 4$.

(b) Evaluate the integral and compare your answers.

Answer:

(a) $73\frac{1}{3}$

(b) $73\frac{1}{3}$

15.2 EXERCISES

In Exercises 1–10, use $n = 4$ to approximate the value of each of the given integrals (a) *by the trapezoidal rule;* (b) *by Simpson's rule.*

1. $\int_0^2 x^2\,dx$

2. $\int_0^2 (2x + 1)\,dx$

3. $\int_0^4 \sqrt{x+1}\,dx$

4. $\int_1^5 \frac{1}{x+1}\,dx$

5. $\int_{-2}^2 (2x^2+1)\,dx$

6. $\int_0^3 (2x^2+1)\,dx$

7. $\int_1^5 \frac{1}{x^2}\,dx$

8. $\int_{-1}^3 \frac{1}{4-x}\,dx$

9. $\int_2^4 \frac{1}{x^3}\,dx$

10. $\int_0^4 x\sqrt{2x-1}\,dx$

APPLIED PROBLEMS

In the following applications, values of $f(x)$ are given.

Natural Science

11. The table below shows the results from a chemical experiment.

Concentration of chemical A, x	1	2	3	4	5	6	7
Rate of formation of chemical B, $f(x)$	12	16	18	21	24	27	32

(a) Plot these points. Connect the points with line segments.
(b) Use the trapezoidal rule to find the area bounded by the broken line of part (a), the x-axis, the line $x = 1$, and the line $x = 7$.
(c) Find the same area using Simpson's rule.

Social Science

12. The results from a research study in psychology were as follows.

Number of hours of study, x	1	2	3	4	5	6	7
Number of extra points earned on a test, $f(x)$	4	7	11	9	15	16	23

Repeat steps (a)–(c) of Exercise 11 for this data.

Management

13. A sales manager presented the following results at a sales meeting.

Year, x	1	2	3	4	5	6	7
Rate of sales, $f(x)$	.4	.6	.9	1.1	1.3	1.4	1.6

Repeat steps (a)–(c) of Exercise 11 for this data to find the total sales over the 7-year period.

14. A company found that its marginal costs in hundreds of dollars were as follows over a 7-year period.

Year, x	1	2	3	4	5	6	7
Marginal cost, $f(x)$	9.0	9.2	9.5	9.4	9.8	10.1	10.5

Repeat steps (a)–(c) of Exercise 11 for this data to find the total cost.

Natural Science

In the study of bioavailability *in pharmacy, a drug is given to a patient. The level of concentration of the drug is then measured periodically, producing* blood level curves *such as the ones shown below. The areas under the curves give the total amount of the drug available to the patient. Use Simpson's rule to find the following areas. Break the problem into two parts: find the area from* 0 *to* 4 *using* 1*-hour intervals, and find the area from* 4 *to* 20 *using* 2*-hour intervals. You will have to estimate the height for many of the values of time. Simpson's rule requires an even number of equal subintervals.* (From D. J. Chodos, M.D., and A. R. DiSantos, Ph.D., *Basics of Bioavailability*, copyright 1974, The Upjohn Company.)

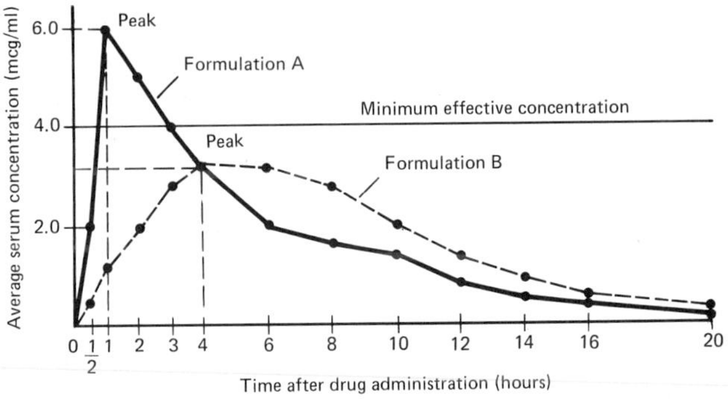

15. Find the area under the curve for Formulation A.

16. Find the area under the curve for Formulation B.

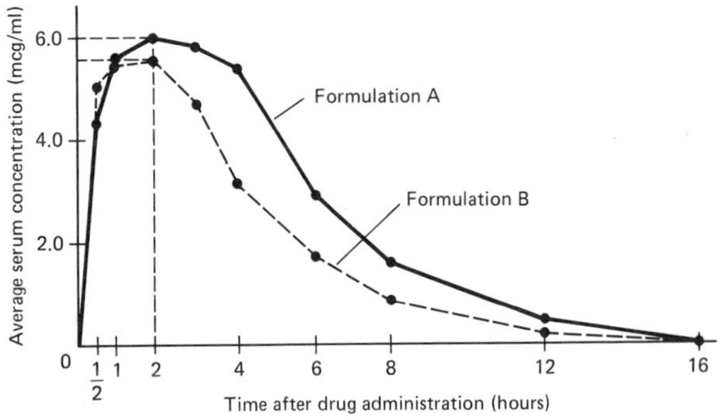

17. Find the area under the curve for Formulation A.

18. Find the area under the curve for Formulation B.

15.3 IMPROPER INTEGRALS

The graph in Figure 15.3(a) shows the area bounded by the curve $f(x) = x^{-3/2}$, the x-axis, and the vertical line $x = 1$. The shaded region could be extended indefinitely to the right. Does this shaded region have an area?

To see if we can find the area of this region, introduce a vertical line at $x = b$, as shown in Figure 15.3(b). With this vertical line we have a region with both upper and lower limits of integration. This region is given by the definite integral

$$\int_1^b x^{-3/2}\,dx.$$

By the fundamental theorem of calculus,

$$\begin{aligned}\int_1^b x^{-3/2}\,dx &= -2x^{-1/2}\Big]_1^b \\ &= -2b^{-1/2} + (2 \cdot 1^{-1/2}) \\ &= -2b^{-1/2} + 2 \\ &= 2 - \frac{2}{b^{1/2}}.\end{aligned}$$

Suppose we now let the vertical line $x = b$ of Figure 15.3(b) move further to the right. That is, suppose we let $b \to \infty$. The expression $-2/b^{1/2}$ would then approach 0, and

$$\lim_{b\to\infty}\left(2 - \frac{2}{b^{1/2}}\right) = 2 - 0 = 2.$$

This limit is defined as the *area* shown in Figure 15.3(a), so that

$$\int_1^\infty x^{-2/3}\,dx = 2.$$

In general, an integral of the form

$$\int_a^\infty f(x)\,dx, \qquad \int_{-\infty}^b f(x)\,dx, \qquad \text{or} \qquad \int_{-\infty}^\infty f(x)\,dx$$

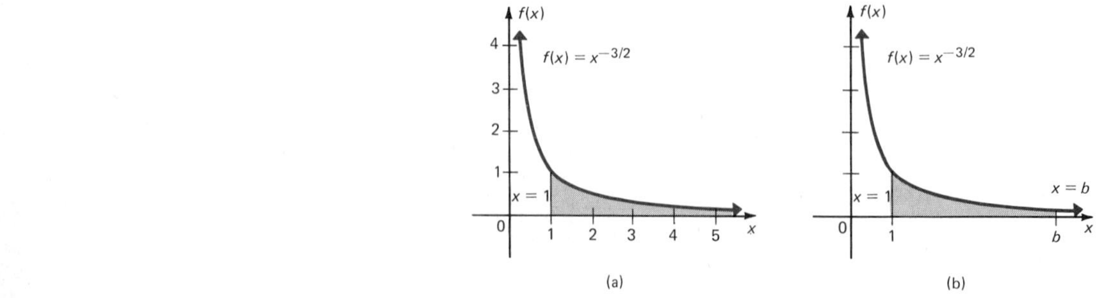

Figure 15.3

is called an improper integral. These **improper integrals** are defined as follows.

$$\int_a^{\infty} f(x)\,dx = \lim_{b\to\infty} \int_a^b f(x)\,dx$$

$$\int_{-\infty}^{b} f(x)\,dx = \lim_{a\to-\infty} \int_a^b f(x)\,dx$$

$$\int_{-\infty}^{\infty} f(x)\,dx = \int_{-\infty}^{c} f(x)\,dx + \int_c^{\infty} f(x)\,dx$$

If the expression on the right side exists, the integrals are **convergent;** otherwise, they are **divergent.**

1. Find the following.

(a) $\int_4^{\infty} x^{-3/2}\,dx$

(b) $\int_9^{\infty} x^{-3/2}\,dx$

Answer:

(a) 1

(b) $\frac{2}{3}$

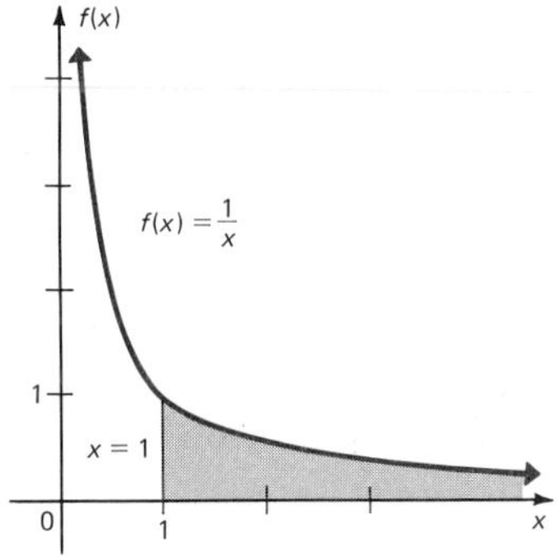

Figure 15.4

Work Problem 1 at the side.

Example 1 Find the following.

(a) $\int_1^{\infty} \frac{dx}{x}$

A graph of this region is shown in Figure 15.4. Use the definition of an improper integral:

$$\int_1^{\infty} \frac{dx}{x} = \lim_{b\to\infty} \int_1^b \frac{dx}{x}.$$

Use the fundamental theorem of calculus to find $\int_1^b \frac{dx}{x}$.

$$\int_1^b \frac{dx}{x} = \ln|x|\Big]_1^b = \ln|b| - \ln|1| = \ln|b| - 0 = \ln|b|$$

Take the limit as $b \to \infty$. We know that $\ln|b|$ gets larger and larger as b gets larger and larger. Therefore,

$$\lim_{b\to\infty} \ln|b| \text{ does not exist.}$$

Since the limit does not exist, $\int_1^{\infty} \frac{dx}{x}$ is divergent.

(b)
$$\int_{-\infty}^{-2} \frac{1}{x^2}\,dx = \lim_{a\to-\infty} \int_a^{-2} \frac{1}{x^2}\,dx$$
$$= \lim_{a\to-\infty} \frac{-1}{x}\Big]_a^{-2}$$
$$= \lim_{a\to-\infty} \left(\frac{1}{2} + \frac{1}{a}\right)$$
$$= \frac{1}{2}$$

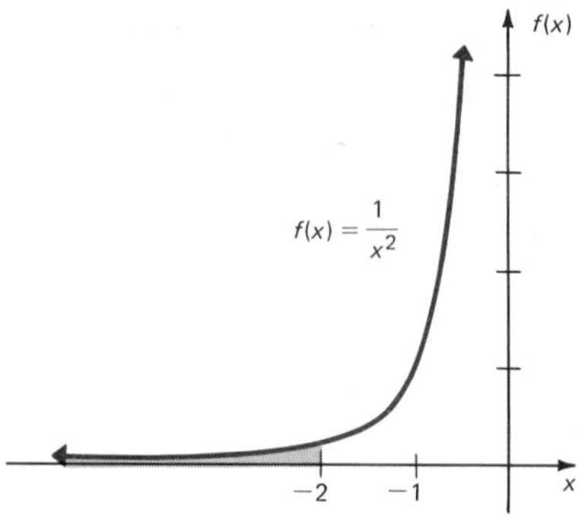

Figure 15.5

A graph of this region is shown in Figure 15.5. This integral converges. ■

2. Find the following.

(a) $\int_{-\infty}^{-2} \frac{1}{x}\,dx$

(b) $\int_{-\infty}^{-3} -4x^{-3}\,dx$

(c) $\int_{1}^{\infty} \sqrt{x}\,dx$

Answer:
(a) divergent
(b) 2/9
(c) divergent

Work Problem 2 at the side.

Example 2 Find $\int_0^{\infty} 4e^{-3x}\,dx$

By definition,

$$\begin{aligned}\int_0^{\infty} 4e^{-3x}\,dx &= \lim_{b\to\infty} \int_0^b 4e^{-3x}\,dx \\ &= \lim_{b\to\infty} -\frac{4}{3}e^{-3x}\Big]_0^b \\ &= \lim_{b\to\infty} \left[-\frac{4}{3}e^{-3b} - \left(-\frac{4}{3}e^{-0}\right)\right] \\ &= \lim_{b\to\infty} \left[\frac{-4}{3e^{3b}} + \frac{4}{3}\right] \\ &= 0 + \frac{4}{3} = \frac{4}{3}. \quad ■\end{aligned}$$

3. Find the following.

(a) $\int_{-\infty}^{0} e^{2x}\,dx$

(b) $\int_0^{\infty} 3e^{-3x}\,dx$

Answer:
(a) $\frac{1}{2}$
(b) 1

Work Problem 3 at the side.

Example 3 A chemical is being released into a small stream. The rate at which the chemical is being released into the stream at time t is given by Pe^{-kt}, where P is the amount of chemical released into the stream initially. Suppose $P = 1000$ and $k = .06$. Find the total amount of the chemical that will be released into the stream into the indefinite future.

We need to find

$$\int_0^{\infty} Pe^{-kt}\,dt = \int_0^{\infty} 1000e^{-.06t}\,dt.$$

Work as above.

$$\int_0^\infty 1000e^{-.06t}\,dt = \lim_{b\to\infty}\int_0^b 1000e^{-.06t}\,dt$$
$$= \lim_{b\to\infty}\left(\frac{1000}{-.06}e^{-.06t}\right)\Bigg]_0^b$$
$$= \lim_{b\to\infty}\left[\frac{1000}{-.06e^{.06b}} - \frac{1000}{-.06}e^0\right]$$
$$= \frac{-1000}{-.06} = 16{,}667$$

total units of the chemical. ■

4. Let R represent the fixed annual rent on a property. The capitalized value of the property is given by $\int_0^b Re^{-kt}\,dt$, where k is the average rate of interest in the economy. Find the capitalized value of property when rent is paid perpetually if
(a) $k = .04$ and $R = 5000$;
(b) $k = .08$ and $R = 12{,}500$.

Answer:

(a) $125,000
(b) $156,250

Work Problem 4 at the side.

15.3 EXERCISES

Find the value of the following which exist. (See Examples 1 and 2.)

1. $\int_2^\infty \frac{1}{x^2}\,dx$

2. $\int_5^\infty \frac{1}{x^2}\,dx$

3. $\int_1^\infty \frac{1}{\sqrt{x}}\,dx$

4. $\int_{16}^\infty \frac{-3}{\sqrt{x}}\,dx$

5. $\int_{-\infty}^{-1} \frac{2}{x^3}\,dx$

6. $\int_{-\infty}^{-4} \frac{3}{x^4}\,dx$

7. $\int_1^\infty \frac{1}{x^{1.001}}\,dx$

8. $\int_1^\infty \frac{1}{x^{.999}}\,dx$

9. $\int_{-\infty}^{-1} x^{-2}\,dx$

10. $\int_{-\infty}^{-4} x^{-2}\,dx$

11. $\int_{-\infty}^{-1} x^{-8/3}\,dx$

12. $\int_{-\infty}^{27} x^{-5/3}\,dx$

13. $\int_0^\infty 4e^{-4x}\,dx$

14. $\int_0^\infty 10e^{-10x}\,dx$

15. $\int_{-\infty}^0 4e^x\,dx$

16. $\int_{-\infty}^0 3e^{4x}\,dx$

17. $\int_{-\infty}^{-1} \ln|x|\,dx$

18. $\int_1^\infty \ln|x|\,dx$

19. $\int_0^\infty \frac{dx}{(x+1)^2}$

20. $\int_0^\infty \frac{dx}{(2x+1)^3}$

21. $\int_{-\infty}^{-1} \frac{2x-1}{x^2-x}\,dx$

22. $\int_0^\infty \frac{2x+3}{x^2+3x}\,dx$

Use the table on page 717 as necessary for the following.

23. $\int_{-\infty}^{1} \frac{2}{3x(2x-7)}\,dx$

24. $\int_1^\infty \frac{7}{2x(5x+1)}\,dx$

25. $\int_1^\infty \frac{4}{9x(x+1)^2}\,dx$

26. $\int_1^5 \frac{5}{4x(x+2)^2}\,dx$

27. $\int_0^\infty xe^{2x}\,dx$

28. $\int_{-\infty}^0 xe^{3x}\,dx$

APPLIED PROBLEMS

Management

Find the capitalized values of the following properties. See Problem 4 in the margin at the end of the text above.

29. A castle for which annual rent of \$60,000 will be paid in perpetuity; the interest rate is 8%

30. A fort on a strategic peninsula in the North Sea, annual rent \$500,000 paid in perpetuity; the interest rate is 6%

Natural Science

Radioactive waste is entering the atmosphere over an area at a decreasing rate. Use the improper integral

$$\int_0^\infty Pe^{-kt}\,dt$$

with $P = 50$ *to find the total amount of the waste that will enter the atmosphere for each of the following values of* k. *(See Example 3.)*

31. $k = .04$

32. $k = .06$

15.4 PROBABILITY DENSITY FUNCTIONS

A bank is interested in improving its services to the public. The manager decides to begin by finding out how much time its tellers spend on each transaction. She decides to time the transactions to the nearest minute. The time for a complete transaction t will then be one of the numbers 1, 2, 3, ... Because the value of the variable t will occur randomly, t is called a **random variable.**

A table of the results of 75 transactions, where $f(t)$ is the relative frequency of a transaction, is shown below. The shortest transaction time was 1 minute, and there were 3 transactions of 1-minute duration. The longest time was 10 minutes. Only one transaction took that long.

t	1	2	3	4	5	6	7	8	9	10
Number of occurrences	3	5	9	12	15	11	10	6	3	1
$f(t)$	$\frac{3}{75}$	$\frac{5}{75}$	$\frac{9}{75}$	$\frac{12}{75}$	$\frac{15}{75}$	$\frac{11}{75}$	$\frac{10}{75}$	$\frac{6}{75}$	$\frac{3}{75}$	$\frac{1}{75}$

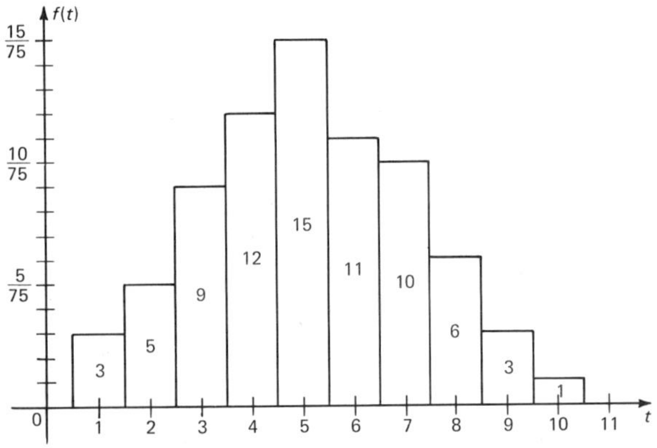

Figure 15.6

A graph of the relative frequency with which the various values of t occurred is shown in Figure 15.6.

The manager could use the results of her survey to get the probability that a transaction would last t minutes. The **probability** of an event can be defined as the relative frequency of occurrence of the event. Thus, the probability that $t = 1$, written $P(t = 1)$, is 3/75. Similarly, $P(t = 2)$ is 5/75 and $P(t = 3)$ is 9/75. To find the probability that the transaction time is no more than 3 minutes, we can add the three probabilities found above. That is,

$$P(t \le 3) = P(t = 1) + P(t = 2) + P(t = 3) = 17/75.$$

1. Find the following using the table of values given above.
(a) $P(t = 5)$
(b) $P(t = 6)$
(c) $P(5 \le t)$

Answer:
(a) 1/5
(b) 11/75
(c) 44/75

Work Problem 1 at the side.

It would have been possible to time the transactions more accurately—to the nearest tenth of a minute or even to the nearest second (or 1/60th of a minute) if desired. Theoretically, at least, the transaction times t could take on any positive real-number value (between, say, 0 and 11 minutes). This allows us to think of the graph of the relative frequency of transaction times as the continuous curve shown in Figure 15.7. As indicated in Figure 15.7, the curve was derived from the graph of Figure 15.6 by connecting the points at the top of the bars and smoothing the resulting polygon into a curve.

Note that the area of the bar above $t = 1$ in Figure 15.6 is 1 times 3/75 or 3/75. Since each bar has a width of 1, its area is equal to the relative frequency for that value of t. The probability that a particular value of t will occur is given by the area of the appropriate bar of the graph. Thus, if we prefer to think of the possible transaction times as all the real numbers between 0 and 11, we can use the area under the curve of Figure 15.7 between any two values of t

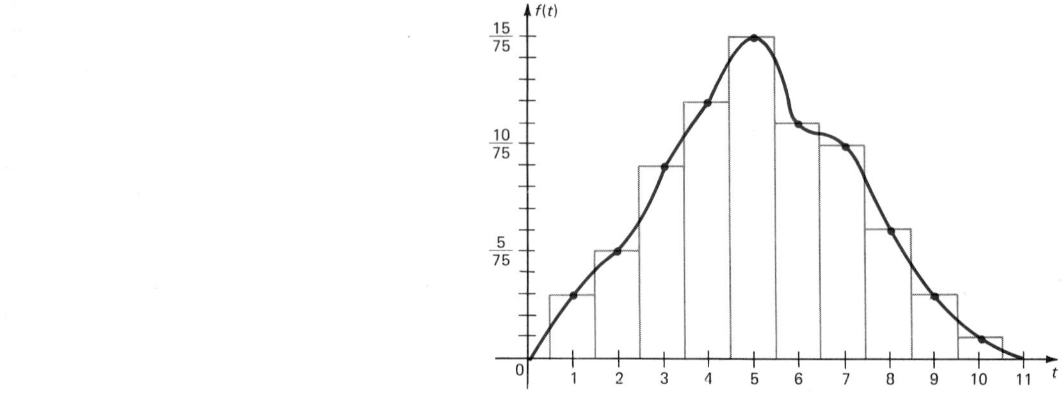

Figure 15.7

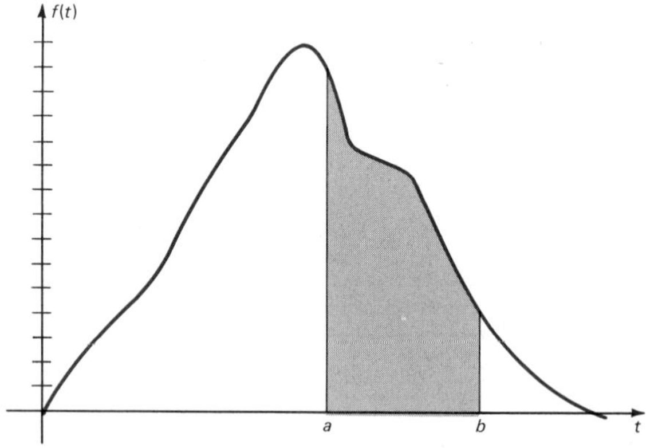

Figure 15.8

to find the probability that a transaction time will be between those numbers. For example, the shaded region in Figure 15.8 corresponds to the probability that t is between a and b, written $P(a \leq t \leq b)$.

In Chapter 14, we saw that the definite integral can be used to find the area under the graph of $f(x)$ from $x = a$ to $x = b$. If we can find a function $f(x)$ to describe a relative frequency curve, we can use the definite integral to find the area under the curve from a to b. Since that area represents the probability that x will be between a and b, we have

$$P(a \leq x \leq b) = \int_a^b f(x)\,dx.$$

A function $f(x)$ which can be used to describe a relative frequency curve is called a **probability density function.** Such a function must satisfy certain conditions. If x is a random variable over the interval $[a, b]$, then $f(x)$ is a probability density function if

1. $\int_a^b f(x)\,dx = 1$, and
2. $f(x) \geq 0$ for all x in the interval $[a, b]$.

Intuitively, property (1) says that the total probability must be 1; *something* must happen. Property (2) says that the probability of a particular event can never be negative.

Example 1 **(a)** Show that $f(x) = \frac{3}{26}x^2$ is a probability density function for the interval $[1, 3]$.

First, we show that condition 1 holds.

$$\int_1^3 \frac{3}{26}x^2\,dx = \frac{3}{26}\left(\frac{x^3}{3}\right)\Bigg]_1^3$$
$$= \frac{3}{26}\left(9 - \frac{1}{3}\right)$$
$$= 1$$

Next, we must show that $f(x) \geq 0$ for the interval $[1, 3]$. Since x^2 is always positive, condition 2 also holds. Thus, $f(x)$ is a probability density function.

(b) Find the probability that x will be between 1 and 2.

The desired probability is given by the area under the graph of $f(x)$ between $x = 1$ and $x = 2$ as shown in Figure 15.9. We find the area by using a definite integral.

$$P(1 \leq x \leq 2) = \int_1^2 \frac{3}{26}x^2\,dx = \frac{3}{26}\left(\frac{x^3}{3}\right)\Bigg]_1^2 = \frac{7}{26} \quad ■$$

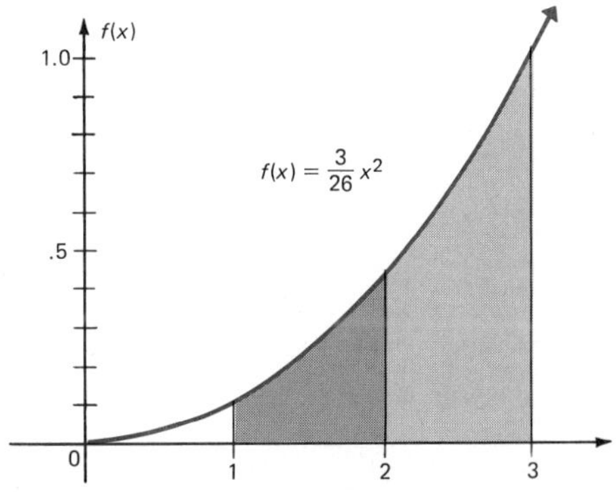

Figure 15.9

Example 2 Is $f(x) = 3x^2$ a probability density function for the interval $[0, 4]$? If not, convert it to one.

We have

$$\int_0^4 3x^2\,dx = x^3\Big]_0^4 = 64.$$

Since the integral is not equal to 1, the function is not a probability density function. To convert it to one, we simply multiply $f(x)$ by $\frac{1}{64}$. The function $P(x) = \frac{3}{64}x^2$ for $[0, 4]$ will be a probability density function since

$$\int_0^4 \frac{3}{64}x^2 = 1,$$

and $P(x) \geq 0$ for all x in $[0, 4]$. ■

2. (a) Is $f(x) = 2x$ a probability density function for the interval $[2, 6]$?
   (b) If your answer to (a) is no, convert $f(x)$ to a probability density function.

Answer:
(a) no
(b) let $f(x) = \dfrac{x}{16}$

Work Problem 2 at the side.

We now discuss some common probability density functions.

Uniform distribution This probability density function is given by

$$f(x) = \frac{1}{b-a} \qquad \text{for } x \text{ in } [a, b],$$

where a and b are real numbers. The graph is shown in Figure 15.10. The uniform distribution requires the assumption that the probability of any value of x between a and b is equally likely.

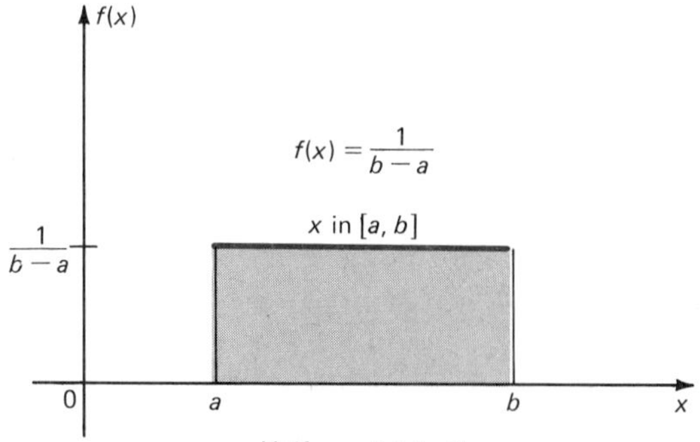

Uniform distribution

Figure 15.10

Example 3 A couple is planning to vacation in San Francisco. They have been told that the maximum daily temperature during the time they plan to be there ranges from 15°C to 27°C. What is the probability that the maximum temperature on the day they arrive will be greater than 24°C?

If we let t be the maximum temperature on a given day, then $f(t) = \frac{1}{12}$ for the interval $[15, 27]$ is the uniform probability density function for t. We have

$$P(t > 24) = \int_{24}^{27} \frac{1}{12}\,dt = \frac{1}{12}t\Big]_{24}^{27} = \frac{1}{4}.$$

By using the uniform probability density function here, we are assuming that the probability of any temperature between 15°C and 27°C is equally likely for any given day during the specified time period. ■

3. A certain species of plant normally grows to a height between 12 cm and 25 cm. Find the probability that a particular plant, selected at random, has a height between 18 and 20 cm. Assume a uniform distribution.

Answer: 2/13

Work Problem 3 at the side.

Exponential distribution The exponential distribution function is given by

$$f(x) = ae^{-ax} \qquad \text{for } 0 \le x < \infty,$$

where a is a positive real number. The graph is shown in Figure 15.11.

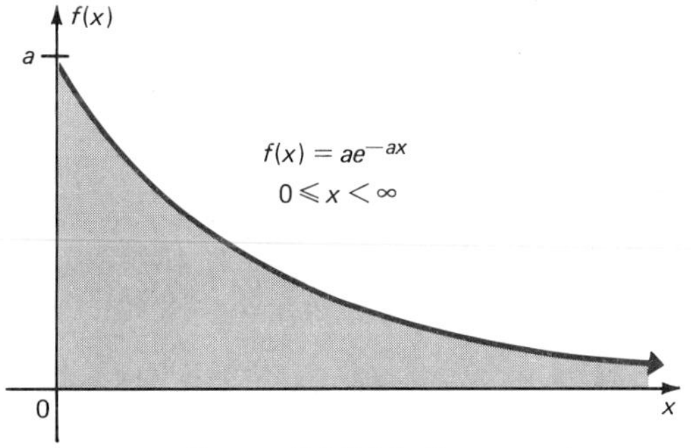

Figure 15.11

Example 4 Suppose the useful life of a flashlight battery is t, in hours, and the probability density function is given by $f(t) = \frac{1}{20}e^{-t/20}$ $(t \ge 0)$. Find the probability that a particular battery, selected at random, has a useful life of less than 100 hours.

The probability is given by

$$\begin{aligned}\int_0^{100} \frac{1}{20}e^{-t/20}\,dt &= \frac{1}{20}\left(-20e^{-t/20}\right)\Big]_0^{100}\\ &= -(e^{-100/20} - e^0)\\ &= -(e^{-5} - 1)\\ &\approx 1 - .0067\\ &= .9933. \quad ■\end{aligned}$$

4. The life span (in days) of a certain insect is a continuous random variable whose probability density function is given by

$$f(x) = .04e^{-.04x}.$$

Find the probability that a particular one of these insects, selected randomly, will die within 10 days of birth.

Answer: .32968

Work Problem 4 at the side.

Normal distribution The normal bell-shaped distribution is undoubtedly the most important probability density function. It is widely used in various applications of statistics. The normal distribution function is given by

$$f(x) = \frac{1}{\sigma\sqrt{2\pi}} e^{-(x-\mu)^2/2\sigma^2} \qquad \text{for } -\infty < x < \infty$$

where μ and σ are real numbers, with $\sigma \geq 0$. Using very advanced techniques, it can be shown that

$$\int_{-\infty}^{\infty} \frac{1}{\sigma\sqrt{2\pi}} e^{-(x-\mu)^2/2\sigma^2} = 1.$$

It would be far too much work to calculate values for this probability distribution for various values of μ and σ. Instead, values are calculated for the **standard normal distribution,** having $\mu = 0$ and $\sigma = 1$. The graph of the standard normal distribution is the bell-shaped curve shown in Figure 15.12.

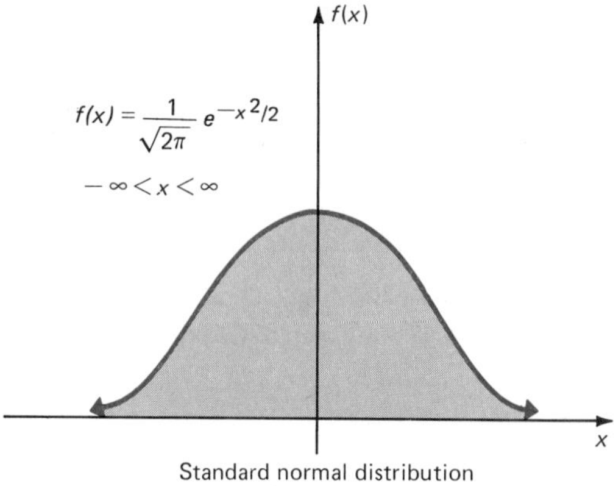

Figure 15.12

To find probabilities for the standard normal distribution we need to evaluate the definite integral

$$\int_a^b \frac{1}{\sqrt{2\pi}} e^{-x^2/2}.$$

As mentioned earlier, $e^{-x^2/2}$ is a function that does not have an antiderivative, so numerical methods must be used to find values of the definite integral.

15.4 EXERCISES

Decide if the following functions are probability density functions. If not, tell why. (See Example 1.)

1. $f(x) = \frac{1}{9}x - \frac{1}{18}, \quad 2 \leq x \leq 5$

2. $f(x) = \frac{1}{3}x - \frac{1}{6}, \quad 3 \leq x \leq 4$

3. $f(x) = \frac{3}{63}x^2, \quad 1 \leq x \leq 4$

4. $f(x) = \frac{3}{98}x^2, \quad 3 \leq x \leq 5$

5. $f(x) = 4x^3, \quad 0 \leq x \leq 3$

6. $f(x) = \frac{x^3}{81}, \quad 0 \leq x \leq 3$

7. $f(x) = \frac{x^2}{16}, \quad -2 \leq x \leq 2$

8. $f(x) = 2x^2, \quad -1 \leq x \leq 1$

9. $f(x) = 2x, \quad -2 \leq x \leq \sqrt{5}$

10. $f(x) = 4x^3, \quad -1 \leq x \leq \sqrt[4]{2}$

For each of the following, find a value of k which will make f(x) a probability density function. (See Example 2.)

11. $f(x) = kx^{1/2}, \quad 1 \leq x \leq 4$

12. $f(x) = kx^{3/2}, \quad 4 \leq x \leq 9$

13. $f(x) = kx^2, \quad 0 \leq x \leq 5$

14. $f(x) = kx^2, \quad -1 \leq x \leq 2$

15. $f(x) = kx, \quad 0 \leq x \leq 3$

16. $f(x) = kx, \quad 2 \leq x \leq 3$

17. $f(x) = kx, \quad 1 \leq x \leq 5$

18. $f(x) = kx, \quad 0 \leq x \leq 4$

19. $f(x) = kx^2, \quad 1 \leq x \leq 3$

20. $f(x) = kx^3, \quad 2 \leq x \leq 4$

APPLIED PROBLEMS

Solve the following problems. (See Examples 3 and 4.)

General

21. The probability density function of a random variable x is

$$f(x) = 1 - \frac{1}{\sqrt{x}} \quad \text{for } 1 \leq x \leq 4.$$

Find the following probabilities.
(a) $P(x \geq 3)$ (b) $P(x \leq 2)$ (c) $P(2 \leq x \leq 3)$

22. The probability density function of a random variable x is

$$f(x) = \frac{1}{11}\left(1 + \frac{3}{\sqrt{x}}\right) \quad \text{for } 4 \leq x \leq 9.$$

Find the following probabilities.
(a) $P(x \geq 6)$ (b) $P(x \leq 5)$ (c) $P(4 \leq x \leq 7)$

23. The probability density function of a random variable x is

$$f(x) = \frac{1}{15}\left(x + \frac{1}{2}\right) \qquad \text{for } 0 \le x \le 5.$$

Find the following probabilities.
(a) $P(x \ge 3)$ (b) $P(x \le 2)$ (c) $P(1 \le x \le 4)$

24. The probability density function of a random variable x is

$$f(x) = \frac{1}{2} \qquad \text{for } 1 \le x \le 3.$$

Find the following probabilities.
(a) $P(x \ge 2)$ (b) $P(x < \frac{3}{2})$ (c) $P(1 \le x \le \frac{5}{2})$

Natural Science

25. The length (in centimeters) of the leaf of a certain plant is a continuous random variable with a probability density function of

$$f(x) = \frac{5}{4}, \qquad 4 \le x \le 4.8.$$

Find the probability that a particular leaf from one of these plants will be between 4.5 and 4.8 cm in length.

Management

26. The price of an item in dollars is a continuous random variable with a probability density function of

$$f(x) = 2, \qquad 1.25 \le x \le 1.75.$$

Find the probability that the price will be less than \$1.35.

Natural Science

27. The length of time, t, in years until a particular radioactive particle decays is a random variable with a probability density function of

$$f(t) = .03e^{-.03t}, \qquad 0 \le t < \infty.$$

Find the probability that the particle decays within 20 years.

28. The length of time, t, in years that a seedling tree survives is a random variable with a probability density function of

$$f(t) = .05e^{-.05t}, \qquad 0 \le t < \infty.$$

Find the probability that a particular tree survives fewer than 10 years.

Social Science

29. The length of time, t, in days required to learn a certain task is a random variable with a probability density function of

$$f(t) = e^{-t}, \qquad 0 \le t < \infty.$$

Find the probability that a certain individual learns the task in 2 to 4 days.

Natural Science

30. The distance, x, in meters that seeds are dispersed from a certain kind of plant is a random variable with a probability density function of

$$f(x) = .1e^{-.1x}, \qquad 0 \le x < \infty.$$

Find the probability that the seeds of a particular plant are dispersed from 1 to 2 meters away.

15.5 SOLUTIONS OF DIFFERENTIAL EQUATIONS

The rate of change of many natural growth processes with respect to time can be expressed as a function of time in the form

$$\frac{dy}{dx} = kx \quad \text{or} \quad \frac{dy}{dx} = ky,$$

where x represents time and k is some constant. Such equations containing a derivative or differential are called **differential equations.** In most differential equations, it is convenient to represent the derivative as dy/dx, rather than y' or $f'(x)$. For a second derivative, the notation d^2y/dx^2 is used. Some examples of differential equations include

$$\frac{dy}{dx} = x^{1/2}, \quad \frac{dy}{dx} = 2y^2, \quad \text{and} \quad \frac{d^2y}{dx^2} = 3x^2 + 4.$$

The time-dating of dairy products depends on the solution of a differential equation. The rate of growth of bacteria in such products increases with time. If y is the number of bacteria in millions present at a time t in days, then the rate of growth of bacteria can be expressed as dy/dt, and we can write

$$\frac{dy}{dt} = kt,$$

where k is an appropriate constant. For simplicity, let us assume that $k = 10$ for a particular product, so that

$$\frac{dy}{dt} = 10t. \tag{1}$$

In Section 13.7, we defined the differential dy as $f'(x)\,dx$, and since $f'(x) = dy/dx$,

$$dy = \frac{dy}{dx}\,dx.$$

Using this idea, we have $dy = (dy/dt)\,dt$, and we can rewrite equation (1) as

$$\frac{dy}{dt}\,dt = 10t\,dt$$

or

$$dy = 10t\,dt.$$

Here we treat dy/dt as a quotient.

If we now integrate both sides of this equation, we get

$$\int dy = \int 10t\,dt.$$

The differential dy on the left indicates that we should integrate there with respect to y, while the dt on the right indicates that we should integrate with respect to t on that side. Performing the two integrations gives

$$y + C_1 = 5t^2 + C_2$$
$$y = 5t^2 + C_2 - C_1.$$

We can replace $C_2 - C_1$ with the single constant C to get

$$y = 5t^2 + C. \qquad \textbf{(2)}$$

Equation (2) is called the **general solution** of differential equation (1). (This process of replacing two or more arbitrary constants with one can be skipped from now on—we shall just include one arbitrary constant and let it go at that.)

Suppose there is a known number of bacteria present at time $t = 0$, say $y = 50$ in millions. Such a condition, called an **initial condition** or **boundary condition,** can be used in the general solution (2) to find C as follows.

$$y = 5t^2 + C$$
$$50 = 5(0)^2 + C$$
$$C = 50$$

With this value of C, equation (2) becomes

$$y = 5t^2 + 50. \qquad \textbf{(3)}$$

Since this solution to the differential equation depends on the particular values of t and y that were given, it is called a **particular solution** of differential equation (1).

After the maximum acceptable value of y is found, the number of days, t, the product is usable can be determined. For example, if the maximum value of y is to be 550 million, from equation (3) we have

$$y = 5t^2 + 50$$
$$550 = 5t^2 + 50$$
$$t^2 = 100$$
$$t = 10.$$

Thus, the product should be dated for 10 days from the date when $t = 0$.

To check the solution we found for equation (1) we differentiate $y = 5t^2 + 50$.

$$y' = \frac{dy}{dt} = 10t$$

Since the result agrees with the given equation we have the correct solution.

Example 1 Find the particular solution to

$$\frac{dy}{dx} = 3x^2 + 4x + 5,$$

given that $y = -1$ when $x = 0$.

We first find the general solution.

$$\frac{dy}{dx} = 3x^2 + 4x + 5$$
$$dy = (3x^2 + 4x + 5)\,dx$$
$$\int dy = \int (3x^2 + 4x + 5)\,dx$$
$$y = x^3 + 2x^2 + 5x + C$$

Then, substituting the given values for x and y, we can find C.

$$-1 = 0^3 + 2(0)^2 + 5(0) + C$$
$$C = -1$$

Thus, this particular solution is

$$y = x^3 + 2x^2 + 5x - 1,$$

as can be verified by differentiating. ■

Work Problem 1 at the side.

1. Find and check the particular solution to

$$\frac{dy}{dx} = 4x^3 - 3x^2 + 2x$$

given that $y = 2$ when $x = 1$.

Answer: $y = x^4 - x^3 + x^2 + 1$

Example 2 Find the particular solution to

$$\frac{d^2y}{dx^2} = 2x + 4, \qquad (4)$$

given that $y = 2$ when $x = 0$ and $y = -4$ when $x = 3$.

Integrating both sides with respect to x, we have

$$\int \frac{d^2y}{dx^2}\,dx = \int (2x + 4)\,dx.$$

The integral of the second derivative becomes

$$\int \frac{d^2y}{dx^2}\,dx = \frac{dy}{dx} + C.$$

Thus, the differential equation can be written as

$$\frac{dy}{dx} = x^2 + 4x + C_1,$$

for some constant C_1. Integrating again, we have

$$y = \frac{x^3}{3} + 2x^2 + C_1 x + C_2 \qquad \textbf{(5)}$$

for some constant C_2. Equation (5) is a general solution of equation (4). To find values of C_1 and C_2, we use the given pairs of values. If $x = 0$, then $y = 2$, so that

$$2 = \frac{0^3}{3} + 2 \cdot 0^2 + C_1 \cdot 0 + C_2$$

$$C_2 = 2.$$

Using $C_2 = 2$, $x = 3$, and $y = -4$, we have

$$-4 = \frac{3^3}{3} + 2 \cdot 3^2 + C_1 \cdot 3 + 2$$

$$-4 = 29 + 3C_1$$

$$C_1 = \frac{-33}{3} = -11.$$

The particular solution is thus

$$y = \frac{x^3}{3} + 2x^2 - 11x + 2.$$

To check this solution, differentiate twice. ■

2. Find and check the particular solution to

$$\frac{d^2y}{dx^2} = x^2 - 4x,$$

given that $y = 5$ when $x = 0$ and $y = 1$ when $x = -1$.

Answer: $y = \frac{x^4}{12} - \frac{2x^3}{3} + \frac{57x}{12} + 5$

Work Problem 2 at the side.

15.5 EXERCISES

Find general solutions for the following differential equations.

1. $\frac{dy}{dx} = x^2$

2. $\frac{dy}{dx} = -x + 2$

3. $\frac{dy}{dx} = -2x + 3x^2$

4. $\frac{dy}{dx} = 6x^2 - 4x$

5. $\frac{dy}{dx} = -4 + 3x^3$

6. $\frac{dy}{dx} = 4x^3 + 3$

7. $\frac{dy}{dx} = e^x$

8. $\frac{dy}{dx} = e^{3x}$

9. $\frac{dy}{dx} = 2e^{-x}$

10. $\frac{dy}{dx} = 3e^{-2x}$

11. $3\frac{dy}{dx} = -4x^2$

12. $3x^3 - 2\frac{dy}{dx} = 0$

13. $4\frac{dy}{dx} - x = 0$

14. $3x^2 - 3\frac{dy}{dx} = 2$

15. $4x - 4 + \frac{dy}{dx} = 6x^2$

16. $-8x^3 - 3x^2 + \frac{1}{2} \cdot \frac{dy}{dx} = 3$

17. $\frac{d^2y}{dx^2} = -8x$

18. $\frac{d^2y}{dx^2} = 4x$

19. $\frac{d^2y}{dx^2} = 5 - 4x$

20. $\frac{d^2y}{dx^2} = 2 + 8x$

21. $\frac{d^2y}{dx^2} = 4x^2 - 2x$

22. $x^2 - \frac{d^2y}{dx^2} = 3x$

23. $5x - 3 + \frac{d^2y}{dx^2} = 0$

24. $\frac{d^2y}{dx^2} = 6 - x$

25. $\frac{d^2y}{dx^2} = e^x$

26. $\frac{d^2y}{dx^2} = 6e^{-x}$

Find particular solutions for the following differential equations. (*See Examples 1 and 2.*)

27. $\frac{dy}{dx} + 2x = 3x^2$; $y = 2$ when $x = 0$

28. $\frac{dy}{dx} = 4x + 3$; $y = -4$ when $x = 0$

29. $\frac{dy}{dx} = 5x + 2$; $y = -3$ when $x = 0$

30. $\frac{dy}{dx} = 6 - 5x$; $y = 6$ when $x = 0$

31. $\frac{dy}{dx} = 3x^2 - 4x + 2$; $y = 3$ when $x = -1$

32. $\frac{dy}{dx} = 4x^3 - 3x^2 + x$; $y = 0$ when $x = 1$

33. $\frac{d^2y}{dx^2} = 2x + 1$; $y = 2$ when $x = 0$; $y = 3$ when $x = -2$

34. $\frac{d^2y}{dx^2} = -3x + 2$; $y = -3$ when $x = 0$; $y = -19$ when $x = 3$

35. $\frac{d^2y}{dx^2} = e^x + 1$; $y = 2$ when $x = 0$; $y = 3/2$ when $x = 1$

36. $\dfrac{d^2y}{dx^2} = -e^x - 2;\quad y = -2$ when $x = 0;\quad y = 3/2$ when $x = 1$

37. $3x^2 - \dfrac{d^2y}{dx^2} = 2;\quad y = 4$ when $x = 0;\quad y = 6$ when $x = 2$

38. $\dfrac{d^2y}{dx^2} + 4x^2 = 1;\quad y = 1/6$ when $x = 1;\quad y = -1$ when $x = 0$

APPLIED PROBLEMS

Natural Science

39. In the time-dating of dairy products example discussed in the text, suppose the number of bacteria present at time $t = 0$ was 250.
(a) Find a particular solution of the differential equation.
(b) Find the number of days the product can be sold if the maximum value of y is 970.

Social Science

40. Suppose the rate at which a rumor spreads—that is, the number of people who have heard the rumor over a period of time—increases with the number of days. If y is the number of people who have heard the rumor, then

$$\frac{dy}{dt} = kt,$$

where t is the time in days.
(a) If y is 0 when $t = 0$, and y is 100 when $t = 2$, find k.
(b) Using the value of k from part (a), find y when $t = 3; 5; 10$.

15.6 SEPARATION OF VARIABLES

The differential equations we have discussed up to this point have been of the form

$$\frac{dy}{dx} = f(x) \qquad \text{or} \qquad \frac{d^2y}{dx^2} = f(x).$$

Another common type of differential equation can be written in the form

$$h(y)\,dy = f(x)\,dx,$$

where all terms involving y (including dy) are on one side of the equation, and all terms involving x (and dx) are on the other side. The process of obtaining a differential equation in this form is called **separating the variables.** When we have a differential equation in which the variables are separated, we can solve the equation by integrating both sides.

Example 1 Find the general solution of

$$y\frac{dy}{dx} = x^2.$$

We begin by separating the variables to get

$$y\,dy = x^2\,dx.$$

The general solution can be found by integrating both sides.

$$\int y\,dy = \int x^2\,dx$$

$$\frac{y^2}{2} = \frac{x^3}{3} + C$$

$$y^2 = \frac{2}{3}x^3 + 2C$$

$$y^2 = \frac{2}{3}x^3 + K \quad ■$$

When the solution of a differential equation leads to a result with y raised to a power, as in Example 1, it is customary to leave the solution in that form, rather than solve explicitly for y.

Work Problem 1 at the side.

Example 2 Find the general solution of

$$2xy\frac{dy}{dx} = 4.$$

We separate the variables as follows.

$$2xy\frac{dy}{dx} = 4$$

$$2y\frac{dy}{dx} = \frac{4}{x} \qquad \text{(assume } x \neq 0\text{)}$$

$$2y\,dy = \frac{4}{x}\,dx$$

Integrating both sides gives the general solution

$$y^2 = 4\ln|x| + C.$$

Verify that this is the correct solution by differentiating implicitly and substituting in the original equation. ■

Work Problem 2 at the side.

1. Find the general solution of

$$3y^2\frac{dy}{dx} = x + 1.$$

Answer: $y^3 = \dfrac{x^2}{2} + x + C$

2. Find the general solution of

$$3x^2\frac{dy}{dx} = 4.$$

Answer: $y = -\dfrac{4}{3x} + C$

Example 3 Find the general solution of $\frac{dy}{dx} = ky$, where k is a constant.

By separating variables we get

$$\frac{1}{y}\,dy = k \cdot dx$$

To solve this equation, we integrate both sides.

$$\int \frac{1}{y}\,dy = \int k \cdot dx$$

$$\ln|y| = kx + C.$$

This general solution can be rewritten, using the definition of natural logarithms, as

$$\begin{aligned} y &= e^{kx+C} \\ &= e^{kx}e^{C} \\ &= Me^{kx}, \end{aligned}$$

where we replaced the constant e^C with the constant M. Since Me^{kx} is never negative, we don't need the absolute value bars around y. ■

In general, in the equation

$$y = Me^{kx},$$

the constant k gives the rate of change of a quantity of size M at time $x = 0$. The equation represents growth when k is positive and decay when k is negative. The function $y = Me^{kx}$ is extremely important because of its many applications in management and the social and behavioral sciences.

Example 4 Find the general solution of

$$x^3 \frac{d^2y}{dx^2} = 3.$$

Since $y' = \frac{dy}{dx}$, we can write $\frac{d^2y}{dx^2}$ as $\frac{dy'}{dx}$.

$$x^3 \frac{dy'}{dx} = 3$$

Now, separate the variables.

$$dy' = \frac{3}{x^3}\,dx$$

If we integrate both sides, we have

$$y' = -\frac{3}{2}x^{-2} + C_1.$$

Substitute $\frac{dy}{dx}$ for y' and integrate again to get the general solution.

$$\frac{dy}{dx} = -\frac{3}{2}x^{-2} + C_1$$

$$dy = \left(-\frac{3}{2}x^{-2} + C_1\right)dx$$

$$y = \frac{3}{2}x^{-1} + C_1x + C_2$$

$$y = \frac{3}{2x} + C_1x + C_2 \quad \blacksquare$$

3. Find the general solution of

$$x^2\frac{d^2y}{dx^2} = 12x^4 + 2x^2.$$

Answer: $y = x^4 + x^2 + C_1x + C_2$.

Work Problem 3 at the side.

In Sections 15.5 and 15.6 we have discussed only the simplest techniques for solving differential equations. It is not always possible to separate the variables in a differential equation. For example, try to separate the variables in

$$\frac{dy}{dx} = x^2 + y^2.$$

For this equation and other types which arise in applications, more advanced methods are required.

15.6 EXERCISES

Find general solutions for the following differential equations. (See Examples 1–4.)

1. $y\frac{dy}{dx} = x$

2. $y^2 \cdot \frac{dy}{dx} = x^2 - 1$

3. $y\frac{dy}{dx} = x^2 - 1$

4. $2x + x^2 - y\frac{dy}{dx} = 0$

5. $\frac{dy}{dx} = 2xy$

6. $\frac{dy}{dx} = x^2y$

7. $\frac{dy}{dx} = 3x^2y - 2xy$

8. $(y^2 - y)\frac{dy}{dx} = x$

9. $\frac{d^2y}{dx^2} - 4x = 0$

10. $\frac{d^2y}{dx^2} - 2x + 3x^2 = 0$

11. $2\frac{d^2y}{dx^2} - 3x^2 = 0$

12. $2\frac{d^2y}{dx^2} = x^2$

Find particular solutions for the following.

13. $\frac{dy}{dx} = \frac{x^2}{y}$; $y = 3$ when $x = 0$

14. $x^2 \frac{dy}{dx} = y$; $y = -1$ when $x = 1$

15. $(2x + 3)y = \frac{dy}{dx}$; $y = 1$ when $x = 0$

16. $x\frac{dy}{dx} - y\sqrt{x} = 0$; $y = 1$ when $x = 0$

17. $\frac{d^2y}{dx^2} + x = 2$; $y = 2$ when $x = 0$ and $y = -2$ when $x = 1$

18. $\frac{4}{x} \cdot \frac{dy^2}{dx^2} = 8x$; $y = -1$ when $x = 0$ and $y = 3$ when $x = -1$

APPLIED PROBLEMS

Natural Science

19. A radioactive substance decays at a rate given by

$$\frac{dy}{dx} = -0.05y,$$

where y represents the amount, in grams, present at time x, in months.
(a) Find the general solution.
(b) Find a particular solution if $y = 90$ when $x = 0$.
(c) Find the amount left when $x = 10$.

Management

20. Extensive experiments have shown that under relatively constant market conditions, sales of a product, in the absence of promotional activities such as advertising, decrease at a constant yearly rate. This rate of sales decline varies considerably from product to product, but it seems to be relatively constant for a particular product. Suppose the yearly rate of sales decrease for a certain company is given by

$$\frac{dy}{dt} = -0.25y,$$

where t is the time in years and y is sales in thousands of dollars.
(a) Find a particular solution if $y = 80$ when $t = 0$.
(b) Find y when $t = 2; 4$.

CHAPTER 15 TEST

Use integration by parts to integrate the following.

[15.1] **1.** $\int x(8 + x)^{3/2}\,dx$ **2.** $\int x\,e^x\,dx$

3. $\int (x+2)e^{-3x}\,dx$

4. $\int (3+x^2)e^{2x}\,dx$

[15.2]

5. Find $\int_2^6 \frac{dx}{x^2-1}$ by using the trapezoidal rule with $n = 4$.

6. Find $\int_2^6 \frac{dx}{x^2-1}$ by using Simpson's rule with $n = 4$.

[15.3]

Find the following improper integrals which are convergent.

7. $\int_{-\infty}^{-2} x^{-2}\,dx$

8. $\int_1^{\infty} x^{-1}\,dx$

9. $\int_0^{\infty} 6e^{-8x}\,dx$

10. $\int_0^{\infty} \frac{dx}{(5x+2)^2}$

[15.4]

11. Is $f(x) = 2x + 4$ for $1 \le x \le 4$ a probability density function? If not, tell why.

12. Find k such that $f(x) = kx + 1$ for $1 \le x \le 3$ is a probability density function.

13. The time t in years until a certain machine requires repairs is a random variable whose density function is

$$f(x) = \frac{5}{112}(1 - x^{-3/2}) \qquad \text{for} \quad 1 \le x \le 25.$$

Find the probability that no repairs are required in the first four years by finding the probability that a repair will be needed in years 4 through 25.

14. The distance in meters that a certain animal moves away from a release point is a random variable x whose probability density function is

$$f(x) = .01e^{-.01x} \qquad \text{for} \quad 0 \le x < \infty.$$

Find the probability that the animal will move no farther than 100 meters away.

[15.5] and [15.6]

Find general solutions for the following.

15. $\frac{dy}{dx} = -3 + 3x^2 + 5x^4$

16. $\frac{dy}{dx} = 4e^{-2x}$

17. $\frac{d^2y}{dx^2} - e^x = 3x^2$

18. $(y + y^2)\frac{dy}{dx} = x^2y$

Find particular solutions for the following.

19. $\sqrt{x} + \frac{dy}{dx} = x;\quad y = 2$ when $x = 0$

20. $\frac{dy}{dx} - 4xy = 0;\quad y = -1$ when $x = 0$

CASE 14 A CROP-PLANTING MODEL*

Many firms in food processing, seed production, and similar industries face a problem every year deciding how many acres of land to plant in each of various crops. Demand for the crop is unknown, as is the actual yield per acre. In this case, we set up a mathematical model for determining the optimum number of acres to plant in a crop.

This model is designed to tell the company the number of acres of seed that it should plant. The model uses the following variables (a simplified version of this model was given in Case 1).

D = number of tons of seed demanded

$f(D)$ = continuous probability density function for the quantity of seed demanded, D

$F(D)$ = cumulative probability distribution for the quantity of seed demanded, D

X = quantity of seed produced per acre of land

Q = quantity of seed carried over in inventory from previous years

S = selling price per ton of seed

C_p = variable costs of production, marketing, etc., per ton of seed

C_c = cost to carry over a ton of seed from previous years

A = number of acres of land to be planted

C_A = variable cost per acre of land contracted

T = total number of tons of seed available for sale

a = lower limit of the domain of $f(D)$

b = upper limit of the domain of $f(D)$

To decide on the optimum number of acres to plant, it is necessary to calculate the *expected value of the profit* from the planting of D acres. Expected value, an idea from probability and statistics, is defined as the product of the profit (or loss) from a certain outcome and the probability of that outcome happening.

Based on the definition of the variables above, the total number of tons of seed that will be available for sale is given by the product of the number of acres planted, A, and the yield per acre, X, added to the carryover, Q. If T represents this total, then

$$T = AX + Q.$$

The variable here is A; we assume X and Q are known and fixed.

The expected profit can be broken down into several parts. The first portion comes from multiplying the profit per ton and the average number of tons demanded. The profit per ton is found by subtracting the variable

---

* Based on work by David P. Rutten, Senior Mathematician, The Upjohn Company, Kalamazoo, Michigan.

cost per ton, C_p, from the selling price per ton, S:

$$\text{profit per ton} = S - C_p.$$

It is shown in more advanced courses that the average number of tons demanded for our interval of concern is given by

$$\int_a^T D \cdot f(D)\, dD.$$

Thus, this portion of the expected profit is

$$(S - C_p) \cdot \int_a^T D \cdot f(D)\, dD. \tag{1}$$

A second portion of expected profit is found by multiplying the profit per ton, $S - C_p$, the total number of tons available, T (recall that this is a variable), and the probability that T or more tons will be demanded by the marketplace.

$$(S - C_p)(T) \int_T^b f(D)\, dD \tag{2}$$

If T is greater than D, there will be costs associated with carrying over the excess seeds. The expected value of these costs is given by the product of the carrying cost per ton, C_c, and the number of tons to be carried over, or

$$-C_c \int_a^T (T - D) f(D)\, dD. \tag{3}$$

The minus sign shows that these costs reduce profit. If $T < D$, this term would be omitted.

Finally, the total cost of producing the seeds is given by the product of the variable cost per acre and the number of acres:

$$-C_A \cdot A. \tag{4}$$

The expected profit is the sum of the expressions in (1)–(4), or

$$\begin{aligned} \text{expected profit} = (S - C_p) \cdot \int_a^T D \cdot f(D)\, dD + (S - C_p)(T) \int_T^b f(D)\, dD \\ - C_c \int_a^T (T - D) f(D)\, dD - C_A \cdot A. \end{aligned} \tag{5}$$

As an example, suppose that

$$\begin{aligned} \text{demand density function} = f(D) &= \frac{1}{1000} \qquad \text{for } 500 \leq D \leq 1500 \text{ tons} \\ a &= 500 \\ b &= 1500 \\ \text{selling price} = S &= \$10{,}000 \text{ per ton} \\ \text{variable cost} = C_p &= \$5000 \text{ per ton} \\ \text{carrying cost} = C_c &= \$3000 \text{ per ton} \\ \text{variable cost per acre} = C_A &= \$100 \\ \text{inventory carryover} = Q &= 200 \text{ tons} \\ \text{yield per acre} = X &= .1 \text{ ton} \\ T = AX + Q &= .1A + 200. \end{aligned}$$

Substitute all this into equation (5).

$$\text{expected profit} = (10{,}000 - 5000)\int_{500}^{.1A+200} D \cdot \frac{1}{1000}\,dD$$
$$+ (10{,}000 - 5000)(.1A + 200)\int_{.1A+200}^{1500} \frac{1}{1000}\,dD$$
$$- 3000\int_{500}^{.1A+200} (.1A + 200 - D)\cdot\frac{1}{1000}\,dD - 100A$$

Simplify all this.

$$\text{expected profit} = \frac{5000}{1000}\cdot\frac{D^2}{2}\Big]_{500}^{.1A+200} + (5000)(.1\text{A} + 200)\frac{D}{1000}\Big]_{.1A+200}^{1500}$$
$$- 3000\left(\frac{.1AD}{1000} + \frac{200D}{1000} - \frac{D^2}{2000}\right)\Big]_{500}^{.1A+200}$$
$$= \frac{5}{2}(.1A + 200)^2 - \frac{5}{2}\cdot 500^2 + 5(.1A + 200)1500$$
$$- 5(.1A + 200)(.1A + 200) - .3A(.1A + 200)$$
$$- 600(.1A + 200) + \frac{3}{2}(.1A + 200)^2 + 150A$$
$$+ 600(500) - \frac{3}{2}(500)^2 - 100A$$
$$= -(.1A + 200)^2 + 6900(.1A + 200) - .3A(.1A + 200)$$
$$+ 50A - 700{,}000$$

To find the maximum expected profit, take the derivative of this function with respect to A and then place it equal to 0.

$$D[\text{expected profit}] = -2(.1A + 200)(.1) + 690 - .06A - 60 + 50$$
$$= -.02A - 40 + 690 - .06A - 10$$
$$= -.08A + 640$$

Place this derivative equal to 0.

$$-.08A + 640 = 0$$
$$640 = .08A$$
$$\frac{640}{.08} = A$$
$$8000 = A$$

If 8000 acres are planted, the maximum profit will be obtained. Compare this with the answer to Case 1.

16
Functions of Several Variables

So far in this book all the functions that we have discussed have used only one independent variable. We have seen functions that give sales in terms of price, or advertising, and functions that give populations in terms of time. In many cases, however, a valid mathematical model must include more than one independent variable.

For example, a mathematical model for the sales of air conditioners would include several variables—the price, the weather, the availability and price of electricity, among other possible variables. A mathematical model for the population of foxes in an area might involve such variables as the population of rabbits, squirrels, gophers, and other food resources.

In this chapter, we shall look at the mathematics of functions of more than one independent variable. We shall see that many of the ideas developed for functions of one variable also apply to functions of more than one variable. In particular, the fundamental idea of a derivative generalizes in a very natural way to functions of more than one variable.

16.1 FUNCTIONS OF SEVERAL VARIABLES

Suppose the value of the variable z depends on the value of x and y according to the relationship

$$z = 4x^2 + 2xy + 3y.$$

For every pair of values of x and y that we might choose we can find exactly one value of z. For example, if $x = 1$ and $y = -4$, we have

$$\begin{aligned} z &= 4x^2 + 2xy + 3y \\ z &= 4(1)^2 + 2(1)(-4) + 3(-4) \\ &= 4 - 8 - 12 \\ z &= -16. \end{aligned}$$

Therefore, if $x = 1$ and $y = -4$, then $z = -16$. To abbreviate this, we let $z = f(x, y)$, so that

$$f(x, y) = 4x^2 + 2xy + 3y.$$

"If $x = 1$ and $y = -4$, then $z = -16$" can now be written as just

$$f(1, -4) = -16.$$

To find $f(3, -2)$, let $x = 3$ and $y = -2$.

$$\begin{aligned} f(x, y) &= 4x^2 + 2xy + 3y \\ f(3, -2) &= 4(3)^2 + 2(3)(-2) + 3(-2) \\ &= 36 - 12 - 6 \\ f(3, -2) &= 18 \end{aligned}$$

Also, $f(0, 0) = 0$ and $f(2, 5) = 51$.

1. Let $f(x, y) = 8x + 9xy - 4y^2 + 1$. Find the following.
(a) $f(2, 1)$
(b) $f(-3, 4)$
(c) $f(0, 0)$

Answer:
(a) 31
(b) -195
(c) 1

Work Problem 1 at the side.

In general, $z = f(x, y)$ is a **function of two variables** whenever each pair of values for x and y leads to just one value of z. The variables x and y are called **independent variables;** z is the **dependent variable.**

Example 1 Let $f(x, y) = 4x\sqrt{x^2 + y^2}$. Find the following.

(a) $f(0, 0)$
Let $x = 0$ and $y = 0$. $f(0, 0) = 4(0)\sqrt{0^2 + 0^2} = 0$.

(b) $f(3, -4)$
$$\begin{aligned} f(3, -4) &= 4(3)\sqrt{3^2 + (-4)^2} = 12\sqrt{9 + 16} = 12\sqrt{25} \\ &= 12(5) = 60 \end{aligned}$$

(c) $f(-5, -7)$
$$f(-5, -7) = 4(-5)\sqrt{(-5)^2 + (-7)^2} = -20\sqrt{74}$$ ■

2. Let $g(x, y) = -4x/\sqrt{x^2 + y^2}$. Find the following, where possible.
(a) $g(3, 0)$
(b) $g(-4, -3)$
(c) $g(0, 0)$

Answer:
(a) -4
(b) $16/5$
(c) function not defined at (0, 0)—can't have a 0 denominator

Work Problem 2 at the side.

Example 2 The amount of money, $M(x, y)$, in dollars, that a twelve-year-old girl will spend in a week on rock-and-roll records depends on her weekly allowance, x, in dollars, and the amount, y, in dollars, that she spends weekly on cosmetics. In a suburb of Carmichael, it was found that $M(x, y)$ can be closely approximated by the mathematical model

$$M(x, y) = 4x - 2y - 9.$$

Find $M(4, 2.5)$.

Let $x = 4$ and $y = 2.5$.

$$M(4, 2.5) = 4(4) - 2(2.5) - 9 = 16 - 5 - 9 = 2$$

The girl spends \$2 on records. This amount, together with the \$2.50 spent on cosmetics, exceeds her allowance by 50¢. ■

Work Problem 3 at the side.

3. In Example 2, how much will be spent on records by a girl whose allowance is \$3 per week and who spends \$1.25 per week on cosmetics?

Answer: 50¢

Example 3 Let x represent the amount of carbon dioxide released by the lungs in one minute; x is measured in cc of carbon dioxide. Let y be the change in the carbon dioxide content of the blood as it leaves the lungs (y is measured in cc of carbon dioxide per 100 cc of blood). Let C be the total output of blood from the heart in one minute (measured in cc). Then

$$C = \frac{100x}{y}.$$

Let $y = 6$ cc of carbon dioxide per 100 cc of blood per minute and find C when $x = 320$ cc of carbon dioxide per minute.

We have

$$C = \frac{100(320)}{6} \approx 5333 \text{ cc of blood per minute}$$

$$\approx 5.3 \text{ liters of blood per minute.}$$ ■

Work Problem 4 at the side.

4. In Example 3, find C when $x = 275$ and $y = 5$.

Answer: 5500 cc or 5.5 liters

While we have defined only a function of two independent variables, similar definitions could be given for functions of three, four, or more independent variables.

Example 4 Let $f(x, y, z) = 4xz - 3yx^2 + 2z^2$. Find the following.

(a) $f(2, -3, 1)$

$$\begin{aligned} f(2, -3, 1) &= 4(2)(1) - 3(-3)(2)^2 + 2(1)^2 \\ &= 8 + 36 + 2 \\ &= 46 \end{aligned}$$

(b) $$\begin{aligned} f(-4, 3, -2) &= 4(-4)(-2) - 3(3)(-4)^2 + 2(-2)^2 \\ &= 32 - 144 + 8 \\ &= -104 \end{aligned}$$ ■

Work Problem 5 at the side.

5. Let $g(x, y, z) = 5x^2 - 4xz + 2yz - 3$. Find the following.

(a) $g(3, 0, 1)$

(b) $g(-2, 1, 4)$

(c) $g(6, -1, 0)$

Answer:

(a) 30

(b) 57

(c) 177

16.1 EXERCISES

Let $f(x, y) = 4x + 5y + 3$. Find the following. (See Example 1.)

1. $f(2, -1)$ **2.** $f(-4, 1)$ **3.** $f(-2, -3)$ **4.** $f(0, 8)$

Let $g(x, y) = -x^2 - 4xy + y^3$. Find the following.

5. $g(-2, 4)$ **6.** $g(-1, -2)$ **7.** $g(-2, 3)$ **8.** $g(5, 1)$

Let $h(x, y) = \sqrt{x^2 + 2y^2}$. Find the following.

9. $h(5, 3)$ **10.** $h(2, 4)$ **11.** $h(-1, -3)$ **12.** $h(-3, -1)$

APPLIED PROBLEMS

Natural Science

The population of cats on a certain farm is approximated by the mathematical model

$$C(x, y) = x^2 + 200y - 1200,$$

where x is the population, in hundreds, of small mice and y is the population, in tens, of large rats. Find the following. (See Examples 2 and 3.)

13. $C(50, 0)$ **14.** $C(30, 4)$

15. How many cats will be present if there are 1400 small mice and 150 large rats?

16. If the farm has 3000 small mice and 200 large rats, how many cats will be present?

Management

The labor charge for assembling a precision camera is given by

$$L(x, y) = 12x + 6y + 2xy + 40,$$

where x is the number of work hours required by a skilled craftsperson and y is the number of hours required by a semiskilled person. Find the following.

17. $L(3, 5)$

18. $L(5, 2)$

19. If a skilled craftsperson requires 7 hours and a semiskilled person needs 9 hours, find the total labor charge.

20. Find the total labor charge if a skilled worker needs 12 hours and a semiskilled worker requires 4 hours.

*Use a calculator with an x^y key, or logarithms, to work the following problems.**

---

* From *Mathematics in Biology* by Duane J. Clow and N. Scott Urquhart. Copyright © 1974 by W. W. Norton & Company, Inc. Used by permission.

Natural Science

The oxygen consumption of a well-insulated mammal which is not sweating is approximated by

$$m = \frac{2.5(T - F)}{w^{.67}},$$

where T is the internal body temperature of the animal (in °C), F is the temperature of the outside of the animal's fur (in °C), and w is the animal's weight in kilograms. Find m for the following data. (See Example 4.)

21. $T = 38°$, $F = 6°$, $w = 32$ kg

22. $T = 40°$, $F = 20°$, $w = 43$ kg

The surface area of a human (in square meters) is approximated by

$$A = 2.02W^{.425}H^{.725},$$

where W is the weight of the person in kilograms and H is the height in meters. Find A for the following data.

23. $W = 72$, $H = 1.78$

24. $W = 65$, $H = 1.40$

25. $W = 70$, $H = 1.60$

26. Find your own surface area.

16.2 GRAPHING FUNCTIONS OF TWO VARIABLES

When we graph functions of one independent variable, we use ordered pairs. We use a similar method to graph functions of two independent variables. For example, if $z = f(x, y) = 4x + 2y$, we might let $x = 2$ and $y = -6$:

$$\begin{aligned} z = f(2, -6) &= 4(2) + 2(-6) \\ &= -4. \end{aligned}$$

If $x = 2$ and $y = -6$, then $z = -4$. This result is written as the **ordered triple** $(2, -6, -4)$. Other ordered triples for the function $f(x, y) = 4x + 2y$ are $(-3, 5, -2)$, $(6, -4, 16)$, $(0, 0, 0)$, and so on. In the ordered triple (x, y, z), we assume that $z = f(x, y)$.

1. Let $z = f(x, y) = -9x + 2y + 5$. Complete the following ordered triples.
(a) $(0, 0, \quad)$
(b) $(-3, 2, \quad)$
(c) $(1, 7, \quad)$

Answer:
(a) $(0, 0, 5)$
(b) $(-3, 2, 36)$
(c) $(1, 7, 10)$

Work Problem 1 at the side.

Ordered pairs are graphed with two axes, an x-axis and a y-axis. Ordered triples are graphed with three axes, an x-axis, a y-axis, and a z-axis. Each of these three axes is perpendicular to the other two. Figure 16.1 shows one possible way to draw the graph of these three axes. In this figure, the plane containing the y-axis

2. Graph the following ordered triples.
(a) (2, 5, −3)
(b) (−1, 6, 3)
(c) (0, 4, −1)
(d) (3, 0, −2)
(e) (2, −5, 0)

Answer:

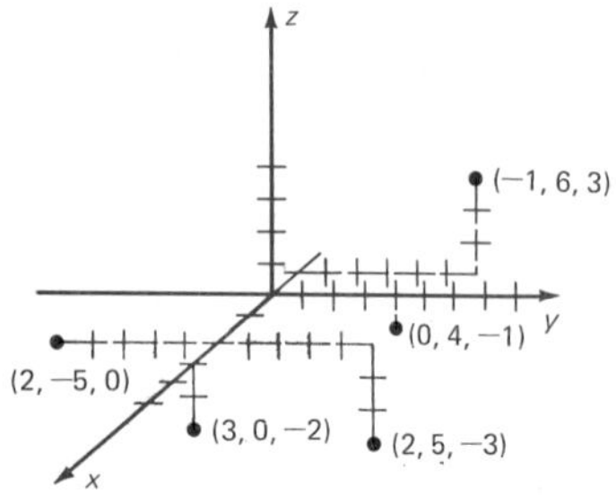

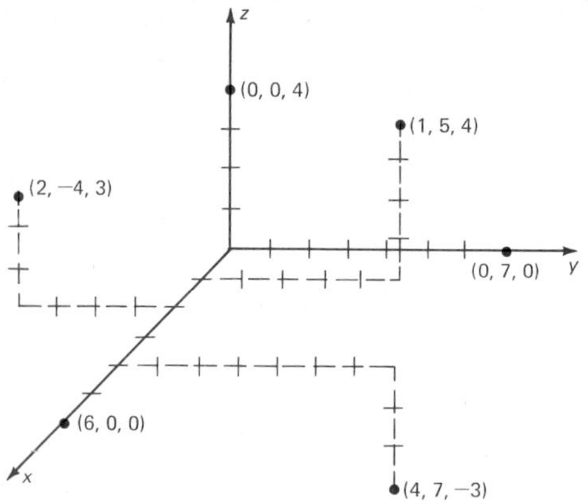

Figure 16.1

and the z-axis is in the plane of the page, while the x-axis is perpendicular to the plane of the page.

To locate the point corresponding to the ordered triple $(2, -4, 3)$, start at the origin and go 2 units along the positive x-axis. Then go 4 units in a negative direction, parallel to the y-axis. Finally, go up 3 units, parallel to the z-axis. The point representing $(2, -4, 3)$ and several other sample points are shown in Figure 16.1. The region of three-dimensional space where all coordinates are positive is called the **first octant.**

Work Problem 2 at the side.

3. Graph the following planes.
(a) $2x + 3y + 4z = 12$
(b) $5x + 2y + 3z = 15$

Answer:

(a)

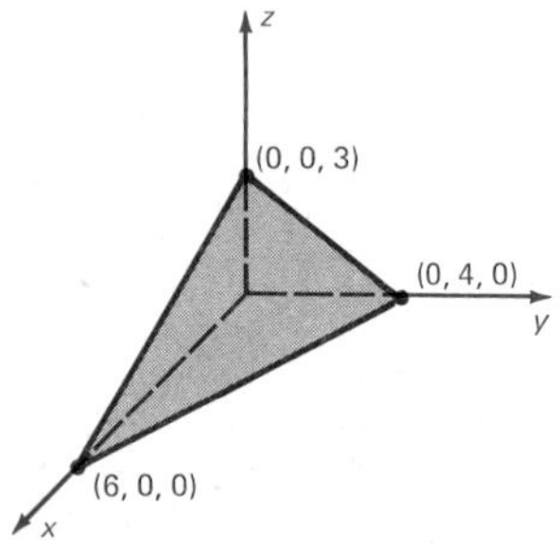

(b)

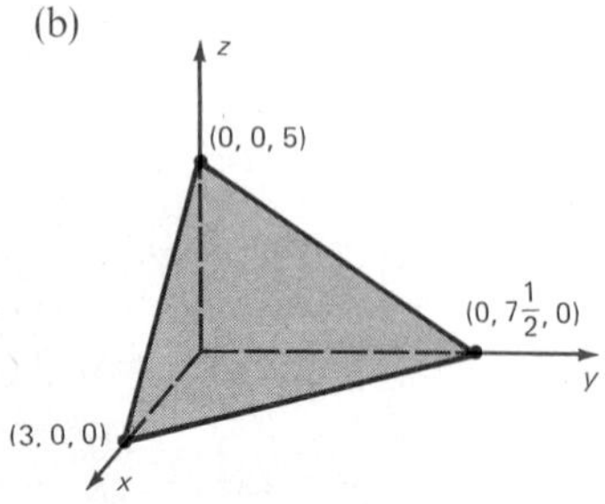

In the rest of this section, we look at various types of functions and the graphs they produce in three-dimensional space.

In two-dimensional space the graph of $ax + by = c$ is a straight line; if not, both a and b are 0. In three-dimensional space the graph of

$$ax + by + cz = d,$$

is a **plane** if a, b, c, and d are real numbers, and not all of a, b, and c are 0.

Example 1 Graph $2x + y + z = 6$.

This equation has a graph which is a plane. To graph the plane, find some typical ordered triples. Some of these ordered triples are $(0, 0, 6)$, $(3, 0, 0)$, and $(0, 6, 0)$. If we plot them we can get the graph shown in Figure 16.2. It is common to show only the portion of the graph which lies in the first octant. ■

Work Problem 3 at the side.

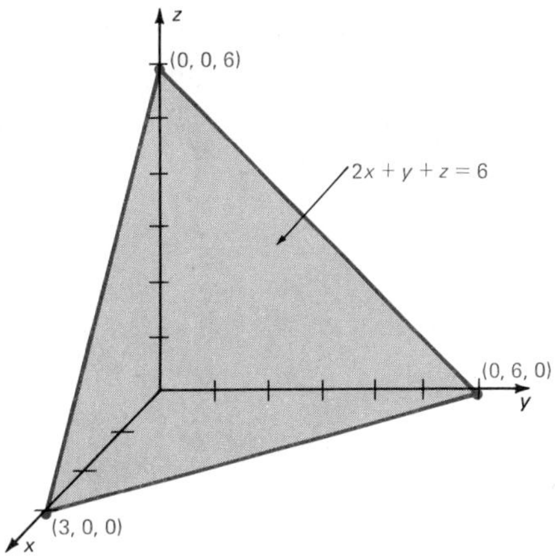

Figure 16.2

Recall from Chapter 3 that the graph of $x = 3$ is a vertical line going through the point $(3, 0)$. The graph is parallel to the y-axis, with y not appearing in the equation $x = 3$. Similar results hold with planes, as shown in the next examples.

4. Graph the following.

(a) $y + z = 7$

(b) $x + 2y = 8$

Answer:

(a)

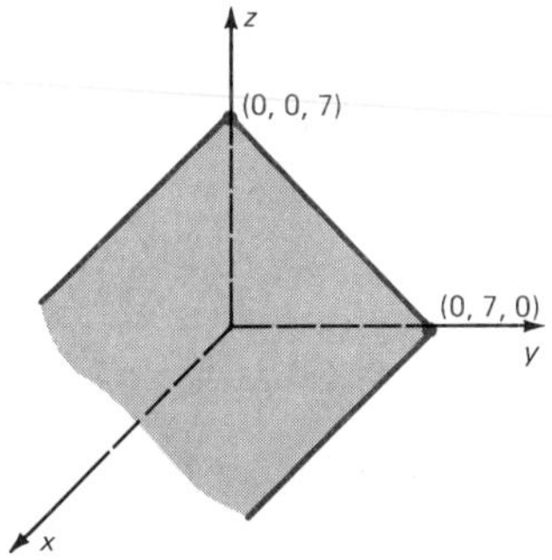

(b)

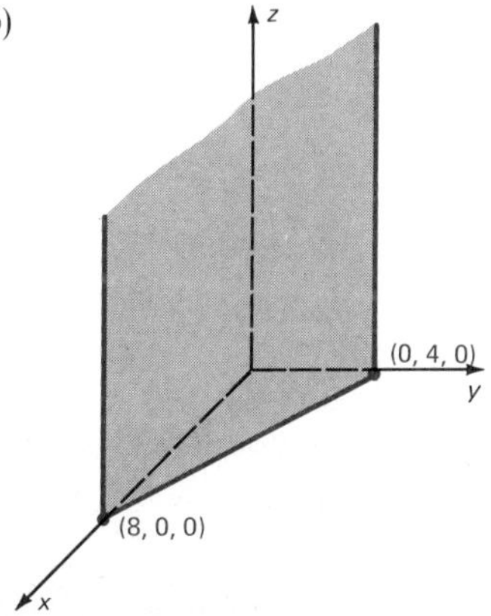

Example 2 Graph $x + z = 6$.

The graph is a plane going through $(6, 0, 0)$, $(0, 0, 6)$, and $(2, 0, 4)$, for example. The plane is parallel to the y-axis (y is missing from the equation $x + z = 6$). The graph is shown in Figure 16.3. ■

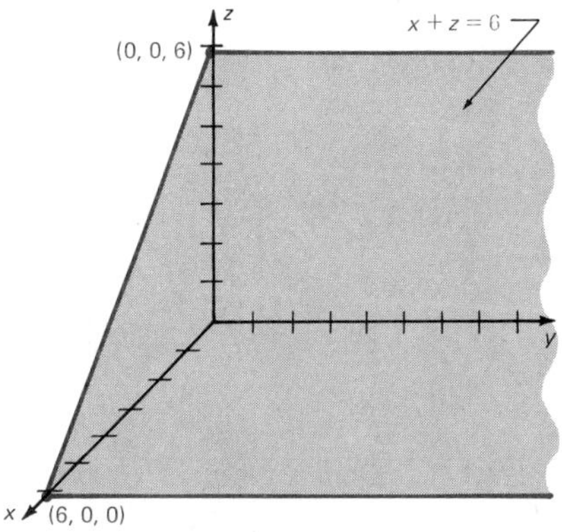

Figure 16.3

Work Problem 4 at the side.

Example 3 Graph $x = 3$.

The plane which is the graph of $x = 3$ goes through $(3, 0, 0)$. It is parallel to both the y-axis and the z-axis. See Figure 16.4. ■

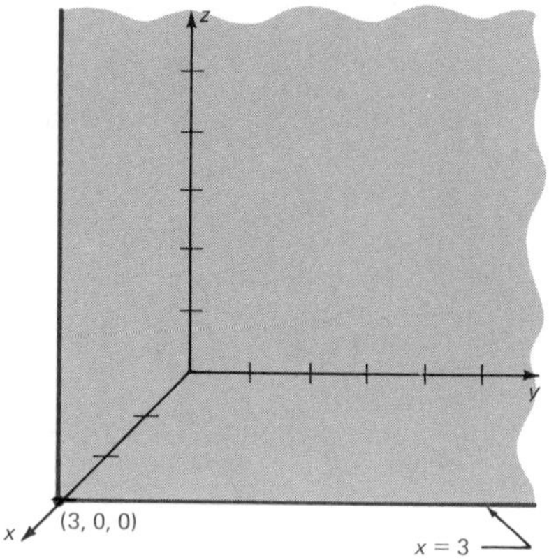

Figure 16.4

Work Problem 5 at the side.

The graph of

$$(x - h)^2 + (y - k)^2 + (z - j)^2 = r^2$$

is a **sphere** with center at (h, k, j) and radius r.

Example 4 Graph $(x - 2)^2 + (y + 3)^2 + (z - 4)^2 = 4$.

This equation represents a sphere of radius 2 with center at $(2, -3, 4)$. See Figure 16.5. Is a sphere the graph of a function? ■

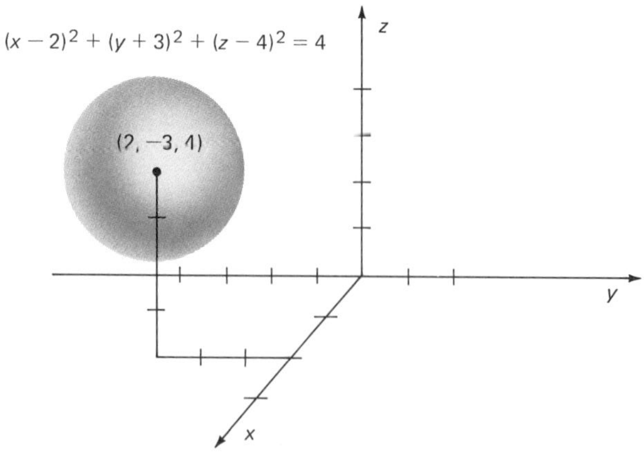

Figure 16.5

5. Graph the following.

(a) $y = 2$

(b) $z = 3$

(c) $y = -4$

Answer:

(a)

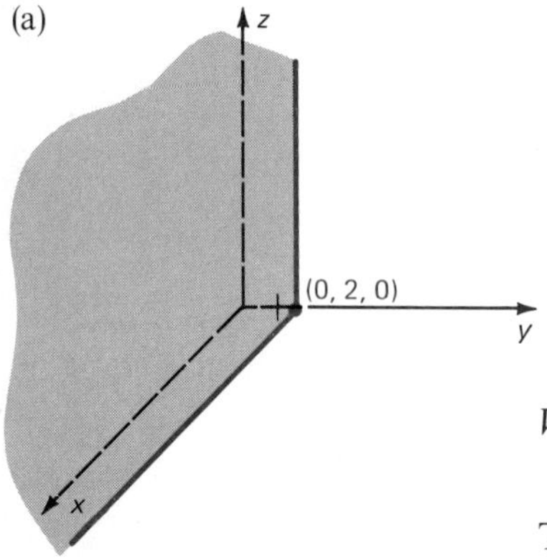

(b)

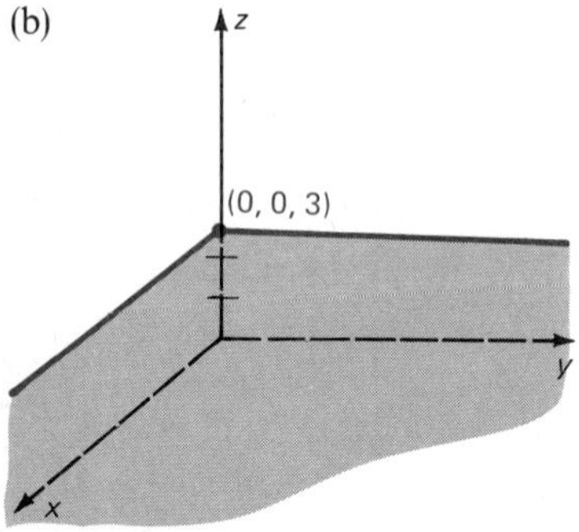

(c)

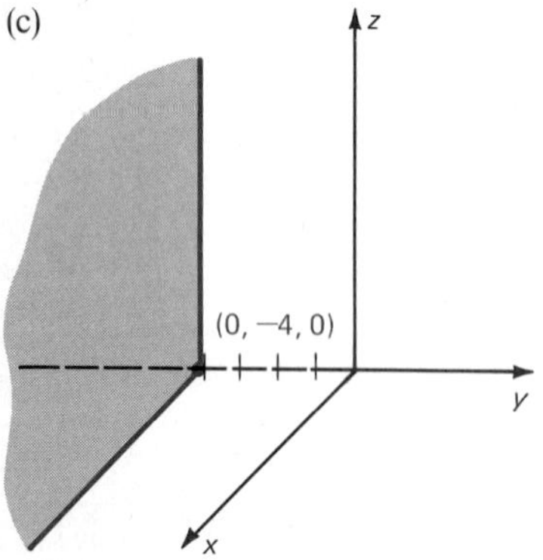

6. Graph the following.
(a) $x^2 + y^2 + z^2 = 9$
(b) $(x - 3)^2 + (y - 4)^2 + (z - 1)^2 = 16$

Answer:

(a)

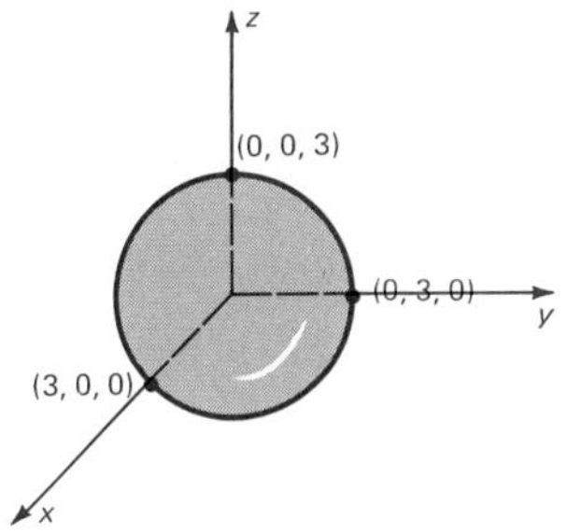

(b)

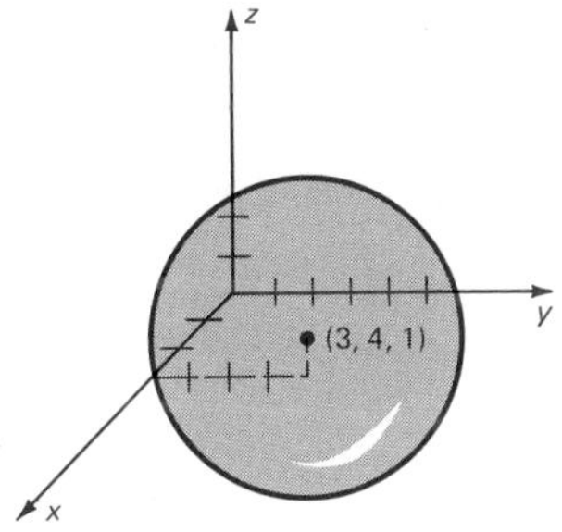

Work Problem 6 at the side.

16.2 EXERCISES

Graph the following planes. (*See Examples 1–3.*)

1. $x + y + z = 6$

2. $x + y + z = 12$

3. $2x + 3y + 4z = 12$

4. $4x + 2y + 3z = 24$

5. $3x - 2y + z = 18$

6. $4x - 3y - z = 12$

7. $x + y = 6$

8. $y + z = 4$

9. $x = 2$

10. $z = -3$

Find the center and radius of the following spheres. Graph each. (*See Example 4.*)

11. $x^2 + y^2 + z^2 = 49$

12. $x^2 + y^2 + z^2 = 16$

13. $(x - 5)^2 + (y - 3)^2 + (z + 4)^2 = 9$

14. $(x + 3)^2 + (y + 2)^2 + (z - 1)^2 = 16$

15. $(x - 2)^2 + (y - 5)^2 + (z + 3)^2 = 25$

16. $(x + 2)^2 + (y - 3)^2 + (z - 2)^2 = 1$

16.3 PARTIAL DERIVATIVES

Just as derivatives are useful in the study of functions of a single variable, partial derivatives are useful with functions of several variables. A *partial derivative* is found by taking the derivative of a function of several variables with respect to one variable at a time.

As an example of this process, suppose that a small firm makes only two products, radios and cassette recorders. The profit of the firm is given by

$$P(x, y) = 40x^2 - 10xy + 5y^2 - 80,$$

where x is the number of radios sold and y is the number of cassette recorders sold. Lately, sales of radios have been steady at 10 units; only the sales of recorders vary. The management would like to know the marginal profit when $y = 12$.

We know that marginal profit is given by the derivative of the profit function. We also know that in our example x is fixed at 10. Using this information, we begin by finding a new function, $f = P(10, y)$. We let $x = 10$ and get

$$\begin{aligned} f = P(10, y) &= 40(10)^2 - 10(10)y + 5y^2 - 80 \\ &= 3920 - 100y + 5y^2. \end{aligned}$$

The function f shows the profit from the sale of y recorders, assuming that x is fixed at 10. To find the marginal profit when $y = 12$, we first find the derivative df/dy.

$$\frac{df}{dy} = -100 + 10y$$

When $y = 12$, we have $-100 + 10(12) = -100 + 120 = 20$, so the marginal profit is 20 when $y = 12$.

1. Assume that x increases to a constant 12.
(a) Find $f = P(12, y)$.
(b) Find the marginal profit under this assumption when $y = 8$.

Answer:
(a) $f = P(12, y) = 5680 - 120y + 5y^2$
(b) -40

Work Problem 1 at the side.

The derivative of the function f that we found above was with respect to y only; we assumed that x was fixed. In general, let $z = f(x, y)$.

> The **partial derivative of f with respect to x** is the derivative of f obtained by treating x as a variable and y as a constant.
>
> The **partial derivative of f with respect to y** is the derivative of f obtained by treating y as a variable and x as a constant.

The symbols f_x (no prime is used) and $\partial f/\partial x$ are used to represent the partial derivative of f with respect to x.

Example 1 Let $f(x, y) = 4x^2 - 9xy + 6y^3$. Find f_x and f_y.
To find f_x, treat y as a constant and x as a variable. This gives

$$f_x = 8x - 9y.$$

If we treat y as a variable and x as a constant, we can find f_y.

$$f_y = -9x + 18y^2 \quad \blacksquare$$

2. Let $f(x, y) = -2x^4 + 6x^3y^3 + 5y^2 - 8$. Find f_x and f_y.

Answer: $f_x = -8x^3 + 18x^2y^3$; $f_y = 18x^3y^2 + 10y$

Work Problem 2 at the side.

Example 2 Let $f(x, y) = \ln(x^2 + y)$. Find f_x and f_y.
Recall the formula for the derivative of a natural logarithm function. [If $g(x) = \ln x$, then $g'(x) = 1/x$.] Using this formula and the chain rule, we have

$$f_x = \frac{2x}{x^2 + y} \quad \text{and} \quad f_y = \frac{1}{x^2 + y}. \quad \blacksquare$$

3. Find f_x and f_y for the following functions.
(a) $f(x, y) = \ln(4x + 9y^2)$
(b) $f(x, y) = xe^y$

Answer:
(a) $f_x = 4/(4x + 9y^2)$; $f_y = 18y/(4x + 9y^2)$
(b) $f_x = e^y$; $f_y = xe^y$

Work Problem 3 at the side.

The notation $f_x(a, b)$ or $(\partial f/\partial x)_{(a, b)}$ represents the value of a partial derivative when $x = a$ and $y = b$, as shown in the next example.

Example 3 Let $f(x, y) = 2x^2 + 3xy^3 + 2y + 5$. Find the following.
(a) $f_x(-1, 2)$
First, find f_x by holding y constant.

$$f_x = 4x + 3y^3$$

Now let $x = -1$ and $y = 2$.

$$f_x(-1, 2) = 4(-1) + 3(2)^3 = -4 + 24 = 20.$$

(b) $f_y(-4, -3)$

$$f_y = 9xy^2 + 2$$
$$f_y(-4, -3) = 9(-4)(-3)^2 + 2$$
$$= 9(-36) + 2 = -322 \quad ■$$

4. Let $f(x, y) = 8x^2 + 5xy^2 - 9y + 4$. Find the following.
(a) $f_x(0, 4)$
(b) $f_x(-1, 2)$
(c) $f_y(-2, 5)$
(d) $f_y(0, 0)$

Answer:
(a) 80
(b) 4
(c) -109
(d) -9

Work Problem 4 at the side.

The derivative of $y = f(x)$ gives the rate of change of y with respect to x. In the same way, if $z = f(x, y)$, then f_x gives the rate of change of z with respect to x, provided that y is held constant.

Example 4 Suppose that the temperature of the water at the point on a river where a nuclear power plant discharges its hot waste water is approximated by

$$T(x, y) = 2x + 5y + xy - 40,$$

where x represents the temperature of the river water in °C before it reaches the power plant and y is the number of megawatts (in hundreds) of electricity being produced annually by the plant.
(a) Find the temperature of the discharge water if the water reaching the plant has a temperature of 8°C and if 300 megawatts of electricity are being produced.
Here $x = 8$ and $y = 3$ (since 300 megawatts of electricity are being produced). Thus,

$$T(8, 3) = 2(8) + 5(3) + 8(3) - 40 = 15.$$

The water at the outlet of the plant is at a temperature of 15°C.

5. Find the temperature of the water at the outlet if the water has an initial temperature of 10°C and if 400 megawatts of electricity are being produced.

Answer: 40°C

Work Problem 5 at the side.

(b) For the function $T(x, y) = 2x + 5y + xy - 40$, find and interpret $T_x(9, 5)$.

$$T_x = 2 + y$$

This partial derivative gives the rate of change of T with respect to x.

$$T_x(9, 5) = 2 + 5 = 7$$

This result, 7, is the approximate change in temperature of the output water if input water temperature changes from $x = 9$ to $x = 9 + 1 = 10$ while y remains constant.

(c) Use the same function to find and interpret $T_y(9, 5)$.

$$T_y = 5 + x$$

This partial derivative gives the rate of change of T with respect to y.

$$T_y(9, 5) = 5 + 9 = 14$$

This result, 14, is the approximate change in temperature resulting from an increase in production of electricity from $y = 5$ to $y = 6$ (600 megawatts) while x remains constant. ■

6. Use the function of Example 4 to find and interpret the following.
(a) $T_x(5, 4)$
(b) $T_y(8, 3)$

Answer:
(a) $T_x(5, 4) = 6$; the approximate increase in temperature if the input temperature increases from 5 to 6 degrees.
(b) $T_y(8, 3) = 13$; the approximate increase in temperature if the production of electricity increases from 300 to 400 megawatts.

Work Problem 6 at the side.

When working with functions of one variable, we found second derivatives by taking the derivative of the first derivative. In much the same way, we can define **second partial derivatives.**

$$\frac{\partial}{\partial x}\left(\frac{\partial z}{\partial x}\right) = \frac{\partial^2 z}{\partial x^2} = f_{xx} \qquad \frac{\partial}{\partial y}\left(\frac{\partial z}{\partial y}\right) = \frac{\partial^2 z}{\partial y^2} = f_{yy}$$

$$\frac{\partial}{\partial y}\left(\frac{\partial z}{\partial x}\right) = \frac{\partial^2 z}{\partial y\,\partial x} = f_{xy} \qquad \frac{\partial}{\partial x}\left(\frac{\partial z}{\partial y}\right) = \frac{\partial^2 z}{\partial x\,\partial y} = f_{yx}$$

Be careful with these last two symbols. If we first find the partial derivative of z with respect to x, and then with respect to y, we have found $(f_x)_y$, or f_{xy}. We have also found

$$\frac{\partial}{\partial y}\left(\frac{\partial z}{\partial x}\right) = \frac{\partial^2 z}{\partial y\,\partial x}, \qquad \text{so that} \qquad \frac{\partial^2 z}{\partial y\,\partial x} = f_{xy}.$$

Note that the order of the symbols x and y is reversed in the last equation.

Example 5 Find all second partial derivatives for

$$f(x, y) = -4x^3 - 3x^2y^3 + 2y^2.$$

First find f_x and f_y.

$$f_x = -12x^2 - 6xy^3 \qquad \text{and} \qquad f_y = -9x^2y^2 + 4y$$

To find f_{xx}, take the partial derivative of f_x with respect to x.

$$f_{xx} = -24x - 6y^3$$

Take the partial derivative of f_y with respect to y; this gives f_{yy}.

$$f_{yy} = -18x^2y + 4$$

We find f_{xy} by starting with f_x. Take the partial derivative of f_x with respect to y.

$$f_{xy} = -18xy^2$$

Finally, find f_{yx} by starting with f_y; take its partial derivative with respect to x.

$$f_{yx} = -18xy^2 \quad \blacksquare$$

7. Let $f(x, y) = 4x^2y^2 - 9xy + 8x^2 - 3y^4$. Find all second partial derivatives.

Answer: $f_{xx} = 8y^2 + 16$;
$f_{yy} = 8x^2 - 36y^2$;
$f_{xy} = 16xy - 9$;
$f_{yx} = 16xy - 9$

Work Problem 7 at the side.

Example 6 Let $f(x, y) = 2e^x - 8x^3y^2$. Find all second partial derivatives.

Here $f_x = 2e^x - 24x^2y^2$ and $f_y = -16x^3y$. [Recall: If $g(x) = e^x$, then $g'(x) = e^x$.] Now find the second partial derivatives.

$$f_{xx} = 2e^x - 48xy^2 \qquad f_{xy} = -48x^2y$$

$$f_{yy} = -16x^3 \qquad f_{yx} = -48x^2y \quad \blacksquare$$

8. Let $f(x, y) = 4e^{x+y} + 2x^3y$. Find all second partial derivatives.

Answer: $f_{xx} = 4e^{x+y} + 12xy$;
$f_{yy} = 4e^{x+y}$;
$f_{xy} = 4e^{x+y} + 6x^2$;
$f_{yx} = 4e^{x+y} + 6x^2$

Work Problem 8 at the side.

In all our examples of second partial derivatives, we found that $f_{xy} = f_{yx}$. This happens for many functions, and certainly for all the ones we shall use in this book.

16.3 EXERCISES

Find f_x and f_y for the following. Then find $f_x(2, -1)$ and $f_y(-4, 3)$. Leave the answers in terms of e in Exercises 7–10. (See Examples 1–4.)

1. $f(x, y) = 12x^2 + 4y^2$
2. $f(x, y) = -6x^2 + 7y^2$
3. $f(x, y) = -2xy + 6y^3 + 2$
4. $f(x, y) = 4x^2y - 9y^2$
5. $f(x, y) = 3x^3y^2$
6. $f(x, y) = -2x^2y^4$
7. $f(x, y) = e^{x+y}$
8. $f(x, y) = 3e^{2x+y}$

9. $f(x, y) = -5e^{3x-4y}$

10. $f(x, y) = 8e^{7x-y}$

Find all second partial derivatives for the following. (See Example 5.)

11. $f(x, y) = 6x^3y - 9y^2 + 2x$

12. $g(x, y) = 5xy^4 + 8x^3 - 3y$

13. $R(x, y) = 4x^2 - 5xy^3 + 12y^2x^2$

14. $h(x, y) = 30y + 5x^2y + 12xy^2$

15. $r(x, y) = \dfrac{4x}{x + y}$

16. $k(x, y) = \dfrac{-5y}{x + 2y}$

17. $z = 4xe^y$

18. $z = -3ye^x$

19. $r = \ln|x + y|$

20. $k = \ln|5x - 7y|$

APPLIED PROBLEMS

Management

Suppose that the manufacturing cost of a precision electronic calculator is approximated by

$$M(x, y) = 40x^2 + 30y^2 - 10xy + 30,$$

where x is the cost of the necessary electronic chips and y is the cost of labor. Find the following.

21. $M_y(4, 2)$

22. $M_x(3, 6)$

23. $(\partial M/\partial x)_{(2,5)}$

24. $(\partial M/\partial y)_{(6,7)}$

Natural Science

The total number of matings per day between individuals of a certain species of grasshoppers is approximated by

$$M(x, y) = 2xy + 10xy^2 + 30y^2 + 20,$$

where x represents the temperature in °C and y represents the number of days since the last rain. Find the following.

25. $(\partial M/\partial x)_{(20,4)}$

26. $(\partial M/\partial y)_{(24,10)}$

27. $M_x(17, 3)$

28. $M_y(21, 8)$

Management

The revenue from the sale of x units of a tranquilizer and y units of an antibiotic is given by

$$R(x, y) = 5x^2 + 9y^2 - 4xy.$$

29. Suppose $x = 9$ and $y = 5$. What is the approximate effect on revenue if x is increased to 10, while y is fixed? (See Example 4.)

30. Suppose $x = 9$ and $y = 5$. What is the approximate effect on revenue if y is increased to 6, while x is fixed?

A production function, $z = f(x, y)$, is a function which gives the quantity of an item produced, z, as a function of two other variables, x and y.

31. The production function z for the United States was once estimated as

$$z = x^{2.1}y^{.3},$$

where x stands for the amount of labor and y stands for the amount of capital. Find the marginal productivity of labor (find $\partial z/\partial x$) and of capital.

32. A similar production function for Canada is

$$z = x^{.4}y^{.6},$$

with x, y, and z as in Exercise 31. Find the marginal productivity of labor and of capital.

Natural Science

(The remaining exercises in this section were first discussed in Section 1 of this chapter.) The oxygen consumption of a well-insulated mammal which is not sweating is approximated by

$$m = m(T, F, w) = \frac{2.5(T - F)}{w^{.67}} = 2.5(T - F)w^{-.67},$$

where T is the internal body temperature of the animal (in °C), F is the temperature of the outside of the animal's fur (in °C), and w is the animal's weight in kilograms.

33. Find m_T.

34. Suppose $T = 38°$, $F = 12°$, and $w = 30$ kg. Find $m_T(38, 12, 30)$.

35. Find m_w.

36. Suppose $T = 40°$, $F = 20°$, and $w = 40$ kg. Find $m_w(40, 20, 40)$.

37. Find m_F.

38. Suppose $T = 36°$, $F = 14°$, and $w = 25$ kg. Find $m_F(36, 14, 25)$.

The surface area of a human, in square meters, is approximated by

$$A(W, H) = 2.02W^{.425}H^{.725},$$

where W is the weight of the person in kilograms and H is the height in meters.

39. Find $\partial A/\partial W$.

40. Suppose $W = 72$ and $H = 1.8$. Find $(\partial A/\partial W)_{(72,1.8)}$.

41. Find $\partial A/\partial H$.

42. Suppose $W = 70$ and $H = 1.6$. Find $(\partial A/\partial H)_{(70,1.6)}$.

In one method of computing the quantity of blood pumped through the lungs in one minute, a researcher first finds each of the following (in milliliters).

b = quantity of oxygen used by body in one minute

a = quantity of oxygen per liter of blood that has just gone through the lungs

v = quantity of oxygen per liter of blood that is about to enter the lungs

In one minute,

amount of oxygen used = amount of oxygen per liter × number of liters of blood pumped.

If C is the number of liters pumped through the blood in one minute, then

$$b = (a - v) \cdot C$$

or

$$C = \frac{b}{a - v}.$$

43. Find C if $a = 160$, $b = 200$, and $v = 125$.

44. Find C if $a = 180$, $b = 260$, and $v = 142$.

Find the following partial derivatives.

45. $\partial C/\partial b$

46. $\partial C/\partial v$

16.4 MAXIMUMS AND MINIMUMS

One of the most important applications of calculus is finding maximums and minimums. Partial derivatives are used to find maximums and minimums of functions of two variables.

To define relative maximums and minimums for a function of two variables, let $z = f(x, y)$ be a function defined for each point in some region of the plane containing the x-axis and y-axis. The function $z = f(x, y)$ has a **relative maximum** at the point (a, b) if there exists a circular region with center at (a, b) such that

$$f(x, y) \leq f(a, b)$$

for every point (x, y) of that region. An example of a relative maximum is shown in Figure 16.6. **Relative minimums** are defined in a similar way. A relative minimum for $z = f(x, y)$ is shown in Figure 16.7.

A function of one variable, such as $y = f(x)$, can have a relative maximum or relative minimum at a point $x = a$ that makes the

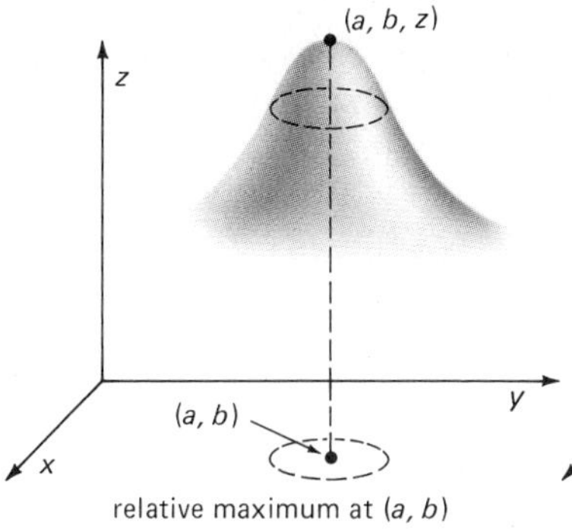

Figure 16.6

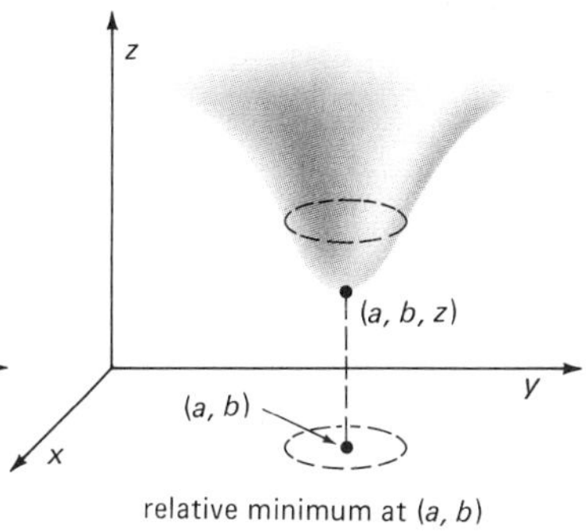

Figure 16.7

derivative equal 0. (But the function does not necessarily have a relative maximum or minimum at $x = a$ just because $f'(a) = 0$.) A similar result holds for functions of two variables, as shown by the next theorem.

Theorem 16.1 Let a function $z = f(x, y)$ have a relative maximum or relative minimum at the point (a, b). Let $f_x(a, b)$ and $f_y(a, b)$ both exist. Then

$$\boldsymbol{f_x(a, b) = 0} \quad \text{and} \quad \boldsymbol{f_y(a, b) = 0.}$$

Once again, the fact that $f_x(a, b) = 0$ and $f_y(a, b) = 0$ is no guarantee that the function has a relative maximum or minimum at (a, b). For example, Figure 16.8 shows the graph of $f(x, y) = z = x^2 - y^2$. Both $f_x(0, 0) = 0$ and $f_y(0, 0) = 0$, and yet $(0, 0)$ leads to neither a relative maximum nor a relative minimum for the function. Here the point $(0, 0, 0)$ is called a **saddle point.** A saddle point is a

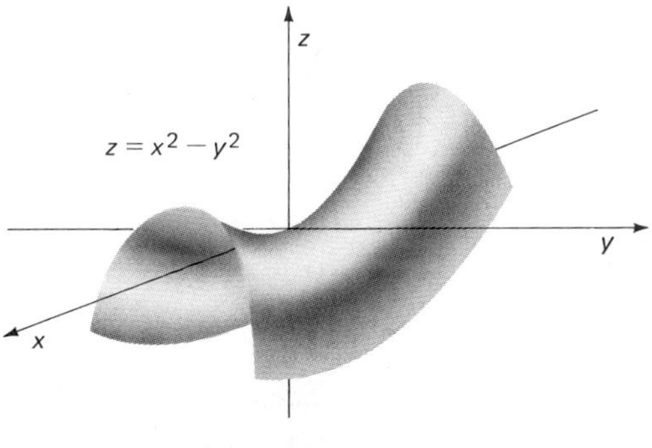

Figure 16.8

minimum when approached from one direction but a maximum when approached from another direction. Thus, a saddle point is neither a maximum nor a minimum.

Figure 16.9 shows the graph of $f(x, y) = x^3$. Check that $f_x(0, 0) = 0$ and $f_y(0, 0) = 0$, but $(0, 0)$ does not lead to a relative maximum, a relative minimum, or a saddle point.

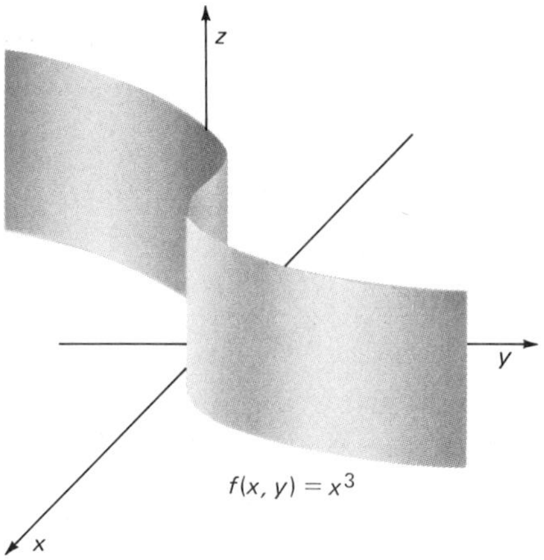

Figure 16.9

Example 1 At what points might the function

$$f(x, y) = 6x^2 + 6y^2 + 6xy + 36x - 5$$

have any relative maximums or minimums?

We begin by finding all points (a, b) such that $f_x(a, b) = 0$ and $f_y(a, b) = 0$; first find f_x and f_y.

$$f_x = 12x + 6y + 36 \quad \text{and} \quad f_y = 12y + 6x$$

Place each of these two partial derivatives equal to 0.

$$12x + 6y + 36 = 0 \quad \text{and} \quad 12y + 6x = 0$$

These two equations make up a system of linear equations. We need to find all values of x and y that make both of the equations true at the same time. To find these values, we rewrite the simplest equation as follows.

$$\begin{aligned} 12y + 6x &= 0 \\ 6x &= -12y \\ x &= -2y \end{aligned}$$

Now we substitute $-2y$ for x in the remaining equation.

$$\begin{aligned} 12x + 6y + 36 &= 0 \\ 12(-2y) + 6y + 36 &= 0 \\ -24y + 6y + 36 &= 0 \\ -18y + 36 &= 0 \\ -18y &= -36 \\ y &= 2 \end{aligned}$$

Since $x = -2y$, we have $x = -2(2) = -4$. The solution of the system of equations is $(-4, 2)$.

Therefore, if the given function $f(x, y) = 6x^2 + 6y^2 + 6xy + 36x - 5$ has any relative maximums or relative minimums, they occur at $(-4, 2)$. To decide whether $(-4, 2)$ leads to a relative maximum, a relative minimum, or neither, we need the results of the next theorem. ■

1. At what point might the function $f(x, y) = x^2 + y^2 + 4x - 6y + 2$ have any relative maximums or minimums?

Answer: $(-2, 3)$

Work Problem 1 at the side.

Theorem 16.2 *M-test for relative maximums and minimums* Let all the following partial derivatives exist for a function $z = f(x, y)$. Let (a, b) be a point for which

$$f_x(a, b) = 0 \qquad \text{and} \qquad f_y(a, b) = 0.$$

Define the number M as

$$M = f_{xx}(a, b) \cdot f_{yy}(a, b) - [f_{xy}(a, b)]^2.$$

Then

(a) $f(a, b)$ is a relative maximum if $M > 0$ and $f_{xx}(a, b) < 0$.

(b) $f(a, b)$ is a relative minimum if $M > 0$ and $f_{xx}(a, b) > 0$.

(c) $f(a, b)$ is a saddle point (neither a maximum nor a minimum) if $M < 0$.

(d) If $M = 0$, the test gives no information.

Example 2 In Example 1 above, we found that if the function

$$f(x, y) = 6x^2 + 6y^2 + 6xy + 36x - 5$$

has any relative maximums or relative minimums, they occur at the point $(-4, 2)$. Does $(-4, 2)$ lead to a relative maximum, a relative minimum, or neither?

We can find out by using the M-test above. We already know that

$$f_x(-4, 2) = 0 \qquad \text{and} \qquad f_y(-4, 2) = 0.$$

We now must find the various second partial derivatives used in finding M. From $f_x = 12x + 6y + 36$ and $f_y = 12y + 6x$, we get

$$f_{xx} = 12, \qquad f_{yy} = 12, \qquad \text{and} \qquad f_{xy} = 6.$$

(If these second partial derivatives had not all been equal to constants, we would have had to evaluate them at the point $(-4, 2)$.) Now we can find M.

$$\begin{aligned} M &= f_{xx}(-4, 2) \cdot f_{yy}(-4, 2) - [f_{xy}(-4, 2)]^2 \\ &= 12 \cdot 12 - 6^2 \\ M &= 108 \end{aligned}$$

Since $M > 0$ and $f_{xx}(-4, 2) = 12 > 0$, part (b) of Theorem 16.2 applies. Thus, $f(x, y) = 6x^2 + 6y^2 + 6xy + 36x - 5$ has a relative minimum at $(-4, 2)$. This relative minimum is given by $f(-4, 2) = -77$. ■

Example 3 Find all points where the function

$$f(x, y) = 50 + 4x - 5y + x^2 + y^2 + xy$$

has any relative maximums or relative minimums.

Here we have

$$f_x = 4 + 2x + y \qquad \text{and} \qquad f_y = -5 + 2y + x.$$

If we place these partial derivatives equal to 0 and rearrange terms, we get

$$2x + y = -4 \qquad \text{and} \qquad x + 2y = 5.$$

One way to solve this system of equations is to multiply both sides of the first equation by -2 and then add the two equations.

$$\begin{array}{r} -4x - 2y = 8 \\ x + 2y = 5 \\ \hline -3x \qquad = 13 \end{array}$$

From this last equation, we get $x = -13/3$. Substitute $-13/3$ for x in either equation to find y. You should find that $y = 14/3$. The only possible point leading to a relative maximum or minimum is $(-13/3, 14/3)$.

In order to use the M-test, we find the following second partial derivatives.

$$f_{xx} = 2, \qquad f_{yy} = 2, \qquad \text{and} \qquad f_{xy} = 1$$

Thus, $M = 2 \cdot 2 - 1^2 = 3.$

Since $M > 0$ *and* $f_{xx}(-13/3, 14/3) = 2$, we see that $f(-13/3, 14/3)$ is a relative minimum. ■

2. Find all points where the following functions have any relative maximums or relative minimums.
(a) $f(x, y) = 4x^2 + 3xy + 2y^2 + 7x - 6y - 6$
(b) $f(x, y) = 4x^2 + 6xy + y^2 + 34x + 8y + 5$

Answer:
(a) $M = 23$, relative minimum at $(-2, 3)$
(b) $M = -20$, saddle point (neither a maximum nor a minimum) at $(1, -7)$

Work Problem 2 at the side.

Example 4 Find all points where the function

$$f(x, y) = 9xy - x^3 - y^3 - 6$$

has any relative maximums or relative minimums.

Here we have

$$f_x = 9y - 3x^2 \qquad \text{and} \qquad f_y = 9x - 3y^2.$$

Place each of these partial derivatives equal to 0.

$$\begin{aligned} f_x &= 0 & f_y &= 0 \\ 9y - 3x^2 &= 0 & 9x - 3y^2 &= 0 \\ 9y &= 3x^2 & 9x &= 3y^2 \\ 3y &= x^2 & 3x &= y^2 \end{aligned}$$

The final equation on the left, $3y = x^2$, can be rewritten as $y = x^2/3$. Substitute this into the final equation on the right.

$$\begin{aligned} 3x &= y^2 \\ 3x &= \left(\frac{x^2}{3}\right)^2 \\ 3x &= \frac{x^4}{9} \\ 27x &= x^4 \end{aligned}$$

Rewrite this equation as

$$\begin{aligned} x^4 - 27x &= 0 \\ x(x^3 - 27) &= 0 \qquad \text{Factor} \\ x = 0 \quad \text{or} \quad x^3 - 27 &= 0 \\ x = 0 \quad \text{or} \quad x^3 &= 27 \\ x = 0 \quad \text{or} \quad x &= 3. \end{aligned}$$

We can use these values of x, along with the equation $3x = y^2$, to find y.

$$\begin{aligned} &\text{If } x = 0, & &\text{If } x = 3, \\ 3x &= y^2 & 3x &= y^2 \\ 3(0) &= y^2 & 3(3) &= y^2 \\ 0 &= y^2 & 9 &= y^2 \\ 0 &= y & \text{or} \quad 3 = y \quad \text{or} \quad -3 &= y. \end{aligned}$$

Since $f_x(3, -3) \neq 0$ and $f_y(3, -3) \neq 0$, the only possible relative maximums or minimums for $f(x, y) = 9xy - x^3 - y^3 - 6$ occur at $(0, 0)$ or $(3, 3)$. To find out which, we use the M-test. Here

$$f_{xx} = -6x, \qquad f_{yy} = -6y, \qquad \text{and} \qquad f_{xy} = 9.$$

Test each of the possible points.

For (0, 0):	For (3, 3):
$f_{xx}(0, 0) = -6(0) = 0$	$f_{xx}(3, 3) = -6(3) = -18$
$f_{yy}(0, 0) = -6(0) = 0$	$f_{yy}(3, 3) = -6(3) = -18$
$f_{xy}(0, 0) = 9$	$f_{xy}(3, 3) = 9$
$M = 0 \cdot 0 - 9^2 = -81.$	$M = -18(-18) - 9^2 = 243.$
Since $M < 0$, we have a saddle point at (0, 0).	Here $M > 0$ and $f_{xx}(3, 3) = -18 < 0$; there is a relative maximum at (3, 3). ■

3. Find all points where the function $f(x, y) = x^3 + y^2 - 6xy + 5$ has any relative maximums or relative minimums.

Answer: at (0, 0), we have $M = -36 < 0$, a saddle point; at (6, 18) we have $M = 36 > 0$ and $f_{xx}(6, 18) = 36 > 0$, a relative minimum

Work Problem 3 at the side.

Example 5 A company is developing a new soft drink. The cost in dollars to produce a batch of the drink is approximated by

$$C(x, y) = 2200 + 27x^3 - 72xy + 8y^2,$$

where x is the number of kilograms of sugar per batch and y is the number of grams of flavoring per batch.

(a) Find the amounts of sugar and flavoring that result in minimum cost for a batch.

We need the following partial derivatives.

$$C_x = 81x^2 - 72y \qquad \text{and} \qquad C_y = -72x + 16y$$

Set each of these equal to 0 and solve for y.

$$81x^2 - 72y = 0 \qquad\qquad -72x + 16y = 0$$
$$-72y = -81x^2 \qquad\qquad 16y = 72x$$
$$y = \frac{9}{8}x^2 \qquad\qquad y = \frac{9}{2}x$$

Since $(9/8)x^2$ and $(9/2)x$ both equal y they are equal. Solve the resulting equation for x.

$$\frac{9}{8}x^2 = \frac{9}{2}x$$
$$9x^2 = 36x$$
$$9x^2 - 36x = 0$$
$$9x(x - 4) = 0$$
$$9x = 0 \quad \text{or} \quad x - 4 = 0$$

The equation $9x = 0$ leads to $x = 0$ which is not a useful answer for our problem. Substitute $x = 4$ into $y = (9/2)x$ to find y.

$$y = \frac{9}{2}x = \frac{9}{2}(4) = 18$$

Now we must check to see if the point (4, 18) leads to a relative minimum. For (4, 18) we have

$C_{xx} = 162x = 162(4) = 648, \qquad C_{yy} = 16, \qquad$ and $\qquad C_{xy} = -72.$

Also, $\qquad M = (648)(16) - (-72)^2 = 5184.$

Since $M > 0$ and $C_{xx}(4, 18) > 0$, the cost at (4, 18) is a minimum.

(b) What is the minimum cost?

To find the minimum cost, we go back to the cost function and evaluate $C(4, 18)$.

$$C(x, y) = 2200 + 27x^3 - 72xy + 8y^2$$
$$C(4, 18) = 2200 + 27(4)^3 - 72(4)(18) + 8(18)^2 = 1336$$

The minimum cost for a batch is \$1336.00. ■

16.4 EXERCISES

Find all points where the following functions have any relative maximums or relative minimums. Identify any saddle points. (See Examples 2–4.)

1. $f(x, y) = xy + x - y$

2. $f(x, y) = 4xy + 8x - 9y$

3. $f(x, y) = x^2 - 2xy + 2y^2 + x - 5$

4. $f(x, y) = x^2 + xy + y^2 - 6x - 3$

5. $f(x, y) = x^2 - xy + y^2 + 2x + 2y + 6$

6. $f(x, y) = x^2 + xy + y^2 + 3x - 3y$

7. $f(x, y) = x^2 + 3xy + 3y^2 - 6x + 3y$

8. $f(x, y) = 5xy - 7x^2 - y^2 + 3x - 6y - 4$

9. $f(x, y) = 4xy - 10x^2 - 4y^2 + 8x + 8y + 9$

10. $f(x, y) = x^2 + xy + 3x + 2y - 6$

11. $f(x, y) = x^2 + xy - 2x - 2y + 2$

12. $f(x, y) = x^2 + xy + y^2 - 3x - 5$

13. $f(x, y) = x^2 - y^2 - 2x + 4y - 7$

14. $f(x, y) = 4x + 2y - x^2 + xy - y^2 + 3$

15. $f(x, y) = 2x^3 + 3y^2 - 12xy + 4$

16. $f(x, y) = 5x^3 + 2y^2 - 60xy - 3$

17. $f(x, y) = x^2 + 4y^3 - 6xy - 1$

18. $f(x, y) = 3x^2 + 7y^3 - 42xy + 5$

APPLIED PROBLEMS

Management

19. Suppose that the profit of a certain firm is approximated by

$$P(x, y) = 1000 + 24x - x^2 + 80y - y^2,$$

where x is the cost of a unit of labor and y is the cost of a unit of goods. Find values of x and y that maximize profit. Find the maximum profit.

20. The labor cost for manufacturing a precision camera can be approximated by

$$L(x, y) = \frac{3}{2}x^2 + y^2 - 2x - 2y - 2xy + 68,$$

where x is the number of hours required by a skilled craftsperson and y is the number of hours required by a semiskilled person. Find values of x and y that minimize the labor charge. Find the minimum labor charge.

21. The number of roosters that can be fed from x pounds of Super-Hen chicken feed and y pounds of Super-Rooster feed is given by

$$R(x, y) = 800 - 2x^3 + 12xy - y^2.$$

Find the number of pounds of each kind of feed that produces the maximum number of roosters.

22. The total profit from one acre of a certain crop depends on the amount spent on fertilizer, x, and hybrid seed, y, according to the model

$$P(x, y) = -x^2 + 3xy + 160x - 5y^2 + 200y + 2{,}600{,}000.$$

Find values of x and y that lead to maximum profit. Find the maximum profit.

23. The total cost to produce x units of electrical tape and y units of packing tape is given by

$$C(x, y) = 2x^2 + 3y^2 - 2xy + 2x - 126y + 3800.$$

Find the number of units of each kind of tape that should be produced so that the total cost is minimum. Find the minimum total cost.

24. The total revenue from the sale of x hot tubs and y solar heaters is approximated by

$$R(x, y) = 12 + 74x + 85y - 3x^2 - 5y^2 - 5xy.$$

Find the number of each that should be sold to produce maximum revenue. Find the maximum revenue.

CHAPTER 16 TEST

[16.1] *Let* $f(x, y) = 9x^2 - 2xy + 4y^2 - 8$. *Find the following.*

1. $f(-4, 8)$
2. $f(1, -6)$

[16.2] *Graph the following. Identify each graph.*

3. $2x + 3y + 4z = 12$
4. $x + 2z = 10$
5. $(x + 1)^2 + (y - 3)^2 + (z - 4)^2 = 9$
6. $x^2 + y^2 + z^2 = 49$

[16.3] *Find* f_x *and* f_y *for the following. Then find* $f_x(-2, 5)$ *and* $f_y(3, 6)$.

7. $f(x, y) = 9x^3 + 2y^2 - 4xy$
8. $f(x, y) = \dfrac{x + 1}{y + 1}$

Find f_{xx}, f_{yy}, *and* f_{xy} *for the following.*

9. $f(x, y) = x^2y^3 + 8xy$
10. $f(x, y) = \dfrac{y}{1 + x}$

[16.4] *Find all points where the following functions have maximums or minimums.*

11. $f(x, y) = x^2 + 5xy - 10x + 3y^2 - 12y$
12. $z = x^3 - 8y^2 + 6xy + 4$

The total cost to manufacture x units of doggy dishes and y units of kitty plates is given by

$$C(x, y) = x^2 + 5y^2 + 4xy - 70x - 164y + 1800.$$

13. Find values of x and y that lead to a minimum cost.
14. What is the minimum cost?

Appendix

LAGRANGE MULTIPLIERS

In the textbook we saw how to find any relative maximums or relative minimums for functions of two variables. In practice, many such functions are given with a secondary condition or **constraint.** For example, we might be asked to find the maximum profit from the sale of x units of small boxes and y units of large boxes, subject to the constraint that the total number of units must be 12 (that is, $x + y = 12$).

Problems with constraints can be worked using the method of **Lagrange multipliers.** This method looks fairly complicated when written down as a theorem, so before we do this we shall look at several examples of the method in use.

Example 1 Find the minimum value of

$$f(x, y) = 5x^2 + 6y^2 - xy,$$

subject to the constraint $x + 2y = 24$.

Go through the following steps.

1. Rewrite the constraint so that it equals 0. Let $g(x, y)$ equal this constraint. In this example,

$$x + 2y = 24 \qquad \text{becomes} \qquad x + 2y - 24 = 0,$$

and finally, $g(x, y) = x + 2y - 24$.

2. Form the **Lagrange function** $F(x, y, \lambda)$. (The symbol λ is the Greek letter lambda.) The Lagrange function is defined as the sum of the function $f(x, y)$ and the product of λ and $g(x, y)$. In this example we have

$$\begin{aligned} F(x, y, \lambda) &= f(x, y) + \lambda \cdot g(x, y) \\ &= 5x^2 + 6y^2 - xy + \lambda(x + 2y - 24) \\ &= 5x^2 + 6y^2 - xy + \lambda x + 2\lambda y - 24\lambda. \end{aligned}$$

3. Find F_x, F_y, and F_λ. Form the system of equations $F_x = 0$, $F_y = 0$, and $F_\lambda = 0$. In our example, we have

$$F_x = 10x - y + \lambda = 0 \qquad \textbf{(1)}$$

$$F_y = 12y - x + 2\lambda = 0 \qquad \textbf{(2)}$$

$$F_\lambda = x + 2y - 24 = 0. \qquad \textbf{(3)}$$

4. Solve the system of equations from Step 3 for x, y, and λ.

To solve the system, multiply both sides of equation (1) by 12 and add the result to equation (2). Doing this, and rearranging terms, we have

$$\begin{array}{r} 120x - 12y + 12\lambda = 0 \\ -x + 12y + 2\lambda = 0 \\ \hline 119x \qquad\quad + 14\lambda = 0. \end{array} \qquad \textbf{(4)}$$

To eliminate y from equations (2) and (3), multiply both sides of equation (3) by -6 and add the result to equation (2). After rearranging terms, we have

$$\begin{array}{rcl} -x + 12y + 2\lambda &=& 0 \\ -6x - 12y \qquad &=& -144 \\ \hline -7x \qquad + 2\lambda &=& -144. \end{array} \qquad \textbf{(5)}$$

We can now find x by solving equations (4) and (5) together. Multiply both sides of equation (5) by -7, and add to equation (4).

$$\begin{array}{rcl} 119x + 14\lambda &=& 0 \\ 49x - 14\lambda &=& 1008 \\ \hline 168x \qquad &=& 1008, \end{array}$$

from which $x = 6$. Substituting 6 for x in equation (3), we have $y = 9$. Also, although we don't use this number, $\lambda = -51$.

Finally, it can be verified that the minimum value for $f(x, y) = 5x^2 + 6y^2 - xy$, subject to the constraint $x + 2y = 24$, is at the point $(6, 9)$. This minimum value is $f(6, 9) = 612$. ■

Work Problem 1 at the side.

Example 2 Find two numbers whose sum is 50 and whose product is a maximum.

If we let x and y represent the two numbers, then we wish to maximize the product

$$f(x, y) = xy,$$

subject to the constraint $x + y = 50$. Go through the four steps presented above.

1. $g(x, y) = x + y - 50$

2. $F(x, y, \lambda) = xy + \lambda(x + y - 50)$

3. $F_x = y + \lambda = 0$ **(6)**

$F_y = x + \lambda = 0$ **(7)**

$F_\lambda = x + y - 50 = 0$ **(8)**

4. From equation (6), we have $\lambda = -y$. Substituting $-y$ for λ in equation (7), we have

$$x + \lambda = 0$$
$$x + (-y) = 0$$
$$x = y.$$

1. Go through the following steps to find a minimum value of $f(x, y) = x^2 + y^2$, subject to the constraint $2x + 3y = 9$.

(a) Put the constraint equal to 0; let $g(x, y)$ equal the constraint

(b) Form the Lagrange function $F(x, y, \lambda)$.

(c) Form the system of equations $F_x = 0$, $F_y = 0$, $F_\lambda = 0$.

(d) Solve the system.

Answer:

(a) $2x + 3y - 9 = 0$;
$g(x, y) = 2x + 3y - 9$

(b) $F(x, y, \lambda) = x^2 + y^2 + \lambda(2x + 3y - 9)$

(c) $F_x = 2x + 2\lambda$;
$F_y = 2y + 3\lambda$;
$F_\lambda = 2x + 3y - 9$

(d) $x = 18/13$, $y = 27/13$, $\lambda = -18/13$

Substituting y for x in equation (8) gives

$$\begin{aligned} x + y - 50 &= 0 \\ y + y - 50 &= 0 \\ 2y &= 50 \\ y &= 25. \end{aligned}$$

Since $x = y$, we have $x = 25$.

Thus, it can be verified that 25 and 25 are the two numbers whose sum is 50 and whose product is a maximum. The maximum product is given by $25 \cdot 25 = 625$. ■

2. Find two numbers x and y whose sum is 12 and such that x^2y is maximized.

Answer: $x = 8$, $y = 4$ leads to the maximum product of $8^2 \cdot 4 = 256$ ($x = 0$, $y = 12$ leads to the *minimum* product of 0.)

Work problem 2 at the side.

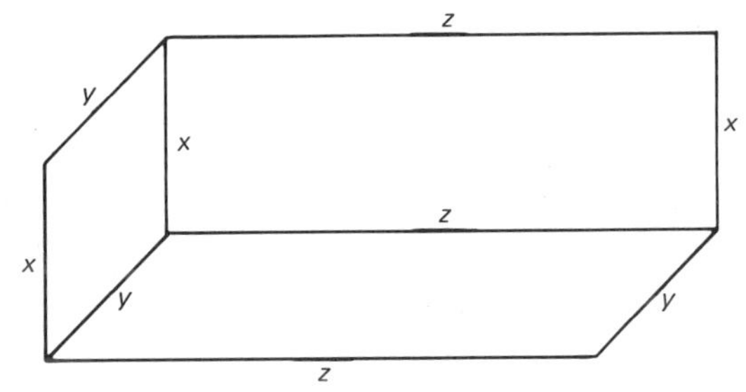

Figure A

Example 3 Find the dimensions of the rectangular box of maximum volume that can be produced from 6 square feet of material.

Let x, y, and z represent the dimensions of the box, as shown in Figure A. The volume of the box is given by

$$f(x, y, z) = xyz.$$

As shown in Figure A, the total amount of material required for the two ends of the box is $2xy$, the total needed for the sides is $2xz$, and the total needed for the top and bottom is $2yz$. Since 6 square feet of material is available, we have

$$2xy + 2xz + 2yz = 6 \qquad \text{or} \qquad xy + xz + yz = 3.$$

In summary, we must maximize $f(x, y, z) = xyz$ subject to the constraint $xy + xz + yz = 3$. This problem involves functions of *three* variables instead of two, but the method of Lagrange multipliers can be used in the same way as before.

1. $g(x, y, z) = xy + xz + yz - 3$
2. $F(x, y, z, \lambda) = xyz + \lambda(xy + xz + yz - 3)$

3. $F_x = yz + \lambda y + \lambda z = 0$
$F_y = xz + \lambda x + \lambda z = 0$
$F_z = xy + \lambda x + \lambda y = 0$
$F_\lambda = xy + xz + yz - 3 = 0$

4. The solution of the system of equations of Step 3 is $x = 1$, $y = 1$, $z = 1$. In other words, the box we want is a cube, 1 foot on a side. ■

Work Problem 3 at the side.

3. A can is to be made that will hold 128π cubic inches of canned salmon. Go through the four steps of the example above to find the dimensions of the can that will produce minimum surface area. The surface area of a can of radius r and height h is given by $S = 2\pi r^2 + 2\pi rh$. The volume is $V = \pi r^2 h$. Remember to treat π as a constant.

Answer:
(Step 1)
$g(r, h) = \pi r^2 h - 128\pi$
(Step 2)
$F(r, h, \lambda) = 2\pi r^2 + 2\pi rh + \lambda(\pi r^2 h - 128\pi)$
(Step 3)
$F_r = 4\pi r + 2\pi h + 2\lambda\pi rh = 0$;
$F_h = 2\pi r + \lambda\pi r^2 = 0$;
$F_\lambda = \pi r^2 h - 128\pi = 0$
(Step 4) From the third equation, $F_\lambda = 0$, we get $\pi r^2 h - 128\pi = 0$, or $\pi r^2 h = 128\pi$, or $r^2 h = 128$, and $h = 128/r^2$. From the second equation, $F_h = 0$, we get $2 + \lambda r = 0$, or $\lambda = -2/r$. From the first equation, we get $2r + h + \lambda rh = 0$. Substitute for h and λ, obtaining $2r + 128/r^2 - 256/r^2 = 0$. Finally, $r = 4$ and $h = 8$.

Now we can state the method of Lagrange multipliers as a theorem. As we said above, the method works with two independent variables, or three, or four, or more. For simplicity, we state the theorem only for two independent variables.

Theorem ***Lagrange multipliers*** Any relative maximum or relative minimum values of the function $z = f(x, y)$, subject to a constraint $g(x, y) = 0$, will be found among those points (x, y) for which there exists a value of λ such that

$$F_x(x, y, \lambda) = 0$$

$$F_y(x, y, \lambda) = 0$$

$$F_\lambda(x, y, \lambda) = 0,$$

where $$F(x, y, \lambda) = f(x, y) + \lambda \cdot g(x, y).$$

We assume that all indicated partial derivatives exist.

One problem with the method of Lagrange multipliers is that it is not easy to tell whether an answer is a relative maximum or a relative minimum—the M-test of the previous section does not apply. (Details on deciding whether a maximum or a minimum has been found are given in Jean E. Draper and Jane S. Klingman *Mathematical Analysis,* second edition, (New York: Harper and Row, 1972), pages 376 and 602. However, we can usually tell from the problem itself—a cost function would probably lead to a minimum, while a problem about profit should lead to maximum.

EXERCISES

Find the following relative maximums or relative minimums. (See Example 1.)

1. Maximum of $f(x, y) = 2xy$, subject to $x + y = 12$.
2. Maximum of $f(x, y) = 4xy + 2$, subject to $x + y = 24$.

3. Maximum of $f(x, y) = x^2y$, subject to $2x + y = 4$.

4. Maximum of $f(x, y) = 4xy^2$, subject to $3x - 2y = 5$.

5. Minimum of $f(x, y) = x^2 + 2y^2 - xy$, subject to $x + y = 8$.

6. Minimum of $f(x, y) = 3x^2 + 4y^2 - xy - 2$, subject to $2x + y = 21$.

7. Maximum of $f(x, y) = x^2 - 10y^2$, subject to $x - y = 18$.

8. Maximum of $f(x, y) = 12xy - x^2 - 3y^2$, subject to $x + y = 16$.

9. Find two numbers whose sum is 20 and whose product is a maximum. See Example 2.

10. Find two numbers whose sum is 100 and whose product is a maximum.

11. Find two numbers x and y such that $x + y = 18$ and xy^2 is maximized.

12. Find two numbers x and y such that $x + y = 36$ and x^2y is maximized.

13. Find three numbers whose sum is 90 and whose product is a maximum.

14. Find three numbers whose sum is 240 and whose product is a maximum.

APPLIED PROBLEMS

Area

15. A farmer has 200 meters of fencing. Find the dimensions of the rectangular field of maximum area that can be enclosed by this amount of fencing.

16. Because of terrain difficulties, two sides of a fence can be built for \$6 a foot, while the other two sides cost \$4 per foot. (See the sketch.) Find the field of maximum area that can be enclosed for \$1200.

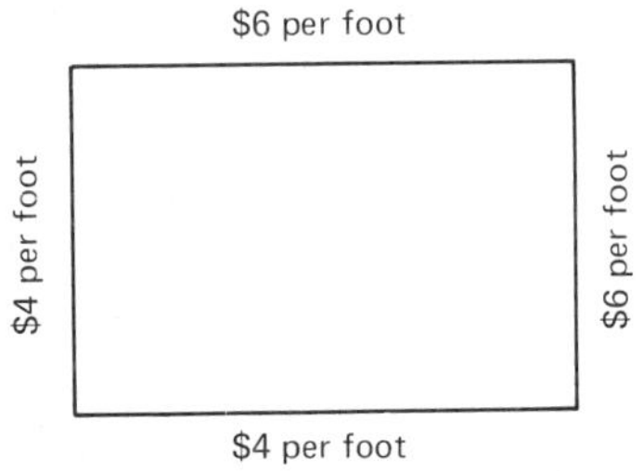

17. Find the area of the largest rectangular field that can be enclosed with 600 meters of fencing. Assume that no fencing is needed along one side of the field.

18. A fence is built against a large building, so that no fencing material is used on that one side. Material for the ends costs \$8 a foot; the side can be built for \$6 per foot. Find the dimensions of the field of maximum area that can be enclosed for \$1200.

Surface area

19. A cylindrical can is to be made that will hold 250π cubic inches of candy. Find the dimensions of the can with minimum surface area.

20. An ordinary 12-ounce beer or soda pop can holds about 25 cubic inches. Use a calculator and find the dimensions of a can with minimum surface area. Measure a can and see how close its dimensions are to the results you found.

Volume

21. A rectangular box with no top is to be built from 500 square meters of material. Find the dimensions of such a box that will enclose the maximum volume.

22. A one-pound soda cracker box has a volume of 185 cubic inches. The end of the box is square. Find the dimensions of such a box that has minimum surface area.

Cost

23. The total cost to produce x large needlepoint kits and y small ones is given by

$$C(x, y) = 2x^2 + 6y^2 + 4xy + 10.$$

If a total of ten kits must be made, how should production be allocated so that total cost is minimized?

Profit

24. The profit from the sale of x units of radiators for automobiles and y units of radiators for generators is given by

$$P(x, y) = -x^2 - y^2 + 4x + 8y.$$

Find values of x and y that lead to a maximum profit if the firm must produce a total of 6 units of radiators.

Production

25. The production of nails depends on the cost of steel and the price of labor. Suppose that the number of units of nails manufactured is given by

$$f(x, y) = -60x + 100y - x^2 - 2y^2 - 3xy + 390{,}000$$

where x is the number of units of steel and y is the number of units of labor. Assume that steel costs \$8 per unit and labor is \$10 per unit. Find the maximum number of units of nails that can be made if the total budget for steel and labor is \$4600.

26. There are two basic ingredients in soft drinks: sugar and carbonated water. If x units of sugar and y units of carbonated water are used, then the total units of soft drinks which can be produced are given by

$$f(x, y) = 120x - 160y - 3x^2 - 2y^2 - 2xy.$$

One unit of sugar costs \$4 and one unit of carbonated water costs \$8. Find the maximum number of units of soft drinks that can be produced with a budget of \$1800 for sugar and carbonated water.

Tables

TABLE 1 SELECTED POWERS OF NUMBERS

n	n^2	n^3	n^4	n^5	n^6
2	4	8	16	32	64
3	9	27	81	243	729
4	16	64	256	1024	4096
5	25	125	625	3125	
6	36	216	1296		
7	49	343	2401		
8	64	512	4096		
9	81	729	6561		
10	100	1000	10,000		

TABLE 2 SQUARES AND SQUARE ROOTS

n	n^2	$\sqrt{n}$	$\sqrt{10n}$	n	n^2	$\sqrt{n}$	$\sqrt{10n}$
1	1	1.000	3.162	51	2601	7.141	22.583
2	4	1.414	4.472	52	2704	7.211	22.804
3	9	1.732	5.477	53	2809	7.280	23.022
4	16	2.000	6.325	54	2916	7.348	23.238
5	25	2.236	7.071	55	3025	7.416	23.452
6	36	2.449	7.746	56	3136	7.483	23.664
7	49	2.646	8.367	57	3249	7.550	23.875
8	64	2.828	8.944	58	3364	7.616	24.083
9	81	3.000	9.487	59	3481	7.681	24.290
10	100	3.162	10.000	60	3600	7.746	24.495
11	121	3.317	10.488	61	3721	7.810	24.698
12	144	3.464	10.954	62	3844	7.874	24.900
13	169	3.606	11.402	63	3969	7.937	25.100
14	196	3.742	11.832	64	4096	8.000	25.298
15	225	3.873	12.247	65	4225	8.062	25.495
16	256	4.000	12.649	66	4356	8.124	25.690
17	289	4.123	13.038	67	4489	8.185	25.884
18	324	4.243	13.416	68	4624	8.246	26.077
19	361	4.359	13.784	69	4761	8.307	26.268
20	400	4.472	14.142	70	4900	8.367	26.458
21	441	4.583	14.491	71	5041	8.426	26.646
22	484	4.690	14.832	72	5184	8.485	26.833
23	529	4.796	15.166	73	5329	8.544	27.019
24	576	4,899	15.492	74	5476	8.602	27.203
25	625	5.000	15.811	75	5625	8.660	27.386
26	676	5.099	16.125	76	5776	8.718	27.568
27	729	5.196	16.432	77	5929	8.775	27.749
28	784	5.292	16.733	78	6084	8.832	27.928
29	841	5.385	17.029	79	6241	8.888	28.107
30	900	5.477	17.321	80	6400	8.944	28.284
31	961	5.568	17.607	81	6561	9.000	28.460
32	1024	5.657	17.889	82	6724	9.055	28.636
33	1089	5.745	18.166	83	6889	9.110	28.810
34	1156	5.831	18.439	84	7056	9.165	28.983
35	1225	5.916	18.708	85	7225	9.220	29.155
36	1296	6.000	18.974	86	7396	9.274	29.326
37	1369	6.083	19.235	87	7569	9.327	29.496
38	1444	6.164	19.494	88	7744	9.381	29.665
39	1521	6.245	19.748	89	7921	9.434	29.833
40	1600	6.325	20.000	90	8100	9.487	30.000
41	1681	6.403	20.248	91	8281	9.539	30.166
42	1764	6.481	20.494	92	8464	9.592	30.332
43	1849	6.557	20.736	93	8649	9.644	30.496
44	1936	6.633	20.976	94	8836	9.695	30.659
45	2025	6.708	21.213	95	9025	9.747	30.822
46	2116	6.782	21.448	96	9216	9.798	30.984
47	2209	6.856	21.679	97	9409	9.849	31.145
48	2304	6.928	21.909	98	9604	9.899	31.305
49	2401	7.000	22.136	99	9801	9.950	31.464
50	2500	7.071	22.361	100	10000	10.000	31.623

TABLE 3 POWERS OF e

x	e^x	e^{-x}
0.00	1.00000	1.00000
0.01	1.01005	0.99004
0.02	1.02020	0.98019
0.03	1.03045	0.97044
0.04	1.04081	0.96078
0.05	1.05127	0.95122
0.06	1.06183	0.94176
0.07	1.07250	0.93239
0.08	1.08328	0.92311
0.09	1.09417	0.91393
0.10	1.10517	0.90483
0.11	1.11628	0.89583
0.12	1.12750	0.88692
0.13	1.13883	0.87810
0.14	1.15027	0.86936
0.15	1.16183	0.86071
0.16	1.17351	0.85214
0.17	1.18530	0.84366
0.18	1.19722	0.83527
0.19	1.20925	0.82696
0.20	1.22140	0.81873
0.30	1.34985	0.74081
0.40	1.49182	0.67032
0.50	1.64872	0.60653
0.60	1.82211	0.54881
0.70	2.01375	0.49658
0.80	2.22554	0.44932
0.90	2.45960	0.40656
1.00	2.71828	0.36787
1.10	3.00416	0.33287
1.20	3.32011	0.30119
1.30	3.66929	0.27253
1.40	4.05519	0.24659
1.50	4.48168	0.22313
1.60	4.95302	0.20189
1.70	5.47394	0.18268
1.80	6.04964	0.16529
1.90	6.68589	0.14956
2.00	7.38905	0.13533
2.10	8.16616	0.12245
2.20	9.02500	0.11080
2.30	9.97417	0.10025
2.40	11.02316	0.09071
2.50	12.18248	0.08208
2.60	13.46372	0.07427
2.70	14.87971	0.06720
2.80	16.44463	0.06081
2.90	18.17412	0.05502
3.00	20.08551	0.04978
3.50	33.11545	0.03020
4.00	54.59815	0.01832
4.50	90.01713	0.01111
5.00	148.41316	0.00674
5.50	224.69193	0.00409
6.00	403.42879	0.00248
6.50	665.14163	0.00150
7.00	1096.63316	0.00091
7.50	1808.04241	0.00055
8.00	2980.95799	0.00034
8.50	4914.76884	0.00020
9.00	8103.08392	0.00012
9.50	13359.72683	0.00007
10.00	22026.46579	0.00005

TABLE 4 NATURAL LOGARITHMS

x	$\ln x$
0.1	7.6974 − 10
0.2	8.3906 − 10
0.3	8.7960 − 10
0.4	9.0837 − 10
0.5	9.3069 − 10
0.6	9.4892 − 10
0.7	9.6433 − 10
0.8	9.7769 − 10
0.9	9.8946 − 10
1.0	0.0000
1.1	0.0953
1.2	0.1823
1.3	0.2624
1.4	0.3365
1.5	0.4055
1.6	0.4700
1.7	0.5306
1.8	0.5878
1.9	0.6419
2.0	0.6931
2.1	0.7419
2.2	0.7885
2.3	0.8329
2.4	0.8755
2.5	0.9163
2.6	0.9555
2.7	0.9933
2.8	1.0296
2.9	1.0647
3.0	1.0986
3.1	1.1314
3.2	1.1632
3.3	1.1939
3.4	1.2238
3.5	1.2528
3.6	1.2809
3.7	1.3083
3.8	1.3350
3.9	1.3610
4.0	1.3863
4.1	1.4110
4.2	1.4351
4.3	1.4586
4.4	1.4816

x	$\ln x$
4.5	1.5041
4.6	1.5261
4.7	1.5476
4.8	1.5686
4.9	1.5892
5.0	1.6094
5.1	1.6292
5.2	1.6487
5.3	1.6677
5.4	1.6864
5.5	1.7047
5.6	1.7228
5.7	1.7405
5.8	1.7579
5.9	1.7750
6.0	1.7918
6.1	1.8083
6.2	1.8245
6.3	1.8405
6.4	1.8563
6.5	1.8718
6.6	1.8871
6.7	1.9021
6.8	1.9169
6.9	1.9315
7.0	1.9459
7.1	1.9601
7.2	1.9741
7.3	1.9879
7.4	2.0015
7.5	2.0149
7.6	2.0281
7.7	2.0412
7.8	2.0541
7.9	2.0669
8.0	2.0794
8.1	2.0919
8.2	2.1041
8.3	2.1163
8.4	2.1281
8.5	2.1401
8.6	2.1518
8.7	2.1633
8.8	2.1748
8.9	2.1861

x	$\ln x$
9.0	2.1972
9.1	2.2083
9.2	2.2192
9.3	2.2300
9.4	2.2407
9.5	2.2513
9.6	2.2618
9.7	2.2721
9.8	2.2824
9.9	2.2925
10	2.3026
11	2.3979
12	2.4849
13	2.5649
14	2.6391
15	2.7081
16	2.7726
17	2.8332
18	2.8904
19	2.9444
20	2.9957
25	3.2189
30	3.4012
35	3.5553
40	3.6889
45	3.8067
50	3.9120
55	4.0073
60	4.0943
65	4.1744
70	4.2485
75	4.3175
80	4.3820
85	4.4427
90	4.4998
95	4.5539
100	4.6052

TABLE 5 COMMON LOGARITHMS

n	0	1	2	3	4	5	6	7	8	9
1.0	.0000	.0043	.0086	.0128	.0170	.0212	.0253	.0294	.0334	.0374
1.1	.0414	.0453	.0492	.0531	.0569	.0607	.0645	.0682	.0719	.0755
1.2	.0792	.0828	.0864	.0899	.0934	.0969	.1004	.1038	.1072	.1106
1.3	.1 139	.1173	.1206	.1239	.1271	.1303	.1335	.1367	.1399	.1430
1.4	.1461	.1492	.1523	.1553	.1584	.1614	.1644	.1673	.1703	.1732
1.5	.1761	.1790	.1818	.1847	.1875	.1903	.1931	.1959	.1987	.2014
1.6	.2041	.2068	.2095	.2122	.2148	.2175	.2201	.2227	.2253	.2279
1.7	.2304	.2330	.2355	.2380	.2405	.2430	.2455	.2480	.2504	.2529
1.8	.2553	.2577	.2601	.2625	.2648	.2672	.2695	.2718	.2742	.2765
1.9	.2788	.2810	.2833	.2856	.2878	.2900	.2923	.2945	.2967	.2989
2.0	.3010	.3032	.3054	.3075	.3096	.3118	.3139	.3160	.3181	.3201
2.1	.3222	,3243	.3263	.3284	.3304	.3324	.3345	.3365	.3385	.3404
2.2	.3424	.3444	.3464	.3483	.3502	.3522	.3541	.3560	.3579	.3598
2.3	.3617	.3636	.3655	.3674	.3692	.3711	.3729	.3747	.3766	.3784
2.4	.3802	.3820	.3838	.3856	.3874	.3892	.3909	.3927	.3945	.3962
2.5	.3979	.3997	.4014	.4031	.4048	.4065	.4082	.4099	.4116	.4133
2.6	.4150	.4166	.4183	.4200	.4216	.4232	.4249	.4265	.4281	.4298
2.7	.4314	.4330	.4346	.4362	.4378	.4393	.4409	.4425	.4440	.4456
2.8	.4472	.4487	.4502	.4518	.4533	.4548	.4564	.4579	.4594	.4609
2.9	.4624	.4639	.4654	.4669	.4683	.4698	.4713	.4728	.4742	.4757
3.0	.4771	.4786	.4800	.4814	.4829	.4843	.4857	.4871	.4886	.4900
3.1	.4914	.4928	.4942	.4955	.4969	.4983	.4997	.5011	.5024	.5038
3.2	.5051	.5065	.5079	.5092	.5105	.5119	.5132	.5145	.5159	.5172
3.3	.5185	.5198	.5211	.5224	.5237	.5250	.5263	.5276	.5289	.5302
3.4	.5315	.5328	.5340	.5353	.5366	.5378	.5391	.5403	.5416	.5428
3.5	.5441	.5453	.5465	.5478	.5490	.5502	.5514	.5527	.5539	.5551
3.6	.5563	.5575	.5587	.5599	.5611	.5623	.5635	.5647	.5658	.5670
3.7	.5682	.5694	.5705	.5717	.5729	.5740	.5752	.5763	.5775	.5786
3.8	.5798	.5809	.5821	.5832	.5843	.5855	.5866	.5877	.5888	.5899
3.9	.5911	.5922	.5933	.5944	.5955	.5966	.5977	.5988	.5999	.6010
4.0	.6021	.6031	.6042	.6053	.6064	.6075	.6085	.6096	.6107	.6117
4.1	.6128	.6138	.6149	.6160	.6170	.6180	.6191	.6201	.6212	.6222
4.2	.6232	.6243	.6253	.6263	.6274	.6284	.6294	.6304	.6314	.6325
4.3	.6335	.6345	.6355	.6365	.6375	.6385	.6395	.6405	.6415	.6425
4.4	.6435	.6444	.6454	.6464	.6474	.6484	.6493	.6503	.6513	.6522
4.5	.6532	.6542	.6551	.6561	.6571	.6580	.6590	.6599	.6609	.6618
4.6	.6628	.6637	.6646	.6656	.6665	.6675	.6684	.6693	.6702	.6712
4.7	.6721	.6730	.6739	.6749	.6758	.6767	.6776	.6785	.6794	.6803
4.8	.6812	.6821	.6830	.6839	.6848	.6857	.6866	.6875	.6884	.6893
4.9	.6902	.6911	.6920	.6928	.6937	.6946	.6955	.6964	.6972	.6981
5.0	.6990	.6998	.7007	.7016	.7024	.7033	.7042	.7050	.7059	.7067
5.1	.7076	.7084	.7093	.7101	.7110	.7118	.7126	.7135	.7143	.7152
5.2	.7160	.7168	.7177	.7185	.7193	.7202	.7210	.7218	.7226	.7235
5.3	.7243	.7251	.7259	.7267	.7275	.7284	.7292	.7300	.7308	.7316
5.4	.7324	.7332	.7340	.7348	.7356	.7364	.7372	.7380	.7388	.7396
n	0	1	2	3	4	5	6	7	8	9

TABLE 5 (CONTINUED)

n	0	1	2	3	4	5	6	7	8	9
5.5	.7404	.7412	.7419	.7427	.7435	.7443	.7451	.7459	.7466	.7474
5.6	.7482	.7490	.7497	.7505	.7513	.7520	.7528	.7536	.7543	.7551
5.7	.7559	.7566	.7574	.7582	.7589	.7597	.7604	.7612	.7619	.7627
5.8	.7634	.7642	.7649	.7657	.7664	.7672	.7679	.7686	.7694	.7701
5.9	.7709	.7716	.7723	.7731	.7738	.7745	.7752	.7760	.7767	.7774
6.0	.7782	.7789	.7796	.7803	.7810	.7818	.7825	.7832	.7839	.7846
6.1	.7853	.7860	.7868	.7875	.7882	.7889	.7896	.7903	.7910	.7917
6.2	.7924	.7931	.7938	.7945	.7952	.7959	.7966	.7973	.7980	.7987
6.3	.7993	.8000	.8007	.8014	.8021	.8028	.8035	.8041	.8048	.8055
6.4	.8062	.8069	.8075	.8082	.8089	.8096	.8102	.8109	.8116	.8122
6.5	.8129	.8136	.8142	.8149	.8156	.8162	.8169	.8176	.8182	.8189
6.6	.8195	.8202	.8209	.8215	.8222	.8228	.8235	.8241	.8248	.8254
6.7	.8261	.8267	.8274	.8280	.8287	.8293	.8299	.8306	.8312	.8319
6.8	.8325	.8331	.8338	.8344	.8351	.8357	.8363	.8370	.8376	.8382
6.9	.8388	.8395	.8401	.8407	.8414	.8420	.8426	.8432	.8439	.8445
7.0	.8451	.8457	.8463	.8470	.8476	.8482	.8488	.8494	.8500	.8506
7.1	.8513	.8519	.8525	.8531	.8537	.8543	.8549	.8555	.8561	.8567
7.2	.8573	.8579	.8585	.8591	.8597	.8603	.8609	.8615	.8621	.8627
7.3	.8633	.8639	.8645	.8651	.8657	.8663	.8669	.8675	.8681	.8686
7.4	.8692	.8698	.8704	.8710	.8716	.8722	.8727	.8733	.8739	.8745
7.5	.8751	.8756	.8762	.8768	.8774	.8779	.8785	.8791	.8797	.8802
7.6	.8808	.8814	.8820	.8825	.8831	.8837	.8842	.8848	.8854	.8859
7.7	.8865	.8871	.8876	.8882	.8887	.8893	.8899	.8904	.8910	.8915
7.8	.8921	.8927	.8932	.8938	.8943	.8949	.8954	.8960	.8965	.8971
7.9	.8976	.8982	.8987	.8993	.8998	.9004	.9009	.9015	.9020	.9025
8.0	.9031	.9036	.9042	.9047	.9053	.9058	.9063	.9069	.9074	.9079
8.1	.9085	.9090	.9096	.9101	.9106	.9112	.9117	.9122	.9128	.9133
8.2	.9138	.9143	.9149	.9154	.9159	.9165	.9170	.9175	.9180	.9186
8.3	.9191	.9196	.9201	.9206	.9212	.9217	.9222	.9227	.9232	.9238
8.4	.9243	.9248	.9253	.9258	.9263	.9269	.9274	.9279	.9284	.9289
8.5	.9294	.9299	.9304	.9309	.9315	.9320	.9325	.9330	.9335	.9340
8.6	.9345	.9350	.9355	.9360	.9365	.9370	.9375	.9380	.9385	.9390
8.7	.9395	.9400	.9405	.9410	.9415	.9420	.9425	.9430	.9435	.9440
8.8	.9445	.9450	.9455	.9460	.9465	.9469	.9474	.9479	.9484	.9489
8.9	.9494	.9499	.9504	.9509	.9513	.9518	.9523	.9528	.9533	.9538
9.0	.9542	.9547	.9552	.9557	.9562	.9566	.9571	.9576	.9581	.9586
9.1	.9590	.9595	.9600	.9605	.9609	.9614	.9619	.9624	.9628	.9633
9.2	.9638	.9643	.9647	.9652	.9657	.9661	.9666	.9671	.9675	.9680
9.3	.9685	.9689	.9694	.9699	.9703	.9708	.9713	.9717	.9722	.9727
9.4	.9731	.9736	.9741	.9745	.9750	.9754	.9759	.9763	.9768	.9773
9.5	.9777	.9782	.9786	.9791	.9795	.9800	.9805	.9809	.9814	.9818
9.6	.9823	.9827	.9832	.9836	.9841	.9845	.9850	.9854	.9859	.9863
9.7	.9868	.9872	.9877	.9881	.9886	.9890	.9894	.9899	.9903	.9908
9.8	.9912	.9917	.9921	.9926	.9930	.9934	.9939	.9943	.9948	.9952
9.9	.9956	.9961	.9965	.9969	.9974	.9978	.9983	.9987	.9991	.9996
n	0	1	2	3	4	5	6	7	8	9

TABLE 6 COMPOUND INTEREST

$(1 + i)^n$

$n \backslash i$	1%	$1\frac{1}{2}$%	2%	3%	4%	5%	6%	8%
1	1.01000	1.01500	1.02000	1.03000	1.04000	1.05000	1.06000	1.08000
2	1.02010	1.03023	1.04040	1.06090	1.08160	1.10250	1.12360	1.16640
3	1.03030	1.04568	1.06121	1.09273	1.12486	1.15763	1.19102	1.25971
4	1.04060	1.06136	1.08243	1.12551	1.16986	1.21551	1.26248	1.36049
5	1.05101	1.07728	1.10408	1.15927	1.21665	1.27628	1.33823	1.46933
6	1.06152	1.09344	1.12616	1.19405	1.26532	1.34010	1.41852	1.58687
7	1.07214	1.10984	1.14869	1.22987	1.31593	1.40710	1.50363	1.71382
8	1.08286	1.12649	1.17166	1.26677	1.36857	1.47746	1.59385	1.85093
9	1.09369	1.14339	1.19509	1.30477	1.42331	1.55133	1.68948	1.99900
10	1.10462	1.16054	1.21899	1.34392	1.48024	1.62889	1.79085	2.15892
11	1.11567	1.17795	1.24337	1.38423	1.53945	1.71034	1.89830	2.33164
12	1.12683	1.19562	1.26824	1.42576	1.60103	1.79586	2.01220	2.51817
13	1.13809	1.21355	1.29361	1.46853	1.66507	1.88565	2.13293	2.71962
14	1.14947	1.23176	1.31948	1.51259	1.73168	1.97993	2.26090	2.93719
15	1.16097	1.25023	1.34587	1.55797	1.80094	2.07893	2.39656	3.17217
16	1.17258	1.26899	1.37279	1.60471	1.87298	2.18287	2.54035	3.42594
17	1.18430	1.28802	1.40024	1.65285	1.94790	2.29202	2.69277	3.70002
18	1.19615	1.30734	1.42825	1.70243	2.02582	2.40662	2.85434	3.99602
19	1.20811	1.32695	1.45681	1.75351	2.10685	2.52695	3.02560	4.31570
20	1.22019	1.34686	1.48595	1.80611	2.19112	2.65330	3.20714	4.66096
21	1.23239	1.36706	1.51567	1.86029	2.27877	2.78596	3.39956	5.03383
22	1.24472	1.38756	1.54598	1.91610	2.36992	2.92526	3.60354	5.43654
23	1.25716	1.40838	1.57690	1.97359	2.46472	3.07152	3.81975	5.87146
24	1.26973	1.42950	1.60844	2.03279	2.56330	3.22510	4.04893	6.34118
25	1.28243	1.45095	1.64061	2.09378	2.66584	3.38635	4.29187	6.84848
26	1.29526	1.47271	1.67342	2.15659	2.77247	3.55567	4.54938	7.39635
27	1.30821	1.49480	1.70689	2.22129	2.88337	3.73346	4.82235	7.98806
28	1.32129	1.51722	1.74102	2.28793	2.99870	3.92013	5.11169	8.62711
29	1.33450	1.53998	1.77584	2.35657	3.11865	4.11614	5.41839	9.31727
30	1.34785	1.56308	1.81136	2.42726	3.24340	4.32194	5.74349	10.06266
31	1.36133	1.58653	1.84759	2.50008	3.37313	4.53804	6.08810	10.86767
32	1.37494	1.61032	1.88454	2.57508	3.50806	4.76494	6.45339	11.73708
33	1.38869	1.63448	1.92223	2.65234	3.64838	5.00319	6.84059	12.67605
34	1.40258	1.65900	1.96068	2.73191	3.79432	5.25335	7.25103	13.69013
35	1.41660	1.68388	1.99989	2.81386	3.94609	5.51602	7.68609	14.78534
36	1.43077	1.70914	2.03989	2.89828	4.10393	5.79182	8.14725	15.96817
37	1.44508	1.73478	2.08069	2.98523	4.26809	6.08141	8.63609	17.24563
38	1.45953	1.76080	2.12230	3.07478	4.43881	6.38548	9.15425	18.62528
39	1.47412	1.78721	2.16474	3.16703	4.61637	6.70475	9.70351	20.11530
40	1.48886	1.81402	2.20804	3.26204	4.80102	7.03999	10.28572	21.72452
41	1.50375	1.84123	2.25220	3.35990	4.99306	7.39199	10.90286	23.46248
42	1.51879	1.86885	2.29724	3.46070	5.19278	7.76159	11.55703	25.33948
43	1.53398	1.89688	2.34319	3.56452	5.40050	8.14967	12.25045	27.36664
44	1.54932	1.92533	2.39005	3.67145	5.61652	8.55715	12.98548	29.55597
45	1.56481	1.95421	2.43785	3.78160	5.84118	8.98501	13.76461	31.92045
46	1.58046	1.98353	2.48661	3.89504	6.07482	9.43426	14.59049	34.47409
47	1.59626	2.01328	2.53634	4.01190	6.31782	9.90597	15.46592	37.23201
48	1.61223	2.04348	2.58707	4.13225	6.57053	10.40127	16.39387	40.21057
49	1.62835	2.07413	2.63881	4.25622	6.83335	10.92133	17.37750	43.42742
50	1.64463	2.10524	2.69159	4.38391	7.10668	11.46740	18.42015	46.90161

TABLE 7 PRESENT VALUE OF A DOLLAR

$(1 + i)^{-n}$

n \ i	1%	$1\frac{1}{2}$%	2%	3%	4%	5%	6%	8%
1	0.99010	0.98522	0.98039	0.97087	0.96154	0.95238	0.94340	0.92593
2	0.98030	0.97066	0.96117	0.94260	0.92456	0.90703	0.89000	0.85734
3	0.97059	0.95632	0.94232	0.91514	0.88900	0.86384	0.83962	0.79383
4	0.96098	0.94218	0.92385	0.88849	0.85480	0.82270	0.79209	0.73503
5	0.95147	0.92826	0.90573	0.86261	0.82193	0.78353	0.74726	0.68058
6	0.94205	0.91454	0.88797	0.83748	0.79031	0.74622	0.70496	0.63017
7	0.93272	0.90103	0.87056	0.81309	0.75992	0.71068	0.66506	0.58349
8	0.92348	0.88771	0.85349	0.78941	0.73069	0.67684	0.62741	0.54027
9	0.91434	0.87459	0.83676	0.76642	0.70259	0.64461	0.59190	0.50025
10	0.90529	0.86167	0.82035	0.74409	0.67556	0.61391	0.55839	0.46319
11	0.89632	0.84893	0.80426	0.72242	0.64958	0.58468	0.52679	0.42888
12	0.88745	0.83639	0.78849	0.70138	0.62460	0.55684	0.49697	0.39711
13	0.87866	0.82403	0.77303	0.68095	0.60057	0.53032	0.46884	0.36770
14	0.86996	0.81185	0.75788	0.66112	0.57748	0.50507	0.44230	0.34046
15	0.86135	0.79985	0.74301	0.64186	0.55526	0.48102	0.41727	0.31524
16	0.85282	0.78803	0.72845	0.62317	0.53391	0.45811	0.39365	0.29189
17	0.84438	0.77639	0.71416	0.60502	0.51337	0.43630	0.37136	0.27027
18	0.83602	0.76491	0.70016	0.58739	0.49363	0.41552	0.35034	0.25025
19	0.82774	0.75361	0.68643	0.57029	0.47464	0.39573	0.33051	0.23171
20	0.81954	0.74247	0.67297	0.55368	0.45639	0.37689	0.31180	0.21455
21	0.81143	0.73150	0.65978	0.53755	0.43883	0.35894	0.29416	0.19866
22	0.80340	0.72069	0.64684	0.52189	0.42196	0.34185	0.27751	0.18394
23	0.79544	0.71004	0.63416	0.50669	0.40573	0.32557	0.26180	0.17032
24	0.78757	0.69954	0.62172	0.49193	0.39012	0.31007	0.24698	0.15770
25	0.77977	0.68921	0.60953	0.47761	0.37512	0.29530	0.23300	0.14602
26	0.77205	0.67902	0.59758	0.46369	0.36069	0.28124	0.21981	0.13520
27	0.76440	0.66899	0.58586	0.45019	0.34682	0.26785	0.20737	0.12519
28	0.75684	0.65910	0.57437	0.43708	0.33348	0.25509	0.19563	0.11591
29	0.74934	0.64936	0.56311	0.42435	0.32065	0.24295	0.18456	0.10733
30	0.74192	0.63976	0.55207	0.41199	0.30832	0.23138	0.17411	0.09938
31	0.73458	0.63031	0.54125	0.39999	0.29646	0.22036	0.16425	0.09202
32	0.72730	0.62099	0.53063	0.38834	0.28506	0.20987	0.15496	0.08520
33	0.72010	0.61182	0.52023	0.37703	0.27409	0.19987	0.14619	0.07889
34	0.71297	0.60277	0.51003	0.36604	0.26355	0.19035	0.13791	0.07305
35	0.70591	0.59387	0.50003	0.35538	0.25342	0.18129	0.13011	0.06763
36	0.69892	0.58509	0.49022	0.34503	0.24367	0.17266	0.12274	0.06262
37	0.69200	0.57644	0.48061	0.33498	0.23430	0.16444	0.11579	0.05799
38	0.68515	0.56792	0.47119	0.32523	0.22529	0.15661	0.10924	0.05369
39	0.67837	0.55953	0.46195	0.31575	0.21662	0.14915	0.10306	0.04971
40	0.67165	0.55126	0.45289	0.30656	0.20829	0.14205	0.09722	0.04603
41	0.66500	0.54312	0.44401	0.29763	0.20028	0.13528	0.09172	0.04262
42	0.65842	0.53509	0.43530	0.28896	0.19257	0.12884	0.08653	0.03946
43	0.65190	0.52718	0.42677	0.28054	0.18517	0.12270	0.08163	0.03654
44	0.64545	0.51939	0.41840	0.27237	0.17805	0.11686	0.07701	0.03383
45	0.63905	0.51171	0.41020	0.26444	0.17120	0.11130	0.07265	0.03133
46	0.63273	0.50415	0.40215	0.25674	0.16461	0.10600	0.06854	0.02901
47	0.62646	0.49670	0.39427	0.24926	0.15828	0.10095	0.06466	0.02686
48	0.62026	0.48936	0.38654	0.24200	0.15219	0.09614	0.06100	0.02487
49	0.61412	0.48213	0.37896	0.23495	0.14634	0.09156	0.05755	0.02303
50	0.60804	0.47500	0.37153	0.22811	0.14071	0.08720	0.05429	0.02132

TABLE 8 AMOUNT OF AN ANNUITY

$$s_{\overline{n}|\,i} = \frac{(1+i)^n - 1}{i}$$

n \ i	1%	$1\frac{1}{2}$%	2%	3%	4%	5%	6%	8%
1	1.00000	1.00000	1.00000	1.00000	1.00000	1.00000	1.00000	1.00000
2	2.01000	2.01500	2.02000	2.03000	2.04000	2.05000	2.06000	2.08000
3	3.03010	3.04523	3.06040	3.09090	3.12160	3.15250	3.18360	3.24640
4	4.06040	4.09090	4.12161	4.18363	4.24646	4.31013	4.37462	4.50611
5	5.10101	5.15227	5.20404	5.30914	5.41632	5.52563	5.63709	5.86660
6	6.15202	6.22955	6.30812	6.46841	6.63298	6.80191	6.97532	7.33593
7	7.21354	7.32299	7.43428	7.66246	7.89829	8.14201	8.39384	8.92280
8	8.28567	8.43284	8.58297	8.89234	9.21423	9.54911	9.89747	10.63663
9	9.36853	9.55933	9.75463	10.15911	10.58280	11.02656	11.49132	12.48756
10	10.46221	10.70272	10.94972	11.46388	12.00611	12.57789	13.18079	14.48656
11	11.56683	11.86326	12.16872	12.80780	13.48635	14.20679	14.97164	16.64549
12	12.68250	13.04121	13.41209	14.19203	15.02581	15.91713	16.86994	18.97713
13	13.80933	14.23683	14.68033	15.61779	16.62684	17.71298	18.88214	21.49530
14	14.94742	15.45038	15.97394	17.08632	18.29191	19.59863	21.01507	24.21492
15	16.09690	16.68214	17.29342	18.59891	20.02359	21.57856	23.27597	27.15211
16	17.25786	17.93237	18.63929	20.15688	21.82453	23.65749	25.67253	30.32428
17	18.43044	19.20136	20.01207	21.76159	23.69751	25.84037	28.21288	33.75023
18	19.61475	20.48938	21.41231	23.41444	25.64541	28.13238	30.90565	37.45024
19	20.81090	21.79672	22.84056	25.11687	27.67123	30.53900	33.75999	41.44626
20	22.01900	23.12367	24.29737	26.87037	29.77808	33.06595	36.78559	45.76196
21	23.23919	24.47052	25.78332	28.67649	31.96920	35.71925	39.99273	50.42292
22	24.47159	25.83758	27.29898	30.53678	34.24797	38.50521	43.39229	55.45676
23	25.71630	27.22514	28.84496	32.45288	36.61789	41.43048	46.99583	60.89330
24	26.97346	28.63352	30.42186	34.42647	39.08260	44.50200	50.81558	66.76476
25	28.24320	30.06302	32.03030	36.45926	41.64591	47.72710	54.86451	73.10594
26	29.52563	31.51397	33.67091	38.55304	44.31174	51.11345	59.15638	79.95442
27	30.82089	32.98668	35.34432	40.70963	47.08421	54.66913	63.70577	87.35077
28	32.12910	34.48148	37.05121	42.93092	49.96758	58.40258	68.52811	95.33883
29	33.45039	35.99870	38.79223	45.21885	52.96629	62.32271	73.63980	103.96594
30	34.78489	37.53868	40.56808	47.57542	56.08494	66.43885	79.05819	113.28321
31	36.13274	39.10176	42.37944	50.00268	59.32834	70.76079	84.80168	123.34587
32	37.49407	40.68829	44.22703	52.50276	62.70147	75.29883	90.88978	134.21354
33	38.86901	42.29861	46.11157	55.07784	66.20953	80.06377	97.34316	145.95062
34	40.25770	43.93309	48.03380	57.73018	69.85791	85.06696	104.18375	158.62667
35	41.66028	45.59209	49.99448	60.46208	73.65222	90.32031	111.43478	172.31680
36	43.07688	47.27597	51.99437	63.27594	77.59831	95.83632	119.12087	187.10215
37	44.50765	48.98511	54.03425	66.17422	81.70225	101.62814	127.26812	203.07032
38	45.95272	50.71989	56.11494	69.15945	85.97034	107.70955	135.90421	220.31595
39	47.41225	52.48068	58.23724	72.23423	90.40915	114.09502	145.05846	238.94122
40	48.88637	54.26789	60.40198	75.40126	95.02552	120.79977	154.76197	259.05652
41	50.37524	56.08191	62.61002	78.66330	99.82654	127.83976	165.04768	280.78104
42	51.87899	57.92314	64.86222	82.02320	104.81960	135.23175	175.95054	304.24352
43	53.39778	59.79199	67.15947	85.48389	110.01238	142.99334	187.50758	329.58301
44	54.93176	61.68887	69.50266	89.04841	115.41288	151.14301	199.75803	356.94965
45	56.48107	63.61420	71.89271	92.71986	121.02939	159.70016	212.74351	386.50562
46	58.04589	65.56841	74.33056	96.50146	126.87057	168.68516	226.50812	418.42607
47	59.62634	67.55194	76.81718	100.39650	132.94539	178.11942	241.09861	452.90015
48	61.22261	69.56522	79.35352	104.40840	139.26321	188.02539	256.56453	490.13216
49	62.83483	71.60870	81.94059	108.54065	145.83373	198.42666	272.95840	530.34274
50	64.46318	73.68283	84.57940	112.79687	152.66708	209.34800	290.33590	573.77016

TABLE 9 PRESENT VALUE OF AN ANNUITY

$$a_{\overline{n}|\,i} = \frac{1-(1+i)^{-n}}{i}$$

n \ i	1%	$1\frac{1}{2}$%	2%	3%	4%	5%	6%	8%
1	0.99010	0.98522	0.98039	0.97087	0.96154	0.95238	0.94340	0.92593
2	1.97040	1.95588	1.94156	1.91347	1.88609	1.85941	1.83339	1.78326
3	2.94099	2.91220	2.88388	2.82861	2.77509	2.72325	2.67301	2.57710
4	3.90197	3.85438	3.80773	3.71710	3.62990	3.54595	3.46511	3.31213
5	4.85343	4.78264	4.71346	4.57971	4.45182	4.32948	4.21236	3.99271
6	5.79548	5.69719	5.60143	5.41719	5.24214	5.07569	4.91732	4.62288
7	6.72819	6.59821	6.47199	6.23028	6.00205	5.78637	5.58238	5.20637
8	7.65168	7.48593	7.32548	7.01969	6.73274	6.46321	6.20979	5.74664
9	8.56602	8.36052	8.16224	7.78611	7.43533	7.10782	6.80169	6.24689
10	9.47130	9.22218	8.98259	8.53020	8.11090	7.72173	7.36009	6.71008
11	10.36763	10.07112	9.78685	9.25262	8.76048	8.30641	7.88687	7.13896
12	11.25508	10.90751	10.57534	9.95400	9.38507	8.86325	8.38384	7.53608
13	12.13374	11.73153	11.34837	10.63496	9.98565	9.39357	8.85268	7.90378
14	13.00370	12.54338	12.10625	11.29607	10.56312	9.89864	9.29498	8.24424
15	13.86505	13.34323	12.84926	11.93794	11.11839	10.37966	9.71225	8.55948
16	14.71787	14.13126	13.57771	12.56110	11.65230	10.83777	10.10590	8.85137
17	15.56225	14.90765	14.29187	13.16612	12.16567	11.27407	10.47726	9.12164
18	16.39827	15.67256	14.99203	13.75351	12.65930	11.68959	10.82760	9.37189
19	17.22601	16.42617	15.67846	14.32380	13.13394	12.08532	11.15812	9.60360
20	18.04555	17.16864	16.35143	14.87747	13.59033	12.46221	11.46992	9.81815
21	18.85698	17.90014	17.01121	15.41502	14.02916	12.82115	11.76408	10.01680
22	19.66038	18.62082	17.65805	15.93692	14.45112	13.16300	12.04158	10.20074
23	20.45582	19.33086	18.29220	16.44361	14.85684	13.48857	12.30338	10.37106
24	21.24339	20.03041	18.91393	16.93554	15.24696	13.79864	12.55036	10.52876
25	22.02316	20.71961	19.52346	17.41315	15.62208	14.09394	12.78336	10.67478
26	22.79520	21.39863	20.12104	17.87684	15.98277	14.37519	13.00317	10.80998
27	23.55961	22.06762	20.70690	18.32703	16.32959	14.64303	13.21053	10.93516
28	24.31644	22.72672	21.28127	18.76411	16.66306	14.89813	13.40616	11.05108
29	25.06579	23.37608	21.84438	19.18845	16.98371	15.14107	13.59072	11.15841
30	25.80771	24.01584	22.39646	19.60044	17.29203	15.37245	13.76483	11.25778
31	26.54229	24.64615	22.93770	20.00043	17.58849	15.59281	13.92909	11.34980
32	27.26959	25.26714	23.46833	20.38877	17.87355	15.80268	14.08404	11.43500
33	27.98969	25.87895	23.98856	20.76579	18.14765	16.00255	14.23023	11.51389
34	28.70267	26.48173	24.49859	21.13184	18.41120	16.19290	14.36814	11.58693
35	29.40858	27.07559	24.99862	21.48722	18.66461	16.37419	14.49825	11.65457
36	30.10751	27.66068	25.48884	21.83225	18.90828	16.54685	14.62099	11.71719
37	30.79951	28.23713	25.96945	22.16724	19.14258	16.71129	14.73678	11.77518
38	31.48466	28.80505	26.44064	22.49246	19.36786	16.86789	14.84602	11.82887
39	32.16303	29.36458	26.90259	22.80822	19.58448	17.01704	14.94907	11.87858
40	32.83469	29.91585	27.35548	23.11477	19.79277	17.15909	15.04630	11.92461
41	33.49969	30.45896	27.79949	23.41240	19.99305	17.29437	15.13802	11.96723
42	34.15811	30.99405	28.23479	23.70136	20.18563	17.42321	15.22454	12.00670
43	34.81001	31.52123	28.66156	23.98190	20.37079	17.54591	15.30617	12.04324
44	35.45545	32.04062	29.07996	24.25427	20.54884	17.66277	15.38318	12.07707
45	36.09451	32.55234	29.49016	24.51871	20.72004	17.77407	15.45583	12.10840
46	36.72724	33.05649	29.89231	24.77545	20.88465	17.88007	15.52437	12.13741
47	37.35370	33.55319	30.28658	25.02471	21.04294	17.98102	15.58903	12.16427
48	37.97396	34.04255	30.67312	25.26671	21.19513	18.07716	15.65003	12.18914
49	38.58808	34.52468	31.05208	25.50166	21.34147	18.16872	15.70757	12.21216
50	39.19612	34.99969	31.42361	25.72976	21.48218	18.25593	15.76186	12.23348

TABLE 10 $1/s_{\overline{n}|i}$

$$\frac{1}{s_{\overline{n}|i}} = \frac{i}{(1+i)^n - 1}$$

n \ i	1%	$1\frac{1}{2}$%	2%	3%	4%	5%	6%	8%
1	1.00000	1.00000	1.00000	1.00000	1.00000	1.00000	1.00000	1.00000
2	0.49751	0.49628	0.49505	0.49261	0.49020	0.48780	0.48544	0.48077
3	0.33002	0.32838	0.32675	0.32353	0.32035	0.31721	0.31411	0.30803
4	0.24628	0.24444	0.24262	0.23903	0.23549	0.23201	0.22859	0.22192
5	0.19604	0.19409	0.19216	0.18835	0.18463	0.18097	0.17740	0.17046
6	0.16255	0.16053	0.15853	0.15460	0.15076	0.14702	0.14336	0.13632
7	0.13863	0.13656	0.13451	0.13051	0.12661	0.12282	0.11914	0.11207
8	0.12069	0.11858	0.11651	0.11246	0.10853	0.10472	0.10104	0.09401
9	0.10674	0.10461	0.10252	0.09843	0.09449	0.09069	0.08702	0.08008
10	0.09558	0.09343	0.09133	0.08723	0.08329	0.07950	0.07587	0.06903
11	0.08645	0.08429	0.08218	0.07808	0.07415	0.07039	0.06679	0.06008
12	0.07885	0.07668	0.07456	0.07046	0.06655	0.06283	0.05928	0.05270
13	0.07241	0.07024	0.06812	0.06403	0.06014	0.05646	0.05296	0.04652
14	0.06690	0.06472	0.06260	0.05853	0.05467	0.05102	0.04758	0.04130
15	0.06212	0.05994	0.05783	0.05377	0.04994	0.04634	0.04296	0.03683
16	0.05794	0.05577	0.05365	0.04961	0.04582	0.04227	0.03895	0.03298
17	0.05426	0.05208	0.04997	0.04595	0.04220	0.03870	0.03544	0.02963
18	0.05098	0.04881	0.04670	0.04271	0.03899	0.03555	0.03236	0.02670
19	0.04805	0.04588	0.04378	0.03981	0.03614	0.03275	0.02962	0.02413
20	0.04542	0.04325	0.04116	0.03722	0.03358	0.03024	0.02718	0.02185
21	0.04303	0.04087	0.03878	0.03487	0.03128	0.02800	0.02500	0.01983
22	0.04086	0.03870	0.03663	0.03275	0.02920	0.02597	0.02305	0.01803
23	0.03889	0.03673	0.03467	0.03081	0.02731	0.02414	0.02128	0.01642
24	0.03707	0.03492	0.03287	0.02905	0.02559	0.02247	0.01968	0.01498
25	0.03541	0.03326	0.03122	0.02743	0.02401	0.02095	0.01823	0.01368
26	0.03387	0.03173	0.02970	0.02594	0.02257	0.01956	0.01690	0.01251
27	0.03245	0.03032	0.02829	0.02456	0.02124	0.01829	0.01570	0.01145
28	0.03112	0.02900	0.02699	0.02329	0.02001	0.01712	0.01459	0.01049
29	0.02990	0.02778	0.02578	0.02211	0.01888	0.01605	0.01358	0.00962
30	0.02875	0.02664	0.02465	0.02102	0.01783	0.01505	0.01265	0.00883
31	0.02768	0.02557	0.02360	0.02000	0.01686	0.01413	0.01179	0.00811
32	0.02667	0.02458	0.02261	0.01905	0.01595	0.01328	0.01100	0.00745
33	0.02573	0.02364	0.02169	0.01816	0.01510	0.01249	0.01027	0.00685
34	0.02484	0.02276	0.02082	0.01732	0.01431	0.01176	0.00960	0.00630
35	0.02400	0.02193	0.02000	0.01654	0.01358	0.01107	0.00897	0.00580
36	0.02321	0.02115	0.01923	0.01580	0.01289	0.01043	0.00839	0.00534
37	0.02247	0.02041	0.01851	0.01511	0.01224	0.00984	0.00786	0.00492
38	0.02176	0.01972	0.01782	0.01446	0.01163	0.00928	0.00736	0.00454
39	0.02109	0.01905	0.01717	0.01384	0.01106	0.00876	0.00689	0.00419
40	0.02046	0.01843	0.01656	0.01326	0.01052	0.00828	0.00646	0.00386
41	0.01985	0.01783	0.01597	0.01271	0.01002	0.00782	0.00606	0.00356
42	0.01928	0.01726	0.01542	0.01219	0.00954	0.00739	0.00568	0.00329
43	0.01873	0.01672	0.01489	0.01170	0.00909	0.00699	0.00533	0.00303
44	0.01820	0.01621	0.01439	0.01123	0.00866	0.00662	0.00501	0.00280
45	0.01771	0.01572	0.01391	0.01079	0.00826	0.00626	0.00470	0.00259
46	0.01723	0.01525	0.01345	0.01036	0.00788	0.00593	0.00441	0.00239
47	0.01677	0.01480	0.01302	0.00996	0.00752	0.00561	0.00415	0.00221
48	0.01633	0.01437	0.01260	0.00958	0.00718	0.00532	0.00390	0.00204
49	0.01591	0.01396	0.01220	0.00921	0.00686	0.00504	0.00366	0.00189
50	0.01551	0.01357	0.01182	0.00887	0.00655	0.00478	0.00344	0.00174

TABLE 11 COMBINATIONS

z	$\binom{n}{0}$	$\binom{n}{1}$	$\binom{n}{2}$	$\binom{n}{3}$	$\binom{n}{4}$	$\binom{n}{5}$	$\binom{n}{6}$	$\binom{n}{7}$	$\binom{n}{8}$	$\binom{n}{9}$	$\binom{n}{10}$
0	1										
1	1	1									
2	1	2	1								
3	1	3	3	1							
4	1	4	6	4	1						
5	1	5	10	10	5	1					
6	1	6	15	20	15	6	1				
7	1	7	21	35	35	21	7	1			
8	1	8	28	56	70	56	28	8	1		
9	1	9	36	84	126	126	84	36	9	1	
10	1	10	45	120	210	252	210	120	45	10	1
11	1	11	55	165	330	462	462	330	165	55	11
12	1	12	66	220	495	792	924	792	495	220	66
13	1	13	78	286	715	1287	1716	1716	1287	715	286
14	1	14	91	364	1001	2002	3003	3432	3003	2002	1001
15	1	15	105	455	1365	3003	5005	6435	6435	5005	3003
16	1	16	120	560	1820	4368	8008	11440	12870	11440	8008
17	1	17	136	680	2380	6188	12376	19448	24310	24310	19448
18	1	18	153	816	3060	8658	18564	31824	43758	48620	43758
19	1	19	171	969	3876	11628	27132	50388	75582	92378	92378
20	1	20	190	1140	4845	15504	38760	77520	125970	167960	184756

For $r > 10$, it may be necessary to use the identity

$$\binom{n}{r} = \binom{n}{n-r}.$$

TABLE 12 AREAS UNDER THE STANDARD NORMAL CURVE

The column under *A* gives the proportion of the area under the entire curve which is between $z = 0$ and a positive value of z.

z	A	z	A	z	A	z	A
.00	.0000	.49	.1879	.98	.3365	1.47	.4292
.01	.0040	.50	.1915	.99	.3389	1.48	.4306
.02	.0080	.51	.1950	1.00	.3413	1.49	.4319
.03	.0120	.52	.1985	1.01	.3438	1.50	.4332
.04	.0160	.53	.2019	1.02	.3461	1.51	.4345
.05	.0199	.54	.2054	1.03	.3485	1.52	.4357
.06	.0239	.55	.2088	1.04	.3508	1.53	.4370
.07	.0279	.56	.2123	1.05	.3531	1.54	.4382
.08	.0319	.57	.2157	1.06	.3554	1.55	.4394
.09	.0359	.58	.2190	1.07	.3577	1.56	.4406
.10	.0398	.59	.2224	1.08	.3599	1.57	.4418
.11	.0438	.60	.2258	1.09	.3621	1.58	.4430
.12	.0478	.61	.2291	1.10	.3643	1.59	.4441
.13	.0517	.62	.2324	1.11	.3665	1.60	.4452
.14	.0557	.63	.2357	1.12	.3686	1.61	.4463
.15	.0596	.64	.2389	1.13	.3708	1.62	.4474
.16	.0636	.65	.2422	1.14	.3729	1.63	.4485
.17	.0675	.66	.2454	1.15	.3749	1.64	.4495
.18	.0714	.67	.2486	1.16	.3770	1.65	.4505
.19	.0754	.68	.2518	1.17	.3790	1.66	.4515
.20	.0793	.69	.2549	1.18	.3810	1.67	.4525
.21	.0832	.70	.2580	1.19	.3830	1.68	.4535
.22	.0871	.71	.2612	1.20	.3849	1.69	.4545
.23	.0910	.72	.2642	1.21	.3869	1.70	.4554
.24	.0948	.73	.2673	1.22	.3888	1.71	.4564
.25	.0987	.74	.2704	1.23	.3907	1.72	.4573
.26	.1026	.75	.2734	1.24	.3925	1.73	.4582
.27	.1064	.76	.2764	1.25	.3944	1.74	.4591
.28	.1103	.77	.2794	1.26	.3962	1.75	.4599
.29	.1141	.78	.2823	1.27	.3980	1.76	.4608
.30	.1179	.79	.2852	1.28	.3997	1.77	.4616
.31	.1217	.80	.2881	1.29	.4015	1.78	.4625
.32	.1255	.81	.2910	1.30	.4032	1.79	.4633
.33	.1293	.82	.2939	1.31	.4049	1.80	.4641
.34	.1331	.83	.2967	1.32	.4066	1.81	.4649
.35	.1368	.84	.2996	1.33	.4082	1.82	.4656
.36	.1406	.85	.3023	1.34	.4099	1.83	.4664
.37	.1443	.86	.3051	1.35	.4115	1.84	.4671
.38	.1480	.87	.3079	1.36	.4131	1.85	.4678
.39	.1517	.88	.3106	1.37	.4147	1.86	.4686
.40	.1554	.89	.3133	1.38	.4162	1.87	.4693
.41	.1591	.90	.3159	1.39	.4177	1.88	.4700
.42	.1628	.91	.3186	1.40	.4192	1.89	.4706
.43	.1664	.92	.3212	1.41	.4207	1.90	.4713
.44	.1700	.93	.3238	1.42	.4222	1.91	.4719
.45	.1736	.94	.3264	1.43	.4236	1.92	.4726
.46	.1772	.95	.3289	1.44	.4251	1.93	.4732
.47	.1808	.96	.3315	1.45	.4265	1.94	.4738
.48	.1844	.97	.3340	1.46	.4279	1.95	.4744

TABLE 12 (CONTINUED)

z	A	z	A	z	A	z	A
1.96	.4750	2.45	.4929	2.94	.4984	3.43	.4997
1.97	.4756	2.46	.4931	2.95	.4984	3.44	.4997
1.98	.4762	2.47	.4932	2.96	.4985	3.45	.4997
1.99	.4767	2.48	.4934	2.97	.4985	3.46	.4997
2.00	.4773	2.49	.4936	2.98	.4986	3.47	.4997
2.01	.4778	2.50	.4938	2.99	.4986	3.48	.4998
2.02	.4783	2.51	.4940	3.00	.4987	3.49	.4998
2.03	.4788	2.52	.4941	3.01	.4987	3.50	.4998
2.04	.4793	2.53	.4943	3.02	.4987	3.51	.4998
2.05	.4798	2.54	.4945	3.03	.4988	3.52	.4998
2.06	.4803	2.55	.4946	3.04	.4988	3.53	.4998
2.07	.4808	2.56	.4948	3.05	.4989	3.54	.4998
2.08	.4812	2.57	.4949	3.06	.4989	3.55	.4998
2.09	.4817	2.58	.4951	3.07	.4989	3.56	.4998
2.10	.4821	2.59	.4952	3.08	.4990	3.57	.4998
2.11	.4826	2.60	.4953	3.09	.4990	3.58	.4998
2.12	.4830	2.61	.4955	3.10	.4990	3.59	.4998
2.13	.4834	2.62	.4956	3.11	.4991	3.60	.4998
2.14	.4838	2.63	.4957	3.12	.4991	3.61	.4999
2.15	.4842	2.64	.4959	3.13	.4991	3.62	.4999
2.16	.4846	2.65	.4960	3.14	.4992	3.63	.4999
2.17	.4850	2.66	.4961	3.15	.4992	3.64	.4999
2.18	.4854	2.67	.4962	3.16	.4992	3.65	.4999
2.19	.4857	2.68	.4963	3.17	.4992	3.66	.4999
2.20	.4861	2.69	.4964	3.18	.4993	3.67	.4999
2.21	.4865	2.70	.4965	3.19	.4993	3.68	.4999
2.22	.4868	2.71	.4966	3.20	.4993	3.69	.4999
2.23	.4871	2.72	.4967	3.21	.4993	3.70	.4999
2.24	.4875	2.73	.4968	3.22	.4994	3.71	.4999
2.25	.4878	2.74	.4969	3.23	.4994	3.72	.4999
2.26	.4881	2.75	.4970	3.24	.4994	3.73	.4999
2.27	.4884	2.76	.4971	3.25	.4994	3.74	.4999
2.28	.4887	2.77	.4972	3.26	.4994	3.75	.4999
2.29	.4890	2.78	.4973	3.27	.4995	3.76	.4999
2.30	.4893	2.79	.4974	3.28	.4995	3.77	.4999
2.31	.4896	2.80	.4974	3.29	.4995	3.78	.4999
2.32	.4898	2.81	.4975	3.30	.4995	3.79	.4999
2.33	.4901	2.82	.4976	3.31	.4995	3.80	.4999
2.34	.4904	2.83	.4977	3.32	.4996	3.81	.4999
2.35	.4906	2.84	.4977	3.33	.4996	3.82	.4999
2.36	.4909	2.85	.4978	3.34	.4996	3.83	.4999
2.37	.4911	2.86	.4979	3.35	.4996	3.84	.4999
2.38	.4913	2.87	.4980	3.36	.4996	3.85	.4999
2.39	.4916	2.88	.4980	3.37	.4996	3.86	.4999
2.40	.4918	2.89	.4981	3.38	.4996	3.87	.5000
2.41	.4920	2.90	.4981	3.39	.4997	3.88	.5000
2.42	.4922	2.91	.4982	3.40	.4997	3.89	.5000
2.43	.4925	2.92	.4983	3.41	.4997		
2.44	.4927	2.93	.4983	3.42	.4997		

TABLE 13 INTEGRALS

C is an arbitrary constant.

1. $\displaystyle\int x^n\,dx = \frac{1}{n+1}x^{n+1} + C \qquad (\text{if } n \neq -1)$

2. $\displaystyle\int e^{kx}\,dx = \frac{1}{k}e^{kx} + C$

3. $\displaystyle\int \frac{a}{x}\,dx = a\ln|x| + C \qquad (a \neq 0)$

4. $\displaystyle\int \ln|ax|\,dx = x(\ln|ax| - 1) + C$

5. $\displaystyle\int \frac{1}{\sqrt{x^2+a^2}}\,dx = \ln\left|\frac{x+\sqrt{x^2+a^2}}{a}\right| + C$

6. $\displaystyle\int \frac{1}{\sqrt{x^2-a^2}}\,dx = \ln\left|\frac{x+\sqrt{x^2-a^2}}{a}\right| + C$

7. $\displaystyle\int \frac{1}{a^2-x^2}\,dx = \frac{1}{2a}\cdot\ln\left|\frac{a+x}{a-x}\right| + C \qquad (x^2 < a^2)$

8. $\displaystyle\int \frac{1}{x^2-a^2}\,dx = \frac{1}{2a}\cdot\ln\left|\frac{x-a}{x+a}\right| + C \qquad (x^2 > a^2)$

9. $\displaystyle\int \frac{1}{x\sqrt{a^2-x^2}}\,dx = -\frac{1}{a}\cdot\ln\left|\frac{a+\sqrt{a^2-x^2}}{x}\right| + C \qquad (0 < x < a)$

10. $\displaystyle\int \frac{1}{x\sqrt{a^2+x^2}}\,dx = -\frac{1}{a}\cdot\ln\left|\frac{a+\sqrt{a^2+x^2}}{x}\right| + C$

11. $\displaystyle\int \frac{x}{ax+b}\,dx = \frac{x}{a} - \frac{b}{a^2}\cdot\ln|ax+b| + C \qquad (a \neq 0)$

12. $\displaystyle\int \frac{x}{(ax+b)^2}\,dx = \frac{b}{a^2(ax+b)} + \frac{1}{a^2}\cdot\ln|ax+b| + C \qquad (a \neq 0)$

13. $\displaystyle\int \frac{1}{x(ax+b)}\,dx = \frac{1}{b}\cdot\ln\left|\frac{x}{ax+b}\right| + C \qquad (b \neq 0)$

14. $\displaystyle\int \frac{1}{x(ax+b)^2}\,dx = \frac{1}{b(ax+b)} + \frac{1}{b^2}\cdot\ln\left|\frac{x}{ax+b}\right| + C \qquad (b \neq 0)$

15. $\displaystyle\int \sqrt{x^2+a^2}\,dx = \frac{x}{2}\sqrt{x^2+a^2} + \frac{a^2}{2}\cdot\ln\left|x+\sqrt{x^2+a^2}\right| + C$

16. $\displaystyle\int x^n\cdot\ln|x|\,dx = x^{n+1}\left[\frac{\ln|x|}{n+1} - \frac{1}{(n+1)^2}\right] + C \qquad (n \neq -1)$

17. $\displaystyle\int x^n e^{ax}\,dx = \frac{x^n e^{ax}}{a} - \frac{n}{a}\cdot\int x^{n-1}e^{ax}\,dx + C \qquad (a \neq 0)$

Answers To Selected Exercises

Chapter 1

Section 1.1 (page 6)

1. Counting number, whole number, integer, rational number, real number **3.** Integer, rational number, real number **5.** Rational number, real number **7.** Irrational number, real number **9.** Irrational number, real number **11.** True **13.** True **15.** False **17.** True

19. 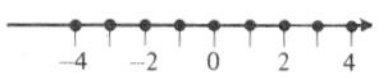

21. 

23. 

25. 

27.

29.

31. 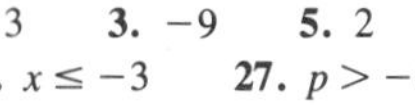

33. 

35. 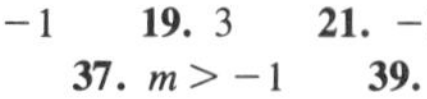

37. 8 **39.** −4 **41.** 2 **43.** −4 **45.** 17 **47.** 4 **49.** −19 **51.** = **53.** < **55.** = **57.** = **59.** = **61.** = **63.** = **65.** Yes **67.** Let x represent the percent of the money paid to the distributor. Then $x \geq 35$. **69.** $x \geq 12{,}600$ **71.** Let x be the manufacturing cost of a product, in percent of the retail price. Then $x \geq 15$. **73.** 100; 0 **75.** 67; 11

Section 1.2 (page 13)

1. 3 **3.** −9 **5.** 2 **7.** 3 **9.** 4 **11.** 2/3 **13.** 7 **15.** −24 **17.** −1 **19.** 3 **21.** −3 **23.** 5 **25.** $x \leq -3$ **27.** $p > -6$ **29.** $a > 0$ **31.** $x \leq 4$ **33.** $p \leq -1$ **35.** $k < 1$ **37.** $m > -1$ **39.** $y \leq 1$ **41.** $p > 1/5$ **43.** 68°F **45.** 15°C **47.** 37.8°C **49.** 104°F **51.** 13% **53.** \$432 **55.** \$1500 **57.** \$205.41 **59.** \$66.50 **61.** $C = 1.2x$ **63.** Profit $= 5x - 1.2x = 3.8x$ **65.** Avis is better if $14(7) > 54 + .07x$, where x is the number of miles driven in a week. They must drive 629 miles for Avis to be a better deal.

Section 1.3 (page 20)

1. $14m + 6$ **3.** $-9k + 5$ **5.** $-x^2 + x + 9$ **7.** $-6y^2 + 3y + 10$ **9.** $-10x^2 + 4x - 2$ **11.** $6p^2 - 15p$ **13.** $-18m^3 - 27m^2 + 9m$ **15.** $12k^2 - 20k + 3$ **17.** $6y^2 + 7y - 5$ **19.** $36y^2 - 4$ **21.** $5(5k + 6)$ **23.** $4(z + 1)$ **25.** $2(4x + 3y + 2z)$ **27.** $2(3r^2 + 2r + 4)$ **29.** $m(m^2 - 9m + 6)$ **31.** $8a(a^2 - 2a + 3)$ **33.** $5p^2(5p^2 - 4p + 20)$ **35.** $(m + 7)(m + 2)$ **37.** $(x + 5)(x - 1)$ **39.** $(z + 5)(z + 4)$ **41.** $(b - 7)(b - 1)$ **43.** Cannot be factored **45.** $(s + 7)(s - 5)$ **47.** $(y - 7)(y + 3)$ **49.** $6(a - 10)(a + 2)$ **51.** $3m(m + 3)(m + 1)$ **53.** $(2x + 1)(x - 3)$ **55.** $(3a + 7)(a + 1)$ **57.** $(2a - 5)(a - 6)$ **59.** $(5y + 2)(3y - 1)$ **61.** Cannot be factored **63.** $(5a + 3)(a - 2)$ **65.** $(7m + 2)(3m + 1)$ **67.** $2a^2(4a - 1)(3a + 2)$ **69.** $4z^3(8z + 3)(z - 1)$ **71.** $(x + 8)(x - 8)$ **73.** $(3m + 5)(3m - 5)$ **75.** $(11a + 10)(11a - 10)$ **77.** Cannot be factored **79.** $(z + 7)^2$ **81.** $(m - 3)^2$ **83.** $(3p - 4)^2$ **85.** $(a - 6)(a^2 + 6a + 36)$ **87.** $(2r - 3)(4r^2 + 6r + 9)$

Section 1.4 (page 26)

1. 5, −4 **3.** −2, −3 **5.** 6, −1 **7.** −8, 3 **9.** 3, −1 **11.** 4 **13.** 5/2, −2 **15.** 4/3, −1/2
17. 5, 2 **19.** 4/3, −4/3 **21.** 0, 1 **23.** $(5 + \sqrt{13})/6 \approx 1.434$; $(5 - \sqrt{13})/6 \approx .232$
25. $(1 + \sqrt{33})/4 \approx 1.686$; $(1 - \sqrt{33})/4 \approx -1.186$ **27.** $(10 + \sqrt{20})/2 \approx 7.236$; $(10 - \sqrt{20})/2 \approx 2.764$
29. $(-12 + \sqrt{104})/4 \approx -.450$; $(-12 - \sqrt{104})/4 \approx -5.550$ **31.** 5/2, 1 **33.** 4/3, 1/2 **35.** −5, 2
37. No real number solutions **39.** 0, −1

41. $-2 < m < 4$

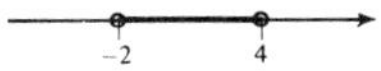

43. $t \leq -6$ or $t \geq 1$

45. $1 < y < 2$

47. $k < -4$ or $k > 1/2$

49. $1 \leq q \leq 6$

51. $m < -1/2$ or $m > 1/3$

53. $-3 \leq y \leq 1/2$

55. $-5 \leq x \leq 5$

57. $54

59. 5/4 months and 6 months

Section 1.5 (page 32)

1. $m/4$ **3.** $z/2$ **5.** $5p/2$ **7.** 8/9 **9.** $3/(t-3)$ **11.** $2(x+2)/x$ **13.** $(m-2)/(m+3)$
15. $(x+4)/(x+1)$ **17.** $(2m+3)/(4m+3)$ **19.** $3k/5$ **21.** $25p^2/9$ **23.** $6/(5p)$ **25.** 2 **27.** 2/9
29. 3/10 **31.** $2(a+4)/(a-3)$ **33.** $(k+2)/(k+3)$ **35.** $(m+6)/(m+3)$ **37.** $(m-3)/(2m-3)$ **39.** $14/r$
41. $19/(6k)$ **43.** $5/(12y)$ **45.** 1 **47.** $(6+p)/(2p)$ **49.** $(8-y)/(4y)$ **51.** $137/(30m)$
53. $(3m-2)/[m(m-1)]$ **55.** $(r-12)/[r(r-2)]$ **57.** $14/[3(a-1)]$ **59.** $23/[20(k-2)]$ **61.** $(x+1)/(x-1)$
63. $-1/(x+1)$

Section 1.6 (page 38)

1. 343 **3.** 1/8 **5.** 1/8 **7.** 1/5 **9.** 1/512 **11.** 27/64 **13.** 25/64 **15.** 8 **17.** 49/4 **19.** 9
21. 3 **23.** 4 **25.** 100 **27.** 4 **29.** −25 **31.** 2/3 **33.** 4/3 **35.** 1/32 **37.** 4/3 **39.** 3/4
41. 3^6 **43.** $1/7^4$ **45.** $1/3^6$ **47.** $1/2^3$ **49.** $1/6^2$ **51.** 4^3 **53.** $1/7^7$ **55.** 8^5 **57.** $1/10^8$ **59.** 5
61. 2^2 **63.** $27^{1/3}$ or 3 **65.** 4^2 **67.** $(\sqrt{7})^3$ **69.** $(\sqrt[5]{28})^3$ **71.** $1/(\sqrt[3]{60})^2$ **73.** $1/(\sqrt[4]{6})^5$ **75.** $12/(\sqrt{x})^3$
77. $8/(\sqrt[5]{m})^4$ **79.** $1/(\sqrt[3]{(3r)})^2$ **81.** $64 **83.** $64,000,000 **85.** About 86 miles **87.** About 211 miles
89. 29 **91.** 177

Chapter 1 Test (page 41)

1. Integer, rational number, real number **2.** Irrational number, real number

3. $x \geq -3$

4. $-4 < x \leq 6$

5. −6 **6.** 4 **7.** 2 **8.** 2/3 **9.** 7 **10.** $k \geq -4$ **11.** $m < 14$ **12.** $84 **13.** $8(y+2)$
14. $(p-7)(p-2)$ **15.** $(k+9)(k-5)$ **16.** $(2a-3)(a+5)$ **17.** $(3m+7)(m-5)$ **18.** $(4p+3)^2$ **19.** −6, 3
20. −2/3, 5/2 **21.** $(4+\sqrt{8})/2 \approx 3.414$; $(4-\sqrt{8})/2 \approx .586$ **22.** $(6+\sqrt{108})/18 \approx .911$; $(6-\sqrt{108})/18 \approx -.244$
23. $-3 \leq x \leq 1$ **24.** $y < -1/3$ or $y > 2$ **25.** $8p^2/5$ **26.** $2r/3$ **27.** 4 **28.** $17/(2r)$ **29.** $119/(72y)$
30. $(5r+3)/[r(r-1)]$ **31.** 1/16 **32.** 36 **33.** 1/8 **34.** 512 **35.** 1/243 **36.** 7/12

Case 1 (page 43)

1. $.1A + 200$ **2.** $.0001A - .3$ **3.** 8000 **4.** 8000 **5.** 800 **6.** 800(10,000) = $8,000,000

Chapter 2

Section 2.1 (page 47)

1. False **3.** True **5.** True **7.** True **9.** False **11.** False **13.** False **15.** True **17.** False **19.** True **21.** True **23.** False **25.** True **27.** True **29.** False **31.** False **33.** False **35.** 8 **37.** 32 **39.** 1 **41.** 32 **43.** $\{e\}$, $\{v\}$

Section 2.2 (page 53)

1. True **3.** False **5.** True **7.** False **9.** True **11.** True **13.** $\{3, 5\}$ **15.** U, or $\{2, 3, 4, 5, 7, 9\}$ **17.** U, or $\{2, 3, 4, 5, 7, 9\}$ **19.** $\{7, 9\}$ **21.** $\varnothing$ **23.** $\varnothing$ **25.** U, or $\{2, 3, 4, 5, 7, 9\}$ **27.** All students in this school not taking this course **29.** All students in this school taking accounting and zoology **31.** All students in this school taking this course or zoology

33.

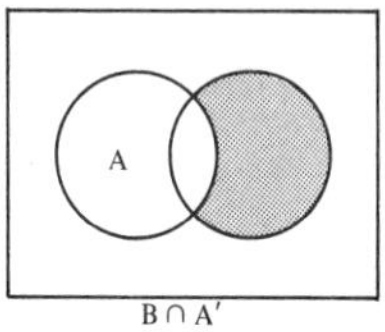

$B \cap A'$

35.

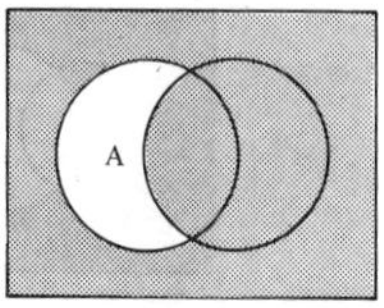

$A' \cup B$

37.

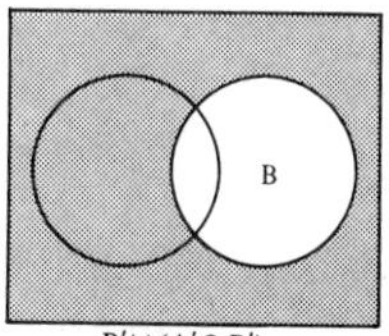

$B' \cup (A' \cap B')$

39. $\varnothing$

41.

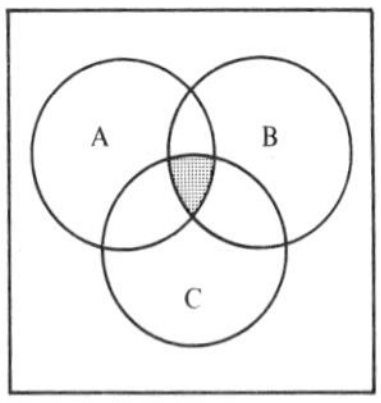

$(A \cap B) \cap C$

43.

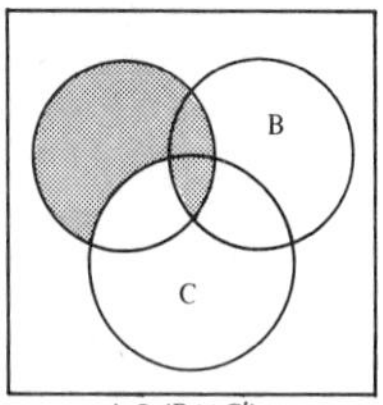

$A \cap (B \cup C')$

45.

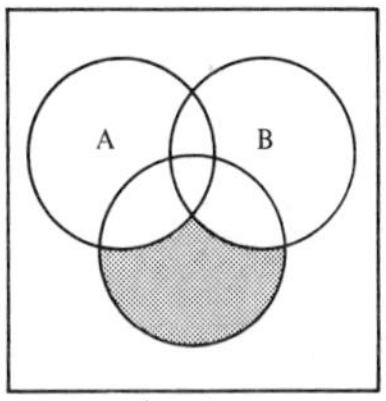

$(A' \cap B') \cap C$

47.

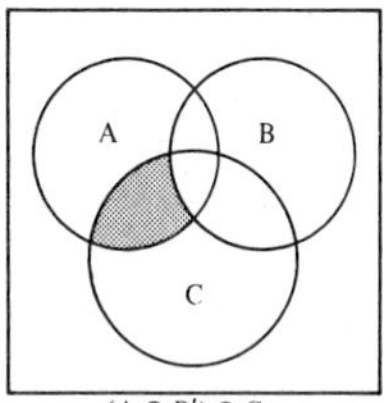

$(A \cap B') \cap C$

49. $\{s, d, c, g, i, m, h\}$ **51.** $\{i, m, h\}$ **53.** $\{s, d, c, g, i, m, h\}$

Section 2.3 (page 56)

1. Yes; his data add up to 142 people **3.** (a) 37 (b) 22 (c) 50 (d) 11 (e) 25 (f) 11 **5.** (a) 50 (b) 2 (c) 14 **7.** (a) 54 (b) 17 (c) 10 (d) 7 (e) 15 (f) 3 (g) 12 (h) 1 **9.** (a) 40 (b) 30 (c) 95 (d) 110 (e) 160 (f) 65

Section 2.4 (page 65)

1. $(-2, -3), (-1, -2), (0, -1), (1, 0), (2, 1), (3, 2)$, range: $\{-3, -2, -1, 0, 1, 2\}$

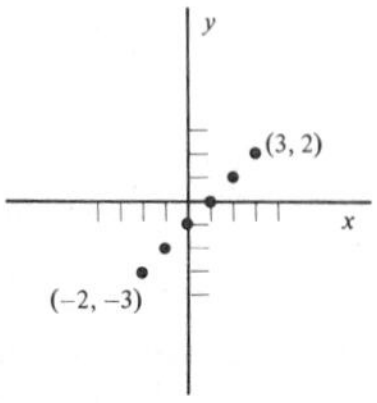

3. $(-2, 17), (-1, 13), (0, 9), (1, 5), (2, 1), (3, -3)$, range: $\{17, 13, 9, 5, 1, -3\}$

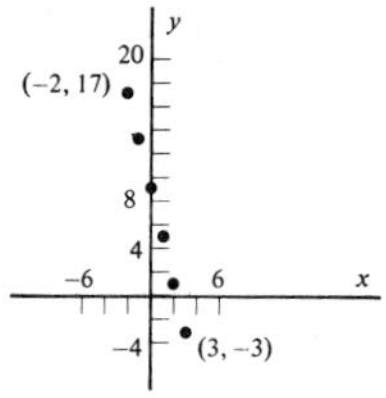

5. $(-2, -3), (-1, -4), (0, -5), (1, -6), (2, -7), (3, -8)$, range: $\{-3, -4, -5, -6, -7, -8\}$

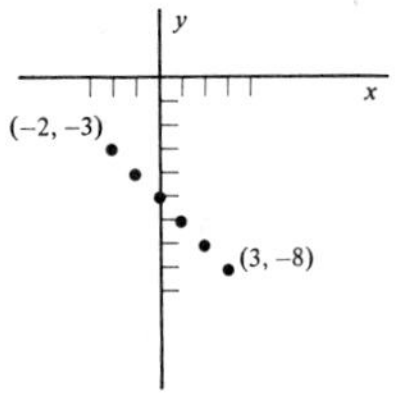

7. $(-2, 13), (-1, 11), (0, 9), (1, 7), (2, 5), (3, 3)$, range: $\{13, 11, 9, 7, 5, 3\}$

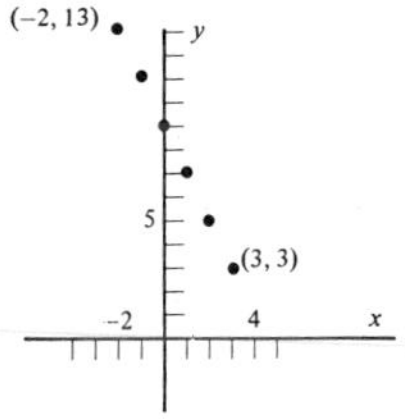

9. $(-2, 3/2), (-1, 2), (0, 5/2), (1, 3), (2, 7/2), (3, 4)$, range: $\{3/2, 2, 5/2, 3, 7/2, 4\}$

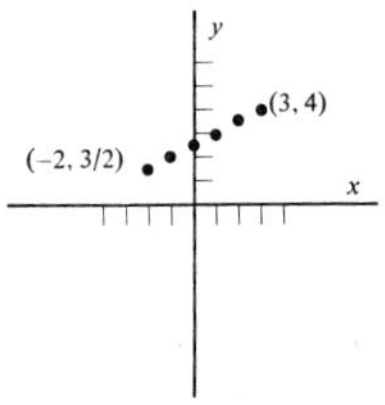

11. $(-2, 2), (-1, 0), (0, 0), (1, 2), (2, 6), (3, 12)$, range: $\{0, 2, 6, 12\}$

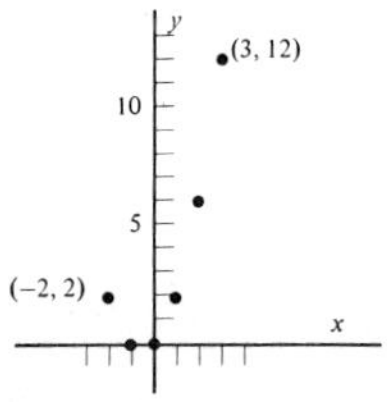

13. $(-2,4),(-1,1),(0,0),(1,1),(2,4),(3,9)$, range: $\{4,1,0,9\}$

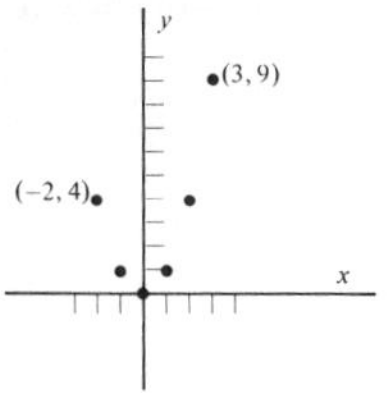

15. $(-2,-13),(-1,-1),(0,3),(1,-1),(2,-13),(3,-33)$, range: $\{-13,-1,3,-33\}$

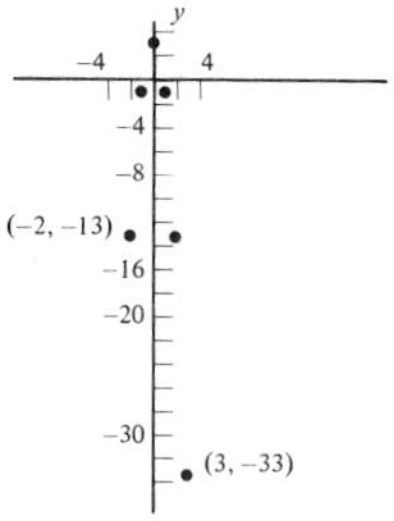

17. $(-2,1),(-1,1/2),(0,1/3),(1,1/4),(2,1/5),(3,1/6)$, range: $\{1,1/2,1/3,1/4,1/5,1/6\}$

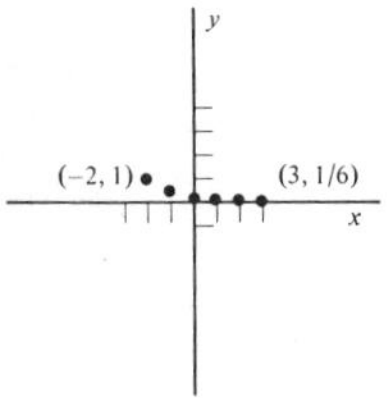

19. $(-2,-3),(-1,-3/2),(0,-3/5),(1,0),(2,3/7),(3,3/4)$, range: $\{-3,-3/2,-3/5,0,3/7,3/4\}$

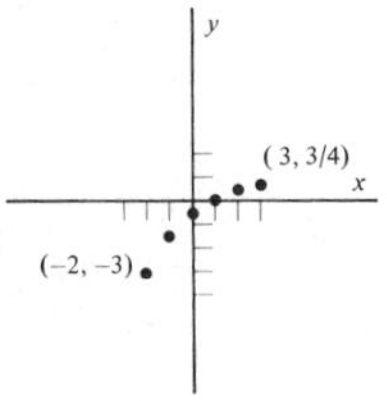

21. $(0,2)$, $(0,3)$, $(0,4)$, $(1,2)$, $(1,3)$, $(1,4)$, $(1,5)$

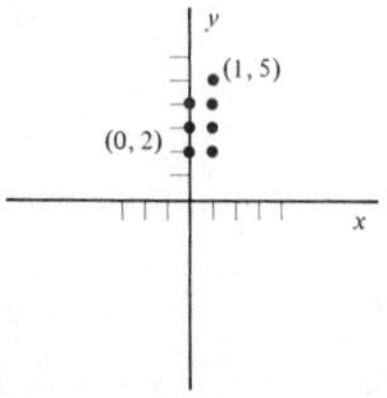

23. (0, 2), (0, 3), (0, 4), (0, 5), (1, 5)

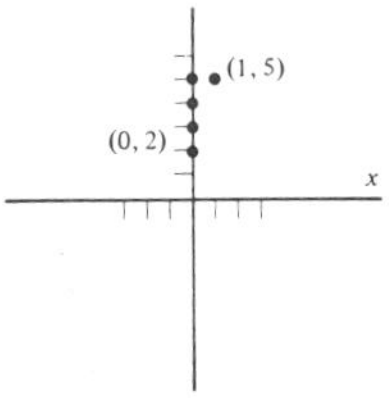

25. Yes **27.** No **29.** Yes **31.** (a) 14 (b) −7 (c) 2 (d) $3a + 2$ **33.** (a) −12 (b) 2 (c) −4 (d) $-2a - 4$ **35.** (a) 48 (b) 6 (c) 0 (d) $2a^2 + 4a$ **37.** (a) 5 (b) −23 (c) 1 (d) $-a^2 + 5a + 1$ **39.** (a) 30 (b) 2 (c) 2 (d) $(a + 1)(a + 2)$ **41.** −3 **43.** −15 **45.** $2a - 3$ **47.** $2m + 3$ **49.** −1 **51.** 1050 thousand dollars **53.** 1200 thousand dollars **55.** $11 **57.** $18 **59.** $32 **61.** $39 **63.** $58 **65.** $58 **67.** $94

Section 2.5 (page 75)

1. 11, 17, 23, 29, 35; arithmetic **3.** −4, −1, 2, 5, 8; arithmetic **5.** −2, −8, −14, −20, −26; arithmetic **7.** 2, 4, 8, 16, 32; geometric **9.** −2, 4, −8, 16, −32; geometric **11.** 6, 12, 24, 48, 96; geometric **13.** 1/3, 3/7, 1/2, 5/9, 3/5; neither **15.** 1/2, 1/3, 1/4, 1/5, 1/6; neither **17.** Arithmetic, $d = 8$ **19.** Arithmetic, $d = 3$ **21.** Geometric, $r = 3$ **23.** Neither **25.** Arithmetic, $d = -3$ **27.** Arithmetic, $d = 3$ **29.** Geometric, $r = -2$ **31.** Neither **33.** 70 **35.** 65 **37.** 78 **39.** −65 **41.** 63 **43.** 678 **45.** 183 **47.** −18 **49.** 48 **51.** −648 **53.** 81 **55.** 64 **57.** 15 **59.** 156/25 **61.** −208 **63.** −15 **65.** 80 **67.** 125 **69.** 134 **71.** 70 **73.** −56 **75.** 125,250 **77.** 90 **79.** 170 **81.** 160 **83.** $15,600 **85.** $1681; $576

Chapter 2 Test (page 78)

1. True **2.** False **3.** True **4.** False **5.** False **6.** True **7.** True **8.** True **9.** {a, e} **10.** {a, c, e, f, g} **11.** {b, d, f, g, h} **12.** {b, d, h}

13.

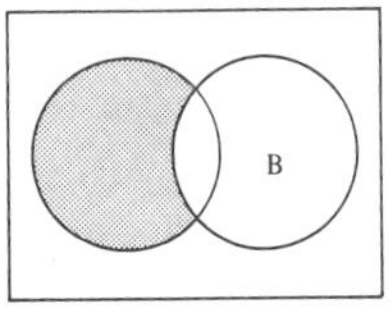

$A \cap B'$

14.

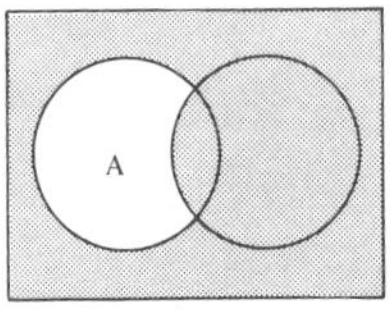

$A' \cup B$

15.

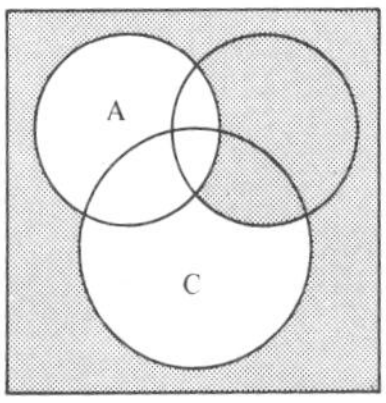

$A' \cap (B \cup C')$

16.

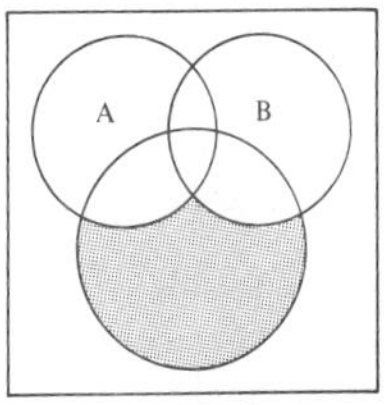

$(A \cup B)' \cap C$

17. 8 **18.** 27 **19.** 28 **20.** 16 **21.** (−2, 8), (−1, 7), (0, 6), (1, 5), (2, 4), (3, 3); domain: {−2, −1, 0, 1, 2, 3}; range: {3, 4, 5, 6, 7, 8} **22.** −13 **23.** −8 **24.** 2 **25.** $-2p^2 + 3p + 1$ **26.** −2, −6, −10, −14, −18; arithmetic **27.** −2, 4, −8, 16, −32; geometric **28.** 1/2, 4/7, 5/8, 2/3, 7/10 **29.** 61 **30.** 1400 **31.** −24 **32.** 5250

Case 2 (page 81)

1. 16 **2.** 30 **3.** 335 **4.** 14 **5.** 2.0% **6.** 13.7% **7.** 52.0% **8.** 97.3%

Case 3 (page 84)

1. $111,504.62 **2.** $89,160.62 **3.** $90,470.62 **4.** $90,658.12 **5.** $90,244.24 **6.** 17,500; yes

Chapter 3

Section 3.1 (page 92)

1.

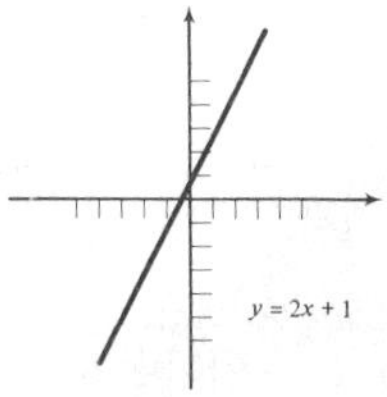

3.

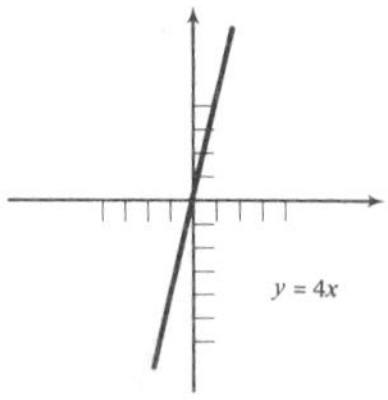

5.

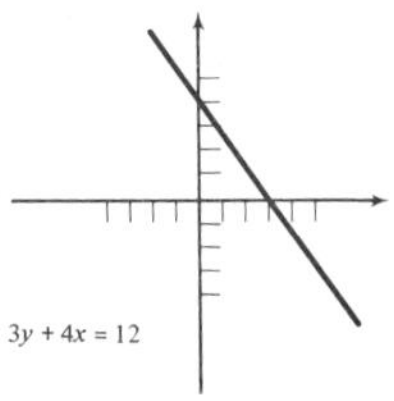

7.

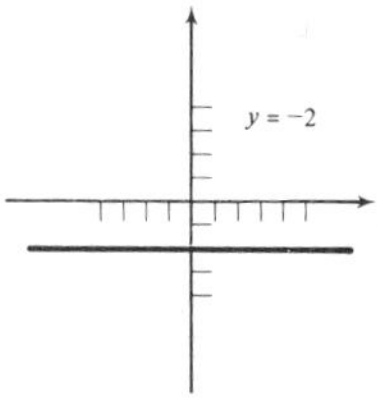

9.

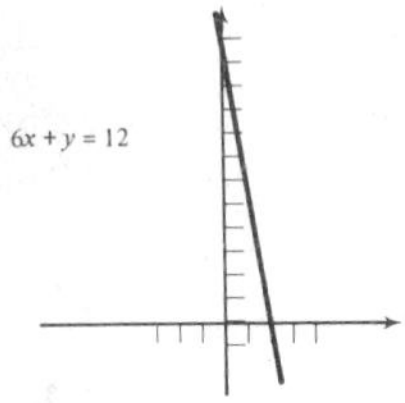

11.

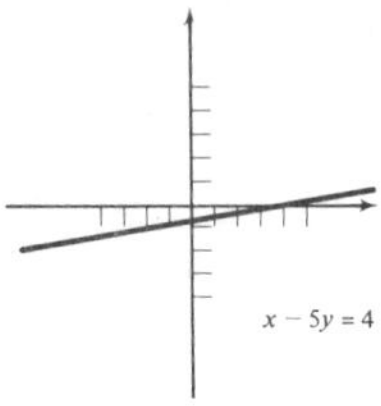

13.

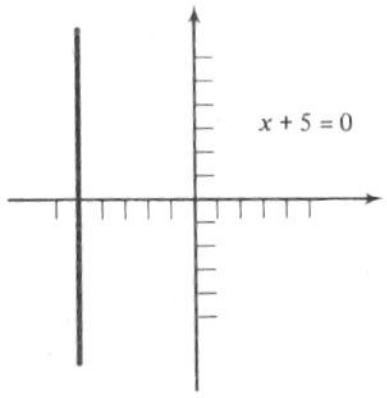

not a function

15.

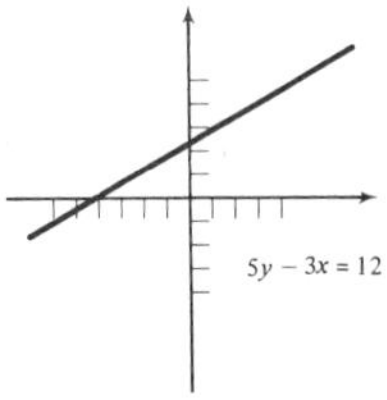

17.

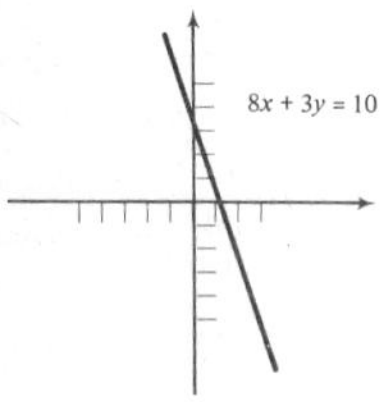

19.

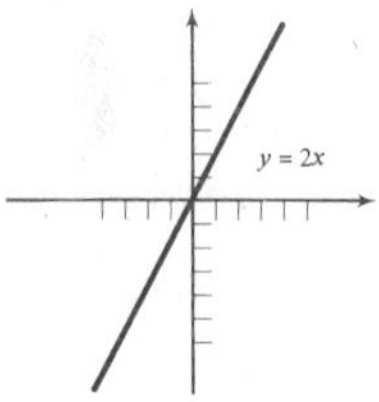

21.

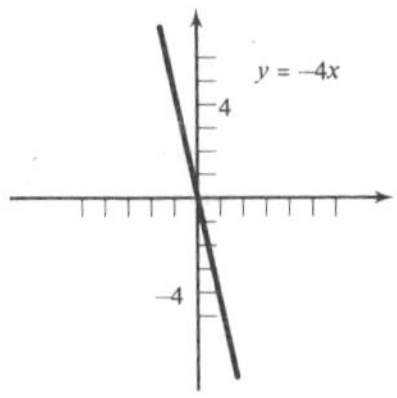

23.

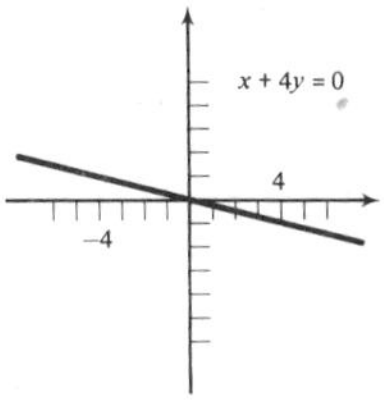

25. 16 **27.** 6 **29.** 4

31.

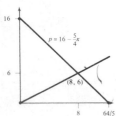

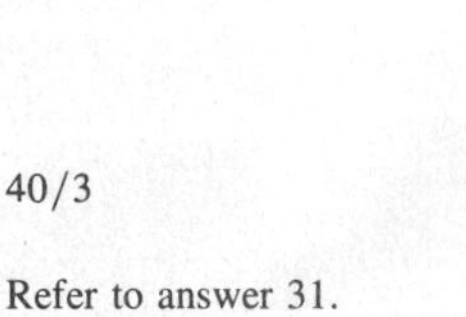

33. 40/3

35. Refer to answer 31.

37. 6 **39.** 36 **41.**

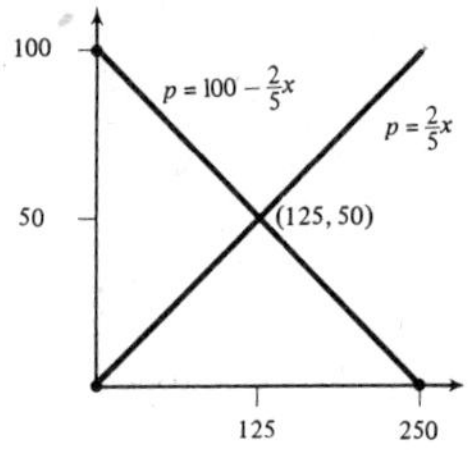

43. 50 **45.** About 1.1 **47.** \$135 **49.** \$275

51.

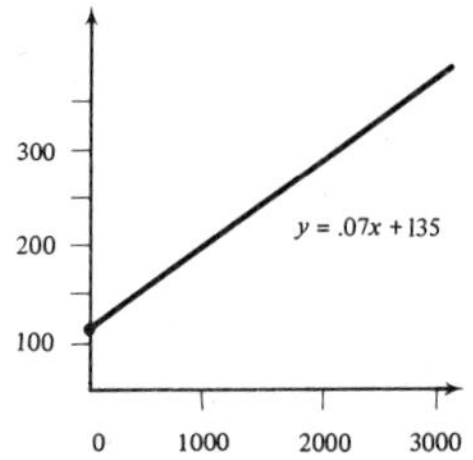

53. \$28.80 **55.**

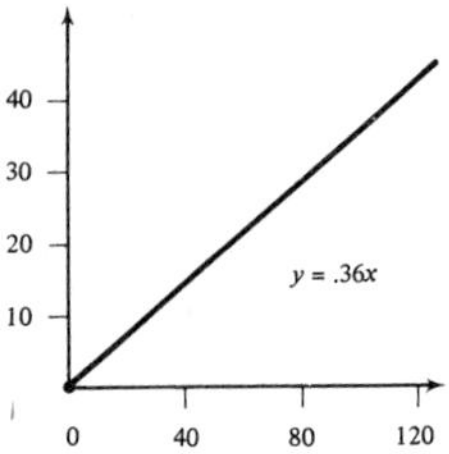

Section 3.2 (page 101)

1. $-1/5$ **3.** $2/3$ **5.** $-3/2$ **7.** No slope **9.** 0 **11.** 3; 4 **13.** -4; 8 **15.** $-3/4$; $5/4$ **17.** -3; 0 **19.** $-2/5$; 0 **21.** 0; 8 **23.** 0; -2 **25.** No slope; no y-intercept

27.

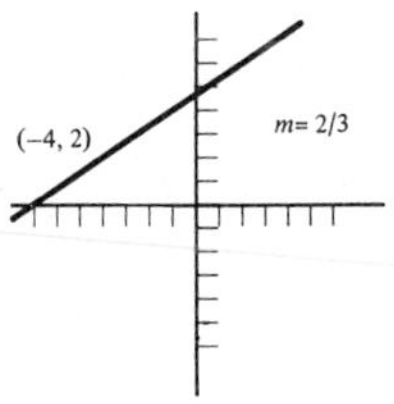

29.

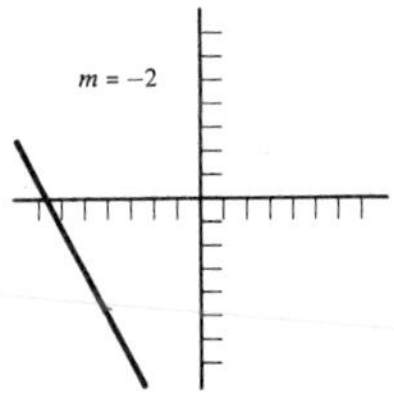

31.

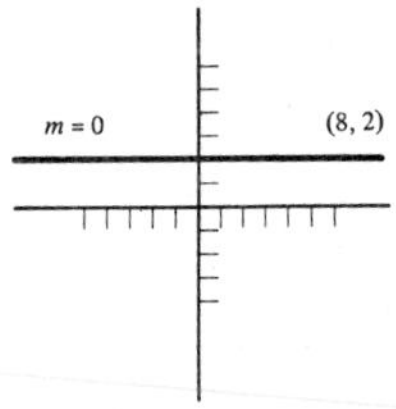

33.

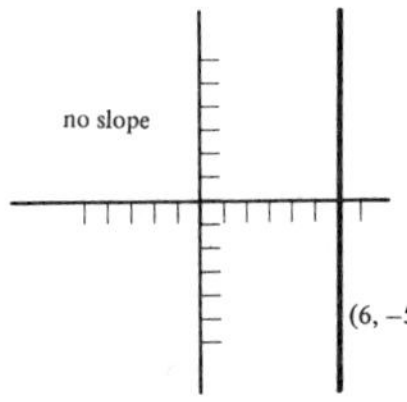

35.

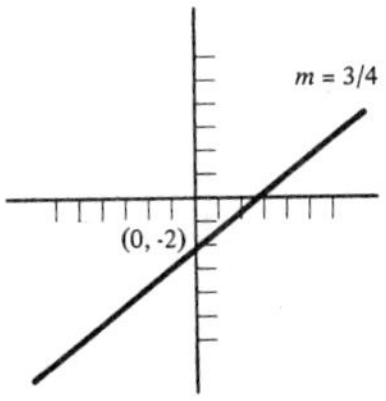

37.

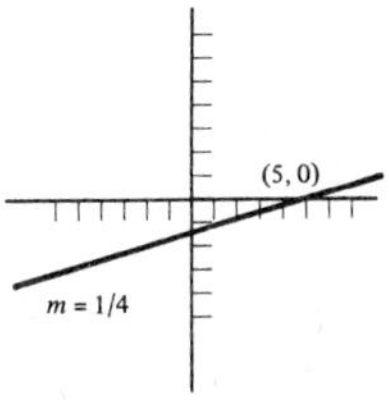

39. $4y = -3x + 16$ **41.** $2y = -x - 4$ **43.** $4y = 6x + 5$ **45.** $y = 2x + 9$ **47.** $y = -3x + 3$ **49.** $4y = x + 5$ **51.** $3y = 4x + 7$ **53.** $3y = -2x$ **55.** $x = -8$ **57.** $y = 3$ **59.** $y = 640x + 1100$; $m = 640$ **61.** $y = -1000x + 40{,}000$; $m = -1000$ **63.** $y = 2.5x - 70$; $m = 2.5$

Section 3.3 (page 111)

1. (a) 2600 (b) 2900 (c) 3200 (d) 2000 (e) 300 **3.** (a) 100 (b) 70 (c) 0 (d) -5; the number is decreasing **5.** (a) $y = 6500x + 85{,}000$ (b) \$137,000 (c) \$169,500 **7.** (a) 480 (b) 360 (c) 120 (d) June 30 (e) -20 **9.** If $C(x)$ is the cost of renting a saw for x hours, then $C(x) = 12 + x$. **11.** If $P(x)$ is the cost (in cents) of parking for x half hours, then $P(x) = 35 + 30x$. **13.** $C(x) = 30x + 100$ **15.** $C(x) = 25x + 1000$ **17.** $C(x) = 50x + 500$ **19.** $C(x) = 90x + 2500$ **21.** \$24 **23.** \$48.048 **25.** \$2000; \$32,000 **27.** \$2000; \$52,000 **29.** \$150,000; \$600,000 **31.** (a) \$20,000 per year (b) \$80,000 **33.** (a) \$4000; \$3000; \$2000; \$1000 (b) \$2500 **35.** \$5090.91; \$3563.64 **37.** \$4171.43; \$2085.71 **39.** 500 units; \$30,000 **41.** Break-even point is 45 units; don't produce

43. Break-even point is −50 units; impossible to make a profit here

45.

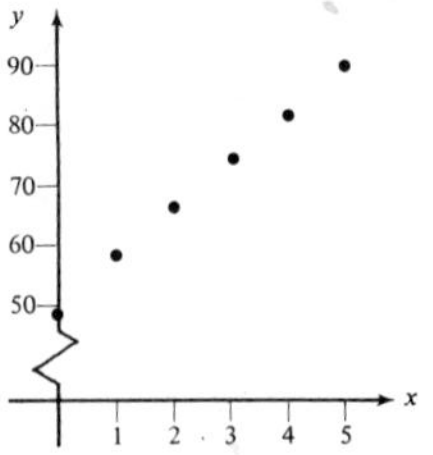

x	y
0	48
1	59
2	66
3	75
4	80
5	90

47. $y = 8x + 50$ **49.** 106 **51.** 32.5 **53.** 145 **55.** 13.95 **57.** 52.2 **59.** About 69

Section 3.4 (page 122)

1.

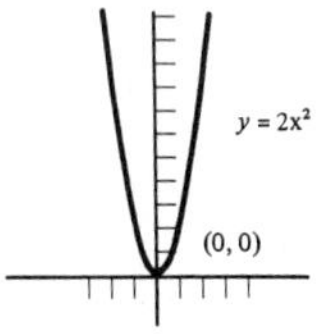

3.

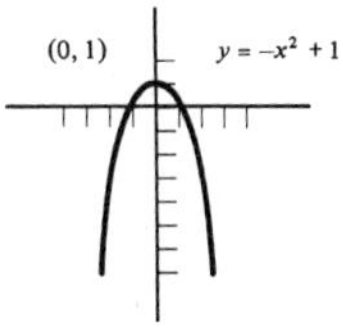

5.

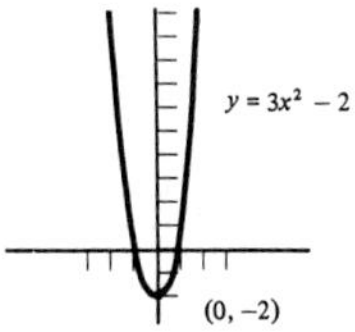

7.

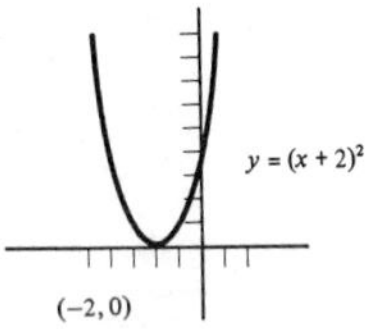

9.

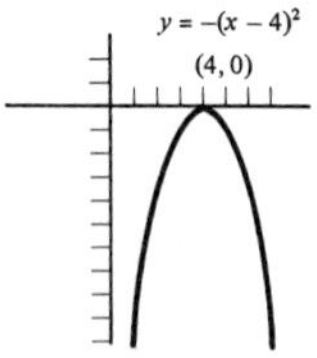

11.

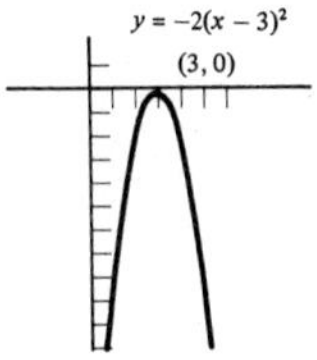

13.

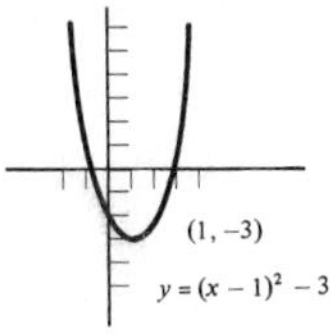

15.

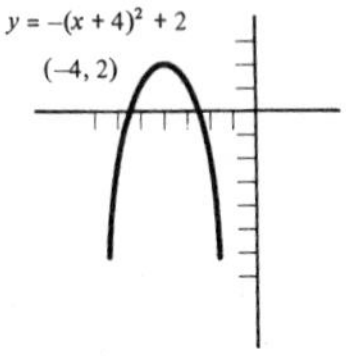

17.

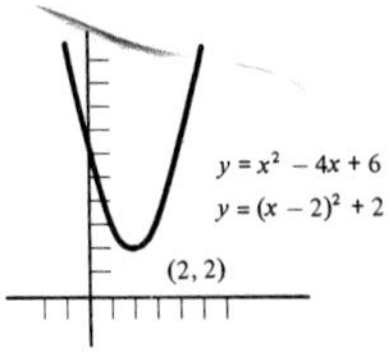

19.

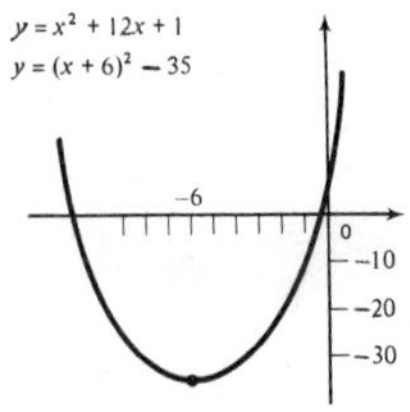

21.

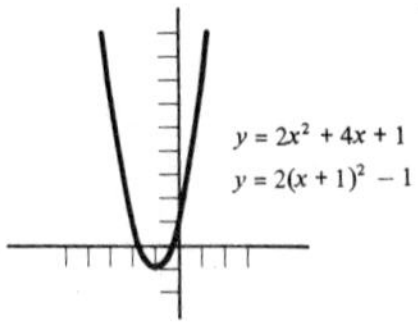

23.

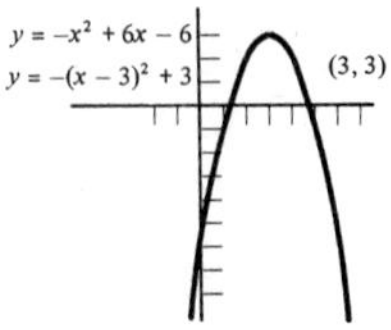

25. (a) $C(x) = (x - 5)^2 + 15$ (b)

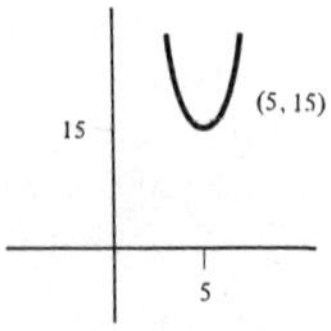

(c) 5 units (d) minimum cost is $15

27. (a) $x = 3$ dimes or 30¢ (b) $p = 8$, or \$800

29. (a) 640 (b) 515 (c) 140 (d)

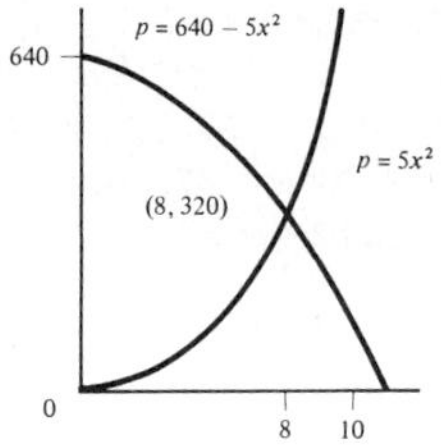

31. (a) About 12 (b) 10 (c) about 7 (d) 0 (e)

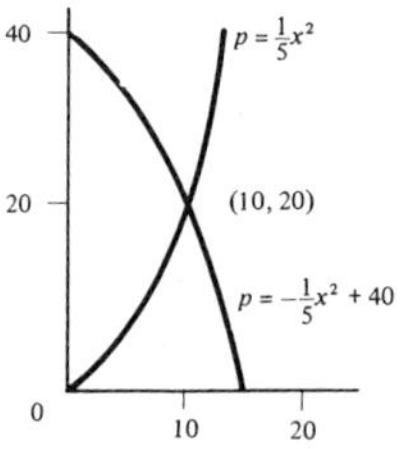

33. 10

35. (a) $(100 - x)(200 + 4x)$ (b)

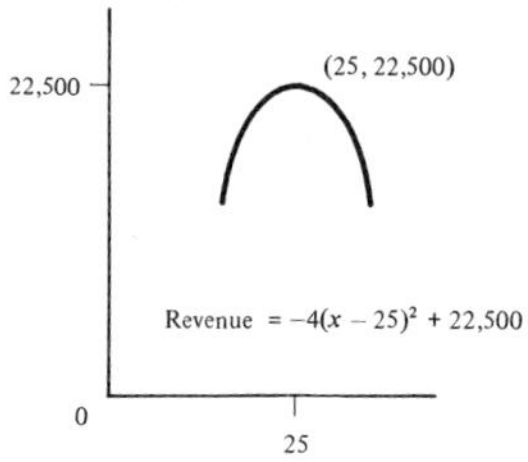

(c) 25 (d) \$22,500

37. (a) 60 (b) 70 (c) 90 (d) 100 (e) 80 (f) 20

Section 3.5 (page 132)

1.

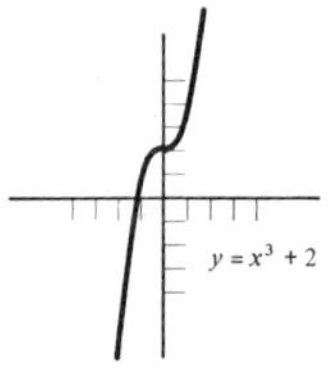

3.

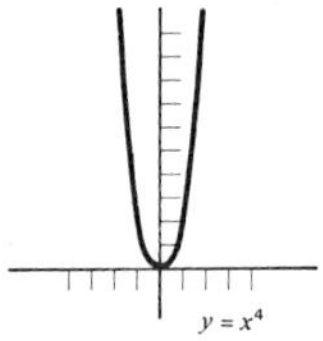

5.

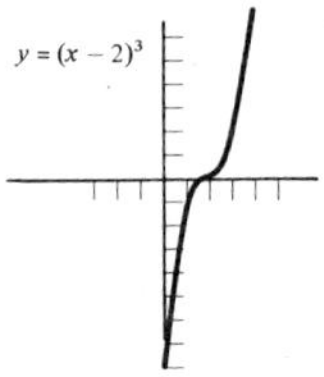

7.

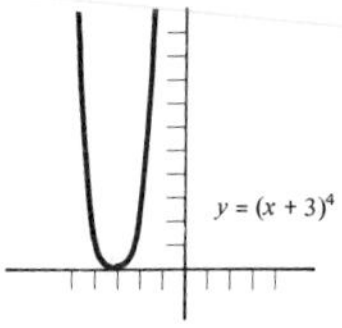

9.

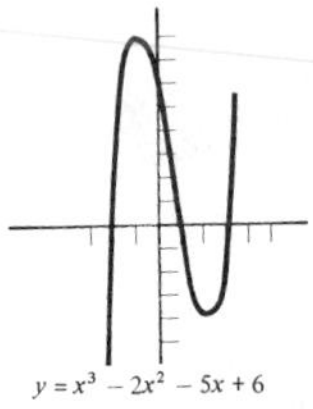

11.

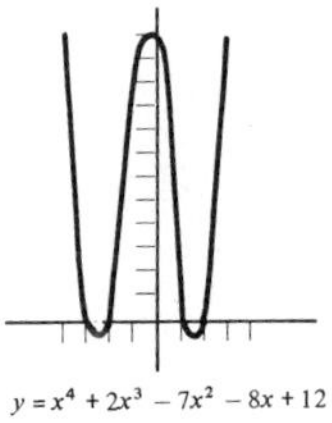

13.

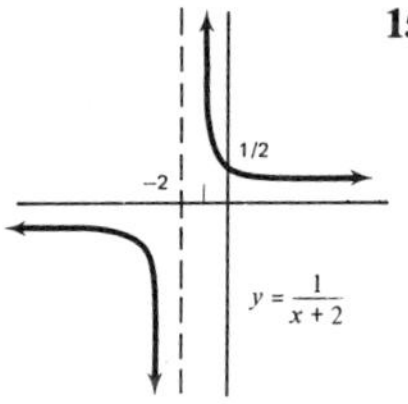

15.

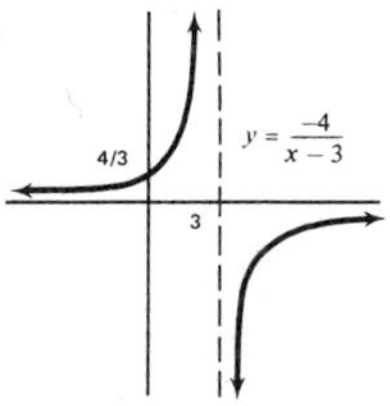

17.

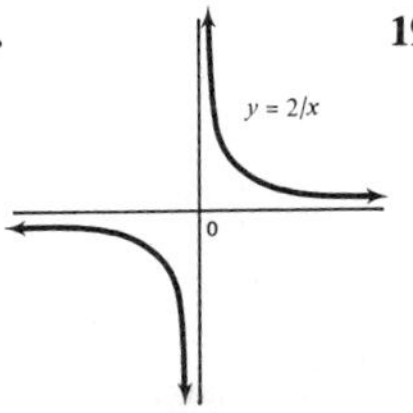

19.

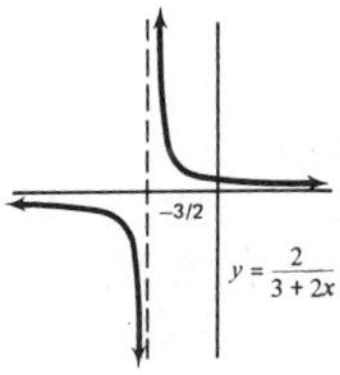

21.

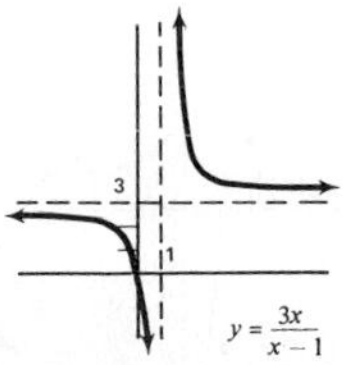

23.

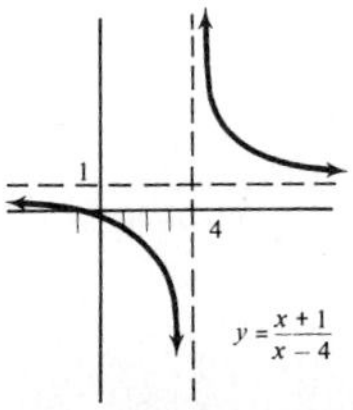

25.

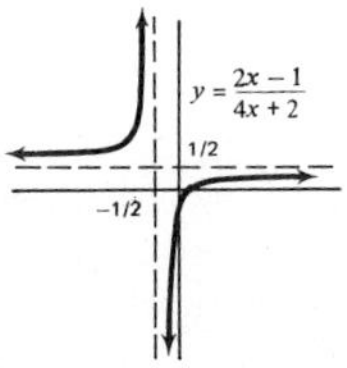

27.

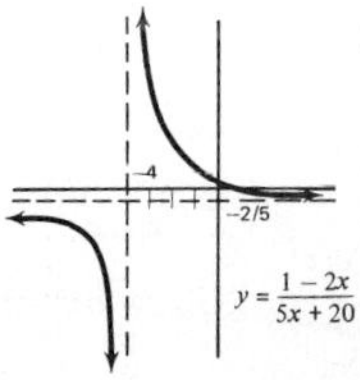

29.

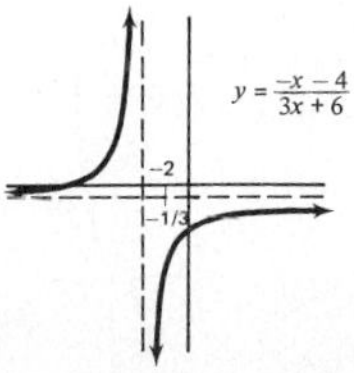

31. 0 **33.** About 2.0 or .20% **35.** About 3.2 or .32% **37.** About 1.0 or .1% **39.** Between 4 and 5 hours, closer to 5 hours **41.** \$12.50 **43.** \$6.25 **45.** \$3.85 **47.** \$440 **49.** About \$232 **51.** \$0 **53.** About \$24,000 **55.** About \$88,000 **57.** \$325,000 **59.** \$6700 **61.** About \$27,000 **63.** About \$77,000 **65.** About \$328,000 **67.** No **69.** About 60% **71.** About \$10,000 **73.** 5000 hundred gallons of gasoline, 1000 hundred gallons of oil

Section 3.6 (page 140)

1. \$60,000 **3.** \$26,667 **5.** \$120,000 **7.** \$76,800

9.

Depreciation in year	Straight-line	Double declining	Sum-of-years' digits
1	\$375	\$933	\$563
2	\$375	\$192*	\$375
3	\$375	\$0	\$188
Totals	\$1125	\$1125	\$1126

*Only \$192 of depreciation may be taken, since the total may not exceed the net cost of \$1125.

11. \$7740 **13.** \$5500 **15.** About 50,900 **17.** About 42,100 **19.**

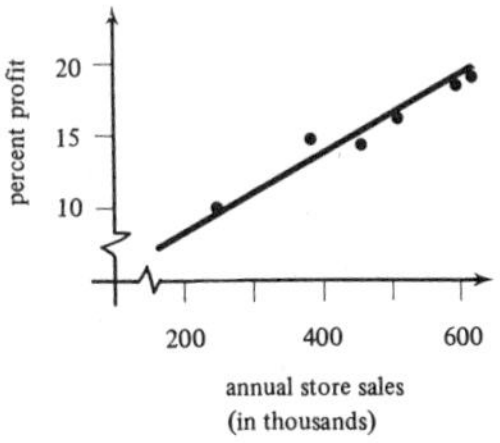

Chapter 3 Test (page 142)

1.

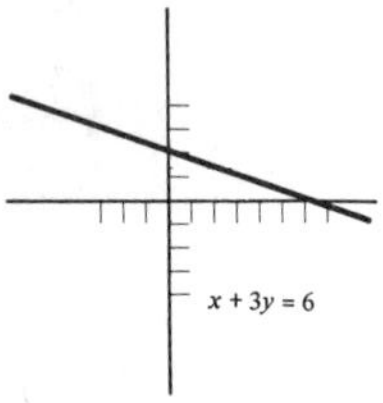

2.

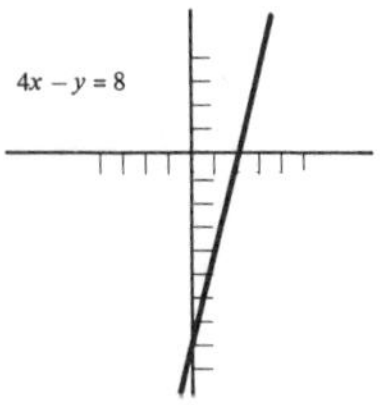

3.

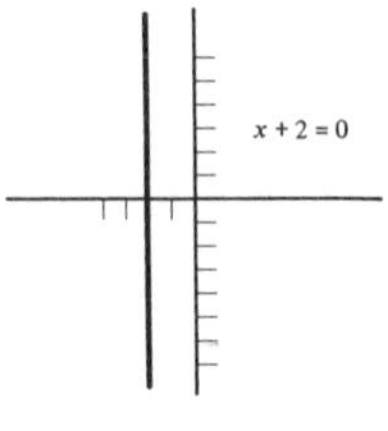

4.

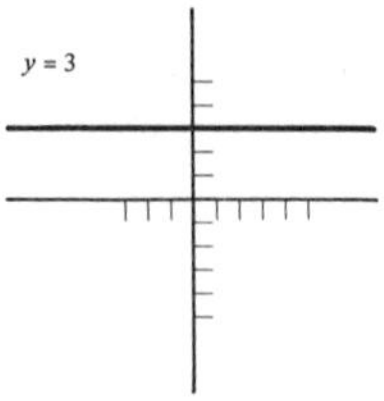

5.

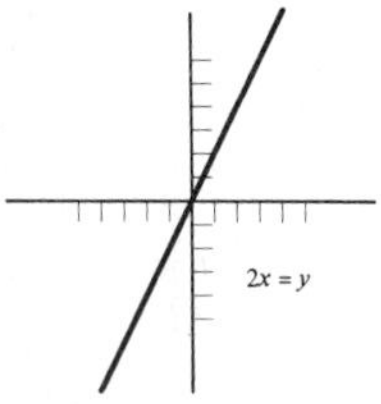

6.

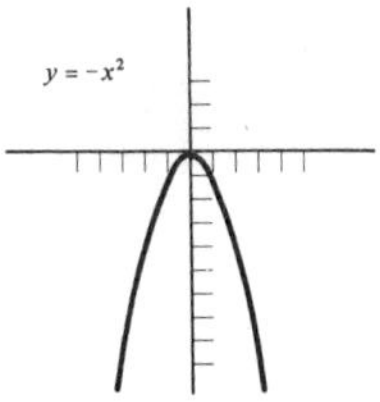

7.

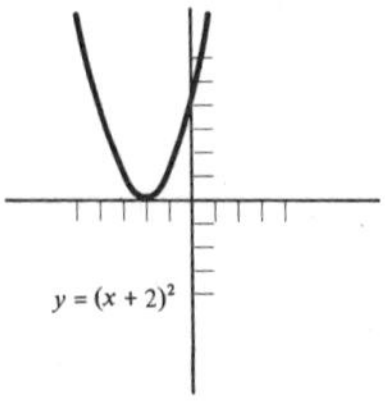

8.

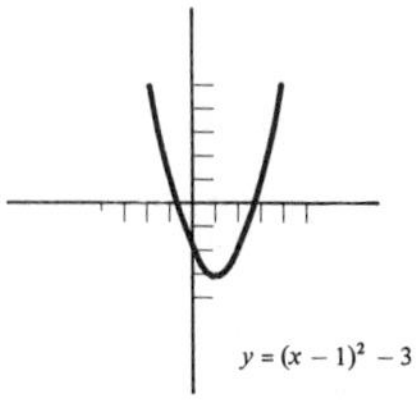

9.

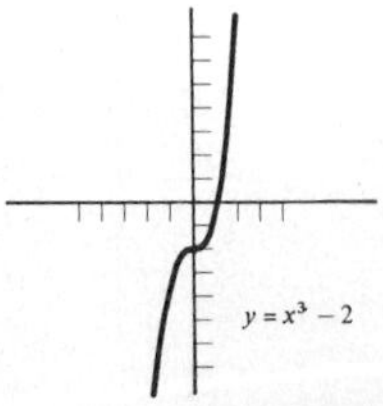

10.

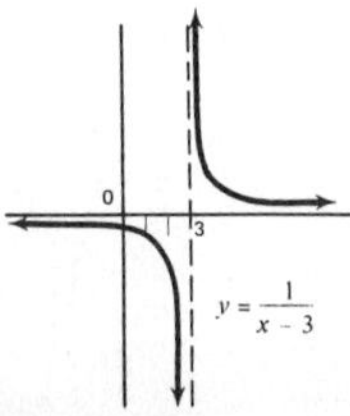

11.

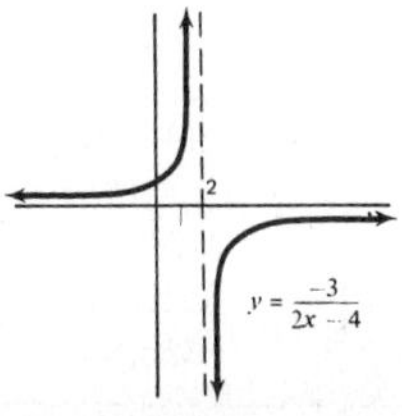

12.

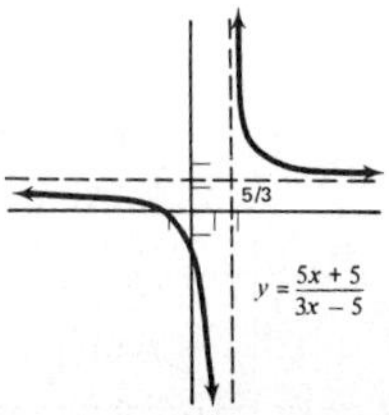

13. 2 **14.** 15 **15.** 1/3 **16.** 8/3 **17.** $2y = x + 8$ **18.** $5y = x - 13$ **19.** $x = 3$ **20.** $C(x) = 30x + 60$ **21.** 5 units **22.** $200 **23.** (3, −4) **24.** (2, −1) **25.** (−3, 10) **26.** 7 **27.** 10 **28.** About $8200 **29.** About $113,000 **30.** About 75% **31.** $3500 **32.** $6000 **33.** $7000

Case 4 (page 145)

1. 4.8 million units **2.** Portion of a straight line going through (3.1, 10.50) and (5.7, 10.67) **3.** In the interval under discussion (3.1 to 5.7 million units), the marginal cost always exceeds the selling price. **4.** (a) 9.87; 10.22 (b) portion of a straight line through (3.1, 9.87) and (5.7, 10.22) (c) .83 million units, which is not in the interval under discussion

Chapter 4

Section 4.1 (page 151)

1. $1368.57 **3.** $14,432.13 **5.** $755.97 **7.** $68,091.04 **9.** $10,730.78 **11.** $10,454.86 **13.** $2864.76 **15.** $34,662.73 **17.** $745.58 **19.** $2663.55 **21.** $2700.58 **23.** $1293.68 **25.** $6946.81 **27.** $1000 now is a little more **29.** 4.04% **31.** 8.16% **33.** 12.36% **35.** About 12 years

Section 4.2 (page 158)

1. 15.91713 **3.** 21.82453 **5.** 22.01900 **7.** .04994 **9.** .02962 **11.** $1200.61 **13.** $305,390.00 **15.** $564,730.23 **17.** $4,180,929.88 **19.** $21,496.30 **21.** $404,601.90 **23.** $6294.79 **25.** $142,840.20 **27.** $12,180.79 **29.** $4798.51 **31.** $628.25 **33.** $1333.54 **35.** $439,414.68 **37.** $12,796.76 **39.** $1047.22 **41.** $444.88

Section 4.3 (page 164)

1. 9.71225 **3.** 12.65930 **5.** 14.71787 **7.** .19702 **9.** .09168 **11.** $7107.82 **13.** $9645.62 **15.** $196,004.40 **17.** $1,483,782.30 **19.** $119,379.40 **21.** $103,796.60 **23.** $14,215.64 **25.** $854.74 **27.** $7494.30 **29.** $6699.00

Section 4.4 (page 168)

1. $147.02 **3.** $5054.93 **5.** $11,727.32 **7.** $274.58 **9.** $354.79 **11.** $732.07 **13.** $2440.29 **15.** $126.91

17.

Payment number	Amount of payment	Interest for period	Portion to principal	Principal at end of period
0	—	—	—	$4000
1	$1154.36	$240	$914.36	$3085.64
2	$1154.36	$185.14	$969.22	$2116.42
3	$1154.36	$126.99	$1027.37	$1089.05
4	$1154.39	$65.34	$1089.05	$0

19.

Payment number	Amount of payment	Interest for period	Portion to principal	Principal at end of period
0	—	—	—	$6000
1	$891.17	$240	$651.17	$5348.83
2	$891.17	$213.95	$677.22	$4671.61
3	$891.17	$186.86	$704.31	$3967.30
4	$891.17	$158.69	$732.48	$3234.82
5	$891.17	$129.39	$761.78	$2473.04
6	$891.17	$98.92	$792.25	$1680.79
7	$891.17	$67.23	$823.94	$856.85
8	$891.12	$34.27	$856.85	$0

21. \$1200

23.

Payment number	Amount of deposit	Interest earned	Total
1	\$3511.58	\$0	\$3511.58
2	\$3511.58	\$105.35	\$7128.51
3	\$3511.58	\$213.86	\$10,853.95
4	\$3511.58	\$325.62	\$14,691.15
5	\$3511.58	\$440.73	\$18,643.46
6	\$3511.58	\$559.30	\$22,714.34
7	\$3511.58	\$681.43	\$26,907.35
8	\$3511.58	\$807.22	\$31,226.15
9	\$3511.58	\$936.78	\$35,674.51
10	\$3511.58	\$1070.24	\$40,256.33
11	\$3511.58	\$1207.69	\$44,975.60
12	\$3511.58	\$1349.27	\$49,836.45
13	\$3511.58	\$1495.09	\$54,843.12
14	\$3511.59	\$1645.29	\$60,000.00

25. \$271.46 **27.** \$3868.05 **29.** \$17,584.58

Section 4.5 (page 173)

1. \$322.00 **3.** \$424.60 **5.** \$207.63 **7.** \$289.50 **9.** \$408.60 **11.** \$278.05 **13.** \$357.38 **15.** 16% **17.** 16% **19.** $15\frac{1}{2}\%$ **21.** $15\frac{1}{2}\%$ **23.** (a) \$5500 (b) 15% **25.** 17%

Chapter 4 Test (page 175)

1. \$1551.33 **2.** \$5057.11 **3.** \$7429.75 **4.** \$3563.10 **5.** \$3353.00 **6.** \$1262.68 **7.** \$885.47 **8.** \$3947.05 **9.** \$10,078.44 **10.** \$22,304.99 **11.** \$15,076.91 **12.** \$11,807.80 **13.** \$3014.06 **14.** \$34,090.08 **15.** \$2940.45 **16.** \$11,255.21 **17.** \$356.24 **18.** \$614.20 **19.** \$1700.24 **20.** \$378.00 **21.** \$392.19 **22.** \$418.58 **23.** $15\frac{1}{2}\%$ **24.** $15\frac{1}{2}\%$

Case 5 (page 179)

1. \$14,038 **2.** \$9511 **3.** \$8837 **4.** \$3968

Chapter 5

Section 5.1 (page 185)

1. False, corresponding elements not all equal **3.** True **5.** True **7.** 2×2, square **9.** 3×4 **11.** 2×1, column **13.** $x = 2, y = 4, z = 8$ **15.** $x = -15, y = 5, k = 3$ **17.** $z = 18, r = 3, s = 3, p = 3, a = 3/4$

19. $\begin{bmatrix} 9 & 12 & 0 & 2 \\ 1 & -1 & 2 & -4 \end{bmatrix}$ **21.** $\begin{bmatrix} 5 & 13 & 0 \\ 3 & 1 & 8 \end{bmatrix}$ **23.** Not possible **25.** $\begin{bmatrix} 1 & 5 & 6 & -9 \\ 5 & 7 & 2 & 1 \\ -7 & 2 & 2 & -7 \end{bmatrix}$

27. $\begin{bmatrix} -12x + 8y & -x + y \\ x & 8x - y \end{bmatrix}$ **29.** $\begin{bmatrix} x & y \\ z & w \end{bmatrix} + \begin{bmatrix} r & s \\ t & u \end{bmatrix} = \begin{bmatrix} x + r & y + s \\ z + t & w + u \end{bmatrix}$ (a 2×2 matrix)

31. $\begin{bmatrix} x + (r + m) & y + (s + n) \\ z + (t + p) & w + (u + q) \end{bmatrix} = \begin{bmatrix} (x + r) + m & (y + s) + n \\ (z + t) + p & (w + u) + q \end{bmatrix}$ **33.** $\begin{bmatrix} m + 0 & n + 0 \\ p + 0 & q + 0 \end{bmatrix} = \begin{bmatrix} m & n \\ p & q \end{bmatrix}$

35. $\begin{bmatrix} 7 & 2 \\ 9 & 0 \\ 8 & 6 \end{bmatrix}; \begin{bmatrix} 7 & 9 & 8 \\ 2 & 0 & 6 \end{bmatrix}$ **37.** (a) $\begin{bmatrix} 2 & 1 & 2 & 1 \\ 3 & 2 & 2 & 1 \\ 4 & 3 & 2 & 1 \end{bmatrix}$ (b) $\begin{bmatrix} 5 & 0 & 7 \\ 0 & 10 & 1 \\ 0 & 15 & 2 \\ 10 & 12 & 8 \end{bmatrix}$ (c) $\begin{bmatrix} 8 \\ 4 \\ 5 \end{bmatrix}$

Section 5.2 (page 195)

1. 2×2; 2×2 **3.** 4×4; 2×2 **5.** 3×2; BA does not exist **7.** AB does not exist; 3×2 **9.** $\begin{bmatrix} -4 & 8 \\ 0 & 6 \end{bmatrix}$

11. $\begin{bmatrix} 24 & -8 \\ -16 & 0 \end{bmatrix}$ **13.** $\begin{bmatrix} -22 & -6 \\ 20 & -12 \end{bmatrix}$ **15.** $\begin{bmatrix} 13 \\ 25 \end{bmatrix}$ **17.** $\begin{bmatrix} -2 & 10 \\ 0 & 8 \end{bmatrix}$ **19.** $\begin{bmatrix} 13 & 5 \\ 25 & 15 \end{bmatrix}$ **21.** $\begin{bmatrix} 13 \\ 29 \end{bmatrix}$

23. $\begin{bmatrix} 11 & -11 \\ 4 & -4 \\ -5 & 5 \end{bmatrix}$ **25.** Not possible **27.** $\begin{bmatrix} 16 & 22 \\ 7 & 19 \end{bmatrix}$ **29.** No; no

31. $(PX)T = \begin{bmatrix} (mx+nz)r + (my+nw)t & (mx+nz)s + (my+nw)u \\ (px+qz)r + (py+qw)t & (px+qz)s + (py+qw)u \end{bmatrix}$

$P(XT)$ is the same, so that $P(XT) = (PX)T$

33. $PX = \begin{bmatrix} mx+nz & my+nw \\ px+qz & py+qw \end{bmatrix}$ (a 2×2 matrix)

35. $k(X+T) = \begin{bmatrix} k(x+r) & k(y+s) \\ k(z+t) & k(w+u) \end{bmatrix}$

$kX + kT = \begin{bmatrix} kx+kr & ky+ks \\ kz+kt & kw+ku \end{bmatrix} = k(X+T)$

37. (a) P, P, X (b) T (c) I maintains the identity of any 2×2 matrix under multiplication.

39. (a) $\begin{bmatrix} 20 & 52 & 27 \\ 25 & 62 & 35 \\ 30 & 72 & 43 \end{bmatrix}$ The rows represent the amounts of fat, carbohydrate, and protein, respectively in each of the daily meals.

(b) $\begin{bmatrix} 75 \\ 45 \\ 70 \\ 168 \end{bmatrix}$ The rows give the number of calories in one exchange of each of the food groups.

Section 5.3 (page 204)

1. $\begin{bmatrix} -1 & -5/4 \\ 4 & 7 \end{bmatrix}$ **3.** $\begin{bmatrix} 1 & 1/5 \\ 0 & 1 \end{bmatrix}$ **5.** $\begin{bmatrix} 1 & 5 & 6 \\ 0 & 13 & 11 \\ 4 & 7 & 0 \end{bmatrix}$ **7.** $\begin{bmatrix} -3 & 1 & -4 \\ 2 & 1 & 3 \\ -17 & 0 & -13 \end{bmatrix}$ **9.** Yes **11.** No **13.** No

15. Yes **17.** $\begin{bmatrix} 0 & 1/2 \\ -1 & 1/2 \end{bmatrix}$ **19.** $\begin{bmatrix} 2 & 1 \\ 5 & 3 \end{bmatrix}$ **21.** None **23.** $\begin{bmatrix} 1 & 0 & 0 \\ 0 & -1 & 0 \\ -1 & 0 & 1 \end{bmatrix}$ **25.** $\begin{bmatrix} 15 & 4 & -5 \\ -12 & -3 & 4 \\ -4 & -1 & 1 \end{bmatrix}$

27. None **29.** $\begin{bmatrix} 7/4 & 5/2 & 3 \\ -1/4 & -1/2 & 0 \\ -1/4 & -1/2 & -1 \end{bmatrix}$ **31.** $\begin{bmatrix} 1/2 & 1/2 & -1/4 & 1/2 \\ -1 & 4 & -1/2 & -2 \\ -1/2 & 5/2 & -1/4 & -3/2 \\ 1/2 & -1/2 & 1/4 & 1/2 \end{bmatrix}$ **33.** $\begin{bmatrix} 1 & 0 \\ 0 & 1 \end{bmatrix} \cdot \begin{bmatrix} a & b \\ c & d \end{bmatrix} = \begin{bmatrix} a & b \\ c & d \end{bmatrix}$

35. $\begin{bmatrix} a & b \\ c & d \end{bmatrix} \cdot \begin{bmatrix} 0 & 0 \\ 0 & 0 \end{bmatrix} = \begin{bmatrix} 0 & 0 \\ 0 & 0 \end{bmatrix}$ **37.** $A^{-1} = \begin{bmatrix} d/(ad-bc) & -b/(ad-bc) \\ -c/(ad-bc) & a/(ad-bc) \end{bmatrix}$

$$A^{-1}A = \begin{bmatrix} d/(ad-bc) & -b/(ad-bc) \\ -c/(ad-bc) & a/(ad-bc) \end{bmatrix} \cdot \begin{bmatrix} a & b \\ c & d \end{bmatrix} = \begin{bmatrix} \dfrac{ad-bc}{ad-bc} & \dfrac{db-bd}{ad-bc} \\ \dfrac{-ca+ac}{ad-bc} & \dfrac{-cb+ad}{ad-bc} \end{bmatrix}$$

$$= \begin{bmatrix} 1 & 0 \\ 0 & 1 \end{bmatrix} = I$$

Section 5.4 (page 212)

1. $\begin{bmatrix} 3 \\ 4 \end{bmatrix}$ **3.** $\begin{bmatrix} 36 & 6 \\ 28 & -2 \end{bmatrix}$ **5.** $\begin{bmatrix} -22 \\ -18 \\ 15 \end{bmatrix}$ **7.** $\begin{bmatrix} 4 & -11 \\ -10 & 11 \\ 4 & -5 \end{bmatrix}$ **9.** $\begin{bmatrix} -6 \\ -14 \end{bmatrix}$

11. $\begin{bmatrix} 35 \\ 92 \end{bmatrix}, \begin{bmatrix} -4 \\ 80 \end{bmatrix}, \begin{bmatrix} 15 \\ 132 \end{bmatrix}, \begin{bmatrix} -9 \\ 99 \end{bmatrix}, \begin{bmatrix} 35 \\ 173 \end{bmatrix}, \begin{bmatrix} 2 \\ 41 \end{bmatrix}, \begin{bmatrix} -8 \\ 61 \end{bmatrix}$

13. $\begin{bmatrix} 76 \\ 77 \\ 96 \end{bmatrix}, \begin{bmatrix} 62 \\ 67 \\ 75 \end{bmatrix}, \begin{bmatrix} 88 \\ 108 \\ 97 \end{bmatrix}, \begin{bmatrix} 141 \\ 160 \\ 168 \end{bmatrix}, \begin{bmatrix} 147 \\ 166 \\ 174 \end{bmatrix}, \begin{bmatrix} 105 \\ 120 \\ 123 \end{bmatrix}, \begin{bmatrix} 111 \\ 131 \\ 119 \end{bmatrix}, \begin{bmatrix} 92 \\ 119 \\ 94 \end{bmatrix}, \begin{bmatrix} 75 \\ 93 \\ 79 \end{bmatrix}, \begin{bmatrix} 181 \\ 208 \\ 208 \end{bmatrix}$

15. [892.9 1178.6] **17.** [1695.65 1565.22 1391.30] **19.** [1538.46 1282.05 1589.74]

21. (a) $\begin{bmatrix} 72 \\ 48 \\ 60 \end{bmatrix}$ (b) $\begin{bmatrix} 2 & 4 & 2 \\ 2 & 1 & 2 \\ 2 & 1 & 3 \end{bmatrix} \begin{bmatrix} x_1 \\ x_2 \\ x_3 \end{bmatrix} = \begin{bmatrix} 72 \\ 48 \\ 60 \end{bmatrix}$ (c) 8, 8, 12

Chapter 5 Test (page 214)

1. 2×2; $a = 2$, $b = 3$, $c = 5$, $q = 9$; both are square matrices **2.** 3×2; $a = 2$, $x = -1$, $y = 4$, $p = 5$, $z = 7$

3. 1×4; $m = 12$, $k = 4$, $z = -8$, $r = -1$; both are row matrices **4.** $\begin{bmatrix} 8 & 8 & 8 \\ 10 & 5 & 9 \\ 7 & 10 & 7 \\ 8 & 9 & 7 \end{bmatrix}$ **5.** $\begin{bmatrix} 9 & 10 \\ -3 & 0 \\ 10 & 16 \end{bmatrix}$ **6.** $\begin{bmatrix} 8 & -6 \\ -10 & -16 \end{bmatrix}$

7. Not possible **8.** $\begin{bmatrix} 26 & 86 \\ -7 & -29 \\ 21 & 87 \end{bmatrix}$ **9.** Not possible **10.** $\begin{bmatrix} 6 & 18 & -24 \\ 1 & 3 & -4 \\ 0 & 0 & 0 \end{bmatrix}$ **11.** $\begin{bmatrix} -7/19 & 4/19 \\ 3/19 & 1/19 \end{bmatrix}$

12. No inverse **13.** $\begin{bmatrix} 18 \\ -7 \end{bmatrix}$ **14.** (a) $\begin{bmatrix} 0 & 1/2 \\ 2/3 & 0 \end{bmatrix}$ (b) [1400 1500]

15. (a) $\begin{bmatrix} 47 \\ 56 \end{bmatrix}, \begin{bmatrix} 0 \\ 130 \end{bmatrix}, \begin{bmatrix} 107 \\ 60 \end{bmatrix}, \begin{bmatrix} 53 \\ 202 \end{bmatrix}, \begin{bmatrix} 72 \\ 88 \end{bmatrix}, \begin{bmatrix} -7 \\ 172 \end{bmatrix}, \begin{bmatrix} 11 \\ 12 \end{bmatrix}, \begin{bmatrix} 83 \\ 74 \end{bmatrix}$ (b) $\begin{bmatrix} 3/13 & 1/26 \\ -1/13 & 2/13 \end{bmatrix}$

Chapter 6

Section 6.1 (page 224)

1. (3, 6) **3.** (−1, 4) **5.** (−2, 0) **7.** (1, 3) **9.** (4, −2) **11.** (2, −2) **13.** ∅ **15.** Same line
17. (12, 6) **19.** (7, −2) **21.** (1, 2, −1) **23.** (2, 0, 3) **25.** ∅ **27.** (0, 2, 4) **29.** (1, 2, 3)
31. (−1, 2, 1) **33.** (4, 1, 2) **35.** x arbitrary, $y = x + 5$, $z = -2x + 1$ **37.** x arbitrary, $y = x + 1$, $z = -x + 3$
39. x arbitrary, $y = 4x - 7$, $z = 3x + 7$, $w = -x - 3$ **41.** (3, −4) **43.** ∅ **45.** ∅ **47.** $a = -3$, $b = -1$
49. $a = 3/4$, $b = 1/4$, $c = -1/2$ **51.** Wife, 40 days; husband, 32 days **53.** 5 model 201 and 8 model 301
55. \$10,000 at 8%, \$7000 at 10%, \$8000 at 9%

Section 6.2 (page 233)

1. $\left[\begin{array}{cc|c} 2 & 3 & 11 \\ 1 & 2 & 8 \end{array}\right]$ **3.** $\left[\begin{array}{cc|c} 1 & 5 & 6 \\ 0 & 1 & 1 \end{array}\right]$ **5.** $\left[\begin{array}{ccc|c} 2 & 1 & 1 & 3 \\ 3 & -4 & 2 & -7 \\ 1 & 1 & 1 & 2 \end{array}\right]$ **7.** $\left[\begin{array}{ccc|c} 1 & 1 & 0 & 2 \\ 0 & 2 & 1 & -4 \\ 0 & 0 & 1 & 2 \end{array}\right]$ **9.** $\left[\begin{array}{ccc|c} 1 & 0 & 0 & 5 \\ 0 & 1 & 0 & -2 \\ 0 & 0 & 1 & 3 \end{array}\right]$

11. $x = 2$, $y = 3$ **13.** $2x + y = 1$, $3x - 2y = -9$ **15.** $x = 2$, $y = 3$, $z = -2$ **17.** (2, 3) **19.** (−3, 0) **21.** (7/2, −1) **23.** (5/2, −1)

25. $\varnothing$ **27.** Same line **29.** $(-2, 1, 3)$ **31.** $(-1, 23, 16)$ **33.** $(3, 2, -4)$ **35.** $\varnothing$ **37.** $(0, 2, -2, 1)$ The answers are given in the order x, y, z, w. **39.** $\left[\begin{array}{cccc|c} 1 & 0 & 0 & 1 & 1000 \\ 1 & 1 & 0 & 0 & 1100 \\ 0 & 1 & 1 & 0 & 700 \\ 0 & 0 & 1 & 1 & 600 \end{array}\right]$; $\left[\begin{array}{cccc|c} 1 & 0 & 0 & 1 & 1000 \\ 0 & 1 & 0 & -1 & 100 \\ 0 & 0 & 1 & 1 & 600 \\ 0 & 0 & 0 & 0 & 0 \end{array}\right]$

41. $x_4 = -x_1 + 1000$; $x_4 = x_2 - 100$; $x_4 = -x_3 + 600$ **43.** 100 **45.** $x_1 = 1000$; $x_2 = 700$; $x_3 = 600$; $x_4 = 600$

Section 6.3 (page 238)

1. $(-1, 4)$ **3.** $(2, 1)$ **5.** $(2, 3)$ **7.** $(4, 8)$ **9.** $(-5/3, 7/3)$ **11.** Same line **13.** $(-8, 6, 1)$ **15.** $(15, -5, -1)$ **17.** $(-31, 24, -4)$ **19.** $(-4, -3, 1)$ **21.** No inverse, $\varnothing$ **23.** $(-7, -34, -19, 7)$ (in the order x, y, z, w)

Section 6.4 (page 243)

1.
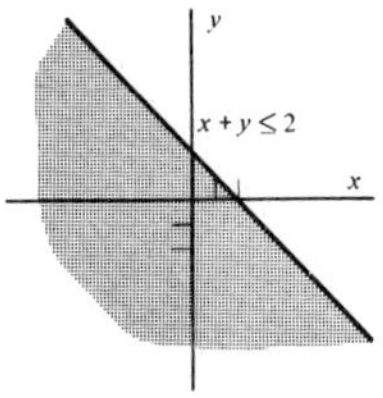

3.
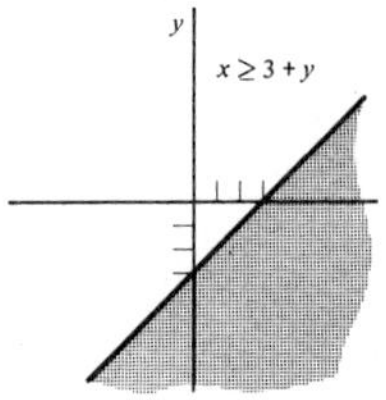

5.
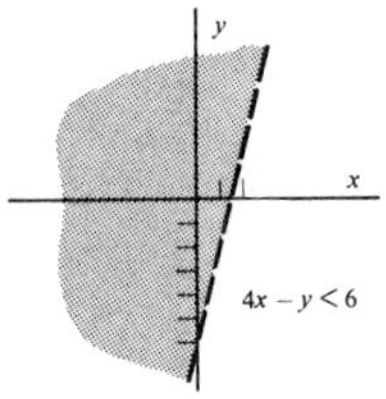

7.
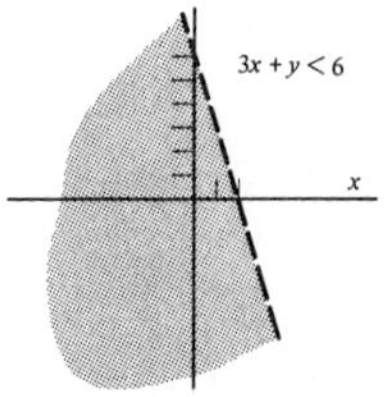

9.
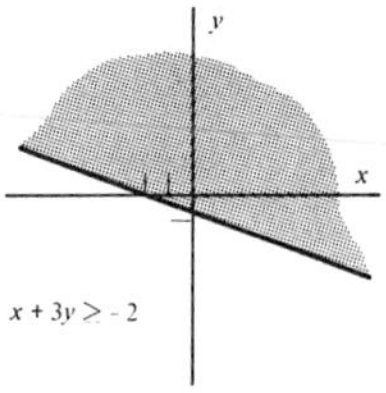

11.
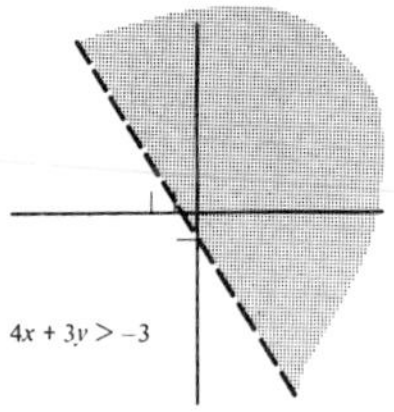

13.
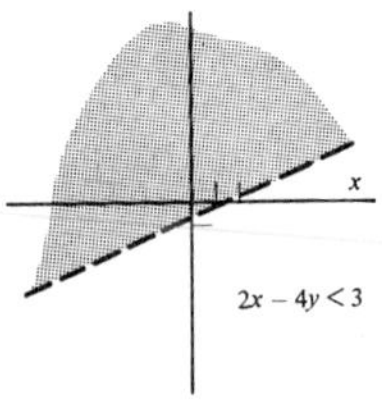

15.
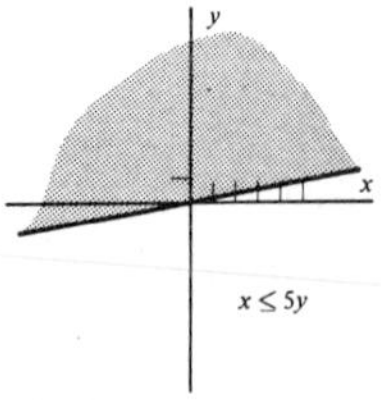

17.
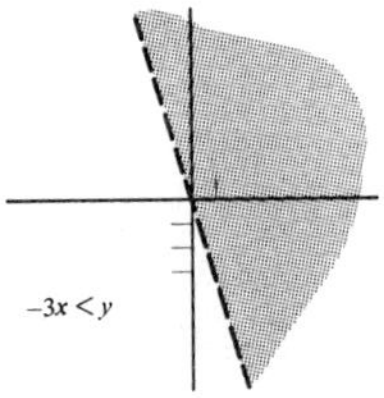

19.
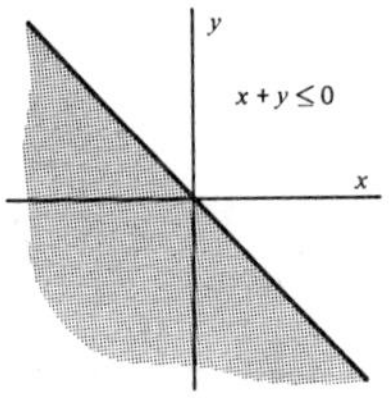

21.
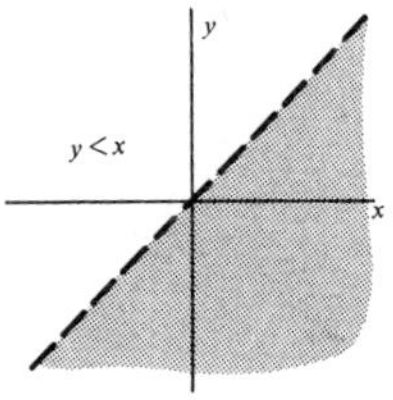

23.
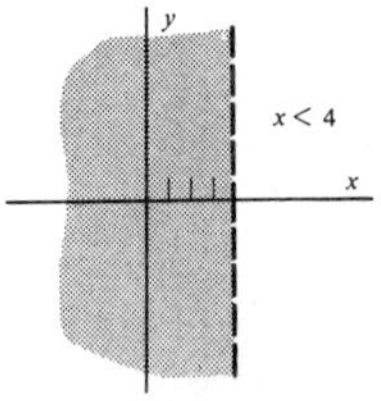

25.
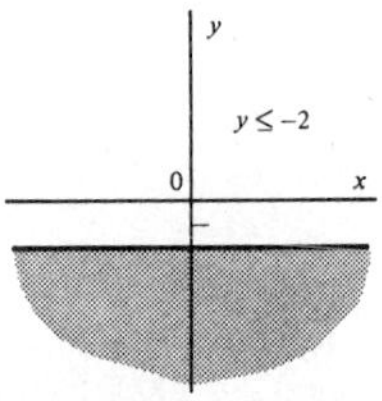

27.
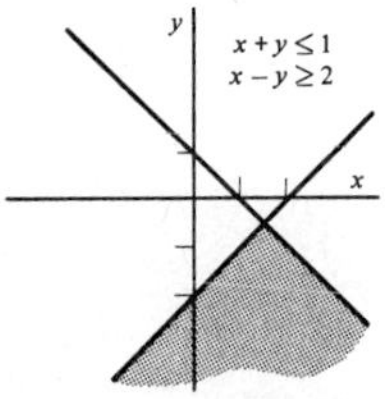

29.
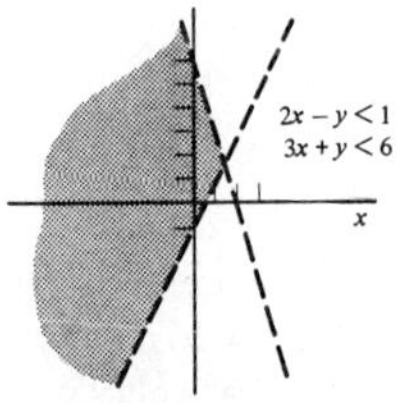

31.
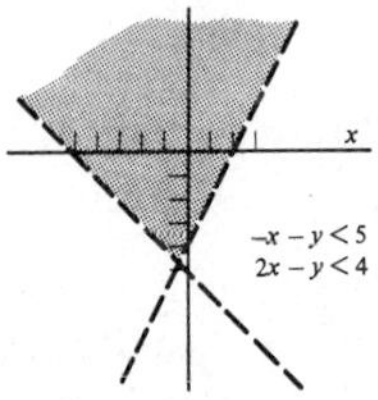

33.
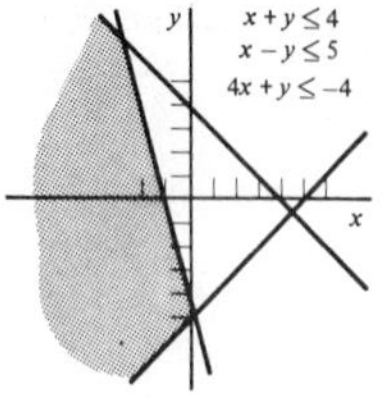

35.
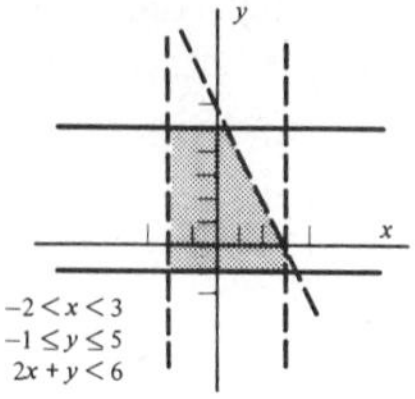

37.
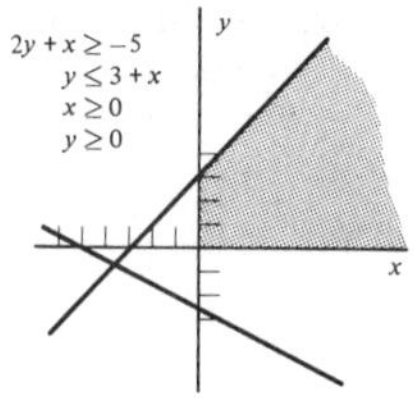

39.
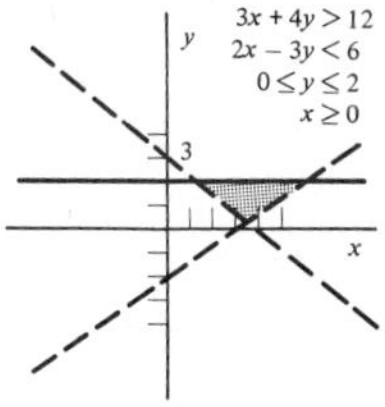

41. (a) $x_1 \geq 1000$; $x_2 \geq 800$; $x_1 + x_2 \leq 2400$; $x_1 \geq 0$; $x_2 \geq 0$ (b)
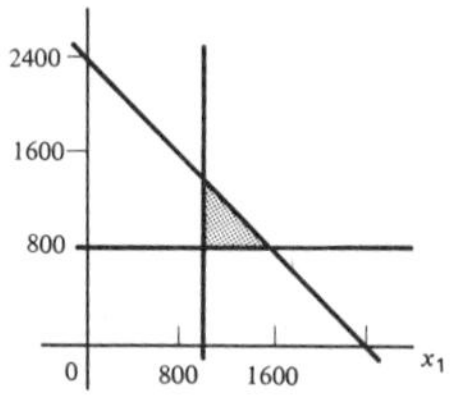

Chapter 6 Test (page 244)

1. $(-4, 6)$ **2.** $(8, -4)$ **3.** $(-1, 2, 3)$ **4.** $a = -4/5$, $b = 12/5$ **5.** 5 blankets, 3 rugs, and 8 skirts **6.** $(-9, 3)$ **7.** $(7, -9, -1)$ **8.** $(2, 1)$ **9.** $(-1, 0, 2)$ **10.**
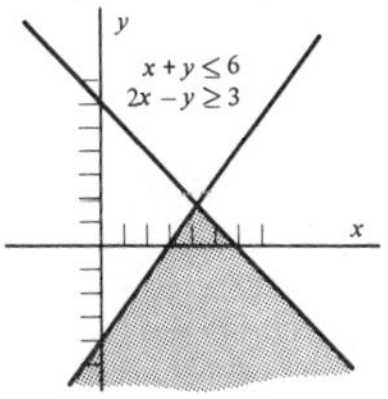

11.
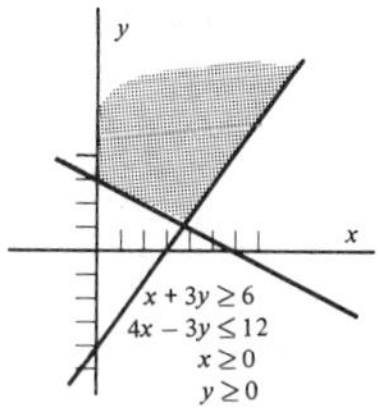

Chapter 7

Section 7.1 (page 253)

1. Maximum of 65 at (5, 10); minimum of 8 at (1, 1) **3.** Maximum of 9 at (0, 12); minimum of 0 at (0, 0) **5.** $x = 6/5$, $y = 6/5$ **7.** $x = 17/3$, $y = 5$ **9.** $x = 105/8$, $y = 25/8$ **11.** (a) Maximum of 204; $x = 18$, $y = 2$ (b) maximum of $117\frac{3}{5}$; $x = 12/5$, $y = 39/5$ (c) maximum of 102; $x = 0$, $y = 17/2$ **13.** 6 pigs, 10 geese, for \$60 maximum profit **15.** 1200 Type 1 and 2400 Type 2 for maximum revenue of \$408 **17.** Ship 20 to A and 80 to B; \$1040 **19.** 150 kilograms half-and-half mix; 75 kilograms other **21.** $3\frac{3}{4}$ servings of A; $1\frac{7}{8}$ servings of B

Section 7.2 (page 259)

1. $x_1 + 2x_2 + x_3 = 6$ **3.** $2x_1 + 4x_2 + 3x_3 + x_4 = 100$ **5.** (a) 3 (b) x_3, x_4, x_5 (c) Find $x_1 \geq 0$, $x_2 \geq 0$, $x_3 \geq 0$, $x_4 \geq 0$, $x_5 \geq 0$ so that $4x_1 + 2x_2 + x_3 = 20$, $5x_1 + x_2 + x_4 = 50$, $2x_1 + 3x_2 + x_5 = 25$ **7.** (a) 2 (b) x_4, x_5 (c) Find $x_1 \geq 0$, $x_2 \geq 0$, $x_3 \geq 0$, $x_4 \geq 0$, $x_5 \geq 0$ so that $7x_1 + 6x_2 + 8x_3 + x_4 = 118$, $4x_1 + 5x_2 + 10x_3 + x_5 = 220$ **9.** (a) 2 (b) x_5, x_6 (c) Find $x_1 \geq 0$, $x_2 \geq 0$, $x_3 \geq 0$, $x_4 \geq 0$, $x_5 \geq 0$, $x_6 \geq 0$, so that $x_1 + x_2 + x_3 + x_4 + x_5 = 50$, $3x_1 + x_2 + 2x_3 + x_4 + x_6 = 100$

11. If x_1 is the number of kilograms of half-and-half mix and x_2 is the number of kilograms of the other mix, find $x_1 \geq 0$, $x_2 \geq 0$, $x_3 \geq 0$, $x_4 \geq 0$ so that $\frac{1}{2}x_1 + \frac{1}{3}x_2 + x_3 = 100$, $\frac{1}{2}x_1 + \frac{2}{3}x_2 + x_4 = 125$, and $z = x_1 + .8x_2$ is maximized.
13. If x_1 is the number of Siamese cats and x_2 is the number of Persian cats raised, find $x_1 \geq 0$, $x_2 \geq 0$, $x_3 \geq 0$, $x_4 \geq 0$, $x_5 \geq 0$, so that $2x_1 + x_2 + x_3 = 90$, $x_1 + 2x_2 + x_4 = 80$, $x_1 + x_2 + x_5 = 50$, and $z = 12x_1 + 10x_2$ is maximized.

Section 7.3 (page 265)

1. (0, 0, 20, 0, 15) **3.** (0, 0, 8, 0, 6, 7) **5.** (a) $\left[\begin{array}{cccccc|c} -1 & 0 & 3 & 1 & -1 & 16 \\ 1 & 1 & 1/2 & 0 & 1/2 & 20 \end{array}\right]$ (b) (0, 20, 0, 16, 0)

7. (a) $\left[\begin{array}{cccccc|c} 2 & 2 & 1 & 1 & 0 & 0 & 12 \\ -5 & -4 & 0 & -3 & 1 & 0 & 9 \\ 1 & -1 & 0 & -1 & 0 & 1 & 8 \end{array}\right]$ (b) (0, 0, 12, 0, 9, 8)

9.
$$\begin{array}{cccc} x_1 & x_2 & x_3 & x_4 \end{array}$$
$$\left[\begin{array}{cccc|c} 2 & 3 & 1 & 0 & 6 \\ 4 & 1 & 0 & 1 & 6 \end{array}\right]$$

11.
$$\begin{array}{ccccc} x_1 & x_2 & x_3 & x_4 & x_5 \end{array}$$
$$\left[\begin{array}{ccccc|c} 1 & 1 & 1 & 0 & 0 & 10 \\ 5 & 2 & 0 & 1 & 0 & 20 \\ 1 & 2 & 0 & 0 & 1 & 0 \end{array}\right]$$

13.
$$\begin{array}{cccc} x_1 & x_2 & x_3 & x_4 \end{array}$$
$$\left[\begin{array}{cccc|c} 3 & 1 & 1 & 0 & 12 \\ 1 & 1 & 0 & 1 & 15 \end{array}\right]$$

15. Maximize $75x_1 + 90x_2 + 100x_3$ where x_1 is the number of prams, x_2 the number of runabouts, and x_3 the number of trimarans.
$$\begin{array}{cccccc} x_1 & x_2 & x_3 & x_4 & x_5 & x_6 \end{array}$$
$$\left[\begin{array}{cccccc|c} 1 & 2 & 3 & 1 & 0 & 0 & 6240 \\ 2 & 5 & 4 & 0 & 1 & 0 & 10800 \\ 1 & 1 & 1 & 0 & 0 & 1 & 3000 \end{array}\right]$$

Section 7.4 (page 273)

1. $x_2 = 4$, $x_5 = 2$; maximum is 20 **3.** $x_1 = 4$, $x_3 = 8$, $x_4 = 2$; maximum is 8 **5.** $x_2 = 4$, $x_5 = 16$, $x_1 = 16$; maximum is 264 **7.** $x_2 = 10$, $x_4 = 40$, $x_5 = 2$; maximum is 120 **9.** $x_1 = 118$, $x_5 = 102$; maximum is 944
11. $x_4 = 50$, $x_6 = 50$; maximum is 250 **13.** 165 kilograms of food P, none of Q, 1090 of R, 146 of S; maximum is 87,600 **15.** 150 kilograms of the half-and-half mix and 75 kilograms of the other; maximum revenue is \$210
17. Make no 1-speed or 3-speed bicycles and make 2700 10-speed; maximum profit is \$59,400

Section 7.5 (page 280)

1. $\begin{bmatrix} 1 & 3 & 1 \\ 2 & 2 & 10 \\ 3 & 1 & 0 \end{bmatrix}$ **3.** $\begin{bmatrix} -1 & 13 & -2 \\ 4 & 25 & -1 \\ 6 & 0 & 11 \\ 12 & 4 & 3 \end{bmatrix}$

5. Minimize $5y_1 + 4y_2 + 15y_3$ subject to $y_1 + y_2 + 2y_3 \geq 4$, $y_1 + y_2 + y_3 \geq 3$, $y_1 + 3y_3 \geq 2$, $y_1 \geq 0$, $y_2 \geq 0$, $y_3 \geq 0$.
7. Maximize $50x_1 + 100x_2$ subject to $x_1 + 3x_2 \leq 1$; $x_1 + x_2 \leq 2$; $x_1 + 2x_2 \leq 1$; $x_1 + x_2 \leq 5$; $x_1 \geq 0$; $x_2 \geq 0$. **9.** $y_1 = 0$; $y_2 = 7$; minimum is 14 **11.** $y_1 = 10$; $y_2 = 0$; minimum is 40 **13.** $y_1 = 0$; $y_2 = 100$; $y_3 = 0$; minimum is 100
15. $3\frac{3}{4}$ servings of A and $1\frac{7}{8}$ servings of B for a minimum cost of \$1.69 **17.** 3 of pill #1, 2 of pill #2 for a minimum cost of 70 cents **19.** Buy 1000 small and 500 large test tubes for a minimum cost of \$210. **21.** $2\frac{2}{3}$ units of I, none of II, and 4 of III for a minimum cost of \$30.67

Chapter 7 Test (page 284)

1. Minimum is 8 at (2, 1); maximum is 40 at (6, 7) **2.** $x = 6$, $y = 0$, maximum is 30

3. (a) Let x_1 = number of units of A she should buy
x_2 = number of units of B
x_3 = number of units of C

(b) $z = 4x_1 + 4x_2 + 5x_3$

(c)
$$\begin{aligned} 5x_1 + 3x_2 + 6x_3 &\leq 1200 \\ x_1 + 2x_2 + 2x_3 &\leq 800 \\ 2x_1 + x_2 + 5x_3 &\leq 500 \end{aligned}$$

4. (a) Let x_1 = amount invested in oil leases
x_2 = amount invested in bonds
x_3 = amount invested in stocks

(b) $z = .15x_1 + .09x_2 + .05x_3$

(c)
$$\begin{aligned} x_1 + x_2 + x_3 &\leq 50{,}000 \\ x_1 + x_2 &\leq (.30)\,50{,}000 \\ x_1 + x_3 &\leq (.5)(50{,}000) \end{aligned}$$

5. (a) $2x_1 + 6x_2 + x_3 = 50$
$x_1 + 3x_2 + x_4 = 25$
$4x_1 + x_2 + x_5 = 18$
$x_1 + x_2 + x_6 = 12$

(b)

$$\begin{array}{cccccc|c} x_1 & x_2 & x_3 & x_4 & x_5 & x_6 & \\ 2 & 6 & 1 & 0 & 0 & 0 & 50 \\ 1 & 3 & 0 & 1 & 0 & 0 & 25 \\ 4 & 1 & 0 & 0 & 1 & 0 & 18 \\ 1 & 1 & 0 & 0 & 0 & 1 & 12 \\ \hline -5 & -3 & 0 & 0 & 0 & 0 & 0 \end{array}$$

(c) pivot is first column, third row entry, 4

6. (a) $x_1 + x_2 + x_3 + x_4 = 100$
$2x_1 + 3x_2 + x_5 = 500$
$x_1 + 2x_3 + x_6 = 350$

(b)

$$\begin{array}{cccccc|c} x_1 & x_2 & x_3 & x_4 & x_5 & x_6 & \\ 1 & 1 & 1 & 1 & 0 & 0 & 100 \\ 2 & 3 & 0 & 0 & 1 & 0 & 500 \\ 1 & 0 & 2 & 0 & 0 & 1 & 350 \\ \hline -2 & -3 & -4 & 0 & 0 & 0 & 0 \end{array}$$

(c) pivot is first row, third column entry, 1

7. Maximum is 82.4; solution is (13.6, 0, 4.8, 0, 0) **8.** (a) $\begin{bmatrix} 2 & -1 & 6 \\ 5 & 4 & -2 \\ 3 & 2 & 5 \end{bmatrix}$ (b) $\begin{bmatrix} 1 & -3 & 0 \\ 7 & 4 & 10 \\ 12 & 8 & -6 \\ 11 & 20 & 9 \end{bmatrix}$

9. Maximize $8x_1 + 10x_2 + 5x_3$ subject to $2x_1 + x_2 + x_3 \le 6$; $2x_1 + 3x_2 \le 4$; $x_1 + x_2 + x_3 \le 1$; $x_1 \ge 0$; $x_2 \ge 0$; $x_3 \ge 0$.
10. (a) $y_1 = 5$; $y_2 = 7$; $y_3 = 3$; minimum is 172 (b) $x_1 = 12$; $x_2 = 8$; $x_3 = 5$; maximum is 172

Case 6 (page 288)

1.

From \ To	A	B	C	D
A	0	10	200	1000
B	20	0	30	15
C	300	0	0	50
D	800	0	75	0

2. 135,552 **3.** 756,970 **4.** 2,756,984 **5.** 5,886,972 **6.** 732,652

Case 7 (page 290)

1. 1: .95; 2: .83; 3: .75; 4: 1.00; 5: .87; 6: .94 **2.** $x_1 = 100$; $x_2 = 0$; $x_3 = 0$; $x_4 = 90$; $x_5 = 210$; $x_6 = 0$

Chapter 8

Section 8.1 (page 296)

1. 1/6 **3.** 2/3 **5.** 1/13 **7.** 1/26 **9.** 1/52 **11.** 1 to 5 **13.** 2 to 1 **15.** $S = \{h\}$
17. $S = \{hhh, hht, hth, htt, thh, tht, tth, ttt\}$ **19.** Let c = correct, w = wrong. Then $S = \{ccc, ccw, cwc, wcc, wwc, wcw, cww, www\}$. **21.** (a) $\{hh, tt\}$; 1/2 (b) $\{hh, ht, th\}$; 3/4 **23.** (a) {2 and 4}; 1/10 (b) {1 and 3, 1 and 5, 3 and 5}; 3/10 (c) $\varnothing$; 0 (d) {1 and 2, 1 and 4, 2 and 3, 2 and 5, 3 and 4, 4 and 5}; 3/5 **25.** 1/5 **27.** 4/15 **29.** 8 to 7
31. 3 to 7 **33.** 4/9 **35.** Not subjective **37.** Subjective **39.** Subjective **41.** Not subjective
43. Not subjective

Section 8.2 (page 302)

1. 1/36 **3.** 1/9 **5.** 5/36 **7.** 1/12 **9.** 5/18 **11.** 11/36 **13.** 5/12 **15.** .08 **17.** .50 **19.** .77
21. 7 to 18 **23.** 7/10 **25.** 3/5 **27.** 7/13 **29.** 3/4 **31.** 4/13 **33.** 25/52 **35.** .9 **37.** .3
39. 1/10 **41.** 9/10 **43.** 11/20 **45.** .951 **47.** .473 **49.** .007 **51.** 3/4 **53.** 1/4 **55.** 1/4
57. .23 **59.** .36 **61.** .75 **63.** .11

Section 8.3 (page 312)

1. 0 **3.** 1 **5.** 1/6 **7.** 4/17 **9.** 25/51 **11.** $\frac{13}{52}\cdot\frac{12}{51}\cdot\frac{11}{50}\cdot\frac{10}{49}\cdot\frac{9}{48}\approx .00050$ **13.** $9/48 = .1875$

15. $4\left(\frac{13}{52}\cdot\frac{12}{51}\cdot\frac{11}{50}\cdot\frac{10}{49}\cdot\frac{9}{48}\right)\approx .00198$ **17.** $(2/3)^3 \approx .296$ **19.** 1/10 **21.** 0 **23.** 2/7 **25.** 2/7

27. The probability of a customer cashing a check given that the customer made a deposit is 5/7. **29.** The probability of a customer not cashing a check given that the customer did not make a deposit is 1/4. **31.** The probability of a customer not both cashing a check and making a deposit is 6/11. **33.** 1/6 **35.** 0 **37.** 2/3 **39.** .06 **41.** 1/4 **43.** 1/4 **45.** 1/7 **47.** .049 **49.** .534 **51.** 42/527 or .080 **53.** Yes **55.** .05 **57.** .25 **59.** not very reasonable

Section 8.4 (page 318)

1. 1/3 **3.** 2/41 **5.** 21/41 **7.** 8/17 **9.** $119/131 \approx .908$ **11.** .824 **13.** 1/6 **15.** $1/176 \approx .006$ **17.** 1/11 **19.** 5/9 **21.** 5/26 **23.** $72/73 \approx .986$ **25.** $165/343 \approx .481$

Section 8.5 (page 326)

1. 12 **3.** 56 **5.** 8 **7.** 24 **9.** 792 **11.** 156 **13.** $\binom{52}{2} = 1326$ **15.** 10 **17.** $5^7 = 78{,}125$

19. $9\cdot 10^4\cdot 1^2 = 90{,}000$ **21.** $9\cdot 9\cdot 8\cdot 7\cdot 6\cdot 5\cdot 4 = 544{,}320$ **23.** $\binom{8}{5}\Big/\binom{19}{5} = 56/11{,}628 = 14/2907$

25. $\binom{8}{3}\cdot\binom{11}{2}\Big/\binom{19}{5} = (56\cdot 55)/11{,}628 = 3080/11{,}628 = 770/2907$ **27.** $\binom{12}{5}\Big/\binom{52}{5} = 792/2{,}598{,}960$

29. $\binom{12}{2}\cdot\binom{4}{3}\Big/\binom{52}{5} = 652{,}080/2{,}598{,}960$ **31.** $\binom{5}{2}\binom{4}{1}\Big/\binom{9}{3} + \binom{5}{3}\Big/\binom{9}{3} = 40/84 + 10/84 = 50/84 = 25/42$

33. $\binom{5}{3}\Big/\binom{9}{3} = 5/42$ **35.** $\binom{3}{3}\Big/\binom{9}{3} = 1/84$ **37.** $\binom{5}{2}\cdot\binom{3}{1}\Big/\binom{9}{3} = 5/14$ **39.** 0 **41.** $5\cdot 3\cdot 2 = 30$

43. (a) $2\cdot 25\cdot 24\cdot 23 = 27{,}600$ (b) $2\cdot 26\cdot 26\cdot 26 = 35{,}152$ (c) $2\cdot 24\cdot 23\cdot 1 = 1104$ **45.** $\binom{4}{2} = 6$

47. (a) $\binom{7}{2} = 21$ (b) $1\cdot\binom{6}{1} = 6$ (c) $\binom{2}{1}\cdot\binom{5}{1} + \binom{2}{2}\cdot\binom{5}{0} = 11$ **49.** (a) $\binom{12}{3} = 220$ (b) $\binom{12}{9} = 220$

51. $6! = 720$ **53.** $5! = 120$ **55.** $2\cdot 26\cdot 25\cdot 24\cdot 10\cdot 9\cdot 8 = 22{,}464{,}000$ **57.** (a) $\binom{9}{3} = 84$ (b) $\binom{5}{3} = 10$ (c) $\binom{5}{2}\cdot\binom{4}{1} = 40$ (d) $1\cdot\binom{8}{2} = 28$

Section 8.6 (page 333)

1. $\binom{5}{2}\left(\frac{1}{2}\right)^2\left(\frac{1}{2}\right)^3 = 5/16$ **3.** $\binom{5}{0}\left(\frac{1}{2}\right)^0\left(\frac{1}{2}\right)^5 = 1/32$ **5.** $\binom{5}{4}\left(\frac{1}{2}\right)^4\left(\frac{1}{2}\right)^1 + \binom{5}{5}\left(\frac{1}{2}\right)^5\left(\frac{1}{2}\right)^0 = 3/16$

7. $\binom{5}{0}\left(\frac{1}{2}\right)^0\left(\frac{1}{2}\right)^5 + \binom{5}{1}\left(\frac{1}{2}\right)^1\left(\frac{1}{2}\right)^4 + \binom{5}{2}\left(\frac{1}{2}\right)^2\left(\frac{1}{2}\right)^3 + \binom{5}{3}\left(\frac{1}{2}\right)^3\left(\frac{1}{2}\right)^2 = 13/16$

9. $\binom{12}{12}\left(\frac{1}{6}\right)^{12}\left(\frac{5}{6}\right)^0 \approx .0000000005$ **11.** $\binom{12}{1}\left(\frac{1}{6}\right)^1\left(\frac{5}{6}\right)^{11} \approx .269$

13. $\binom{12}{0}\left(\frac{1}{6}\right)^0\left(\frac{5}{6}\right)^{12} + \binom{12}{1}\left(\frac{1}{6}\right)^1\left(\frac{5}{6}\right)^{11} + \binom{12}{2}\left(\frac{1}{6}\right)^2\left(\frac{5}{6}\right)^{10} + \binom{12}{3}\left(\frac{1}{6}\right)^3\left(\frac{5}{6}\right)^9 \approx .875$

15. $\binom{5}{5}\left(\frac{1}{2}\right)^5\left(\frac{1}{2}\right)^0 = 1/32$ **17.** $\binom{5}{0}\left(\frac{1}{2}\right)^0\left(\frac{1}{2}\right)^5 + \binom{5}{1}\left(\frac{1}{2}\right)^1\left(\frac{1}{2}\right)^4 + \binom{5}{2}\left(\frac{1}{2}\right)^2\left(\frac{1}{2}\right)^3 + \binom{5}{3}\left(\frac{1}{2}\right)^3\left(\frac{1}{2}\right)^2 = 13/16$

19. $\binom{20}{0}(.05)^0(.95)^{20} \approx .358$ **21.** $\binom{6}{2}\left(\frac{1}{5}\right)^2\left(\frac{4}{5}\right)^4 \approx .246$

23. $\binom{6}{4}\left(\frac{1}{5}\right)^4\left(\frac{4}{5}\right)^2 + \binom{6}{5}\left(\frac{1}{5}\right)^5\left(\frac{4}{5}\right)^1 + \binom{6}{6}\left(\frac{1}{5}\right)^6\left(\frac{4}{5}\right)^0 \approx .017$ **25.** $\binom{3}{1}\left(\frac{5}{50}\right)^1\left(\frac{45}{50}\right)^2 \approx .243$

27. $\binom{10}{5}(.20)^5(.80)^5 \approx .026$ **29.** .879 **31.** $\binom{20}{17}(.70)^{17}(.30)^3 \approx .072$ **33.** .035 **35.** $\binom{10}{7}\left(\frac{1}{5}\right)^7\left(\frac{4}{5}\right)^3 \approx .00079$

37. $\approx .999922$ **39.** $\binom{3}{1}(.80)^1(.20)^2 = .096$ **41.** $\binom{3}{0}(.80)^0(.20)^3 + \binom{3}{1}(.80)^1(.20)^2 = .104$ **43.** .185

45. .256 **47.** $1 - (.99999975)^{10,000} \approx .0025$ **49.** $\binom{12}{4}(.2)^4(.8)^8 \approx .133$ **51.** .795

Section 8.7 (page 338)

1. Yes **3.** Yes **5.** No **7.** No **9.** Yes **11.** Yes **13.** No

15.

	Small	Large
Small	.80	.20
Large	.60	.40

17. [.74896 .25104]

19.

	G_0	G_1	G_2
G_0	.85	.10	.05
G_1	0	.80	.20
G_2	0	0	1

21. [36,125 8250 5625]

23. [26,100 11,241 12,659] **25.** [42,500 5000 2500] **27.** [33,281 11,513 5206] **29.** [.257 .597 .146]

31. [.251 .584 .165] **33.**

	Agricultural	Urban	Idle
Agricultural	.80	.15	.05
Urban	0	.90	.10
Idle	.10	.20	.70

35. [.252 .492 .256]

37.

	Single	Multiple
Single	.90	.10
Multiple	.05	.95

39. 51.8% in single family dwellings and 48.2% in multiple family dwellings

41.

	Liberal	Conservative	Independent
Liberal	.80	.15	.05
Conservative	.20	.70	.10
Independent	.20	.20	.60

43. [44% 40.5% 15.5%]

45. [47.84% 36.705% 15.455%]

Chapter 8 Test (page 341)

1. 1/4 **2.** 1/26 **3.** 3/13 **4.** 8/13 **5.** 1/2 **6.** 1/3 **7.** 1 **8.** 1 to 3 **9.** 1 to 25 **10.** 2 to 11

11.

N_1N_2	N_1T_2
T_1N_2	T_1T_2

12. 1/4 **13.** 1/2 **14.** 1/4 **15.** 0 **16.** 1/6 **17.** 1/3 **18.** 1/7 **19.** 4/9

20. $\binom{20}{4}(.01)^4(.99)^{16} \approx .00004$ **21.** $\binom{20}{3}(.01)^3(.99)^{17} \approx .00096$

22. .81791 + .16523 + .01586 + .00096 + .00004 = 1.00000

23. $\binom{20}{12}(.01)^{12}(.99)^8 + \binom{20}{13}(.01)^{13}(.99)^7 + \cdots + \binom{20}{20}(.01)^{20}(.99)^0$ **24.** (a) [.54 .46] (b) [.6464 .3536]

25. 2/3 of the market

Case 8 (page 343)

1. $1/2000 = .0005$ **2.** $1999/2000 = .9995$ **3.** $(1999/2000)^a$ **4.** $1 - (1999/2000)^a$ **5.** $(1999/2000)^{Nc}$
6. $1 - (1999/2000)^{Nc}$ **7.** $1 - .741 = .259$ **8.** $1 - .882 = .118$

Case 9 (page 347)

$32,150 at a cutoff of 55

Chapter 9

Section 9.1 (page 352)

1. 3.6 **3.** 14.64 **5.** −$.64 **7.** 9/7 or 1.3 **9.** (a) 5/3 or 1.67 (b) 4/3 or 1.33 **11.** 1
13. No; the expected value is about −21¢ **15.** −2.7¢ **17.** −20¢ **19.** (a) Yes; the probability of a match is still 1/2 (b) 40¢ (c) −40¢ **21.** 2.5 **23.** $4500 **25.** 118 **27.** 3.51 **29.** (a) $50,000 (b) $65,000 **31.** 61
33. 76 **35.** 51; on the average, a man with a blood pressure of 145 has a life expectancy of 51 years. This is an average which may or may not apply to you.

Section 9.2 (page 357)

1. (a) Buy speculative (b) buy blue-chip (c) buy speculative; $24,300 (d) buy blue-chip **3.** (a) Set up in the stadium (b) set up in the gym (c) set up both; $1010

5. (a)

	New product better	Not better
Market new product	50,000	−25,000
Don't	−40,000	−10,000

(b) $5000 if they market new product, −$22,000 if they don't: market the new product

7. (a)

	Strike	None
Bid $30,000	−5500	4500
Bid $40,000	4500	0

(b) bid $40,000

9. Environment; 14.25

Section 9.3 (page 365)

1. $E_1 = 18.61M$, $E_2 = 2.898M - 42.055$, $E_3 = .56M - 48.6$ **3.** $E_1 = 19M$, $E_2 = 1.7M - 45$, $E_3 = .6M - 48.5$

Section 9.4 (page 369)

1. $6 from B to A **3.** $2 from A to B **5.** $1 from A to B **7.** Yes
9. $\begin{bmatrix} -2 & 8 \\ -1 & -9 \end{bmatrix}$ **11.** $\begin{bmatrix} 4 & -1 \\ 3 & 5 \end{bmatrix}$ **13.** $\begin{bmatrix} 8 & -7 \\ -2 & 4 \end{bmatrix}$
15. (1, I); 3; strictly determined **17.** No saddle point; not strictly determined **19.** (3, I); 3; strictly determined
21. (1, III); 1; strictly determined **23.** No saddle point; not strictly determined **25.** (2, III); 6

Section 9.5 (page 373)

1. 1: 1/5; 2: 4/5; I: 3/5; II: 2/5; 17/5 **3.** 1: 7/9; 2: 2/9; I: 4/9; II: 5/9; −8/9 **5.** 1: 8/15; 2: 7/15; I: 2/3; II: 1/3; 5/3 **7.** 1: 6/11; 2: 5/11; I: 7/11; II: 4/11; −12/11 **9.** Strictly determined; saddle point at (2, II); value of the game is −5/12 **11.** Allied should select strategy 1 with probability 10/27 and strategy 2 with probability 17/27.

The value of the game is 1/18, which represents increased sales of \$55,556. **13.** The doctor should prescribe medicine 1 about 5/6 of the time and medicine 2 about 1/6 of the time. The effectiveness will be about 50%.

15. (a)

		Number of fingers 0	2
Number of fingers	0	0	−2
	2	−2	4

(b) For both players A and B: choose 1 (or I) with probability 3/4 and 2 (or II) with probability 1/4. The value of the game is −1/2. **17.** He should invest in rainy day goods about 5/9 of the time and sunny day goods about 4/9 of the time for a steady profit of \$72.22.

Chapter 9 Test (page 377)

1. −5/6 dollar **2.** \$8500 **3.** 2.06 **4.** Hostile **5.** Friendly **6.** Hostile; \$785 **7.** Friendly—payoff is \$680 **8.** (1, I); −2 **9.** (2, III); 3 **10.** (2, II); 0; fair game **11.** $\begin{bmatrix} -11 & 6 \\ -10 & -12 \end{bmatrix}$ **12.** $\begin{bmatrix} -1 & 9 & 0 \\ 8 & -6 & 7 \end{bmatrix}$
13. 1: 5/6, 2: 1/6; I: 1/2, II: 1/2; value is 1/2 **14.** 1: 8/13, 2: 5/13; I: 8/13, II: 5/13; value is 1/13
15. 1: 1/9, 2: 8/9; I: 5/9, II: 4/9; value is 5/9 **16.** 1: 8/19, 2: 11/19; I: 5/19, II: 14/19; value is −2/19
17. 1: 1/5, 2: 4/5; I: 3/5, II: 0, III: 2/5; value is −12/5 **18.** 1: 1/4, 2: 3/4; I: 0, II: 1/2, III: 0, IV: 1/2; value is −3/2

Case 10 (page 382)

1. $\begin{bmatrix} 2 & 2 \\ 1 & 3 \end{bmatrix}$ **2.** Both (1, I) and (1, II) are saddle points.

Chapter 10

Section 10.1 (page 388)

1. (a)

0–24	4
25–49	3
50–74	6
75–99	3
100–124	5
125–149	9

(b)

Interval	Frequency	Cumulative frequency
0–24	4	4
25–49	3	7
50–74	6	13
75–99	3	16
100–124	5	21
125–149	9	30

(c)–(d)

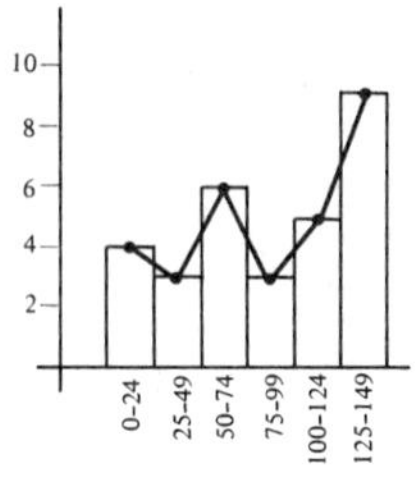

(e)

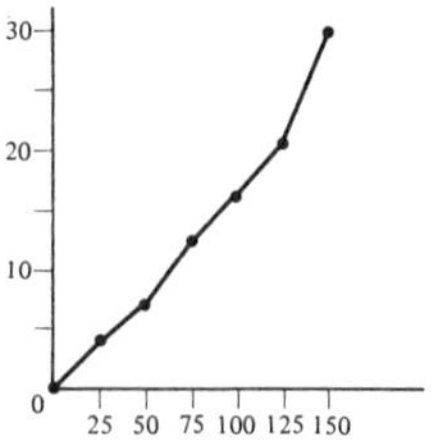

3. (a)–(b)

Interval	Frequency	Cumulative frequency
70–74	2	2
75–79	1	3
80–84	3	6
85–89	2	8
90–94	6	14
95–99	5	19
100–104	6	25
105–109	4	29
110–114	2	31

(c)–(d)

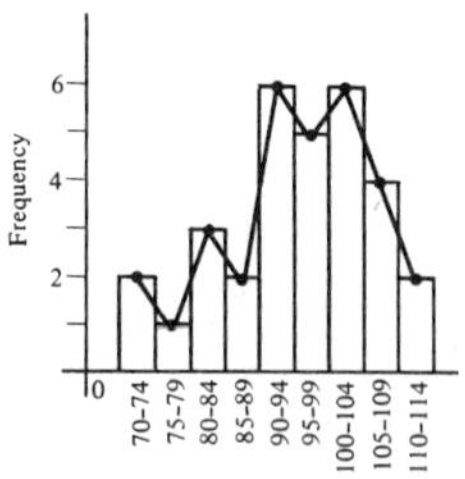

(e)

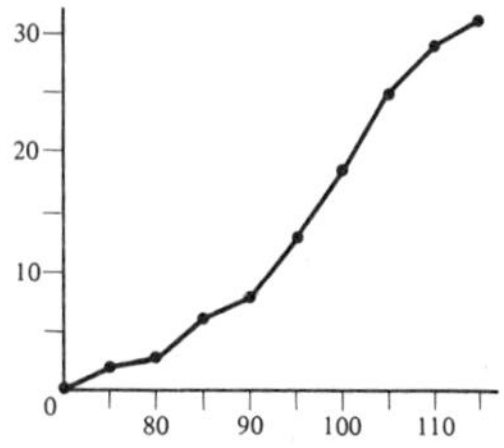

5. E: 18.1%; T: 8.8%; A: 6.2%; O: 4.4%; N: 7.5%; I: 7.1%; R: 5.8%; S: 5.8%; H: 5.8%; D: 2.2%; L: 5.3%; C: 4.9%; U: 4.0%; M: 1.8%; F: 4.0%; P: 1.8%; Y: 1.3%; W: 1.8%; G: 4.0%; B: 0%; V: .4%; K: 0%; X: 0%; J: 0%; Q: 1.3%; Z: 0%

9.

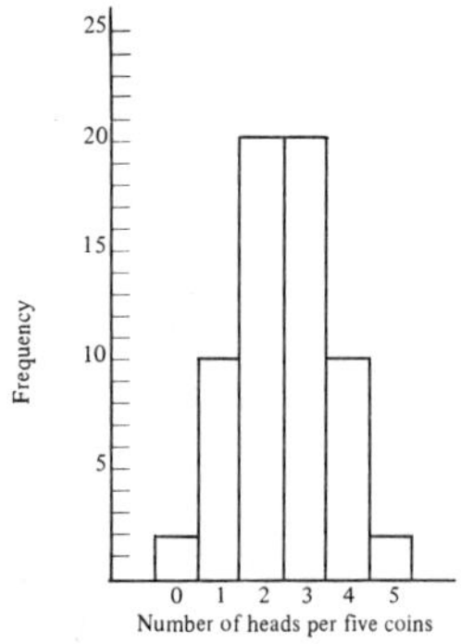

Section 10.2 (page 396)

1. 16 **3.** 150.3 **5.** 27,955 **7.** 7.7 **9.** .1 **11.** 6.7 **13.** 17.2 **15.** 118.8 **17.** 51 **19.** 130 **21.** 49 **23.** 562 **25.** 29.1 **27.** 9 **29.** 64 **31.** 68 and 74 **33.** No mode **35.** 6.1 and 6.3 **37.** Mean = 86.2, modal class 125–149 **39.** Mean = 69.4, modal class 60–69 **41.** $300,000 **43.** About $335,000 **45.** $41,000 **47.** Mean = 1.9, median = 1, mode = 1 **49.** Mean = 17.2, median = 17.5, no mode

Section 10.3 (page 404)

1. 6; $\sqrt{4} = 2$ **3.** 12; $\sqrt{18.3} = 4.3$ **5.** 53; $\sqrt{407.4} = 20.2$ **7.** 46; $\sqrt{215.3} = 14.7$ **9.** 24; $\sqrt{59.4} = 7.7$ **11.** 30; $\sqrt{111.8} = 10.6$ **13.** About 44 **15.** 14.5 **17.** 3/4 **19.** 24/25 **21.** At least 88.9% **23.** At least 96% **25.** No more than 11.1% **27.** Forever Power, $\bar{x} = 26.2$, $s = 4.1$; Brand X, $\bar{x} = 25.5$, $s = 6.9$ (a) Forever Power (b) Forever Power **29.** 12.5 **31.** 4.3 **33.** 15.5

Section 10.4 (page 412)

1. 49.38% **3.** 17.36% **5.** 45.64% **7.** 49.91% **9.** 7.7% **11.** 47.35% **13.** 92.42% **15.** 32.56% **17.** 1.64 or 1.65 **19.** −1.04 **21.** 5000 **23.** 4332 **25.** 642 **27.** 9918 **29.** 19 **31.** 15.87% **33.** .62% **35.** 84.13% **37.** 37.79% **39.** 2.27% **41.** 99.38% **43.** 189 **45.** .0062 **47.** .4325 **49.** $38.62 and $25.89 **51.** .0823 **53.** 82 **55.** 70

Section 10.5 (page 419)

1. (a)

x	0	1	2	3	4	5	6
$P(x)$	.335	.402	.201	.054	.008	.001	.000

(b) 1.00 (c) .91

3. (a)

x	0	1	2	3
$P(x)$	.941	.058	.001	.000

(b) .06 (c) .24

5. (a)

x	0	1	2	3	4
$P(x)$	.0081	.0756	.2646	.4116	.2401

(b) 2.8 (c) .92

7. 12.5; 3.49 **9.** 51.2; 3.2 **11.** .1974 **13.** .1210 **15.** .0240 **17.** .9032 **19.** .8665 **21.** .0956 **23.** .0760 **25.** .6443 **27.** .0146 **29.** .1974 **31.** .0092 **33.** .0001

Section 10.6 (page 426)

1. (a) $y' = .3x + 1.5$ (b) $r = .20$ (c) 2.4 **3.** (a) $y' = 3.35x - 78.2$ (b) 123 (c) 156 (d) $r = .66$
5. (a) 26,920; 23,340; 19,770 (b) 29,370; 25,790; 22,210 (c) \$42 **7.** $y' = 8.06x + 49.52$; $r = .996$
9. (a) $y' = 1.02x - 135$ (b) $r = .74$ (c) \$375,000
11. (a)–(b) $y' = .16x - .89$

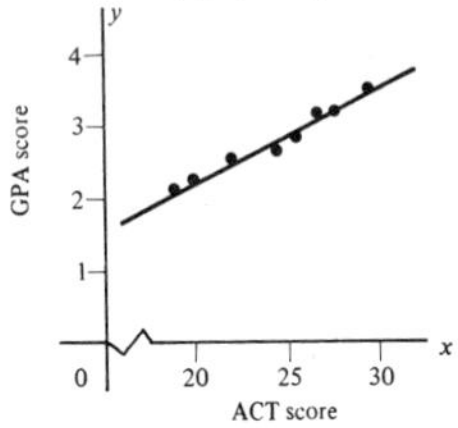

(c) 3.6 (d) $r = .95$

Chapter 10 Test (page 431)

1.

Interval	Frequency
450–474	5
475–499	6
500–524	5
525–549	2
550–574	2

2.

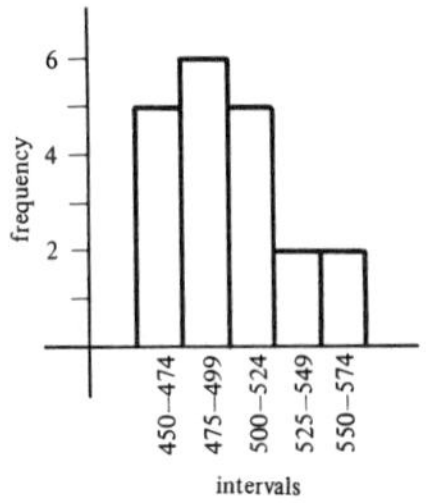

3. 73 **4.** 34.9 **5.** Median = 44; mode = 46 **6.** Median = 38; mode is 36 or 38 **7.** 30–39
8. 18; $\sqrt{38.8} = 6.2$ **9.** 67; $\sqrt{512.2} = 22.6$ **10.** 12.5 **11.** 2.27% **12.** 15.87% **13.** 68.26% **14.** .3980
15. .9521 **16.** .3409 **17.** 25.14% **18.** 28.10% **19.** 22.92% **20.** 56.25% **21.** .1977 **22.** .0718
23. $y = .97x + 31.9$ **24.** About 216 **25.** $r = .93$

Case 11 (page 435)

1. $r = -.677$; $y = -.557x + 6.45$ **2.** $r = -.714$; $y = -.115x + 6.3$ **3.** $r = -.699$; $y = -.012x + 6.55$
4. 4 **5.** 4 **6.** 5

Chapter 11

Section 11.1 (page 439)

1.

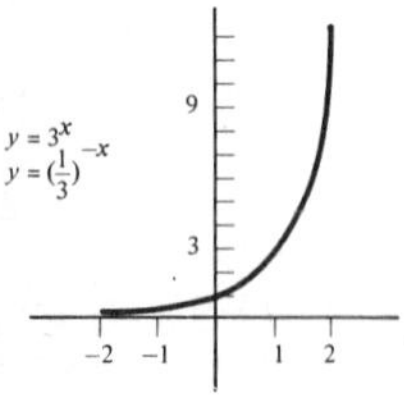

3.

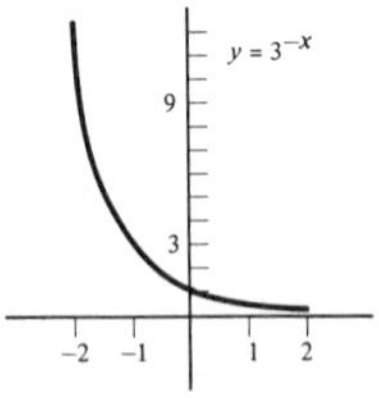

5.

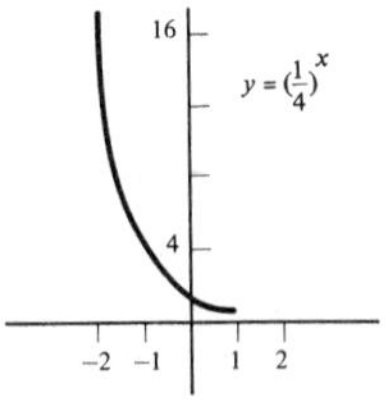

7. Refer to answer 1.

9.

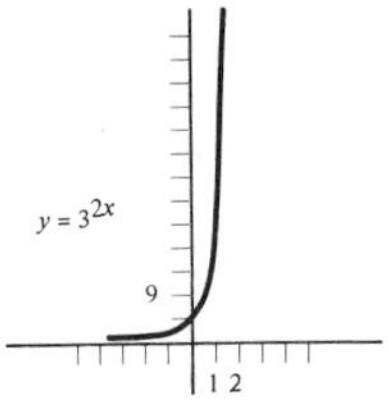

11.

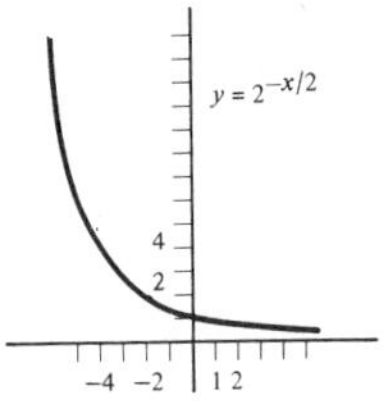

13.

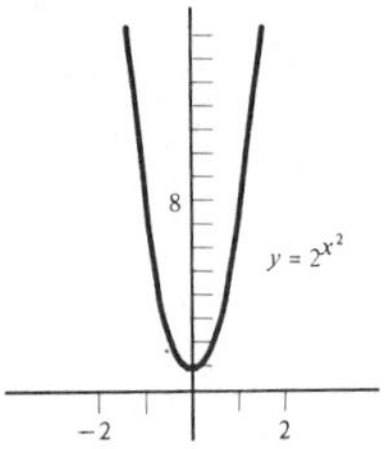

15.

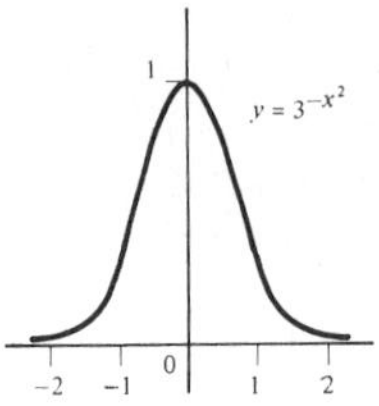

17.

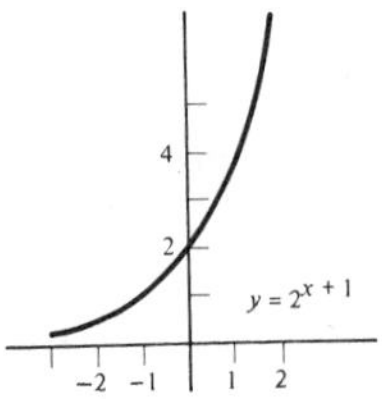

19.

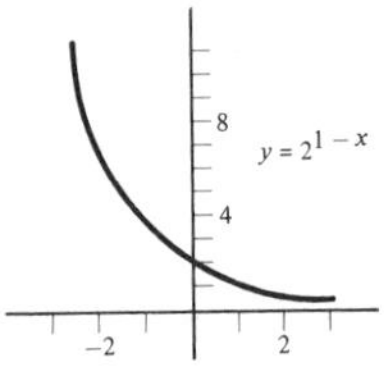

21. $x = 2$ **23.** $x = -3$ **25.** $x = 3/2$ **27.** $x = 3/2$ **29.** $x = -2$ **31.** $x = 3$ **33.** $x = -5/4$
35. $x = 1/9$ **37.** 1.06; 1.12; 1.19; 1.26; 1.42; 1.50; 1.59; 1.69 **39.** .92; .85; .78; .72; .61; .56; .51; .47
41. About $116,000 **43.** 12; 24; 48; 96; 384; 768; 1536; 3072 **45.** About 1,120,000 **47.** About 1,410,000
49. 2,000,000 **51.** 8,000,000

Section 11.2 (page 446)

1. 1,000,000 **3.** About 1,080,000 **5.** 100 **7.** 182 **9.** 500 g **11.** 335 g **13.** 10 g **15.** About 18.2 g
17. 25,000 **19.** About 37,300 **21.** 50,000 **23.** About 40,900 **25.** 12 **27.** About 27 **29.** $21,665.60
31. $44,510.80 **33.** 1000 **35.** About 4460 **37.** Almost 5000 **39.** 30 **41.** Almost 60 **43.** 0
45. About 432 **47.** Almost 500 **49.** (a) About 45,200 (b) about 37,000 **51.** .13 **53.** 40 **55.** 11 **57.** 3

Section 11.3 (page 453)

1. $\log_2 8 = 3$ **3.** $\log_3 81 = 4$ **5.** $\log_{1/3} 9 = -2$ **7.** $2^3 = 8$ **9.** $10^2 = 100$ **11.** $10^4 = 10{,}000$ **13.** 4
15. 3 **17.** −2 **19.** 1/9; 1/3; 1; 3; 27

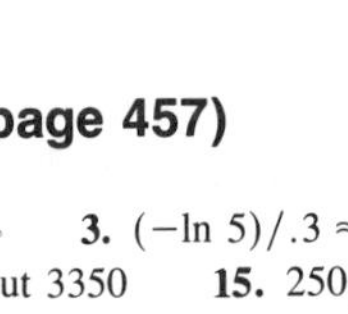

21. 1.8 **23.** .4 **25.** 4.4 **27.** .07 **29.** −.22
31. 2.9 **33.** 1 **35.** 1.7 **37.** 2.9957 **39.** 4.0943 **41.** 6.6847 **43.** 6.2767 **45.** 6.6439
47. 10.9770

Section 11.4 (page 457)

1. $\ln 6/(2 \ln 5) \approx .56$ **3.** $(-\ln 5)/.3 \approx -5.4$ **5.** $(\ln 10)/.02 \approx 115$ **7.** $(\ln 4) - 2 \approx -.61$ **9.** $(5 \ln 8)/2 \approx 5.2$
11. 5000 **13.** About 3350 **15.** 250,000 **17.** About 112,000 **19.** $-(\ln .1)/.04 \approx 57.6$ hours

21. (5600 ln 2.5)/ln 2 ≈ 7400 years **23.** (5600 ln 100)/ln 2 ≈ 37,200 years **25.** About 37% **27.** About 46 days
29. About 3500 years **31.** About 1800 years **33.** (ln 2)/.05 ≈ 13.9 years

Chapter 11 Test (page 459)

1. 4 **2.** −1 **3.** 1/2 **4.** (ln 7)/(5 ln 2) ≈ .56

5.

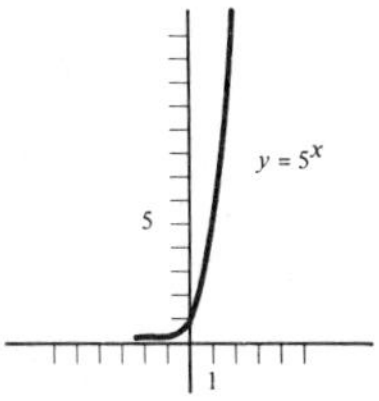

6. and 7.

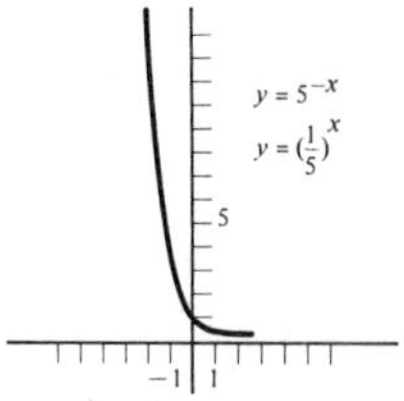

8.

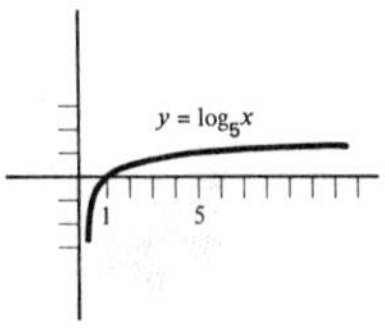

9. $\log_2 64 = 6$ **10.** $\log_3 \sqrt{3} = 1/2$ **11.** $\log_{1000} .001 = -1$ **12.** $2^5 = 32$ **13.** $10^2 = 100$ **14.** $27^{1/3} = 3$
15. 2 **16.** 1/2 **17.** 1.79 **18.** −.18 **19.** .645 **20.** 1.8245 **21.** 6.5511 **22.** 6.1799
23. (a) 100,000 (b) about 149,000 (c) about 272,000 **24.** (a) 5000 (b) about 3350 (c) −(ln .5)/.2 ≈ 3.5 days
25. (a) 0 (b) about 55 (c) about 98 (d) about 100

Case 12 (page 460)

1. .0315 **2.** λy_0 is about 118,000; painting is a forgery **3.** 163,000; forgery **4.** 24,000; cannot be modern forgery
5. 142,000; forgery **6.** 134,000; forgery

Chapter 12

Section 12.1 (page 463)

1. 3 **3.** Does not exist **5.** 2 **7.** 0 **9.** −11 **11.** 9 **13.** No limit **15.** 10 **17.** No limit **19.** 12
21. 29 **23.** 41 **25.** 9/7 **27.** −1 **29.** 6 **31.** −5 **33.** −1 **35.** $\sqrt{5}$ **37.** 1/10 **39.** 500
41. $1500 **43.** 3/5 **45.** 1/2 **47.** 1/2 **49.** 0 **51.** No limit **53.** .57 **55.** .52

Section 12.2 (page 476)

1. −1 **3.** −1 **5.** 0; 2 **7.** 1 **9.** Yes; no **11.** No; no; yes **13.** Yes; no; no **15.** Yes; no; yes
17. Yes; no; yes **19.** Yes; yes; yes **21.** $96

23.

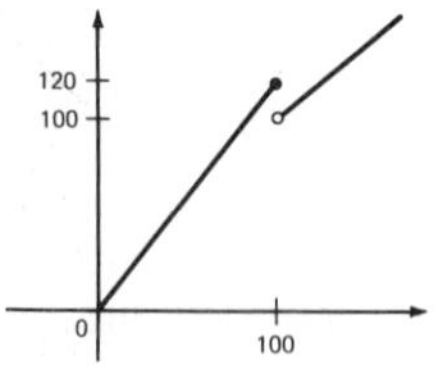

25. $260 **27.** $525 **29.**

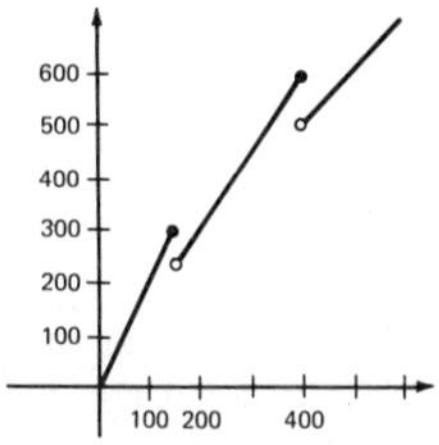

31. *m* **33.** [−9, 15] **35.** (10, ∞) **37.** [−8, 0] **39.** (12, 20) **41.** (−∞, 3) **43.** (−10, ∞)
45. (−∞, −6); (−6, 0); (0, 4); (4, ∞) **47.** (−∞, 0); (0, ∞)

Section 12.3 (page 483)

1. 1 **3.** $6/10$ or $3/5$ **5.** 2 million **7.** $-.8$ million **9.** -2.2 million **11.** 3 **13.** 1 **15.** -2 **17.** -3 **19.** About \$5 million **21.** About \$35 million **23.** About $-\$150/7 \approx -\$21$ million **25.** 5 **27.** 8 **29.** $1/3$ **31.** 6 **33.** $-1/3$ **35.** 7 **37.** 3.02; 3.002; 3.0002 **39.** 3 **41.** 15 **43.** 13.01; 13.001; 13.0001 **45.** 17

Section 12.4 (page 492)

1. $f'(x) = 4x$; 8; 0; -12 **3.** $f'(x) = -8x + 11$; -5; 11; 35 **5.** $f'(x) = -9$; -9; -9; -9 **7.** $f'(x) = 8$; 8; 8; 8 **9.** $f'(x) = 3x^2 + 3$; 15; 3; 30 **11.** $f'(x) = 2/x^2$; $1/2$; does not exist; $2/9$ **13.** $f'(x) = -4/(x-1)^2$; -4; -4; $-1/4$ **15.** $f'(x) = 1/(2\sqrt{x})$; $1/(2\sqrt{2})$; does not exist; does not exist **17.** -4 **19.** -6 **21.** Has a derivative everywhere **23.** 0 **25.** -2 **27.** -1; 2 **29.** -4; 0 **31.** -8 **33.** 0; no **35.** -16; no **37.** 30 **39.** 10 **41.** -10

Section 12.5 (page 501)

1. $y' = 10x$ **3.** $y' = -30x^4$ **5.** $y' = 30x$ **7.** $y' = 18x - 8$ **9.** $y' = 6x - 4$ **11.** $y' = -12x^2 + 4x$ **13.** $y' = 30x^2 - 18x + 6$ **15.** $y' = 4x^3 - 15x^2 + 18x$ **17.** $y' = 9x^{.5}$ **19.** $y' = -48x^{2.2}$ **21.** $y' = -27x^{1/2}$ **23.** $y' = 16x^{-1/2}$ or $16/x^{1/2}$ **25.** $y' = 75x^{1/2} - 3x^{-1/2}$ or $75x^{1/2} - 3/x^{1/2}$ **27.** $y' = 4x^{-1/2} + 6$ or $4/x^{1/2} + 6$ **29.** $y' = -30x^{-6}$ or $-30/x^6$ **31.** $y' = 8x^{-3} - 3x^{-2}$ or $8/x^3 - 3/x^2$ **33.** $y' = -20x^{-3} - 12x^{-5} - 6$ or $-20/x^3 - 12/x^5 - 6$ **35.** $y' = -6x^{-2} + 16x^{-3}$ or $-6/x^2 + 16/x^3$ **37.** $y' = -6x^{-3/2}$ or $-6/x^{3/2}$ **39.** $y' = 5x^{-3/2} - 12x^{-5/2}$ or $5/x^{3/2} - 12/x^{5/2}$ **41.** $x = 5/3$ **43.** $x = 2/3$; $x = 1$ **45.** $x = 5/2$; $x = -1$ **47.** $x = \sqrt[3]{4}$ **49.** 10; increasing **51.** 1 **53.** 1% **55.** 450 **57.** The blood sugar level is decreasing at a rate of 4 points per unit of insulin. **59.** $dV/dr = 160\pi r$ **61.** 960π cubic mm **63.** \$15 **65.** \$207 **67.** \$8 **69.** \$428 **71.** 4 **73.** 2 2/3 or $8/3$ **75.** (a) $v(t) = 6$ (b) 6; 6; 6 **77.** (a) $v(t) = 22t + 4$ (b) 4; 114; 224 **79.** (a) $v(t) = 12t^2 + 16t$ (b) 0; 380; 1360

Section 12.6 (page 508)

1. $y' = 12x - 15$ **3.** $y' = 4x + 3$ **5.** $y' = 48x + 66$ **7.** $y' = 18x^2 - 6x + 4$ **9.** $y' = 9x^2 - 4x - 5$ **11.** $y' = 36x^3 + 21x^2 - 18x - 7$ **13.** $y' = 4(2x - 5)$ or $8x - 20$ **15.** $y' = 4x(x^2 - 1)$ or $4x^3 - 4x$ **17.** $y' = 3x^{1/2}/2 + x^{-1/2}/2 + 2$ or $3\sqrt{x}/2 + 1/(2\sqrt{x}) + 2$ **19.** $y' = 10 + 3x^{-1/2}/2$ or $10 + 3/(2\sqrt{x})$ **21.** $y' = -3/(2x - 1)^2$ **23.** $y' = 53/(3x + 8)^2$ **25.** $y' = -6/(3x - 5)^2$ **27.** $y' = -17/(4 + x)^2$ **29.** $y' = (x^2 - 2x - 1)/(x - 1)^2$ **31.** $y' = (-x^2 + 4x + 1)/(x^2 + 1)^2$ **33.** $y' = (-2x^2 - 6x - 1)/(2x^2 - 1)^2$ **35.** $y' = (x^2 + 6x - 14)/(x + 3)^2$ **37.** $y' = [-\sqrt{x}/2 - 1/(2\sqrt{x})]/(x - 1)^2$ or $(-x - 1)/[2\sqrt{x}(x - 1)^2]$ **39.** $y' = (5\sqrt{x}/2 - 3/\sqrt{x})/x$ or $(5x - 6)/(2x\sqrt{x})$ **41.** 86.8 **43.** $A(x) = 9x - 4 + 8/x$ **45.** \$72.75 **47.** $A(x) = 10x - 5 - 18/x$ **49.** $G'(20) = -1/200$; go faster **51.** $N'(t) = 6t^2 - 80t + 200$ **53.** 46 **55.** $s'(x) = m/(m + nx)^2$ **57.** $5000/3600 \approx 1.39$

Section 12.7 (page 514)

1. $y' = 4(2x + 9)$ **3.** $y' = 90(5x - 1)^2$ **5.** $y' = -144x(12x^2 + 4)^2$ **7.** $y' = 36(2x + 5)(x^2 + 5x)^3$ **9.** $y' = 36(2x + 5)^{1/2}$ **11.** $y' = -21(8x + 9)(4x^2 + 9x)^{1/2}/2$ **13.** $y' = 16(4x + 7)^{-1/2}$ or $16/\sqrt{4x + 7}$ **15.** $y' = -(2x + 4)(x^2 + 4x)^{-1/2}$ or $-(2x + 4)/\sqrt{x^2 + 4x}$ **17.** $y' = 16x(2x + 3) + 4(2x + 3)^2$ or $12(2x + 3)(2x + 1)$ **19.** $y' = 2(x + 2)(x - 1) + (x - 1)^2 = 3(x - 1)(x + 1)$ **21.** $y' = 10(x + 3)^2(x - 1) + 10(x - 1)^2(x + 3)$ or $10(x + 3)(x - 1)(2x + 2)$ or $20(x + 3)(x - 1)(x + 1)$ **23.** $y' = (x + 1)^2 \cdot x^{-1/2}/2 + 2x^{1/2}(x + 1)$ or $x^{-1/2}(x + 1)(5x + 1)/2$ **25.** $y' = -2(x - 4)^{-3}$ or $-2/(x - 4)^3$ **27.** $y' = (x^2 - 2x - 15)/(x - 1)^2$ **29.** $y' = 8/(x + 2)^3$ **31.** -3.2 **33.** -2.4 **35.** $A(x) = 1000/x - 4 + x/250$ **37.** 6 **39.** $138/7 \approx 19.71$ **41.** $-1/2 = -.5$ **43.** $-.011$ **45.** $R'(Q) = -Q/[6\sqrt{C - Q/3}] + \sqrt{C - Q/3}$

Chapter 12 Test (page 520)

1. $17/3$ **2.** 8 **3.** -13 **4.** $1/6$ **5.** $1/2$ **6.** 0 **7.** None **8.** -4; 2 **9.** $[-4, 2]$ **10.** $(4, \infty)$ **11.** 30 **12.** -60 **13.** $9/77$ **14.** $y' = 4$ **15.** $y' = 10x + 6$ **16.** 315 **17.** 1015 **18.** 1515

19. $55/3 = 18\ 1/3$ **20.** $65/4 = 16\ 1/4$ **21.** 15 **22.** (a) $v(t) = 60$ (b) 60; 60; 60 **23.** (a) $v(t) = 24t^2 - 8t$ (b) 0; 1872; 9440 **24.** $y' = 3x^2 - 8x$ **25.** $y' = 21x^{3/4}/2$ **26.** $y' = 36x + 27$
27. $y' = 2(x+2)/\sqrt{x} + 4\sqrt{x}$ or $6x^{1/2} + 4x^{-1/2}$ **28.** $y' = 16/(2x+1)^2$ **29.** $y' = (x^2 - 2x)/(x-1)^2$
30. $y' = 60x(x^2+2)^4$ **31.** $y' = 3(6x-11)^{-1/2}$ or $3/\sqrt{6x-11}$
32. $y' = 45x(x+1)^4 + 9(x+1)^5$ or $9(6x+1)(x+1)^4$ **33.** $y' = (-2x+10)/(x+5)^3$

Chapter 13

Section 13.1 (page 522)

1. Minimum of -4 at 1 **3.** Maximum of 3 at -2 **5.** Maximum of 3 at -4; minimum of 1 at -2
7. Maximum of 3 at -4; minimum of -2 at -7 and -2 **9.** Maximum of 2 at -4 and 5 at 2; minimum of -3 at -2
11. Minimum of -44 at -6 **13.** Minimum of -46 at 7 **15.** Maximum of 22 at -4 **17.** Maximum of 17 at -3
19. Maximum of -8 at -3; minimum of -12 at -1 **21.** None **23.** Maximum of 57 at 2; minimum of 30 at 5
25. Minimum of $-377/6$ at -5; Maximum of $827/96$ at $-1/4$ **27.** Maximum of -4 at 0; minimum of -85 at 3 and -3
29. $(-6, -42)$ **31.** $(2, -10)$ **33.** $(-1, 2)$ **35.** $(4/3, -4/3)$ **37.** Increasing on $(-6, \infty)$; decreasing on $(-\infty, -6)$
39. Increasing on $(-\infty, 4/3)$; decreasing on $(4/3, \infty)$ **41.** Increasing on $(-\infty, -3)$ and $(4, \infty)$; decreasing on $(-3, 4)$
43. Increasing on $(-\infty, -3/2)$ and $(4, \infty)$; decreasing on $(-3/2, 4)$ **45.** Always decreasing **47.** 30; 1080 **49.** 12°
51. $(0, 7/3)$; $(5, \infty)$ **53.** (a) 6 (b) \$1604 **55.** (a) 10 (b) 500

Section 13.2 (page 534)

1. Absolute maximum at x_3; no absolute minimum **3.** Neither **5.** Absolute minimum at x_1; no absolute maximum
7. Absolute maximum at x_1; absolute minimum at x_2 **9.** Absolute maximum at x_2; absolute minimum at x_3
11. Absolute maximum at 0; absolute minimum at -3 **13.** Absolute maximum at -3; absolute minimum at $-3/4$
15. Absolute maximum at -1; absolute minimum at -5 **17.** Absolute maximum at -2; absolute minimum at 4
19. Absolute maximum at -2; absolute minimum at 3 **21.** Absolute maximum at 4; absolute minimum at -1
23. Absolute maximum at 6; absolute minimum at -4 and 4 **25.** About 34.4; 60 **27.** Use all the wire to make a circle
29. Absolute maximum is about 3.4, at about $x = 1.3$; absolute minimum is -7 at $x = 0$ **31.** Absolute maximum on $[-5, -4]$ is about 60.4 at -4.5 or -4.4; absolute minimum is 58, at -5; absolute maximum on $[0, 1]$ is 4 at 1; absolute minimum is about 1.6 at about .4 or .5

Section 13.3 (page 540)

1. $f''(x) = 6x + 8$; 8; 20; -10 **3.** $f''(x) = -12x^2 + 12x - 2$; -2; -26; -146 **5.** $f''(x) = 16$; 16; 16; 16
7. $f''(x) = 6(x-2)$; -12; 0; -30 **9.** $f''(x) = 4/(x-1)^3$; -4; 4; $-1/16$ **11.** $f'''(x) = 48x - 18$; $f^{(4)}(x) = 48$
13. $f'''(x) = 180x^2 - 24x + 12$; $f^{(4)}(x) = 360x - 24$ **15.** $f'''(x) = -6x^{-4} = -6/x^4$; $f^{(4)}(x) = 24x^{-5} = 24/x^5$
17. $f'''(x) = 24(2x+1)^{-4} = 24/(2x+1)^4$; $f^{(4)}(x) = -192(2x+1)^{-5} = -192/(2x+1)^5$
19. Relative minimum at 6 **21.** Relative minimum at 1 **23.** Relative maximum at 0; relative minimum at $4/3$
25. Relative maximum at 2; relative minimum at -5 **27.** Relative maximum at -1; relative minimum at 6
29. Critical value at 0, but neither a maximum nor a minimum there **31.** Relative maximum at 0; relative minimum at -2 and 2
33. Relative maximum at -2; relative minimum at 2 **35.** Relative maximum at -2; relative minimum at 2
37. No critical values; no maximums or minimums **39.** Always concave up; no points of inflection
41. Always concave downward; no points of inflection **43.** Concave downward on $(-\infty, 1/2)$; concave upward on $(1/2, \infty)$; point of inflection when $x = 1/2$ **45.** Concave downward on $(-4, \infty)$; concave upward on $(-\infty, -4)$; point of inflection when $x = -1/4$ **47.** Concave downward on $(-\infty, -1)$; concave upward on $(-1, \infty)$; no point of inflection
49. To the left of trash compactors **51.** Color tv's, room air conditioners; rate of growth of sales will now decline
53. Maximum at 8; minimum at $4/3$ **55.** 50 **57.** (a) After 3 hours (b) 2/9% **59.** $v(t) = -6t - 6$; $a(t) = -6$; -6; -30; -6; -6 **61.** $v(t) = 9t^2 - 8t + 8$; $a(t) = 18t - 8$; 8; 120; -8; 64 **63.** $v(t) = -(t+3)^{-2}$; $a(t) = 2(t+3)^{-3}$; $-1/9$; $-1/49$; $2/27$; $2/343$ **65.** -96 feet per second **67.** -256 feet per second **69.** 260 **71.** 32

Section 13.4 (page 551)

1. (a) $y = 100 - x$ (b) $P = x(100 - x)$ (c) $P' = 100 - 2x$; $x = 50$ (d) 50 and 50 (e) $50 \cdot 50 = 2500$ **3.** 100; 100 **5.** 100; 50 **7.** 5; -5 **9.** (a) $R(x) = 100{,}000x - 100x^2$ (b) 500 (c) 25,000,000 cents **11.** (a) $\sqrt{3200} \approx 56.6$ mph (b) \$22.63 **13.** (a) $1200 - 2x$ (b) $A(x) = 1200x - 2x^2$ (c) 300 m (d) 180,000 sq m **15.** 405,000 sq m **17.** 200 feet on the \$3 sides; 100 feet on the \$6 sides **19.** (a) $40 - 2x$ (b) $100 + 5x$ (c) $R(x) = 4000 - 10x^2$ (d) pick now (e) \$40 per tree **21.** (a) 90 (b) \$405 **23.** 4 by 4 by 2 **25.** 3 by 6 by 2 **27.** 50.0 mph gives the minimum cost of \$400. **29.** 8 miles from A

Section 13.5 (page 561)

1. 10 **3.** 60 **5.** 5

Section 13.6 (page 565)

1. $-4x/(3y)$ **3.** $-y/(y + x)$ **5.** $2/y$ **7.** $-3y^2/(6xy - 4)$ **9.** $(-y - x)/x$ **11.** $(-6x - 4y)/(4x + y)$ **13.** $3x^2/(2y)$ **15.** $-2y/x$ **17.** $-y/x$ **19.** $-2xy/(x^2 + 3y^2)$ **21.** $-y^{1/2}/x^{1/2}$ **23.** $-y^{1/2}x^{-1/2}/(x^{1/2}y^{-1/2} + 2) = -y/(x + 2x^{1/2}y^{1/2})$ **25.** $4y = 3x + 25$ **27.** $y = x + 2$ **29.** $y = -24x + 57$ **31.** $4y = -x + 12$ **33.** $-7/6$ feet per minute **35.** 6π square feet per minute **37.** 50π cubic inches per minute

Section 13.7 (page 571)

1. $dy = 12x\,dx$ **3.** $dy = x^{-1/2}dx$ **5.** $dy = (14x - 9)dx$ **7.** $dy = [-22/(x - 3)^2]dx$ **9.** $\Delta y = .6$; $dy = .6$ **11.** $\Delta y = -1.52$; $dy = -1.6$ **13.** $\Delta y = .109$; $dy = .1$ **15.** $\Delta y = .125$ (rounded); $dy = .130$ (rounded) **17.** $\Delta y = -.024$ (rounded); $dy = -.023$ (rounded) **19.** 3 1/6 **21.** 3 7/8 **23.** 11 1/11 **25.** 1 11/12 **27.** 4 1/48 **29.** 2 1/32 **31.** 2.005 **33.** 1 119/120 $\approx$ 1.992 **35.** 12.8π cubic centimeters **37.** $.48\pi$ square miles

Section 13.8 (page 575)

1. $y' = 4e^{4x}$ **3.** $y' = 12e^{-2x}$ **5.** $y' = -16e^{2x}$ **7.** $y' = -16e^{x+1}$ **9.** $y' = 2xe^{x^2}$ **11.** $y' = 12xe^{2x^2}$ **13.** $y' = 16xe^{2x^2-4}$ **15.** $y' = xe^x + e^x = e^x(x + 1)$ **17.** $y' = 2(x - 3)(x - 2)e^{2x}$ **19.** $y' = -1/(3 - x)$ or $1/(x - 3)$ **21.** $y' = (4x - 7)/(2x^2 - 7x)$ **23.** $y' = 1/[2(x + 5)]$ **25.** $y' = (2xe^x - x^2e^x)/e^{2x} = x(2 - x)/e^x$ **27.** $y' = x + 2x \ln |x| = x(1 + 2 \ln |x|)$ **29.** $y' = (4x + 7 - 4x \ln |x|)/[x(4x + 7)^2]$ **31.** $y' = [x(\ln |x|)e^x - e^x]/[x(\ln |x|)^2]$ or $e^x[x \ln |x| - 1]/[x(\ln |x|)^2]$ **33.** $y' = [x(e^x - e^{-x}) - (e^x + e^{-x})]/x^2$ **35.** $y' = -20{,}000e^{.4x}/(1 + 10e^{.4x})^2$ **37.** $y' = 8000e^{-.2x}/(9 + 4e^{-.2x})^2$ **39.** $y' = 1/(x \ln |x|)$ **41.** $200e^{.4} \approx 298$; $200e^{1.6} \approx 991$ **43.** (a) $e^{-.02} \approx .98$ (b) $e^{-.2} \approx .82$ (c) $e^{-2} \approx .14$ (d) $-.02e^{-2} \approx -.0027$; the rate of change in the proportion wearable when $x = 100$ **45.** (a) .005 (b) .0007 (c) .000014 (d) $-.022$ (e) $-.0029$ (f) $-.000054$ **47.** A maximum of $1/e$ at $x = -1$ **49.** A minimum of 0 at $x = 0$; a maximum of $4/e^2$ at $x = 2$ **51.** A minimum of 2 at $x = 0$ **53.** $C'(t) = [k/(b - a)](-ae^{-at} + be^{-bt})$ **55.** $b/a = e^{-at}/e^{-bt}$ **57.** $(b - a)t = \ln b/a$ **59.** .04

Chapter 13 Test (page 582)

1. Relative maximum at 2; maximum is -4 **2.** Relative maximum of $-151/54$ at $1/3$; relative minimum of $-27/8$ at $-1/2$ **3.** None **4.** Increasing on $(5/2, \infty)$; decreasing on $(-\infty, 5/2)$ **5.** Increasing on $(-4, 2/3)$; decreasing on $(-\infty, -4)$ and $(2/3, \infty)$ **6.** Absolute maximum of 29/4 at 5/2; absolute minimum of 5 at 1 and 4 **7.** Absolute maximum of 39 at -3; absolute minimum of $-319/27$ at 5/3 **8.** $v(t) = 18t - 7$; $a(t) = 18$; -7; 47; 18; 18 **9.** $v(t) = -6t^2 + 8t - 6$; $a(t) = -12t + 8$; -6; -36; 8; -28 **10.** $f''(x) = 36x - 18$; -18; -198 **11.** $f''(x) = -68(2x + 3)^{-3}$ or $-68/(2x + 3)^3$; $-68/27$; $68/343$ **12.** $f'''(x) = 54$; $f^{(4)}(x) = 0$ **13.** $f'''(x) = 12(x - 4)^{-4} = 12/(x - 4)^4$; $f^{(4)}(x) = -48(x - 4)^{-5} = -48/(x - 4)^5$ **14.** 6 hundred boxes **15.** \$178 **16.** 225 by 450 **17.** 40

18. $(-4y - 2xy^3)/(3x^2y^2 + 4x)$ **19.** $(y - 3y^2)/(4y^2 + x)$ **20.** $16y = -23x + 94$ **21.** $dy = (24x^2 - 4x)dx$ **22.** $dy = 24x(x^2 - 1)^2dx$ **23.** 12 23/26 **24.** $y' = 12xe^{6x^2}$ **25.** $y' = 3/(3x - 2)$ **26.** $y' = -1/[2(3 - x)]$ or $1/[2(x - 3)]$ **27.** $y' = e^{2x}(2x - 1)/(5x^2)$ **28.** $y' = e^x(2/x + \ln x^2)$ or $e^x(2 + x \ln x^2)/x$ **29.** $y' = (6x \ln |x| - 6x - 4)/[x(\ln |x|)^2]$ **30.** a relative maximum at $-1/10$

Case 13 (page 584)

1. $-C_1/m^2 + DC_3/2$ **2.** $m = \sqrt{2C_1/(DC_3)}$ **3.** About 3.33 **4.** $m^+ = 4$ and $m^- = 3$ **5.** $Z(m^+) = Z(4) = \$11,400$; $Z(m^-) = Z(3) = \$11,300$ **6.** 3 months; 9 trainees per batch

Chapter 14

Section 14.1 (page 586)

1. $5x^3/3 + C$ **3.** $6x + C$ **5.** $x^2 + 3x + C$ **7.** $x^3/3 - 2x^2 + 5x + C$ **9.** $x^4 + x^3 + x^2 - 6x + C$ **11.** $2x^{3/2}/3 + C$ **13.** $2x^{3/2}/3 + 2x^{5/2}/5 + C$ **15.** $4x^{5/2} - 4x^{7/2} + C$ **17.** $6x^{3/2} + 4x + C$ **19.** $-x^{-1} + C$ or $-1/x + C$ **21.** $-x^{-2}/2 - 2x^{1/2} + C$ or $-1/(2x^2) - 2x^{1/2} + C$ **23.** $-4e^x + C$ **25.** $30e^x + 5x^3/3 + C$ **27.** $2 \ln |x| + C$ **29.** $3 \ln |x| + e^x + C$ **31.** $C(x) = 2x^2 - 5x + 8$ **33.** $C(x) = .07x^3 + 10$ **35.** $C(x) = 2x^{3/2}/3 - 8/3$ **37.** $C(x) = x^3/3 - x^2 + 3x + 6$ **39.** $P(x) = -x^2 + 20x - 50$ **41.** $y = 2x^3 - 2x^2 + 3x + 1$ **43.** $v(t) = t^3/3 + t + 6$ **45.** $s(t) = -16t^2 + 6400$; 20 seconds

Section 14.2 (page 593)

1. 32; 38 **3.** 15; 31/2 **5.** 20; 30 **7.** 16; 14 **9.** 12.8; 27.20 **11.** 2.67; 2.08 **13.** (a) 15.656 (b) 18.258 (c) 21 **15.** About 10,000,000 **17.** We get about 20 **19.** A rough answer is about 2900 feet **21.** 30 **23.** 65 **25.** 357 **27.** 20

Section 14.3 (page 602)

1. 35/2 **3.** 44 **5.** 24 **7.** 42 **9.** 12 **11.** 124 **13.** 46 **15.** 36 **17.** 768 **19.** 14/3 **21.** $3(\ln 5 - \ln 2) \approx 2.75$ **23.** $e^2 - e \approx 4.67$ **25.** $e^3 - e + 4 \approx 21.37$ **27.** $3/4 + \ln 4 \approx 2.14$ **29.** 161 **31.** 21 **33.** 81/4 **35.** $\ln 2 \approx .69$

Section 14.4 (page 607)

1. \$280; \$11.67 **3.** \$31; \$216 **5.** No **7.** 220 **9.** 49.5 days **11.** 850 **13.** 8700 **15.** 6960 **17.** 1.36 **19.** (a) 75 (b) 100 **21.** (a) 14.26 (b) 3.55

Section 14.5 (page 613)

1. 15 **3.** 4 **5.** 40 **7.** 1735/6 **9.** 5/3 **11.** (a) 8 years (b) about 148 (c) about 771 **13.** 19.6 days **15.** \$807.52 hundreds or \$80,752 **17.** 5733.33 **19.** 83.33 **21.** (a)

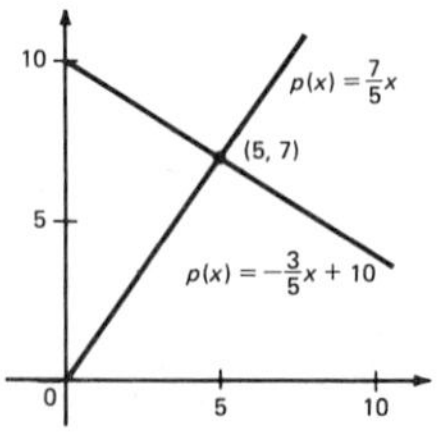

(b) $x^* = 5$ (c) 7.50 (d) 17.50 **23.** $I(.1) = .019$; the lower 10% of the income producers earn 1.9% of the total income of the population. **25.** $I(.6) = .384$; the lower 60% of the income producers earn 38.4% of the total income of the population.

27.

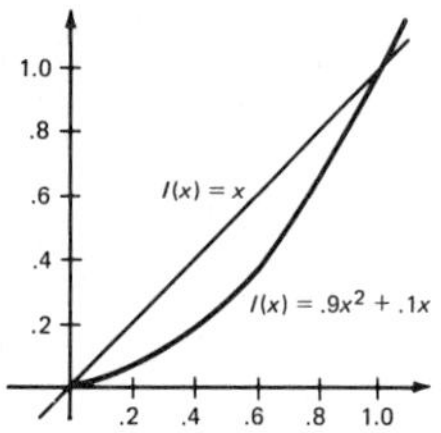

Section 14.6 (page 623)

1. $2(2x + 3)^5/5 + C$ **3.** $-2(x - 2)^{-2} + C$ or $-2/(x - 2)^2 + C$ **5.** $2(x - 4)^{3/2}/3 + C$ **7.** $(x^2 + 1)^4/4 + C$
9. $2(x^2 + 12x)^{3/2}/3 + C$ **11.** $e^{2x}/2 + C$ **13.** $-2e^{2x} + C$ **15.** $e^{2x^3}/2 + C$ **17.** $-8 \ln |1 + x| + C$
19. $(\ln |2x + 1|)/2 + C$ **21.** $\ln |x^2 + 4x| + C$ **23.** $-(3x^2 + 2)^{-3}/3 + C$ or $-1/[3(3x^2 + 2)^3] + C$
25. $-(2x^2 - 4x)^{-1}/4 + C$ or $-1/[4(2x^2 - 4x)] + C$ **27.** $[(1/x) + x]^2/2 + C$ **29.** $14/3$ **31.** $15/4$
33. $1024/3$ **35.** 1 **37.** $(e^2 - 1)/2 \approx 3.19$ **39.** $5(e^{-.4} - e^{-.8}) \approx 1.105$ **41.** About \$1350 **43.** $c(t) = 1.2e^{.04t}$
45. 14.75 **47.** .5 **49.** 15 years

Section 14.7 (page 629)

1. $x \ln 4 + x(\ln |x| - 1) + C$ **3.** $-4 \ln |(x + \sqrt{x^2 + 36})/6| + C$ **5.** $\ln |(x - 3)/(x + 3)| + C, \quad (x^2 > 9)$
7. $(4/3) \ln [(3 + \sqrt{9 - x^2})/x] + C, \quad (0 < x < 3)$ **9.** $-2x/3 + 2 \ln |3x + 1|/9 + C$ **11.** $(-2/15) \ln |x/(3x - 5)| + C$
13. $\ln |(2x - 1)/(2x + 1)| + C$ **15.** $-3 \ln |(1 + \sqrt{1 - 9x^2})/(3x)| + C$ **17.** $x^5(\ln |x/5| - 1/25) + C$
19. $(1/x)(-\ln |x| - 1) + C$ **21.** $-xe^{-2x}/2 - e^{-2x}/4 + C$ **23.** $-x^2e^{-2x} - xe^{-2x} - e^{-2x}/2 + C$
25. $x^3e^x - 3x^2e^x + 6xe^x - 6e^x + C$

Chapter 14 Test (page 632)

1. $x^4/2 + C$ **2.** $x^3/3 + 5x^2/2 + C$ **3.** $2x^{5/2}/5 + C$ **4.** $-x^{-3}/3 + C$ or $-1/(3x^3) + C$ **5.** $3e^x + C$
6. $5 \ln |x| + C$ **7.** 24 **8.** 28 **9.** 24 **10.** $\ln 3 \approx 1.10$ **11.** $-2(e^4 - 1) \approx -107.2$ **12.** 36,000
13. 1000 **14.** $64/3$ **15.** 2.5 years; about \$99,000 **16.** $(5x^2 + 6)^{3/2}/15 + C$ **17.** $(x^2 - 5x)^5/5 + C$
18. $-e^{-4x}/4 + C$ **19.** $12 \ln |x^2 + 9x + 1| + C$ **20.** 243 **21.** $\ln |(x + \sqrt{x^2 - 64})/8| + C$
22. $-\ln |(5 + \sqrt{25 + x^2})/x| + C$ **23.** $-(1/2) \ln |x/(3x + 4)| + C$
24. $x\sqrt{x^2 + 49}/2 + 49 \ln |(x + \sqrt{x^2 + 49})/2| + C$ **25.** $3x^3[(\ln |x|)/3 - 1/9] + C$ or $3x^3[3 \ln |x| - 1]/9$

Chapter 15

Section 15.1 (page 635)

1. $x(x + 1)^6/6 - (x + 1)^7/42 + C$ **3.** $2x^2(1 + x^2)^{5/4}/5 - 8(1 + x^2)^{9/4}/45 + C$ **5.** $2x(x - 1)^{1/2} - 4(x - 1)^{3/2}\ 3 + C$
7. $-2x(x + 5)^{-5}/5 - (x + 5)^{-4}/10 + C$ **9.** $-x^2(3 - x^2)^{1/2} - 2(3 - x^2)^{3/2}/3 + C$ **11.** $xe^x - e^x + C$
13. $-(5x - 9)e^{-3x}/3 - 5e^{-3x}/9 + C$ **15.** $(x^2 + 1)e^x - 2xe^x + 2e^x + C$ **17.** $(1 - x^2)e^{2x}/2 + xe^{2x}/2 - e^{2x}/4 + C$
19. $x \ln |x| - x + C$ **21.** $(x^2 \ln |x|)/2 - x^2/4 + C$ **23.** $(x^2 - x) \ln |3x| - x^2/2 + x + C$ **25.** $e^4 + e^2 \approx 61.9872$

Section 15.2 (page 639)

1. (a) 2.75 (b) 2.67 **3.** (a) 6.76 (b) 6.79 **5.** (a) 16 (b) 14.67 **7.** (a) .94 (b) .84 **9.** (a) .10 (b) .10

11. (a)

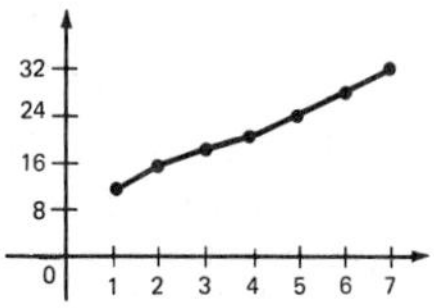

(b) 128 (c) 128

13. (a)

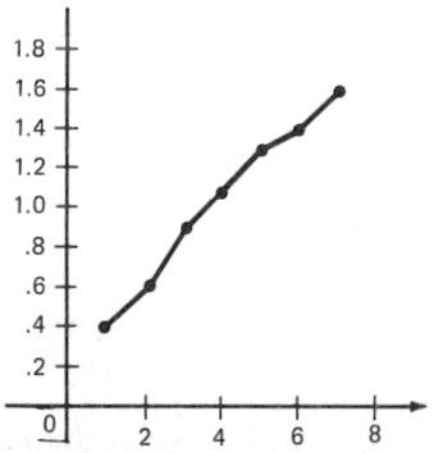

(b) 6.3 (c) 6.27

15. About 34 **17.** About 33

Section 15.3 (page 644)

1. $1/2$ **3.** Divergent **5.** -1 **7.** 1000 **9.** 1 **11.** $3/5$ **13.** 1 **15.** 4 **17.** Divergent **19.** 1
21. Divergent **23.** $(2 \ln 2.5)/21 \approx .087$ **25.** $-4(1/2 + \ln 1/2)/9 \approx .086$ **27.** Divergent **29.** \$750,000
31. 1250

Section 15.4 (page 648)

1. Yes **3.** Yes **5.** No; $\int_0^3 4x^3 dx \neq 1$ **7.** No; $\int_{-2}^{2} (x^2/16)\, dx \neq 1$ **9.** No; $2x \leq 0$ in the interval $-2 \leq x \leq 0$
11. $3/14$ **13.** $3/125$ **15.** $2/9$ **17.** $1/12$ **19.** $3/26$ **21.** (a) .46 (b) .17 (c) .37
23. (a) $3/5$ (b) $1/5$ (c) $3/5$ **25.** .38 **27.** .45 **29.** .12

Section 15.5 (page 657)

1. $y = x^3/3 + C$ **3.** $y = -x^2 + x^3 + C$ **5.** $y = -4x + 3x^4/4 + C$ **7.** $y = e^x + C$ **9.** $y = -2e^{-x} + C$
11. $y = -4x^3/9 + C$ **13.** $y = x^2/8 + C$ **15.** $y = 2x^3 - 2x^2 + 4x + C$ **17.** $y = -4x^3/3 + C_1x + C_2$
19. $y = 5x^2/2 - 2x^3/3 + C_1x + C_2$ **21.** $y = x^4/3 - x^3/3 + C_1x + C_2$ **23.** $y = -5x^3/6 + 3x^2/2 + C_1x + C_2$
25. $y = e^x + C_1x + C_2$ **27.** $y = x^3 - x^2 + 2$ **29.** $y = 5x^2/2 + 2x - 3$ **31.** $y = x^3 - 2x^2 + 2x + 8$
33. $y = x^3/3 + x^2/2 - 5x/6 + 2$ **35.** $y = e^x + x^2/2 - ex + 1$ **37.** $y = x^4/4 - x^2 + x + 4$
39. (a) $y = 5t^2 + 250$ (b) 12 days

Section 15.6 (page 662)

1. $y^2 = x^2 + C$ **3.** $y^2 = 2x^3/3 - 2x + C$ **5.** $y = ke^{x^2}$ **7.** $y = ke^{(x^3 - x^2)}$ **9.** $y = 2x^3/3 + C_1x + C_2$
11. $y = x^4/8 + C_1x + C_2$ **13.** $y^2 = 2x^3/3 + 9$ **15.** $y = e^{(x^2 + 3x)}$ **17.** $y = x^2 - x^3/6 - 29x/6 + 2$
19. (a) $y = ke^{-.05x}$ (b) $y = 90e^{-.05x}$ (c) About 55 grams

Chapter 15 Test (page 666)

1. $2x(8 + x)^{5/2}/5 - 4(8 + x)^{7/2}/35 + C$ **2.** $xe^x - e^x + C$ **3.** $-(x + 2)e^{-3x}/3 - e^{-3x}/9 + C$
4. $(3 + x^2)e^{2x}/2 - xe^{2x}/2 + e^{2x}/4 + C$ **5.** .414 **6.** .387 **7.** $1/2$ **8.** Divergent **9.** $3/4$ **10.** $1/10$
11. No; $\int_1^4 (2x + 4)\, dx \neq 1$ **12.** $1/4$ **13.** .91 **14.** .632 **15.** $y = -3x + x^3 + x^5 + C$ **16.** $y = -2e^{-2x} + C$
17. $y = x^4/4 + e^x + C_1x + C_2$ **18.** $y + y^2/2 = x^3/3 + C$ **19.** $y = x^2/2 - 2x^{3/2}/3 + 2$ **20.** $y = -e^{2x^2}$

Chapter 16

Section 16.1 (page 671)

1. 6 **3.** -20 **5.** 92 **7.** 47 **9.** $\sqrt{43}$ **11.** $\sqrt{19}$ **13.** 1300 **15.** 1996 **17.** 136 **19.** 304
21. 7.85 **23.** 18.9 **25.** 17.3

Section 16.2 (page 675)

1.

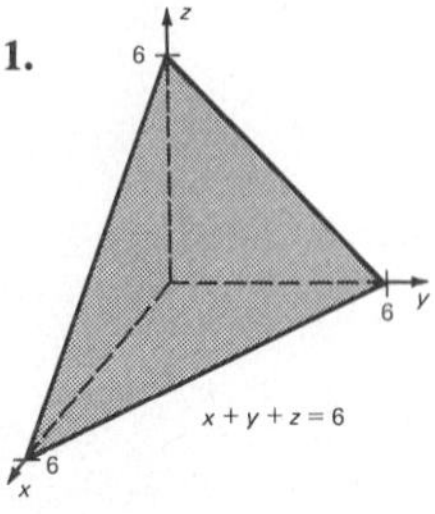

3.

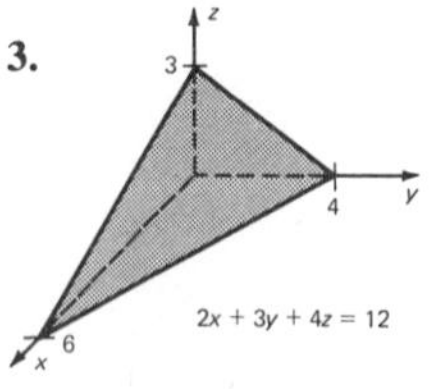

5.

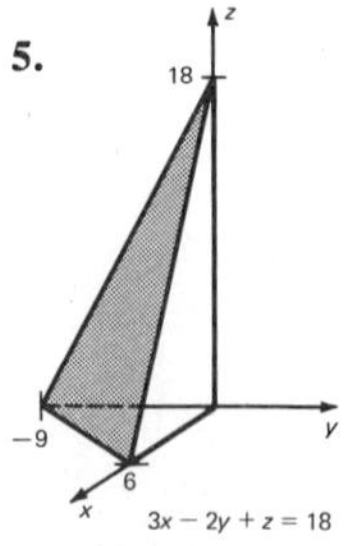

7.

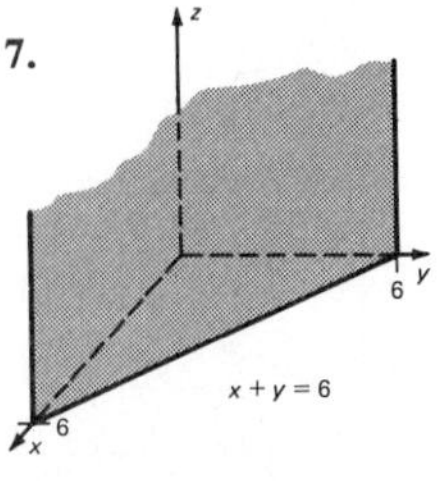

9.

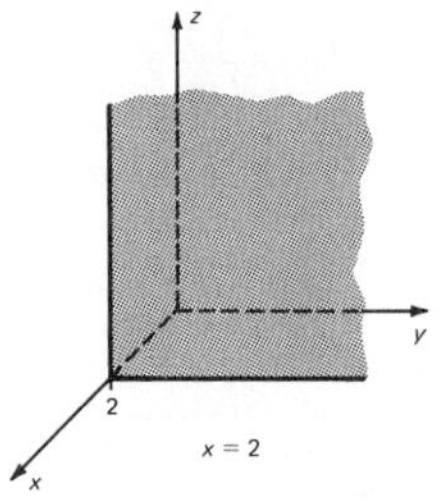

11. Center: (0, 0, 0); radius 7

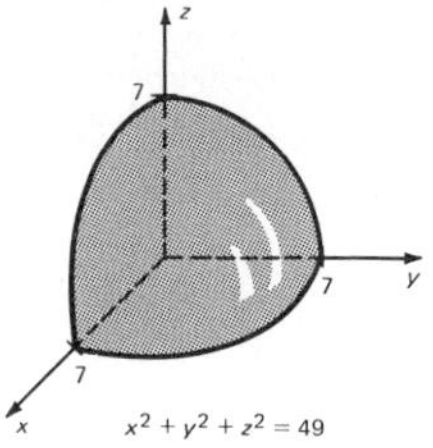

13. Center: (5, 3, −4); radius 3

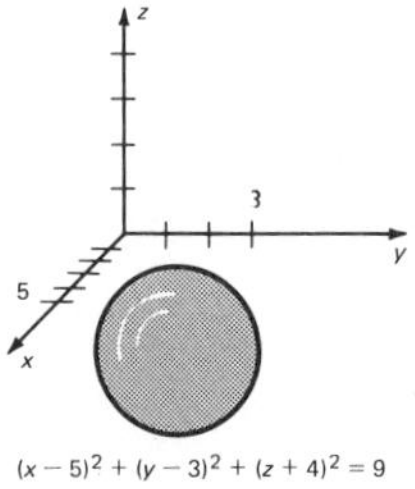

15. Center: (2, 5, −3); radius 5

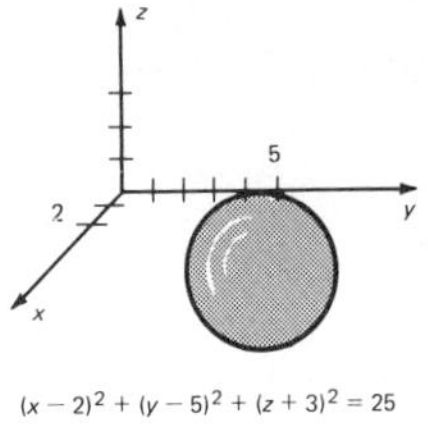

Section 16.3 (page 679)

1. $f_x = 24x$; $f_y = 8y$; 48; 24 **3.** $f_x = -2y$; $f_y = -2x + 18y^2$; 2; 170 **5.** $f_x = 9x^2y^2$; $f_y = 6x^3y$; 36; −1152 **7.** $f_x = e^{x+y}$; $f_y = e^{x+y}$; e^1 or e; e^{-1} **9.** $f_x = -15e^{3x-4y}$; $f_y = 20e^{3x-4y}$; $-15e^{10}$; $20e^{-24}$ **11.** $f_{xx} = 36xy$; $f_{yy} = -18$; $f_{xy} = f_{yx} = 18x^2$ **13.** $R_{xx} = 8 + 24y^2$; $R_{yy} = -30xy + 24x^2$; $R_{xy} = R_{yx} = -15y^2 + 48yx$ **15.** $r_{xx} = -8y/(x+y)^3$; $r_{yy} = 8x/(x+y)^3$; $r_{xy} = r_{yx} = (4x - 4y)/(x+y)^3$ **17.** $z_{xx} = 0$; $z_{yy} = 4xe^y$; $z_{xy} = z_{yx} = 4e^y$ **19.** $r_{xx} = -1/(x+y)^2$; $r_{yy} = -1/(x+y)^2$; $r_{xy} = r_{yx} = -1/(x+y)^2$ **21.** 80 **23.** 110 **25.** 168 **27.** 96 **29.** Increase of $70 **31.** $\partial z/\partial x = 2.1x^{1.1}y^{.3}$; $\partial z/\partial y = .3x^{2.1}y^{-.7}$ **33.** $2.5w^{-.67}$ **35.** $-1.675(T - F)w^{-1.67}$ **37.** $-2.5w^{-.67}$ **39.** $.8585W^{-.575}H^{.725}$ **41.** $1.4645W^{.425}H^{-.275}$ **43.** About 5.71 **45.** $1/(a - v)$

Section 16.4 (page 686)

1. Saddle point at (1, −1) **3.** Relative minimum at (−1, −1/2) **5.** Relative minimum at (−2, −2) **7.** Relative minimum at (15, −8) **9.** Relative maximum at (2/3, 4/3) **11.** Saddle point at (2, −2) **13.** Saddle point at (1, 2) **15.** Saddle point at (0, 0); relative minimum at (4, 8) **17.** Saddle point at (0, 0); relative minimum at (9/2, 3/2) **19.** $P(12, 40) = 2744$ **21.** $x = 12$; $y = 72$ **23.** $C(12, 25) = 2237$

Chapter 16 Test (page 695)

1. 456 **2.** 157 **3.**

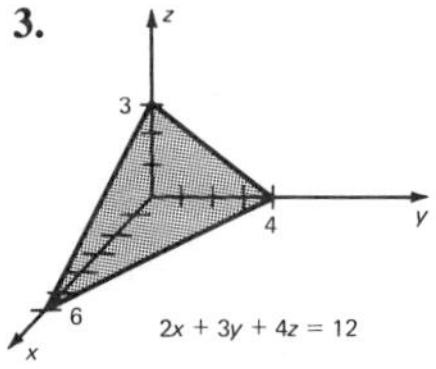

4.

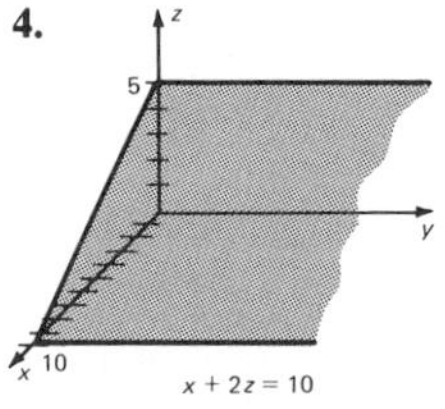

5.

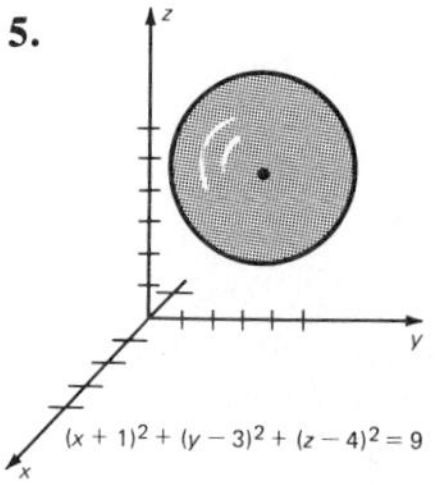

6. sphere, center at (0, 0, 0), radius 7

7. $f_x = 27x^2 - 4y$; $f_y = 4y - 4x$; 88;12 **8.** $f_x = 1/(y+1)$; $f_y = -(x+1)/(y+1)^2$; 1/6; −4/49 **9.** $f_{xx} = 2y^3$; $f_{yy} = 6x^2y$; $f_{xy} = f_{yx} = 6xy^2 + 8$ **10.** $f_{xx} = 2y/(1+x)^3$; $f_{yy} = 0$; $f_{xy} = f_{yx} = -1/(1+x)^2$ **11.** Saddle point at (0, 2) **12.** Saddle point at (0, 0); relative maximum at (−3/4, −9/32) **13.** $x = 11$; $y = 12$ **14.** $C(11, 12) = 431$

Index

What do you think of *Mathematics and Calculus with Applications?*

We would appreciate it if you would take a few minutes to answer these questions. Then cut the page out, fold it, seal it, and mail it. No postage is required.

Which chapters did you cover?
(circle) 1 2 3 4 5 6 7 8 9 10 11 12 13 14 15 16 All

Which helped most?
Explanations ______ Examples ______ Exercises ______
All three ______

Does the book have enough worked-out examples? Yes ______ No ______

Does the book have enough exercises? Yes ______ No ______

Were the problems in the margin helpful? Yes ______ No ______

Were the answers in the back of the book helpful? Yes ______ No ______

Did you have trouble getting the form of the answer in the back of the book?
Often ______ Sometimes ______ Never ______

Were the cases in the text helpful? Yes ______ No ______

Did your instructor cover the cases in class? Yes ______ No ______

Did you use the *Study Guide with Computer Problems?*
Yes ______ No ______ Did not know of it ______

If yes, was the *Study Guide* helpful?
Yes ______ For some topics ______ No ______

For you, was the course elective ______
required by ______

Do you plan to take more mathematics courses? Yes ______ No ______

If yes, which ones?
Statistics ______ College Algebra ______
Analytic geometry ______ Trigonometry ______
Calculus (engineering and physics) ______ Other ______

How much algebra did you have before this course?
Years in high school (circle) 0 ½ 1 1½ 2 more
Courses in college 0 1 2 3

If you had algebra before, how long ago?
Last 2 years ______ 3–5 years ______ 5 years or more

What is your major or your career goal? ______ Your age? ______

We would appreciate knowing of any errors you found in the book. (Please supply page numbers.)

What did you like most about the book?

FOLD HERE

What did you like least about the book?

College ______________________ State ______________

FOLD HERE

NO POSTAGE NECESSARY IF MAILED IN THE UNITED STATES

BUSINESS REPLY MAIL

FIRST CLASS PERMIT NO. 31 GLENVIEW, IL

Postage will be paid by
SCOTT, FORESMAN AND COMPANY
College Division Attn: Lial/Miller
1900 East Lake Avenue LM-1
Glenview, Illinois 60025